Edison is seen here in later years practicing on his telegraph key. He is said to have been the fastest telegraph operator in the country during his early years. He obtained over 100 patents for telegraph equipment and systems.

Experiencing Electricity and Electronics

Electron Flow Version

Second Edition

Mark E. Hazen

Florida Advanced Technology Center
Brevard Community College

Saunders College Publishing

Harcourt Brace Jovanovich College Publishing

Fort Worth   Philadelphia   San Diego   New York   Orlando   Austin
San Antonio   Toronto   Montreal   London   Sydney   Tokyo

Text Typeface: Times Roman
Compositor: Waldman Graphics
Acquisitions Editor: Emily Barrosse
Senior Developmental Editor: Lloyd Black
Managing Editor: Carol Field
Senior Project Manager: Marc Sherman
Project Editor: Janet Nuciforo
Copy Editor: Andy Potter
Manager of Art and Design: Carol Bleistine
Art Director: Anne Muldrow
Art and Design Coordinator: Caroline McGowen
Text Designer: Rebecca Lemna
Cover Designer: Lawrence Didona
Text Artwork: Grafacon
Layout Artwork: York Production Services
Director of EDP: Tim Frelick
Production Manager: Joanne Cassetti
Marketing Manager: Monica Wilson

The following chapter opener photographs were provided by Mark Hazen, courtesy of Edison-Ford Winter Estates, Fort Myers, Florida: Chapters 1 through 12, 14, 16, 18, 19, and 22.

Chapter opener photographs for Chapters 13, 15, and 21 from The Bettmann Archive.

Chapter opener photographs for Chapters 17 and 20 courtesy of Edison-Ford Winter Estates, Fort Myers, Florida.

Photographs for Part Openers, I and III, courtesy of Edison-Ford Winter Estates, Fort Myers, Florida; Part opener photographs for Parts II, IV, and V from The Bettmann Archive.

Endsheet photographs courtesy of Edison-Ford Winter Estates, Fort Myers, Florida.

Printed in the United States of America

EXPERIENCING ELECTRICITY & ELECTRONICS, 2nd edition
Electron Flow Version

ISBN 0-03-076691-5

Library of Congress Catalog Card Number: 92-056722

3 4 5 6 032 9 8 7 6 5 4 3 2 1

Dedicated to the memory of
my father Edward L. Hazen (1917–1982)
and
my mother Elizabeth E. Hazen (Gibbs, 1920–1992).

The Saunders College Publishing Series in Electronics Technology

Preface

Experiencing Electricity and Electronics, Second Edition, is intended for the introductory first-year course in DC and AC theory, analysis, and design. The text was written to meet the vigorous requirements for the electronics technician and technology curriculum. Improvements made in this edition are the result of an extensive survey of electronics instructors from across the country.

A Special Edition

Experiencing Electricity and Electronics, Second Edition, is a special edition for two reasons. First, this revision incorporates the suggestions of many of its faithful users, including improvements such as the coverage of all DC components, advanced DC circuit-analysis theorems, and resonance topics, respectively, in discrete chapters. Further, a new chapter covering advanced AC circuit analysis has been added. Also, the use of E notation in problem solutions has been virtually eliminated in favor of the preferred use of standard prefixes.

The second reason for this being a very special edition is the spotlight on the life and inventions of Thomas Alva Edison. Starting with ''The Age of Edison'' (following the Table of Contents), the reader is introduced to this amazing man who had a photographic memory and accumulated 1,093 different patents during his lifetime. Throughout this book, readers will be fascinated and inspired by photographs of Edison's inventions displayed in part openers and chapter openers accompanied by brief narratives. Students will get a sense of Edison's genius and creativity, which he attributed to ''99% perspiration and 1% inspiration.'' The Edison theme will appear in this edition only, making it a valued book to keep.

Subject Development and Approach

The word ''experience'' in the text title is appropriate because DC/AC fundamentals are consistently related to their applications in the world of electronics. The thorough presentation and use of laboratory test equipment throughout the text and color insert sections enhance the student's experiences with electricity and electronics. Students are further motivated through the now familiar Need To Know scenarios and practical applications, guided by the use of sectional Self-Checks and worked examples, and introduced to new content by lucid writing and informative illustrations. Students have praised the text's understandable writing style and problem-solving techniques as a welcome relief from the usual conceptual jungle that exists in many of their textbooks.

Professors find this text of great benefit not only because of its value to the student, but because of its modular approach in handling subjects. First, the text is divided into six parts: Part I: Getting Acquainted with Electricity and Electronics; Part II: DC Circuits and Theorems; Part III: Magnetism and Magnetic Devices; Part IV: Inductors, Capacitors, and Pulse Response; Part V: AC Theory and Circuit Analysis; Part VI: Appendices, Answers to Odd-Numbered Problems, Glossary, and Index. Next, chapters are divided into sections containing their own Self-Check, questions, and problems. Instructors do not need to rearrange their order of presenting topics to fit the text. Rather, instructors can use the text to accommodate their individual course outlines.

Chapter Organization

Each chapter is organized with the following pedagogical and instructional elements:

- Edison photograph(s) with caption(s)
- chapter outline
- objectives
- Need To Know scenario
- introduction
- two-color text format
- numbered sections and accompanying Self-Checks
- numerous worked examples and illustrations
- Design Notes
- boldfaced special terms
- four-part summary: formulas, concepts, procedures, and a listing of special terms
- Need To Know solution
- questions grouped by section
- problems grouped by section
- answers to Self-Checks
- Suggested Projects

Innovative Features

- *Focus on Edison.* Coverage of the life and inventions of Thomas Alva Edison can be found in the special section "The Age of Edison 1847–1931" (following the Table of Contents), in the part and chapter openers, and on the end papers.
- *Modular Chapter Organization.* Each chapter is divided into several sections, and each section concludes with a Self-Check. At the end of each chapter, answers to Self-Checks are identified by section. Also, questions and problems are arranged by section. This modular flexibility enables the text to conform to any instructional program—the program, or instructor, does not have to conform to the text.

- *Need To Know.* Each chapter begins with a motivational section that establishes the student's need to know. The "Need to Know" presents students with an interesting real world problem that they can solve after learning the material covered in the chapter. A Need To Know solution is always presented at the end of the chapter.
- *Design Notes.* Design Notes are special features that spotlight circuit theory, analysis, and design through the use of BASIC programs. Procedures are outlined in the BASIC program steps.
- *Example Groupings.* Very often examples are presented in related groups so the student can easily compare important effects of parameter changes.
- *Problem/Text Compatibility.* All questions and problems relate directly to the material presented in each section. Careful attention has been paid to insuring that students do not become confused because of problems not related to the topics covered.
- *Color Inserts.* This text contains color photograph sections highlighting the use of test equipment. The instructor and student should take time to examine these photographs when alerted to do so in the margins of the text.
- *Biographical Sketches.* In addition to the focus on the life of Edison, photographs and short biographies of other pioneers in electricity are integrated throughout the book.
- *Suggested Projects.* Projects are suggested at the end of each chapter for the student's enrichment.
- *Icons.* A variety of icons have been added throughout to alert the reader to such things as Need To Know, troubleshooting, Self-Checks, Self-Check solutions, and suggested projects.

End-of-Book Material

The following helps are provided at the end of this book:

- *Appendices.* English/Metric Conversion, The Periodic Table, Algebraic Operations, Basic Trigonometry, Simultaneous Equations and Determinants, Standard Resistor Values.
- *Answers to Odd-Numbered Problems.* Answers to all odd-numbered problems are provided at the end of the text to provide the student with immediate feedback.
- *Glossary.* This extensive glossary offers students definitions to all important technical terms used in the text.
- *Index.* A detailed index is provided.

Ancillary Package

The following supplements are available to adopters of *Experiencing Electricity and Electronics*:

- *Instructor's Manual with Transparency Masters.* Written by the text author, this instructor's manual contains solutions to all end-of-chapter problems and a set of 100 transparency masters to aid class lectures.
- *Laboratory Manual.* Written by the text author, this laboratory manual contains 36 multipart experiments, sufficient to satisfy a two-semester course.
- *Instructor's Laboratory Manual.* Written by the text author, this supplement contains suggested laboratory results.

- *Parts Kits.* Low-cost parts kits that complement the laboratory experiments are available through MEH LABS, P.O. Box 100004, Palm Bay, FL 32910–0004.
- *Test Bank.* Written by the text author, the test bank includes 700 multiple choice questions and problems for class use.
- *ExaMaster*™ *Computerized Test Bank.* The printed test bank questions are presented in an easy-to-use computerized format for IBM. Users can select questions, edit questions, add questions, and print out assorted versions of the same test.
- *RequesTest*™. Instructors without access to a personal computer may contact Saunders Software Support Department at (800) 447-9457 to request tests prepared from the computerized test bank. The test will be mailed or faxed to the instructor within 48 hours.
- *Study Guide.* Written by the text author, the study guide includes a chapter-by-chapter review of terminology, concepts, and additional practice problems. Questions include multiple choice, completion, and circuit-analysis problems.
- *Design Note Software.* All circuit analysis and design programs featured in the Design Note sections of the text are available on IBM PC® diskette. This software is free to all adopters and may be copied by students enrolled in a course that uses this textbook. Instructors may obtain a copy of this software directly from the author. See the order form on page 2 of the Instructor's Manual or call (407) 723-8197.

Acknowledgments

I first wish to thank my wife Sharon, my son Jonathan, and my daughter Valerie for their patience, sacrifice, and understanding during the development and revision of this book. The author is not the only one who must make sizable sacrifices when a project of this magnitude is undertaken.

Second, I wish to express my appreciation to the hardworking professional team at Saunders College Publishing: Emily Barrosse (Senior Acquisitions Editor), Laura Shur (Assistant Editor), Lloyd Black (Senior Developmental Editor), Marc Sherman (Senior Project Manager), and Janet Nuciforo (Project Editor).

Third, I want to thank the reviewers of and contributors to my work for their dedication and mutual concern for a quality and thorough text. The following professionals have enhanced the quality of this project with their insights, suggestions, and information.

James L. Alward, El Camino College
Larry D. Anderson, Skagit Valley College
Robert Bales, Northeast Wisconsin Technical College
Bruce Barnes, Central Washington University
Ronald C. Boyer, Jefferson College
M.L. Bratley, Skagit Valley College
John Conforti, Mount San Antonio College
Gerald G. Cottrell, Mid-State Technical College
Gary Crossman, Chippewa Valley Technical College
Michael D. Dotson, Southern Illinois University
Robert Doyle, Wilkes Community College
John Fitzen, Idaho State University
Daniel Fox, Lakeshore Technical College
Joseph E. Francis, Spoon River College

Deborah Jo Greathouse, Southern Illinois University
Allen Grommet, East Arkansas Community College
Bill Hames, Midlands Technical College
Kenneth J. Hanson, Cerritos College
Steve Harsany, Mount San Antonio College
David Hilse, Central Missouri State University
Gary L. Hobbie, Oklahoma State University
Ron Jones, Salt Lake Community College
Walter Kahan, El Camino College
Cecil A. Legg, Central Piedmont Community College
Dave Longobardi, Antelope Valley College
Robert Ludeman, Andrews University
Dennis D. Meyer, Mid-State Technical College
Wallace Niebel, College of the Redwoods
David L. Olejniczak, Northeast Wisconsin Technical Center
Joe F. Pedersen, Skagit Valley College
Calli Pindrus, Chippewa Valley Technical College
Jerry Place, Mid-State Technical College
James Predko, Lansing Community College
Marcus S. Rasco, DeVry Institute of Technology, Dallas
Ernie Renner, Central Oregon Community College
Hal Sappington, Central Missouri State University
James J. Schreiber, DeVry Institute of Technology, Phoenix
Richard Shields, Winona State University
Rex Sinclair, College of the Redwoods
Harry M. Smith, Mount San Antonio College
Ames L. Stewart, Central Missouri State University
Michael Szymkewicz, Olympic College
John Thomsen, Moorpark College
Thomas G. Tinsley, St. Charles County Community College
Andrew J. Van Camp II, East Central College
Neal Voke, Triton College
Jack Warfield, ITT Technical Institute, Fort Wayne
Thomas E. Warner, University of Hartford
Davis Watson, Arizona Western College
Richard Zoladz, Triton College

Finally, I want to offer a very special thank you to all of the fine folks at the Edison-Ford Winter Estates in Fort Meyer, Florida: Robert P. Halgrim (Manager), Judy Surprise (Assistant Manager), Dr. Leslie H. Marietta (Historian), Robert Beeson (Resource Person), Richard Gainer (Resource Person), James Niccum (Resource Person). These people were very kind and generous in their help with photographs and information.

A Special Note to the Student

The experiences you are about to begin here in this text are far more exciting than you can imagine. Through these pages, I will have the pleasure of introducing you to the wonderful world of electricity with many applications to electronics. You will gain new knowledge and make exciting discoveries. Your experiences here will not be of words, diagrams, and pages but of theory and components that make up the high-tech world in which we live. Your natural sense of creativity and curiosity will feed on every concept, theory, and application. And that is where the strength of this text lies, in the real application of theory, components, and circuits. To help you in your experience with electricity and electronics, I have included the following special features:

1. As with my other books, I begin each chapter with a Need To Know scenario. These brief sorties are designed to help you prepare mentally for the material covered in the chapter. Their main purpose is to establish a need to know in your mind. Please take time to read these and give them some thought before beginning each chapter.

2. Every chapter is divided into sections with a short Self-Check involving questions and problems at the end of each section. All answers to these Self-Checks are provided at the end of the chapter. The Self-Check is intended to help you see if you are understanding the material.

3. At the end of each chapter, questions and problems are divided up according to section to help you focus on a smaller area at one time. Answers to all odd-numbered problems are given at the end of the book.

4. Virtually all circuits in examples and figures are labeled with standard component values so you can build and test any circuit.

5. In some chapters, I have included what I call Design Notes. These are special emphasis features that include diagrams, circuits, formulas, and BASIC computer programs that you can use to analyze or design circuits. The programs will run on any IBM® compatible PC.

6. Color photographs have been included that, for the most part, demonstrate the use of test equipment. You will be directed to particular photographs by notes in the margin at key points throughout the book.

7. The appendices, glossary, and answers to odd-numbered problems at the end of the book are designed to assist you in your learning experience. Take a few moments to familiarize yourself with these.

8. Try not to overlook the suggested projects at the end of each chapter.

Oh, by the way, I highly recommend that you keep this text in your personal technical library. You will find it to be very helpful as you take higher-level courses and as a reference as you begin your career.

Take time to read ''The Age of Edison'' just following the Table of Contents and study the many photographs of Edison and his inventions throughout this book. It is my hope that Edison's life will be an inspiration to you. He was truly a remarkable man who never gave up in the face of continual failure (for example, the invention of the light bulb).

About the Author

You might be wondering who I am. I often wondered who the authors of some of my textbooks were. So, let me tell you just a little bit about myself. As I write this, I am in my 43rd year. Looking back, I have had a very exciting career in electronics that started in the U.S. Air Force. I taught electronics a little over twenty years ago at Kessler Air Force Base in Biloxi, Mississippi. Between then and now, I have been all over the world, not to mention the United States. Among my adventures, I installed a control room in a recording studio in Monte Carlo, a 400,000 watt AM broadcast transmitter in Sri Lanka, a 100,000 watt shortwave transmitter on Guam, and wired a professional recording studio complex in Hong Kong. I taught electronics at DeVry Institute of Technology in Phoenix, Arizona for nearly six years, then left Phoenix for Florida to complete two additional books—*Fundamentals of DC and AC Circuits* and *Exploring Electronic Devices*. I am back in the classroom now teaching electronics at the Florida Advanced Technology Center of Brevard Community College near the Cape. We have much in common you and I— we are both learning. That's what makes me so excited about electronics—there is always something new to challenge and stimulate the imagination. You're not alone in your learning experiences. Enjoy!

Mark Edward Hazen
Palm Bay, Florida
December 1992

Contents

The Age of Edison: 1847–1931

Thomas Alva Edison was born to Samuel and Nancy Edison on February 11, 1847, in Milan, Ohio. The world had no reason to suspect that with the birth of Thomas a new age of discovery was also born. By the time Edison died in 1931 at the age of 84, he had amassed 1,093 patents. His accomplishments were so astounding that Henry Ford proposed that the period of Edison's life be proclaimed ''The Age of Edison.'' Edison's life and accomplishments have been and continue to be an inspiration to all of us, appealing to our curious and inventive nature. It is my hope that Edison's life and inventions described here and throughout this book will be an inspiration and challenge to you. This was Edison's wish as well.

> *If I have spurred men on to greater effort, and if our work has widened the horizon of men's understanding even a little and given a measure of happiness in the world, I am content.—Thomas Alva Edison*

Edison the Boy

Like many children Edison was curious about nearly everything around him. He was constantly asking questions. Though his mother had once been a schoolteacher, she could not begin to answer all of his questions. He wanted to know such things as how birds fly, how chicken eggs hatch, and why water can extinguish fires. Often he sought answers to these questions on his own through experimentation. Once he sat on a nest of eggs to see if he could hatch them. Learning that balloons fly because they are filled with gas, he once persuaded a friend to take a triple dose of Seidlitz powders. These powders produced a carbonated gas when dissolved in water. He was expecting his friend to swell and begin to float in the air. Instead, his friend swelled and lay in distress on the ground.

When Al, as his family called him, was 7, he and his family moved to Port Huron, Michigan. There, he entered the public school system and immediately gained the recognition of his teachers—though not as you might think. Al wearied his teachers and schoolmaster with an endless flow of questions. It is told that Al received whippings with a belt

Thomas Alva Edison at age 12. (Photograph from The Bettmann Archive.)

from the schoolmaster because of his questions. Unfortunately, or fortunately, Al overheard his schoolmaster tell the district school superintendent that he believed Edison was "addled." Naturally, this outraged Al's parents and they took him out of the public school system. Al's formal public education lasted only three months. From that point on, his mother taught him at home. At the age of 9, his mother bought him a chemistry book which he absorbed voraciously. He accumulated nearly 100 different chemicals and tested every experiment in the book.

When he was 12, he became a "candy butcher" on the Port Huron to Detroit stretch of the Grand Trunk Railway. His primary duty was to sell candy, fruit, and newspapers to the passengers. He also began to print and publish a newspaper called the *Weekly Herald* from the train. Al could not bear to be apart from his experimenting very long, so he set up a chemical laboratory in the baggage car of the train. Unfortunately for Edison, a stick of phosphorus burst into flames one day and set the baggage car on fire. The conductor responded by boxing Al's ears and throwing him off the train with all of his wares.

In his closing years, Edison became almost completely deaf. Some have thought that this ear-boxing experience contributed to Al's deafness. Edison is reported to have told of an event later than this in which he tried to board a moving train. The conductor of that train saw that he was having difficulty boarding, so he reached out and grabbed him by the ears to hoist him into the train. It was then that something snapped in his head and his deafness began. However, some historians believe that these are just stories and that his deafness began with his chronic childhood illnesses such as scarlet fever. In any case, his deafness grew worse with time but he considered it a blessing since he was able to shut out the outside world and concentrate on his work.

At the age of 16, Al became a telegraph operator for the Grand Trunk Railway in Ontario, Canada. Later, he would become one of the fastest telegraph operators in the United States. He learned the Morse code from a grateful station agent at the railway station in Mt. Clemens, Michigan, after rescuing the agent's son from certain death under the wheels of a rolling freight car. Nearly 150 of Edison's patents dealt with innovations in telegraphy equipment and circuits.

Edison the Inventor

Edison was always creating new things or improving upon old ideas. He was driven by the excitement of discovery and creation. At the age of 21, he signed his first patent—for his Electrographic Vote Recorder (October 13, 1868) which he attempted to sell to Congress in Washington, DC. His machine was rejected. This experience led him to proclaim, ''I will never again invent anything which nobody wants!'' From that time on, Edison dedicated himself to what he termed ''the desperate needs of the world.''

In 1869, he moved to New York financially broke. While visiting the Gold Indicator Company, he convinced an employee to allow him to sleep in an office. While there, he carefully studied the stock ticker machine used to transmit the price of gold over wire to stockbrokers. A couple of days later the stock ticker broke down and no one seemed to be able to fix it. Edison asked if he could try and in a short time did fix the machine. The supervisor rewarded Edison with a full-time job at $300 per month. That was a lot of money. Edison soon made many significant improvements to stock tickers and obtained patents. In early 1870, Edison received the glorious sum of $40,000 from the Gold and Stock Telegraph Company for the patent of a greatly improved stock ticker. Edison was hoping for $5,000 and was willing to take less. He was now able to finance his own dreams so he moved to Newark, New Jersey, and opened his first workshop/laboratory.

The following is a chronological summary of Edison's life and inventions beginning with the improved stock ticker of 1870 when he was 23. This chronology is by no means complete.

1870 Edison received $40,000 for the patent of his stock ticker. He opened his first workshop and laboratory in Newark, New Jersey, where be began manufacturing his improved stock ticker and other telegraph equipment.

1871 Edison's mother Nancy died on April 9. On December 25, Edison married Mary Stilwell.

1872 to 1876 manufactured telegraph equipment for Western Union Telegraph Company and Automatic Telegraph Company, opened many shops in Newark, and invented the motograph; automatic telegraph system; duplex, quadraplex, and multiplex telegraph systems; wax paper; and carbon rheostats. He also improved the typewriter, which had been slower than handwriting, by replacing all wooden parts with metal parts.

1876 moved to Menlo Park, New Jersey, to open a new shop and laboratory. On August 8, he was granted a patent for the electric pen used to make stencils for his mimeograph machine. These were later licensed to the A.B. Dick Company of Chicago. Also in this year he invented the carbon microphone, which greatly improved Bell's telephone system.

1877 recorded and played back ''Mary had a little lamb'' on his first cylindrical phonograph on December 6. The invention worked immediately, with only an initial false start. In the fall of this year, he developed the carbon button telephone microphone.

1878 first phonograph patent was granted on February 19. He became interested in the possibility of electric lighting. On November 15, he held an incorporation meeting for the Edison Electric Lighting Company.

1879 installed an experimental electrical generating station aboard the S.S. *Jeannette* for the George Washington De Long expedition to the Arctic. He invented the first practical incandescent light bulb using a carbon filament. His first successful bulb burned for more than 40 hours starting the evening of October 19. In this same year, he made significant improvements to dynamos (electrical generators). He increased their energy conversion efficiency from 40 to 50 percent to 90 percent. He also invented electrical distribution systems, regulators, energy consumption meters, sockets, switches, fuses, and more. On December 31, he gave a public demonstration of his electric lighting system to Menlo Park. He became known as "The Wizard of Menlo Park."

1880 to 1887   obtained well over 300 patents that related to electrical lighting and power systems.

1880   invented the magnetic iron ore separator. In February he discovered what became known as the "Edison effect," electrical current through a vacuum from a heated filament. The first commercial incandescent light bulb was manufactured by Edison at his Menlo Park Edison Lamp Works on October 1.

1881   opened the Edison Machine Works in New York.

1882   opened the first commercial lighting and power plant at Holborn Viaduct, London, England, on January 12. On May 1, he moved his Edison Lamp Works from Menlo Park to Harrison, New Jersey. He established factories to manufacture dynamos and related electrical apparatus in Harrison. The first American central commercial lighting station was set in operation on September 4 at 257 Pearl Street in New York City.

1883   opened the first three-wire electric lighting system on July 4 at Sunbury, Pennsylvania. On November 15, he filed for a patent on an "electric indicator" which made use of the "Edison effect." This was the very beginning of the age of vacuum tubes.

1884   Edison's wife Mary died on August 9 at Menlo Park.

1885   executed a patent on March 27 for a wireless telegraph system to be used by moving trains. On May 14, he applied for a patent on a ship-to-shore wireless telegraph system.

1886   married Mina Miller of Akron, Ohio, on February 24 and moved to Florida to establish a winter home in Fort Myers. Their entire estate with houses, laboratory, and beautiful gardens can be seen today.

Edison listening to one of his cylindrical phonographs (age 41). (Courtesy of Edison-Ford Winter Estates, Fort Myers, Florida.)

Edison standing in a section of his Menlo Park, New Jersey, laboratory (in his 50s). (Courtesy of Edison-Ford Winter Estates, Fort Myers, Florida.)

1887   moved his laboratory from Menlo Park to West Orange, New Jersey, where he established a larger and more modern facility. Here, over a four-year period, he made significant improvements to his cylindrical phonographs, taking out over 80 patents. He also established an extensive business in manufacturing and selling phonographs.

1889   first projection of an experimental motion picture on October 6 at his West Orange laboratory. The pictures were synchronized with sound from a phonograph.

1891   applied for a patent on the first motion picture camera (projector) on August 24. He continued to improve upon the kinetoscope.

1893   Edison-Lalande primary cells used to supply power for railroad signals near Phillipsburg, New Jersey.

1894   first commercial showing of motion pictures on April 14 at the kinetoscope parlor on Broadway in New York.

1896   developed the fluoroscope, the forerunner of the X-ray machine, but did not patent it. He chose to leave it to the public domain for the good of mankind. On February 26, Edison's father died in Norwalk, Ohio. On May 16, Edison applied for a patent on the fluorescent light.

1900   began ten years of research and experimentation to develop the Edison nickel-iron-alkaline storage cell. Alkaline cells are commonly used today. (Edison received 141 battery patents in all.)

1901   opened the Edison Cement Plant in New Jersey.

1902   improved the copper oxide primary cell.

1903   applied for patent on long rotary kilns used in the production of cement.

1907   invented the universal electric motor (series-wound motor), which runs on AC or DC.

1910   began work on improving disk-type phonographs. He developed the ''Diamond Disk'' instrument, which produced high-fidelity records.

1913   introduced the Kinetophone for talking motion pictures.

Edison examining some of his culture dishes in his Fort Myers Laboratory (in his late 70s or early 80s). (Courtesy of Edison-Ford Winter Estates, Fort Myers, Florida.)

1914   patent for electric safety lanterns for miners. He also developed a process to produce synthetic carbolic acid. He built a plant in less than a month that would produce more than a ton per day to help with the shortage caused by World War I. The Telescribe was invented, which combined the telephone and business phonograph to record and play back telephone messages.

1915   established manufacturing facilities for coal tar derivatives. On October 7, Edison became president of the Naval Consulting Board to assist with national defense.

1927   began a search for a domestic source of natural rubber. He collected and tested over 17,000 plants and eventually produced a new strain of goldenrod that grew 12 to 14 feet tall in one growing season and yielded a significant amount of latex rubber. Harvey Firestone made him a set of tires for his 1907 Model T Ford from this source of rubber.

1931   Thomas Alva Edison died on October 18 at Llewellyn Park in West Orange. He was survived by sons Thomas Alva, Jr., and William Leslie and daughter Marion from his first wife, and sons Charles and Theodore and daughter Madeleine from his second wife. Charles Edison became secretary of the navy in 1939 and governor of New Jersey in 1941.

Edison the Man

Edison has been characterized by those who knew him as dedicated to his work, relentless, somewhat of a loner, not the ideal family man, driven, and independent. He loved practical jokes and, to say the least, was inventive and creative. Edison was blessed with a photographic memory and could read an entire book in about 15 minutes. In later years, he would entertain his grandchildren by memorizing names and numbers in the phone book and reciting them on demand. His relentlessness and optimism were demonstrated during his quest for the nickel-iron-alkaline storage battery. He had performed over 9,000 exper-

Edison enjoying the company of his close friends Henry Ford (left) and Harvey
Firestone (right) at his winter home in Fort Myers, Florida. (Courtesy of Edison-Ford
Winter Estates, Fort Myers, Florida.)

iments with no success. When confronted by an assistant with the verdict of failure he
merely pointed out that they had discovered 9,000 things that will not work—9,000 things
that would not have to be tried again. It took 41,000 more experiments before he was
indeed successful in the development of the nickel-iron-alkaline cell.

Henry Ford and Harvey Firestone were two of Edison's close friends. Ford purchased
an estate next to Edison's in Fort Myers where they could be nearby during the cold
northern winters. One of Edison's most prized possessions was a 1907 Model T Ford given
to him by his dear friend. Every year, Ford wanted to exchange the old Model T for a new
model. Edison always refused. Can you guess what the T stood for in Model T? That's
right, Thomas. Toward the end of his life, Edison discovered natural rubber in goldenrod.
Harvey Firestone took the first harvest of rubber from Edison and made him a complete
set of tires for his beloved Model T.

Edison has always been thought to be a genius and a gift to mankind. When Edison
was once approached with the idea of his genius he said it was ''1 percent inspiration and
99 percent perspiration.'' Genius or not, he knew the meaning and value of hard work—
a method he proved over and over again. ◆

The author would like to thank the following people of the Edison-Ford Winter Estates
of Fort Myers, Florida, for their generous supply of information for this biography of
Edison:

• Robert P. Halgrim, Manager
• Judy Surprise, Assistant Manager
• Dr. Leslie H. Marietta, Historian
• Robert Beeson, Resource Person
• Richard Gainer, Resource Person
• James Niccum, Resource Person

Part I

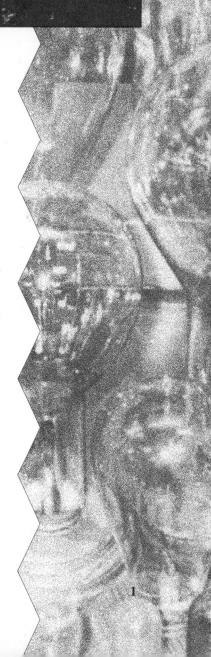

Getting Acquainted

Thomas A. Edison's laboratory at his winter home in Fort Myers, Florida is shown above and to the left. The laboratory can be seen today just as Edison left it back in 1931.

The main room of the laboratory is composed of a machine shop and an extensive chemical laboratory. It is here that Edison developed a new strain of goldenrod from which rubber could be produced. The new strain would grow 12 to 14 feet tall in one growing season and yield a generous quantity of rubber. In 1927, Edison sent samples of this new rubber to his friend Harvey Firestone, who used the rubber to make a new set of tires for Edison's 1907 Model T Ford. The Model T was a treasured gift from another famous friend, Henry Ford. The Model T and tires can be seen today at the Thomas Alva Edison Home in Fort Myers, Florida.

Edison's office adjoining his laboratory is preserved just as he left it (see left). Of special interest is the cot where Edison would catnap. He claimed that he did not need a full night of sleep. A short twenty-minute nap was sufficient for another four or five hours of work. He accepted self-imposed challenges with all of his being. (Photographs courtesy of Edison-Ford Winter Estates, Ft. Myers, FL 33901)

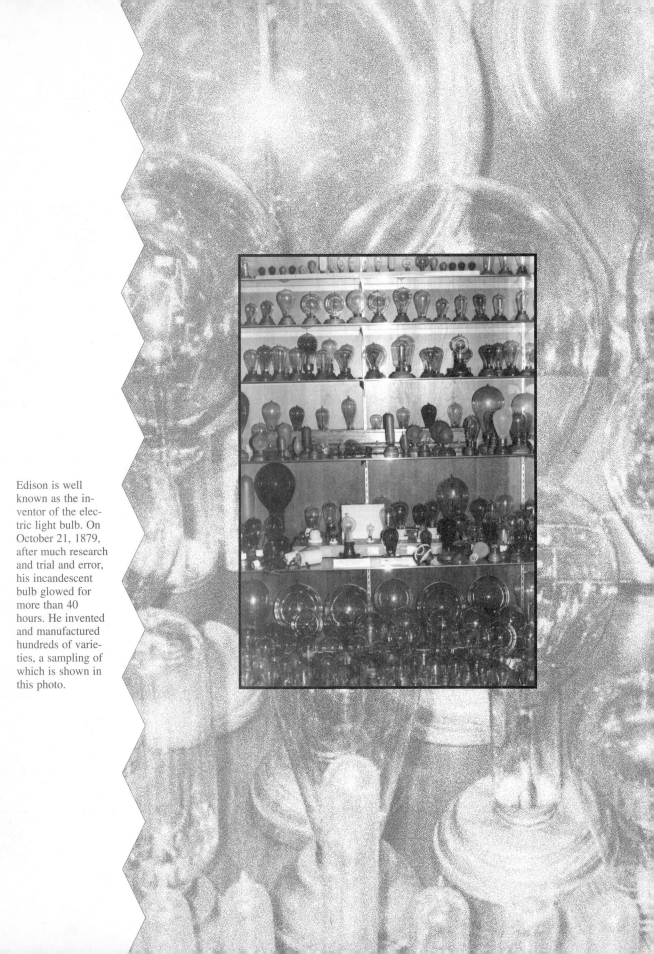

Edison is well known as the inventor of the electric light bulb. On October 21, 1879, after much research and trial and error, his incandescent bulb glowed for more than 40 hours. He invented and manufactured hundreds of varieties, a sampling of which is shown in this photo.

Chapter 1

Getting Started

OBJECTIVES

After studying this chapter, you will be able to
• use SI symbols and prefixes to describe
  electrical values.

• manipulate numbers in mathematical
  operations using scientific notation,
  engineering notation, and E notation to
  aid in mathematical circuit analysis in
  later chapters.

INTRODUCTION

Getting started in your study of electricity and electronics is exactly what this chapter is all about. There are some things you simply need to know before we begin analyzing circuits and chasing electrons. This chapter will help prepare you for chapters to come. We will discuss: an International System of units (SI) used as a standard in electrical measurement, scientific notation, mathematical operations involving powers of ten, and the use of prefixes to modify units. Some of this material may be a review for you, especially if you have had a strong math background. Whether it be review or new, the material is significant, forming a foundation for further study.

NEED TO KNOW

In the field of electronics, as in any science or discipline, you must master certain basic fundamentals to such a degree that their use requires little effort. There are many such fundamentals in electronics, and you will learn them in a logical and practical way. Perhaps the most fundamental of these will be your ability to manipulate very large and very small numbers. These numbers can, and will, be expressed in many different ways. For example, five and one-half million volts may be expressed as 5,500,000 V, or 5,500 kV, or 5.5 MV, or $5.5 \cdot 10^6$ V, or even as 5.5 E6 V. The point is, you must be able to recognize all of these expressions and quickly convert from one form to another.

Suppose you were given this short test right now. Could you express each of these quantities in at least two other forms? If not, don't worry. You will be able to do so by the end of this chapter.

Short Test

Express the following in two different forms.

1. Twenty milliseconds =
2. 3.3 kilovolts =
3. 4.5 μS =
4. 1.9 E6 watts =
5. $1.5 \cdot 10^6 \ \Omega$ =

1-1 SI Units

In your study of electricity and electronics, as in any science, you will apply theory, collect data, calculate unknowns, measure, analyze, categorize, and compare values and information. Your results must be expressed in some standard form, or units, readily understood and usable by someone else. You must either use standard, mutually understood, units of measurement or be able to quickly and easily convert from your units to someone else's and vice versa. For example, people in the United States find themselves using units of measurement from two different systems—the English system and the metric system. A distance may be expressed in miles or kilometers. If you want to express a given number of miles in terms of kilometers, you must use a conversion factor that will translate the miles to kilometers. In this case, there are 1.609 kilometers for every mile. So, if the distance is 5 miles, it can be expressed as 8.045 kilometers. (For more information on system conversion factors, see Appendix A, "Metric/English Conversion.")

As you can see, it would be much less confusing if everyone simply used the same units of measurement. Fortunately such a system of weights and measures does exist and has been adopted by most scientists and engineers around the world. That system is the **Système International d'Unités** (International System of Units) or simply **SI units**. This system, based on the metric system, was compiled in its present form and adopted in France in 1960. Since then it has gained wide acceptance the world over. The basic SI units are the **Meter** for length, the **Kilogram** for weight, the **Second** for time, and the **Ampere** for electric current. It is for this reason that the SI is often referred to as the **MKSA** system.

Table 1-1 is a summary of SI units and symbols. You will learn to use many of these as you progress in your study of electricity and electronics. Take time to study the table. Do not try to memorize the quantities, units, and symbols at this time. You will do that in a very natural way as you progress in your study of electricity and electronics. The various units and symbols will be introduced to you one at a time. For now, try Self-Check 1-1.

TABLE 1-1   SI Units

Quantity	Unit	Symbol
Length	meter	m
	kilometer	km
	centimeter	cm
	millimeter	mm
Mass	kilogram	kg
	gram	g
	milligram	mg
Time	second	s
	millisecond	ms
	microsecond	μs
Electrical		
Admittance (Y)	siemen	S
Capacitance (C)	farad	F
Charge (Q)	coulomb	C
Conductance (G)	siemen	S
Current (I)	ampere	A
Electromotive force (E or EMF)	volt	V
Energy (E)	joule	J
Frequency (f)	hertz	Hz
Impedance (Z)	ohm	Ω
Inductance (L)	henry	H
Power (P)	watt	W
Reactance (X)	ohm	Ω
Resistance (R)	ohm	Ω
Susceptance (B)	siemen	S
Wavelength (λ)	meter	m

SELF-CHECK 1-1

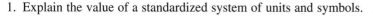

1. Explain the value of a standardized system of units and symbols.

2. What does MKSA stand for?

3. What does SI stand for?

1-2 Scientific and Engineering Notation

Scientific Notation

The study of electricity and electronics is very quantitative. That is, the elements we deal with have definite quantity, size, or amount. This means that nearly everything we talk about will be expressed as having a number value. We will be using very large numbers like 25,346,000,000 and very small numbers like 0.000000000005. Often, it will be nec-

TABLE 1-2   Powers of 10

	1,000,000,000.	=	$1 \cdot 10^9$
	100,000,000.	=	$1 \cdot 10^8$
	10,000,000.	=	$1 \cdot 10^7$
	1,000,000.	=	$1 \cdot 10^6$
	100,000.	=	$1 \cdot 10^5$
>1	10,000.	=	$1 \cdot 10^4$
	1,000.	=	$1 \cdot 10^3$
	100.	=	$1 \cdot 10^2$
Positive	10.	=	$1 \cdot 10^1$
Exponents	1.	=	$1 \cdot 10^0$

$1 \cdot 10^{-1}$	=	.1
$1 \cdot 10^{-2}$	=	.01
$1 \cdot 10^{-3}$	=	.001
$1 \cdot 10^{-4}$	=	.0001
$1 \cdot 10^{-5}$	=	.00001
$1 \cdot 10^{-6}$	=	.000001
$1 \cdot 10^{-7}$	=	.0000001
$1 \cdot 10^{-8}$	=	.00000001
$1 \cdot 10^{-9}$	=	.000000001
$1 \cdot 10^{-10}$	=	.0000000001
$1 \cdot 10^{-11}$	=	.00000000001
$1 \cdot 10^{-12}$	=	.000000000001

Negative Exponents

<1

essary to multiply or divide numbers such as these. Naturally, we can expect some pretty wild answers using very large or very small numbers. Fortunately, we can simplify the way we express numbers and obtain answers through the use of **scientific notation**.

Scientific notation lets us express very large or very small numbers in a simplified form by using powers of ten. For example, the number 25,346,000,000 can be expressed as $2.5346 \cdot 10^{10}$, and the number 0.000000000005 can be expressed as $5 \cdot 10^{-12}$. Do you see what I did to convert these numbers to scientific notation? Simply **express the number as a number between 1 and 10 times a power of ten**. Take time to study Table 1-2. It will help you understand powers of ten.

Let's take a closer look at what we actually do to convert a number to scientific notation. First, the number must be expressed as a number between 1 and 10. This means that the decimal point must be shifted left or right. For example:

25,346,000,000. ← SHIFT LEFT to get 2.5346

As you shift left, count the number of decimal places. The number of decimal places will be the exponent, or power of ten, that you will use. In this case, you shifted the decimal 10 places to the left. Therefore, the fully converted number is $2.5346 \cdot 10^{10}$. You may also express this number in the form 2.5346 E10 or 2.5346 E + 10, where the E stands for **E**xponent. We call this **E notation**. This method of expressing powers of ten is common to the way large numbers, or very small numbers, are entered into computers and scientific calculators. You will find both methods very useful. Let's look at our other example.

0.000000000005

In order to express this as a number between 1 and 10, we must shift the decimal point to the right.

SHIFT RIGHT → 0.000000000005 to get 5.0

Now, how many places to the right did we shift the decimal point? That's correct, twelve. Therefore, the exponent is 12. However, since this is a number less than 1, the exponent is negative. So, our fully converted number is $5.0 \cdot 10^{-12}$ or 5.0 E − 12.

RULE

When you shift the decimal to the *left*, the exponent will be *greater*. When you shift the decimal to the *right*, the exponent will be *less*.

Take time to study Example Set 1-1.

EXAMPLE SET   1-1

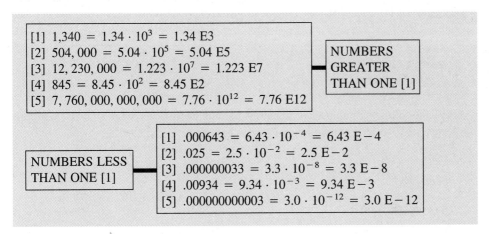

[1] $1{,}340 = 1.34 \cdot 10^3 = 1.34$ E3
[2] $504{,}000 = 5.04 \cdot 10^5 = 5.04$ E5
[3] $12{,}230{,}000 = 1.223 \cdot 10^7 = 1.223$ E7
[4] $845 = 8.45 \cdot 10^2 = 8.45$ E2
[5] $7{,}760{,}000{,}000{,}000 = 7.76 \cdot 10^{12} = 7.76$ E12

NUMBERS GREATER THAN ONE [1]

NUMBERS LESS THAN ONE [1]

[1] $.000643 = 6.43 \cdot 10^{-4} = 6.43$ E − 4
[2] $.025 = 2.5 \cdot 10^{-2} = 2.5$ E − 2
[3] $.000000033 = 3.3 \cdot 10^{-8} = 3.3$ E − 8
[4] $.00934 = 9.34 \cdot 10^{-3} = 9.34$ E − 3
[5] $.000000000003 = 3.0 \cdot 10^{-12} = 3.0$ E − 12

Engineering Notation

Engineering notation is much like scientific notation. The difference is that the number is expressed as a number times a power of ten that is a multiple of three (10^3 and E3, 10^6 and E6, 10^9 and E9, 10^{12} and E12, 10^{-3} and E − 3, 10^{-6} and E − 6, 10^{-9} and E − 9, 10^{-12} and E − 12). This means that the number cannot always be expressed as a number between 1 and 10 times a power of ten, as in strict scientific notation. To demonstrate what I have said, let's convert the numbers of Example Set 1-1 to engineering notation.

Numbers Greater Than One [1]

1. 1.34 E3 is already in engineering notation. The exponent is 3.
2. 5.04 E5 = 504 E3 = 0.504 E6. Either form is proper engineering notation. Notice that when the decimal is moved to the right two places, the exponent is decreased by

two (504.0 E3). When the decimal is moved to the left one place, the exponent is increased by one. When the number is made smaller, the exponent must be made larger and vice versa.

3. $1.223 \text{ E7} = 1.223 \cdot 10^7 = 12.23 \cdot 10^6$ or 12.23 E6. Both forms are engineering notation.

4. $8.45 \text{ E2} = 0.845 \text{ E3}$ or $0.845 \cdot 10^3$

5. 7.76 E12 or $7.76 \cdot 10^{12}$ is already in engineering notation.

Numbers Less Than One [1]

1. $6.43 \text{ E}-4 = 0.643 \text{ E}-3 = 643 \text{ E}-6$

2. $2.5 \text{ E}-2 = 25 \text{ E}-3$ or $25 \cdot 10^{-3}$

3. $3.3 \text{ E}-8 = 0.033 \text{ E}-6 = 33 \text{ E}-9$ or $33 \cdot 10^{-9}$

4. $9.34 \text{ E}-3$ is already in engineering notation.

5. $3.0 \text{ E}-12$ is already in engineering notation.

It is important for you to realize that scientific notation and engineering notation are different and that E notation can be used to express both scientific and engineering notation.

Practice using scientific notation and engineering notation by converting the numbers of Self-Check 1-2.

S E L F - C H E C K   1 - 2

Express the following numbers in scientific and engineering notation. Show each number in powers of ten and E notation.

1. 5,600

2. 23,000,000

3. 9,543,000,000,000

4. 0.000342

5. 0.000000000951

1-3 Mathematical Operations with Powers of 10

Now that you are familiar with scientific notation, we must examine its use in mathematical operations. It is one thing to express a number in scientific form and another to actually use that form in a formula to find an unknown quantity. We shall examine six fundamental operations: addition, subtraction, multiplication, division, square, and square root. Let's explore them one at a time, stating the rules and using examples.

Addition

> **RULE**
>
> Terms expressed in powers of ten must be expressed in the *same* power before they can be added.
>
> $$A \cdot 10^N + B \cdot 10^N = (A + B) \cdot 10^N$$

4.5 E3 + 2.756 E3 = 7.256 E3. No problem here, we simply add the numbers, 4.5 + 2.756 = 7.256, and adopt E3 as the power of 10.   EXAMPLE 1

1250 + 6.54 E2 = 12.50 E2 + 6.54 E2 = 19.04 E2 = 1.904 E3. Notice that we must   EXAMPLE 2
express 1250 as a number times 10 to the second power, E2. Then the numbers are added, and E2 is adopted as the power of 10. The answer may then be converted to scientific notation.

9.3 + 1.82 E−2 = 9.3 + 0.0182 = 9.3182. Here, we must convert 1.82 E−2 to a   EXAMPLE 3
regular number, then simply add it to 9.3. You always have the option to convert back to regular numbers before adding. However, this can be cumbersome with very large or very small numbers.

66.32 E7 + 1.59 E4 = 66320.0 E4 + 1.59 E4 = 66321.59 E4 or 66.32 E7 + 0.00159   EXAMPLE 4
E7 = 66.32159 E7. As you can see, you may express both numbers times 10^4 or 10^7. Notice that when we convert 66.32 E7 to 66320.0 E4, the number part becomes larger [66.32 → 66320.0] and the exponent becomes smaller [E7 → E4] as the decimal point is shifted to the right. This, of course, is because every decimal position is a step up or down in power. If we change the position of the decimal point in the number, we must change the exponent to compensate and maintain the number at its original value. When we change 1.59 E4 to 0.00159 E7, it is necessary to decrease the number part by shifting the decimal point three places to the left because the exponent is increased from E4 to E7.

4.256 E−3 + 6.8 E−5 = 4.256 E−3 + 0.068 E−3 = 4.324 E−3 or 425.6 E−5 +   EXAMPLE 5
6.8 E−5 = 432.4 E−5. We may express our terms as either E−3 or E−5. Again, we are adjusting the exponent and compensating by shifting the decimal point. When we convert 6.8 E−5 to 0.068 E−3 the exponent is *increased* from −5 to −3 and the number becomes smaller by shifting the decimal point two places to the left. If we choose to convert 4.256 E−3 to 425.6 E−5, the exponent is *decreased* from −3 to −5 and the number is made larger by shifting the decimal point two places to the right.

Subtraction

> **RULE**
>
> Terms expressed in powers of ten must be expressed in the same power before one is subtracted from the other.
>
> $$A \cdot 10^N - B \cdot 10^N = (A - B) \cdot 10^N$$

The procedure for subtracting numbers expressed with powers of ten is virtually the same as for adding them. Here are some examples.

EXAMPLE 1   4.59 E−3 − 1.25 E−3 = 3.34 E−3. No problem here. Both terms are already expressed in the same power of ten.

EXAMPLE 2 $12.43 \cdot 10^{-2} - 34.5 \cdot 10^{-3} = 12.43 \cdot 10^{-2} - 3.45 \cdot 10^{-2} = 8.98$ E−2. In this case we decide to convert the smaller term (3.45 E−3) to match the exponent of the larger term. Since we are increasing the exponent from −3 to −2, we must decrease the number by one decimal place (34.5 to 3.45). Of course, both terms could be converted to regular numbers and subtracted in normal fashion (0.1243 − 0.0345 = 0.0898).

Multiplication

> **RULE**
>
> When multiplying terms expressed in powers of ten, multiply the numerical part of the terms and add their exponents.
>
> $$(A \cdot 10^N)(B \cdot 10^M) = (A \cdot B) \cdot 10^{(N+M)}$$

EXAMPLE 1 $84.23 \cdot 10^3 \cdot 21.2 \cdot 10^2 = 1785.676 \cdot 10^5 = 1.785676 \cdot 10^8$

EXAMPLE 2   1.954 E6 · 4.75 E4 = 9.2815 E10

EXAMPLE 3   456.324 E9 · 123.5 E−3 = 56356.014 E6 = 5.6356014 E10

EXAMPLE 4   100 E−3 · 55 E−6 = 5500 E−9 = 5.5 E−6

Division

RULE

When dividing terms expressed in powers of ten, divide the numerical part of the terms and subtract their exponents.

$$\frac{A \cdot 10^N}{B \cdot 10^M} = \left(\frac{A}{B}\right) \cdot 10^{(N-M)}$$

$5.34 \cdot 10^3/2.3 \cdot 10^2 = 2.3217 \cdot 10^1 = 23.217$ E X A M P L E   1

$7.12 \text{ E}6/9.9 \text{ E}8 = 0.72 \text{ E}-2 = 7.2 \text{ E}-3$ E X A M P L E   2

$56.34 \text{ E}-6/21.4 \text{ E}-3 = 2.633 \text{ E}-3$ E X A M P L E   3

$94.1 \text{ E}-2/7.7 \text{ E}-4 = 12.22 \text{ E}2 = 1.222 \text{ E}3$ E X A M P L E   4

Square

RULE

To square a term expressed in powers of ten, square the number part of the term and double the exponent.

$$(A \cdot 10^N)^M = A^M \cdot 10^{MN}$$

If $M = 2$, then $(A \cdot 10^N)^2 = A^2 \cdot 10^{2N}$

$(12 \text{ E}3)^2 = 12 \text{ E}3 \cdot 12 \text{ E}3 = 144 \text{ E}6 = 1.44 \text{ E}8$ E X A M P L E   1

$(5.34 \cdot 10^4)^2 = (5.34 \cdot 10^4)(5.34 \cdot 10^4) = 28.516 \cdot 10^8 = 2.8516 \text{ E}9$ E X A M P L E   2

Square Root

RULE

To find the square root of a term expressed in powers of ten, take the square root of the number part of the term and divide the exponent by 2.

$$\sqrt{A \cdot 10^N} = \sqrt{A} \cdot 10^{N/2}$$

EXAMPLE   1 $\sqrt{49\ E6} = \sqrt{49} \cdot E6/2 = 7\ E3 = 7 \cdot 10^3$

EXAMPLE   2 $\sqrt{64\ E7} = ?$

Now what do we do? Exponents should be expressed as whole numbers. If we divide the 7 by 2, we get 3.5 for an exponent. No good! So, we may either convert 64 E7 to a regular number—640,000,000—or shift the decimal point to give us an even exponent such as 6 or 8.

$$\sqrt{64\ E7} = \sqrt{6.4\ E8} = \sqrt{6.4} \cdot E8/2 = 2.53\ E4 = 2.53 \cdot 10^4$$

Time for a Self-Check.

SELF-CHECK   1-3

1. $78.23 \cdot 10^3 + 3.1 \cdot 10^5 =$

2. $9.523\ E4 - 14.2\ E2 =$

3. $34.3\ E7 \cdot 66.0\ E3 =$

4. $17.4 \cdot 10^{-5}/2.6 \cdot 10^{-3} =$

5. $(8.34\ E4)^2 =$

6. $\sqrt{38\ E9} =$

1-4 The Prefix Game

As was shown in the previous two sections, extremely large and small numbers may be expressed and manipulated much more easily using powers of ten, or E notation. In this section, we will go one step further and express very large and very small numbers using prefixes. The use of prefixes is extremely common in electronics. Voltages, currents, frequencies, resistances, capacitances, inductances, and more are all labeled using standard prefixes. Table 1-3 is a collection of the most common prefixes and their meanings. Take time to study the table. You will want to commit these to memory as soon as you can.

TABLE   1-3   Engineering Prefixes

Prefix	Symbol	Meaning
tera-	T	10^{12}
giga-	G	10^9
mega-	M	10^6
kilo-	k	10^3
milli-	m	10^{-3}
micro-	μ	10^{-6}
nano-	n	10^{-9}
pico-	p or $\mu\mu$	10^{-12}

Using Prefixes

Let's take a few moments to see how these prefixes are used. Naturally, we will not be able to cover every application of prefixes here. However, we will look at several examples to give you the general idea. As you consider these examples, notice that the number is first converted to engineering notation and then the corresponding prefix is assigned.

100,000 meters can be expressed as 100 E3 m = 100 km.   E X A M P L E   1

1,000,000 ohms (unit of resistance) may be written as 1,000 E3 Ω = 1,000 kΩ or 1 E6   E X A M P L E   2
Ω = 1 MΩ.

0.0000022 amps (unit of current) is more easily expressed as 2.2 μA (2.2 E−6 A) or   E X A M P L E   3
2,200 nA (2,200 E−9 A).

0.000000000035 watts (unit of power) = $35 \cdot 10^{-12}$ W = 35 pW.   E X A M P L E   4

As you can see, prefixes are merely a substitute for certain powers of ten used in engineering notation.

Prefix Conversion

Now, let's take a close look at how we convert from one prefix to another. You might call this "The Prefix Game." Again, we shall examine several examples to establish a method for these conversions.

Suppose we want to express 10,000 nA as μA. Notice that we will be going from nA   E X A M P L E   1
(10^{-9} amps) to μA (10^{-6} amps). Is the new prefix larger or smaller? Yes, μ is larger than n. Therefore, the number part of our expression must be decreased to compensate and maintain our expression at its original value. Since the prefix increases by 10^3, the number must be reduced by three decimal places. So, 10,000 nA = 10 μA. As you can see, working with prefixes is just like working with powers of ten.

Convert 46 mΩ to kΩ. First identify the prefixes: m = milli = 10^{-3}; k = kilo = 10^3.   E X A M P L E   2
Are we increasing or decreasing the prefix when we go from milli to kilo? Increasing, of course. How much? Notice that kilo is 10^6 above milli. So, since the prefix is being increased, we must decrease the number by six decimal places. Therefore, 46 mΩ = 0.000046 kΩ.

Convert 25 MW (megawatts) to mW (milliwatts). In this example, the prefix is being   E X A M P L E   3
decreased from M (10^6) to m (10^{-3}). The difference between M and m is 10^9. Therefore, we must increase the number by nine decimal places. So, 25 MW = 25,000,000,000 mW.

Now try Self-Check 1-4.

SELF-CHECK 1-4

Convert:

1. 1,230,000 Ω to MΩ

2. 0.00000015 A to μA

3. 0.001 μA to nA

4. 13.5 GHz (gigahertz) to kHz (kilohertz)

5. 200 pA to μA

Summary

CONCEPTS

- Numbers greater than one, expressed in powers of ten, have positive exponents.
- Numbers less than one, expressed in powers of ten, have negative exponents.
- When a number is converted to scientific notation, the decimal point is shifted left or right. When the decimal point is shifted to the left, the exponent is made greater. When the decimal point is shifted to the right, the exponent is made less.
- When a quantity is expressed as a number times a certain power of ten and you want to express it as a number times a different power of ten (by increasing or decreasing the exponent), you must also decrease or increase the number by shifting the decimal point. If the exponent is increased, the number is decreased by shifting the decimal point to the left. If the exponent is decreased, the number is increased by shifting the decimal point to the right.
- Engineering notation uses powers of ten that are a multiple of 3 and accommodate standard prefixes. [10^3 = kilo (k), 10^6 = mega (M), 10^9 = giga (G), 10^{12} = tera (T), 10^{-3} = milli (m), 10^{-6} = micro (μ), 10^{-9} = nano (n), 10^{-12} = pico (p)]
- Two terms, or quantities, must be expressed in the same power of ten before they can be added or subtracted.
- When multiplying terms expressed in powers of ten, multiply the numerical part of the terms and add the exponents.
- When dividing terms expressed in powers of ten, divide the numerical part of the terms and subtract their exponents.
- To square a term expressed in powers of ten, square the number part of the term and double the exponent.
- To find the square root of a term expressed in powers of ten, take the square root of the number part of the term and divide the exponent by two (even exponents only).

SPECIAL TERMS

- SI
- MKSA

- Scientific notation, engineering notation
- E notation
- Exponent
- Tera-, giga-, mega-, kilo-
- Centi-, milli-, micro-, nano-, pico-

Need to Know Solution

Answers to Short Test:

1. Twenty milliseconds = 20 ms = 0.02 s
2. 3.3 kilovolts = 3.3 kV = 3,300 V
3. 4.5 μS = 4.5 microsiemens = 0.0045 mS
4. 1.9 E6 watts = 1.9 megawatts = 1,900 kW
5. $1.5 \cdot 10^6 \ \Omega$ = 1.5 MΩ = 1,500,000 Ω

Questions

1-1 SI Units

1. What are SI units?
2. What system is the SI based on?
3. What are four of the basic units of the SI?
4. What is the international symbol for resistance?
5. What is the international symbol for power?

1-2 Scientific and Engineering Notation

6. What is scientific notation?
7. What is E notation?
8. Numbers less than one expressed in scientific notation have an exponent that is (a) positive, or (b) negative.
9. A number is made larger by shifting its decimal point to the (a) left, or (b) right.
10. What is the difference between scientific notation and engineering notation?

1-3 Mathematical Operations with Powers of 10

11. What must you do before you add these two terms?

    1.23 E6   and   4.5 E3

12. In what way is subtracting powers of ten like adding powers of ten?
13. Explain how you multiply powers of ten, such as $10^3 \cdot 10^5$.

1-4 The Prefix Game

14. What are the prefixes for (a) 10^{-3}, (b) 10^6, and (c) E $-$ 12?

Problems

1-2 Scientific and Engineering Notation

Express the following in scientific and engineering notation:

1. 155,000
2. 2,330,000,000
3. 0.0000045
4. 0.00108
5. 946.5

1-3 Mathematical Operations with Powers of 10

6. $4{,}600 + 2.34 \text{ E2} =$
7. $21.8 \text{ E12} + 167.4 \text{ E7} =$
8. $0.001 \text{ E}-3 + 0.022 \text{ E}-6 =$
9. $3.24 \cdot 10^9 + 7.65 \cdot 10^7 =$
10. $0.000006 + 0.098 \cdot 10^{-4} =$
11. $7.64 \text{ E6} - 9.8 \text{ E4} =$
12. $1.2 \text{ E2} - 22 =$
13. $3 \cdot 10^0 - 0.2 \text{ E}-2 =$
14. $0.000092 - 0.001 \text{ E}-4 =$
15. $1.8 \cdot 10^7 - 66.7 \cdot 10^3 =$
16. $5.4 \text{ E3} \cdot 6.8 \text{ E4} =$
17. $0.001 \text{ E}-3 \cdot 0.02 \text{ E}-3 =$
18. $4{,}320 \cdot 2.3 \text{ E4} =$
19. $5.5 \cdot 10^4 \cdot 3.33 \cdot 10^2 =$
20. $65 \text{ E4} \cdot 0.03 \text{ E}-2 =$
21. $2.54 \text{ E3}/3.8 \text{ E2} =$
22. $98.65 \cdot 10^{-4}/0.628 \cdot 10^{-2} =$
23. $0.0002 \text{ E}-6/0.01 \text{ E}-12 =$
24. $56.2 \text{ E7}/33 \text{ E9} =$
25. $854 \text{ E6}/423 \text{ E6} =$
26. $(33.3 \cdot 10^4)^2 =$
27. $(0.024 \text{ E}-3)^2 =$
28. $\sqrt{45 \cdot 10^4} =$
29. $\sqrt{0.024 \text{ E3}} =$

1-4 The Prefix Game

Express the following in prefix form:

30. 100,000 W
31. 0.000002 F
32. 45,000,000,000 Hz
33. 0.000000000004 F
34. 1.5 E6 Ω

Express the following in engineering notation. Use E notation to express the engineering notation.

35. 150 MΩ
36. 24 pF
37. 12.5 GHz
38. 0.01 nA
39. 47 kΩ

Solve the following:

40. 1,000 pF = _____ nF
41. 0.01 μF = _____ pF
42. 4,000 kHz = _____ MHz
43. 0.023 mA = _____ μA
44. 2.2 kΩ = _____ mΩ

Answers to Self-Checks

Self-Check 1-1

1. Using standard mutually understood units eliminates having to convert from one system to another. Obviously this saves time and effort.
2. Meter, kilogram, second, ampere
3. Système International

Self-Check 1-2

1. $5.6 \cdot 10^3$ or 5.6 E3
2. $2.3 \cdot 10^7$ or 2.3 E7 = 23 E6
3. $9.543 \cdot 10^{12}$ or 9.543 E12
4. $3.42 \cdot 10^{-4}$ or 3.42 E−4 = 342 E−6 = 0.342 E−3
5. $9.51 \cdot 10^{-10}$ or 9.51 E−10 = 0.951 E−9 = 951 E−12

Self-Check 1-3

1. 3.8823 E5
2. 9.381 E4
3. 2.2638 E13
4. 6.6923 E−2
5. 6.95556 E9
6. 1.9494 E5

Self-Check 1-4

1. 1.23 MΩ
2. 0.15 μA
3. 1 nA
4. 13,500,000 kHz
5. 0.0002 μA

SUGGESTED PROJECTS

Electricity and Electronics Notebook

Start your own Electricity and Electronics Notebook. This should be neat and well organized. You might want to include tables and diagrams with brief explanations. You will want something that you can add to easily, such as a three-ring binder. Later in your studies you will want to develop your own design notes and include them in your notebook. If you are a student, you may want to include information from your class notes. This notebook can be a valuable source of reference for you in the years ahead.

(1) This is just one of many of Edison's dynamos (DC generators), a common source of electromotive force (voltage). More than 300 of Edison's 1,093 patents dealt with electric lighting and power distribution.

(2) This is a later and improved Edison nickel-iron-alkaline storage battery.

Chapter 2

The Nature
of Electricity

OBJECTIVES

After studying this chapter, you will be able to
- describe basic atomic structure.
- understand and explain the significance of
  valence electrons.
- explain the difference between conductors,
  semiconductors, and insulators in
  terms of valence.
- apply the Law of Attraction and Repulsion to
  atomic structure and electron flow.
- understand and use Coulomb's Law.

- describe electrical current in terms of
  coulombs.
- describe the difference between electron
  flow and conventional flow.
- understand electromotive force and its
  unit of measure, the volt.
- understand the concepts and units of measure
  for resistance and conductance and convert
  from one to the other.

INTRODUCTION

The nature of electricity is best explained and understood through a short course in atomic structure and theory. In this chapter, we will peer deeply into the atomic organization and structure of a variety of elements to see what electricity actually is. You will learn why some materials are good for conducting electricity and others are not. We will examine what actually causes electrical current and what works against the current to limit it. It is important that we study these concepts now, before we begin analyzing electrical circuits, because it is in this chapter that we answer the hows and whys of the basic nature of electricity.

19

NEED TO KNOW

What is electricity? Good question, huh? We all know what electricity does. But what is it, really? How is it created? How does it travel in a wire? Why doesn't electricity leave the wire and flow into the plastic insulation that surrounds the wire? How is electricity controlled? What are the different ways we measure electricity? Suddenly, our first question has grown into a whole list of questions! All of them are very important.

The answers to these questions are definitely need-to-know items, since they form the foundation for all the electronics you will learn. All electricity and electronic theory is based on the electricity theory presented in this chapter. So keep these questions in mind as you study. I am sure you will be able to answer them satisfactorily in your own words as you finish the chapter. A brief need-to-know discussion is also provided at the chapter's end.

2-1 Understanding the Atom

A fundamental understanding of the atom is basic to understanding the nature of electricity. In this section we go beyond our knowledge of household uses of electricity to the very nature of electricity. To do this, we must study the atom. Atoms contribute the very small particles needed to create electricity. In our study of the atom, we will discover what these particles are and the part they play in making electricity.

Atomic Particles

The simplest and most basic substances in the universe, of which all things are made, are called **elements**. The smallest part of an element that still retains the characteristics of that element is the **atom**. Let us begin our investigation of the atom with the simplest of all atoms, the hydrogen atom. As you can see from Figure 2-1, the hydrogen atom is made of a central area called the **nucleus** and a small particle, called an **electron**, that actually orbits the nucleus.

In the hydrogen (H) atom, the nucleus consists of a single particle called a **proton**, a rather heavy little particle that has a mass 1836 times greater than an electron. However, hydrogen is the simplest of all the elements known to mankind and is actually missing a nuclear particle common to all other elements. That particle is the **neutron**, which has a

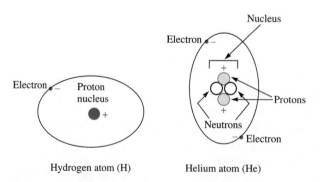

Hydrogen atom (H)   Helium atom (He)

Figure 2-1 The atomic structure of hydrogen (H) and helium (He) atoms.

mass slightly greater than the proton. So, in the atoms of the remainder of the 103 known natural elements, the nucleus is composed of protons and neutrons. The helium atom (also shown in Figure 2-1), which consists of 2 neutrons, 2 protons, and 2 electrons, illustrates this fact.

All elements are composed of atoms, which contain different numbers of protons, neutrons, and electrons. In fact, elements are listed by the number of protons in the nucleus of their atoms in a special table known as the **Periodic Table**. The number of protons in the nucleus of an atom is known as the **atomic number**. Hydrogen has 1 proton in its nucleus. So its atomic number is 1. Helium (He) is next in the Periodic Table because it has an atomic number of 2 (2 protons in the nucleus). Likewise, the atomic number for carbon is 6 because it has 6 protons. (See Appendix B, ''The Periodic Table.'')

Atomic Structure

Atomic structure is most often described as being **planetary** in nature, a model first proposed in 1913 by Niels Bohr. That is, the electrons orbit the nucleus as planets orbit a star in a solar system. In actual fact, the structure of an atom is three-dimensional and is composed of energy levels or regions in which certain electrons might be found. Textbook models are two-dimensional to illustrate the various energy levels and number of electrons contained in each level. Each electron in any energy level has a path that it follows around the nucleus. The electron's orbital velocity and mass create an outward force that prevents the electron from being pulled into the nucleus, just as the Earth is spared from the Sun. But what prevents the electron from leaving its orbit? We know that the Earth is held in its orbit by a mutual gravitational force between the Earth and the Sun. Is this how electrons are held in their orbits? Physicists tell us that electrons and protons are charged particles. Years ago the proton was assigned a **positive charge** and the electron a **negative charge**. Each charge is equal in magnitude yet opposite in sign or polarity. These opposite charges on the proton and electron create a force of attraction. Furthermore, the number of protons in the nucleus is balanced by an equal number of electrons in orbit. Therefore, the overall charge on the atom is neutral, just as the neutrons in the nucleus are neutral. However, the number of neutrons in the nucleus does not have to match the number of protons or electrons; i.e., an atom of gold contains 79 electrons and 79 protons, but 118 neutrons.

Shells

It has long been known that electrons orbit the nucleus at different distances and energy levels. Electrons orbiting at greater distances from the nucleus are said to exist at higher energy levels. These orbits, or energy levels, are called **shells**. As many as seven shells may surround the nucleus of an atom. As shown in Figure 2-2, they are labeled alphabetically starting with K as the first or innermost shell.

Figure 2-3 shows the carbon (C) and copper (Cu) atoms. The carbon atom has two shells, K and L. Notice that the K shell, closest to the nucleus, has only 2 electrons and the outer shell has 4 electrons. Why are there not 4 electrons in the inner shell and 2 in the outer shell, or perhaps some other combination? Scientists have discovered that there is a maximum number of electrons permitted in each shell. The first shell is full when 2 electrons are present, and the second shell is full when 8 electrons are present. As you can see, the outer shell of the carbon atom is not full, nor does it have to be.

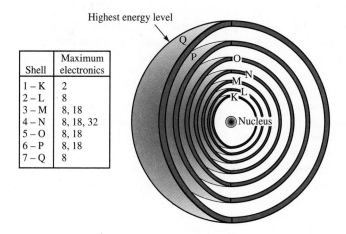

Shell	Maximum electronics
1 – K	2
2 – L	8
3 – M	8, 18
4 – N	8, 18, 32
5 – O	8, 18
6 – P	8, 18
7 – Q	8

Figure 2-2   Electrons orbit the nucleus of an atom at different energy levels, called shells. The shells represented here merely demonstrate the areas in which electrons can exist. (Diagram courtesy of William Reed, DeVry, Kansas City.)

As a general rule of thumb, the number of electrons permitted in any given shell is determined by the following formula:

$$\#\text{Electrons per shell} = 2N^2 \qquad (N \text{ is the shell number}) \qquad (2.1)$$

Understand that this is general rule that does yield to exceptions. Some elements have a maximum of 8, 18, or 32 electrons in higher energy level shells (M through Q). The outermost shell of an atom of any element never contains more than 8 electrons. If the outer shell of an atom is full, it will have 8 electrons (except for helium which has 2) and

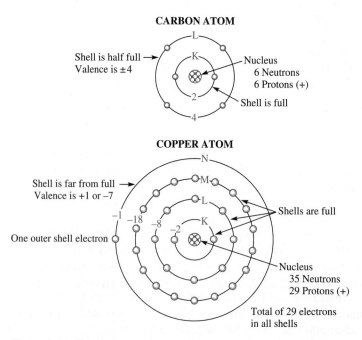

Figure 2-3   The atomic structure of carbon (C) and copper (Cu) atoms.

the element is said to be very **stable**. This means that it will not react chemically with other elements. Inert gases (such as helium, neon, argon, etc.) are full in the outer shell and are very stable elements.

Valence

The **valence** of an atom is a positive or negative number that indicates the number of electrons contained in the outer shell (also called the valence shell). If an element, such as carbon, has 4 electrons in the valence shell, we say the valence is $+4$. We might also say that carbon has a valence of -4. The -4 indicates that the valence shell is 4 electrons short of being full (remember 8 is full). As shown in Figure 2-3, copper has one electron in its outer shell. Therefore, it has a valence of $+1$. We may also say that copper has a valence of -7 because its valence shell is 7 electrons short of being full. The valence indicates how stable an element is or how readily an element might enter into a chemical reaction. As you can see, copper is more chemically reactive than carbon and carbon more so than an inert gas.

Free Electrons

Free Electrons in Conductors

In order for an element to conduct electricity, there must be an ample supply of free electrons. **Free electrons** are electrons that normally exist in the valence shell of an atom but are able to break free under the influence of an external source of energy. Metals such as silver (Ag), copper (Cu), gold (Au), aluminum (Al), and iron (Fe) are good **conductors** because they have many free electrons. Silver is the best conductor, followed by copper, gold, aluminum, and iron. Valence electrons in these elements are loosely held to the atom and are easily dislodged by heat, voltage potentials, and magnetic fields. Even at room temperature a conductor has trillions of free electrons moving from the valence shell of one atom to another in a random manner. At absolute zero ($-273°C$) this activity ceases and the electrons remain in their valence shells until motivated by a non-heat energy source, such as a voltage source. When a voltage is applied to a circuit, vast quantities of valence electrons are freed and migrate, or flow, in an organized manner from atom to atom, thus producing a current.

Generally speaking, conductors have less than three valence electrons and are willing to give them up in a chemical reaction or as free electrons for electrical current. In other materials, valence electrons are not as easily freed as they are in conductors. We may classify these materials as *semiconductors* and *insulators*.

Free Electrons in Semiconductors

Semiconductors usually have four valence electrons in each atom and are not as willing to give them up as free electrons. Semiconductors are poor conductors. Nevertheless, they serve valuable purposes in electronics. Carbon, the most common semiconductor, is widely used in electronic components called *resistors* (more on resistance and resistors later). Germanium (Ge) and silicon (Si) are semiconductor elements used in devices such as diodes, transistors, and integrated circuits.

Free Electrons in Insulators

Insulators are normally chemically stable compounds (combined elements) such as air, glass, ceramic, plastics, resins, wax, and rubber. These substances have no free electrons. All valence electrons are bonded tightly in small atomic structures called **molecules**, which are formed when valence electrons are shared between atoms of one element or between two or more different elements. Good insulators hold these valence electrons securely in each molecule. As a result, insulators conduct very little electrical current. We are all aware of at least some of the many applications for insulators in electronics—such as wire insulation, packaging transistors and integrated circuits and other components, and safety in isolating dangerous high-voltage sources such as those found in television sets.

We have really only just begun to explore the wonders of the atom in this short section. However, what we have covered is adequate for a basic understanding of the nature of electricity. This brief investigation has made you aware of the structure of the atom and the importance of the electrons that orbit its nucleus. Basic atomic theory will help you understand the sections that follow. However, before you continue, take time to do Self-Check 2-1.

SELF-CHECK 2-1

1. The smallest part of an element that still retains the characteristics of that element is a(n) _____. (choose one: electron, molecule, atom)

2. The center of an atom is called the _____. (choose one: valence shell, nucleus, centrus, proton)

3. The various energy levels of an atom are called _____. (choose one: degrees, heights, shells)

4. The outermost energy level of an atom is known as the _____. (choose one: valence shell, regis belt, centripetal ring)

5. What electrons become free electrons when an external energy source dislodges them from their orbits?

6. An element with four valence electrons, such as carbon, is classified as a(n) _____. (choose one: conductor, semiconductor, insulator)

2-2 Charged Bodies and the Coulomb Connection

In this section, we shall explore the nature of electricity a little further by examining the principles of electrical charge. We shall consider the nature, characteristics, and measure of electrical charge as we study charged bodies and the coulomb.

Charged Bodies

In our study of charged bodies, we must once again consider the atom. As you now know, an atom, in its normal neutral state, is composed of an equal number of protons in the nucleus and electrons in the shells. Each electron has a negative charge and each proton has an equally strong positive charge. The charges are said to be **polarized** (meaning: of two opposite states or positions). The proton is positive and the electron is negative. Each forms a **pole**—one a positive pole, the other a negative one. We say that the electron and proton are opposite in **polarity**. However, the overall charge on the atom is neutral because of the balance between the negative and positive charges.

If we add or remove electrons to or from the valence shell of an atom, the balance in charge is disrupted and the atom is no longer neutral. Removing electrons makes the atom positive and adding electrons makes the atom negative in charge. In such a case, we have created what is known as an **ion**. A **positive ion** is an atom that has lost electrons, and a **negative ion** is an atom that has gained electrons. The process involved in creating ions is known as **ionization**. In order for ionization to take place, some form of energy must act on the valence electrons of the atom. When an electron is removed, energy is added to the electron to enable it to break away from the atom. That source of energy might be heat caused by friction, light, radiation, or a voltage source. The atmosphere high above us is ionized by radiation from the sun. Thus, it is named the *ionosphere*. Air masses become ionized due to friction as one mass of lower temperature and higher density moves in against another. Extremely large quantities of charge are evidenced in dramatic displays of lightning.

Atoms are not the only things that may become charged. Most likely you have experienced the sudden and surprising jolt from a doorknob or someone's mischievous finger after a short stroll across a carpet (a charged human body as it were). Or how about the many times you took your clothes out of the dryer and your socks clung desperately to your shirts, skirts, slacks, etc. The charges acquired in these instances are called **static charges**. Static charges are the result of large quantities of electrons being removed or added to an object or objects. The word *static* means "stationary." In other words, the charge accumulates or builds up and has nowhere to go—until, of course, a path is provided from the charged object to a neutral or oppositely charged body (an unsuspecting victim).

In our example of the socks clinging to other items of clothing, a very important law of charged bodies is illustrated. This law is known as the **Law of Attraction and Repulsion** and is basically stated as follows: **Like charges repel and unlike charges attract**. The friction between the items of clothing and the inside of the dryer causes some pieces to become negatively charged and others relatively positively charged. The result is the familiar "static cling" because **unlike charges attract**. (You will recall that this is how the atom itself is held together—the attraction of the negative electrons to the positive protons.)

The repulsion of like charges can also be demonstrated in everyday life. If you have ever lived in a cold climate and have needed to wear a warm pullover sweater, you probably remember removing it over your head only to discover your hair standing on end afterward. Each hair on your head had become charged by the passing sweater. Since each hair had acquired the same charge, they were making every effort to get away from each other because **like charges repel**.

There is a more scientific way to demonstrate the Law of Attraction and Repulsion. This is done with a very simple device known as the **electroscope**. Figure 2-4a shows an electroscope with its parts labeled. The electroscope is used to detect and measure (at least in a relative way) small **electrostatic**, or static, charges.

In order to demonstrate the Law of Attraction and Repulsion, we must have a source of static electricity. The glass rod and ebonite rod (hard rubber rod) have traditionally been used for this purpose. The glass rod becomes positively charged when rubbed with a silk

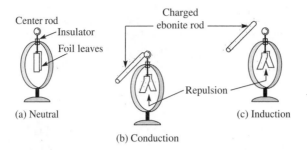

(a) Neutral

(b) Conduction

(c) Induction

Figure 2-4   The electroscope can be used to demonstrate charge characteristics.

cloth because electrons are removed from the glass and collected on the silk cloth. The ebonite rod becomes negatively charged when rubbed with animal fur because electrons are readily given up by the fur atoms and collected by the rod. Either rod may be used as a source of static electricity in demonstrating the electroscope.

Figure 2-4b shows a charged ebonite rod delivering a negative charge to the insulated metal center rod. This method of delivering a charge is known as **conduction**. Electrons are directly transferred (conducted) from the ebonite rod to the metal center rod by *physical contact*. As a result, the foil leaves become negatively charged and repel each other. When the ebonite rod is removed, the metal leaves remain charged and separated. The electrons must be drained off by providing a conductive path before the metal foils will collapse and return to a neutral or vertical position.

The principle of **induction** is demonstrated in Figure 2-4c. Notice that this time the ebonite rod is *not actually touching* the metal center rod on the electroscope. In this case a positive charge has been induced on the top of the electroscope center rod. Since like charges repel, a generous supply of free electrons has been coerced to migrate to the metal leaves. The top of the metal center rod is left positive while the bottom tips of the foil leaves are very negative. Again, the foil leaves separate because like charges repel. This time, when the ebonite rod is removed, the foil leaves will return to their vertical, neutral position. This is because the electrons migrate back through the center rod to the positive area and neutralize. The original normal distribution of free electrons is restored.

The Coulomb Connection

Charles Coulomb experimented extensively with charged bodies. In 1785 he developed a mathematical formula for calculating the actual force between two charged objects, taking

---

Charles Augustin de Coulomb (1736–1806)

Charles Coulomb was an early pioneer in electricity and mechanics. He was a noted French scientist and inventor who made significant contributions in the areas of friction, static electricity, and magnetism. Because of his contributions, the unit of electrical charge was named in his honor. One of his most significant contributions was a mathematical formula used to calculate the force between two charged bodies. This formula is known today as Coulomb's Law. (Photo courtesy of The Bettmann Archive.)

---

into account the amount of charge on each object and the distance between them. This formula is known, naturally, as Coulomb's Law.

Coulomb's Law

$$F = \frac{k \cdot Q_1 \cdot Q_2}{r^2} \qquad (2.2)$$

Where: F is force measured in newtons (N). One newton is the force required to accelerate a 1 kg mass at the rate of 1 m/s/s. Q stands for the quantity of electrical charge and is measured in coulombs (C). One coulomb is $6.25 \cdot 10^{18}$ electrons; k is a constant equal to $9 \cdot 10^9$ Nm^2/C^2; and r is the distance between the centers of the two charged bodies measured in meters (m).

The quantity of charge on any given body is expressed in coulombs. A **coulomb** is the quantity of charge created by a rather large number of electrons. There are $6.25 \cdot 10^{18}$ electrons in 1 coulomb of charge. This is an important unit that will be used to describe or define other electrical units as we progress in our study of electricity. If one coulomb contains 6.25 E18 electrons, one electron must possess a charge of approximately 1.6 E−19 C (coulombs).

1 C (coulomb) = 6.25 E18 e (electron charge)

Therefore: $|1\ e| = 1$ C/6.25 E18 = 1.6 E−19 C

The charge on one electron is expressed as $-e$, and the charge on a proton is expressed as $+e$. Thus, the charge on an electron is − 1.6 E−19 C and the charge on a proton is + 1.6 E−19 C. Remember, the electron and proton have the same quantity of charge but are opposite in polarity. When a body is negatively charged, it is because the body contains a surplus of electrons. Therefore, the charge on the body is expressed as a negative coulomb quantity, such as −0.56 C of charge. Also, when a body is positively charged due to a lack of electrons, we express the charge as a positive coulomb quantity, such as +0.67 C of charge. With this in mind, take time to study the following examples and design note of the application of Coulomb's Law.

EXAMPLE 2-1

Given: Two charged bodies whose centers are separated by 4 cm.
Body #1 has a charge of +50 μC and body #2 has a charge of −75 μC.
Find: The force of attraction or repulsion in newtons.

Coulomb's Law: $F = \dfrac{k \cdot Q_1 \cdot Q_2}{r^2}$

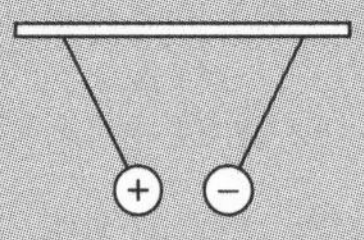

$$F = \frac{9\ \text{E}9 \cdot 50\ \text{E}-6 \cdot -75\ \text{E}-6}{(4\ \text{E}-2)^2}$$

Figure 2-5

$$= -33.75/1.6\ \text{E}-3 = -21,094\ \text{N (strong attraction)}$$

└──(indicates attraction)

EXAMPLE   2-2

Given: Two charged bodies whose centers are separated by 8 cm.
Body #1 has a charge of $+0.4$ μC and body #2 has a charge of $+0.12$ μC.
Find: The force of attraction or repulsion in newtons.

Coulomb's Law: $F = \dfrac{k \cdot Q_1 \cdot Q_2}{r^2}$

$$F = \frac{9\ E9 \cdot 0.4\ E-6 \cdot 0.12\ E-6}{(8\ E-2)^2}$$

$= 4.32\ E-4/6.4\ E-3 = +.0675$ N (repulsion)

$\quad\quad\quad\quad\quad\quad\quad\llcorner$(indicates repulsion)

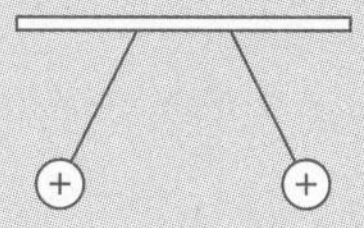

Figure 2-6

EXAMPLE   2-3

Given: One electron spaced a distance of $20\ E-10$ m from a nucleus containing three protons.
Find: The force of attraction or repulsion in newtons.

Coulomb's Law: $F = \dfrac{k \cdot Q_1 \cdot Q_2}{r^2}$

$$F = \frac{9\ E9 \cdot -1.6\ E-19 \cdot (3 \cdot 1.6\ E-19)}{(20\ E-10)^2}$$

$= -6.912\ E-28/4\ E-18 = -1.728\ E-10$ N (attraction)

$\quad\quad\quad\quad\quad\quad\llcorner$(indicates attraction)

Figure 2-7

In this section you learned that there is a definite force of attraction between charged bodies of opposite polarity and a force of repulsion between charged bodies with the same polarity. By using Coulomb's Law, you are now able to calculate the strength of that force in newtons. The Law of Attraction and Repulsion is a very important one as it is basic to the concepts of current and electromotive force, which we will discuss in the next two sections. Before going on, take time to do Self-Check 2-2.

SELF-CHECK   2-2

1. What is the meaning of the term *polarized*?

2. What is an ion?

3. What is the meaning of the term *static*, as in "static charges"?

4. Write the Law of Attraction and Repulsion.

5. What is a coulomb?

6. What is the charge on a proton expressed in coulombs?

7. Calculate the force between two bodies whose centers are spaced at a distance of 0.024 m if the charge on body #1 is -0.33 μC while the charge on body #2 is $+0.12$ μC? Is this a force of attraction or repulsion?

DESIGN NOTE 2-1:   Coulomb's Law

Coulomb's Law is used to determine the force of attraction or repulsion between two charged bodies.

Coulomb's Law: $F = \dfrac{k \cdot Q_1 \cdot Q_2}{r^2}$

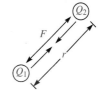

F is force in newtons (N)
k is a constant: 9 E9 Nm^2/C^2
Q is charge on body in Coulombs (C)
r is the distance from center of body #1 to center of body #2

Example: $Q_1 = 0.02$ μC, $Q_2 = 0.32$ μC, $r = 0.06$ m. Find: F

CALCULATOR SOLUTION	"BASIC" PROGRAM SOLUTION
START	10 CLS
	20 PRINT "COULOMB'S LAW"
k 9 EXP 9	30 PRINT ""
	40 PRINT "F = K * Q1 * Q2 / r^2"
×	50 PRINT ""
	60 PRINT "ENTER CHARGE IN COULOMBS ON BODY
Q_1	#1 USING E NOTATION."
. 0 2 EXP +/− 6	70 INPUT Q1
	80 PRINT "ENTER CHARGE IN COULOMBS ON BODY
×	#2 USING E NOTATION."
Q_2	90 INPUT Q2
. 3 2 EXP +/− 6	100 PRINT "ENTER DISTANCE BETWEEN CENTERS OF
	BODIES IN METERS."
÷	110 INPUT R
	120 F = 9.000001E+09 * Q1 * Q2 / R^2
r . 0 6	130 PRINT "THE FORCE IS ";F; " NEWTONS."
	140 INPUT "CALCULATE ANOTHER? (Y/N) ";A$
X²	150 IF A$ = "Y" THEN GOTO 30
	160 CLS:END
=	
F 0.016	
newtons	

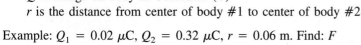

2-3 Electrical Current

Electron Flow

In Section 2-1, we briefly discussed the concept of electron flow in conductors in comparison to semiconductors and insulators. We said that, in order to have electron flow, there must be an ample supply of free electrons and the electrons must be influenced to flow in an organized manner from atom to atom. In this section, we will define electron flow mathematically in terms of coulombs and identify its unit of measurement.

> *Electron flow is the organized migration of electrons from atom to atom in a conductor, or semiconductor, under the influence of an electrical pressure created by a difference in charge.*

Usually that source of energy is a voltage (electrical pressure) created by chemical reaction in a battery or a magnetic field in a generator. The source of energy might also be light applied to photovoltaic cells, or heat applied to thermocouples (more on this in Section 2-4). Figure 2-8 shows a section of a copper wire with electrons flowing from left to right in an organized manner. The reason the electrons are flowing is because of the influence of the negative polarity on the left and the positive polarity on the right. The law of attraction and repulsion is clearly at work here. The electrons, being negatively charged, are very much attracted to the positive end of the wire and repelled from the negative end. This electron flow is also referred to as electrical **current**.

Electrical current is symbolized with the letter *I* for *I*ntensity of flow. Water flow intensity, as for water released from a dam or water flowing in a pipe, is measured in gallons, or liters, per minute. This is an important measurement for city engineers because they must keep track of the rate of use of the water supply and determine pipe sizes in water distribution systems, etc. Likewise, in electronics we are interested in the rate and quantity of electrical current. Instead of the minute, we use the second (s) as the unit of time (*t*). Naturally we do not use gallons or liters to measure the quantity of electrons but, rather, coulombs (C). Recall that one coulomb is $6.25 \cdot 10^{18}$ electrons. Therefore, electrical

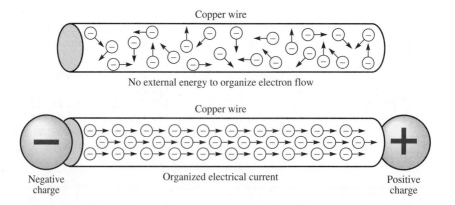

Figure 2-8   Electrons flow from a negative potential to a positive potential through a conductor. This is known as electron flow.

André-Marie Ampère (1775–1836)

André Ampère was a French mathematician and physicist. He is credited with the discovery of electromagnetism. In the 1820s he formulated basic electromagnetic principles. He invented the first electromagnet and the first ammeter used to measure electrical current. Ampère was a professor at the Ecole Polytechnique in Paris and published his *Mathematical Theory of Electromag-* *netic Phenomena* in 1827. The unit of measure for electrical current was named in his honor. (Photo courtesy of The Bettmann Archive.)

current (I) is mathematically defined as a certain quantity (Q) of electrons flowing past a certain point in an electrical circuit in a certain amount of time (t).

$$I = Q/t \qquad (2.3)$$

where Q is **Quantity** of charge in coulombs and t is **time** in seconds (s).

The electrical unit for current is the **ampere (A)**, named in honor of André-Marie Ampère. One ampere is a rate of electrical current defined as *one coulomb of electrons flowing past a given point every second.*

1 A = 1 C/1 s [6.25 E18 electrons per second]

Thus,

$$\# \text{ of amperes of current} = \frac{\# \text{ of coulombs}}{\# \text{ of seconds}}$$

$$\# \text{ A} = \frac{\# \text{ C}}{\# \text{ s}} \qquad (2.4)$$

Study the following examples.

EXAMPLE   2-4

Given: 23 coulombs of electrons pass a point
in a conductor every 14 seconds.
Find: The amount of current measured in amps (amperes).

$I = Q/t$ therefore: $I = 23 \text{ C}/14\text{s} = 1.643 \text{ A}$

The rate or intensity of flow is 1.643 coulombs per second (amperes or amps).

Calculator Solution:

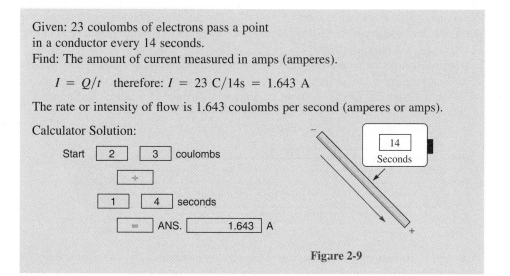

Figure 2-9

EXAMPLE   2-5

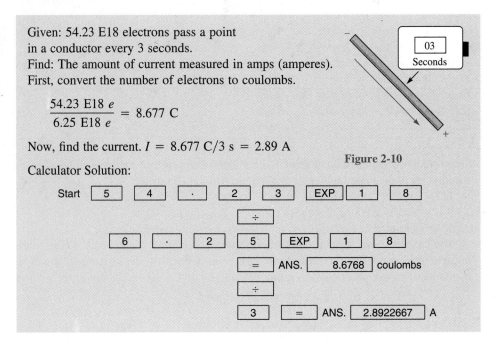

Given: 54.23 E18 electrons pass a point
in a conductor every 3 seconds.
Find: The amount of current measured in amps (amperes).
First, convert the number of electrons to coulombs.

$$\frac{54.23 \text{ E18 } e}{6.25 \text{ E18 } e} = 8.677 \text{ C}$$

Now, find the current. $I = 8.677 \text{ C}/3 \text{ s} = 2.89 \text{ A}$

Figure 2-10

Calculator Solution:

Start [5] [4] [.] [2] [3] [EXP] [1] [8]

[÷]

[6] [.] [2] [5] [EXP] [1] [8]

[=] ANS. [8.6768] coulombs

[÷]

[3] [=] ANS. [2.8922667] A

Conventional Flow

Scientists have not always known that electrons flow in a circuit from a negative potential to a positive potential. In the early days of electricity discovery, it was thought that positively charged particles flowed from the positive terminal of a battery, through the circuit, to the negative side of the battery. In those days, it was decided to establish a convention, or agreement, for the direction of flow of electrical current. So electrical current was described as an organized flow of positive particles from a positive electrode, through a circuit, to a negative electrode (see Figure 2-11). The positive electrode was thought of as the source and the negative electrode as the return. This description of electrical current is known as **conventional flow**.

Many engineering books today use conventional flow instead of electron flow because of its direct compatibility with semiconductor symbology. For example, conventional current is in the direction of the arrow in the schematic symbols for diodes and transistors.

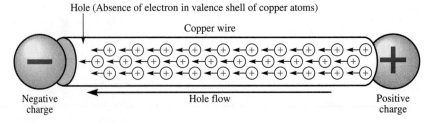

Figure 2-11   Holes appear to migrate toward the negative potential as electrons flow to the positive potential. This positive hole flow is known as conventional current.

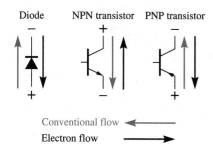

Diode   NPN transistor   PNP transistor

Conventional flow ←
Electron flow →

Figure 2-12   The direction of conventional current conforms to the direction of the arrow in the schematic symbols of electronic devices.

Electron flow goes against the arrow (in the direction opposite to that indicated by the arrow). This is illustrated in Figure 2-12.

How can conventional flow be reconciled with electron flow? As electrons flow through a conductor, or semiconductor, they migrate from atom to atom. An electron that leaves the valence shell of an atom migrates toward a positive charge and away from a negative charge. An electron that escapes the valence shell of an atom leaves behind a ''hole'' (the absence of an electron, creating a positive ion) and leaves the atom positively charged, due to the imbalance between electrons and protons. Another migrating electron may temporarily fall into that hole after having just left a previous atom. Thus, as the electrons migrate from a negative charge, through a conductor, to a positive charge, there is a relative migration of holes in the opposite direction (from positive to negative). This relative migration is sometimes referred to as **hole flow** (which, in fact, is conventional flow).

Regardless of whether the electrical current is considered to be electron flow or conventional flow, the quantity of charge is still measured in coulombs and the rate of flow is still measured in amperes. For every forward motion of each electron there is a corresponding and relative backward motion of a hole. The rate and quantity of electron or hole migration is the same.

This textbook is the **electron flow version** of *Experiencing Electricity and Electronics.* As such, all future references to electrical current imply electron flow, unless otherwise specified. All arrows used to indicate the direction of current in a circuit shall indicate electron flow.

Thus far, you have learned that electrical current flows in a conductor, or semiconductor, if a force or source of energy is provided. You have also seen that the electrons move away from a negative charge and toward a positive charge as they migrate through the conductor. Also, conventional flow is the relative migration of holes from a positive charge, through a conductor, to a negative charge. Conventional flow is in the direction opposite to electron flow. In the next part, we will expand on these concepts as you study the **electromotive force** that causes electrical current to flow. Make sure you understand this section by completing Self-Check 2-3.

S E L F - C H E C K   2 - 3

1. What is the symbol for electrical current?

2. What is the unit of measurement for electrical current?

3. How many coulombs is 3.125 E18 electrons?

4. How many amperes of current are flowing if 6 coulombs of electrons pass a given point in a circuit every 2 seconds?

5. How many amperes of current are flowing if 25 E18 electrons pass a given point every half second?

6. Briefly compare conventional flow to electron flow.

2-4 Voltage—The Electrical Pressure

Electromotive Force (EMF)

In this section, we shall examine the force, or energy, that causes charges in a wire to flow or conduct in an organized manner to produce current. Electrons are able to carry energy and do work, such as causing lights to glow and motors to turn. However, electrons do not generate this energy on their own. The energy is acquired by the electrons from a source. Some force must impart energy to the electrons to free them from the valence shells of atoms and influence them to flow in an organized manner. That force is called the **electromotive force** (**EMF** or **E**)—the force that motivates electrons, if you like. We shall examine this force and five of its common sources.

The electromotive force or EMF is a **pressure** that forces electrons to flow. The water-pipe analogy is once again useful here. See Figure 2-13. As you know, water will not flow in the pipes that contain it unless there is pressure forcing it to do so. Often this pressure is created by the water source being higher in altitude than its end users. A reservoir in the mountain, or a high water tower and a tank are examples of this. The force, or pressure, is created by a difference in height.

Electromotive force is also created by a difference. In this case, it is a difference in charge, or potential, on each end of a conductor or electrical circuit. In Figure 2-14, the car battery has a positive terminal and a negative terminal. A chemical reaction inside the battery has caused a large number of electrons to leave the positive terminal and collect on the negative terminal. The positive terminal is left with a large accumulation of positive ions. This great difference in charge is a source of EMF. You should recall your study of

Count Alessandro Volta (1745–1827)

Alessandro Volta was born into a noble family in Como, Italy. In the first thirty years of his life, he firmly established himself in the scientific community and became noted for his research in electricity. Volta is credited for the invention of the battery and his experiments with electrolysis. He also invented the capacitor. The unit of potential difference, the volt, was named in his honor. (Photo courtesy of The Bettmann Archive.)

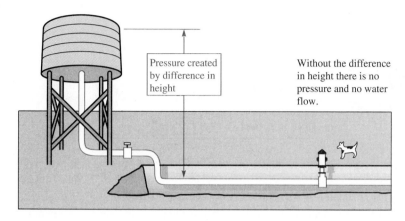

Figure 2-13   The amount of water pressure is a function of the height of the water level above the pipe lines. Water pressure is analogous to voltage–electrical pressure.

current earlier. When a negative charge, or potential, is applied to one end of a conductor and a positive charge to the other, we have the difference we need for current. This difference of potential is the electromotive force, whose unit of measure is the **volt (V)**, named in honor of Italian scientist Alessandro Volta. I am sure you are at least familiar with the term *volt*, since it is commonly used in reference to household electricity, automobile batteries, flashlight cells, etc.

Electrically, the volt is defined as the potential difference that exists when one **joule (J)** of energy is needed to move one coulomb of charge between two points. The joule is an SI unit for work and energy, defined as the energy required to move one coulomb of charge between two points having a difference of potential of one volt. Physicists describe the joule as the work done by one newton of force acting over a distance of one meter. Mathematically, one volt is one joule of energy available per coulomb of charge.

1 volt = 1 joule per 1 coulomb   or   1 V = 1 J/1 C

More voltage means more electrical pressure or a greater electromotive force. A twelve-volt (12 V) battery has a greater difference of potential than a six-volt (6 V) battery, and is therefore a stronger EMF.

Automobile battery

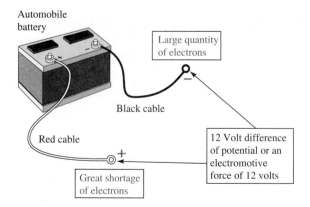

Large quantity of electrons

Black cable

Red cable

12 Volt difference of potential or an electromotive force of 12 volts

Great shortage of electrons

Figure 2-14   Electromotive force (EMF) is a difference of electrical potential such as exists between the terminals of a battery.

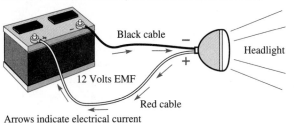

Electromotive force created by chemical reaction in the battery

Black cable

Headlight

12 Volts EMF

Red cable

Arrows indicate electrical current

Figure 2-15   Current flows when a path is provided. Voltage is the pressure that causes current to flow.

Keep in mind that voltage and current are two different things. It is important that you understand that voltage is an electromotive force or potential difference, and is not to be confused with the actual current that it causes. Voltage is pressure and electrical current is the result! A voltage is the electromotive force or difference of potential that exists between two terminals of an electrical energy source such as a battery or laboratory power supply. As shown in Figure 2-15, current results when a device is connected between those two terminals. Voltage is always present and ready to work. Current is only present when a device is connected to that voltage, providing a path for the current.

Sources of EMF

Electrochemical Source of EMF

An **electrochemical** source of EMF is one in which a chemical reaction is taking place, with free electrons being produced as a byproduct of molecular bonding. Batteries are an electrochemical source of EMF. In the case of a common lead–acid battery, sulfuric acid reacts with spongy lead (negative plates) to form lead sulfate. The chemical combination of the lead and the sulfate releases unneeded electrons as free electrons, which accumulate on the negative plate (the sulfate comes from the breakdown of the sulfuric acid). Positive hydrogen ions ($+H$) are a byproduct of this reaction. These positive ions are attracted to the positive plate, where they combine with negative oxygen ions ($-O$) that are being released as the sulfuric acid reacts with the lead peroxide on the positive plate. The $+H$ ions combine with the $-O$ ions to form water (H_2O). See Figure 2-16.

The buildup of electrons on the negative plate creates a difference of potential, or EMF, with positive ions on the positive plate. The source of energy used to create the difference of potential is the energy contained in every atom and molecule. The chemical reaction is actually a release of this energy during the reorganization of the molecular structure as new chemical compounds are formed (such as the lead sulfate and water).

Electromagnetic Source of EMF

An **electromagnetic** source of EMF is one in which the mechanical motion of a conductor, moving perpendicular to and through a magnetic field, causes electrons to flow in the conductor. The flow is organized, creating a difference of potential, or EMF. This source of EMF is often called **electromechanical**, since the apparatus involved is definitely mechanical. It is, without any doubt, the single most important and most common source of EMF. Enormous generators are used to supply our communities with electricity. We also

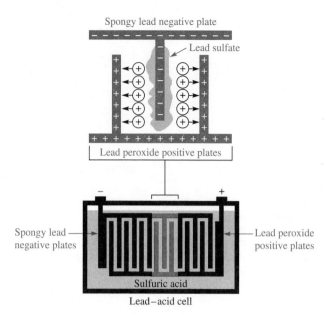

Figure 2-16   A car battery is an electrochemical source of EMF.

make use of the electromagnetic principle in automobiles where alternating current (AC) generators, called *alternators*, are used to maintain the charge in our batteries. (The AC is converted to DC to charge the battery.) Figure 2-17 illustrates the operation of a simple generator, showing the three basic conditions that must be met in order to produce an EMF electromagnetically.

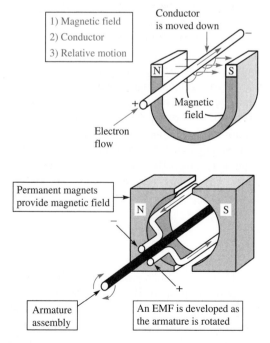

Figure 2-17   Generators are an electromagnetic source of EMF.

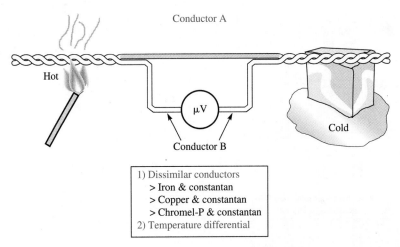

Figure 2-18   Physical contact of dissimilar metals creates a low-power source of EMF called a thermocouple.

Thermoelectric Source of EMF

A **thermoelectric** source of EMF is one in which electrons are forced to flow due to the presence of heat. In 1821 a scientist by the name of Thomas Johann Seebeck discovered that when two conductors made of different materials are joined on each end to form a loop, and the temperature of one end is maintained hotter than the other, an EMF is produced. This arrangement (shown in Figure 2-18) is known as a **thermocouple**. Direct current (DC) is produced as one end of the thermocouple is heated while the other end is cooled. It sounds too good, or too simple, to be true. The problem is that the voltage produced is in the range of microvolts at an extremely low power level. One of the most practical applications for the thermocouple is as a temperature probe for temperature measurement devices.

   As with any source of EMF, thermocouples may be arranged in series and parallel to obtain higher voltage and current capability. An arrangement such as this is called a **thermopile**. A thermopile combined with a source of energy is referred to as a **thermoelectric generator**. It is interesting to note that in 1969 the American Apollo 12 astronauts left a small nuclear-powered thermoelectric generator on the moon to power scientific experiments. In recent times, some manufacturers have begun selling small thermopiles for the home, to be heated by fireplaces. Power output is less than 100 watts at a very high initial cost. The thermopile may be heated by almost any means: burning wood, coal, natural gas, petroleum products, or nuclear reaction. Again, the greatest limiting factor is very poor efficiency.

Photoelectric Source of EMF

A **photoelectric** source of EMF is one in which light forces electrons to migrate in a material, causing a difference of potential. This source of EMF is in wide use today, with **solar cells** converting solar energy to electrical power for remote radio, telephone, and television relay stations, satellites, irrigation systems, individual homes, and entire communities. Many people in remote areas enjoy the comforts of electricity produced entirely from the sun's energy.

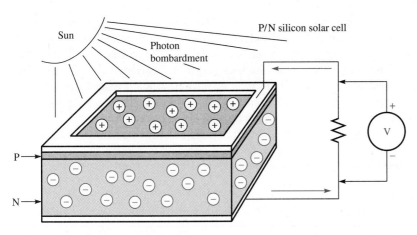

Figure 2-19 Semiconductor materials are used to create the PV cell, a photovoltaic source of EMF.

The key component of the photoelectric system is the individual solar cell, known as a **photovoltaic (PV) cell**. Figure 2-19 illustrates the construction and operation of the photovoltaic cell. As you can see, a very thin layer of *p*-type semiconductor material is layered on a much thicker *n*-type material. (*p*-type material has many holes and *n*-type material has many extra electrons.) Photons carrying light energy penetrate the thin *p*-layer, colliding with and liberating electrons. The liberated electrons leave holes behind in the *p* material. (The hole is the absence of a valence electron.) The liberated electrons then move into the *n* semiconductor material and are collected by the plated contact for transfer to the load device. The *n* material already has an abundance of free electrons and is very willing to give them up. The continual pressure from the light energy creates a steady supply of free electrons and an organized current.

Piezoelectric Source of EMF

A **piezoelectric** source of EMF is one in which a physical stress on a crystalline material produces a difference of potential. In 1880, two French scientists, Pierre and Jacques Curie, discovered that certain crystalline minerals would produce a voltage, or difference of potential, on their surfaces when they were flexed or compressed along an axis (see Figure 2-20). A few of these crystalline minerals are quartz, rochelle salt, and tourmaline. This

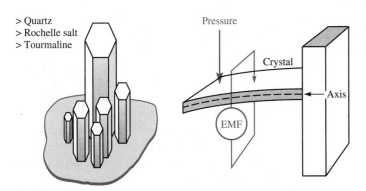

Figure 2-20 Crystals under stress produce voltage. This is known as the piezoelectric effect.

phenomenon is known as the **piezoelectric effect**. The use of piezoelectric-generated EMF is very common throughout the world today. No, it is not commonly used to generate electricity for our homes, even though research is being done in that area. Instead, the piezoelectric effect is being used in a multitude of gadgets and devices that enhance modern living. Some of these devices are the transducer on the tone arm of a record player, crystal microphones, ultrasonic transducers for sonar and medicine, and quartz crystals that serve as frequency-determining elements for radio transmitters and receivers.

Later, we will express voltage and current in more practical mathematical terms. For now, it is important that you understand the basic concepts of voltage and current. Test yourself with Self-Check 2-4 before continuing.

SELF-CHECK 2-4

1. What is the force called that causes electrons to flow, or migrate, from one end of a conductor to the other in an organized manner?

2. What does EMF stand for?

3. What is the unit of measure for EMF?

4. Describe this unit of EMF in terms of joules and coulombs.

5. Explain the difference between voltage and current.

6. List the five common sources of EMF.

2-5 Resistance and Conductance

Thus far, you have learned that voltage is the pressure that forces electrical current to flow. But what actually limits the amount of electrical current that flows in a circuit? Surely the amount of current in a circuit is not unlimited. How can the amount of current that flows in a circuit be controlled? In this section, these questions are answered as you explore the concepts of resistance and conductance.

Resistance (*R*)

Resistance Explained

Resistance is the *opposition* to current. It is a characteristic of all conductors and electrical or electronic devices. Resistance is the characteristic that *limits* or *restricts* and therefore *controls* the amount of current. Resistance is desirable in some electronic components, while in conductors it is generally undesirable. Without resistance in a circuit, flashlight and automobile batteries would be drained dead almost instantly. It would be like the bottom falling out of the water tank in the water pressure analogy (Figure 2-13).

In the water tank analogy, it is the size of the pipe and control valve setting that limits the amount of water flow. The amount of pressure is also a factor in determining how much water will flow. The resistance of the pipes and valve are always working against the water pressure to control the flow, and so it is with electrical circuits: The resistance in a circuit is working against the electromotive force to control current. If the EMF is increased, more current will flow. If the resistance is increased, less current will flow.

Thus, we can adjust the amount of current in a circuit by adjusting the voltage, the resistance, or both.

Conductors have much less resistance than semiconductors. Why? The reason is once again explained using atomic theory. Conductors are materials that usually have only one or two valence electrons. These electrons are easily liberated by an energy source, be it a potential difference, a magnetic field, or simply room-temperature heat. The nuclei of the atoms do not offer much opposition to the removal of the electrons. Thus, electron flow is fairly easy to accomplish. There is very little resistance.

In semiconductors, the electrons are held more tightly in the valence shells of the atoms. More energy is now required to free the electrons for current. When an electron is freed, a hole is left in the valence shell. This hole causes the atom to be positively charged. In other words, the atom is now short one negative charge and the proton/electron balance is disrupted. This positive atom is now very interested in capturing an electron to restore the balance. Almost as quickly as the atom lost one valence electron, it will capture another. The captured electron releases its energy in the form of heat. Then the electromotive force must apply more energy to remove the electron once again and continue current. As you can see, semiconductor atoms offer much opposition to current. There is much resistance. Therefore, at a given energy level, or EMF, far fewer electrons will actually break free and flow.

The Ohm (Ω)

The **ohm** is the basic unit of measurement for resistance. It is symbolized with the Greek letter omega (Ω). As you can see, from our previous discussion, resistance, voltage, and current are all conceptually related. Thus, the ohm is related to the volt and the ampere. One ohm (1 Ω) is defined as the amount of resistance (R) present in a circuit that will allow one ampere (1 A) of current to flow when one volt (1 V) of EMF is applied. Mathematically, $1\ \Omega = 1\ \text{V}/1\ \text{A}$ (1 V per 1 A).

The resistance of conductors is measured in milliohms ($1\ \text{m}\Omega = 1 \cdot 10^{-3}\ \Omega$) and microhms ($1\ \mu\Omega = 1 \cdot 10^{-6}\ \Omega$), while semiconductors have resistances in ohms, kilohms ($1\ \text{k}\Omega = 1 \cdot 10^{3}\ \Omega$), megohms ($1\ \text{M}\Omega = 1 \cdot 10^{6}\ \Omega$), and gigohms ($1\ \text{G}\Omega = 1 \cdot 10^{9}\ \Omega$). Carbon is a semiconductor material used widely for very common electronic components appropriately named resistors. We will look at a variety of resistors later.

Conductance (G) $G = 1/R$

Conductance Defined

Conductance is a characteristic of a circuit that is the inverse or reciprocal of resistance. In other words, conductance is $1/R$. The symbol for conductance is the capital letter G. Therefore, $G = 1/R$. If resistance is low in a circuit, conductance will be high, and if resistance is high, conductance is very low.

The Siemen (S)

For many years the unit of measure for conductance was the **mho** (ohm spelled in reverse). This unusual name did serve a purpose, since it was a reminder that conductance is mathematically the reciprocal of resistance. However, today the **siemen (S)**, named in honor of Werner von Siemens (1816–1892), a famous German inventor, is used as the standard SI unit of conductance.

To convert ohms (Ω) to siemens (S), you simply take the reciprocal of ohms. Likewise, to convert siemens to ohms, you take the reciprocal of siemens. For example, to convert 100 Ω to siemens, you take the reciprocal of 100 Ω. $1/(100\ \Omega) = 0.01$ S or 10 mS. Thus, 100 Ω of resistance is equal to 10 mS of conductance. To convert back to ohms, take the reciprocal of 10 mS. $1/(10\ \text{mS}) = 100\ \Omega$. The unit of conductance is at times more convenient to use than resistance. The analysis of some circuits is easier to accomplish when calculations are done in siemens. However, most of your analysis and design work will be done in ohms.

Take a moment to do Self-Check 2-5.

SELF-CHECK 2-5

1. Define resistance in terms of voltage and current.

2. What is the basic unit of resistance?

3. Materials with four valence electrons per atom are _____ . (Choose one: good conductors, poor conductors.)

4. Resistance offers what to current flow?

5. The reciprocal of resistance is termed what?

Summary

FORMULAS

(2.1) #Electrons per shell $= 2N^2$
   (N is the shell number)

(2.2) Coulomb's Law: $F = \dfrac{k \cdot Q_1 \cdot Q_2}{r^2}$

   F is force in newtons
   k is a constant, 9 E9 Nm²/C²
   Q is quantity of charge in coulombs
   r is the distance between centers of charged bodies

(2.3) $I = Q/t$
   I is the intensity of current measured in amperes
   Q is quantity of electrons in coulombs
   t is time in seconds

(2.4) #A $=$ #C/#s
   Number of amperes is equal to the number of coulombs divided by the number of seconds.

CONCEPTS

- Electrons are negatively charged and protons are positively charged.
- Electrons orbit the nucleus at various energy levels, called shells.

- Valence electrons become free electrons when acted upon by an external source of energy, such as heat, a magnetic field, or a chemical reaction.
- Conductors have a large number of free electrons.
- Semiconductors have few free electrons.
- Insulators have no free electrons.
- The absence of a valence electron leaves a hole.
- The Law of Attraction and Repulsion states that like charges repel and opposite charges attract.
- Electron flow is the organized migration of electrons from atom to atom in a conductor, or semiconductor, under the influence of some source of energy.
- One ampere (A or amp) of current (I) is one coulomb ($1\ C = 6.25 \cdot 10^{18}\ e$) of electrons flowing past a given point in a circuit every second.
- Conventional flow (hole flow) is in the direction opposite to electron flow. Conventional flow is from positive to negative.
- One volt (V) is one joule (J) of energy per coulomb (C) of electrons.
- Voltage doesn't flow; electrical current does.
- Resistance is opposition to electrical current.
- Resistance (R) is measured in ohms (Ω).
- Conductance (G) is a mathematical concept which is the reciprocal of resistance. $G = 1/R$.
- Conductance is measured in siemens (S).

SPECIAL TERMS

- Elements
- Atom, nucleus
- Neutron, proton, electron
- Positive charge, negative charge
- Shells, valence, free electrons
- Conductors, semiconductors, insulators
- Pole, polarity, polarized
- Negative and positive ions
- Electroscope, conduction, induction
- Coulomb (C), current (I), ampere (A)
- Electron flow, conventional flow
- Electromotive force (EMF)
- Volt (V)
- Electrochemical, electromagnetic, thermoelectric
- Photovoltaic, piezoelectric
- Resistance (R), ohm (Ω), conductance (G), siemen (S)

Need to Know Solution

Electricity can be thought of as a force called *electromotive force*, which is measured in volts. It may also be thought of in terms of electrical current measured in amperes. The electromotive force is the pressure that causes the current to flow. Electrical current is the organized migration of electrons from atom to atom in a conductor as a result of a potential difference (electromotive force measured in volts). Electrical current is also thought of conventionally as hole flow in the opposite direction of electron flow. An EMF can be created by a chemical reaction (battery), a magnetic field (generator), or even light (solar cells).

Conductors conduct electricity because their valence electrons are easily liberated as free electrons for current. Insulators do not conduct electricity because the valence electrons are held very securely to their atoms and are not easily liberated. This is why electron flow is confined to a wire conductor and not its insulation. Electrical current is controlled by the amount of resistance a conductor or semiconductor has and the amount of electrical pressure (voltage) that is applied.

Questions

2-1 Understanding the Atom

1. Protons orbit the nucleus of an atom. (T/F)
2. The proton is equal in mass to the electron. (T/F)
3. Atoms have an equal number of electrons and protons under normal conditions. (T/F)
4. What does it mean when we say that atomic structure is planetary in nature?
5. What do scientists call the energy levels surrounding the nucleus of an atom?
6. The outer shell of an atom is considered to be full with how many electrons?
7. What is the maximum number of electrons permitted in the third shell of a five-shell atom?
8. What is the meaning of *valence*?
9. What are free electrons?
10. Describe an insulator in terms of atomic structure and free electrons.

2-2 Charged Bodies and the Coulomb Connection

11. What is a positive ion?
12. What does the word *static* mean?
13. Give an example of household static electricity.
14. Write out the Law of Attraction and Repulsion.
15. Describe the process of charging by *conduction*.
16. Describe the process of charging by *induction*.
17. One coulomb is equal to how many electrons?
18. If the distance between charged bodies decreases, the force between them will (a) increase, or (b) decrease.

2-3 Electrical Current

19. Describe electron flow in your own words.
20. What is the symbol for electrical current and its unit of measure?
21. Describe current in terms of coulombs and time.
22. Define the unit for current.
23. Do electrons flow from negative to positive or positive to negative?
24. What is a hole?
25. Is conventional flow from negative to positive or positive to negative?

2-4 Voltage—The Electrical Pressure

26. What is electromotive force?
27. What is the unit of measure for EMF?
28. Define the unit for EMF in terms of joules and coulombs.
29. Which has the larger EMF: a 1.5 volt flashlight cell or a 9 volt battery for a radio?
30. Explain the difference between voltage and current.

2-5 Resistance and Conductance

31. What is the purpose for resistance in a circuit?
32. Why do semiconductors have more resistance than conductors? Explain in terms of atomic structure.
33. What is the unit of measure for resistance and its symbol?
34. Define the unit of resistance in terms of current and voltage.
35. Define conductance in terms of resistance.
36. What is the unit of measure for conductance?

Problems

2-2 Charged Bodies and the Coulomb Connection

1. If body #1 has $+0.034$ μC of charge and body #2 has $+0.80$ μC of charge and the distance between their centers is 5 cm, what is the force between them in newtons? Is it a force of attraction or of repulsion?
2. If body #1 has a charge of $+1.5$ μC and body #2 has a charge of -0.6 μC and the distance between centers is 0.7 m, what is the force between them in newtons? Is it a force of attraction or of repulsion?

2-3 Electrical Current

3. If 10 C of electrons pass a certain point in a conductor every 3 seconds, how much current is flowing?
4. If 2 C of electrons pass a certain point in a conductor every 0.25 seconds, how much current is flowing?
5. If 32 E18 electrons pass a certain point in a conductor every 1.5 seconds, how much current is flowing?
6. If 2.65 E18 electrons pass a certain point in a conductor every 0.5 seconds, how much current is flowing?

2-5 Resistance and Conductance

Calculate conductance for the following:

7. 150 Ω =
8. 4.7 kΩ =
9. 0.5 Ω =
10. 1 MΩ =

Answers to Self-Checks

Self-Check 2-1

1. Atom
2. Nucleus
3. Shells
4. Valence shell
5. Valence
6. Semiconductor

Self-Check 2-2

1. Of two opposite states or positions. In electronics it means opposite in electrical charge, positive and negative.
2. An atom that has gained or lost an electron
3. Stationary, as in charges that do not move or conduct
4. Like charges repel, and opposite charges attract.
5. A quantity of electrical charge equal to 6.25 E18 electrons
6. $+1.6\,E-19\,C$
7. $-0.6188\,N$ of attraction

Self-Check 2-3

1. I
2. Amperes (A or amps)
3. ½ C
4. 3 A
5. 8 A
6. Conventional flow is relative hole flow from positive to negative, while electron flow is the migration of electrons from negative to positive.

Self-Check 2-4

1. Electromotive force (E or EMF)
2. Electromotive force
3. Volts (V)
4. One volt is the electromotive force available when 1 joule of energy is available per 1 coulomb of charge.
5. Voltage is only the force or pressure, while current is the actual electron flow. You can have a voltage source with nothing connected to it and have no current, but you cannot have current without voltage.
6. Electrochemical, electromagnetic, thermoelectric, photoelectric, piezoelectric

Self-Check 2-5

1. Resistance is the opposition to current that is being moved by voltage. 1 Ω of resistance will allow a current of 1 A to flow under a pressure of 1 V.
2. The ohm (Ω)
3. Poor conductors and classified as semiconductors
4. Opposition
5. Conductance (G), measured in siemens (S)

1. Add the formulas and important concepts from this chapter to your Electricity and Electronics Notebook. You might also want to transfer the Design Note in this chapter to your personal notebook.

2. Build and experiment with an electroscope. You may want to use a glass canning jar and some aluminum foil leaves suspended by thin wires from the lid.

3. You may want to start your own electronics lab or workshop at this point in time. Start by collecting basic tools such as a soldering pencil, solder, wire strippers/cutters, small heat sinks, small screwdrivers, and pliers. Digital multimeters in the $40.00 to $80.00 price range are usually more than adequate. Have fun!

EDISON PRIMARY BATTERY M-1002
Multiple Element 1000 Amp. Hrs.
Price $10.25 1921

Renewal Element $5.75
Caustic Soda 85¢
Battery Oil 15¢

(1)

(1) One of Edison's glass-jar primary batteries. The small jar to the right contained the battery oil used to cover the surface of the caustic soda solution and prevent its evaporation. Notice the 1,000 ampere-hour capacity of the battery. (2) Three of Edison's glass-jar batteries with ceramic lids. These were used to power electric fans and early radio equipment.

(2)

Chapter 3

The DC Circuit, Components and Symbols

OBJECTIVES

After studying this chapter, you will be able to

- list the four essentials of a practical circuit.
- understand basic circuit documentation and use schematic symbols.
- define basic battery terminology and characteristics.
- describe the proper arrangement of cells to obtain a desired voltage and current capacity.
- use the multimeter to measure battery voltages and currents.
- understand resistivity of conductors and calculate the total resistance of a length of wire.

- describe the use of a variety of switches as control devices.
- explain the use of a fuse or circuit breaker as a circuit protection device.
- describe a wide variety of resistors and identify schematic symbols.
- use the color code to determine resistor values.

INTRODUCTION

In this chapter, you will be introduced to the DC circuit and the basic essentials necessary for making it a practical one. We shall examine: **batteries**, which provide an important source of DC; **conductors**, which provide an interconnecting path for current; **switches**, which provide a means of circuit control; and circuit protection devices such as **fuses** and **circuit breakers**. The most common and most often used piece of test equipment, the **multimeter**, will also be demonstrated in measuring voltage and current. You will find this chapter filled with useful information that will strengthen and enhance your experience in electricity and electronics.

NEED TO KNOW

Here is an assignment for you.
Design and document a DC-motor control circuit from the following specifications:

- It is known that the motor will have a starting current of 7 A for ¼ second, then drop to a running current of 1 A.
- The power supply will be alkaline D cells, with no more than ¼ A being drawn from any one cell during normal motor operation.
- The total supply voltage shall be 4.5 V.
- A switch must be provided to turn the motor on and off.
- A switch must be provided for changing motor direction (forward and reverse).
- The batteries must be protected against sudden discharge due to a shorted circuit.

Can you do it? If not, there is much that you need to know. Study this chapter carefully and see if you are able to accomplish this assignment later.

3-1 Circuit Essentials and Documentation

In your study and work in electronics, you will design, construct, test, and troubleshoot a great variety of practical circuits. But, what is a practical circuit? What do you use to make a practical electronic circuit? In other words, what elements are essential or necessary to construct a practical circuit? In this section, we shall examine the basic essentials needed for a practical DC circuit. We will also discuss circuit documentation in the form of circuit diagrams and component symbols.

The DC Circuit

The **DC circuit** is a circuit like Figure 3-1. It is powered by a DC source of voltage. **DC** stands for **D**irect **C**urrent. Direct current is current that always *flows in one direction*. Electrons originate at the negative terminal of the DC supply and migrate through the circuit to return to the positive terminal. The DC source is the source of EMF needed to force the electrons to flow and do work for us. In a DC circuit, the current never changes direction because the polarity of the DC supply does not change. The electron flow is always negative to positive.

The opposite of direct current is **Alternating Current (AC)**. This is the type of current we use in our homes and factories. Alternating current changes direction very frequently.

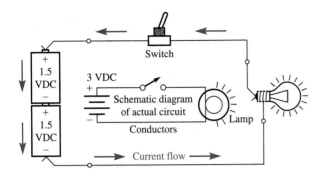

Figure 3-1   A practical circuit has a source of EMF, a load device, and a continuous path for current. The switch serves as a control device.

In the United States household current changes direction 60 times every second, in Europe and Asia it alternates 50 times per second. In this chapter, we will concentrate on DC circuits and save the topic of AC for later.

Essentials

So, what are the essential requirements for a practical DC circuit? Figure 3-1 illustrates the basic circuit essentials. A practical DC circuit will meet the following requirements:

1. a **DC source** of EMF (voltage)
2. **conductors** to carry the current (wires)
3. a **practical load** device (a device that makes use of the energy contained in the electrons—such as a lamp, motor, transistor radio, or a tape player)
4. **continuity** (the switch must be on and there must be a continuous unbroken path for current to flow from the negative side of the DC supply through the conductors, the load device, and back to the positive side)

If any of the essentials listed above is missing, the circuit will not be practical or operational. You must have a source of EMF, a conductive path for current flow, a practical load, and continuity.

Documentation

Circuit documentation is also essential. An accurate diagram of a circuit can be invaluable when it comes to repairing that circuit. The design engineer will spend dozens, even hundreds, of hours working on circuit documentation before any actual circuit assembly begins. He will make modifications or changes to his circuit and his documentation as he tests and modifies the new device or circuit. The final documentation is marketed along with the circuit, or piece of equipment, and is of tremendous value to the technician.

The documentation will contain a written description of the circuit along with the circuit diagram. The circuit diagram is referred to as a **schematic**. The schematic is a map that clearly shows circuit values and the interconnection of all circuit components. Figure 3-1 shows a schematic of our practical circuit. Notice that each part of the actual circuit is symbolized for simplification. Every device or component that is used in an electrical

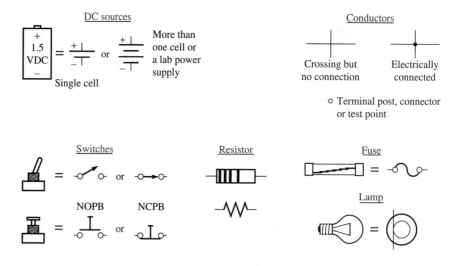

Figure 3-2   A variety of common schematic symbols.

or electronic circuit will have a corresponding schematic symbol. Figure 3-2 illustrates some of the more common symbols. This simple list of schematic symbols will be greatly expanded as you progress in your study of electricity and electronics.

Review the important points of this section. Then test your knowledge with Self-Check 3-1.

SELF-CHECK 3-1

1. What does DC stand for?

2. How is DC different from AC?

3. What is continuity and why is it essential to a practical circuit?

4. How would you indicate that two wires are electrically connected together in a schematic diagram?

5. Draw the schematic symbol for a fuse.

3-2 Batteries

Starting with this section, we will take a closer look at some of the more common components that make a practical circuit. We begin with batteries because these devices are an essential source of EMF needed in many practical circuits.

There are many questions we might ask about battery sizes, types, and applications. The very fact that there is a large variety of batteries indicates that some batteries are better suited than others for certain applications. So, which battery is for which application? Which batteries are rechargeable? What is the shelf life for various battery types? How do you arrange batteries for more voltage or more current? How are batteries rated? These

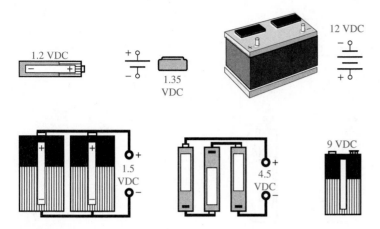

Figure 3-3

are only a few of the questions we might ask. However, it is not our purpose to cover all there is to know about batteries. We will discuss batteries in a very practical way, covering battery terminology, battery types, discharging and charging, battery applications and selection, and series and parallel combinations of batteries.

Terminology and Characteristics

Battery or Cell?

What is a battery? Often the hefty little object that we call a battery really isn't a battery. A flashlight "battery," for example, is not really a battery in the strictest sense of the term. Technically, the common 1.5 V flashlight battery should be called a flashlight *cell*. A cell is what a flashlight battery really is. It is a single voltaic or galvanic cell (chemical unit) that produces a certain amount of voltage due to a chemical reaction. So, if a battery isn't a cell, what is a battery? Technically, a **battery** is a combination of two or more cells. For example, a 9 V transistor radio battery is a battery because it is a combination of six 1.5 volt cells in one little package. An automobile battery is a battery because it is a combination of six 2 volt cells. So, you ask, is it wrong to call flashlight cell a battery? Technically, yes. Practically speaking, no! Even the manufacturers call individual cells batteries and label them as such. The only reason I am making such a point of the terminology here is to help you understand that actual batteries are made of individual voltaic units, called *cells*, that produce a definite amount of voltage. The actual voltage of the battery naturally depends on the individual cell voltage, since the battery voltage will be some multiple of the cell voltage, depending on the number of cells and the chemical composition of each cell.

The Voltaic Cell

As stated earlier, the individual cell is referred to as a voltaic cell or galvanic cell, named after Luigi Galvani (1737–1798). The **voltaic** or **galvanic cell** consists of two different conductive materials that interact chemically with a special mixture called an electrolyte. See Figure 3-4. The **electrolyte** is either a solid, a semisolid, or a liquid (usually an aqueous

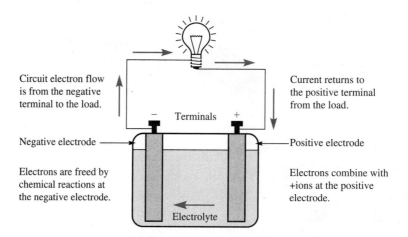

Figure 3-4   The voltaic cell produces an EMF.

salt solution) that permits ionic conduction, or current, within the cell between the positive and negative **electrodes** (conductive materials that create the charges in the chemical reaction).

Dry Cell and Wet Cell

A **dry cell** is a voltaic cell that contains an electrolyte that is not a liquid. The electrolyte in a dry cell is more like a paste. The dry cell exists in much greater variety than the wet cell. However, the wet cell is not uncommon, since it is the type of cell used in the construction of automobile batteries. A 12 V automobile battery is made up of six of these wet cells in a series arrangement (each cell contributing a little over 2 volts). The electrolyte of a **wet cell** is definitely a liquid and in many cases will leak out if the battery is not maintained in an upright position.

Primary and Secondary Cells

A **primary cell** is a disposable single-use cell. Common flashlight batteries of the carbon–zinc or alkaline systems are considered primary cells and cannot or should not be recharged. They are primarily single-use batteries. **Secondary cells** can be recharged and are advertised as such. Nickel–cadmium (Ni–Cad) and automotive or motorcycle batteries are examples of secondary batteries.

Charging

In rechargeable batteries, the chemical reaction is safely reversed. The electrodes, which have deteriorated in chemical reaction during discharge, are actually redeposited or rebuilt through this reverse chemical reaction during charge. The energy required for this reverse process is obtained from a battery charger that forces current in the reverse direction. The positive terminal on the charger is connected to the positive terminal of the battery and the negative to the negative. To insure that current flows in the reverse direction, the charger voltage must be higher than the battery voltage. As shown in Figure 3-5, automobile battery chargers are in the neighborhood of 13.5 to 15 volts.

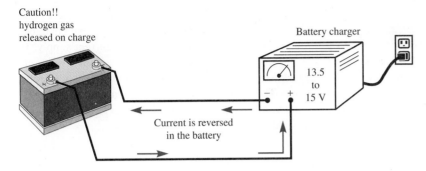

Caution!!
hydrogen gas
released on charge

Battery charger

13.5
to
15 V

Current is reversed
in the battery

Figure 3-5   Current must be reversed in the battery to recharge it.

The charging rate should always be kept within the manufacturer's recommendations. The **charging rate** is simply the amount of reverse current applied to the battery over a period of time. For example, a nickel–cadmium cell might be charged at the rate of 100 mA for a 15-hour period. A higher rate would be 200 mA for 8 hours, and a lower rate would be 50 mA for 25 hours. Each time the cell, or battery is discharged, a charge/discharge **cycle** is completed. The number of cycles a secondary cell can survive is called the **cycle life**. Ni–Cads, for example, have a cycle life of 1,000 or more.

Capacity

The **capacity** is an indication or measure of the amount of service you can expect to get *from* a battery. This capacity is expressed in **ampere hours (AH)**. If a battery is rated at 100 AH, as an automobile battery might be, it simply means that the battery can potentially deliver 100 A of current to a load for 1 hour, or it could deliver 50 A for 2 hours, or 10 A for 10 hours, or 1 A for 100 hours, etc. As you can see, the **AH rating** is simply the amount of current delivered to a load multiplied by the time before the battery voltage has dropped to an unusable level.

$$\text{Life (Hours)} = \frac{\text{Capacity}}{\text{Current}} = \frac{\text{AH rating}}{\text{\# of amperes}} \tag{3.1}$$

Shelf Life

The **shelf life** for a particular battery type is simply the length of time the battery can sit idle before it is considered unusable. As a battery ages, it very slowly discharges itself internally. Also, the electrolyte may dry out very slowly. After a period of time, the battery is considered useless because its capacity and voltage have been severely reduced.

Open and Closed Circuit Voltage

The **open circuit voltage** of a battery is the voltage or difference of potential available at the battery's terminals with no circuit connected. In other words, it is the voltage present with no current being drawn from the battery. The **closed circuit voltage** is the actual voltage the battery is able to supply to a circuit that is "on" and demanding current. The closed circuit voltage is usually less than the open circuit voltage. The amount of drop in voltage from open to closed circuit conditions depends upon the ambient (environmental) temperature, the amount of current demanded, and the age of the battery. The chemical reaction in the battery is what produces the difference of potential and current supply. If the operating temperature is too low or too high and/or the battery is old and/or the current demand is very high, the chemical reaction will not occur rapidly enough to maintain the difference of potential at its rated value. This may be referred to as an **internal resistance** (r_i). We will develop this concept of internal resistance to a greater degree in a later chapter where practical voltage sources are covered.

Battery Systems

Carbon–Zinc (Primary Cells)

The **carbon–zinc** family of batteries is the least expensive system of batteries for general purpose applications. Carbon–zinc is really a general name for two specific systems of batteries: LeClanché and zinc chloride. The **LeClanché battery** is a carbon–zinc cell that has an electrolyte composed of ammonium chloride and zinc chloride in water mixed with manganese dioxide and carbon forming a paste. The **zinc chloride battery** is similar to the LeClanché battery except its electrolyte is composed mainly of zinc chloride in water.

The zinc chloride version of the carbon–zinc battery has many advantages over the LeClanché. The output capacity (AH) decreases less at increased drain rates. Low-temperature performance is better. Also, the closed circuit voltage, or voltage under load, is higher. In other words, the drop in voltage from a no-load to full-load condition is less with zinc chloride batteries than with LeClanché batteries.

Alkaline–Manganese Dioxide (Primary Cells)

The **alkaline** battery is higher in unit cost than a carbon–zinc battery. However, it is more economical in terms of cost per hour of use on high current drains. The alkaline cell can last 200% to 700% longer than its carbon–zinc counterpart, depending on the rate of current drain (the heavier the current drain, the greater the percentage). It has a better low-temperature performance and longer shelf life than carbon–zinc batteries. The construction of the alkaline cell is shown in Figure 3-6.

Nickel–Cadmium (Secondary Cells)

The nickel–cadmium (Ni–Cad) cell is an extremely durable cell that can be recharged hundreds and hundreds of times. Its terminal voltage remains flat or constant until the cell is almost completely drained. A flashlight powered by Ni–Cads will be very bright and suddenly go dim. You experience a long period of full service then, very suddenly, the battery is drained and must be recharged. Figure 3-7 demonstrates the difference in discharge characteristics between carbon–zinc, alkaline, and Ni–Cad cells.

One of the few disadvantages of the Ni–Cad cell is that it has a lower terminal voltage (an average of about 1.2 to 1.25 volts closed circuit) than common primary cells. If Ni–Cads are used in place of carbon–zinc or alkaline cells, the total series voltage will be less. For example, 8 alkaline cells in series produce 12 VDC while 8 Ni–Cad cells in

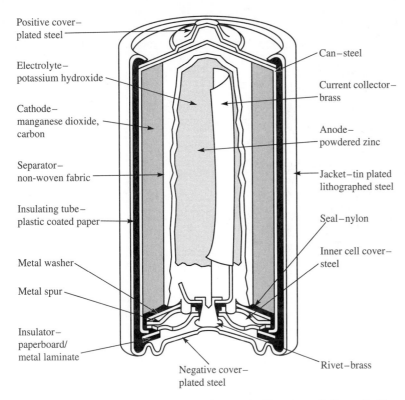

Positive cover–
plated steel

Electrolyte–
potassium hydroxide

Cathode–
manganese dioxide,
carbon

Separator–
non-woven fabric

Insulating tube–
plastic coated paper

Metal washer

Metal spur

Insulator–
paperboard/
metal laminate

Can–steel

Current collector–
brass

Anode–
powdered zinc

Jacket–tin plated
lithographed steel

Seal–nylon

Inner cell cover–
steel

Negative cover–
plated steel

Rivet–brass

Figure 3-6   This is a cutaway view of the popular Energizer alkaline cell. (Courtesy of Union Carbide—Eveready.)

series produce only about 10 VDC. An additional one or two Ni–Cad cells are needed to make up the difference.

Figure 3-8 illustrates the assembly of a cylindrical Ni–Cad cell. Notice the "jelly roll" construction. This method of construction offers the largest surface area for chemical reaction between the cathode (positive electrode mixture) and the anode (negative electrode mixture). A resealable safety vent is incorporated in the design to allow gas to safely escape during excessive overcharging.

Chart 3-1 summarizes the battery systems thus far discussed.

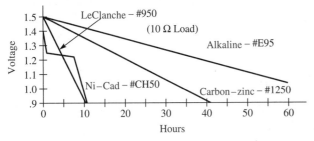

Figure 3-7   These graphs demonstrate the relative discharge rate of various electrochemical cells through a 10 Ω load resistance.

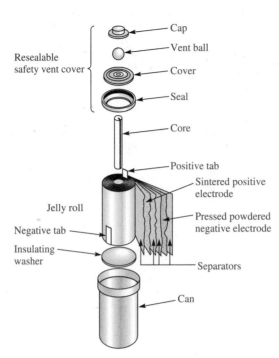

Cap

Vent ball

Cover

Seal

Resealable
safety vent cover

Core

Positive tab

Sintered positive
electrode

Jelly roll

Pressed powdered
negative electrode

Negative tab

Insulating
washer

Separators

Can

Figure 3-8   The Ni–Cad cell is very
different from the alkaline cell.
(Courtesy of Union Carbide—
Eveready.)

Lithium Cell (Primary or Secondary Cell Depending on Construction)

Lithium cells are in a category all their own. They have a very high power density and an extremely long shelf life—over ten years at room temperature. They exhibit exceptionally long service life when properly matched to the equipment. Initially, these cells are a little more costly. However, they are very, very reliable over long service periods, making them an excellent choice for watches, cameras, computer memory power backup, medical equipment, smoke detectors, satellites, and electric vehicles.

Figure 3-9 shows a rechargeable lithium cell made by Molienergy Limited of Canada. Most lithium cells are not rechargeable. However, the Molicel™ is an exception, making it a direct competitor with the Ni–Cad cells. It is a very high power density lithium–molybdenum disulfide system (Li/MoS_2). The individual cell voltage is slightly over 2 V (fully charged) gradually sloping to about 1.3 V on discharge before recharging is necessary. The power to weight ratio for this technology is about three times that of even the powerful lead–acid batteries. Most lithium cells are the primary type and are manufactured as three basic systems: lithium–iron sulfide (Li/FeS_2), lithium–manganese dioxide (Li/MnO_2), and lithium–thionyl chloride ($Li/SOCL_2$). They exhibit a nominal output voltage per cell of 1.5, 3.0, and 3.6 volts respectively.

Figure 3-9   One of the first rechargable lithium
cells. (Courtesy of Molienergy Limited of
Mississauga, Ontario.)

Usual Name	LeClanché	Zinc Chloride	Alkaline–Manganese Dioxide (MnO_2)	Mercuric Oxide	Silver Oxide	Nickel–Cadmium
Electrochemical System	Zinc–Manganese Dioxide	Zinc–Manganese Dioxide	Zinc–Alkaline Manganese dioxide	Zinc–Mercuric Oxide	Zinc–Silver Oxide	Nickel–Cadmium
Voltage Per Cell	1.5	1.5	1.5	1.35	1.5 (monovalent)	1.2
Negative Electrode	Zinc	Zinc	Zinc	Zinc	Zinc	Cadmium
Positive Electrode	Manganese Dioxide	Manganese Dioxide	Manganese Dioxide	Mercuric Oxide	Monovalent Silver Oxide	Nickelic Hydroxide
Electrolyte	Aqueous solution of ammonium chloride and zinc chloride	Aqueous solution of zinc chloride (may contain some ammonium chloride)	Aqueous solution of potassium hydroxide	Aqueous solution of potassium hydroxide or sodium hydroxide	Aqueous solution of potassium hydroxide or sodium hydroxide	Aqueous solution of potassium hydroxide
Type	Primary	Primary	Primary	Primary	Primary	Rechargeable
Rechargeability	Not Recommended	Not Recommended	Not Recommended	Not Recommended	Not Recommended	Yes
Number of Cycles						300 to 1,000
Input if Rechargeable						Minimum of 140% of energy withdrawn
Overall Equations of Reaction	$2MnO_2 + 2NH_4Cl + Zn \rightarrow ZnCl_2 \cdot 2NH_3 + Mn_2O_3 \cdot H_2O$	$8MnO_2 + 4Zn + ZnCl_2 + 9H_2O \rightarrow 8MnOOH + ZnCl_2 \cdot 4ZnO \cdot 5H_2O$	$2Zn + 2KOH + 3MnO_2 \rightarrow 2ZnO + 2KOH + Mn_3O_4$	$Zn + HgO + KOH \rightarrow ZnO + Hg + KOH$	$Zn + Ag_2O + KOH \rightarrow ZnO + 2Ag + KOH$	$Cd + 2NiOOH + KOH + 2H_2O \rightleftarrows Cd(OH)_2 + 2Ni(OH)_2 + KOH$
Typical Commercial Service Capacities	60 mAH to 30 AH	Several hundred mAH to 18 AH	30 mAH to 45 AH	45 mAH to 14 AH	15 mAH to 210 mAH	1.50 mAH to 4 AH
Energy Density (Commercial) Watt-Hour/Lb	20	40	20–45	50	50	12–16
Energy Density (Commercial) Watt-Hour/Cubic Inch	1–2	2–2.5	2.5–4.5	8	8	1.2–1.5
Practical Current Drain Rates Pulse	Yes	Yes	Yes	Yes	Yes	Yes
Practical Current Drain Rates High (More Than 50mA)	100 mA/square inch of zinc area (''D'' cell)	150 mA/square inch of zinc area (''D'' cell)	200 mA/square inch of separator area (''D'' cell)	No	No	8 10 A ≈ 1 amp/sq. in. of electrode
Practical Current Drain Rates Low (Less Than 50mA)	Yes	Yes	Yes	Yes	Yes	Yes
Discharge Curve (Shape)	Sloping	Sloping	Sloping	Flat	Flat	Flat
Temperature Range (Storage)	−40°C to 50°C (−40°F to 120°F)	−40°C to 50°C (−40°F to 120°F)	−40°C to 50°C (−40°F to 120°F)	−40°C to 60°C (−40°F to 140°F)	−40°C to 60°C (−40°F to 140°F)	−40°C to 60°C (−40°F to 140°F)
Temperature Range (Operating)	−5°C to 55°C (20°F to 130°F)	−20°C to 55°C (0°F to 130°F)	−30°C to 55°C (−20°F to 130°F)	−10°C to 55°C (14°F to 130°F)	−10°C to 55°C (14°F to 130°F)	Discharge −20°C to 45°C (−4°F to 113°F) Charge CH Types 0°C to 45°C (32°F to 113°F) Charge CF type 15°C to 45°C (60°F to 113°F)
Effect of Temperature on Service Capacity	Poor at low temperatures	Good at low temperatures relative to carbon–zinc	Good at low temperatures	Good at high temperatures; low-temperature performance depends upon construction	Low-temperature performance depends upon construction	Very good at low temperatures
Impedance	Moderate	Low	Very low	Low	Low	Very low
Leakage	Low to medium under abusive conditions	Low	Rare	Some salting	Some salting	No
Gassing	Medium	Higher than carbon zinc	Low	Very low	Very low	Low
Reliability (Lack of Duds, 95% Confidence Level)	99% at 2 years	99% at 2 years	99% at 2 years	99% at 2 years	99% at 2 years	99% at 2 years
Shock Resistance	Fair to good	Good	Good	Good	Good	Good
Cost Initial	Low	Low to Medium	Medium Plus	High	High	High
Cost Operating	Low	Low to Medium	Medium to high at high power requirements	High	High	Low

Chart 3-1   Battery comparison. (Courtesy of Union Carbide—Eveready.)

Lead–Acid Battery

The **lead–acid battery** is very common and well known to almost everyone. It is the hard-working giant that resides under the hood of our automobiles and faithfully starts our engines at the twist of a key. On demand, it will deliver 300 to 400 A of current to the starter motor, which we hope will start the engine very soon. The lead–acid battery is often thought of as a ''transportation battery'' simply because it is the most commonly used battery system in vehicles of every kind. The popularity of these batteries is due to the simple fact that no other system can deliver similar amounts of energy at a competitive price. The capacity of these batteries can be from 20 to 40 AH for motorcycles, to 100 AH for automobiles, and to 1,000 AH or more for electric vehicles. Each cell yields about 2 to 2.2 volts due to chemical reaction. (See the treatment of electrochemical sources of EMF in Section 2-4.)

Battery Selection

While battery selection for a particular application should not be done in a random manner, it need not be a difficult task either. A battery should be matched to the application just as an engine is matched to a particular automobile or truck. Chart 3-2 is an aid to battery selection for many common applications.

System	Type	Features	Recommended Applications
Carbon–Zinc (LeClanché)	Primary	Low cost; sloping discharge; decrease in efficiency at high current drains; poor low temperature performance.	Radios, barricade flashers, marine depth finders, toys, lighting systems, signaling circuits, novelties, flashlights, paging, laboratory instruments.
Zinc Chloride	Primary	Sloping discharge; low temperature performance and capacity better than LeClanché; decrease in efficiency at high current drain but better than LeClanché; less variety of shapes and sizes than LeClanché.	Cassette players and recorders, calculators, motor driven toys, radios, clocks, video games.
Alkaline	Primary	Sloping discharge; better low temperature performance than LeClanché or Zinc Chloride and higher capacity; cost per hour advantage over Carbon–Zinc types at moderate to high current drains.	Radios (particularly high current drain), shavers, electronic flash, movie cameras, tape recorders, TV sets, walkie talkies, calculators, toys, any high drain use.
Mercuric Oxide	Primary	Higher AH per cc than Silver Oxide; higher energy density than Alkaline or Carbon–Zinc; better service maintenance than Carbon–Zinc.	Secondary voltage standard, walkie talkies, paging, radiation detection, test equipment, hearing aids, watches, calculators, microphones, cameras.
Silver Oxide	Primary	Relatively flat discharge; higher operating voltage and service maintenance than Mercury; energy density equal to Mercury.	Hearing aids, reference voltage source, cameras, instruments, watches, calculators.
Nickel–Cadmium	Secondary	Sealed maintenance free construction; relatively flat discharge; good high temp. performance; cost per hour of operation is very low; long cycle life; high initial cost; fair charge retention.	Portable hand tools, shavers, toothbrushes, photoflash equipment, dictating machines, movie cameras, instruments, portable communications equipment, tape recorders, radios, TV sets, calculators, RC models.
Lithium	Primary/Secondary	High energy density; high voltage; very long shelf life; wide temp. range for operation and storage; ideal standby power source; reliable in emergency or medical equipment.	Audio equipment, calculators, cameras, light meters, data acquisition systems, communications devices, electronic games, watches and clocks, hearing aids, medical equipment, memory retention, military electronics, security devices.

Chart 3-2   Battery selection. (Courtesy of Union Carbide—Eveready.)

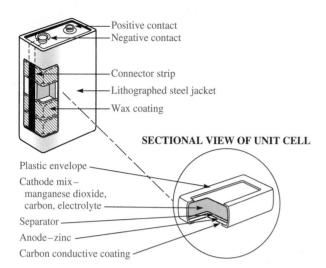

Positive contact
Negative contact

Connector strip
Lithographed steel jacket
Wax coating

SECTIONAL VIEW OF UNIT CELL

Plastic envelope
Cathode mix—
  manganese dioxide,
  carbon, electrolyte
Separator
Anode—zinc
Carbon conductive coating

Figure 3-10 Cutaway view of a general purpose 9 V transistor radio battery. (Courtesy of Union Carbide—Eveready.)

Connecting Cells for Higher Voltage

When it is desired to obtain a higher DC voltage from voltaic cells, two or more cells may be connected in a series arrangement. The overall resulting voltage is the individual cell voltage times the number of cells connected in series. All of the cells must be of the same size, system, and age. The weakest battery in the series will determine the overall capacity (AH rating) of your DC supply. If the AH rating for each cell is 0.5 AH, then the overall capacity for the series arrangement will be the same, 0.5 AH.

The series arrangement of cells is very common. A flashlight that uses two 1.5 volt batteries will provide 3 volts to the bulb. A portable cassette player that runs on 6 volts will require four 1.5 volt cells. We have already seen that an automobile battery is a series arrangement of six 2 volt cells, yielding a total of 12 volts. Figure 3-10 shows the assembly of a common 9 volt transistor radio battery. As you can see, this battery is actually a series stack of six small 1.5 volt cells. Cells connected in series are mated negative to positive, negative to positive, etc., as shown in Figure 3-11. In this way, the difference of potential is increased and the current is through each consecutive cell.

Figure 3-11 also demonstrates the use of the voltmeter in measuring battery voltage. Multimeters have a function switch which you would preset to DC VOLTS. There may also be a range switch, which must be preset to a range that is just greater than the highest voltage you intend to measure. As you connect the test leads, make sure you match polarity—the red test probe to the positive side and the black probe to the negative side of the voltage source being measured. Figure 3-11 also makes use of the schematic symbol for the voltmeter in demonstrating its proper use.

Connecting Cells for Higher Capacity

If it is not possible or practical to obtain a single cell with a sufficient capacity (AH rating) to drive a particular load, you can connect two or more smaller cells in parallel. Figure 3-12 demonstrates this method. As you can see, all negative terminals are connected together electrically as are all positive terminals. Each cell will share the responsibility of supplying current to the load. If the load current demanded is 600 mA, each cell will contribute 200 mA. This will more than triple the service life as compared to a single cell,

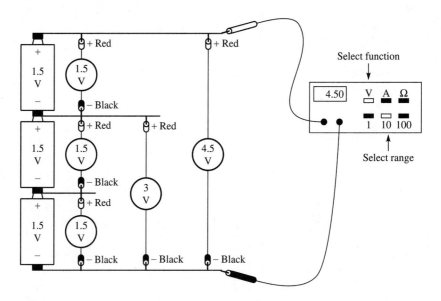

Figure 3-11   Connecting cells in series produces a higher voltage.

since each cell is operating at a lower drain rate. Recall that capacity is generally reduced at higher discharge rates. Again, as in series, the individual cells must be matched as closely as possible so the load current can be shared equally among them.

As you can see, in Figure 3-12 ammeters are being used to demonstrate the total current and the amount of current being contributed to that total by each individual battery. When using your multimeter, be sure to select the DC AMPS function and a range high enough to handle the current you are measuring. At this point you may not have enough technical experience to estimate the amount of current or voltage you are measuring in a circuit. Thus, you may have difficulty in selecting a range on your multimeter. To be safe, simply start with the highest range and switch down a range at a time until you get a reasonable reading on your meter. Make sure the negative lead (black) of the ammeter is facing toward the negative voltage source terminal and the positive ammeter lead (red) is toward the positive voltage source terminal.

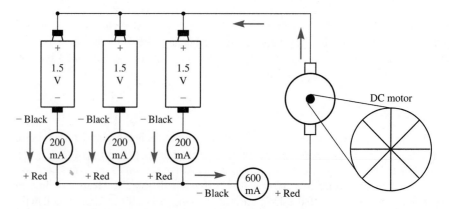

Figure 3-12   Connecting cells in parallel produces a higher capacity.

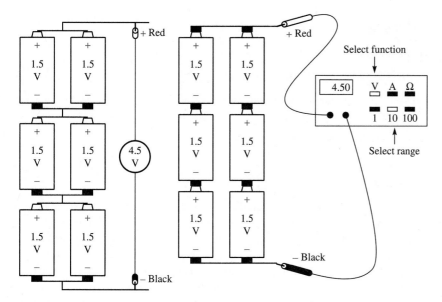

Figure 3-13   Series-parallel combinations of cells yield both higher voltage and capacity.

NOTE: It is extremely important that the ammeter be connected in series with the voltage source and the load as shown. The ammeter has very little internal resistance and is meant to be placed in series with the load to measure load current. Placing the ammeter in parallel with a voltage source will short out and possibly damage the voltage source and the ammeter!!

Connecting Cells for Higher Voltage and Capacity

If it is desired to obtain both a higher voltage and increased capacity, cells may be arranged in a series/parallel combination. This is not as common as straight series or parallel but it can be done. Parallel groups of cells may be chained in series or series chains of cells may be placed in parallel. Figure 3-13 demonstrates these possibilities. Again, it is very important that all cells be equally matched.

Take time to review this section by answering the questions in Self-Check 3-2.

SELF-CHECK   3-2

1. What is the technical difference between a battery and a cell?

2. What is an electrolyte?

3. What is the difference between a dry cell and a wet cell?

4. What is a secondary cell?

5. What is the unit of measure for cell capacity?

6. What is meant by *shelf life*?

7. What battery system offers the lowest initial cost but a very high cost per hour of use?

8. Is a Ni–Cad of the primary or secondary type?

9. List two advantages of lithium cells.

10. Which battery is thought of as the "transportation battery"?

11. How can you connect four 1.5 V cells to produce 3 volts output at twice the capacity of a single cell?

3-3 Conductors

As you have already learned, a practical circuit must have continuity. It is not enough to simply have a source of voltage and a practical device. Obviously, the device must be connected to that source of EMF, or voltage, if it is to do work for you. The continuity of the circuit (completed path for electron flow) is accomplished by the use of conductors like those shown in Figure 3-14. Conductors are a very important part of any practical circuit. In this section, you will explore common materials used for conductors, standard wire sizes, and conductor resistivity. You will learn how wire size and material relates to the amount of total resistance in a given length of wire.

Conductor Materials

Some of the more common materials used as conductors are silver, copper, gold, aluminum, and iron. **Silver** is often used as plating on copper wire or as tubing in radio frequency applications such as coils in transmitters and receivers. The silver plating improves conductivity on the surface of the conductors where most radio frequency current travels. **Gold** is also used as a plating on conductors. Since gold is very resistant to corrosion, it is used to plate connectors, plugs, jacks, relay contacts, and leads, or legs, on electronic components. This will insure a good electrical connection for many years.

Copper and **aluminum** are good conductors and are less expensive than silver or gold. Copper is by far most common material used for general wiring and as cladding on printed circuit boards. It costs more than aluminum but it has some properties that aluminum just can't match. One very practical difference is that copper can be soldered

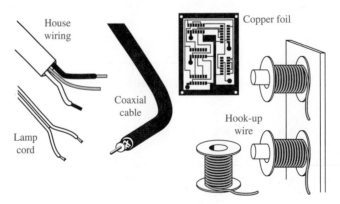

Figure 3-14

for a solid mechanical and electrical connection. Aluminum cannot be soldered, or at least not without tremendous difficulty. Also, copper conducts electricity with less resistance than aluminum, which means a thinner copper wire will do the same job as a thicker aluminum wire. Finally, copper is physically more stable and has a higher melting temperature than aluminum. In spite of its disadvantages, aluminum is used in many applications to save on cost. High-voltage distribution lines, Cable TV coaxial lines, and some house wiring are areas where it is practical to use aluminum.

Iron is a poorer conductor than those already mentioned. However, it does have one property that the others do not. It is very strong. For many years, aboveground telephone lines were made of copper-clad (plated) steel wire. The steel core of the lines gave great strength to withstand the tension and weight of long spans between poles or towers, while the copper plating improved electrical conductivity. Also, iron, or steel, is a common conductor used in automobiles. The steel chassis and body act as a common conductor for the negative side of the battery supplying current to all of the various circuits in the automobile. In this case, the poor conductivity of iron is overcome by size. Current flow is unhindered because the chassis is very large, making the overall resistance very small.

Resistivity (Specific Resistance)

One of the most important characteristics used in comparing conductor material is resistivity or specific resistance (ρ). The **resistivity** of a material is the amount of resistance in a one-foot length of wire made of that material, with a cross-sectional area of one circular mil (CM). The **circular mil (CM)** is the (very small) cross-sectional area of a wire that is 0.001 inch (1 mil) in diameter. This cross-sectional area is found by simply squaring the diameter expressed in mils. See Figure 3-15 for illustration and examples.

Naturally, if a material has a low resistivity, it will be a better conductor. Table 3-1 illustrates how these common materials compare. The values shown are for a room temperature of 20°C.

As you can see, silver is the best conductor, with only 9.8 CM · Ω/ft. Iron is by far the poorest of those listed.

$$A \ (\text{CM}) \ = \ d^2 \ (\text{mils}) \tag{3.2}$$

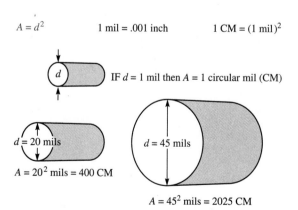

$A = d^2$ 1 mil = .001 inch   1 CM = (1 mil)^2

IF d = 1 mil then A = 1 circular mil (CM)

d = 20 mils

$A = 20^2$ mils = 400 CM

d = 45 mils

$A = 45^2$ mils = 2025 CM

Figure 3-15   Cross-sectional area is measured in circular mils (CM), which is equal to the diameter in mils squared.

TABLE   3-1   Resistivity

Material	Resistivity (ρ) CM · Ω/ft @ 20°C
Silver	9.8
Copper	10.4
Gold	14
Aluminum	17
Iron	58

Wire Resistance

If the resistivity of a wire is known and the length and cross-sectional area are known, the overall resistance for that section of wire can easily be found by the following formula:

$$R = \rho \cdot L/A \tag{3.3}$$

where ρ is the resistivity in CM · Ω/ft
   L is the length of wire in feet
   A is the cross-sectional area in CM ($A = d^2$)

Take time to study Example Set 3-1.

EXAMPLE SET   3-1   WIRE RESISTANCE

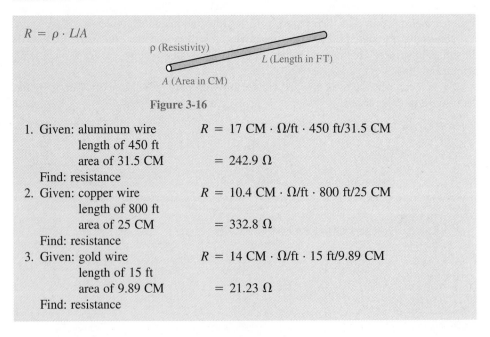

$R = \rho \cdot L/A$

ρ (Resistivity)

L (Length in FT)

A (Area in CM)

Figure 3-16

1. Given: aluminum wire
      length of 450 ft
      area of 31.5 CM
   Find: resistance

   $R = 17\ \text{CM} \cdot \Omega/\text{ft} \cdot 450\ \text{ft}/31.5\ \text{CM}$
   $= 242.9\ \Omega$

2. Given: copper wire
      length of 800 ft
      area of 25 CM
   Find: resistance

   $R = 10.4\ \text{CM} \cdot \Omega/\text{ft} \cdot 800\ \text{ft}/25\ \text{CM}$
   $= 332.8\ \Omega$

3. Given: gold wire
      length of 15 ft
      area of 9.89 CM
   Find: resistance

   $R = 14\ \text{CM} \cdot \Omega/\text{ft} \cdot 15\ \text{ft}/9.89\ \text{CM}$
   $= 21.23\ \Omega$

Wire resistance may also be determined by using a table. Table 3-2 shows the American standard or system for wire sizing and resistance. This system is based on the cross sectional area of wire expressed in CM (which is based on 0.001 inch) and is not accepted worldwide. Many countries base their wire sizing, or gauging, on the millimeter (mm). If you live in a country where the American system is not used, similar tables based on that country's gauging system can be obtained from the local government Bureau of Standards or a public library.

T A B L E   3-2   American Wire Gauge System (AWG) for Solid Round Copper Wire

AWG#	Area (CM)	Ω/1000 ft @ 20°C
0000 (4/0)	211,600	0.0490
000 (3/0)	167,810	0.0618
00 (2/0)	133,080	0.0780
0	105,530	0.0983
1	83,694	0.1240
2	66,373	0.1563
3	52,634	0.1970
4	41,742	0.2485
5	33,102	0.3133
6	26,250	0.3951
7	20,816	0.4982
8	16,509	0.6282
9	13,094	0.7921
10	10,381	0.9989
11	8,234.0	1.260
12	6,529.0	1.588
13	5,178.4	2.003
14	4,106.8	2.525
15	3,256.7	3.184
16	2,582.9	4.016
17	2,048.2	5.064
18	1,624.3	6.385
19	1,288.1	8.051
20	1,021.5	10.15
21	810.10	12.80
22	642.40	16.14
23	509.45	20.36
24	404.01	25.67
25	320.40	32.37
26	254.10	40.81
27	201.50	51.47
28	159.79	64.90
29	126.72	81.83
30	100.5	103.2
31	79.70	130.1
32	63.21	164.1
33	50.13	206.9
34	39.75	260.9
35	31.52	329.0
36	25.00	414.8
37	19.83	523.1
38	15.72	659.6
39	12.47	831.8
40	9.89	1049.0

The American system is referred to as the **American Wire Gauge system (AWG)**. It starts with an AWG# of 0000 (4-ought) and continues to 40. Notice that as the AWG# increases, the size of the wire decreases. Also notice that the resistance per 1,000 ft increases with higher AWG numbers because the wire size is decreasing. The smaller the wire, the higher the resistance!

Table 3-2 is created for round, solid copper wire. If aluminum wire is substituted for copper, the aluminum wire must be at least two sizes larger (the AWG# must be two steps lower) to make up for the higher resistivity of aluminum (1.635 times that of copper). For example, #10 aluminum wire may be substituted for #12 copper wire. We will explore the significance of wire resistance mathematically when we progress in the study of DC circuit analysis. Our biggest concern will be to minimize the resistance contributed by the conductors in a practical circuit. The purpose of the conductor is to deliver current to a load without significantly hindering current itself.

Table 3-2 is meant to be very practical and useful. If you were an electrician, you would be interested in knowing the overall resistance in a long run (length) of copper wire. The table indicates the amount of resistance per 1,000 ft for all standard sizes. If you have a 2,500-ft run of #14 copper wire, the overall resistance in the wire will be 2,500 ft times 2.525 Ω/1,000 ft, or 6.313 Ω ($R = $ 2,500 ft $\cdot$ 2.525 Ω/1,000 ft). If you have 10 ft of #40 wire, the resistance is 10 ft $\cdot$ 1,049.0 Ω/1,000 ft $=$ 10.49 Ω. Note that wire resistance doubles as the AWG# is increased by three—i.e., $R = $ 0.9989 for AWG#10 and $R = $ 2.003 for AWG#13.

Study Design Note 3-1. You may find the BASIC program very useful in determining the resistance of a length of wire.

In this section we have discussed many important things about conductors. We have examined the different materials commonly used for conductors and discussed their particular merits. Also, you are now aware of the resistance in a wire determined by material used, wire size and length. Take a few moments now to review this section by doing Self-Check 3-3.

SELF-CHECK 3-3

1. What is the purpose of the conductors in a practical circuit?

2. List the five most common materials used as conductors.

3. Which conductor material has the least resistance, or highest conductivity?

4. Explain resistivity.

5. What is the resistivity of gold?

6. Name at least two factors that affect the resistance of a wire.

7. How much resistance is in 300 ft of #28 solid copper wire?

8. Name at least two advantages of copper wire over aluminum wire.

DESIGN NOTE 3-1: Wire Resistance

The resistance of a wire is dependent on the material the wire is made of, the cross-sectional area in circular mils, and the overall length of the wire.

$$R = \rho \cdot L/A$$

R is overall resistance in ohms
ρ is the resistivity in CM $\cdot$ Ω/ft
L is the length in feet
A is the area in circular mils (CM)

CALCULATOR SOLUTION	"BASIC" PROGRAM SOLUTION
EXAMPLE: $\rho = 58$, $L = 80$ ft, $A = 159.8$ CM. Find: R START ρ [5] [8] [×] L [8] [0] [÷] A [1] [5] [9] [.] [8] [=] R 29.036295 ohms	10 CLS 20 PRINT"THIS PROGRAM WILL CALCULATE THE RESISTANCE IN A COPPER OR ALUMINUM" 30 PRINT"WIRE OF ANY SIZE OR LENGTH.":PRINT" " 40 INPUT"SELECT (1) COPPER OR (2) ALUMINUM WIRE. (1 OR 2)";M 50 P = 17 60 IF M = 1 THEN P = 10.4 70 PRINT" ":INPUT"ENTER LENGTH IN FEET. ";L 80 PRINT"DO YOU WISH TO CALCULATE RESISTANCE BY (1) ENTERING DIAMETER IN MILS" 90 PRINT"OR (2) ENTERING AREA IN CIRCULAR MILS?" 100 INPUT"SELECT 1 OR 2";C 110 IF C = 2 THEN GOTO 300 120 PRINT" " 130 INPUT"ENTER DIAMETER OF WIRE IN MILS. ";D 140 R = P * L / D^2 150 GOTO 330 300 PRINT" " 310 INPUT"ENTER CROSS SECTIONAL AREA IN CIRCULAR MILS. ";A 320 R = P * L / A 330 PRINT" ":PRINT"THE RESISTANCE OF THIS WIRE IS ";R;" OHMS." 340 PRINT" " 350 PRINT"CALCULATE ANOTHER? (Y/N)";A$ 360 IF A$ = "Y" THEN GOTO 40 370 CLS:END

3-4 Switches

Although we may not go so far as to say that the switch is absolutely essential to a practical circuit, we certainly can argue that the switch is extremely important. The switch allows us to control the circuit operation. It is the most common of all the human–electronic

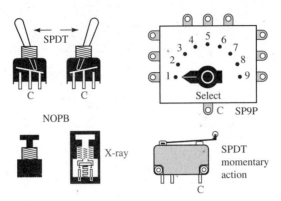

Figure 3-17

interfaces that permit man to safely control a light or any one of a million other electrical or electronic devices. We open a circuit by turning the switch off and we close a circuit by turning a switch on. Control is what we are seeking as we employ a seemingly endless variety of these important devices to operate the inventions of this world. In this section, we will explore switch kinds and types, operational states, and schematic symbols.

Kinds of Switches

Here, we will quite literally look at a good sampling of **kinds** of switches. When we talk about kinds of switches, we are talking about the switch's mechanical or physical characteristics. Figures 3-17 and 3-18 show a good variety of common switches.

Figure 3-18   Here is a sampling of switch kinds: (a) pushbutton (b) multiple pushbutton array (c) Lever-lite illuminated switch (d) key switch for security (e) rotary switch (f) slide and rocker switches (g) micro switches (h) DIP switches. (Photographs a through d courtesy of Switchcraft, Inc., a Raytheon Company.)

The ABC's of Careers in Electronics

Careers in Electronics: Automation

REQUIRED SKILLS AND KNOWLEDGE:
- General Mechanics
- Electrical Wiring
- Basic Electronics (circuit theory and devices)
- Digital Circuitry
- Computer Basics
- Computer Interfacing • Computer Programming
- Industrial Circuitry and Controls
- Use of Specialized Test Equipment for Setup and Trouble-shooting

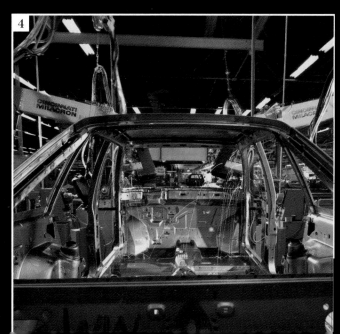

1) Automobile chassis assembly accomplished with automated assembly lines and robot welders.

2) Precision welding performed by a computer controlled robot arm. (Courtesy of Owens/Corning Fiberglass)

3) A fully automated manufacturing line designed to machine both car and truck water pumps. Part of Ford Motor Company's "factory of the future," this line runs four times faster than traditional machining lines. (Courtesy of Ford Motor Company)

4) Automobile body assembly on a fully automated robot assembly line. (Courtesy of Cincinnati Milacron)

Careers in Electronics: Broadcasting

REQUIRED SKILLS AND KNOWLEDGE:
- Basic Electricity and Electronics
- Audio Circuitry
- Digital Circuitry
- Radio Frequency Circuitry
- Transmission Lines and Waveguides
- Transmitters and Receivers
- Antenna Systems
- Microwave Systems
- Fiber Optics
- Computer Programming and Interfacing
- Use of Common and Specialized Test Equipment for Setup and Troubleshooting

5) Satellite systems are vital to the modern world of broadcasting and communications. Here two astronauts use the space shuttle's remote manipulator system (RMS) to retrieve the Westar VI satellite from orbit. (Courtesy of NASA)

6) The 4,444 pound RCA SATCOM K-1 communications satellite launched from space shuttle *Atlantis* late in November 1985. (Courtesy of NASA)

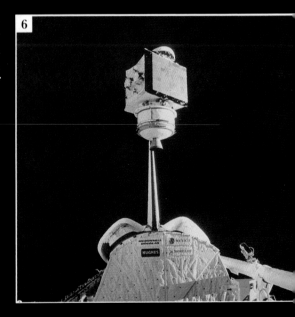

7) Line of sight Microwave Systems used for communications and broadcast relay. (Courtesy of J. Pickerell/FPG)

Careers in Electronics: Broadcasting

8) A mountaintop VHF and UHF relay sight for public service communications such as police, fire, and paramedics. (Terry Qing/FPG)

9) Technicians busy at work inside a modern mobi television production broadcast van. (Tom Campbe FPG)

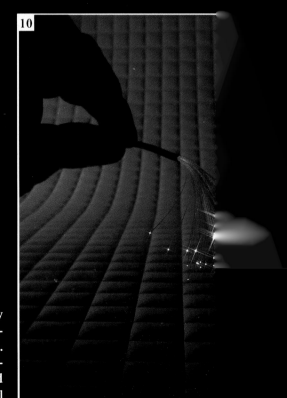

10) Tiny glass fibers can carry thousands of telephone and television signals all at the same time. Fiber optics has taken a prominent role in communications and broadcasting. (Courtesy of Michael Stuckey/Comstock Inc.)

Careers in Electronics: Computers

REQUIRED SKILLS AND KNOWLEDGE
- Basic Electricity and Electronics
- Digital Circuitry
- Microprocessor Theory
- Microprocessor Support Chips
- Microprocessor Programming
- Computer Interfacing
- Computer Memory Devices and Systems
- Computer Control Applications
- Computer Peripheral Systems for Input, Output, and Data Storage
- Use of Common and Specialized Equipment for System Development and Troubleshooting

11) Critical manufacturing processes are controlled and monitored by computer. (Courtesy of Texas Instruments)

12) Sophisticated design work is accomplished on high resolution color computer systems. (Courtesy of Hewlett-Packard Company)

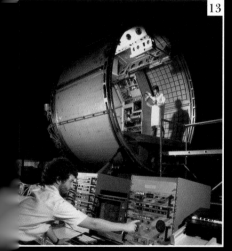

13) Computer simulation and testing of systems in the European space lab. Courtesy of McDonnell Douglas Corp.)

integrated circuits (VLSI).

Careers in Electronics: Computers

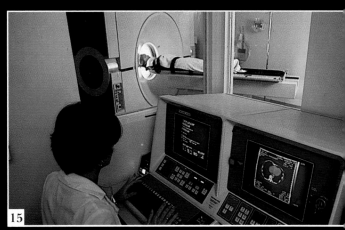

15) Here, a CAT scan is performed under computer control. Computers are an integral component in nearly all areas of medicine. (Larry Mulvehill/Photo Researchers)

16) Automated and computer con-trolled mass storage and inventory systems. (Courtesy of IBM)

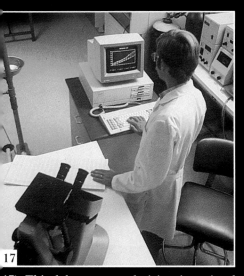

18) Virtually all space shuttle functions are under computer control. (Courtesy of NASA)

17) This laboratory technician is assisted with diagnostics by a powerful desktop per-sonal computer. (Courtesy of Sperry Corpo-ration)

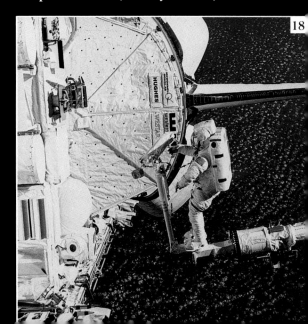

Code Practice

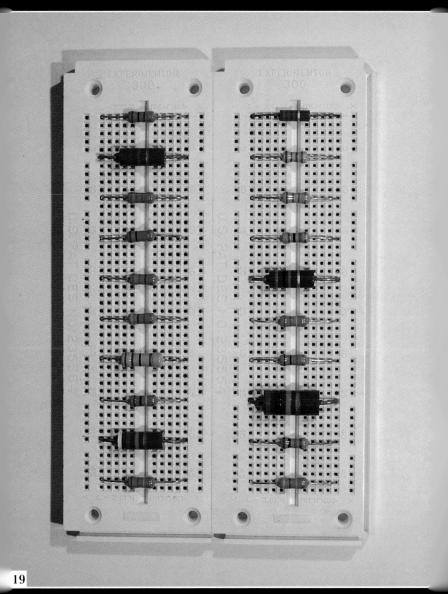

19

9) Identify the value of each resistor in the two columns shown. (Section 3-6

SWERS:
LUMN 1: 150 Ω 5%, 6.8 kΩ 10%, 2.2 kΩ 5%, 15 kΩ 5%, 330 Ω 5%, 4.7 kΩ
Ω 5%, 100 kΩ 5%, 91 kΩ 5%, 120 kΩ 5%
LUMN 2: 120 Ω 10%, 1.5 MΩ 5%, 22 kΩ 5%, 1 kΩ 5%, 100 Ω 10%, 220 kΩ

Proving Ohm's Law

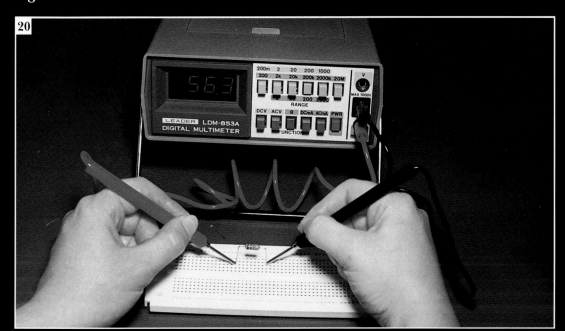

20) Step One: Use the multimeter to measure the actual resistance of the resistor. This resistor is found to be 56.3 kΩ. (Section 4-1) (test instruments courtesy of Leader Instruments Corp., Hauppauge, NY)

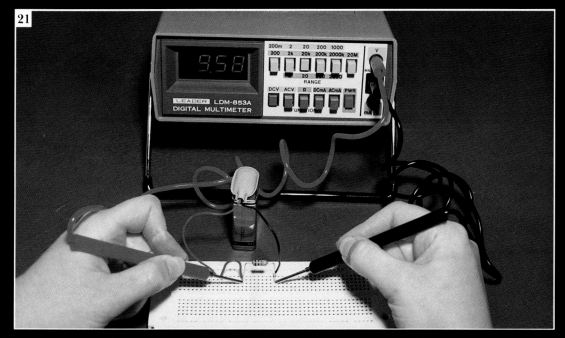

21) Step Two: Use the multimeter to measure an applied voltage. Be sure to observe proper polarity with your test leads. Also, make sure your fingers do not touch the metal tip and collar of the probes. Carelessness here could be a shocking experience someday. This fresh transistor radio battery has an EMF of 9.58 V. (Section 4-1) (test instruments courtesy of Leader Instruments Corp., Hauppauge, NY)

Proving OHM's Law

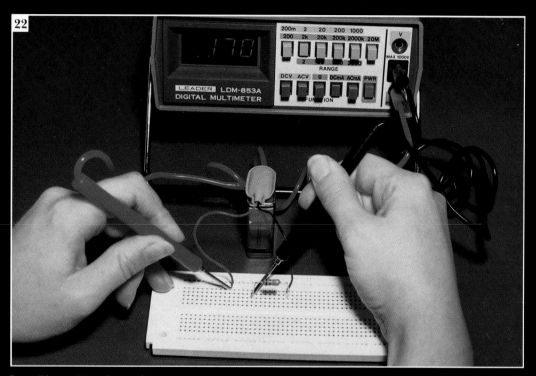

22) Step Three: Use the multimeter to measure the actual current flowing through the resistor. Notice the test leads and meter are placed in series with the resistor and the battery. Again, observe polarity and do not touch metal probe tips. The meter is set to the 2mA range and indicates a current of 0.170 mA (170 μA). According to Ohm's Law, I = E/R. In this case, I = 9.58 V/56.3 kΩ = 0.170 mA. Thus, Ohm's Law is proven experimentally to be true. (Section 4-1) (test instruments courtesy of Leader Instruments Corp., Hauppauge, NY)

23) This photo shows a power supply set at 20 V and supplying 111 mA of current to a 180 Ω resistor. Ohm's Law: I = E/R = 20 V/180Ω = 111 mA. (equipment courtesy of Tektronix, Inc.)

Pushbutton Switch

The **pushbutton switch** is manufactured with a great number of case styles and operational characteristics. The push of the button may turn a circuit on, turn it off, or do both. Pushbutton switches that turn a circuit on with the first push, then turn the circuit off with the second push are sometimes referred to as *push-push* switches. Many pushbutton switches are momentary in operation. That is, the switch performs an open circuit (OFF) or closed circuit (ON) operation only as long as you press the button. If the switch is normally closed (NC), pushing the button will open the circuit, or turn it off. If the switch is normally open (NO), pushing the button will turn the circuit on.

Multiple Pushbutton Array

The **multiple pushbutton array** is very useful in controlling a piece of equipment that has many functions or for selecting from a variety of signal inputs or outputs. These switch arrays are commonly used on machinery or electronic equipment such as tape recorders, audio mixers, and audio routing panels in studios. The individual pushbutton switches in the array may be *interlocked* or *independent*. If they are designed to be interlocking, only one switch may be activated at a time. When a button is pushed, the one previously pushed will automatically be released. This eliminates having to release a button before another is chosen. If the pushbutton switches are independent, they may be operated individually with no interaction between them. Any number may be activated at one time. Notice, in the photo, that these switches can be quite elaborate with many terminals and internal contacts.

Toggle Switch

The **toggle switch** is one of the most common switches used in electronics. The word *toggle* means "to switch back and forth between two points." These switches have a small handle that can be flipped between two positions to determine if a circuit is on or off. The handle might be baseball-bat shaped, paddle shaped, or a simple flat lever. In some cases, a toggle switch is designed to control more than one circuit. Additional terminals and internal contacts are added for this purpose. The toggle switch may also have a **center off (CO)** position. This is a neutral position in which no circuits are on. Moving the lever left will activate circuit #1, and moving the lever right will activate circuit #2. The lever switch shown in the photo is a very elaborate one. It has many sets of terminals and contacts, as shown in the schematic diagram on the side of the switch, and the lever itself is illuminated internally.

Rotary Switch

The **rotary switch** is often used for channel selection on television sets and communication equipment. This switch is sometimes called a *wafer switch* because the main shaft passes through the center of one or more ceramic, fiberglass, or phenolic wafers on which the terminals and contacts are mounted. The center section of each wafer can be rotated, passing a common contact to one of the many stationary contacts mounted around the wafer's perimeter. In this way, a particular circuit or channel is selected. Using more than one wafer or set of contacts on the same mechanical shaft is called **gang switching**. In this way, all sets of contacts are switched at the same time. Often, the contacts are plated with silver or gold to ensure good contact and reliable operation.

Slide and Rocker Switches

The **slide switch** is often used as a mode switch to select a certain mode of operation such as HIGH or LOW, FAST or SLOW, 50 Hz or 60 Hz operation, and FORWARD or REVERSE. Slide switches can be very simple, with one set of contacts, or very elaborate, with many sets in a straight-line arrangement. At the heart of the slide switch is a plastic or Bakelite slide with electrical contacts embedded in its underside. As the slide position is changed, the contacts slide and mate with a new set of stationary contacts.

The **rocker switch** is often simply a modified slide switch. Pressing on one side of the rocker-arm mechanism will cause the slide to be forced in the other direction. However, some rocker switches are very different from the slide switch. The switching contacts are actually on a metallic teeter-totter (rocker arm) that is tripped into action by pressing one side or the other of the plastic rocker arm on the outside of the switch.

Micro Switch

The **micro switch** is a short-travel, light-pressure switch. It is normally used to monitor the movement of a mechanical device. Micro switches are sometimes referred to as *limit*, *range limit*, or *end stop* switches. They are commonly used in machinery to tell the motor when a process has finished, or a cam or arm has reached the limit of travel.

DIP Switches

DIP switches are small switch assemblies designed for mounting on printed circuit boards. **DIP** stands for **D**ual **I**n-line **P**ackage. The pins or terminals on the bottom of the DIP Switch are the same size and spacing as an 8-, 14-, or 16-pin Dual In-line Package integrated circuit. They look right at home on a crowded printed circuit board. The individual switches may be of the toggle, rocker, or slide kind. They are mainly used to preset a complex circuit to certain conditions. Computers and computer equipment interfaces make generous use of these switches.

Types of Switches

When we talk about different **types** of switches, we are talking about the electrical arrangement of the switch's terminals and contacts. The type description will tell you immediately if a switch is suitable for a particular application by indicating how many poles and how many positions a switch has. A **pole** on a switch is a single common terminal that is switched to one or more other terminals. The type description may also indicate the operational state of the switch—such as being normally ON or normally OFF. Let's now examine some of the most common types of switches one at a time. Then we will summarize and clarify what has been discussed by using schematic symbols and sample circuits.

Single Pole Single Throw (SPST)

The **SPST** is one of the simplest of all switch types. It is a two-terminal switch that is strictly for ON/OFF operations. The common terminal, or pole, is switched in contact with the auxiliary terminal for the ON state.

Single Pole Double Throw (SPDT)

The **SPDT** switch has three terminals, one of which is common to the other two. The common single pole can be switched to either of the auxiliary terminals. This would allow a single voltage source to be switched to one or the other of two circuits. Also, two voltage sources connected to the two auxiliary terminals may be switched to the common pole. Some double throw switches have center OFF (CO) positions. This is a neutral center position in which the common pole is disconnected from both auxiliary poles.

Double Pole Single Throw (DPST)

The **DPST** switch has two sets of terminals. Each set has one common pole and one auxiliary terminal. This switch is basically an ON/OFF switch. It may be used to switch both poles of a battery into a circuit at the same time or to connect two different voltage sources to two separate circuits, thus turning both circuits on and off at the same time.

Double Pole Double Throw (DPDT)

The **DPDT** switch has two sets of terminals. Each set has one common pole and two auxiliary terminals. This switch is very versatile and may be used for all applications previously mentioned. It is frequently used as a FORWARD/REVERSE switch and a two-line switch, but it has many other applications as well.

Triple Throw (TT)

Single and double pole switches are also manufactured as **triple throw** types. This type of switch is usually the slide kind and has three auxiliary terminals for every common pole. Therefore, the common pole can be switched to one of three auxiliary terminals.

Rotary Switch Type Descriptions

Rotary switches may have many poles and many auxiliary terminals and switch positions. Normally, each wafer section of a multisection rotary switch represents one pole. Each wafer may have any number of auxiliary terminals. These switches are described as having a certain number of poles and a certain number of positions. Some examples are: SP10P = single pole, 10 positions; 4P23P = 4 poles, 23 positions; TP6P = triple pole, 6 positions.

Normally Open Pushbutton (NOPB) and Normally Closed Pushbutton (NCPB)

The **NOPB** and **NCPB** are momentary-contact pushbuttons. As long as your finger is on the button, the circuit will be open (off), when using a NCPB, or closed (on), when using a NOPB. NCPBs are frequently used as reset buttons to open and restart a circuit such as a burglar alarm. NOPBs are used in calculator and computer keyboards.

Study Figure 3-19. It will illustrate and clarify what has been discussed.

In this section, you have been introduced to various kinds and types of switches along with schematic symbols. You will discover how very practical this information is as you continue your studies and gain experience in electricity and electronics. Test your knowledge with Self-Check 3-4 before continuing.

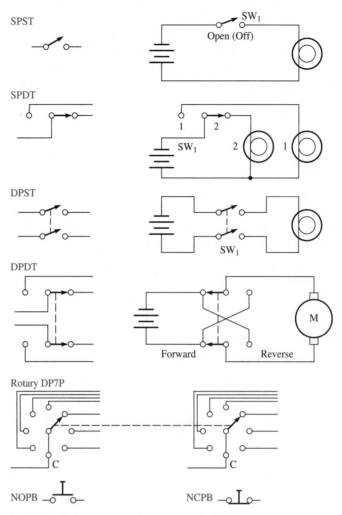

Figure 3-19 A sampling of switch types and applications.

SELF-CHECK 3-4

1. Describe the operation of a normally open pushbutton switch.

2. What is an interlocked pushbutton array?

3. What is a toggle switch?

4. Describe a 5P12P rotary switch.

5. For what are micro switches used?

6. What does SPDT stand for?

3-5 Fuses and Circuit Breakers

Thus far, we have discussed the basic practical DC circuit and the various elements essential or important to its operation. We have examined: batteries, a source of EMF; conductors, which establish continuity; and switches, which provide control. All of these elements are either essential or important to provide, deliver, and control current to our practical load.

However, there is yet another element that should not be overlooked in the design and implementation of a practical circuit. That element is a *circuit-protection device* such as a fuse or a circuit breaker. A circuit-protection device is extremely important in minimizing excessive damage to the circuit in case of overload or component failure. As you know, resistance limits the amount of current flowing in a circuit. If that resistance is bypassed (shorted out) in such a manner as to eliminate it or reduce it to zero ohms, there would no longer be a resistance to limit the current. Therefore, the current would be very high. This excessive amount of current could be very damaging to the voltage source, the wiring, the switch, or any other device in the path of the current. However, if a circuit-protection device, such as a fuse or circuit breaker, is included in the circuit design, little or no damage will result to the rest of the circuit because the fuse will burn open, or the breaker will trip.

Fuses

Types and Sizes of Fuses

As can be seen in Figures 3-20 and 3-21, fuses are manufactured in a variety of sizes and styles. Most fuses used in electronic equipment in the United States are sized according to an AG or AB number. Table 3-3 (p. 77) gives the various dimensions for these fuse sizes. The **AG** used in the sizing nomenclature stands for **A**utomobile **G**lass (because the early sizes were used mainly in the automotive industry). The AB designation indicates that the fuse body is made of some insulator other than glass such as Bakelite, fiber, or ceramic.

Fuse Ratings

The **voltage rating** of a fuse is usually marked on the fuse. It is the highest voltage at which that particular fuse was designed to safely interrupt the current. When a fuse burns open, due to a fault condition, an electrical arc will occur within the fuse. The fuse must

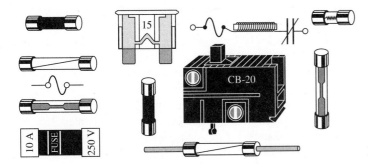

Figure 3-20

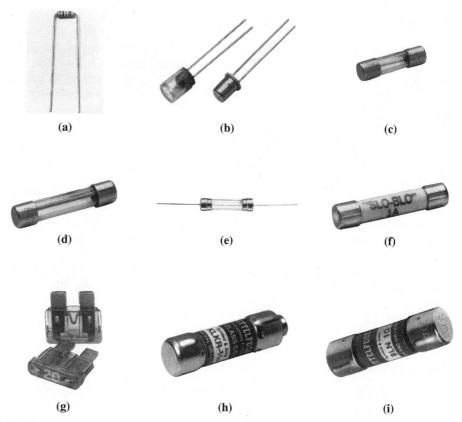

(a) **(b)** **(c)**

(d) **(e)** **(f)**

(g) **(h)** **(i)**

Figure 3-21 A variety of fuse types: (a) PICO II® fast-acting color-coded fuse for PC boards (b) fast-acting Microfuse for very low current capability (c) fast-acting metric-sized glass-body fuse (d) glass-body Slo-Blo® fuse (e) glass fuse with wire leads for PC board mounting (f) 3AB ceramic body Slo-Blo® fuse (g) plug in Autofuse® (h, i) fibre body ''Midget'' fuses for very high current interruption (photos reproduced courtesy of Littelfuse, Inc.)

be designed to contain and extinguish the arc. A fuse may be used at any voltage at or below the rated voltage.

Fault-condition current is the current rating, usually marked on the fuse, that indicates how much current the fuse will sustain without fault or failure. When that amount is exceeded, the fuse will begin to heat up and start to deteriorate. How long it takes for the fuse to burn open will be determined by how much the circuit current has exceeded the rating. For example, if the circuit current is 110% of the rating it may take four hours or more for the fuse to fail; if the current is 150% of the rating it may take 20 or 30 minutes to fail; and if the current is 200% of the rating it may take 5, 10, or 15 seconds to fail. The fault current rating is also affected by ambient temperature (temperature of the environment immediately surrounding the fuse). Most fuses are rated at 25°C. According to LITTELFUSE®, a leading manufacturer of fuses in the United States, in choosing a fuse for a particular application, the current rating of the fuse should be derated 25% at 25°C to avoid nuisance blowing. For example, a 10 A fuse should not be operated at a circuit current of more than 7.5 A at 25°C.

TABLE 3-3 Fuse Sizes

Size	Diameter	Length
1AG	1/4″	5/8″
3AG	1/4″	1-1/4″
4AG	9/32″	1-1/4″
5AG	13/32″	1-1/2″
7AG	1/4″	7/8″
8AG	1/4″	1″

The **short circuit rating** of the fuse is also very important. This is also referred to as the *interrupting current*, since it is the maximum current the fuse can safely interrupt at a rated voltage. Many fuses have short circuit ratings of 35 A, 1,500 A, or even 200,000 A. If the power source can only deliver a maximum of 10 A, then the 35 A short circuit rating would be fine. However, if the power supply could deliver 100 A on demand, the 35 A short-circuit-rated fuse would be inadequate.

Fuse Response Time

Fuses are characterized according to response time. Some fuses are fast acting, to protect exceptionally delicate circuitry. Their response time to a 200% overload may be a fraction of a second. Other fuses are classified as medium acting and may take a few seconds to blow at 200% overload. Finally, slow acting fuses may take 10 or 15 seconds to blow. Slow acting fuses are used in circuits, such as motor circuits, that have a large amount of surge current (initial starting current). The circuit may require a certain amount of time to reach a normal level of operating current, and a fast or medium acting fuse would blow too quickly.

Fuse Placement

Fuse placement in a circuit is almost as important as choosing the proper fuse for the circuit. The fuse should be placed as close to the source of voltage as is practical or possible. Figure 3-22 illustrates the proper placement for a fuse in various circuits. Notice that the fuse is placed before the switch, as close to the power source as possible. The idea is to protect as much of the circuit as possible.

Circuit Breakers

Circuit breakers, like fuses, come in a wide variety of shapes, sizes, and capacities for an equally wide variety of applications. They range from small, low-current "Button Breakers" (used to replace fuses in low-current, low-voltage DC supplies) to 1,000 A or 2,000 A heavy-duty industrial circuit breakers. We cannot possibly begin to discuss all

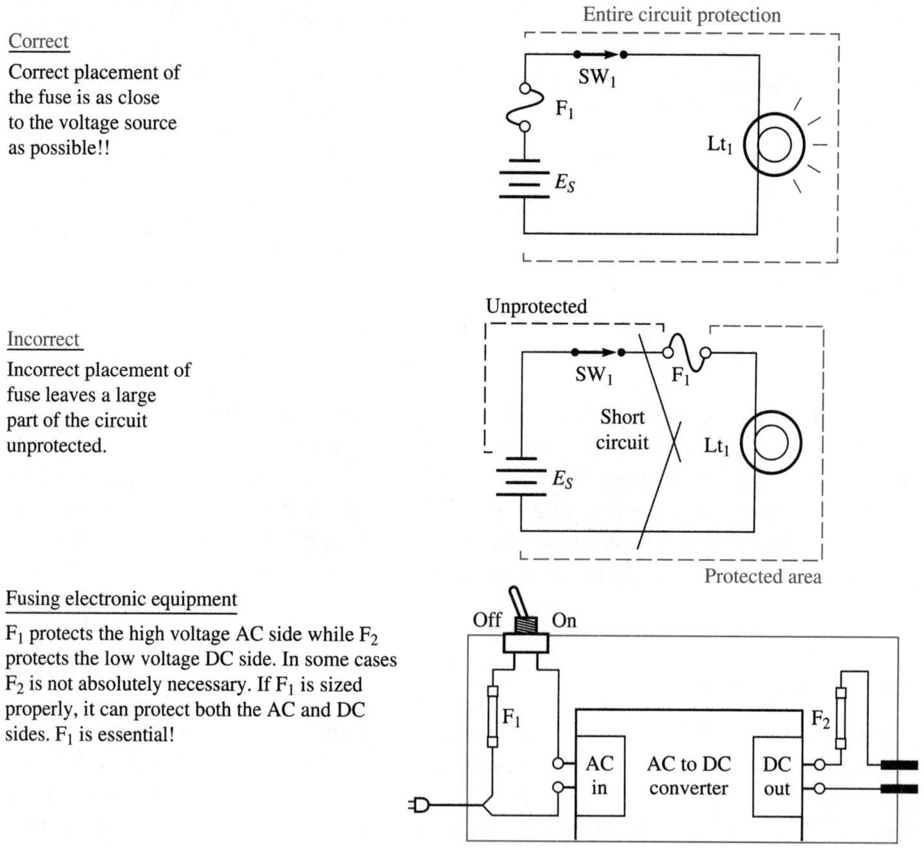
Correct
Correct placement of the fuse is as close to the voltage source as possible!!
Entire circuit protection
SW1
F1
Lt1
ES
Incorrect
Incorrect placement of fuse leaves a large part of the circuit unprotected.
Unprotected
SW1
F1
Short circuit
Lt1
ES
Protected area
Fusing electronic equipment
F1 protects the high voltage AC side while F2 protects the low voltage DC side. In some cases F2 is not absolutely necessary. If F1 is sized properly, it can protect both the AC and DC sides. F1 is essential!
Off
On
F1
F2
AC in
AC to DC converter
DC out

Figure 3-22 Fuse placement is extremely important for full-circuit protection.

there is to know about these devices. However, we shall examine their operational characteristics and a typical application.

Operational Characteristics

Circuit breakers have certain operational characteristics that may or may not make them more desirable than fuses as circuit-protection devices. Most notably, the circuit breaker can be reset and therefore reused, whereas the fuse must be replaced. However, circuit breakers tend to be medium or slow acting devices. This is not desirable in circuits with sensitive components that cannot tolerate relatively long-term overloads.

Mechanical Description

The operational characteristics of circuit breakers can best be understood through a brief mechanical description. Circuit breakers have a pair of spring-loaded contacts and some type of trip mechanism. The trip mechanism is similar to the notched trip mechanism of a mouse trap. When the mechanical catch slips out of the notch, the trap is sprung (tripped).

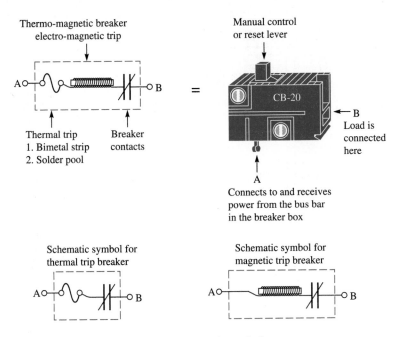

Figure 3-23   Circuit breaker and schematic symbols.

In the case of a circuit breaker, the contacts are quickly pulled apart when the breaker is tripped. The breaker may then be reset with a mechanical reset lever. The trip mechanism is some type of current sensing device. It could be a **thermal trip mechanism**, an **electromagnetic trip mechanism**, or a combination of both. The schematic symbols for these types of trip mechanisms and breakers are shown in Figure 3-23.

Typical Application

Standard household circuit breakers are a combination of the bimetal-strip thermal trip and electromagnetic trip mechanisms. The electromagnetic trip is fast acting and will respond very quickly to a dangerous overload, such as a short circuit in the wiring. The thermal trip will protect the circuit against sustained moderate overloads, as when many appliances are connected to the circuit and operating at one time. Figure 3-24 shows a typical house wiring circuit that uses these breakers. The AC current and voltage is supplied to the house from the utility company via the power meter, which is normally mounted on the side of the house. The main lines from the power meter are brought into a metal breaker-panel box to the heavy bus bars. (Usually there are two, in some cases three, "hot" bus bars in the breaker panel even though only one is shown in the diagram.) The circuit breakers are designed to snap or plug into the bus bars in a side-by-side arrangement. Each breaker is for a separate circuit in the home. This diagram is intended only for the purpose of illustrating the use of circuit breakers in a domestic situation. Local and national building and electrical codes should be consulted before extensive house wiring is attempted.

In this section, we have discussed the importance of and various types of fuses and circuit breakers. Take a few minutes now to answer the questions in Self-Check 3-5.

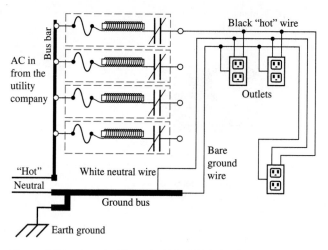

Figure 3-24   A typical household breaker circuit.

SELF-CHECK 3-5

1. What is the purpose of a fuse or circuit breaker in a practical circuit?

2. What is the fault-condition current rating of a fuse?

3. What is meant by fuse response time?

4. What is the proper placement for a fuse in a circuit?

3-6 Resistors

Among other things, circuit analysis involves the proper identification of component types, values, and operating characteristics. One of the most common components found in an electronic circuit is the resistor, which comes in a wide variety of sizes and types for various functions. In this section, we will take time to investigate a wide variety of resistors. As can be seen in Figure 3-25, you will soon discover that resistors vary greatly in size, shape, design, and power handling capability.

Fixed Resistors

Carbon Composition Resistors

The **carbon composition resistor** was once the most commonly used resistor in electronic circuitry. Carbon composition resistors are manufactured in a variety of standard values and power ratings. Figure 3-26 illustrates the standard power ratings and resistor construction. The ohmic value of these resistors is marked on each resistor using a color code system. The various colored ring combinations represent different values of resistance (in ohms).

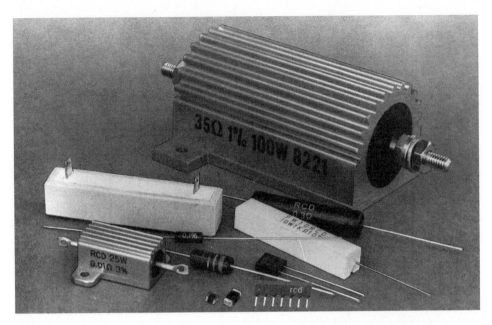

Figure 3-25 A variety of resistors: wirewound power, wirewound precision, carbon and metal film, SIP resistor network, small chip resistors. (Courtesy of RCD Components, Inc.)

Carbon Film Resistors

Shown in Figure 3-27, **carbon film resistors** are low-power resistors that are formed by depositing a thin carbon film on a solid insulator. The film deposit is then cut in a spiral fashion to form the resistive element, called the *helix*. Usually the insulator is tubular and metal end caps are used as contacts or terminals to which the component leads are attached. The manufacturing process for this type of resistor can be precisely controlled, making it easy to keep resistor tolerance within $\pm 5\%$ and even as low as $\pm 2\%$.

Metal Film Resistors

Metal film resistors are similar in appearance to carbon film resistors. Instead of a carbon film, a thin metal film is deposited, then cut in a spiral fashion around the insulator from

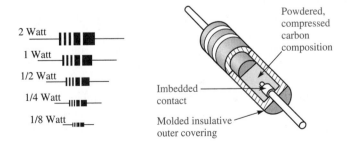

Figure 3-26 Carbon composition resistors and sizing for various power levels.

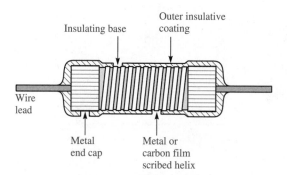

Figure 3-27   Cutaway view of tiny film resistors.

one end to the other. Once again, Figure 3-27 illustrates the construction. The width, thickness, and length of the metal film determines the component's resistance. These resistors are very much more precise in value than the common carbon composition or carbon film resistors, typically ±1%, ±0.5%, ±0.1%. Naturally, they cost significantly more than the other types.

The metal film resistors not only have the advantage of closer tolerance, they are also less sensitive to temperature changes, are affected very little by aging, and generate much less noise in sensitive circuits. Temperature can and does affect the value of a resistor. A wide temperature variation can change the value of a carbon composition resistor by as much as 5% or more, while very little change will occur in the value of a metal film resistor. Carbon resistors that normally operate at high temperatures will gradually increase their resistance over a period of time until they are far out of tolerance. Metal film resistors age very little. Carbon resistors are a generous source of noise that can be amplified in very sensitive circuits. The atomic structure of the metal film is responsible for greatly reducing this type of unwanted noise. These advantages qualify the metal film resistor as a prime candidate for use in sensitive test equipment where long-term calibration, accuracy, and low noise are essential.

Wirewound Resistors

Wirewound resistors are constructed by winding resistance wire, such as constantan or nichrome, on an insulator form, such as ceramic. The manufacturing process for these resistors is very precise, allowing consistent production of close-tolerance precision resistors (±1% and less). See Figure 3-28.

Wirewound resistors are used not only where precision is required, but where high power dissipation is required as well. These resistors are generally used in applications where the power dissipation of the resistor will exceed two watts. Power ratings for these resistors range from a couple to several hundred watts. Along with the high power capability, these resistors are very temperature stable, with very little change in resistance due to large temperature changes.

Figure 3-29 illustrates three common implementations of wirewound construction for resistors. **Fixed** wirewound resistors are the most common, with two terminals or leads for circuit connection. **Tapped** wirewound resistors let you choose a variety of fixed resistances from the same resistor simply by selecting the appropriate pair of terminals. **Adjustable** wirewound resistors have a sliding metal collar that you can loosen and relocate on the resistor to acquire exactly the amount of resistance desired from one end or the other. These are referred to as *adjustable but not variable*, since the resistance is not

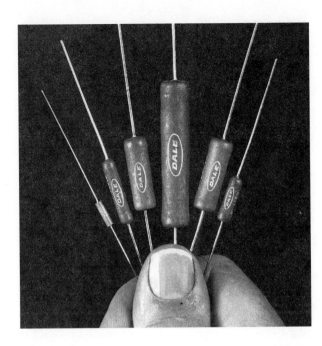

Figure 3-28   Precision wirewound resistors. (Courtesy of Dale Electronics, Inc., Columbus, Nebraska.)

quickly or easily varied with a knob. The tapped and adjustable resistors are often used as voltage dividers, allowing the user to choose a desired voltage. The voltage applied to the resistor is distributed, or divided, along the body of the resistor, allowing selection of a voltage lower than that which is applied. Voltage dividers are very interesting and useful. Naturally, we will discuss them in detail at the appropriate time.

Resistor Networks

Figure 3-30 shows a variety of **resistor networks**, or **arrays**, used on high-density printed circuit boards due to their compactness. Two common styles are the SIP (single in-line package) and the DIP (dual in-line package). The resistors in each package are close-tolerance ($\pm 1\%$ to $\pm 0.05\%$) film resistors with their value marked on each package (the value may be included in the part number marked on the package). These networks are commonly used in modern mass production and offer the benefits of higher quality, higher parts density, higher productivity, and lower cost.

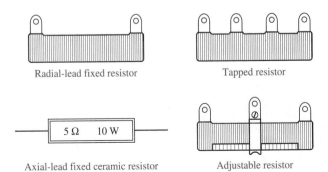

Radial-lead fixed resistor

Tapped resistor

5 Ω   10 W

Axial-lead fixed ceramic resistor

Adjustable resistor

Figure 3-29   Common wirewound power resistors.

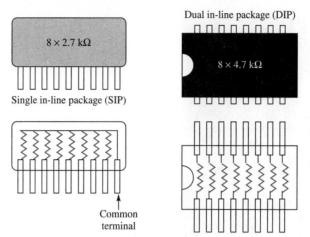

Figure 3-30   SIP and DIP resistor networks.

Chip Resistors

Chip resistors are small, high-quality, extremely stable resistors manufactured on small, rectangular, ceramic chips. See Figure 3-31. A typical chip size is $1.6 \times 3.2 \times 0.6$ mm, and new, smaller standards are being established every day. Power dissipation is typically ⅛ to ¼ watt. A thick carbon film is deposited on the ceramic chip and wraparound metal end terminals are attached for easy surface mounting to a printed circuit board. Chip resistors are rugged, temperature-stable, discrete components available in tolerances of ±1% or ±5%. They are commonly used as discrete (separate, distinct, or individual) components along with integrated circuits on a small PC (printed circuit) board to form a hybrid circuit. Often, these hybrid circuits are sealed in a resin compound or small casing with pins protruding for connection to the outside world. Chip resistors were among the first components used in the more modern **surface mount technology (SMT)** of today. Instead of components having leads that are bent and stuck through mounting holes in a

Figure 3-31   Chip resistors for surface-mount PC boards. (Courtesy of Allen-Bradley Co.)

printed circuit board, many of today's mass-production components are leadless, mounted directly to the surface of the PC board.

The Resistor Color Code

Why Have a Color Code?

Reading the color code, like almost anything else, is very easy if you know how. Like any other skill, it takes practice. You might wonder why anyone would bother using a color code when all you need to do is print the value on the component. That's a good question. Many components, including resistors, do simply have the value clearly marked. However, this can be more of a problem than reading a color code, because the numbers are often rubbed off or the component is mounted on a printed circuit board in such a way that you cannot read the value.

See color photo insert page A6.

Determining Resistor Value

Figure 3-32 demonstrates how the color code relates to the resistor's color bands. Notice that there are four (4) color bands, or markings, of concern. The first two color bands represent the two significant digits of the value. The third band, or color dot, is called the *multiplier* because the two significant digits are multiplied by the power of ten represented by this band, or dot.

It is easy to remember the multiplier band if you remember it this way: The digit represented by the color of the third color band is the exponent to the base 10 that is multiplied by the significant digits. Essentially, the third band tells you how many zeros (0) to add to the two significant digits. For example: if the first three bands are red, orange, black, the two significant digits are 2 (red) and 3 (orange) while the multiplier band is 0 (black, which stands for $10^0 = 1$). Since the multiplier band is black, or 0, you would add

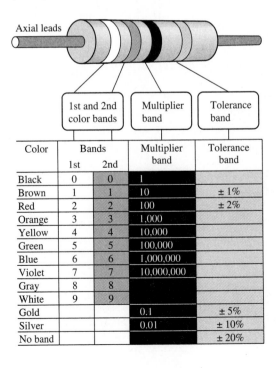

Color	Bands		Multiplier band	Tolerance band
	1st	2nd		
Black	0	0	1	
Brown	1	1	10	± 1%
Red	2	2	100	± 2%
Orange	3	3	1,000	
Yellow	4	4	10,000	
Green	5	5	100,000	
Blue	6	6	1,000,000	
Violet	7	7	10,000,000	
Gray	8	8		
White	9	9		
Gold			0.1	± 5%
Silver			0.01	± 10%
No band				± 20%

Figure 3-32 The resistor color code (also used for five band coded resistors).

no zeros and the value would be 23 Ω (23 · 1 = 23). If the third band is brown, you add one zero and the value is 230 Ω (23 · 10^1 = 23 · 10 = 230), etc. Notice that the number represented by the color of the third band is the exponent of the multiplier.

Determining Tolerance

The fourth band is the tolerance band. It indicates that the actual value of the resistor is within plus or minus so many percent of the value indicated by the first three bands. There are three possible colors for the fourth (tolerance) band: red (±2%), gold (±5%), and silver (±10%). If the fourth band is missing, the tolerance of the resistor is ±20%. The actual resistance of the resistor will be somewhere within the range of tolerance. To determine the range of tolerance for a particular resistor, first determine the color code value from the first three bands, then calculate the indicated percentage. The calculated percentage is then added to and subtracted from the color coded value to find the range of tolerance. For example: a 2,700 Ω resistor that has a 5% tolerance will have a range of tolerance from 2,565 Ω to 2,835 Ω (2,700 · 0.05 = 135 Ω, so 2,700 − 135 = 2,656 Ω and 2,700 + 135 = 2,835 Ω). If, for example, a 10 kΩ resistor is manufactured with a tolerance of ±10%, the actual value of the resistor could be anywhere between 9 kΩ and 11 kΩ (10 kΩ · 0.1 = 1 kΩ, therefore the actual value is in the range of 10 kΩ − 1 kΩ to 10 kΩ + 1 kΩ). The manufacturer does not guarantee that the actual value of the resistor will match the marked value. The manufacturer only guarantees that the actual value will be within the range of tolerance. Common tolerances are ±1% for precision wirewound and film resistors to ±5%, ±10%, and ±20% for carbon composition and carbon film resistors. The 20% tolerance resistors are now virtually nonexistent.

EXAMPLE SET   3-2   USING THE COLOR CODE

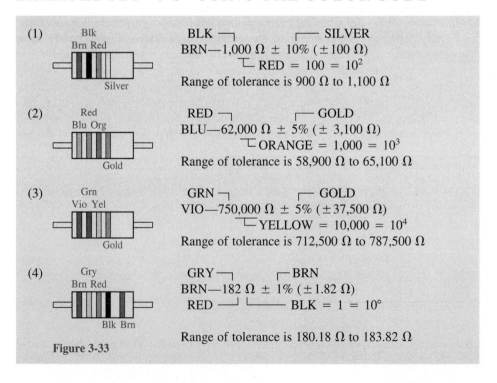

(1)   Blk
      Brn Red

      Silver

BLK ⌐   ⌐ SILVER
BRN—1,000 Ω ± 10% (±100 Ω)
            ⌐ RED = 100 = 10^2
Range of tolerance is 900 Ω to 1,100 Ω

(2)   Red
      Blu Org

      Gold

RED ⌐   ⌐ GOLD
BLU—62,000 Ω ± 5% (± 3,100 Ω)
            ⌐ ORANGE = 1,000 = 10^3
Range of tolerance is 58,900 Ω to 65,100 Ω

(3)   Grn
      Vio Yel

      Gold

GRN ⌐   ⌐ GOLD
VIO—750,000 Ω ± 5% (±37,500 Ω)
            ⌐ YELLOW = 10,000 = 10^4
Range of tolerance is 712,500 Ω to 787,500 Ω

(4)   Gry
      Brn Red

      Blk Brn

GRY ⌐   ⌐ BRN
BRN—182 Ω ± 1% (±1.82 Ω)
RED ⌐   ⌐ BLK = 1 = 10^0

Range of tolerance is 180.18 Ω to 183.82 Ω

Figure 3-33

Example Set 3-2 will help you get started using the color code. You may want to purchase a grab bag (random assortment) of resistors and go through them, determining their color code values for practice. Also, it is good to have an assortment on hand for experiments and projects ($\frac{1}{2}$-, $\frac{1}{4}$-, and $\frac{1}{8}$-watt sizes are the sizes most commonly used). See Appendix F for tables of standard values. Also, a color code tutorial program is part of the circuit analysis software available free to users of this textbook.

The Five Band Code

Precision 1% and 2% film resistors are labeled with a five band color code. The first three bands indicate three significant digits, the fourth band is the multiplier, and the fifth band indicates the percentage tolerance. Color and multiplier values of Figure 3-32 are also used for the five band code. An example is given in Example Set 3-2. See Appendix F for tables of standard values.

Remembering the Code

I do *not* recommend that you purchase a clever pocket color code calculator or that you carry a color code chart in your wallet or purse for the rest of your life. This will only create a dependency on the chart and, in the long run, make the color code uncomfortable for you to use. It is far better to remember the code by mental association. I have heard many little rhymes over the years that are designed to help you remember the code in a natural way, some of which I would not repeat. However, I will offer you one of my own.

Black	Berries	Ripen	On	Yonder	Gates	But	Violets	Grow	Wild
Black	Brown	Red	Orange	Yellow	Green	Blue	Violet	Gray	White
0	1	2	3	4	5	6	7	8	9

Better	Run	Get	Some	Now
Brown	Red	Gold	Silver	No Color
1%	2%	5%	10%	20%

Variable Resistors

Variable resistors are designed with a sliding contact, or wiper, that can easily be moved from one end of the resistance element to the other for the purpose of selecting a different amount of resistance measured from a given end. Like fixed resistors, variable resistors are manufactured in a large variety of sizes and shapes with a variety of materials. We shall take time here to examine the major types and uses.

Potentiometers (Pots)

Potentiometers are generally low-power variable resistors used to adjust the amplitude, or level, of an AC or DC voltage. The word *potentiometer* is coined from the words *potential* and *meter*. The idea is that the potential (voltage) can be metered out (varied).

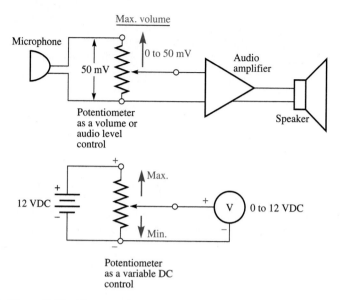

Figure 3-34   The potentiometer is used to vary voltage as in audio volume control or variable voltage control.

These devices are generally used as volume or tone controls on audio equipment and level controls on test equipment. Some potentiometers have a built-in switch that allows a piece of equipment to be turned on as the knob is first rotated to increase the volume. The resistive element in the potentiometer may be carbon on a phenolic base, CERMET (a mixture of metal powders and binders), conductive plastic, or wirewound. The schematic symbol for the potentiometer and examples of its use are shown in Figure 3-34.

Rotary or Slide

Potentiometers are manufactured in two basic mechanical designs: rotary and slide. The **rotary pot** (potentiometer) has a wiper arm connected to a shaft that is rotated with a knob or screwdriver. The **slide pot**, or **slider pot**, has a straight resistive element and a wiper contact that is slid along its length. Figure 3-35 illustrates these two mechanical designs.

Trim Pots

Potentiometers come in all sizes and shapes. Figure 3-36 includes a modest selection of potentiometers. The very small thumbwheel- or screwdriver-adjusted potentiometers are usually mounted on a printed circuit board to allow for calibration of a sensitive circuit. These miniature pots are normally found inside a piece of equipment, and therefore are not frequently adjusted by the operator. Often, these miniature potentiometers are called **trim pots** or **trimmer pots**, because they are used to fine-tune, or calibrate, a circuit. For very delicate fine adjustments, multiturn potentiometers are used. A ten-turn (10T) trim pot, for example, will have a threaded shaft that the sliding contact will follow. The shaft must be rotated completely ten times in order for the wiper to move from one end of the resistive element to the other. This allows for very fine adjustment.

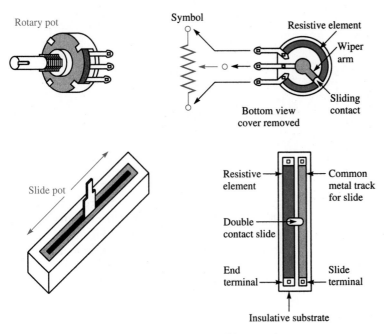

Figure 3-35   The very common rotary and slide potentiometers.

Taper

The **taper** of a potentiometer refers to the distribution of resistance along the resistive element. The carbon element can be made to have more resistance in one half than the other by controlling the cross-sectional area of the carbon or its density. If the resistance of the element is evenly distributed, the potentiometer is said to be *linear*, or have a **linear**

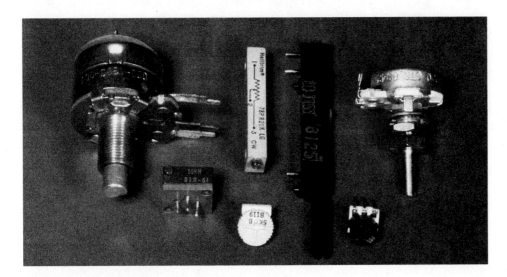

Figure 3-36   Common chassis-mount and PC board-mount potentiometers.

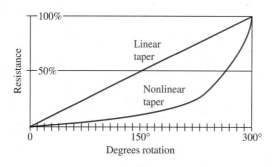

Figure 3-37   The resistance element of potentiometers is manufactured to provide a linear or nonlinear change in resistance known as taper.

taper. If the distribution of resistance increases as the wiper arm is moved from one end to the other, the potentiometer is said to be nonlinear, or have a **nonlinear taper**. Single-turn rotary pots have about 300° of rotation from one end to the other. Figure 3-37 graphically demonstrates the difference between a linear taper and a nonlinear taper potentiometer by comparing degrees of rotation to resistance. The nonlinear graph may be logarithmic or exponential in nature. Audio volume controls are of the nonlinear type, allowing greater control of loudness at normal or low listening levels due to the slow change in resistance in the first 150° of rotation.

Rheostats

The **rheostat** is a close relative of the potentiometer in that it is constructed in similar fashion. In fact, a potentiometer can be used as a rheostat as long as its power rating is not exceeded. Usually, however, a rheostat has a wirewound resistive element that can handle higher power levels than a potentiometer.

The greatest distinction between a potentiometer and a rheostat is found in their purpose or use in a circuit. The potentiometer is used to control voltage, while the rheostat is used to control current. Figure 3-38 demonstrates the use of a rheostat and includes the schematic symbols. Note that the rheostat is functionally a two-terminal device, whereas the potentiometer is a three-terminal device. When a potentiometer is used as a rheostat, one of the end terminals is left unused or tied to the wiper terminal.

Thermistors

Thermistors are a very interesting and useful family of devices. They are, in fact, *therm*al res*istors*, from which the name *thermistor* is coined. The thermistor is a two-lead device whose resistance changes with any change in temperature. Most thermistors are specified as having a certain number of ohms of resistance at 25°C. The thermistor's value may increase or decrease as the temperature of the thermistor is raised above the 25°C reference temperature. Thermistors are also referred to as resistive temperature devices (RTDs).

If the resistance of the thermistor increases as temperature increases, it has a **P**ositive **T**emperature **C**oefficient (**PTC**). If the resistance decreases with an increase in temperature, the thermistor has a **N**egative **T**emperature **C**oefficient (**NTC**). Most thermistors are of the NTC type. The temperature coefficient tells you the amount of change in resistance for every degree Celsius change in temperature. It is sometimes called the thermistor *sensitivity* and is expressed as a certain percentage change in resistance per degree Celsius. For example, a thermistor may have a sensitivity of 6%/°C (@25°C). An NTC thermistor

RHEOSTAT USED AS LIGHT DIMMER

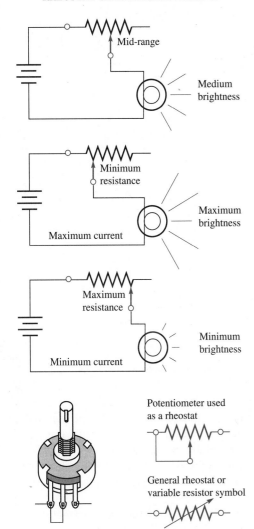

Potentiometer used
as a rheostat

General rheostat or
variable resistor symbol

Figure 3-38 The rheostat is used to vary
and control the amount of current in a
circuit. A potentiometer can be wired to
serve as a rheostat.

with a 6%/°C sensitivity will drop in resistance by 6% for every 1°C increase in temperature. As you might imagine, every succeeding 6% drop in resistance is actually fewer
ohms because the overall resistance is getting less each time. In other words, 6% of 100
Ω is 6 Ω and 6% of 94 Ω is 5.64 Ω and 6% of 88.36 Ω is 5.3 Ω. See what I mean? This
demonstrates the nonlinearity of these devices. The change in resistance versus change in
temperature is not a straight-line function (does not graph as a straight line). In addition,
the sensitivity itself changes with temperature. As the temperature increases, the sensitivity
decreases, and vice versa. This compounds the nonlinearity. The actual graph is more
exponential, as is shown in Figure 3-39.

Because of the negative temperature effect, the NTC thermistor is an excellent choice
for suppressing current surges when a circuit is first energized. An incandescent light bulb
with a tungsten filament has a positive temperature coefficient, and therefore has a very
low initial resistance when the power is first applied. As the filament increases in temper-

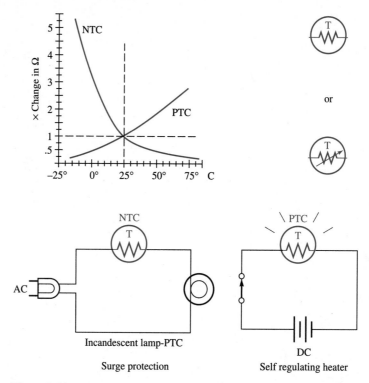

Figure 3-39   Thermistors have either a positive or a negative temperature coefficient.

ature, the resistance of the filament increases greatly. However, the initial surge current is quite large and destructive. If an NTC thermistor is placed in series with the light bulb, the high initial resistance of the thermistor will counteract the very low initial resistance of the filament and prevent the surge current. Naturally, this greatly increases the life of the bulb. Special disc-type NTC thermistors can be placed in a light socket and the bulb can then be screwed in. The NTC thermistor is used in many applications for the purpose of surge suppression.

Figure 3-39 also shows a typical graph for a silicon PTC thermistor. Note that silicon PTC thermistors are very linear compared to the NTC type. This makes them attractive choices for remote temperature sensors in analog and digital thermometers and computer-controlled systems. Also, PTC thermistors make excellent miniature self-regulating heaters. The current flowing through the thermistor heats the thermistor, which causes the resistance of the device to increase. This in turn decreases current, which controls the amount of heat. In other words, the device reaches equilibrium at some temperature and current. The PTC thermistor heaters are often used to heat LCDs (Liquid Crystal Displays) to operating temperature.

Like many other components, thermistors are manufactured in a great variety of shapes, sizes, and values for various applications. Whatever their shape, thermistors are basically a mixture of semiconducting materials. Powdered metal oxides such as nickel and manganese oxides are mixed with water and various binders to form a paste, which is baked at very high temperatures. A metal deposit, usually silver, is placed on opposite sides of the thermistor chunk to which leads are connected. The specific mixture of oxides and binder will determine the range of resistance for the thermistor.

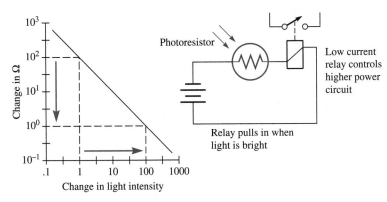

Figure 3-40   The resistance of a photoresistor changes with impinging light intensity.

Photoresistors

Photoresistors, often called photoconductive cells, are special devices that change resistance with changes in light levels. Light falling on the surface of the cell will actually cause the resistance of the device to change. If the light is bright, the resistance will be very small, and if the cell is totally hidden from light the resistance will be very high. A certain photoresistor, for example, may have 1000 Ω of resistance in normal room lighting, 100 Ω of resistance in sunlight, and 1 MΩ in the dark. Figure 3-40 shows a graph of resistance versus light intensity for a typical photoresistor as well as a simple light-sensitive circuit.

The most common photoresistors are cadmium sulfide, referred to as CdS cells. This semiconductor material is very active in the presence of light releasing many free electrons. The external light energy frees the electrons, making them available for electron flow.

You have probably already thought of several applications for this interesting device. Some of those applications might be: to automatically turn on security lighting when it starts to get dark or turn street lights on at dusk; as sensors in light-beam burglar systems or light meters for cameras; and the list definitely goes on.

Take time now to review this section and test your knowledge with Self-Check 3-6.

S E L F - C H E C K   3 - 6

1. What is the difference between a chip resistor and an axial lead film resistor?

2. What is the purpose for the fourth band of the four-band color code for carbon resistors?

3. Determine the value and range of tolerance for a carbon resistor marked as follows: red, yellow, orange, gold.

4. Determine the value and range of tolerance for a metal film resistor marked as follows: yellow, blue, yellow, red, red.

5. Describe the purpose or use for (a) a potentiometer, and (b) a rheostat.

Summary

FORMULAS

(3.1) Life (Hours) $= \dfrac{\text{Capacity}}{\text{Current}} = \dfrac{\text{AH rating}}{\#\ \text{of amperes}}$

(3.2) A (CM) $= d^2$ (mils)

(3.3) $R = \rho \cdot L/A$

R is resistance in ohms.
ρ is resistivity in CM $\cdot$ Ω/ft.
L is length of wire in ft.
A is cross-sectional area in CM [$A = d^2$ (d is diameter in mils.)]

CONCEPTS

- In a **DC** circuit, current flows only in one direction.
- A practical circuit is one that has a source of voltage, conductors, a practical load, and continuity.
- Schematic symbols are used to document a circuit.
- A cell is a single unit that produces a certain amount of voltage due to a chemical reaction.
- A battery is a combination of two or more cells forming a single unit.
- Secondary cells are rechargeable, primary cells are not.
- The capacity of a battery is expressed in ampere hours (AH).
- Cells are connected in series for higher voltages.
- Cells are connected in parallel for higher capacity.
- Voltmeters are connected in parallel across the terminals of the voltage source to measure voltage.
- Ammeters are connected in series and actually become part of the circuit in order to measure current flowing in the circuit.
- The resistivity or specific resistance of a material is the amount of resistance in a one-foot length of wire, made of that material, with a cross-sectional area of one circular mil (CM).
- The higher the AWG number, the smaller the wire size.
- Switches are used in electrical circuits as control devices.
- Fuses and circuit breakers are circuit-protection devices. They are used to protect circuit components, not people.
- Metal film resistors are manufactured to closer tolerance, are less affected by temperature change, and generate less noise in a sensitive circuit than carbon resistors do.
- Wirewound resistors are close tolerance and are used in high-power applications.
- Chip resistors are used for surface mount (SMT) applications.
- Potentiometers are used to vary or control voltage, whereas rheostats are used to control current.
- Thermistors change in resistance as temperature changes (PTC if resistance increases with temperature and NTC otherwise).
- Photoresistors change resistance with changes in light intensity: the more light, the less resistance.

SPECIAL TERMS

- DC, AC
- Continuity
- Schematic, schematic symbol
- Battery, voltaic cell, electrolyte
- Dry cell, wet cell, primary cell, secondary cell
- Capacity, ampere hours (AH)
- Shelf life
- Open circuit voltage, closed circuit voltage
- Battery system
- Charge and discharge rate
- Resistivity, specific resistance (ρ)
- Circular mil (CM), cross-sectional area, AWG
- Normally closed (NC), normally open (NO)
- Interlocked, independent
- Pole, throw, SPST, SPDT, DPST, DPDT, TT
- Fuse voltage rating, fault-condition current
- Short circuit rating, fuse response time
- Thermal trip, electromagnetic trip
- Carbon composition resistor, film resistor
- Adjustable resistor, variable resistor
- Potentiometer, rheostat, taper
- Thermistor, NTC, PTC
- Photoresistor

Need to Know Solution

Each of the four series strings of D cells will contribute $\frac{1}{4}$ A to the total 1 A of current. The three cells in series produce 4.5 V. The fuse must be slow acting to prevent blowing

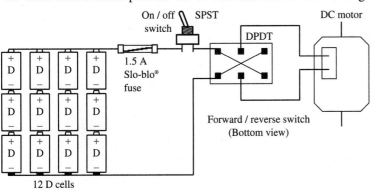

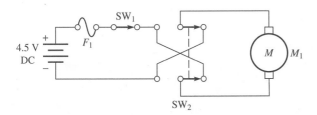

every time the motor is turned on. The DPDT switch provides FORWARD/REVERSE operation by changing the voltage polarity applied to the motor.

It is interesting to note that if the mechanical load on the motor is increased, more current will be drawn from the batteries. If the current exceeds 1.5 A, the fuse will eventually burn open and the motor, of course, will stop. This is good, since the blown fuse prevents the motor from burning up due to overload.

Questions

3-1 Circuit Essentials and Documentation

1. List the four essentials of a practical DC circuit.
2. What is the meaning of the term *continuity*?
3. What does it mean to document a circuit?
4. What is a schematic?
5. How do you indicate that two crossing conductors in a schematic are not connected?

3-2 Batteries

6. What is the difference between a cell and a battery?
7. What is a secondary cell?
8. What are two examples of secondary cells?
9. What is the unit of measure for battery capacity?
10. What characteristic makes the Ni–Cad very different from the alkaline cell?
11. Which probe on a multimeter, the red or the black, should you connect to the positive terminal on the battery when measuring voltage?
12. Describe in detail how an ammeter is connected in a DC circuit to measure current. How does the use of the ammeter differ from the use of the voltmeter?

3-3 Conductors

13. Which of the materials discussed in this chapter is the best conductor?
14. What is specific resistance?
15. Define the circular mil.
16. What is the resistivity of copper?
17. List at least three factors that determine the overall resistance of a piece of wire.

3-4 Switches

18. What is the purpose for switches in electrical circuits?
19. If a switch is normally closed (NC), what does that mean?
20. What is SPDT?
21. What is DPTT?
22. Describe this rotary switch type—4P12P.
23. What does NOPB stand for?

3-5 Fuses and Circuit Breakers

24. Explain how a fuse can protect a circuit.
25. Define fuse voltage rating.

26. What are slow acting fuses and when are they used?
27. Describe the proper placement of a fuse in a circuit.
28. What is one advantage of the circuit breaker over the fuse as a circuit-protection device?

3-6 Resistors

29. Carbon composition resistors are manufactured to what standard tolerances?
30. Describe the construction of a film resistor.
31. What are three advantages of the metal film resistor over the carbon resistor?
32. What is a tapped wirewound resistor?
33. Describe a chip resistor.
34. What is the meaning of SMT?
35. What is the purpose of the fourth color band on the carbon composition or carbon film resistor?
36. Describe the color code for 1% and 2% film resistors.
37. What is the purpose or function of a potentiometer?
38. For what are trim pots used?
39. What is meant by the taper of a potentiometer?
40. What is a rheostat used to control?
41. Explain how a potentiometer can be used as a rheostat.
42. What is a PTC thermistor?
43. Describe the operation of a photoresistor in relation to light intensity.
44. What is CdS?

Problems

3-2 Batteries

1. How many hours of service can you expect to get from a 5 AH battery that is supplying ¼ A of current to a load?
2. How many hours of service can you expect to get from a 20 AH battery that is supplying 2 A of current to a load?
3. If six 1.5 V cells that each have a capacity of 4 AH are placed in series, what is the overall capacity of the series arrangement? What is the overall voltage?
4. How would you arrange eight 1.5 V cells to produce 6 V?
5. How many 0.5 AH cells will you need to make a 6 AH supply? What arrangement will you use to connect them together?

3-3 Conductors

6. If the diameter of a wire is 0.008 inches, what is its diameter expressed in mils? What is its cross-sectional area in CM?
7. What is the overall resistance of an aluminum wire that has a diameter of 0.002 inches and a length of 1,200 ft?
8. What is the overall resistance of a copper wire that has a diameter of 0.014 inches and a length of 200 ft?
9. What is the overall resistance of a 3,000-ft length of AWG#18 copper wire?
10. What is the overall resistance of a 12,000-ft length of AWG#24 copper wire?

3-6 Resistors

Sharpen your color code skills on the following:

	1st Band	2nd Band	3rd Band	4th Band	5th Band	Value	Range of Tolerance
11.	brn	red	org	gold	none	?	? to ?
12.	red	vio	yel	none	none	?	? to ?
13.	wht	brn	org	sil	none	?	? to ?
14.	blu	red	red	gold	none	?	? to ?
15.	org	blk	grn	sil	none	?	? to ?
16.	red	red	red	gold	none	?	? to ?
17.	brn	yel	blk	brn	brn	?	? to ?
18.	red	yel	wht	org	red	?	? to ?
19.	grn	brn	brn	blk	brn	?	? to ?
20.	blu	yel	wht	red	brn	?	? to ?

Answers to Self-Checks

Self-Check 3-1

1. Direct current
2. Direct current is current that flows continually in one direction, while alternating current is current that continually reverses direction at a certain rate or frequency.
3. Continuity means a continuous, unbroken path for current, forming a complete circuit. It is essential because no practical circuit will operate if there is no complete path for current.
4. A heavy black dot over the intersection of the wires.
5. See the symbol given in Figure 3-2.

Self-Check 3-2

1. A cell is a single voltaic unit that produces a definite voltage according to the chemical reaction that takes place within. A battery is formed by combining two or more cells for higher capacity or voltage.
2. A solid, semisolid, or liquid inside the cell that permits ionic conduction, or current, between the electrodes
3. A dry cell has a solid or semisolid electrolyte and a wet cell has a liquid electrolyte.
4. A rechargeable cell
5. Ampere hour (AH)
6. The length of time a cell can sit idle before it is considered useless
7. LeClanché cell
8. Secondary
9. Very long shelf life and high capacity
10. Lead–acid
11. Two series strings of two cells in parallel; a 2 × 2 arrangement

Self-Check 3-3

1. To connect the load to the voltage source and provide continuity
2. Silver, copper, gold, aluminum, iron
3. Silver
4. The amount of resistance in a one-foot length of wire, made of a particular material, with a cross-sectional area of one circular mil
5. 14 CM · Ω/ft @ 20°C
6. Length, cross-sectional area, temperature, type of material
7. 19.47 Ω
8. Copper can be soldered and has less resistance.

Self-Check 3-4

1. The circuit is off until the button is pushed.
2. The buttons mechanically interact with each other. When one is depressed, the one that was previously depressed is automatically released.
3. A kind of switch that usually has two positions, one or the other of which is chosen.
4. 5 poles and 12 positions.
5. To monitor movement of a mechanical device.
6. **S**ingle **P**ole **D**ouble **T**hrow

Self-Check 3-5

1. Circuit protection in case of a short or overload.
2. The current labeled on the fuse. The fuse will blow when this current level is exceeded.
3. How long it takes for the fuse to blow when the fault-condition is exceeded.
4. As close to the voltage source as possible.

Self-Check 3-6

1. The chip resistor has no leads, only metal end caps. The chip resistor is mounted directly on the surface of the PC board (SMT). Axial lead resistors are hole mounted.
2. The fourth band is the tolerance band.
3. 24,000 Ω ±5%, 22,800 Ω to 25,200 Ω
4. 46,400 Ω ±2%, 45,472 Ω to 47,328 Ω
5. a. A potentiometer is used to vary voltage.
   b. A rheostat is used to vary current.

SUGGESTED PROJECTS

1. Update your personal Electricity and Electronics Notebook by adding what you consider to be the more important or practical information from this chapter.

2. Purchase a variety of C or D cells. Test their capacity by connecting them to a fixed load and timing them while measuring their voltage until it drops to 1 V. Draw a discharge curve for each type for comparison. I suggest a 10 Ω, 10 watt, wirewound resistor available at most electronic parts stores.

3. Obtain an SPST toggle switch and a DPDT toggle switch. Experiment with the circuit in the ''Need To Know'' part of this chapter. You don't need to use many batteries; two or three in series will make a sufficient DC source.

Edison's light bulbs ranged from very tiny low-power bulbs to large high-power bulbs such as those shown here. The large incandescent bulb on the left in this photograph is rated at 50,000 W and was manufactured in 1929 for The Observance of Lights Golden Jubilee, a 50th anniversary celebration honoring Edison and his wondrous invention. The bulb on the right is rated at 75,000 W and was built by General Electric and Corning Glass for the 1954 celebration of The Observance of Lights Diamond Jubilee. The bulbs can be seen at the Edison-Ford Winter Estates in Fort Myers, Florida.

Chapter 4

Introduction to Basic Circuit Analysis

CHAPTER OUTLINE

4-1 Ohm's Law
4-2 Power

4-3 Circuit Analysis Procedures

OBJECTIVES

After studying this chapter, you will be able to

- use Ohm's Law to solve for current, resistance, or voltage when two of these three parameters are known.
- calculate available power, power dissipation, or power loss when any two of the three main parameters (current, voltage, resistance) are known.
- explain the concepts of power rating and derating.

- understand and read the watthour meter and calculate the cost of electricity when the electric rate and total usage for a given period of time are known.
- explain the difference between an open and a short in terms of resistance and current.
- explain the difference between theoretical and experimental circuit analysis.
- use the multimeter in experimental circuit analysis.

INTRODUCTION

Now that you have become familiar with the basic elements of a practical circuit, you are ready for an **introduction to basic circuit analysis**. In this chapter, you will be introduced to the mainstay of electronic theory and analysis, Ohm's Law. You will soon discover how valuable Ohm's Law is in analyzing the relationship between current, voltage, and resistance in a practical circuit. Electrical power will also be discussed, along with power calculations. You will learn how to perform theoretical and experimental circuit analysis, comparing calculated and measured circuit values. Naturally, basic troubleshooting will be discussed as a practical application of your circuit analysis skills. This chapter is an introduction to the more detailed circuit analysis to be presented in later chapters. A thorough understanding of this material will prove invaluable as you continue your experiences in electricity and electronics.

NEED TO KNOW

Suppose you have a small power supply and you want to add a resistor in series with one of the power supply's terminal posts to limit the current to a certain maximum value in the event the terminals are shorted together. The questions you must answer are: What value resistor should I choose? What power rating should it be? There are two things you need to know: how to calculate the needed resistance when voltage and current are known, and how to calculate the power dissipation of the resistor. Let us suppose that the voltage is 12 V and the maximum current the resistor will permit to flow during short circuit is 1 A. What value, power rating, and type should the resistor be? Not sure how to proceed? Then you definitely have a need to know.

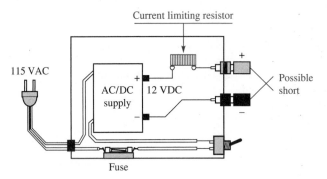

4-1 Ohm's Law

The Law

There is a simple mathematical relationship that demonstrates that the amount, or intensity, of current in a circuit is determined by the electromotive force applied to the circuit and by the amount of resistance in the circuit. This mathematical relationship is known world-

---

Georg Simon Ohm (1787 – 1854)

Georg Simon Ohm was born in Erlangen, Germany. He graduated from the University of Erlangen and later became a professor of physics at the University of Nuremburg. In 1826, Ohm discovered the mathematical relationships between current, voltage, and resistance. This discovery became known as Ohm's Law, forming the basic foundation for all electrical circuit analysis. The

unit of resistance, the ohm (Ω), was named in his honor. His contributions earned him a permanent place of recognition in electronics for all time. (Photo courtesy of The Bettmann Archive.)

---

wide as Ohm's Law in honor of its discoverer, Georg Simon Ohm. The relationship is a simple one and is stated as follows:

The amount of current in a circuit is directly proportional to the amount of voltage applied to the circuit and inversely proportional to the amount of resistance in the circuit.

Mathematically, Ohm's Law is expressed like this:

Ohm's Law

$$I = E/R \qquad\qquad (4.1)$$

Where I is the Intensity of current measured in amperes (A), E is the electromotive force measured in volts (V); R is the resistance measured in ohms (Ω).

Ohm's Law is the most fundamental of all mathematical tools used in circuit analysis. You will probably use Ohm's Law more frequently than any other electrical or electronic formula throughout your entire career. It will become an old friend, and its use will become as natural as breathing. In the next few subsections, we shall examine the practical use of Ohm's Law to solve for an unknown value of current, voltage, or resistance when two of these three circuit parameters are known.

Practicing the Current Law—$I = E/R$

As you can see, in Figure 4-1, the amount of current in a circuit is determined by the amount of applied voltage divided by the resistance in the circuit. The pie-shaped chart may be of help to you in remembering and using Ohm's Law. We can clearly see that if the voltage is increased, the current in the circuit will increase, and if the voltage is decreased, current will decrease. Changing resistance has the opposite effect. If the circuit resistance is increased, the current will decrease and vice versa. We say that current varies directly with voltage and inversely with resistance. Study the examples shown in Example Set 4-1.

See color photo insert pages A7 and A8.

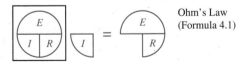

Ohm's Law
(Formula 4.1)

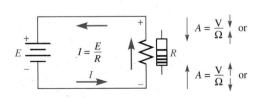

Figure 4-1   Current is a function of voltage and resistance: $I = E/R$.

EXAMPLE SET 4-1: SOLVING FOR CURRENT

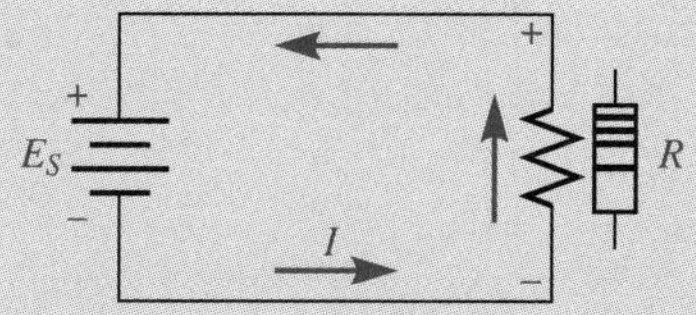

$$I = E/R$$

Figure 4-2

(1) Given: $E_s = 10$ V, $R = 10 \ \Omega$
$I = 10$ V$/10 \ \Omega = 1$ A
(2) Given: $E_s = 115$ V, $R = 1$ kΩ
$I = 115$ V$/1$ k$\Omega = 115$ mA
(3) Given: $E_s = 35$ V, $R = 2.7$ kΩ
$I = 35$ V$/2.7$ k$\Omega = 0.013$ A $= 13$ mA
(4) Given: $E_s = 9$ V, $R = 33$ kΩ
$I = 9$ V$/33$ k$\Omega = 0.273$ mA $= 273 \ \mu$A
(5) Given: $E_s = 6$ kV, $R = 1.5$ MΩ
$I = 6$ kV$/1.5$ M$\Omega = 4$ mA

Practicing the Voltage Law—$E = I \cdot R$

Figure 4-3 shows you how to find the amount of applied voltage if only circuit current and circuit resistance are known. Ohm's Law is simply rearranged with the unknown voltage parameter on the left and the known values of current and resistance on the right of the equation. This form of Ohm's Law tells us that current flowing through a resistance will leave a difference of potential (voltage) across the resistance. This is an important concept and we will develop it more thoroughly as we progress in circuit analysis. Consider the examples shown in Example Set 4-2.

$$E = I \cdot R \tag{4.2}$$

Practicing the Resistance Law—$R = E/I$

Here, as shown in Figure 4-5, the value of an unknown resistance can be found if current and voltage are known. The current through a resistance is divided into the voltage across the resistance to find the value of the resistance. This is an important basic circuit analysis

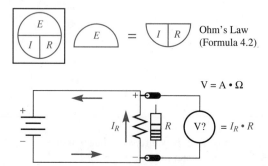

Ohm's Law
(Formula 4.2)

$V = A \cdot \Omega$

$= I_R \cdot R$

Figure 4-3 Resistor voltage is a function of current and resistance: $V = I \cdot R$.

EXAMPLE SET 4-2:   SOLVING FOR VOLTAGE

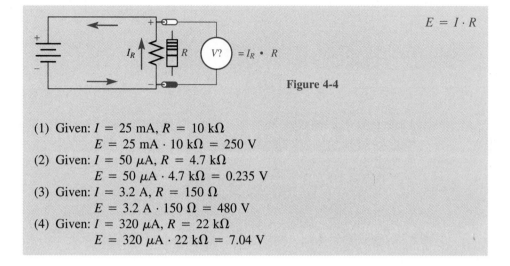

Figure 4-4

(1) Given: $I = 25$ mA, $R = 10$ kΩ
$E = 25$ mA · 10 kΩ $= 250$ V
(2) Given: $I = 50$ μA, $R = 4.7$ kΩ
$E = 50$ μA · 4.7 kΩ $= 0.235$ V
(3) Given: $I = 3.2$ A, $R = 150$ Ω
$E = 3.2$ A · 150 Ω $= 480$ V
(4) Given: $I = 320$ μA, $R = 22$ kΩ
$E = 320$ μA · 22 kΩ $= 7.04$ V

technique. You can calculate any unknown value of resistance using this form of Ohm's Law when current and voltage are known or obtained by direct measurement. Example Set 4-3 illustrates the use of this form of Ohm's Law.

$$R = E/I \qquad\qquad (4.3)$$

Using Conductance—(G)

You may recall from Chapter 2 that conductance (G) is the reciprocal of resistance (R). That is: $G = 1/R$. You learned that ohms divided into 1 equals siemens (S). (Expressed like this: # S $= 1/\#$ Ω.) At times, it may be convenient for you to convert resistances to conductances in solving particular problems. In a formula such as Ohm's Law, $1/R$ may be replaced with G or R may be replaced with $1/G$. For example, the basic Ohm's Law formula for current may be modified by substituting G for $1/R$. Therefore, current can be found by multiplying voltage by conductance. Mathematically it looks like this:

$$I = E/R = E \cdot (1/R) = E \cdot G \text{ by substitution}$$

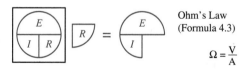

Ohm's Law
(Formula 4.3)

$$\Omega = \frac{V}{A}$$

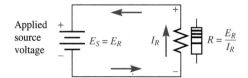

Figure 4-5 Circuit resistance can be found when voltage and current are known: $R = E/I$.

EXAMPLE SET   4-3:   SOLVING FOR RESISTANCE

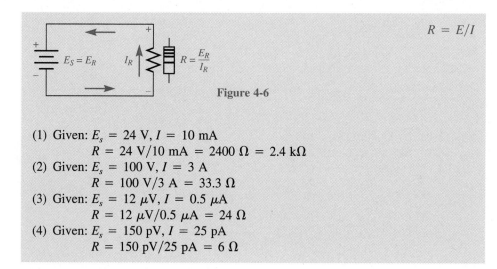

Figure 4-6

$R = E/I$

(1) Given: $E_s = 24$ V, $I = 10$ mA
$\qquad R = 24$ V/10 mA $= 2400$ $\Omega = 2.4$ kΩ

(2) Given: $E_s = 100$ V, $I = 3$ A
$\qquad R = 100$ V/3 A $= 33.3$ Ω

(3) Given: $E_s = 12$ μV, $I = 0.5$ μA
$\qquad R = 12$ μV/0.5 μA $= 24$ Ω

(4) Given: $E_s = 150$ pV, $I = 25$ pA
$\qquad R = 150$ pV/25 pA $= 6$ Ω

So, Ohm's Law may be written in conductance form like this:

$$I = EG \quad \text{or} \quad \text{amperes} = \text{volts} \cdot \text{siemens} \quad (\# \text{ A} = \# \text{ V} \cdot \# \text{ S}) \tag{4.4}$$

We can do the same thing with the voltage form of Ohm's Law: $1/G$ may be substituted for R.

$$E = IR = I \cdot (1/G) = I/G \text{ by substitution} \tag{4.5}$$

$$E = I/G \quad \text{or} \quad \text{volts} = \text{amperes/siemens} \quad (\# \text{ V} = \# \text{ A}/\# \text{ S})$$

Finally, we can show that the conductance of a circuit can be found by dividing the current of the circuit by the voltage applied to the circuit.

$$R = E/I \quad \text{therefore} \quad 1/G = E/I \text{ (by substituting } 1/G \text{ for } R).$$

EXAMPLE SET   4-4:   USING CONDUCTANCE

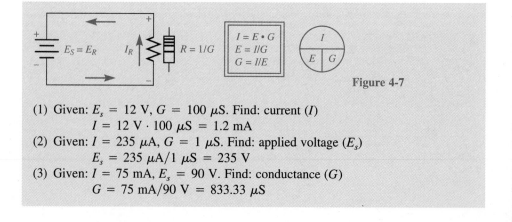

Figure 4-7

(1) Given: $E_s = 12$ V, $G = 100$ μS. Find: current (I)
$\qquad I = 12$ V $\cdot$ 100 μS $= 1.2$ mA

(2) Given: $I = 235$ μA, $G = 1$ μS. Find: applied voltage (E_s)
$\qquad E_s = 235$ μA/1 μS $= 235$ V

(3) Given: $I = 75$ mA, $E_s = 90$ V. Find: conductance (G)
$\qquad G = 75$ mA/90 V $= 833.33$ μS

If $1/G = E/I$ then $G/1 = I/E$, because it is legal to invert both sides of the equation. Therefore:

$G = I/E$ or siemens = amperes/volts (# S = # A/# V) $\qquad$ (4.6)

Do not try to memorize the conductance forms of Ohm's Law. Simply remember that resistance and conductance are reciprocals of each other and substitute G (for $1/R$) or $1/G$ (for R) in the Ohm's Law formulas. Example Set 4-4 will help you understand the use of conductance in basic circuit analysis. Also, take time to study Design Note 4-1.

DESIGN NOTE 4-1:   Ohm's Law

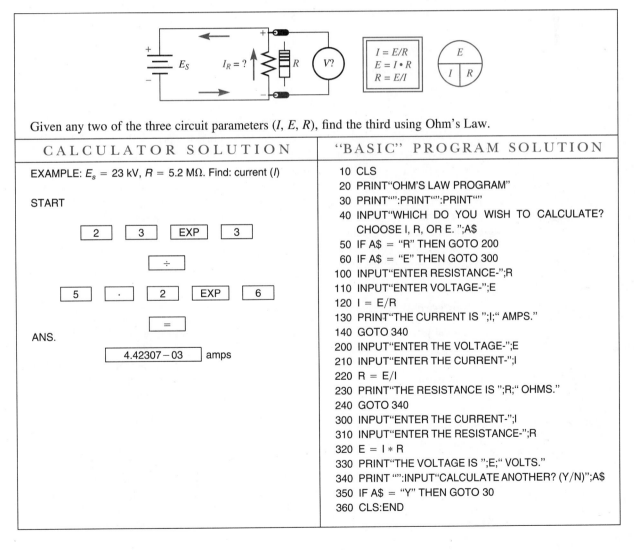

Given any two of the three circuit parameters (I, E, R), find the third using Ohm's Law.

CALCULATOR SOLUTION	"BASIC" PROGRAM SOLUTION
EXAMPLE: $E_s = 23$ kV, $R = 5.2$ MΩ. Find: current (I) START [2] [3] [EXP] [3] [÷] [5] [.] [2] [EXP] [6] [=] ANS. [4.42307 − 03] amps	10 CLS 20 PRINT"OHM'S LAW PROGRAM" 30 PRINT"":PRINT"":PRINT"" 40 INPUT"WHICH DO YOU WISH TO CALCULATE? CHOOSE I, R, OR E. ";A$ 50 IF A$ = "R" THEN GOTO 200 60 IF A$ = "E" THEN GOTO 300 100 INPUT"ENTER RESISTANCE-";R 110 INPUT"ENTER VOLTAGE-";E 120 I = E/R 130 PRINT"THE CURRENT IS ";I;" AMPS." 140 GOTO 340 200 INPUT"ENTER THE VOLTAGE-";E 210 INPUT"ENTER THE CURRENT-";I 220 R = E/I 230 PRINT"THE RESISTANCE IS ";R;" OHMS." 240 GOTO 340 300 INPUT"ENTER THE CURRENT-";I 310 INPUT"ENTER THE RESISTANCE-";R 320 E = I ∗ R 330 PRINT"THE VOLTAGE IS ";E;" VOLTS." 340 PRINT "":INPUT"CALCULATE ANOTHER? (Y/N)";A$ 350 IF A$ = "Y" THEN GOTO 30 360 CLS:END

SELF-CHECK 4-1

1. Increasing the voltage applied to a circuit will cause the circuit current to do what?

2. Increasing resistance in a circuit will cause the circuit current to do what?

3. What two factors determine the amount of current in a circuit?

4. How much current will result in a circuit if 30 V is applied and the circuit has a resistance of 10 Ω?

5. What is the resistance of a circuit that has 150 V applied and 100 mA of current flowing?

6. What is the amount of current flowing in a circuit that has a conductance of 45.45 μS and an applied voltage of 75 V?

4-2 Power

Electrical Power

Power Explained

Current, voltage, and resistance are not the only circuit parameters that you will be concerned with in the design, analysis, or troubleshooting of a practical circuit. You will also be very concerned with the amount of electrical power a circuit is consuming or dissipating. The amount of power tells you how fast, or at what rate, electrical energy is being converted to heat, accoustical energy, mechanical energy, or electromagnetic energy. The concept of power is probably not unfamiliar to you. In everyday conversation we talk about horsepower, manpower, power plants, hydroelectric power, etc. Simply stated, **power** is the rate at which a certain amount of work is done (or the rate at which an amount of energy is expended). Stated another way, power equals a certain quantity of work per unit of time (power = work/time).

Power Formula Derivation—$P = I \cdot E$

Electrical power uses the second(s) as the unit of time and the **joule (J)** as the basic unit of energy or work. One joule of energy expended every second of time is one **watt (W)** of electrical power (1 watt = 1 joule/1 second). Thus the watt is the basic unit of measure for electrical power. It is from this basic definition of the watt (1 W = 1 J/1 s) that we are able to derive a very useful formula for calculating electrical power. You may recall, from Chapter 2, that we defined the volt as one joule of energy available per coulomb of charge (1 volt = 1 joule/1 coulomb). By rearranging this relationship, we see that: 1 joule = 1 volt · 1 coulomb (1 J = 1 V · C). Since 1 watt = 1 joule/1 second, we can show that 1 watt = 1 volt · 1 coulomb/1 second by substituting 1 volt · 1 coulomb for 1 joule. We also know that 1 coulomb/1 second is 1 ampere (1 A = 1 C/1 s). If we substitute 1 ampere for 1 coulomb/1 second we see that 1 watt = 1 volt · 1 ampere (1 W = 1 V · 1 A = 1 volt · 1 ampere). Thus, from the basic definition of electrical power, we have derived a very useful and practical formula for calculating power when the amount of voltage and current are known. We have shown that power = current (I) times the electromotive force (E).

$P = I \cdot E$ (watts = amperes · volts)   (4.7)

(This formula is often referred to as the PIE power formula for obvious reasons.)

Other Power Formulas—$P = \dfrac{E^2}{R}$ and $P = I^2 \cdot R$

We can use our basic PIE power formula and Ohm's Law to derive two more very useful power formulas. All we need to do is substitute the Ohm's Law equivalents for the I in the PIE formula, then for the E. Here's what I mean. Ohm's Law says that $I = E/R$. So, let's substitute the E/R for the I in the PIE power formula.

$P = I \cdot E$ and $I = E/R$ so $P = (E/R) \cdot E$ (by substitution)

Therefore:

$P = E^2/R$ (W = V^2/Ω)   (4.8)

We also know that $E = I \cdot R$, and we can substitute $I \cdot R$ for the E in the PIE power formula.

$P = I \cdot E$ and $E = I \cdot R$ so $P = I \cdot I \cdot R$ (by substitution).

Therefore:

$P = I^2 \cdot R$ (W = A^2 · Ω)   (4.9)

Thus, we have two additional power formulas that we may use in analyzing the power demand or usage of an electrical or electronic circuit. In the following subsections, we will make use of these formulas in practical circuit analysis. Take a few minutes now to study Design Note 4-2.

Available Power

Available power is the amount of power the power source is capable of supplying. Usually, a power supply is rated at a certain voltage and current. For example, the specifications for a small AC to DC converter power supply might state that the supply can deliver a maximum of 3 A of current at 24 V. Therefore, the maximum available power is 3 A · 24 V or 72 W. Naturally, you would not choose this supply for a circuit that would demand more than 72 W of power. If you did, the power supply would overheat and components would fail (unless, of course, the power supply was properly fused to avoid such a problem). As you can see, available power is easily determined by using the PIE power formula.

Power Dissipation

Power dissipation is the amount of power that a circuit, or component in the circuit, is actually using or converting to another form of energy (heat, mechanical motion, etc.). Resistors will convert electrical power into heat. Therefore, it is very important that we

DESIGN NOTE 4-2:   Power Calculation

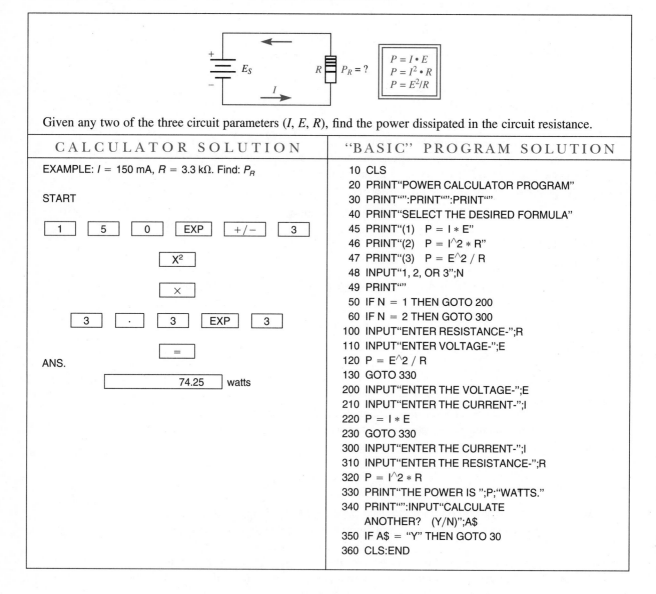

Given any two of the three circuit parameters (I, E, R), find the power dissipated in the circuit resistance.

CALCULATOR SOLUTION	"BASIC" PROGRAM SOLUTION
EXAMPLE: I = 150 mA, R = 3.3 kΩ. Find: P_R	10 CLS
	20 PRINT"POWER CALCULATOR PROGRAM"
START	30 PRINT"":PRINT"":PRINT""
	40 PRINT"SELECT THE DESIRED FORMULA"
▢1 ▢5 ▢0 ▢EXP ▢+/− ▢3	45 PRINT"(1) P = I * E"
	46 PRINT"(2) P = I^2 * R"
▢X²	47 PRINT"(3) P = E^2 / R"
	48 INPUT"1, 2, OR 3";N
▢×	49 PRINT""
	50 IF N = 1 THEN GOTO 200
▢3 ▢. ▢3 ▢EXP ▢3	60 IF N = 2 THEN GOTO 300
	100 INPUT"ENTER RESISTANCE-";R
▢=	110 INPUT"ENTER VOLTAGE-";E
ANS.	120 P = E^2 / R
	130 GOTO 330
▢ 74.25 ▢ watts	200 INPUT"ENTER THE VOLTAGE-";E
	210 INPUT"ENTER THE CURRENT-";I
	220 P = I * E
	230 GOTO 330
	300 INPUT"ENTER THE CURRENT-";I
	310 INPUT"ENTER THE RESISTANCE-";R
	320 P = I^2 * R
	330 PRINT"THE POWER IS ";P;"WATTS."
	340 PRINT"":INPUT"CALCULATE ANOTHER? (Y/N)";A$
	350 IF A$ = "Y" THEN GOTO 30
	360 CLS:END

do not overdissipate such components with too much electrical power. (Heat is one of the worst enemies of electrical and electronic circuits.)

To calculate the amount of power dissipation in an entire circuit or an individual component, we must know at least two out of three of the main circuit parameters (I, E, R). If we want to determine the amount of power dissipated by the entire circuit, we can multiply the applied source voltage (E_s) by the total current (I_t), using the PIE power formula ($P_t = I_t \cdot E_s$). If you know the total resistance (R_t) of the circuit, you can use it along with total current or source voltage to determine total power dissipation (P_t). Figure 4-8 illustrates these approaches.

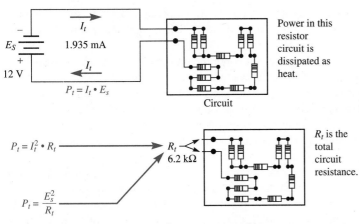

$$P_t = 1.935 \text{ mA} \cdot 12 \text{ V} = (1.935 \text{ mA})^2 \cdot 6.2 \text{ k}\Omega = (12 \text{ V})^2 / 6.2 \text{ k}\Omega = 23.22 \text{ mW}$$

Figure 4-8   The total power dissipated by a circuit is the sum of all individual power dissipations in the circuit and can be determined using total resistance, total current, and the applied voltage.

The power dissipation of an individual component, such as a resistor, is determined in similar fashion. However, you must be careful to use the parameter values that apply specifically to that component. The current through the component and the voltage directly across the component (or the resistance of the component with its current or voltage) must be known. The current and voltage are measured or calculated values, while resistance is usually marked in some fashion on the resistor (or it can be measured with the ohmmeter for accuracy). See Figure 4-9.

Power Rating and Derating

Power rating and derating of various components will be an important area of concern for you as a technician or engineer. Most components (resistors, transformers, and transistors, to name a few) are manufactured in a variety of sizes to handle different amounts of power. The manufacturer **rates** a certain size component according to the maximum amount of power the component can safely handle under normal operating conditions.

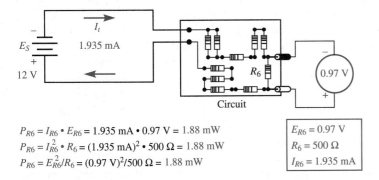

$$P_{R6} = I_{R6} \cdot E_{R6} = 1.935 \text{ mA} \cdot 0.97 \text{ V} = 1.88 \text{ mW}$$
$$P_{R6} = I_{R6}^2 \cdot R_6 = (1.935 \text{ mA})^2 \cdot 500 \text{ }\Omega = 1.88 \text{ mW}$$
$$P_{R6} = E_{R6}^2/R_6 = (0.97 \text{ V})^2/500 \text{ }\Omega = 1.88 \text{ mW}$$

$E_{R6} = 0.97$ V
$R_6 = 500$ Ω
$I_{R6} = 1.935$ mA

Figure 4-9   The power dissipated by any component depends on the current through and voltage across that component.

Normal operating conditions may mean an ambient temperature of 25°C and a specified maximum altitude above sea level (due to air density). The manufacturer may also specify air flow requirements.

As an engineer or technician, you must honor the rquirements for operating conditions for the component as specified by the manufacturer if you want good, reliable service from the component. In order to insure long, reliable service, the power rating of the component should be **derated**. Derating a component means that you will consider the component to have a much lower power-handling capability than its actual rating. This will allow the component to survive operating in an environment whose temperature is well beyond the standard 25°C (77°F).

By how much should we derate a component? The answer to this question is provided by the component's manufacturer. In the case of a transistor, the manufacturer will specify a rating adjustment of so many mW or W per degree centigrade variance from 25°C. For example: a transistor is rated as 2 W at 25°C and has a derating factor of 50 mW/1°C. If this transistor is operated at an ambient temperature of 40°C, it must be derated by 0.75 W (15°C · 50 mW/1°C = 0.75 W). As you can see, the amount of derating is dependent upon the number of degrees above 25°C and the derating factor itself. In this example, under actual operating conditions, this transistor should not be caused to dissipate any more than 1.25 W, well below the manufacturer's rating of 2 W.

Derating a resistor is a little more straightforward. As a general rule of thumb, it is considered safe and reasonable to derate almost any component by 50%. In other words, a 2 W resistor is considered as a 1 W resistor and therefore is not permitted to dissipate more than 1 W of power. This insures a reasonable margin of safety for most applications. However, this rule of thumb should not be used as an excuse to ignore manufacturer's specifications and recommendations for a component. Also, if you are involved in R&D (research & development) for the government or a private company, you may be required to follow very strict guidelines for component selection, not an arbitrary 50% rule of thumb.

Power Loss

It is usually desirable to deliver all of the intended power to the load device that will convert the power to some other form of energy. A speaker connected to your home stereo system is supposed to convert the electrical power to accoustical power or sound pressure. An electric water pump is to convert its electrical power to mechanical motion and water pressure. A light bulb will convert its electrical power to heat and light. If some power is lost somewhere before it gets to the devices just named, power will be wasted and the efficiency of the system will be less than it could or should be.

Power loss is definitely to be avoided. In the examples just listed, the most common cause or place for power loss is in the wiring that is delivering the power to the load. This loss is due to the resistance in the wiring. Recall that $P = I^2R$. If, for example, the resistance of the wiring itself is 1 Ω and the wire is delivering 10 A to a load, the power loss in the wiring will be 100 W ($10^2 \cdot 1 = 100$ W). See Figure 4-10. As you can see, the resistance of the wiring must be reduced significantly to avoid this very large power loss. You may recall that a larger diameter wire or a shorter length of wire will reduce this resistance. We will discuss power loss further as you progress in circuit analysis techniques.

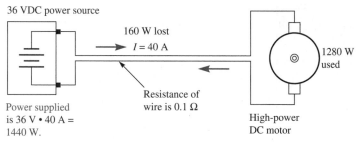

36 VDC power source

160 W lost
$I = 40$ A

Resistance of
wire is 0.1 Ω

1280 W
used

Power supplied
is 36 V • 40 A =
1440 W.

High-power
DC motor

Power loss in wire is found by using the formula

$$P = I^2 \cdot R_{WIRE} = 40^2 \cdot 0.1 = 160 \text{ W}$$

Actual power delivered to the DC motor equals
power supplied – power lost
    1440 W – 160 W = 1280 W

Figure 4-10   Significant power can be lost in a long run of wire.

Measuring Power

The first thing most people think of when we talk about measuring power is the power meter on the side of the house (Figure 4-11). This little glass-faced gadget seems to be among the most reliable of all devices in the world—never requiring service and, unfortunately, never breaking down. This type of meter is very interesting because it is actually an analog computer. It contains two electromagnetic coils; one is energized by the voltage

Figure 4-11   The residential power meter is a watthour meter used to measure total energy usage. (Courtesy of Edison Winter Home, Fort Myers, Florida.)

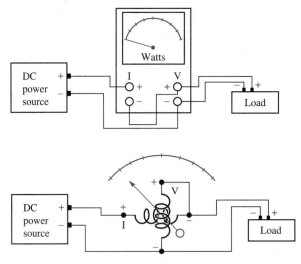

The interaction of the magnetic fields of the two coils causes the meter needle to deflect.

Figure 4-12   The wattmeter computes and displays power from current and voltage.

delivered to the home and the other is energized by the current delivered to the home. The interaction of these two coils determines the amount of accumulated power use (energy use), which is faithfully recorded on the little rotating number dials. This analog computer is continually calculating and recording current times voltage every hour of the day. This is why these meters are called **watthour meters**. The accumulation of power is measured in kilowatthours, which we will discuss more thoroughly momentarily.

Figure 4-12 illustrates a typical wattmeter with an analog meter movement calibrated in watts on its meter face plate. This type of wattmeter is similar to the watthour meter previously discussed in that it also has two electromagnetic coils, which in this case determine needle deflection on the meter scale. Again, one coil is energized by load current and the other by applied voltage. Thus $I \cdot E$ is continually computed and displayed as needle deflection. You will learn much about meter movements in a later chapter.

A simple ammeter or voltmeter may also be used to measure power, since power is a function of current and voltage. The face plate of the ammeter or voltmeter can be calibrated in watts instead of amperes or volts. However, if an ammeter or voltmeter is used, the load resistance must remain fixed and constant in value, or the calibration of this type of wattmeter will be inaccurate. If an ammeter is used, the face-plate calibration will be done according to this formula (using a specific value of load resistance) $P = I^2 \cdot R$. As you can see, the meter will respond to changes in current, and calibration is dependent upon R remaining fixed. See Figure 4-13.

Figure 4-13 also shows a voltmeter used as a wattmeter. Again, it is very important that the load resistance remain fixed at the specific calibration value. When a voltmeter is used, power calibration is determined by the formula $P = E^2/R$. Here, power is a function of voltage with resistance held constant.

The Kilowatthour (kWh)

The **kilowatthour** is actually a unit of measure for accumulated power usage. It is power (kW) times time (h) (# kWh = # kW · # h). For example, an electric iron that consumes 1 kW of power (115 V · 8.7 A = 1 kW) for a period of 1 hour has used 1 kWh of power.

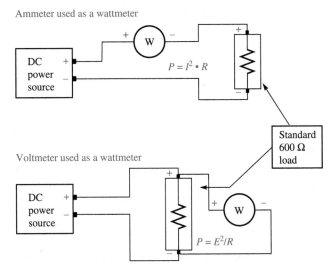

Ammeter used as a wattmeter

$P = I^2 \cdot R$

Standard 600 Ω load

Voltmeter used as a wattmeter

$P = E^2/R$

In these examples, the meters are calibrated to a 600-Ω load.

Figure 4-13   Ammeters and voltmeters can be used as wattmeters when a fixed-value load resistance is used.

If the local cost of electricity is 10 cents per kilowatthour, it will cost 10 cents to operate the iron for 1 hour. If the iron is left on for 5 hours, 5 kWh of electricity will be used, for a total cost of 50 cents. Likewise, if the iron is only used for half an hour, the power used will be 0.5 kWh (1 kW · 0.5 h = 0.5 kWh), for an operating cost of 5 cents (0.5 kWh · $0.10/kWh = $0.05). Figure 4-14 offers another practical example.

To summarize, total consumption in kilowatthours is found by

$$\# \text{ kWh} = \# \text{ kW} \cdot \# \text{ hours} \tag{4.10}$$

and total operating cost is found by

$$\text{Cost} = \# \text{ kWh} \cdot \text{rate (rate is the cost per kWh)} \tag{4.11}$$

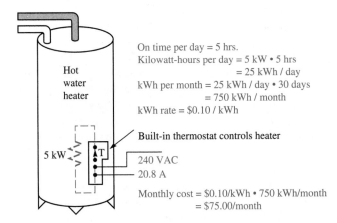

Hot water heater

On time per day = 5 hrs.
Kilowatt-hours per day = 5 kW • 5 hrs
= 25 kWh / day
kWh per month = 25 kWh / day • 30 days
= 750 kWh / month
kWh rate = $0.10 / kWh

Built-in thermostat controls heater

5 kW

240 VAC
20.8 A

Monthly cost = $0.10/kWh • 750 kWh/month
= $75.00/month

Figure 4-14   Total cost to operate a typical hot water heater for one month.

Take a few minutes to review what you have studied by skimming this important section over again. Then test your understanding by answering the questions in Self-Check 4-2.

SELF-CHECK 4-2

1. The basic unit of electrical power is the _____ .

2. If current and resistance are known, what power formula should you use to calculate power?

3. How much power is there available from a power supply that can deliver 5 A at 36 V?

4. If a 100 Ω resistor has 25 mA of current flowing through it, how much power is it dissipating?

5. If a component is operated at 30°C and is rated at 5 W at 25°C and has a derating factor of 20 mW/1°C, what is the derated power rating of the component?

6. How much power loss is there in a wire carrying 20 A of current if the resistance of the wire is 0.05 Ω?

7. Explain how an ammeter can be used as a wattmeter.

8. What is your operating cost if you operate a 1.5 kW electric frying pan for 2 hours and the electric rate is $0.12/kWh?

4-3 Circuit Analysis Procedures

Circuit analysis is essential and very basic to successful troubleshooting. You will undoubtedly use Ohm's Law more than any other formula in your circuit analysis and troubleshooting. Knowing how a circuit *should* behave as compared to how it *is* behaving is a basic key to unlocking the mysteries of a troubled circuit. Ohm's Law will provide you with a means of determining how the circuit should behave, while actual circuit measurements will reveal how the circuit is behaving.

The Open and Short

Your circuit measurements may reveal an abnormally high current through, or voltage across, a circuit component. This may indicate a short somewhere in the circuit. Or your measurements may indicate an abnormally low current possibly caused by an open somewhere in the circuit. As you gain actual troubleshooting experience, you will discover that most circuit troubles are either opens or shorts. Figure 4-15 illustrates the open and short conditions.

Short circuits, or **shorted circuits**, are circuits in which a resistor or other component has been bypassed by a very low-resistance path either internally or externally. This path may be inside the device itself and may have been created by a sudden surge in current or voltage. An arc may have occurred inside the component, causing a low-resistance carbon path through the component. Or, excessive current flowing through the component

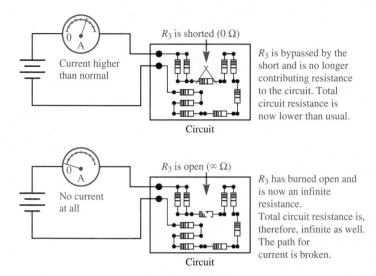

Figure 4-15 Opens and shorts produce totally opposite circuit symptoms.

over a period of time may have overheated it, causing it to break down internally. This kind of damage cannot be repaired and the component must be replaced.

Sometimes circuit components are bypassed externally and there is nothing wrong with the component itself. Some examples might be: a solder bridge between foils on a printed circuit board, a metal chassis coming in contact with a component lead, a bent component lead coming in contact with a neighboring component lead, and worn insulation on wiring.

Earlier it was stated that abnormally high circuit current is generally the result of a short in the circuit. Ohm's Law clearly demonstrates that a sharp decrease in circuit resistance, due to the short, will cause a correspondingly sharp increase in circuit current ($\uparrow I = E/R\downarrow$).

Open circuits are circuits in which current has been interrupted due to a break in the normal current path. The open may be due to a hairline crack in a copper foil, a fatigued and broken wire, a very poor solder connection, a blown fuse, a faulty switch, a faulty connector or plug, an overloaded and burned component, etc. Resistors will burn open when too much current flows through them (possibly the result of a lead touching a metal chassis or a short occurring in another component). Solid-state semiconductor components, such as diodes and transistors, will short internally and then, in some cases, burn open due to excessive voltage and/or current. In any case, the open will block the flow of DC, acting as an *infinite* resistance. Ohm's Law can be used to demonstrate what happens to current when the open occurs: $\downarrow I = E/R\uparrow\infty$. Since the open is an infinite resistance (∞ Ω), the current permitted to flow is 0 A.

To summarize, the short is a very low- or no-resistance circuit trouble that will generally cause an increase in circuit current. The open is a very high- or infinite-resistance circuit trouble that will block the flow of current.

Theoretical Circuit Analysis

One of the first questions you should ask yourself about a circuit that you are analyzing or troubleshooting is, How should this circuit be acting? In other words, what are the

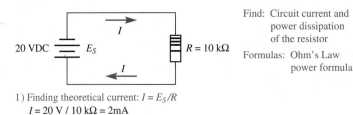

Find: Circuit current and
        power dissipation
        of the resistor

Formulas: Ohm's Law
              power formula

1) Finding theoretical current: $I = E_S/R$
   $I = 20 \text{ V} / 10 \text{ k}\Omega = 2\text{mA}$

2) Finding theoretical power dissipation: $P = I^2 \cdot R$
   $P = (2 \text{ mA})^2 \cdot 10 \text{ k}\Omega = 0.04 \text{ W} = 40\text{mW}$

Figure 4-16   Theoretical circuit analysis using formulas.

normal values of current and voltage? The answer to this question is usually determined theoretically. That is, the circuit is analyzed by using electronic theory and formulas. Ohm's Law will definitely play a large role here.

Usually, theoretical circuit analysis is done from a schematic of the circuit. Applied source voltage and component values are marked on the circuit diagram. By applying circuit theory and formulas, such as Ohm's Law, you can determine what the circuit currents and voltages should be. In this chapter, we have discussed the use of Ohm's Law and the power formulas in analyzing very simple single resistor circuits. In later chapters, you will learn how to analyze much more complex circuits using Ohm's Law and other formulas.

For now, it is important that you fully understand the use of Ohm's Law and the power formulas to analyze a simple DC circuit. Take time to study Figure 4-16 carefully.

Experimental Circuit Analysis

The process of experimental circuit analysis involves actual circuit measurement. You are collecting information about how the circuit is actually operating. If these measured values are too far from the theoretical values, you know you have a problem. Opens and shorts will definitely cause measured values to be way out of line with theoretical ones. However, some circuit troubles are caused by certain components, such as resistors, changing value due to aging or heat effects. A resistor may be marked as 10 kΩ and actually be 14 kΩ or 15 kΩ. This is soon discovered through experimental circuit analysis.

During experimental analysis, you will be using your multimeter to measure various parameters in the actual circuit. The ammeter and voltmeter functions of your multimeter will be used with circuit voltage applied while the *ohmmeter is only used with the circuit turned off—power removed.*

Figure 4-17 demonstrates the use of the multimeter in measuring current, voltage, and resistance for verification of theoretical values in a simple circuit. You should recall that the DC ammeter and voltmeter functions of your multimeter are polarized. It is therefore very important to match polarity with that of the circuit (the positive test lead, red, toward positive and the negative lead, black, toward negative). Since the ammeter must be connected within the circuit, temporarily becoming part of the circuit, it is necessary to remove power from the circuit until all connections are made.

The ohmmeter function of your multimeter makes use of the battery inside the multimeter as a power source for measuring resistance. It is therefore neither necessary nor

The ammeter is placed in series
with the load to measure the current.

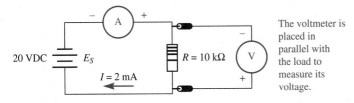

The voltmeter is
placed in
parallel with
the load to
measure its
voltage.

Voltage is disconnected here by removing wire

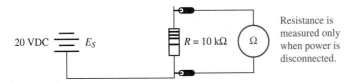

Resistance is
measured only
when power is
disconnected.

Figure 4-17   Experimental circuit analysis using test equipment.

desirable for the circuit power to be on while you measure the value of a resistor. When
you are using the ohmmeter, circuit power must be off and at least one lead of the resistor
should be disconnected from the circuit. Disconnecting a resistor lead will insure that the
ohmmeter will measure only the resistance of the resistor and nothing else. In a very simple
circuit, like those of Figure 4-17, nothing is really gained by disconnecting a lead of the
resistor, since the resistor is practically all there is to the circuit. However, you can be sure
the circuits will become more complex as we go along. Therefore, it is good to develop
proper analysis techniques early. Simply make a habit of isolating the resistor by disconnecting one of its leads from the circuit before measuring it with the ohmmeter. (If you
are using an analog multimeter, make sure you short the test leads and zero the meter
before taking measurements.)

Take time now to test your knowledge with Self-Check 4-3.

SELF-CHECK   4-3

1. Describe a short in terms of resistance.

2. Does an open generally cause an increase in circuit current or a decrease?

3. What might cause a component to open?

4. List two examples of a short external to a component.

5. Describe the proper use of the ohmmeter in checking the value of a resistor that is part
   of a circuit.

Summary

FORMULAS

(4.1) Ohm's Law: $I = E/R$ (A = V/Ω)

(4.2) Ohm's Voltage Law: $E = I \cdot R$ (V = A $\cdot$ Ω)

(4.3) Ohm's Resistance Law: $R = E/I$ (Ω = V/A)

(4.4) $I = E \cdot G$ (A = V $\cdot$ S)

$\quad$ G is conductance measured in Siemens (S)

(4.5) $E = I/G$ (V = A/S)

(4.6) $G = I/E$ (S = A/V)

(4.7) Power Formula #1: $P = I \cdot E$ (W = A $\cdot$ V)

(4.8) Power Formula #2: $P = E^2/R$ (W = V^2/Ω)

(4.9) Power Formula #3: $P = I^2 \cdot R$ (W = A^2 $\cdot$ Ω)

(4.10) Total Energy Usage: # kWh = # kW $\cdot$ # hours

(4.11) Total Operating Cost: Cost = # kWh $\cdot$ rate

$\quad$ where rate is the cost per kWh

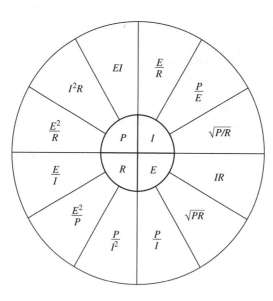

Figure 4-18   Memory wheel for solving problems.

CONCEPTS

- Ohm's Law: The amount of current flowing in a circuit is directly proportional to the amount of voltage applied to the circuit and inversely proportional to the amount of resistance in the circuit.
- Power is the rate at which energy is used.
- Electrical power is measured in watts.
- Available power is the maximum amount of power a power source is capable of delivering to a load.

- Dissipated power is the amount of power actually demanded by a circuit and converted to another form of energy.
- Power rating is the amount of power, specified by the manufacturer, that a component is capable of safely dissipating without damage.
- Power derating is the method used to adjust the power rating of a component to match other than normal operating conditions.
- Power loss is wasted power due to conductor resistance and is calculated using I^2R.
- The watthour meter measures the accumulated power consumption, or energy usage, over a period of time. Its basic unit of measure is the kilowatthour.
- The wattmeter is used to measure the rate of energy usage, or power. It is an analog computer electromechanically multiplying current and voltage while displaying the answer in watts on its meter face.
- The ammeter or voltmeter may be calibrated as a wattmeter if the load resistance is always the same.
- An open has infinite resistance and will not allow current to pass.
- A short has no resistance and does not hinder the flow of current at all.

SPECIAL TERMS

- Power (P), watt (W), joule (J)
- Available power, power dissipation, power loss
- Power rating, derate, derating factor
- Watthour (Wh), kilowatthour (kWh)
- Open circuit, short circuit
- Theoretical circuit analysis
- Experimental circuit analysis

Need to Know Solution

- **Value of the Resistor.**   The DC voltage is 12 V and the maximum current must not exceed more than 1 A in the event of a short circuit. Ohm's Law is used to determine the value of the resistor needed to limit the current to 1 A.

$$R = E/I = 12 \text{ V}/1 \text{ A} = 12 \ \Omega$$

A 12 Ω resistor will insure the current limit.

- **Size or Power Rating.**   The power rating for the resistor must be determined from the maximum power dissipation of the resistor during a short circuit condition. When a short circuit occurs, 1 A of current will be flowing through the resistor's 12 Ω resistance. The power dissipation will be as follows:

$$P_D = I^2R = (1 \text{ A})^2 \cdot 12 \ \Omega = 12 \text{ W}$$

We can assume that any short that might occur will only be temporary and the resistor will not be dissipating 12 W of power continually. Therefore, a 12 W, or larger, resistor will be adequate (more than likely a 20 W standard value).

- **Type of Resistor.**   The resistor will be wirewound, since they are designed for high-power applications.

Questions

4-1 Ohm's Law

1. Is the amount of current in a circuit directly or inversely proportional to the amount of applied voltage?
2. Is the amount of current in a circuit directly or inversely proportional to the amount of resistance in the circuit?

4-2 Power

3. Define the watt in terms of joules.
4. What is a generally safe and accepted rule of thumb for component derating?
5. Explain the basic function of a wattmeter. How is it like an analog computer?
6. Under what restriction is it possible to use a voltmeter as a wattmeter?
7. What is the difference between a watthour meter and a wattmeter? Which type do we find mounted on the power panels of our homes?

4-3 Circuit Analysis Procedures

8. Describe a short circuit in terms of current.
9. Is a short no resistance or an infinite amount of resistance?
10. Describe an open in terms of circuit current.
11. What circuit condition will indicate an open?
12. Describe an open in terms of resistance.
13. What is the difference between theoretical and experimental circuit analysis?
14. What precautions must you take when using an ohmmeter?
15. Does an ohmmeter require an external power source for operation? Explain.
16. What is one precaution that you must observe when using an ammeter or a voltmeter?
17. When connecting an ammeter into a circuit, should the power be (a) on or (b) off until all connections are made? Explain.

Problems

4-1 Ohm's Law

Use Ohm's Law to solve for the unknowns in each of the following:

	Resistance Ω	Applied Voltage	Current A	Conductance S
1.	?	50 mV	100 μA	?
2.	2.7 kΩ	18 V	?	?
3.	?	24 V	100 mA	?
4.	?	10 V	?	133.33 μS
5.	130 kΩ	?	50 μA	?
6.	?	?	1500 μA	4.286 μS
7.	?	23 kV	12 A	?
8.	11 MΩ	1.5 kV	?	?
9.	?	100 mV	?	21.277 mS
10.	?	?	670 mA	45.455 μS

4-2 Power

11. Five joules of energy expended every two seconds equals how many watts of power?
12. One hundred joules of energy expended every five seconds equals how many watts of power?

Solve for the unknowns in the following:

	Power W	Resistance Ω	Current A	Voltage V
13.	?	?	100 mA	250 V
14.	?	100 kΩ	35 mA	?
15.	?	22 kΩ	?	24 V
16.	1200 W	?	3 A	?
17.	90 μW	1.5 MΩ	?	?
18.	?	560 Ω	250 mA	?
19.	?	33 kΩ	?	9 V
20.	100 W	?	?	300 V
21.	?	?	760 μA	250 V
22.	50 μW	?	?	25 mV

23. If a DC power supply can deliver 12 A of current at 24 V, what is the maximum available power?
24. The voltage measured directly across a 10 kΩ resistor is found to be 5.8 V. How much power is the resistor dissipating?
25. The voltage measured directly across a 33 kΩ resistor is found to be 18 V. How much power is the resistor dissipating?
26. Would a ⅛ watt resistor be adequate for Problem 24?
27. Would a ⅛ watt resistor be adequate for Problem 25?
28. If a certain component has a derating factor of 25 mW/°C, and is rated at 1.5 W (at 25°C), what is its derated value at an ambient temperature of 40°C?
29. If a certain component has a derating factor of 60 mW/°C, and is rated at 10 W (at 25°C), what is its derated value at an ambient temperature of 60°C?
30. If a load device is demanding 30 A of current from a distant power source and the total resistance of the conductors is 0.01 Ω, what is the amount of power loss due to the resistance in the conductors?
31. If a load device is demanding 12 A of current from a distant power source and the total resistance of the conductors is 0.32 Ω, what is the amount of power loss due to the resistance in the conductors?
32. How much energy (in kWh) will a 500 W appliance use if it is operating for 30 minutes?
33. How much energy (in kWh) will a 2.5 kW appliance use if it is operating for 5 hours?
34. How much energy will a 4 kW electric hot water heater consume in a month's time (30 days) if it comes on an average of 6 hours per day?
35. If the electric rate is $0.12/kWh, what is the monthly cost of heating water in Problem 34?

Answers to Self-Checks

Self-Check 4-1

1. Increase
2. Decrease
3. Voltage and resistance
4. 3 A
5. 1,500 Ω
6. 3.41 mA

Self-Check 4-2

1. Watt
2. $P = I^2R$
3. 180 W
4. 62.5 mW
5. 4.9 W
6. 20 W
7. The ammeter can be calibrated in watts and used in series with a specific resistance load. If the load is changed to a different value of resistance, the calibration will be inaccurate.
8. $0.36

Self-Check 4-3

1. A short has no resistance (0 Ω) or a resistance that is very much lower than the component that is being shorted.
2. Decrease
3. Too much current or voltage, which generally causes the component to first short internally, then burn open.
4. The lead or leads of a compartment touching a metal chassis or a solder bridge between copper foils on a printed circuit board.
5. Make sure circuit power is removed. Isolate the resistor by disconnecting at least one of its leads from the circuit. Set the multimeter to the proper range and ohmmeter function. If you are using an analog multimeter, short the test leads and zero the meter. Connect the test leads across the resistor and measure.

SUGGESTED PROJECTS

1. Add the formulas and important concepts from this chapter to your personal Electricity and Electronics Notebook. You might also want to transfer the Design Notes from this chapter to your notebook.

2. Prove Ohm's Law to yourself by constructing simple DC circuits using a single resistor and a 9 V battery. Use your multimeter as an ammeter. Try a variety of resistors (1 kΩ or greater). Your test circuit should be like that of Figures 4-16 and 4-17. Do not forget to observe proper polarity of your test leads.

EDISON EMBOSSING TRANSLATING
TELEGRAPH

Part II

EDISON TYPEWRITING
TELEGRAPH

DC Circuits and Theorems

PART OUTLINE

Edison stands proudly next to his electric carriage. Notice the New Jersey license plate. Most of his inventing was done in his New Jersey laboratories. He is holding one of his improved nickel-iron-alkaline storage batteries, which power his vehicle. In addition to inventing the batteries, he greatly improved the DC motor used to power everything from fans to electric vehicles. His batteries were used extensively in mining vehicles, train cars, and railroad signaling applications, to name only a few. (Photograph from The Bettmann Archive.)

In the openings of the Part II chapters that follow, you will discover that Edison invented the mimeograph machine, ticker tape machines, automatic telegraph machines, and the forerunner of the fax machine. There seemed to be no end to his creativity.

(1) This "Electric Pen" was invented by Edison to produce stencils for his mimeograph machine. His batteries supplied current to a coil which vibrated much like a doorbell causing a pin-pointed needle to rapidly project from the bottom of the pen piercing holes in the paper as a design or lettering was traced. A U.S. patent was granted on August 8, 1876. A license agreement was later granted to the A. B. Dick Company for the manufacture of the mimeograph machine. (2) Edison invented the mimeograph machine in 1875 while in Newark, New Jersey.

EDISON ELECTRIC PEN

(1)

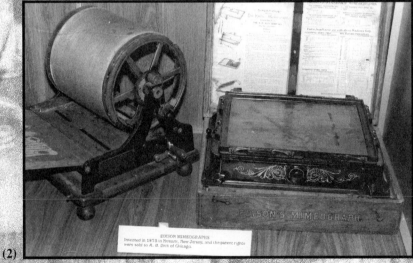

EDISON MIMEOGRAPHS
Invented in 1875 in Newark, New Jersey, and the patent rights were sold to A. B. Dick of Chicago.

(2)

Chapter 5

Series DC Circuits

CHAPTER OUTLINE

OBJECTIVES

After studying this chapter, you will be able to
- calculate total resistance in a series resistive circuit.
- calculate current, voltage drops, and power dissipation in a series resistive circuit.
- discuss and apply Kirchhoff's Voltage Law to series circuits.
- apply the concepts of relative polarity and common ground.
- troubleshoot a series DC circuit with an ohmmeter and a voltmeter.

INTRODUCTION

In this chapter, you will learn the important characteristics of a series DC circuit. You will learn how to calculate values of current, voltage, and power. The important concepts of relative polarity and common ground will be introduced. All of these topics will continue to build upon the foundation that was laid in previous chapters. The coverage of series circuit theory and calculation will prepare you for the sections on troubleshooting with the ohmmeter and the ammeter. By the time you finish this chapter, you will have gained much practical and real-world knowledge to reinforce your experiences in electricity and electronics.

NEED TO KNOW

Some time ago, people would decorate their Christmas trees with strings of lights that were wired in series. They used to describe these by saying, "when one goes out, they all go out." When one of the bulbs burned out, they had to start at one end of the string and replace each bulb one at a time to find the bad bulb. This was a long and aggravating process. Can you explain why the entire string went out when one went out?

Today, most people are using so-called "improved" series strings of lights on their Christmas trees. One wire is interrupted with lights, forming a complete loop back to the plug. But, when one goes out, they don't all go out, as they used to. Why? Also, when one goes out, the others get a little brighter, though this is not always noticeable. To make things worse, the remaining lights begin to burn out more quickly. In fact, each burned-out light accelerates the burnout process. Why? What does one burned-out light have to do with the other good lights? The answer to these questions has everything to do with series circuit theory and good troubleshooting sense. Do you have a need to know?

5-1 Series Resistors

In this section, we will examine series resistor circuits. You will learn about **total resistance** (R_t) and how it is affected when resistive devices are connected in a series arrangement as shown in Figure 5-1. You will learn how to calculate and measure total resistance in a series resistor circuit.

See color photo insert page B1.

First, it must be understood that any power source connected to terminals A and B of Figure 5-1 would have but one path for current. The current resulting from the applied EMF would flow from A to B, or B to A, through R_1, R_2, and R_3. Since the current is forced to flow through all these resistors, the total resistance (total opposition to current) is the sum of the individual resistors. In other words, each resistor adds to, and increases, the total resistance offered by the circuit to the EMF. Therefore:

$$R_t = R_1 + R_2 + R_3 + \cdots \tag{5.1}$$

The values of any additional resistors placed in series with the three resistors shown would have to be added to the present total to show the resulting increase in total resistance. Example 5-1 is a practical application of Formula 5.1.

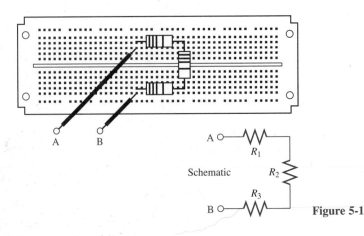

Schematic

Figure 5-1

EXAMPLE   5-1

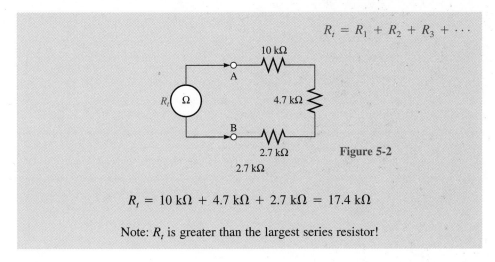

$$R_t = R_1 + R_2 + R_3 + \cdots$$

10 kΩ
A
R_t Ω   4.7 kΩ
B
2.7 kΩ   **Figure 5-2**
2.7 kΩ

$$R_t = 10 \text{ k}\Omega + 4.7 \text{ k}\Omega + 2.7 \text{ k}\Omega = 17.4 \text{ k}\Omega$$

Note: R_t is greater than the largest series resistor!

If you were to construct the circuit of Example 5-1, you would be able to verify your calculations by using a multimeter. You would set your meter to the OHMs function and select the appropriate range. By connecting the test leads across each resistor, you can read and record its actual value of resistance. Finally, the total resistance can be measured, as is shown in Figure 5-3, and compared to the calculations that you obtained from marked values. You should be aware that, due to manufacturer's tolerances, your measured total resistance will probably not be exactly the same as the total resistance you calculated using the marked values.

Remember this rule:

The total resistance in a series circuit is the sum of all individual resistances.

Now, let's take a look at Example 5-2.

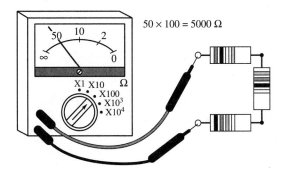

$50 \times 100 = 5000 \ \Omega$

Figure 5-3   Using the ohmmeter to measure total resistance of a series circuit.

EXAMPLE   5-2

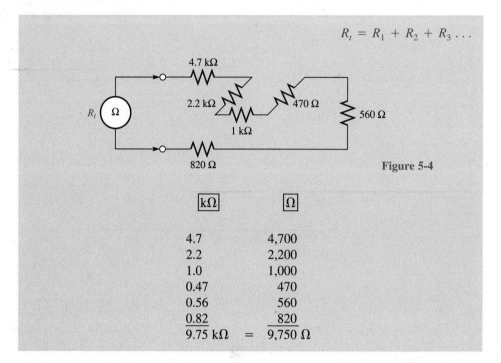

$$R_t = R_1 + R_2 + R_3 \ldots$$

Figure 5-4

kΩ		Ω
4.7		4,700
2.2		2,200
1.0		1,000
0.47		470
0.56		560
0.82		820
9.75 kΩ	=	9,750 Ω

Figure 5-5 demonstrates the use of a popular plug-in type of experiment board in setting up a series resistive circuit and measuring total resistance. This type of experiment board is arranged in two sets of vertical contact strips, each strip having five component access holes. Above and below are single horizontal contact strips that have many access holes. These horizontal strips are useful to jumper (positive or negative) DC connections to any and all circuits that you have laid out on the board. We will use this type of experiment board throughout this book to illustrate circuits and laboratory techniques. You may presently own one of these boards and find it useful in verifying and reinforcing what you learn. If you do not own such a board, you will want to invest in one.

Take a few moments to examine Design Note 5-1 on page 133. Then try your new skills on Self-Check 5-1.

SELF-CHECK   5-1

1. Write the formula for total resistance in a series circuit. $R_t =$

2. Four resistors are in series: $R_1 = 10 \ \Omega$, $R_2 = 5.6 \ \Omega$, $R_3 = 8.2 \ \Omega$, $R_4 = 6.2 \ \Omega$. Find R_t

3. Six resistors are in series: $R_1 = 360 \ \Omega$, $R_2 = 470 \ \Omega$, $R_3 = 680 \ \Omega$, $R_4 = 910 \ \Omega$, $R_5 = 1.1 \ k\Omega$, $R_6 = 2.4 \ k\Omega$. Find R_t

DESIGN NOTE 5-1:   Total Resistance in Series

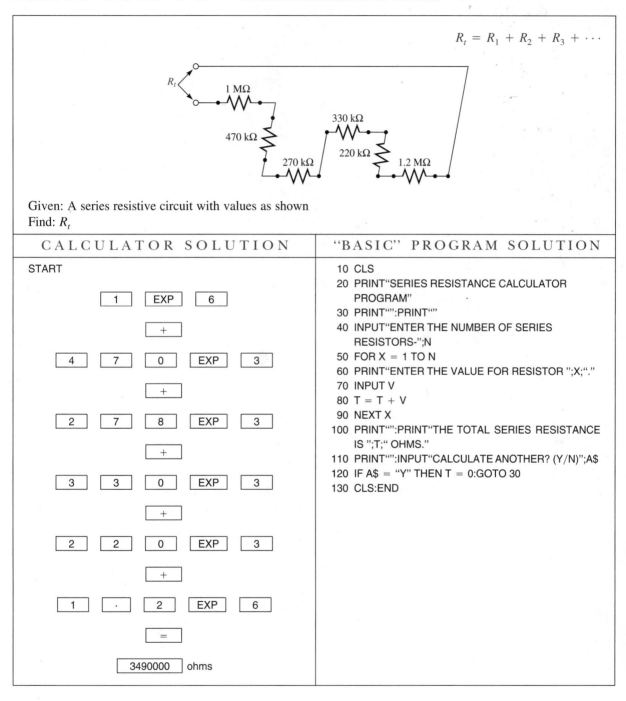

$$R_t = R_1 + R_2 + R_3 + \cdots$$

Given: A series resistive circuit with values as shown
Find: R_t

CALCULATOR SOLUTION	"BASIC" PROGRAM SOLUTION
START	10 CLS

CALCULATOR SOLUTION:

START

| 1 | EXP | 6 |

| + |

| 4 | 7 | 0 | EXP | 3 |

| + |

| 2 | 7 | 8 | EXP | 3 |

| + |

| 3 | 3 | 0 | EXP | 3 |

| + |

| 2 | 2 | 0 | EXP | 3 |

| + |

| 1 | . | 2 | EXP | 6 |

| = |

| 3490000 | ohms

"BASIC" PROGRAM SOLUTION:

```
10 CLS
20 PRINT"SERIES RESISTANCE CALCULATOR
   PROGRAM"
30 PRINT"":PRINT""
40 INPUT"ENTER THE NUMBER OF SERIES
   RESISTORS-";N
50 FOR X = 1 TO N
60 PRINT"ENTER THE VALUE FOR RESISTOR ";X;"."
70 INPUT V
80 T = T + V
90 NEXT X
100 PRINT"":PRINT"THE TOTAL SERIES RESISTANCE
    IS ";T;" OHMS."
110 PRINT"":INPUT"CALCULATE ANOTHER? (Y/N)";A$
120 IF A$ = "Y" THEN T = 0:GOTO 30
130 CLS:END
```

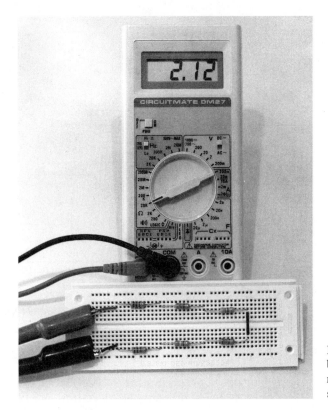

Figure 5-5 A typical DMM being used to measure the total resistance of a bread-boarded series circuit.

5-2 One Path for Current

In this section, we emphasize the one characteristic that makes the simple series circuit unique from other circuit configurations: The series circuit has only **one path for current**. The electromotive force, or difference of potential, causes electrons to flow in an organized manner from the negative source terminal through the conductor and consecutive components, and finally return to the positive source terminal. The amount of current allowed to flow is determined by the EMF (source voltage) and the total resistance in the circuit, as is shown in Figure 5-6. The total current, therefore, is calculated using Ohm's Law for

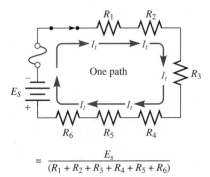

$$= \frac{E_s}{(R_1 + R_2 + R_3 + R_4 + R_5 + R_6)}$$

Figure 5-6 The series circuit has one path for current.

current ($I = E/R$). More specifically, total current (I_t) equals the source voltage (E_S) divided by the total resistance (R_t).

$$I_t = E_S/R_t \quad \text{(where } R_t = R_1 + R_2 + R_3 + \cdots\text{)} \tag{5.2}$$

Many examples of solving for total current in a series circuit are shown in Example Set 5-3. Take time to study them carefully.

EXAMPLE SET 5-3

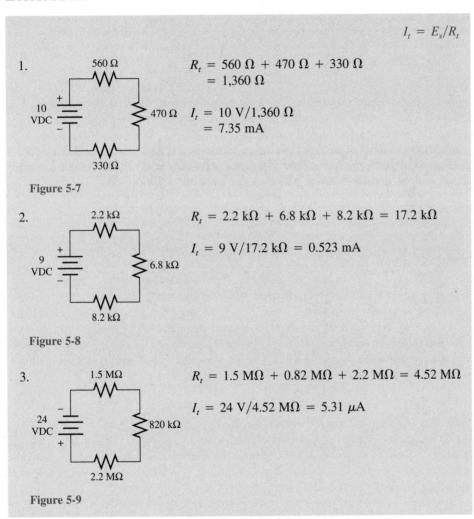

$$I_t = E_S/R_t$$

1.   560 Ω

$R_t = 560\ \Omega + 470\ \Omega + 330\ \Omega$
$\quad = 1{,}360\ \Omega$

10 VDC

470 Ω $I_t = 10\ \text{V}/1{,}360\ \Omega$
$\quad = 7.35\ \text{mA}$

330 Ω

Figure 5-7

2.   2.2 kΩ

$R_t = 2.2\ \text{k}\Omega + 6.8\ \text{k}\Omega + 8.2\ \text{k}\Omega = 17.2\ \text{k}\Omega$

9 VDC

6.8 kΩ $I_t = 9\ \text{V}/17.2\ \text{k}\Omega = 0.523\ \text{mA}$

8.2 kΩ

Figure 5-8

3.   1.5 MΩ

$R_t = 1.5\ \text{M}\Omega + 0.82\ \text{M}\Omega + 2.2\ \text{M}\Omega = 4.52\ \text{M}\Omega$

24 VDC

820 kΩ $I_t = 24\ \text{V}/4.52\ \text{M}\Omega = 5.31\ \mu\text{A}$

2.2 MΩ

Figure 5-9

Before continuing, test your skills on Self-Check 5-2.

1. A series circuit has three resistors: 10 kΩ, 33 kΩ, and 27 kΩ. The source voltage is 24 VDC. Find R_t and I_t

2. A series circuit has four resistors: 5.6 kΩ, 2.7 kΩ, 4.7 kΩ, and 3.3 kΩ. The source voltage is 9 VDC. Find R_t and I_t

3. A series circuit has three resistors: 1.5 MΩ, 9.1 MΩ, and 2.4 MΩ. The source voltage is 75 VDC. Find R_t and I_t

5-3 Series Voltage Drops

See color photo insert page B2.

Another unique characteristic of the series circuit is the individual voltage drops that occur across each resistor in the circuit. A **voltage drop** is simply the amount of voltage each resistor has across itself as a result of current flowing through its resistance. This is easily proven and demonstrated using Ohm's Voltage Law ($E = I \cdot R$). Ohm's Voltage Law tells us that a current flowing through a resistance develops a voltage across the resistance. As an example, if 10 A of current flows through a resistance of 20 Ω, the resulting voltage drop across the resistor is 200 V (Ohm's Law: 10 A $\cdot$ 20 Ω $=$ 200 V).

> As long as you know the value of a resistor and the current flowing through that resistor, you can calculate the voltage across that resistor.

In a series circuit containing more than one resistor, you can quickly and easily calculate the voltage drop on each resistor. To do this, you must know the value of each resistor and the current flowing through the resistors (total current, I_t). The voltage drop on each resistor is then calculated using the total current (I_t) and the value of each resistor. Therefore, the voltage on any single resistor is found using the following formula from Ohm's Law:

$$V_R = I_t \cdot R \tag{5.3}$$

where R is any single resistor in a series circuit.

Example Set 5-4 illustrates the use of Formula 5-3 to calculate the individual voltage drops on resistors in series circuits. Notice that the total current must first be calculated in order to solve for individual voltage drops. Naturally, you must solve for total resistance (R_t) before you can find the total current.

Try your new knowledge on Self-Check 5-3 before continuing.

Find the individual voltage drops for the resistors in the problems of Self-Check 5-2.

EXAMPLE SET   5-4

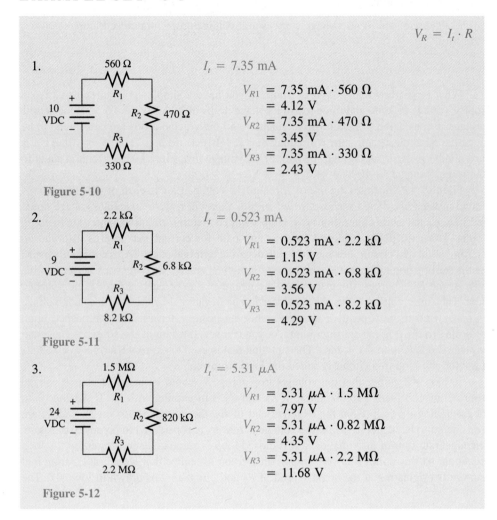

$$V_R = I_t \cdot R$$

1.

$I_t = 7.35 \text{ mA}$

$V_{R1} = 7.35 \text{ mA} \cdot 560 \ \Omega$
$= 4.12 \text{ V}$
$V_{R2} = 7.35 \text{ mA} \cdot 470 \ \Omega$
$= 3.45 \text{ V}$
$V_{R3} = 7.35 \text{ mA} \cdot 330 \ \Omega$
$= 2.43 \text{ V}$

Figure 5-10

2.

$I_t = 0.523 \text{ mA}$

$V_{R1} = 0.523 \text{ mA} \cdot 2.2 \text{ k}\Omega$
$= 1.15 \text{ V}$
$V_{R2} = 0.523 \text{ mA} \cdot 6.8 \text{ k}\Omega$
$= 3.56 \text{ V}$
$V_{R3} = 0.523 \text{ mA} \cdot 8.2 \text{ k}\Omega$
$= 4.29 \text{ V}$

Figure 5-11

3.

$I_t = 5.31 \ \mu\text{A}$

$V_{R1} = 5.31 \ \mu\text{A} \cdot 1.5 \text{ M}\Omega$
$= 7.97 \text{ V}$
$V_{R2} = 5.31 \ \mu\text{A} \cdot 0.82 \text{ M}\Omega$
$= 4.35 \text{ V}$
$V_{R3} = 5.31 \ \mu\text{A} \cdot 2.2 \text{ M}\Omega$
$= 11.68 \text{ V}$

Figure 5-12

5-4 Kirchhoff's Voltage Law and the Voltage Divider Rule

Kirchhoff's Voltage Law

You may have discovered a very important and useful circuit analysis law while studying Example Set 5-4 and Self-Check 5-3. That important law is commonly known as **Kirchhoff's Voltage Law** and is stated in *simplified* form as follows. The sum of all resistor voltage drops in a series circuit is equal to the source voltage: $E_s = V_{R1} + V_{R2} + V_{R3} + \cdots$. This law establishes the rule for voltage distribution in any *closed loop* of a circuit. It is a basic concept that will become useful to you as you begin to analyze more complex circuits. The voltage law is stated as follows:

See color photo insert page B2.

Kirchhoff's Voltage Law

The algebraic sum of all voltage sources and voltage drops in a closed loop is equal to zero.

What is this law really saying? Suppose you have a simple series circuit that has a source voltage of 10 V and three resistor voltage drops of 2 V, 3 V, and 5 V. If the source voltage is considered to be a positive quantity and the voltage drops are considered to be negative quantities, their sum will equal zero [$(+10) + (-2 - 3 - 5) = 0$]. This is just another way of saying that the sum of all voltage drops in a series circuit is equal to the source voltage ($+10$ V $= +2$ V $+ 3$ V $+ 5$ V).

Figure 5-13 illustrates the use of Kirchhoff's Voltage Law in each of the closed loops found in what is called a *series-parallel circuit*. You will learn to calculate voltage drops in circuits such as this later. For now, simply notice that this circuit has three such closed loops. Thus, we define a **closed loop** as any path for current that returns to a chosen starting point in a circuit. For loop #1, the negative terminal of the DC source was chosen as the starting point. Current is then traced through R_2 and R_1, arriving back at the positive side of the DC source. Along the way, the negative side of each resistor is approached first. Thus their voltage drops are considered to be negative quantities. The source voltage is considered to be a positive quantity, since its positive terminal is approached upon returning to the original starting point. As you can see, the sum of the source voltage and voltage drops in loop #1 is zero. The equation that is used to express Kirchhoff's Voltage Law for any loop in a circuit is called the **loop equation**.

In loop #2, notice that the voltage drop across R_2 could be considered as a voltage source, since the series combination of R_3 and R_4 is in parallel with R_2. If we choose to begin tracing the loop from the negative end of R_2, the voltage drops on R_2 and R_3 are considered to be negative, while the voltage across R_2 is considered to be positive. Again, the algebraic sum of all voltages in this closed loop is zero, as demonstrated by the loop equation. At this point, it is very important for you to realize that we are not stating that current is originating at the negative end of R_2 and circulating as drawn in loop #2. The

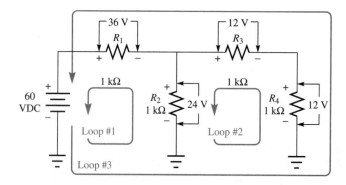

Loop #1: $-12\,\text{V} -36\,\text{V} + 60\,\text{V} = 0$

Loop #2: $-12\,\text{V} -12\,\text{V} + 24\,\text{V} = 0$

Loop #3: $-12\,\text{V} -12\,\text{V} -36\,\text{V} + 60\,\text{V} = 0$

Figure 5-13   Kirchhoff's Voltage Law

Gustav Robert Kirchhoff (1824 – 1887)

Gustav Kirchhoff was a famous nineteenth-century German physicist. He is well known for his current and voltage laws, which he introduced in 1845, and for his work in the spectral analysis of elements. In 1854, Kirchhoff was appointed a professor of physics at the University of Heidelberg. He and German chemist Robert Benson are credited with discovering cesium in 1860 and rubidium in 1861. He is well known today among chemists, physicists, and electronics technicians and engineers. (Photo courtesy of The Bettmann Archive.)

---

loop arrow does not indicate current, rather it shows a complete circuit loop of voltage drops.

Loop #3 is the largest loop. It in fact does indicate the proper current direction. As you can see, R_2 is not considered as being part of loop #3. Only R_4, R_3, and R_1 fall in the path of this loop. Again, the negative end of each resistor is approached as the loop is traced. Thus, the resistor voltage drops are considered to be negative quantities and the source voltage is positive.

In this particular illustration, all loop voltages were traced in a counterclockwise direction. However, it was not necessary to do so. We could just as easily have traced each loop in a clockwise direction. Doing so would make voltage drops positive, since the positive side of each resistor would be approached first, and the source voltage negative, since its negative terminal would be approached first upon returning to the original starting point. Regardless of the direction in which you trace a loop, the sum of all voltages in the loop is zero. If the sum is not equal to zero, it is because an error has been made in determining voltage drops in the loop. Thus, Kirchhoff's Voltage Law becomes a valuable tool for checking calculated voltage drops found during circuit analysis.

The Voltage Divider Rule

The resistors in a series circuit share, or distribute, the source voltage among themselves according to their relative sizes. The larger-value resistors have larger voltage drops than the lower-value resistors. You may have already noticed this from Self-Check 5-3. If not, go back and re-examine Self-Check 5-3. Since it is true that larger-value resistors have greater voltage drops than lower-value resistors, we can calculate voltage drops by proportion using the following voltage divider rule:

> **Voltage Divider Rule**
>
> The voltage drop across any resistor in a series circuit is proportional to the value of the resistor as compared to the total resistance.

Since the current through any resistor in a series circuit is equal to the total current, we can make the following mathematical rationalization:

$$I_t = E_s/R_t = V_R/R$$

therefore

$$E_s \cdot R = V_R \cdot R_t$$

and

$$(E_s \cdot R)/R_t = V_R$$

Thus, we arrive at the following proportional formula, which is the mathematical expression of the voltage divider rule:

$$V_R = (R/R_t) \cdot E_s \tag{5.4}$$

where R is a single resistor in a series circuit and the voltage divider ratio is R/R_t.

Formula 5.4 tells us: If 20 VDC is applied to a series circuit that has a total resistance of 20 kΩ and the circuit is composed of twenty 1 kΩ resistors, each resistor will drop 1 VDC [$V_R = (1 \text{ k}\Omega/20 \text{ k}\Omega) \cdot 20 \text{ V} = 1 \text{ V}$]. Take time to study Example Set 5-5. This will help you become familiar with the proportion formula and voltage divider rule.

EXAMPLE SET   5-5

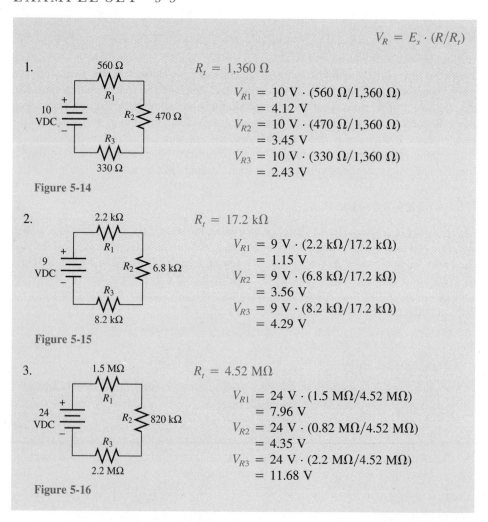

$$V_R = E_s \cdot (R/R_t)$$

1.   560 Ω $R_t = 1,360 \ \Omega$

R_1

10 VDC

R_2 470 Ω

R_3

330 Ω

$$V_{R1} = 10 \text{ V} \cdot (560 \ \Omega/1,360 \ \Omega)$$
$$= 4.12 \text{ V}$$
$$V_{R2} = 10 \text{ V} \cdot (470 \ \Omega/1,360 \ \Omega)$$
$$= 3.45 \text{ V}$$
$$V_{R3} = 10 \text{ V} \cdot (330 \ \Omega/1,360 \ \Omega)$$
$$= 2.43 \text{ V}$$

Figure 5-14

2.   2.2 kΩ $R_t = 17.2 \text{ k}\Omega$

R_1

9 VDC

R_2 6.8 kΩ

R_3

8.2 kΩ

$$V_{R1} = 9 \text{ V} \cdot (2.2 \text{ k}\Omega/17.2 \text{ k}\Omega)$$
$$= 1.15 \text{ V}$$
$$V_{R2} = 9 \text{ V} \cdot (6.8 \text{ k}\Omega/17.2 \text{ k}\Omega)$$
$$= 3.56 \text{ V}$$
$$V_{R3} = 9 \text{ V} \cdot (8.2 \text{ k}\Omega/17.2 \text{ k}\Omega)$$
$$= 4.29 \text{ V}$$

Figure 5-15

3.   1.5 MΩ $R_t = 4.52 \text{ M}\Omega$

R_1

24 VDC

R_2 820 kΩ

R_3

2.2 MΩ

$$V_{R1} = 24 \text{ V} \cdot (1.5 \text{ M}\Omega/4.52 \text{ M}\Omega)$$
$$= 7.96 \text{ V}$$
$$V_{R2} = 24 \text{ V} \cdot (0.82 \text{ M}\Omega/4.52 \text{ M}\Omega)$$
$$= 4.35 \text{ V}$$
$$V_{R3} = 24 \text{ V} \cdot (2.2 \text{ M}\Omega/4.52 \text{ M}\Omega)$$
$$= 11.68 \text{ V}$$

Figure 5-16

Let's take a few minutes to review the important characteristics of series circuits.

1. *A series circuit has one current and one path for current. The current in every resistor is the same and is the same value as the total current:* $I_t = E_s/R_t$.

2. *Each resistor voltage drop can be found using Ohm's Law:* $V_R = I_t \cdot R$.

3. *All individual voltage drops added together equal the source voltage (Kirchhoff's Voltage Law).*

4. *The amount of voltage a series resistor drops is proportionate to its resistive value compared to the total resistance (Voltage Divider Rule):* $V_R = (R/R_t) \cdot E_s$.

Study Design Note 5-2. Then, solve the problems in Self-Check 5-4 using the proportion formula. You may check your answers using Ohm's Law and Kirchhoff's Voltage Law.

S E L F - C H E C K   5 - 4

1.
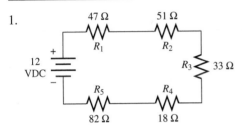

Find:

$V_{R1} =$
$V_{R2} =$
$V_{R3} =$
$V_{R4} =$
$V_{R5} =$

Figure 5-17

2.
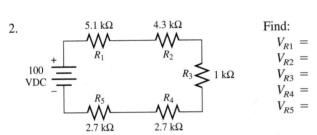

Find:

$V_{R1} =$
$V_{R2} =$
$V_{R3} =$
$V_{R4} =$
$V_{R5} =$

Figure 5-18

3.
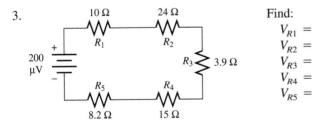

Find:

$V_{R1} =$
$V_{R2} =$
$V_{R3} =$
$V_{R4} =$
$V_{R5} =$

Figure 5-19

4. Write a loop equation for each of the three circuits above.

DESIGN NOTE 5-2: Voltage Drops

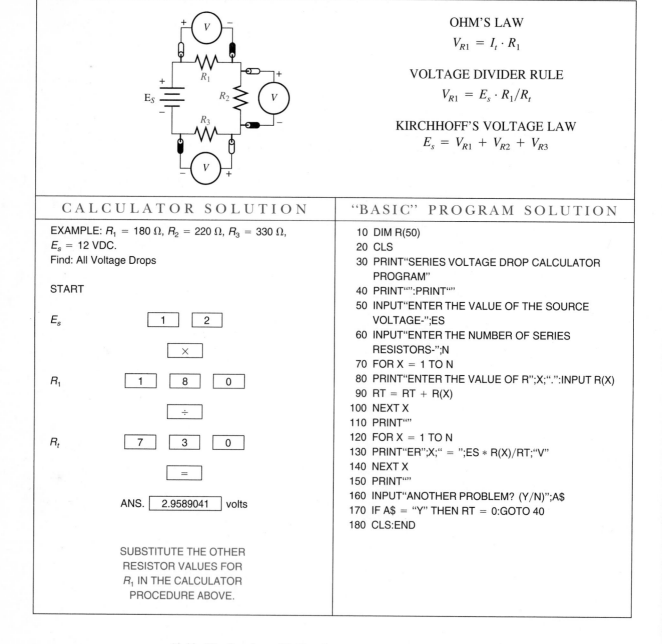

OHM'S LAW

$$V_{R1} = I_t \cdot R_1$$

VOLTAGE DIVIDER RULE

$$V_{R1} = E_s \cdot R_1/R_t$$

KIRCHHOFF'S VOLTAGE LAW

$$E_s = V_{R1} + V_{R2} + V_{R3}$$

CALCULATOR SOLUTION	"BASIC" PROGRAM SOLUTION

EXAMPLE: R_1 = 180 Ω, R_2 = 220 Ω, R_3 = 330 Ω, E_s = 12 VDC.
Find: All Voltage Drops

START

E_s [1]   [2]

[×]

R_1 [1]   [8]   [0]

[÷]

R_t [7]   [3]   [0]

[=]

ANS. [2.9589041] volts

SUBSTITUTE THE OTHER
RESISTOR VALUES FOR
R_1 IN THE CALCULATOR
PROCEDURE ABOVE.

```
10 DIM R(50)
20 CLS
30 PRINT"SERIES VOLTAGE DROP CALCULATOR
   PROGRAM"
40 PRINT"":PRINT""
50 INPUT"ENTER THE VALUE OF THE SOURCE
   VOLTAGE-";ES
60 INPUT"ENTER THE NUMBER OF SERIES
   RESISTORS-";N
70 FOR X = 1 TO N
80 PRINT"ENTER THE VALUE OF R";X;".":INPUT R(X)
90 RT = RT + R(X)
100 NEXT X
110 PRINT""
120 FOR X = 1 TO N
130 PRINT"ER";X;" = ";ES * R(X)/RT;"V"
140 NEXT X
150 PRINT""
160 INPUT"ANOTHER PROBLEM? (Y/N)";A$
170 IF A$ = "Y" THEN RT = 0:GOTO 40
180 CLS:END
```

5-5 Relative Polarity

See color photo insert page B2.

You have already learned that each resistor in a series circuit receives a voltage drop in proportion to its relative size. We are able to prove this concept using Ohm's Law or the proportion formula (voltage divider rule). However, we have not yet discussed the polarity of the voltage drops on each resistor. It is very important that we fully understand polarity of voltage drops since proper circuit analysis, design, and troubleshooting depend on it.

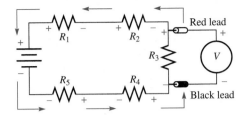

Figure 5-20   The polarity of voltage drops depends on the polarity of the source.

Figure 5-20 illustrates the polarity of individual voltage drops in a typical series circuit. It is very important that you notice the + and − on *each* resistor. Notice also that the − polarity of each resistor voltage drop is toward the − side of the DC supply and the + polarity of each resistor voltage drop is toward the + side of the DC supply. Remember, each resistor has its own voltage drop and therefore its own polarity. Electrons are traveling from the negative side of the DC supply through each series resistor. Each resistor converts a certain amount of energy, carried by the electrons, into heat. This drop in energy is seen as a drop in voltage on each resistor. The end of the resistor at which electrons enter (−) has a higher negative potential than the end at which electrons exit (+).

Do not be concerned that the wires connecting resistors are labeled with a − and a +. You are correct in thinking that the wire cannot be charged negatively and positively at the same time. The − and + symbols merely represent the relative polarity of voltage drop on each resistor, not the interconnecting wires.

Figure 5-21 illustrates the concept of **relative polarity**. Notice that test point A is the most positive since it is connected to the positive side of the voltage source and test point C is the most negative since it is connected to the negative side of the source. In other words, A is very positive relative to C and C is very negative relative to A. But what about test point B? Is test point B negative, positive, or neutral? The answer is: It's relative! Test point B is negative relative to test point A, since B is closer to the negative side of the source than A is. At the same time, test point B is positive relative to test point C, since B is closer to the positive side of the source than C is. Another way to state the relative polarity of these three test points is that *A is more positive than B*, while *C is more negative than B*. If a common bus wire is connected to test point B, you have a dual-polarity voltage source. You have a choice of a negative or a positive voltage relative to bus wire B. This will come in handy later when you learn about certain integrated circuits that require a plus and minus (dual-polarity) supply.

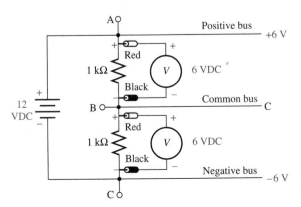

Figure 5-21   The polarity of an output is relative to a chosen reference.

SELF-CHECK 5-5

    Given: This circuit

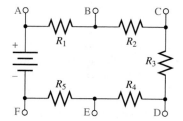

Figure 5-22

1. Mark the polarity of each resistor above with a '+' and a '−' symbol.

2. What is the polarity of test point B relative to point F?

3. What is the polarity of test point C relative to point A?

4. What is the polarity of test point E relative to point C?

5. What is the polarity of test point D relative to point F?

5-6 Common Ground

See color photo insert page B3.

Up to this point, all of the circuits that we have used in figures and examples have had two main wires carrying current in and out of the circuit. This has been fine and proper for these relatively small circuits. However, most pieces of electrical or electronic equipment are far more complicated or extensive. A schematic diagram of a television receiver or a computer can be quite large and detailed. The wires that carry current to and from the power source throughout the various circuits in a piece of equipment can really add to the confusion when you read a complicated schematic. **Common-ground** symbols are used in schematics to help reduce confusion created by too many wires drawn on a diagram. Figure 5-23 illustrates two common-ground return symbols. The pitchfork symbol is considered by some authorities to be the most appropriate symbol for metal chassis or copper foil common grounds on PC boards, while the inverted pyramid symbolizes earth ground (such as from the chassis to the third prong on the AC power plug connection). The inverted

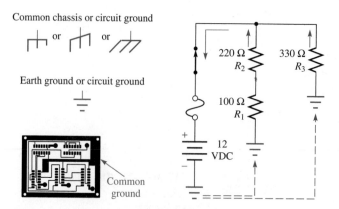

Figure 5-23   Symbols used to represent common ground.

pyramid is also used to represent a large metallic mass such as the chassis of a vehicle or the metal enclosure of a piece of equipment. The particular use for both symbols in schematic diagrams is at best inconsistent, with most electrical or electronics draftsmen preferring the inverted pyramid for most applications.

The common-ground return symbols are placed in a schematic to represent the ground foil on a PC board or a metal chassis. In an automobile, the entire metal body, frame, and engine block is the common ground. That is, the negative side of the battery is connected to the metallic mass of the vehicle, which acts as a large current bus. The positive side of the battery is connected to a fuse panel from which insulated wires meander throughout the vehicle to various circuits. A schematic of the wiring in an automobile has lines drawn for the positive wires and uses common-ground symbols to represent the metallic mass of the vehicle.

Positive or Negative Ground?

A ground is considered negative if the negative side of the supply is connected to the common metallic mass. Most vehicles have a negative ground, as do most pieces of electronics equipment. However, some pieces of equipment have a positive ground for one reason or another. Then again, there are some circuits in which the ground serves as both positive and negative. Such a case is a circuit that requires a $+$ and $-$ supply with respect to a common ground. Some amplifier circuits and computer chips require this.

As you can see in Figure 5-24, if a circuit has a negative ground, all voltages in the circuit are positive with respect to ground. This means you can clip the negative lead of your voltmeter to a ground point, such as a chassis, and the positive lead can be used to measure voltages throughout the circuit. Alternately, if a circuit has a positive ground, all voltages in the circuit are negative with respect to ground and the positive lead of your voltmeter is clipped to ground, while the negative lead is used for circuit measurements.

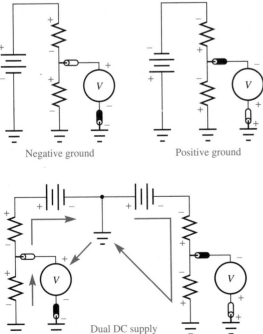

Negative ground   Positive ground

Dual DC supply

Figure 5-24 Grounds can be positive or negative.

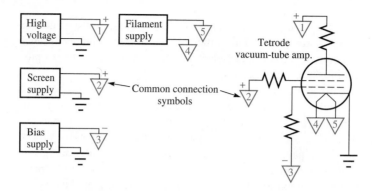

Figure 5-25   The small triangle can be used as a common-connection symbol in a complex schematic diagram.

If a dual ($\pm$) supply is used, both positive and negative voltages are available with respect to ground. In this case, if you are using an analog voltmeter, the leads of your meter will have to be reversed when measuring positive and negative voltages with respect to ground. Most digital voltmeters handle any polarity without damage (negative voltages are indicated with a negative sign on the display).

Common-Connection Symbols

The common-connection symbol is sometimes used as a ground-return symbol but is basically intended to indicate a common connection in a very complicated circuit or piece of equipment. For example, most schematics for large broadcast transmitters make generous use of these common-connection symbols to show that certain portions of a circuit are connected to a particular power supply. As shown in Figure 5-25, a large transmitter may have five or six different power supplies, all using the same common ground. Therefore, one side of each supply in the schematic is drawn with the ground-return symbol, while the other side of each supply is shown with a numbered common-connection symbol. This eliminates having to draw many long interconnecting wires on the schematic.

S E L F - C H E C K   5 - 6

1. What is the purpose for a common-ground symbol in a schematic?

2. What is the difference between a positive ground and a negative ground?

3. Could a ground be both positive and negative?

4. What is the purpose of a common-connection symbol in a schematic?

5-7   The Voltage Divider

You will recall that resistors in a series circuit each receive a certain voltage drop according to their relative ohmic values (the voltage divider rule). You also learned that the sum of all voltage drops in a series circuit equals the source voltage (Kirchhoff's Voltage Law). In other words, the source voltage is divided among the series resistors in the circuit. This concept is not only very basic, it is very useful and practical. Voltage dividers are extremely

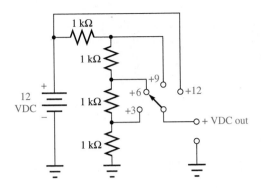

Figure 5-26 A series circuit can be used as a voltage divider.

common circuits in electricity and electronics. They are used to bias transistors and integrated circuits for proper operation, establish preset voltage levels in power supplies or voltage regulators, calibrate test equipment, and in many other applications.

We will begin with analysis, then discuss the design of the voltage divider. Figure 5-26 illustrates a typical voltage divider that provides a choice of four different positive voltages with respect to ground. In the *analysis* of a voltage divider, you will be concerned with determining the voltages available from the resistors with respect to ground. When *designing* a voltage divider, you will already know the voltages you want and will have to determine the value of each resistor needed to obtain those voltages.

Analysis First

Analyzing a voltage divider is very similar to analyzing a simple series circuit. First you need to determine the total resistance of the series circuit. Then the individual voltage drops across each resistor must be found. To do this, you may want to use Ohm's Voltage Law ($E = I \cdot R$), after having found the total current. Or, you could use the proportion formula to find individual voltage drops (voltage divider rule). Finally, you will be able to solve for output voltages from your voltage divider by adding all voltage drops that lie between any voltage divider output and ground. Now, let's examine this procedure by carefully studying Example 5-6.

EXAMPLE   5-6

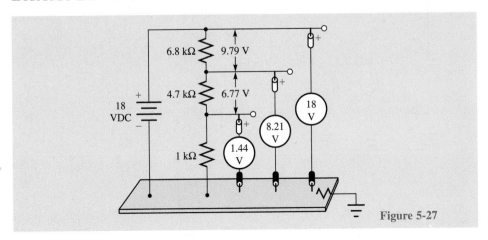

Figure 5-27

Notice that the voltage available at each terminal is actually the sum of all voltage drops between that terminal and ground. Can you think of a slightly more direct approach in solving for the various terminal voltages? Suppose you want to determine a particular terminal voltage in a large voltage divider without having to calculate every individual voltage drop. Example 5-7 illustrates a direct solution.

EXAMPLE   5-7

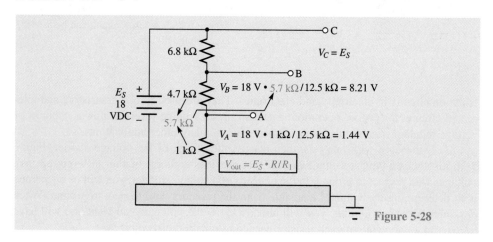

Figure 5-28

Voltage Divider Design

Voltage divider design may seem a little more interesting since it tends to stir up your creative juices. You start with a given set of specifications and design a circuit that will fulfill those specs. One of the most common voltage dividers with which you will work throughout your career is the very simple two-resistor series circuit. So this is where we start, and we'll make things tougher later.

EXAMPLE   5-8

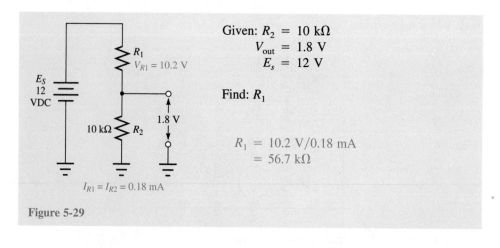

Figure 5-29

A First Approach

As you can see in Example 5-8, we want to obtain 1.8 volts from a 12 V supply. The resistance from the 1.8 V terminal to ground is given as 10 kΩ. In this particular approach to voltage divider design, three important circuit parameters (values, characteristics, or specifications) have been given: source voltage, desired output voltage, and terminal resistance to ground. With this information, you are able to solve for the value of R_1 (the series dropping resistor). To solve for R_1, you must know the amount of current through R_1 and the voltage across R_1. Now, here is where you can put some of your circuit theory to work for you. Since R_1 and R_2 form a simple series circuit, the current through R_1 is the same as the current through R_2. Thus, I_{R2} can be used to solve for R_1. Because of Kirchhoff's Voltage Law, you are also aware that the voltage across R_1 is simply the difference between the source voltage and the terminal voltage (voltage across R_2). V_{R1} is therefore 10.2 V. The value of R_1 is then found using Ohm's Law.

EXAMPLE   5-9

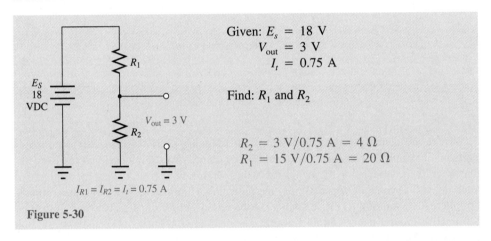

Given: $E_s = 18$ V
$V_{out} = 3$ V
$I_t = 0.75$ A

Find: R_1 and R_2

$R_2 = 3\text{ V}/0.75\text{ A} = 4\ \Omega$
$R_1 = 15\text{ V}/0.75\text{ A} = 20\ \Omega$

$I_{R1} = I_{R2} = I_t = 0.75$ A

Figure 5-30

A Second Approach

In another approach to voltage divider design, the total current may be stated as a required specification. In other words, you may be given the source voltage, the desired output voltage, and the total current. You then have to determine R_1 and R_2, keeping in mind that $R_1 + R_2$ must equal the total resistance that limits current to the specified level. Example 5-9 is such a case. You'll be delighted to see how simple the solution to this type of design problem really is. First, find total resistance. This is easy since the source voltage and total current are given ($R_t = E_s/I_t$). Now solve for the terminal resistance (R_2)—again, by using Ohm's Law ($R_2 = V_{out}/I_t$). Finally, the dropping resistor (R_1) is found by simply subtracting R_2 from the total resistance ($R_1 = R_t - R_2$) or by using Ohm's Law.

Getting Tougher!

Here is a chance to really flex your muscles. How would you design a voltage divider with more than one output or terminal voltage? Where would you begin? The answer to this last question depends upon what information has been given or specified. Normally, for a large voltage divider with many terminals, you know the source voltage, the desired output voltages, and the total current. Now what? Well, it won't be so tough after all. Look at Example 5-10.

EXAMPLE 5-10

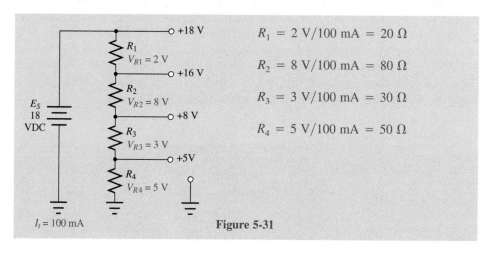

$R_1 = 2 \text{ V}/100 \text{ mA} = 20 \ \Omega$

$R_2 = 8 \text{ V}/100 \text{ mA} = 80 \ \Omega$

$R_3 = 3 \text{ V}/100 \text{ mA} = 30 \ \Omega$

$R_4 = 5 \text{ V}/100 \text{ mA} = 50 \ \Omega$

Figure 5-31

There are actually several approaches to designing such a voltage divider. In this particular example, each resistor is found by first identifying each individual resistor voltage drop, then solving for the resistance using Ohm's Law ($R_X = V_{RX}/I_t$). Notice that each individual voltage drop is simply the potential difference between output terminals. An alternative solution to this example would be to find terminal resistances from each terminal to ground using Ohm's Law ($R_{\text{term.}} = V_{\text{term.}}/I_t$). Then, the resistance of each resistor can be determined by the difference in resistance between terminals.

Let us summarize a useful general approach to designing a voltage divider.

Voltage Divider Design

1. *Total current, source voltage, and all terminal voltages must be known.*

2. *Determine individual resistor voltage drops by taking the difference between terminal voltages.*

3. *Solve for the value of each resistor using Ohm's Law. $R_X = V_{RX}/I_t$.*

Power Dissipation

In the design of any circuit, you will want to consider the power dissipation of the components. As you already know, the power dissipation of a resistor in a circuit is determined by the amount of current through the resistor and the amount of voltage across the resistor. **The total power dissipation in any DC circuit is always the sum of individual dissipations or the product of total current and source voltage.** Example 5-11 is an analysis of power dissipation in the voltage divider of Example 5-10.

EXAMPLE   5-11

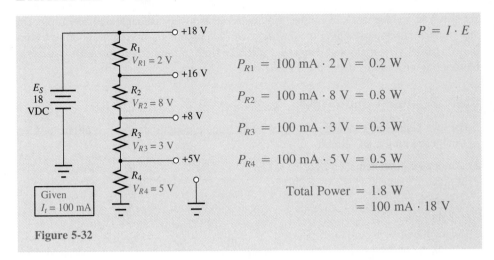

$$P = I \cdot E$$

$$P_{R1} = 100 \text{ mA} \cdot 2 \text{ V} = 0.2 \text{ W}$$

$$P_{R2} = 100 \text{ mA} \cdot 8 \text{ V} = 0.8 \text{ W}$$

$$P_{R3} = 100 \text{ mA} \cdot 3 \text{ V} = 0.3 \text{ W}$$

$$P_{R4} = 100 \text{ mA} \cdot 5 \text{ V} = \underline{0.5 \text{ W}}$$

$$\text{Total Power} = 1.8 \text{ W}$$
$$= 100 \text{ mA} \cdot 18 \text{ V}$$

Figure 5-32

Now it's your turn to do some voltage divider design. Test your skills on Self-Check 5-7.

SELF-CHECK   5-7

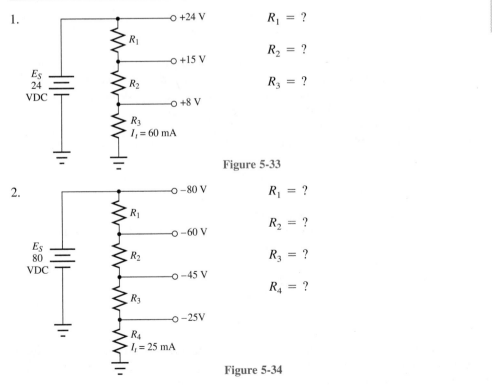

1.

$R_1 = ?$

$R_2 = ?$

$R_3 = ?$

Figure 5-33

2.

$R_1 = ?$

$R_2 = ?$

$R_3 = ?$

$R_4 = ?$

Figure 5-34

3. Calculate the power dissipated by each resistor in problem #2.

5-8 Troubleshooting with an Ohmmeter

By now you are probably aware that the ohmmeter is a very valuable and useful piece of test equipment. Understanding the proper use of the ohmmeter in troubleshooting a complex circuit is very important. Here, we will take time to familiarize you with basic troubleshooting procedures using the ohmmeter.

Basic Trouble Conditions

Before we begin, let's identify the three basic trouble conditions that we might look for in troubleshooting a DC circuit.

1. An open component such as a
   - fuse
   - resistor
   - light bulb
   - DC motor
   - broken conductor

2. A shorted component such as a
   - resistor
   - circuit (conductors)
   - DC motor

3. An out-of-tolerance component such as a
   - resistor

The Open

As you know, an open is a break in the normal current path. A copper foil on a printed circuit board may have a hairline crack. A wire may be fatigued and broken inside its insulation. A fuse may have burned open due to excessive current demand. A resistor may have overheated and burned open. These are only a few examples in which an open may occur. The result is a DC current path that is now infinite in resistance. Placing the test leads of your ohmmeter across any one of these examples would result in an infinite meter indication.

The Short

A short is the opposite of an open. A shorted component would result in a zero-ohms (0 Ω) reading (or close to it) on your ohmmeter. A typical short might be a solder bridge between copper foils on a printed circuit board, or a section of a metal chassis touching the leads of a component. An overloaded DC motor may develop shorts within its windings due to enamel insulation burning away and bare wires touching. Many of these conditions, and more, may be located with the proper use of the ohmmeter.

Out of Tolerance

Out-of-tolerance resistors can be located as well. Ordinary carbon-composition resistors are known to increase in resistance due to overheating and aging. There may be no visible evidence of this but a quick check with the ohmmeter will reveal the resistor's true value.

Methods for Troubleshooting

Now that we know what types of troubles to expect, we should consider some example circuits and develop useful methods for troubleshooting.

Troubleshooting a Series Circuit

Consider Figure 5-35. Notice that this is a series circuit and has a fuse to protect the wiring and other components should a short occur somewhere in the circuit. As you know, the fuse is placed as close as possible to the DC source to protect as much of the circuit as possible. You will recall that a good fuse has very little resistance (close to $0\,\Omega$) and will look like a short when measured with the ohmmeter.

Let us say that the circuit has a problem, and as a result, the light bulb will not light when the switch is closed. The question is: What is the actual trouble that is preventing the bulb from lighting? More important, what procedure shall we follow in locating this trouble?

The first step in the troubleshooting process is:

1. *Clearly identify the symptoms, problem, or abnormal condition.*

In this example, we have already stated that the bulb will not light when the switch is closed. The next step in the troubleshooting process is:

2. *Make a list of possible troubles that would cause the identified symptom or symptoms.*

In this example we would make the following list:

Possible Troubles

1. light bulb burned out
2. fuse burned open
3. wire or connection broken
4. resistor burned open
5. switch is bad

(We will assume that the battery is good and has already been tested.)

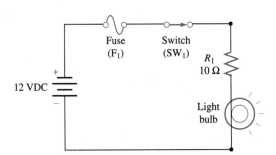

Figure 5-35   Troubleshooting a series circuit.

The next step in our troubleshooting process is to:

3. *Make a visual inspection of the circuit and components with the power safely disconnected.*

A visual circuit inspection is a good step but it is not always revealing. You may not see anything that is broken or burned. Broken wires covered with plastic insulation are almost impossible to find visually. Hairline cracks in copper foils are very difficult to find visually. If the fuse casing is clear glass a careful inspection will reveal if it is good or open in most cases. Always double-check with the ohmmeter. A burned-open resistor is very obvious to the eye, as is a burned-open light bulb under careful inspection. A bad switch is not obvious to the eye and the ohmmeter must be used.

Our final step in troubleshooting will be to:

4. *Isolate each suspected component and measure with the ohmmeter.*

We must, of course, make sure the power is off. In this example, the fuse can be removed and/or the batteries can be removed. If the DC source is an AC/DC converter (DC power supply), it must be unplugged. As a general rule, when the ohmmeter is used, each component should be isolated from the other components to prevent false readings (more on this in Chapter 6). In a simple series circuit that has been broken (switch opened and/or fuse removed), each component can be measured without removing it from the circuit.

The main suspects in our investigation are the light bulb, the fuse, the resistor, the wiring, and the switch. Normally, you would begin isolating and testing components in the order of greatest probability. Figure 5-36 illustrates the proper way to test each com-

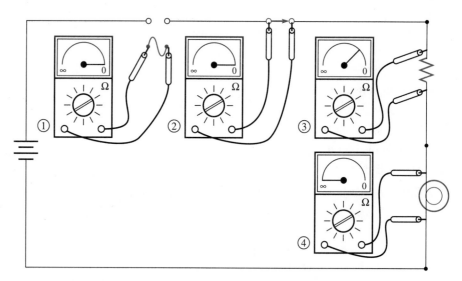

TEST RESULTS

Component	Should be	Measured as	Condition
1. Fuse	0 ohms	0 ohms	Good
2. Switch (On)	0 ohms	0 ohms	Good
3. Resistor	10 ohms	10 ohms	Good
4. Light bulb	2 ohms (Cold)	Infinite	Bad (Open)

Figure 5-36   The ohmmeter is used to troubleshoot the series circuit with the circuit power disconnected.

ponent. Notice that the power has been completely disconnected by removing the fuse so there is no chance of damaging the ohmmeter. The wiring may be checked by testing each interconnecting length of wire for continuity from one component lead to another. In this case it is discovered that the bulb is burned open.

Before we continue, let's review this basic troubleshooting procedure.

TROUBLESHOOTING PROCEDURE

1. Identify the symptoms or abnormal conditions.
2. Make a list of possible causes. A schematic or circuit diagram is normally essential for this.
3. With the power disconnected, make a visual inspection of the circuit and its components.
4. Isolate each suspected component and measure with the ohmmeter.

Test your memory with Self-Check 5-8.

SELF-CHECK 5-8

1. Try to list the four troubleshooting steps from memory.

2. In order to prevent the ohmmeter from being damaged, what precaution must you take before measuring circuit commponents?

3. What does it mean to isolate a component?

5-9 Troubleshooting with a Voltmeter

Now that you have an understanding of series circuit theory and basic troubleshooting techniques, you are ready to learn to troubleshoot such a circuit using a voltmeter (or your multimeter as a voltmeter). In this section, we will discuss the use of the voltmeter in detecting opens and shorts in a troubled series circuit. Figure 5-37 shows a typical analog multimeter that you might use in troubleshooting a circuit. Notice the various scale range markings on the meter face. These correspond to various multimeter functions and range selections. When using a multimeter as a DC voltmeter, select a DC voltage range that is higher than the source voltage to begin taking measurements. Switch down to a lower range as needed.

See color photo insert page B2.

Detecting Opens

The first trouble we will discuss is the open. We already know that an open in a circuit is a break in the flow of current. It may actually be a broken wire, a hairline crack in a foil on a PC board, a bad fuse, a burned light bulb, a burned resistor, etc. Whatever the cause, we know that an open will not allow current to flow. An open can also be thought of as an infinite resistance (a resistance so large that no current can flow). Obviously, an open

Figure 5-37 A handy analog multimeter. (Courtesy of Amprobe Instruments, a div. of Core Industries, Inc.)

is the largest resistance in the circuit (∞ Ω). Since the open is the largest resistance in the series circuit, all of the source voltage will appear across it. Remember what we said about determining voltage drops by proportion? That's correct. The largest resistance gets the largest voltage drop. If the largest resistance is infinite (open), then it will get all of the available source voltage, since any other resistances will look insignificant! The other circuit resistances will have no voltage drop at all, since there is no current due to the open!

We have just examined three very important troubleshooting facts regarding an open:

1. *When an open occurs in a series circuit, the circuit current drops to zero (0 A).*

2. *The entire source voltage is dropped across an open in a series circuit.*

3. *All voltages across other circuit components drop to zero volts (0 V).*

We will be interested in facts 2 and 3 when we troubleshoot a series circuit with a voltmeter. Figure 5-38 illustrates the use of the voltmeter in troubleshooting a circuit that contains an open. Notice that when the open is found, the voltmeter will indicate maximum voltage (the source voltage).

Detecting Shorts

The short is the opposite of an open. Instead of dropping to zero, as it does when an open occurs, the current will increase. The increase in current is naturally due to the reduction in resistance caused by the short. You will recall that a shorted component is one whose resistance is being bypassed because of accidental contact with a metal chassis or some internal damage to the component. In any case, the resistance of the component now looks like zero ohms (no resistance at all, or very little as compared to what it should be).

Since a short has no resistance, it cannot have a voltage drop. In other words, if R drops to zero, due to a short, the voltage must also drop to zero. This is easy to prove with Ohm's Voltage Law ($E = I \cdot R$). If R is zero, then 0 Ω $\cdot$ A = 0 V. The current can be any value and the voltage will still be zero volts across the short because any value multiplied by zero is zero.

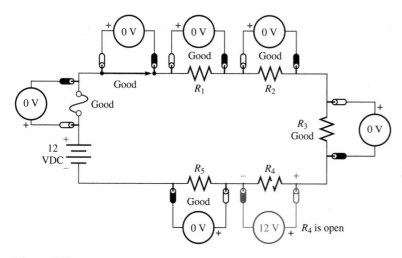

Figure 5-38 The source voltage will be measured across an open in a series circuit.

Finally, when a short occurs, the voltage that was across the shorted component will be distributed to the other components. The voltage across the other circuit components will increase. Again, this is easily proven with Ohm's Voltage Law. The total current in the circuit has increased, thus increasing voltage drops around the circuit ($\uparrow V_R = \uparrow I_t \cdot R$).

Thus, we have described three important troubleshooting facts regarding a short:

1. *Total circuit current increases when a short occurs.*

2. *The voltage across a component drops to zero when it is shorted.*

3. *The voltages across other components increase when the short occurs.*

Again, we will be interested in facts 2 and 3 when we troubleshoot a series circuit with a voltmeter. Figure 5-39 illustrates the use of the voltmeter in troubleshooting a circuit that contains a short. Notice that when the short is found, the voltmeter will indicate zero volts (0 V).

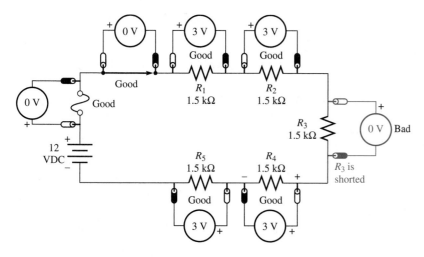

Figure 5-39 Zero volts (0 V) will be measured across a dead short in a series circuit.

Take a few moments to review this section by testing your skills on Self-Check 5-9 before continuing.

SELF-CHECK 5-9

1. When an open occurs in a series circuit, will the total current increase or decrease?

2. What will your voltmeter indicate when an open is found in a series circuit?

3. List three troubleshooting facts about a short in a series circuit.

Summary

FORMULAS

(5.1) $R_t = R_1 + R_2 + R_3 + \cdots$

   (total resistance for series resistors)

(5.2) $I_t = E_s/R_t$

   (to find total current in a resistive circuit)

(5.3) $V_R = I_t \cdot R$

   (to find the voltage drop across a series resistor)

(5.4) $V_R = (R/R_t) \cdot E_s = \dfrac{E_s R}{R_t}$

   (proportion formula or voltage divider rule for series circuits)

CONCEPTS

- Adding resistors to a series resistive circuit increases total resistance.
- The voltage drop on any resistor in a series circuit is proportional to the value of the resistor as compared to the total resistance.
- The algebraic sum of all voltages in a closed loop is equal to 0 volts (Kirchhoff's Voltage Law).
- The end of a component that is electrically closer to the negative side of the voltage source is considered to be negative with respect to its other end or the positive side of the source.
- The total source voltage is measured across an open in a simple series circuit.
- The total current is zero amps when an open occurs in a simple series circuit.
- Zero volts is always measured across a short.
- Total circuit current increases when a short occurs in any circuit.
- If a resistor in a simple series circuit becomes shorted, the voltage drops on other resistors will increase.

PROCEDURES

Troubleshooting Procedure

1. Identify the symptoms or abnormal conditions.

2. List all possible causes.

3. With the power off, make a component and wiring inspection.

4. Isolate each suspected component and measure it with the ohmmeter.

Troubleshooting with a Voltmeter

1. Set the voltmeter to a range higher than the source voltage to begin your measurements.

2. Observe polarity. (Be careful around lethal voltages.)

SPECIAL TERMS

- Series circuit
- Total resistance (R_t), equivalent resistance (R_{eq})
- One path, voltage drop
- Kirchhoff's Voltage Law
- Voltage divider rule
- Relative polarity
- Common ground, common connection
- Positive ground, negative ground

Need to Know Solution

Why did the entire string go out when one went out in the early Christmas lights? Because the lights were in series and a burned-out bulb broke the path for current. The modern series-wired Christmas lights use special bulbs that form a bridging contact inside when the element burns out. In this way, the current path is unbroken. When one bulb burns open, all of the voltage is distributed to the remaining bulbs. Also, there is increased current because there is less total resistance (less total bulb-element resistance). Therefore, the amount of power each remaining bulb must dissipate increases with every bulb that burns out. This makes the remaining bulbs burn open sooner and sooner. Neat, huh?!

Questions

5-1 Series Resistors

1. What happens to the total resistance of a series circuit if another resistor is placed in series with the circuit?

2. What happens to the total resistance of a series circuit if one of its resistors becomes shorted?

5-2 One Path for Current

3. Name one important characteristic of a series circuit.

4. What two things determine the amount of current in a series circuit?

5-3 Series Voltage Drops

5. What two things must you know about a resistor in order to calculate the voltage across the resistor in a series circuit?

6. Write the formula you would use to calculate individual voltage drops in a series circuit.

5-4 Kirchhoff's Voltage Law and the Voltage Divider Rule

7. In a series circuit, the sum of all voltage drops must equal what?
8. The voltage drop across any resistor in a series circuit is proportional to its resistance compared to what?

5-5 Relative Polarity

9. What does it mean to say a point in a series circuit is more negative than another point?
10. Explain how a single DC source and two resistors can be used to make a dual-polarity DC supply.

5-6 Common Ground

11. Draw a series DC circuit using a common-ground symbol for the DC source and part of the circuit.
12. Explain the use of a common-connection symbol.

5-8 Troubleshooting with an Ohmmeter

13. Should circuit power be (1) on, or (2) off when using the ohmmeter to troubleshoot? Explain.
14. What is the first step in using the ohmmeter to troubleshoot a circuit?
15. What does it mean to isolate a component for measurement?

5-9 Troubleshooting with a Voltmeter

16. If a resistor in a series circuit is shorted, what reading would you expect to see when placing your voltmeter leads across the resistor?
17. If the voltage is measured across an open resistor in a series circuit, what reading would you expect to see?
18. If a fuse is good and there are no opens in the circuit, what voltage reading would you expect to see across the fuse?
19. If one of three series resistors is shorted, what has happened to the voltage drops across the other two resistors?
20. Explain how you would use your voltmeter to find an open in a simple series circuit.

Problems

5-1 Series Resistors

1. Find the total resistance between terminals A and B in Figure 5-40.

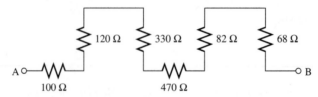

Figure 5-40

2. Find the total resistance between terminals A and B in Figure 5-41.

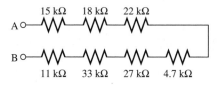

Figure 5-41

3. Will adding another resistor to Figure 5-41 (1) increase, or (2) decrease total resistance?
4. Where should you place your ohmmeter leads to measure the total resistance of Figure 5-42?

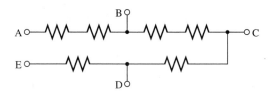

Figure 5-42

5-2 One Path for Current

5. If a series circuit has a total resistance of 22 kΩ and a source voltage of 28 VDC, what is the total current?
6. A 24 VDC source is connected to three series resistors (27 kΩ, 18 kΩ, and 39 kΩ). Find R_t and total current.

5-3 Series Voltage Drops

7. If total current in a series circuit is 150 μA, what is the voltage drop across a 220 kΩ resistor in that circuit?
8. Find total resistance, total current, and all voltage drops in Figure 5-43.

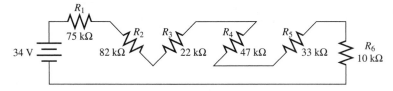

Figure 5-43

5-4 Kirchhoff's Voltage Law and the Voltage Divider Rule

9. Use the voltage divider formula to solve for the voltage across a 10 kΩ resistor when the source voltage is 48 V and R_t is 120 kΩ. Show all work.
10. In a series circuit that has a 22 kΩ resistor and a 4.7 kΩ resistor, prove mathematically that the lower value resistor has the lower voltage drop. ($E_s = 28$ VDC)
11. Write a loop equation for Figure 5-43.

12. Write a loop equation for Figure 5-44.

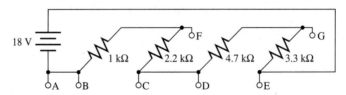

Figure 5-44

5-5 Relative Polarity

13. Mark the correct voltage-drop polarity on each resistor in Figure 5-44.
14. In Figure 5-44, is point D more positive or more negative than point B?
15. In Figure 5-44, if point F is considered a reference point for all voltages in the circuit, list all points that are relatively negative and all points that are relatively positive.

5-7 The Voltage Divider

16. Solve for output voltages and power dissipations of the divider resistors in Figure 5-45.

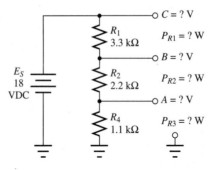

Figure 5-45

17. Solve for output voltages and power dissipations of the divider resistors in Figure 5-46.

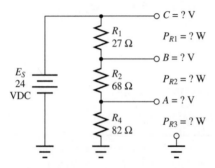

Figure 5-46

18. Design a voltage divider that has a source voltage of 36 V and two output voltages of +8 V and +28 V. The total resistance of the voltage divider shall be 150 Ω. Also, determine the proper power rating for each of the divider resistors. (*Hint:* Three divider resistors are needed.)

19. Design a voltage divider that has a source voltage of 12 V and two output voltages of −4 V and −8 V. The total resistance of the voltage divider shall be 20 kΩ. Also, determine the proper power rating for each of the divider resistors. (*Hint:* Three divider resistors are needed.)

Answers to Self-Checks

Self-Check 5-1

1. $R_t = R_1 + R_2 + R_3 + \cdots$
2. $R_t = 30 \; \Omega$
3. $R_t = 5,920 \; \Omega = 5.92 \; k\Omega$

Self-Check 5-2

1. 70 kΩ and 0.343 mA
2. 16.3 kΩ and 0.552 mA
3. 13 MΩ and 5.77 μA

Self-Check 5-3

1. 3.43 V, 11.3 V, 9.26 V
2. 3.1 V, 1.5 V, 2.6 V, 1.82 V
3. 8.66 V, 52.5 V, 13.85 V

Self-Check 5-4

1. $V_{R1} = 2.44$ V; $V_{R2} = 2.65$ V; $V_{R3} = 1.71$ V, $V_{R4} = 0.94$ V; $V_{R5} = 4.26$ V
2. $V_{R1} = 32.3$ V; $V_{R2} = 27.2$ V; $V_{R3} = 6.33$ V; $V_{R4} = 17.1$ V; $V_{R5} = 17.1$ V
3. $V_{R1} = 32.7 \; \mu$V; $V_{R2} = 78.6 \; \mu$V; $V_{R3} = 12.8 \; \mu$V; $V_{R4} = 49.1 \; \mu$V; $V_{R5} = 26.8 \; \mu$V
4. 1. -4.26 V $- 0.94$ V $- 1.71$ V $- 2.65$ V $- 2.44$ V $+ 12$ V $= 0$ V
   2. -17.1 V $- 17.1$ V $- 6.33$ V $- 27.2$ V $- 32.3$ V $+ 100$ V $= 0$ V
   3. $-26.8 \; \mu$V $- 49.1 \; \mu$V $- 12.8 \; \mu$V $- 78.6 \; \mu$V $- 32.7 \; \mu$V $+ 200 \; \mu$V $= 0 \; \mu$V

Self-Check 5-5

1. See Figure 5-47.

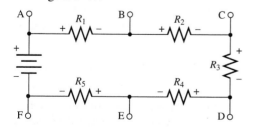

Figure 5-47

2. Positive
3. Negative
4. Negative
5. Positive

Self-Check 5-6

1. The common-ground symbol represents a common metallic mass and reduces the number of wires that must be drawn on a schematic.
2. A negative ground is created when the negative side of the supply is connected to the common metallic mass or common bus. A positive ground is created when the positive side of the supply is connected to the common metallic mass or common bus.
3. Yes—as in a dual power supply.
4. The common-connection symbol is used to show a connection between two points in a circuit without having to draw a wire from one point to the other on the schematic.

Self-Check 5-7

1. $R_1 = 150\ \Omega$; $R_2 = 117\ \Omega$; $R_3 = 133\ \Omega$
2. $R_1 = 800\ \Omega$; $R_2 = 600\ \Omega$; $R_3 = 800\ \Omega$; $R_4 = 1\ k\Omega$
3. $P_{R1} = 0.5\ W$; $P_{R2} = 0.375\ W$; $P_{R3} = 0.5\ W$; $P_{R4} = 0.625\ W$

Self-Check 5-8

1. (1) Identify symptoms, (2) list possible causes, (3) inspect visually, (4) isolate and measure with ohmmeter
2. To prevent damage to the ohmmeter, first remove power (unplug power supply, remove batteries, remove fuse, turn off switch).
3. "Isolate the component" means to disconnect it, if necessary, to make sure that only the component itself is being measured with the ohmmeter.

Self-Check 5-9

1. Decrease
2. The source voltage
3. (1) Total current increases; (2) the voltage on the shorted component will be zero; and (3) voltage drops on all other components will increase.

SUGGESTED PROJECTS

1. Add the formulas and important concepts from this chapter to your Electricity and Electronics Notebook. You might also want to transfer the Design Notes in this chapter to your personal notebook.

2. Build a few of the circuits shown in this chapter and verify your calculations with a multimeter.

3. Practice your troubleshooting skills by constructing a circuit and simulating shorts and opens. Use 3 or 4 resistors in series. Place a jumper wire across a resistor to simulate a short. Remove the resistor to simulate an open. Be careful not to short out your power supply (there should be a fuse).

This is just one of Edison's many inventions used to transmit and receive stock quotations and other information. Notice the semicircular keyboard arrangement for transmitting characters. Incoming information was printed on ribbon paper that was fed from a spool mounted at the top (paper not shown). Edison invented his first stock ticker machine late in 1869 (''Universal Stock Ticker''). He also invented the ''Unison'', a device that automatically synchronized all stock tickers on the same circuit. In 1870, he received an astounding $40,000 from the Gold and Stock Telegraph Company for his stock ticker.

EDISON UNIVERSAL PRINTER
THIS 1871 MODEL, EDISON'S FIRST
COMMERCIAL INVENTION WAS MANUFACTURED
AT NEWARK, NEW JERSEY, FACTORY AND SOLD
TO THE GOLD AND STOCK TELEGRAPH
COMPANY.

Chapter 6

Parallel DC Circuits

CHAPTER OUTLINE

OBJECTIVES

After studying this chapter, you will be able to
- calculate total resistance in parallel circuits.
- calculate branch currents and find the total current in a parallel circuit.
- calculate branch currents using a resistance ratio.
- explain Kirchhoff's Current Law and basic parallel-circuit theory.

- design practical current dividers.
- calculate power and determine power ratings for resistors in parallel circuits.
- use the ohmmeter and the ammeter in troubleshooting a parallel circuit.

INTRODUCTION

As time goes on and you progress in your study of electricity and electronics, you will eventually be able to read large schematics of radios, televisions, and a great variety of other pieces of electronic equipment. You will be able to follow DC current throughout various parts of a complex circuit and be able to analytically determine voltages across various components. The use of the multimeter, as a valuable troubleshooting tool, will become second nature to you as you explore and repair these interesting circuits. You will be methodical, confident, and deliberate in your analysis and troubleshooting approach, but it will take time and effort to reach that level of expertise.

These skills are developed over time in a step-by-step process. In Chapter 4 you were introduced to Ohm's Law with basic circuit analysis and troubleshooting. Then, in Chapter 5, you learned how to analyze, design, and troubleshoot series circuits. In this chapter, you will increase your troubleshooting skills and learn to analyze and design very practical parallel circuits. Don't forget, with each chapter, we continue to build upon the knowledge and skills presented in previous chapters. Study this chapter well. The information contained here is vital to succeeding chapters and your career.

Suppose you live in a part of the world where the electricity supplied to your home is not reliable (particularly in the evening when you really need it). You decide that you want to build an emergency lighting system using 12 V light bulbs powered from your car battery. Once you decide how many lights you need, you will want to calculate the total current drain on your battery. How can you determine the total current mathematically and experimentally? How can you determine the power dissipated by each and by all of the lights? Can you estimate how long the lights will burn brightly from your car battery? What things do you need to know before you can begin your analysis? Keep this problem in mind as you study parallel circuits. As usual, a solution to this problem is discussed at the end of the chapter. (No fair looking ahead!!)

6-1 Parallel Resistors

Finding the total resistance of a circuit made up of parallel resistors is quite different from the series circuits we have already examined. Consider Figure 6-1. This is a parallel resistor circuit where $R_1 \parallel R_2 \parallel R_3$. The two vertical lines $\parallel$ mean "in parallel with." Notice that terminal A is connected to the top lead of all three resistors through a common wire (sometimes referred to as a **bus wire**). The bottom leads of the three resistors are bus-wired to terminal B. Therefore, current supplied to this circuit, through bus A, will separate and flow through each of the three branches, provided by the three parallel resistors, and recombine on bus B. Since the current has three separate branches it may follow, instead of only one as in a series circuit, the total opposition or resistance is reduced! In fact, the more parallel branches you add, the lower the total resistance becomes!! This is the exact opposite of a series circuit. Also, the total resistance for a parallel circuit is always less than any single branch resistance.

See color photo insert page B3.

Remember this basic rule for total resistance in parallel circuits:

The total resistance of a parallel circuit is always less than the lowest branch resistance.

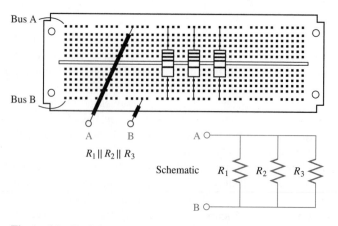

Figure 6-1   Resistors arranged to form a parallel circuit.

Also,

> Adding parallel branches to a parallel circuit always decreases resistance and increases conductance ($\uparrow G = 1/R\downarrow$).

Parallel Analysis Using Ohm's Law

Now, let's take a few moments and prove what has been said about parallel circuits. To do this, we shall use Figure 6-2.

In order to prove our parallel rules, we shall analyze this simple circuit using our old friend, Ohm's Law ($I = E/R$). You will recall that current is determined by the amount of resistance and the applied EMF. Also, if the source voltage (applied voltage) and the total current are known, the total resistance can easily be determined using the Ohm's Law resistance formula ($R = E/I$). Therefore, a formula that may be used to calculate total resistance in any circuit, series or parallel, is

See color photo insert pages B3 and B4.

$$R_t = E_s/I_t = \frac{\text{source voltage}}{\text{total current}} \qquad (6.1)$$

In Figure 6-2, E_s is 1.5 V. This voltage is the same across each resistor because of the bus wires. Therefore, the 1.5 V must be used to calculate the individual currents through R_1, R_2, and R_3. These branch currents may then be added together to find the total current. Once the total current is determined, we can solve for the total resistance using Formula 6.1.

Now, let's do what we have just discussed. First, we will solve for individual branch currents.

$$I_{R1} = 1.5\ \text{V}/10\ \Omega = 0.15\ \text{A}$$
$$I_{R2} = 1.5\ \text{V}/5\ \Omega = 0.3\ \text{A}$$
$$I_{R3} = 1.5\ \text{V}/3\ \Omega = 0.5\ \text{A}$$

Therefore:

$$I_t = I_{R1} + I_{R2} + I_{R3} + \cdots \qquad (6.2)$$
$$I_t = 0.15\ \text{A} + 0.30\ \text{A} + 0.5\ \text{A} = 0.95\ \text{A}$$

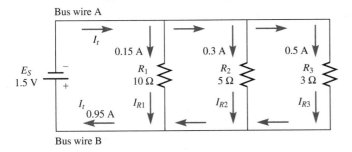

Figure 6-2 A circuit demonstrating parallel circuit theory.

Now that we know the amount of total current, we can use Formula 6.1 to solve for R_t.

$$R_t = E_s/I_t = 1.5 \text{ V}/0.95 \text{ A} = 1.58 \text{ } \Omega$$

Notice that the calculated value of total resistance is, in fact, less than the resistance of any one branch (less than 3 Ω). You can see that if we add another parallel branch, there will be more current and less total resistance, or opposition, to total current flow supplied by the source. In other words, the total conductance of the circuit increases with each additional path (branch) for current. Thus our parallel rules hold true.

The Reciprocal Formula

Ohm's Law is very useful in proving parallel-circuit theory, as we have just done. However, it is not the best way to calculate total resistance. There are three other formulas that are far more useful in determining total resistance. The first formula is known as the *reciprocal formula* (or *conductance formula*). This formula is the reciprocal of the sum of all branch conductances. Remember, resistance is the reciprocal of conductance.

$$R_t = \frac{1}{1/R_1 + 1/R_2 + 1/R_3 + \cdots} = \frac{1}{G_t} = \frac{1}{G_1 + G_2 + G_3 + \cdots} \quad (6.3)$$

Ohm's Law can be used to prove Formula 6.3 as follows:

$$I_t = I_{R1} + I_{R2} + I_{R3} + \cdots$$

Since $I = E/R$, we may substitute E/R for I like this:

$$\frac{E_s}{R_t} = \frac{E_s}{R_1} + \frac{E_s}{R_2} + \frac{E_s}{R_3}$$

The rules of algebra allow us to rearrange the equation like this:

$$E_s \cdot 1/R_t = E_s \cdot (1/R_1 + 1/R_2 + 1/R_3)$$

We may now divide both sides of the equation by E_s leaving:

$$1/R_t = (1/R_1 + 1/R_2 + 1/R_3) = G_t = G_1 + G_2 + G_3$$

Taking the reciprocal of both sides of the equation yields:

$$R_t = \frac{1}{1/R_1 + 1/R_2 + 1/R_3 + \cdots}$$

We can now prove the reciprocal formula by simply inserting the values from Figure 6-2. If the formula holds true, the answer will be 1.58 Ω.

$$R_t = \frac{1}{1/R_1 + 1/R_2 + 1/R_3}$$

so

$$R_t = \frac{1}{1/10 \text{ } \Omega + 1/5 \text{ } \Omega + 1/3 \text{ } \Omega}$$
$$= 1/(G_1 + G_2 + G_3) = 1/G_t$$
$$= 1/(0.1 + 0.2 + 0.333) = 1/0.633 = 1.58 \text{ } \Omega$$

As you can see, this matches our earlier calculation using Ohm's Law.

The reciprocal formula may be used to solve for total resistance for any number of parallel resistors, or branches. Example 6-1 has four branches. Study it carefully. Note that the four resistors are in parallel even though the wiring does look a bit odd. No rule says that parallel resistors must actually be soldered physically aligned and parallel with each other. Electrically they are in parallel! Many copper-clad printed-circuit boards have such a confusing appearance. In time, you will learn to translate a meandering printed circuit into a neatly organized schematic.

EXAMPLE   6-1

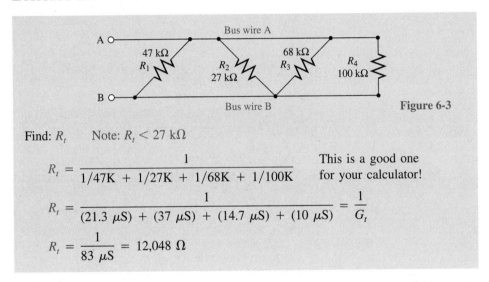

Bus wire A

A ○

47 kΩ
R_1

R_2
27 kΩ

68 kΩ
R_3

R_4
100 kΩ

B ○

Bus wire B

Figure 6-3

Find: R_t Note: $R_t < 27$ kΩ

$$R_t = \frac{1}{1/47K + 1/27K + 1/68K + 1/100K}$$ This is a good one for your calculator!

$$R_t = \frac{1}{(21.3 \ \mu S) + (37 \ \mu S) + (14.7 \ \mu S) + (10 \ \mu S)} = \frac{1}{G_t}$$

$$R_t = \frac{1}{83 \ \mu S} = 12{,}048 \ \Omega$$

In a moment, we shall take a look at two other formulas used to solve for total resistance in parallel circuits. First, take time to study Design Note 6-1.

The next two formulas for total resistance in parallel circuits may seem a bit easier to use. They are special-case formulas. That is, you may use these formulas only under certain conditions.

The Product over the Sum Formula

The first special-case formula is the product over the sum formula.

$$R_t = \frac{R_1 \cdot R_2}{R_1 + R_2} \tag{6.4}$$

This formula is used when you want to find total resistance for only two parallel resistors at a time. Suppose you have a 4.7 kΩ resistor in parallel with a 2.7 kΩ resistor. Simply insert these values into the product over the sum formula like this:

$$R_t = \frac{4{,}700 \ \Omega \cdot 2{,}700 \ \Omega}{4{,}700 \ \Omega + 2{,}700 \ \Omega} = \frac{12{,}690{,}000}{7{,}400} = 1{,}715 \ \Omega$$

DESIGN NOTE 6-1:   Total Resistance in Parallel

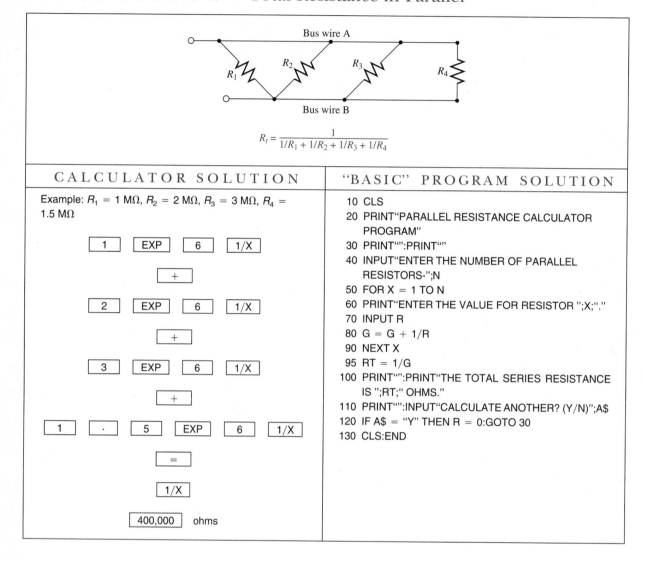

CALCULATOR SOLUTION	"BASIC" PROGRAM SOLUTION
Example: $R_1 = 1\,M\Omega$, $R_2 = 2\,M\Omega$, $R_3 = 3\,M\Omega$, $R_4 = 1.5\,M\Omega$	10 CLS 20 PRINT"PARALLEL RESISTANCE CALCULATOR PROGRAM" 30 PRINT"":PRINT"" 40 INPUT"ENTER THE NUMBER OF PARALLEL RESISTORS-";N 50 FOR X = 1 TO N 60 PRINT"ENTER THE VALUE FOR RESISTOR ";X;"." 70 INPUT R 80 G = G + 1/R 90 NEXT X 95 RT = 1/G 100 PRINT"":PRINT"THE TOTAL SERIES RESISTANCE IS ";RT;" OHMS." 110 PRINT"":INPUT"CALCULATE ANOTHER? (Y/N)";A$ 120 IF A$ = "Y" THEN R = 0:GOTO 30 130 CLS:END

Of course, you may still use the reciprocal formula to solve this problem. However, since you want the total resistance for just two parallel resistors, you may choose to use the product over the sum formula. In actuality, the product over the sum formula is derived from the reciprocal formula as follows:

$$R_t = \frac{1}{1/R_1 + 1/R_2}$$

First, we need to find a common denominator for $1/R_1$ and $1/R_2$. Of course that would be $R_1 \cdot R_2$. Using the common denominator, our formula looks like this:

$$R_t = \cfrac{1}{\cfrac{R_2}{R_1 \cdot R_2} + \cfrac{R_1}{R_1 \cdot R_2}} = \cfrac{1 \text{ (numerator)}}{\cfrac{R_2 + R_1}{R_1 \cdot R_2} \text{ (denominator)}}$$

To simplify, we may now invert the fractional denominator and multiply it by the numerator. This yields the product over the sum formula for two parallel resistors.

$$R_1 = \frac{R_1 \cdot R_2}{R_1 + R_2}$$

If your skills in algebra are a little weak, you might consider studying Appendix C, ''Algebraic Operations.''

The Same Values Formula

The second special-case formula for determining total resistance in a parallel resistive circuit is the same values formula. This formula is the simplest of all. It may be used for any resistors having the same value. You simply take the value of one resistor and divide it by the number of resistors having that same value.

$$R_t = \frac{R}{N} \tag{6.5}$$

where R = value of one resistor and N = the number of like resistors.

Consider a parallel circuit containing five 10 Ω resistors. The total resistance is simply $10 \ \Omega/5 = 2 \ \Omega$. The five parallel resistors are equal to a single 2 Ω resistor.

Use Them All

On occasion, you will want to use more than one formula to solve for total resistance. Example 6-2 is a case in point.

E X A M P L E   6 - 2

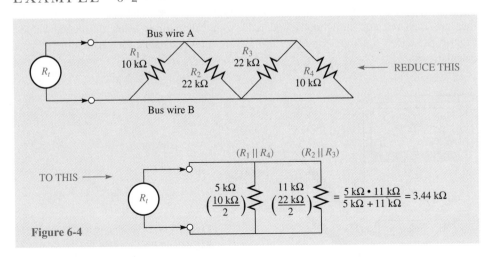

Figure 6-4

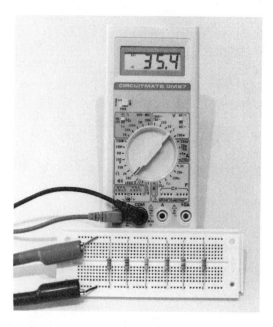

Figure 6-5   Here, the DM27 DMM by Beckman Industrial is being used to measure the total resistance of a parallel resistor circuit.

Naturally, the reciprocal formula may be used to solve for total resistance in this example. However, we may first simplify this circuit by using the same values formula. Notice that R_1 and R_4 are both 10 kΩ. These resistors quickly reduce to an equivalent resistance of 5 kΩ by use of the same values formula (10 kΩ/2 = 5 kΩ). Resistors R_2 and R_3 are also the same value and reduce to an equivalent 11 kΩ resistor. We now have a simplified two-resistor circuit that we can further solve using the product over the sum formula. The total resistance for this circuit is found to be 3.44 kΩ (55 kΩ/16 kΩ = 3.44 kΩ). Note that 3.44 kΩ is a reasonable answer because it is less than the smallest branch resistance (5 kΩ in the redrawn and reduced version).

As before, you can verify your calculations by actually constructing the circuit in Example 6-2. Remember that the marked values on the resistor body and the actual values are slightly different due to manufacturer's tolerance. For extreme accuracy, you should measure each resistor individually and redo your calculations using the actual values. Figures 6-5 and 6-6 demonstrate how to measure total resistance for verification of your calculations.

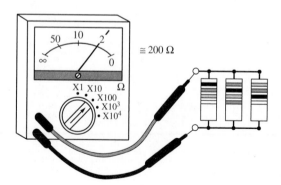

Figure 6-6   The ohmmeter can be used to measure parallel total resistance when circuit power is disconnected.

Now take time to test your skills on Self-Check 6-1.

1. Write the reciprocal formula.

   $R_t =$

2. Write the product over the sum formula.

   $R_t =$

3. Under what conditions can you use the product over the sum formula?

4. Write the same values formula.

   $R_t =$

5. When should you use the same values formula?

6. Four resistors (470 Ω, 270 Ω, 270 Ω, and 510 Ω) are arranged in parallel. Find the total conductance (G_t) and total resistance (R_t).

7. Four resistors (3.3 MΩ, 1.8 MΩ, 1.3 MΩ, and 2.7 MΩ) are arranged in parallel. Find the total conductance (G_t) and total resistance (R_t).

6-2 Kirchhoff's Current Law and Distributed Current

Kirchhoff's Current Law

Formula 6-2 ($I_t = I_{R1} + I_{R2} + I_{R3} + \cdots$) is a mathematical expression of what is commonly known as Kirchhoff's Current Law. As applied to a simple parallel circuit, Kirchhoff's Current Law can be stated as follows: **The sum of all branch currents in a parallel circuit must equal the total source current.**

See color photo insert page B5.

    Like Kirchhoff's Voltage Law, the current law is a useful tool for circuit analysis. If all branch currents of a parallel circuit are added and the total does not equal the known total current, an error has occurred. The individual branch currents and the total currents can be recalculated until the error is found.

    Kirchhoff's Current Law is often stated in a more formal way:

> **Kirchhoff's Current Law**
>
> The algebraic sum of all currents entering and leaving any point in a circuit must equal zero.

    This simply means that if currents entering a junction are considered positive quantities and currents leaving the junction are considered negative quantities, the sum of the currents entering and leaving will always be zero. This is something that you already knew

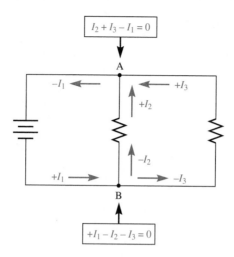

Figure 6-7 Kirchhoff's Current Law.

and assumed to be true, even though you thought about it in a different way. If 10 A of current flows down a wire and separates into two branches at a junction, the sum of the two branch currents must equal 10 A. Therefore, 10 A into the junction equals 10 A out of the junction (10 A$_{in}$ = 10 A$_{out}$). We may also express this equality in this form: 10 A$_{in}$ − 10 A$_{out}$ = 0. Thus, we have the mathematical expression of Kirchhoff's Current Law: (Sum of all currents in) − (Sum of all currents out) = 0.

The circuit shown in Figure 6-7 illustrates Kirchhoff's Current Law. Notice that currents entering a junction, specifically junctions A and B, are considered to be positive quantities, and currents leaving a junction are considered to be negative quantities. Since the positive and negative quantities at any junction are equal in magnitude and opposite in sign, their sum is always zero.

Distributed Current

Most electrical and electronic circuits are of the parallel type. The power supply provides an EMF and current to a number of branches through common bus wires. The wiring in automobiles, the wiring in homes, and the overall circuit layout of virtually any piece of electronic equipment are all examples of parallel circuitry.

In parallel circuits, current provided by the power source is distributed to a number of branches. It is very similar to water entering your home through one main line and branching out through the plumbing to the kitchen, washing machine, and bathrooms. The total water flow can be measured in the main line supplying your home, while each branch within your home has only a portion of this total flow. Thus, any parallel electrical circuit has a main supply and return line with many branches in between. Each branch receives an amount of current that depends upon the amount of resistance in that branch and the amount of voltage applied to the branch. The amount of voltage applied to each branch is the same for every branch and equal to the supply voltage, since the same main supply and return lines are connected to every branch. Example 6-3 illustrates these concepts.

We can use Ohm's Law to calculate the total current in a parallel circuit using two different approaches. First, if you know the resistance in each branch, you can simply solve for total resistance, then divide it into the source voltage ($I_t = E_s/R_t$). The second approach is to calculate each branch current using Ohm's Law ($I_B = E_s/R_B$). Then, the total current is found by simply adding all branch currents (Kirchhoff's Current Law).

$$I_t = I_{B1} + I_{B2} + I_{B3} + \cdots \quad \text{or} \quad I_t - I_{B1} - I_{B2} - I_{B3} = 0 \qquad (6.6)$$

where $B = $ branch.

EXAMPLE 6-3

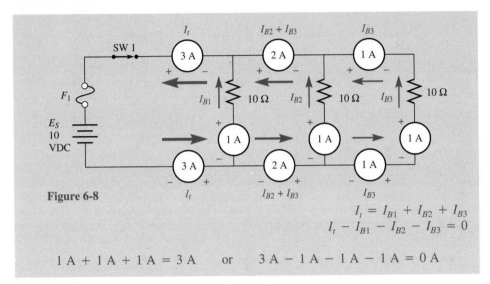

Figure 6-8

$$I_t = I_{B1} + I_{B2} + I_{B3}$$
$$I_t - I_{B1} - I_{B2} - I_{B3} = 0$$

$$1\,A + 1\,A + 1\,A = 3\,A \quad \text{or} \quad 3\,A - 1\,A - 1\,A - 1\,A = 0\,A$$

As you can see from Formula 6.6, adding more branches to a parallel circuit increases the total current demanded from the power supply through the bus wires. This is why only a certain number of branches can be connected to each circuit breaker in a house wiring or industrial wiring circuit. The various branches in household electrical circuits are usually lights and electrical outlets (wall sockets). A typical household circuit may be designed to handle 15 A of current with a 15 A circuit breaker. If too many electrical devices are plugged into a particular circuit, the total current demand may exceed the 15 A limit, thus tripping the breaker. Each device is a separate parallel branch that demands its current from a pair of common wires delivering current throughout the circuit. The last device plugged in and turned on may be the straw that breaks the camel's back.

Important Parallel Circuit Characteristics to Remember

1. *The source voltage is applied to every branch.*

   $$E_s = E_{B1} = E_{B2} = E_{B3} = \cdots$$

2. *The total current supplied to the parallel circuit depends on the source voltage and the total resistance of the circuit.*

   $$I_t = E_s/R_t$$

3. *The total circuit current is the sum of all individual branch currents. (Kirchhoff's Current Law)*

$$I_t = I_{B1} + I_{B2} + I_{B3} + \cdots \quad or \quad I_t - I_{B1} - I_{B2} - I_{B3} = 0$$

4. *Adding a new branch increases total current and removing a branch decreases total current.*

Design Note 6-2 and Self-Check 6-2 will help you focus in on these concepts.

DESIGN NOTE 6-2:   Parallel Analysis

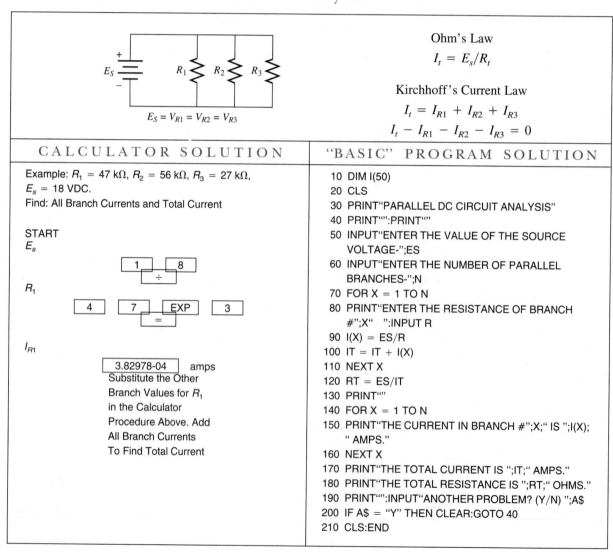

Ohm's Law

$$I_t = E_s/R_t$$

Kirchhoff's Current Law

$$I_t = I_{R1} + I_{R2} + I_{R3}$$
$$I_t - I_{R1} - I_{R2} - I_{R3} = 0$$

CALCULATOR SOLUTION

Example: $R_1 = 47\ k\Omega$, $R_2 = 56\ k\Omega$, $R_3 = 27\ k\Omega$, $E_s = 18\ VDC$.
Find: All Branch Currents and Total Current

START
E_s

R_1

| 1 | ÷ | 8 |

| 4 | 7 | EXP | 3 |
| | = | |

I_{R1}

| 3.82978-04 | amps
Substitute the Other
Branch Values for R_1
in the Calculator
Procedure Above. Add
All Branch Currents
To Find Total Current

"BASIC" PROGRAM SOLUTION

```
10  DIM I(50)
20  CLS
30  PRINT"PARALLEL DC CIRCUIT ANALYSIS"
40  PRINT"":PRINT""
50  INPUT"ENTER THE VALUE OF THE SOURCE
      VOLTAGE-";ES
60  INPUT"ENTER THE NUMBER OF PARALLEL
      BRANCHES-";N
70  FOR X = 1 TO N
80  PRINT"ENTER THE RESISTANCE OF BRANCH
      #";X"  ":INPUT R
90  I(X) = ES/R
100 IT = IT + I(X)
110 NEXT X
120 RT = ES/IT
130 PRINT""
140 FOR X = 1 TO N
150 PRINT"THE CURRENT IN BRANCH #";X;" IS ";I(X);
      " AMPS."
160 NEXT X
170 PRINT"THE TOTAL CURRENT IS ";IT;" AMPS."
180 PRINT"THE TOTAL RESISTANCE IS ";RT;" OHMS."
190 PRINT"":INPUT"ANOTHER PROBLEM? (Y/N) ";A$
200 IF A$ = "Y" THEN CLEAR:GOTO 40
210 CLS:END
```

1.

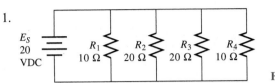

Figure 6-9

Find: I_{R1}, I_{R2}, I_{R3}, I_{R4}, I_t, R_t. Write a Kirchhoff's Current Law expression.

2.

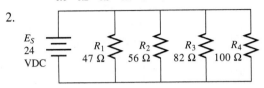

Figure 6-10

Find: I_{R1}, I_{R2}, I_{R3}, I_{R4}, I_t, R_t. Write a Kirchhoff's Current Law expression.

3.

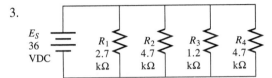

Figure 6-11

Find: I_{R1}, I_{R2}, I_{R3}, I_{R4}, I_t, R_t. Write a Kirchhoff's Current Law expression.

6-3 The Current Divider Rule

What type of circuit comes to your mind when you think of dividing current? Of course, the parallel circuit. The parallel circuit provides many branches in which current may flow. Thus, the total current in a parallel circuit is divided among its branches. But, how is the current divided? "That's easy!" you say. "All you have to do is solve for current in each branch by dividing branch resistance into the source voltage across the branch." Yes, that is true. However, in this section we will take another look at parallel circuits acting as current dividers and examine a slightly different approach to analyzing and designing them.

A Different Approach to Analysis?

Here, we consider a different approach to solving branch currents, one that emphasizes the branch resistance as compared to the total resistance in the circuit. You will see that the amount of current a particular branch receives depends upon the inverse relationship between the branch's resistance and the total resistance of the parallel circuit. Thus, the portion of total current that is distributed to any branch can be found as follows:

$$I_B = I_t \cdot R_t/R_B = \frac{I_t \cdot R_t}{R_B} \tag{6.7}$$

where I_B is the branch current and R_B is the branch resistance.

Formula 6.7 expresses the current divider rule, which can be stated as follows:

The Current Divider Rule

Total current is distributed among parallel branches in accordance with the ratio of total resistance to individual branch resistance.

Before we put Formula 6.7 to work, let's take a moment to see how it was derived. First, we start with the Ohm's Law solution to branch current: $I_B = E_s/R_B$. Since $E_s = I_t \cdot R_t$, we can substitute $I_t \cdot R_t$ for E_s. Our formula now looks like this: $I_B = I_t \cdot R_t/R_B$. If the total circuit current and total circuit resistance are known, the current in any branch can be solved by substituting the various branch resistances into this formula. Consider Example 6-4.

EXAMPLE   6-4

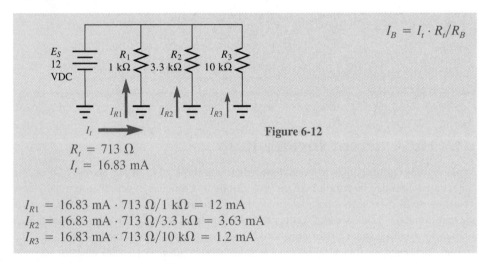

$$I_B = I_t \cdot R_t/R_B$$

Figure 6-12

$R_t = 713\ \Omega$
$I_t = 16.83\ \text{mA}$

$I_{R1} = 16.83\ \text{mA} \cdot 713\ \Omega/1\ \text{k}\Omega = 12\ \text{mA}$
$I_{R2} = 16.83\ \text{mA} \cdot 713\ \Omega/3.3\ \text{k}\Omega = 3.63\ \text{mA}$
$I_{R3} = 16.83\ \text{mA} \cdot 713\ \Omega/10\ \text{k}\Omega = 1.2\ \text{mA}$

By expanding Ohm's Law to Formula 6.7, we are able to see, and better understand, the distribution of total current according to total resistance and individual branch resistances. We can see that current division is very much dependent upon the branch resistances of a parallel circuit. We can also see that if the total current is increased (by increasing E_s), each branch will receive a proportionate increase in current while maintaining a proportionate distribution of current among branches. This is true because increasing total current does not change the ratio of total resistance to each individual branch resistance. This is illustrated in Example 6-5. Example 6-5 uses the same circuit as Example 6-4. The only difference is the increased total current, which causes proportionately increased branch currents.

EXAMPLE   6-5

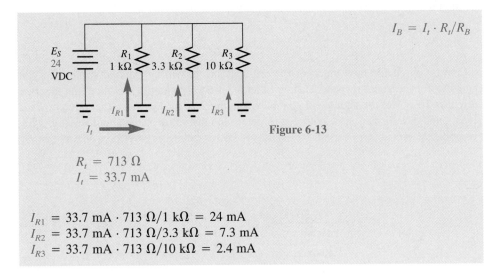

$$I_B = I_t \cdot R_t / R_B$$

Figure 6-13

$R_t = 713 \ \Omega$
$I_t = 33.7 \ \text{mA}$

$I_{R1} = 33.7 \ \text{mA} \cdot 713 \ \Omega / 1 \ \text{k}\Omega = 24 \ \text{mA}$
$I_{R2} = 33.7 \ \text{mA} \cdot 713 \ \Omega / 3.3 \ \text{k}\Omega = 7.3 \ \text{mA}$
$I_{R3} = 33.7 \ \text{mA} \cdot 713 \ \Omega / 10 \ \text{k}\Omega = 2.4 \ \text{mA}$

Designing a Practical Current Divider

Suppose you have a resistor in a circuit and you want to bypass some of the current around the resistor so not all of the current will go through the resistor. In other words, you want to divide the available current between two resistors instead of just one. The second resistor will allow a specific amount of current to bypass the first resistor. This bypass resistor is known as a **shunt** because it will *shunt* current around a particular component or device (the word ''shunt'' means to place in parallel or bypass). Figure 6-14 illustrates the use of a shunt to bypass a specific amount of current around a resistor.

In order to design a shunt for a particular application, you must know the total current and how it is to be distributed between the shunt and the shunted resistor. For sake of discussion, we will call the shunted resistor R_1 and the shunt resistance R_{sh}. The total current (I_t) is divided between R_1 (I_{R1}) and the shunt (I_{sh}): $I_t = I_{R1} + I_{sh}$, (Kirchhoff's Current Law). Since the voltage across R_1 (V_{R1}) is the same as the voltage across the shunt (V_{sh}), the R_1 current (I_{R1}) times R_1 will equal the shunt current (I_{sh}) times the shunt resis-

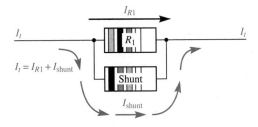

$$R_{shunt} = I_{R1} \cdot R_1 / I_{shunt}$$

Figure 6-14   Current division using a shunt.

tance (R_{sh}). In other words: $V_{sh} = V_{R1}$, therefore $I_{sh} \cdot R_{sh} = I_{R1} \cdot R_1$. Rearranging this equation to solve for the shunt resistance yields:

$$R_{sh} = \frac{I_{R1} \cdot R_1}{I_{sh}} \qquad (6.8)$$

Once you decide how much of the original current will be shunted around R_1, you can solve for the shunt resistor. Follow Example 6-6 very carefully. Notice that the total current is 10 A and the shunt will take 8 A, leaving 2 A for R_1. Since the value of R_1 is given, enough information is available to determine the value of the shunt resistor using Formula 6.8.

EXAMPLE   6-6

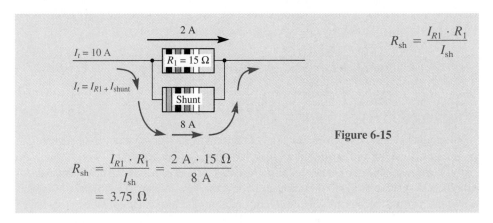

Figure 6-15

Now that you have a basic understanding of the shunt as a current divider, we can consider a very real and practical example. One very common application is a shunt for an ammeter. The ammeter, by itself, may be rated at 1 mA of current for a **Full S**cale needle **D**eflection (1 mA FSD). In other words, when the current through the ammeter reaches 1 mA, the needle will have reached the top end of the scale indicating maximum current. But, suppose you need to measure a current in the range of 100 mA? The scale, on the meter face, can be marked for a range of 0 to 100 mA but it is still a 1 mA meter. What do we do with the extra 99 mA? The extra 99 mA must be *shunted* around the meter. The 1 mA (maximum) through the ammeter will *represent* the entire 100 mA. If 50 mA is being measured, 0.5 mA will go through the ammeter and 49.5 mA will be shunted around the ammeter. Thus, the needle will deflect to half scale, indicating 50 mA.

Example 6-7 illustrates the ammeter and shunt. In order to calculate the resistance of the shunt, we must know the internal resistance of the ammeter. Then, we can apply the meter-resistance and circuit currents to Formula 6.8. A pushbutton or rotary-selector switch could be used with several different shunt resistors providing many different ranges for the ammeter. Notice that the meter resistance combined with the shunt resistance is very low. Therefore, the voltage drop across the ammeter circuit is very small and insignificant. The low ammeter circuit resistance is typical for ammeters in general and is the reason for using the ammeter *only in series* with circuit resistance and *never in parallel* with the source voltage.

EXAMPLE 6-7

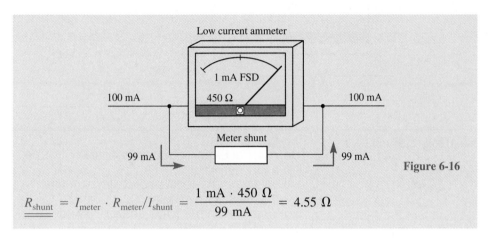

Figure 6-16

$$R_{shunt} = I_{meter} \cdot R_{meter}/I_{shunt} = \frac{1 \text{ mA} \cdot 450 \text{ } \Omega}{99 \text{ mA}} = 4.55 \text{ } \Omega$$

Shunt Power Dissipation

It is important not to ignore the power dissipation of the shunt itself, since it is usually handling most of the current. Recall that power is equal to current squared times resistance ($P = I^2 \cdot R$). To insure reliability, the power rating of the shunt should be twice its maximum dissipation. In Example 6-7, a ⅛ W or ¼ W, 4.55 Ω precision wirewound resistor would be very adequate. The actual power dissipation of the shunt is only 0.045 W at maximum current [$(99 \text{ mA})^2 \cdot 4.55 \text{ } \Omega = 0.045 \text{ W}$].

Design Note 6-3 and Self-Check 6-3 will give you an opportunity to review and reinforce what we have discussed in this section.

SELF-CHECK 6-3

1. What ratio determines the amount of current in a particular branch of a parallel circuit?

2. If the total current of a parallel circuit is 200 mA and the total resistance of the circuit is 500 Ω, how much current would a 3.3 kΩ branch receive?

3. If the total current of a parallel circuit is 1.5 A and the total resistance of the circuit is 6 Ω, how much current would a 22 Ω branch receive?

4. What do you call a resistor used to bypass current around a component or device?

5. If a basic ammeter has an internal meter resistance of 1.5 kΩ and is rated at 250 μA for full scale deflection, what size shunt would you need to enable the meter to read 10 mA full scale?

DESIGN NOTE 6-3:   The Shunt

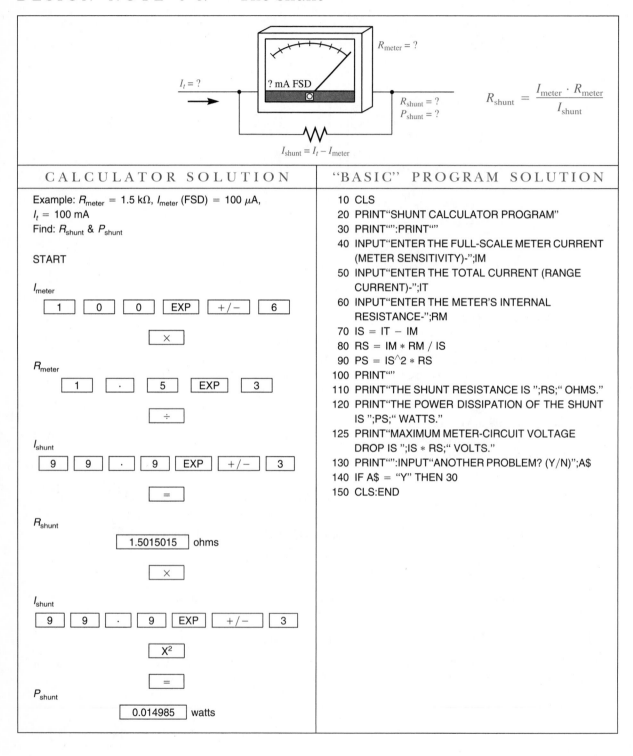

$$R_{shunt} = \frac{I_{meter} \cdot R_{meter}}{I_{shunt}}$$

$I_{shunt} = I_t - I_{meter}$

CALCULATOR SOLUTION

Example: $R_{meter} = 1.5\ k\Omega$, I_{meter} (FSD) $= 100\ \mu A$,
$I_t = 100\ mA$
Find: R_{shunt} & P_{shunt}

START

I_{meter}

| 1 | 0 | 0 | EXP | +/− | 6 |

| × |

R_{meter}

| 1 | . | 5 | EXP | 3 |

| ÷ |

I_{shunt}

| 9 | 9 | . | 9 | EXP | +/− | 3 |

| = |

R_{shunt}

| 1.5015015 | ohms |

| × |

I_{shunt}

| 9 | 9 | . | 9 | EXP | +/− | 3 |

| X² |

| = |

P_{shunt}

| 0.014985 | watts |

"BASIC" PROGRAM SOLUTION

```
10  CLS
20  PRINT"SHUNT CALCULATOR PROGRAM"
30  PRINT"":PRINT""
40  INPUT"ENTER THE FULL-SCALE METER CURRENT
    (METER SENSITIVITY)-";IM
50  INPUT"ENTER THE TOTAL CURRENT (RANGE
    CURRENT)-";IT
60  INPUT"ENTER THE METER'S INTERNAL
    RESISTANCE-";RM
70  IS = IT − IM
80  RS = IM ∗ RM / IS
90  PS = IS^2 ∗ RS
100 PRINT""
110 PRINT"THE SHUNT RESISTANCE IS ";RS;" OHMS."
120 PRINT"THE POWER DISSIPATION OF THE SHUNT
    IS ";PS;" WATTS."
125 PRINT"MAXIMUM METER-CIRCUIT VOLTAGE
    DROP IS ";IS ∗ RS;" VOLTS."
130 PRINT"":INPUT"ANOTHER PROBLEM? (Y/N)";A$
140 IF A$ = "Y" THEN 30
150 CLS:END
```

6-4 Troubleshooting with an Ohmmeter

The ohmmeter can be a valuable tool in troubleshooting parallel circuits just as it is for series circuits. In this section, you will see how the ohmmeter is used to troubleshoot parallel circuits with a strong emphasis on the importance of isolating components for measurement.

Recall the basic troubleshooting procedure:

TROUBLESHOOTING   PROCEDURE

1. Identify the symptoms or abnormal conditions.
2. Make a list of possible causes. A schematic or circuit diagram is normally essential for this.
3. With the power disconnected, make a visual inspection of the circuit and its components.
4. Isolate each suspected component and measure with the ohmmeter.

Troubleshooting a Parallel Circuit

Notice that step four of our troubleshooting procedure says to isolate each suspected component. Figure 6-17 illustrates the importance of isolating a parallel component from the rest of the circuit. As you can see, R_1 has not been isolated as R_3 has been. At least one lead of the component must be removed from the circuit, as is done for R_3. This insures that only R_3 is being measured by the ohmmeter. Since R_1 is not isolated, both R_1 and R_2 are being measured at the same time. The meter will display the equivalent resistance of the two ($R_1 \parallel R_2$). If you know what the equivalent resistance should be for $R_1 \parallel R_2$, this

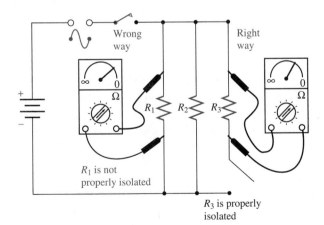

Figure 6-17   The ohmmeter can be used to troubleshoot a parallel circuit as long as circuit power is disconnected and each component is isolated before it is measured.

may not be such a problem. However, when checking components on a complicated and scrambled printed circuit board, you may not be aware of other components that may affect your measurements. As a general rule:

> Isolating a component before measuring it with your ohmmeter is a good habit!!

As we discussed in Section 5-8, you are using the ohmmeter to determine if the component is good, shorted, open, or out of tolerance. It should be noted that resistors are rarely internally shorted. However, the resistor can become shorted by surrounding metallic surfaces or wiring.

Time for a brief Self-Check.

SELF-CHECK 6-4

1. Explain why it is so very important to isolate a component in a parallel circuit before measuring it with an ohmmeter.

2. Which of these conditions will a resistor rarely exhibit—(1) out of tolerance, (2) shorted, (3) open?

6-5 Troubleshooting with an Ammeter

See color photo insert page B5.

The ammeter can be a useful troubleshooting tool, since it can be inserted in a conductive path to measure current that is being supplied to a circuit or a particular section of a circuit. Here, we shall investigate the proper use of the ammeter and consider typical meter indications that result from common circuit troubles. Figure 6-18 shows a typical digital multimeter that can be used as an ammeter in troubleshooting a circuit. This particular multimeter has six DC current ranges—from 200 μA to 10 A.

A Few Words of Caution

As we begin our discussion of the use of the ammeter in troubleshooting, it is very important that we take time to recall a few basic facts illustrated in Figure 6-19. First of all, *it is important that you observe polarity when connecting the ammeter in a circuit.* The ammeter is a polarized instrument. That is, the ammeter has a positive and a negative test lead, or terminal. If the ammeter is an analog type (moving needle) and it is placed incorrectly in the circuit, the needle will remain pinned against the left end of the scale, making a reading impossible. If the ammeter is digital, you need not be concerned about meter damage due to polarity reversal. The value will simply be displayed with a negative sign ($-$) preceding the digits in the display.

Second, *never place the ammeter in parallel with a voltage source or in parallel with a resistor, as you would a voltmeter.* It is important for you to realize that the ammeter is a very low-resistance instrument. For this reason, the ammeter can take the place of a conductor in a circuit for the purpose of measuring current. The ammeter will not add a

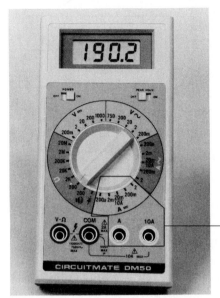

DC current ranges
200 μA to 10 A

Figure 6-18 A DMM can be used to measure current in a DC circuit. (Courtesy of Beckman Industrial Corp., a subsidiary of Emerson Electric Co.)

significant amount of resistance to the circuit, so current measurements will be accurate. Since the ammeter has no significant resistance of its own, it depends on circuit resistance to limit current. Thus, the ammeter is always placed in series with circuit resistances. If the ammeter is placed in parallel with the voltage source, severe damage will almost instantly occur to the ammeter (unless it is internally fused).

When using the ammeter function of a multimeter, it is a good practice to *start on the highest range and switch down to lower ranges until a reasonable in-range reading is obtained.* However, many multimeters are autoranging—that is, the proper range is selected automatically by the electronic circuitry of the multimeter.

One final word of caution. *Always remove power before you attempt to connect or disconnect the ammeter.* This will insure your safety and prevent any accidental shorts from occurring. Check your work carefully before turning power on.

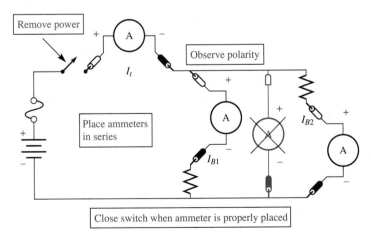

Figure 6-19 Always place the ammeter in series with circuit components, never in parallel.

TROUBLESHOOTING WITH AN AMMETER

See color photo insert
page B5.

1. Remove power before connecting the ammeter.
2. Be careful to observe proper polarity of the ammeter.
3. Always place the ammeter in series in the circuit.
4. For manual ranging multimeters, start on the highest range and switch down to lower ranges until a reasonable in-range reading is obtained.
5. Remove power before disconnecting the ammeter and restoring the circuit.

Detecting Opens

Again, you will recall that an open is a break in a current path. The open will not allow current to flow. If a resistor in a branch of a parallel circuit is open, that branch will have 0 A of current. Naturally, an ammeter placed in that branch will indicate no current (0 A). Figure 6-20 illustrates this. Furthermore, an open in a branch will result in a lower total current. Notice that the ammeter marked I_t is placed in the circuit at a point where total current can be measured. The current indicated will be lower than it should be, since a branch is open. If branch resistances are known, it is possible to determine the open branch using Ohm's Law. Simply calculate individual branch currents and determine which combination of currents will equal the total indicated current. If all branch resistances are the same, this method will not work. The ammeter would then have to be inserted in each branch to find the open.

Detecting Shorts

An ammeter can also be somewhat helpful in detecting a short in a circuit. If a short has occurred in a circuit, the total current will be higher than what it should be. Notice the circuit shown in Figure 6-21. The total current indicated by the ammeter is higher than the normal amount of current for that circuit. In cases where total current is higher than normal, you will have to finish your circuit investigation with a voltmeter or an ohmmeter.

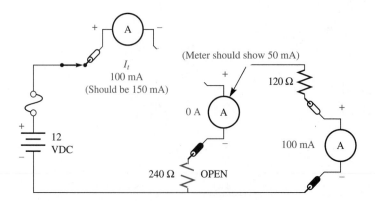

Figure 6-20   The ammeter will show zero amps when an open is discovered.

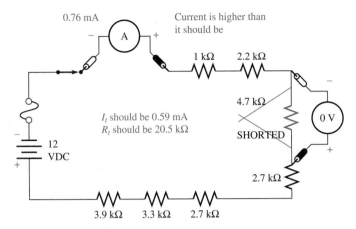

Figure 6-21   The ammeter will indicate a higher-than-normal current when a short exists in a circuit. The voltmeter clearly identifies the shorted component.

Recall that 0 volts will be indicated on the voltmeter when the meter's leads are placed across a short. Later we will discuss how the ammeter can further be used to detect a short in more complex circuits. However, you will discover that in practice, the ammeter is not used in extensive troubleshooting. It is far more convenient to use the voltmeter with power on or the ohmmeter with power off.

Test your skills by troubleshooting the following circuits in Self-Check 6-5.

SELF - CHECK   6 - 5

1.

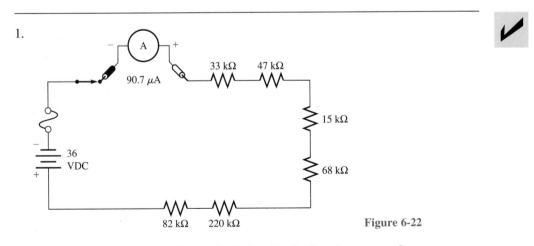

Figure 6-22

a. What should the total resistance be in the circuit above? _____ Ω
b. What should the total current be in the circuit above? _____ μA
c. Is the current indicated on the meter correct? If not, what is wrong?
d. If a voltmeter were placed across the 82 kΩ resistor, what voltage would it read? _____ V

2.

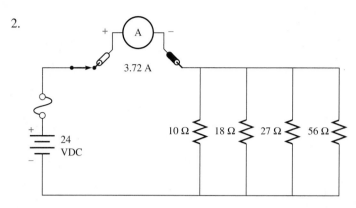

Figure 6-23

a. What should the total resistance be in the circuit above? _____ Ω
b. What should the total current be in the circuit above? _____ A
c. Is the current indicated on the meter correct? If not, what is wrong?
d. If a voltmeter were placed across the 10 Ω resistor, what voltage would it read? _____ V

Summary

FORMULAS

(6.1) $R_t = E_s/I_t$

(applies to series or parallel circuits)

(6.2) $I_t = I_{R1} + I_{R2} + I_{R3} + \cdots$

(total current is the sum of all parallel resistor currents in a simple parallel DC circuit)

(6.3) $R_t = \dfrac{1}{1/R_1 + 1/R_2 + 1/R_3 + \cdots} = \dfrac{1}{G_t} = \dfrac{1}{G_1 + G_2 + G_3 + \cdots}$

(total resistance for any parallel resistive DC circuit)

(6.4) $R_t = \dfrac{R_1 \cdot R_2}{R_1 + R_2}$

(product over the sum formula for any two parallel resistors)

(6.5) $R_t = R/N$

(same values formula for parallel resistors)

(6.6) $I_t = I_{B1} + I_{B2} + I_{B3} + \cdots$ or $I_t - I_{B1} - I_{B2} - I_{B3} = 0$

(total current is the sum of all branch currents in any DC, and some AC, parallel circuits regardless of how complex)

(6.7) $I_B = I_t \cdot R_t/R_B = \dfrac{I_t \cdot R_t}{R_B}$

(branch current, current divider formula)

$$(6.8) \quad R_{sh} = \frac{I_{R1} \cdot R_1}{I_{sh}}$$

(to determine the value of a current bypass resistor—a shunt)

CONCEPTS

- Adding resistors in parallel decreases total resistance and increases conductance.
- Total resistance for a parallel circuit is always less than the lowest resistance branch.
- The source voltage is applied to every branch of a parallel circuit.
- The sum of all branch currents is equal to the total current in a parallel circuit (Kirchhoff's Current Law).
- Adding a new branch to a parallel circuit increases total current, while removing a branch decreases total current.
- Total current in a parallel circuit is distributed to its branches, and depends upon the ratio of total resistance to the branch's resistance
- A shunt is used to bypass current around some other component or device.

PROCEDURES

Troubleshooting Procedure

1. Identify the symptoms or abnormal conditions.
2. List all possible causes.
3. With the power off, make a component and wiring inspection.
4. Isolate each suspected component and measure it with the ohmmeter.

Troubleshooting with an Ammeter

1. Remove power before connecting the ammeter.
2. Observe polarity.
3. Always place the ammeter in series.
4. Start on the highest range and switch down until a reasonable in-range reading is obtained.
5. Remove power before disconnecting the ammeter and restoring the circuit.

SPECIAL TERMS

- Parallel circuit
- Bus, bus wire, branch
- Kirchhoff's Current Law, distributed current
- Shunt, full scale deflection (FSD)

Need to Know Solution

To build your lighting system, you will need to answer the following:

1. How many lights do I need?

2. What amount of current will each light demand? The answer to this will depend on the type of bulbs used. You may have to set up a test circuit to measure the amount of current a bulb requires with the battery connected. Measuring the bulb with an ohm-meter to determine its resistance, then using Ohm's Law to determine bulb current, will not work. For example, the resistance of a taillight bulb such as the GE 1141 is about 1 Ω when it is cold and about 10 Ω when it is hot (on). The filament has a positive temperature coefficient.

3. What is the ampere hour (AH) capacity of my car battery? If it is not written on the battery, call the dealer and find out.

4. How long will my system operate before the battery must be recharged?

$$\text{Life (Hours)} = \text{capacity/current demand} = \#\text{AH}/I_t$$

The total current is the sum of all individual bulb currents, since all bulbs are wired in parallel with the battery.

Questions

6-1 Parallel Resistors

1. When can the reciprocal formula be used?
2. When would you use the product over the sum parallel resistance formula?
3. Is the total resistance of a simple parallel circuit greater than or less than the lowest-value parallel resistor?
4. What is a bus wire?
5. Explain why a measured value of total resistance might be somewhat different from the calculated value.
6. Will adding another parallel resistor to a simple parallel circuit (a) increase, or (b) decrease the total resistance?

6-2 Kirchhoff's Current Law and Distributed Current

7. List two examples of parallel circuitry.
8. In a series circuit, voltage is divided, while in a parallel circuit, _____ is divided.
9. Name the two factors that determine the current in each branch of a parallel circuit.
10. Name two ways in which the total current of a parallel circuit can be calculated.

6-3 The Current Divider Rule

11. What does the total resistance of parallel resistors and individual branch resistance have to do with the way current is divided among the branches?
12. What is a shunt?
13. Name one very practical application for a shunt resistor.
14. Write the current divider formula for shunt resistors.

6-4 Troubleshooting with an Ohmmeter

15. Given a circuit consisting of a battery, a switch, and two parallel resistors, describe how you would correctly use the ohmmeter to check each resistor.

16. Explain why it is important to isolate a component before checking it with an ohmmeter.

6-5 Troubleshooting with an Ammeter

17. List the four cautions that must be observed when using the ammeter for troubleshooting.
18. If an ammeter is placed in a parallel circuit in such a way as to measure total current, and an open occurs in one branch of the circuit, what will happen to the amount of current indicated by the ammeter?
19. The internal resistance of an ammeter is (a) very high, or (b) very low.
20. If the circuit of Question 18 is not protected by a fuse, what might happen to the ammeter if one of the branches becomes shorted?

Problems

6-1 Parallel Resistors

1. How many 100 Ω resistors would you need to place in parallel to give you a total resistance of 20 Ω?
2. Calculate the total resistance and conductance appearing between terminals A and B in Figure 6-24.

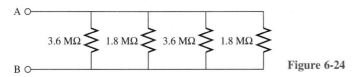

Figure 6-24

3. Calculate the total resistance and conductance appearing between terminals A and B in Figure 6-25.

Figure 6-25

6-2 Kirchhoff's Current Law and Distributed Current

4. Solve for total conductance, total resistance, total current, and individual branch currents in Figure 6-26.

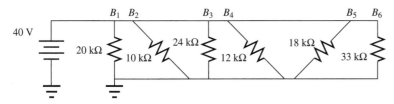

Figure 6-26

5. Solve for total conductance, total resistance, total current, and individual branch currents in Figure 6-27.

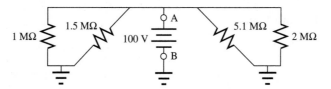

Figure 6-27

6. Solve for total resistance, total current, and source voltage in Figure 6-28. Also, find the current through R_1 and R_2.

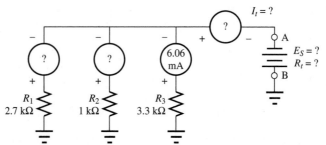

Figure 6-28

6-3 The Current Divider Rule

7. Given the partial circuit of Figure 6-29, find the amount of current in each branch.

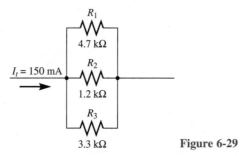

Figure 6-29

8. Given the partial circuit of Figure 6-30, find the amount of current in each branch.

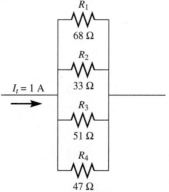

Figure 6-30

9. Given the partial circuit of Figure 6-31, find the resistance and power dissipation of the shunt.

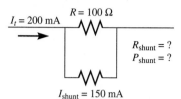

$I_t = 200$ mA $R = 100 \, \Omega$

$R_{shunt} = ?$
$P_{shunt} = ?$

$I_{shunt} = 150$ mA   **Figure 6-31**

10. Given the partial circuit of Figure 6-32, find the resistance and power dissipation of the shunt.

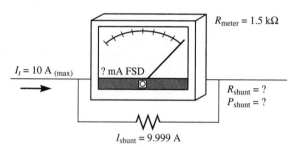

$R_{meter} = 1.5 \, k\Omega$

$I_t = 10$ A $_{(max)}$? mA FSD

$R_{shunt} = ?$
$P_{shunt} = ?$

$I_{shunt} = 9.999$ A   **Figure 6-32**

Answers to Self-Checks

Self-Check 6-1

1. $R_t = 1/(1/R_1 + 1/R_2 + 1/R_3 + \cdots)$
2. $R_t = (R_1 \cdot R_2)/(R_1 + R_2)$
3. For two different valued parallel resistors
4. $R_t = R/N$
5. For any number of parallel resistors of the same value
6. $G_t = 0.0115$ S $= 11.5$ mS; $R_t = 87 \, \Omega$
7. $G_t = 2 \, \mu$S; $R_t = 0.5 \, M\Omega = 500 \, k\Omega$

Self-Check 6-2

1. $I_{R1} = 2$ A; $I_{R2} = 1$ A; $I_{R3} = 1$ A; $I_{R4} = 2$ A; $I_t = 6$ A; $R_t = 3.33 \, \Omega$
2. $I_{R1} = 0.51$ A; $I_{R2} = 0.429$ A; $I_{R3} = 0.293$ A; $I_{R4} = 0.24$ A; $I_t = 1.47$ A; $R_t = 16.3 \, \Omega$
3. $I_{R1} = 13.3$ mA; $I_{R2} = 7.66$ mA; $I_{R3} = 30$ mA; $I_{R4} = 7.66$ mA; $I_t = 58.6$ mA; $R_t = 614 \, \Omega$

Self-Check 6-3

1. Total resistance/branch resistance
2. 30.3 mA
3. 0.41 A
4. Shunt
5. 38.5 Ω

Self-Check 6-4

1. Isolation of a component prevents measuring other components along with the one you are measuring
2. (2) shorted

Self-Check 6-5

1. (a) 465 kΩ; (b) 77.4 μA; (c) No, the 68 kΩ resistor is shorted; (d) 7.44 V
2. (a) 4.75 Ω; (b) 5.1 A; (c) The 18 Ω resistor is open; (d) 24 V

SUGGESTED PROJECTS

1. Add the formulas and important concepts from this chapter to your Electricity and Electronics Notebook. You might also want to transfer the Design Notes in this chapter to your personal notebook.

2. Build a few of the circuits shown in this chapter and verify your calculations with a multimeter. Measure the actual resistance of each resistor to check tolerance. Use the ammeter function to measure branch currents and total current. Review the ohmmeter and ammeter troubleshooting procedures before you start.

This is the Edison Typewriting Telegraph machine invented in 1871. Edison improved upon the designs of others and originated some of his own. The circular arrangement of keys was used to transmit Morse code automatically and the pointer in the center rotated to indicate characters as they were received.

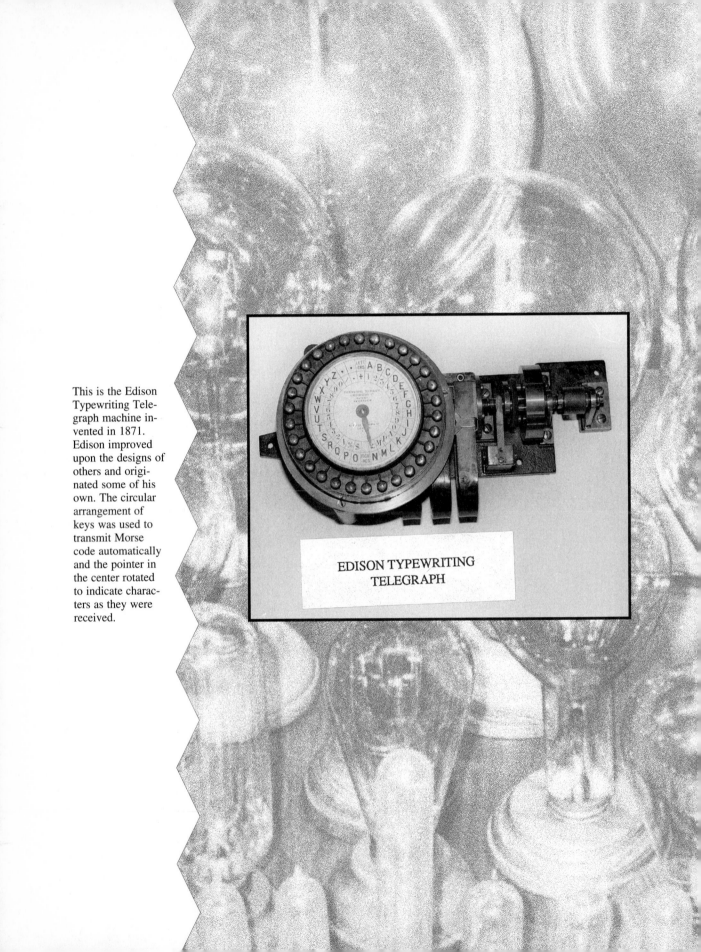

EDISON TYPEWRITING
TELEGRAPH

Chapter 7

Combined Series and Parallel DC Circuits

OBJECTIVES

After studying this chapter, you will be able to

- calculate total resistance for overall series and overall parallel circuits.
- solve for voltage drops, currents, and power dissipation in overall series and overall parallel circuits.
- analyze the Wheatstone bridge and discuss some of its applications.

- calculate current and voltage drops in loaded voltage dividers.
- explain the effects of different loads on a given voltage divider.
- troubleshoot loaded voltage dividers with a multimeter.
- explain the loading effect of a voltmeter.

INTRODUCTION

Among the most common and practical circuits found in electronics are the complex overall parallel circuit, the loaded voltage divider, and the Wheatstone bridge. With these circuits, we are able to perform the essential task of distributing current and/or voltage as needed in a complex circuit. Complex electronic systems ranging from transistor radios to computers are essentially overall parallel circuits in which a common DC supply provides voltage and current to many branches. Many of those branches are overall series circuits, or loaded voltage dividers. These voltage dividers are used to provide correct voltage levels for transistors and integrated circuits. The Wheatstone bridge is an important circuit that is simple yet extremely useful in a multitude of applications. This circuit is used for precision measurement of resistance, temperature, pressure, liquid level, weight, physical balance of mechanical systems, automatic positioning, automatic tuning, and much more. As you can see, there are many interesting and practical things to be learned in this chapter. Have fun and learn much.

NEED TO KNOW

Suppose you are being interviewed for the first job of your electronics career and the circuit shown below is placed in front of you as a test of your basic circuit analysis skills. (Don't laugh, it might really happen!) Can you determine the amount of current flowing through R_4 or the amount of current through R_5? Can you calculate the voltage across R_3? Can you determine the correct polarity of the voltage drops across each resistor? Do you know how to use an ammeter to measure total current or the current through a certain resistor? If R_1 were open, how much voltage would you measure across R_1? If R_6 were externally shorted, how much voltage would you measure across it with a voltmeter? Can you answer all of these questions confidently and correctly? No? Then at least you are clearly aware of a need to know. The circuit analysis and troubleshooting procedures that you will learn in this chapter are extremely important since they are skills that you will apply throughout your entire electronics career.

As always, a need-to-know discussion is provided at the close of the chapter.

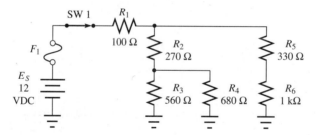

7-1 Overall Series Circuits

See color photo insert
page B6. Photograph 34.

In this section, you will become familiar with, and learn to calculate total resistance for, overall series circuits as shown in Figure 7-1. This arrangement is often referred to as a series/parallel circuit where the overall circuit is a series circuit.

Identifying Characteristics

It is helpful to recognize that there are two important identifying characteristics for overall series circuits. Understanding these characteristics will help you solve for total resistance in very complex overall series circuits.

First, if a voltage source were provided at terminals A and B, the resulting total current would have to flow through R_1. The total current would then be divided between R_2 and

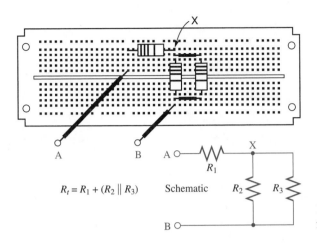

$R_t = R_1 + (R_2 \| R_3)$ Schematic

Figure 7-1   Resistors arranged to form an overall series circuit.

R_3 according to their relative values (see Section 6-3). The point here is that in an overall series circuit, total current flows through a common junction, wire, or component *within* the circuit. In this case, total current must flow through R_1 and junction X.

The second identifying characteristic is that the overall series circuit can be simplified, or reduced, to a simple series circuit. In Figure 7-1, R_2 and R_3 can be combined into one equivalent resistance. The product over the sum formula can be used for this simplification. The circuit can then be redrawn as a simple two-resistor series circuit. The total resistance is then $R_t = R_1 + (R_2 \| R_3)$.

Figure 7-2 shows another overall series circuit. Notice that wire X is a common wire within the circuit through which total current must flow. Also, you can reduce this circuit to a simple series circuit by using the 'product over the sum' formula for $(R_1 \| R_2)$ and $R_3 \| R_4)$. Therefore, the total resistance for this circuit is found as follows: $R_t = (R_1 \| R_2) + (R_3 \| R_4)$. Take time to study Examples 7-1 and 7-2.

Overall Series (Series-Parallel)

1. Total current flows through a common junction, wire, or component within the circuit.
2. The circuit can be reduced to a simple series circuit.

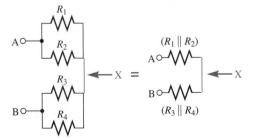

Figure 7-2   The overall series, or series-parallel, circuit has a conductor or component within the circuit through which total current must flow.

EXAMPLE   7-1

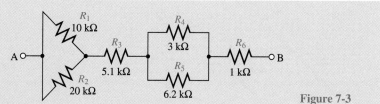

Figure 7-3

$$R_t = (R_1 \parallel R_2) + R_3 + (R_4 \parallel R_5) + R_6$$

• First solve the two parallel sections.

$$(R_1 \parallel R_2) = \frac{10\ k\Omega \cdot 20\ k\Omega}{10\ k\Omega + 20\ k\Omega} = 6.67\ k\Omega$$

$$(R_4 \parallel R_5) = \frac{3\ k\Omega \cdot 6.2\ k\Omega}{3\ k\Omega + 6.2\ k\Omega} = 2.02\ k\Omega$$

• Now solve for R_t

$$R_t = 6.67\ k\Omega + 5.1\ k\Omega + 2.02\ k\Omega + 1\ k\Omega = 14.79\ k\Omega$$

EXAMPLE   7-2

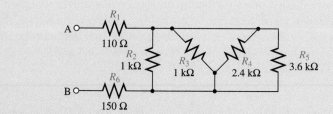

Figure 7-4

$$R_t = R_1 + (R_2 \parallel R_3 \parallel R_4 \parallel R_5) + R_6$$

• First solve the parallel section

$$(R_2 \parallel R_3 \parallel R_4 \parallel R_5) = \frac{1}{(1/1\ k\Omega) + (1/1\ k\Omega) + (1/2.4\ k\Omega) + (1/3.6\ k\Omega)}$$
$$= 371\ \Omega$$

• Now solve for R_t

$$R_t = 110\ \Omega + 371\ \Omega + 150\ \Omega = 631\ \Omega$$

To solve for total resistance in the previous examples, we merely applied the skills learned in Sections 5-1 and 6-1. You will find that the solution to any complex problem is simply the application of basic skills in a step-by-step process. Here are the steps we follow when solving for total resistance in an overall series circuit:

Calculating R_t for Overall Series Circuits

1. Reduce any parallel sections to their equivalent resistances using one or more parallel R_t formulas.
2. Use the series R_t formula to find the total resistance of the simplified circuit. In other words, add all individual and equivalent resistances in the series chain.

Let's look at one more example of an overall series circuit. Follow Example 7-3 carefully.

EXAMPLE   7-3

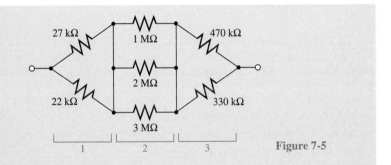

Figure 7-5

This circuit has three parallel sections! It may help to redraw it like this:

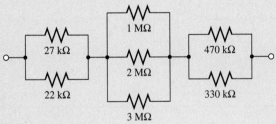

Figure 7-6

Section 1 $= \dfrac{27 \text{ k}\Omega \times 22 \text{ k}\Omega}{27 \text{ k}\Omega + 22 \text{ k}\Omega} = 12.12 \text{ k}\Omega$

Section 2 $= \dfrac{1}{\dfrac{1}{1 \text{ M}\Omega} + \dfrac{1}{2 \text{ M}\Omega} + \dfrac{1}{3 \text{ M}\Omega}} = 545 \text{ k}\Omega$

Section 3 $= \dfrac{470 \text{ k}\Omega \times 330 \text{ k}\Omega}{470 \text{ k}\Omega + 330 \text{ k}\Omega} = 193.88 \text{ k}\Omega$

$R_t = 12.12 \text{ k}\Omega + 545 \text{ k}\Omega + 193.88 \text{ k}\Omega = 751 \text{ k}\Omega$

Before continuing, test your skills on Self-Check 7-1.

SELF-CHECK 7-1

1. Find R_t between A and B in Figure 7-7.

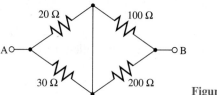

Figure 7-7

Hint: This circuit has two parallel sections in series.

2. Find R_t between A and B in Figure 7-8.

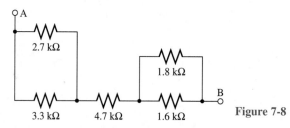

Figure 7-8

3. Find R_t between A and B in Figure 7-9.

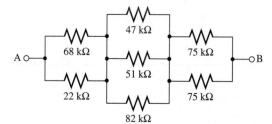

Figure 7-9

7-2 Voltage, Current, and Power in Overall Series Circuits

In this section, you will learn almost nothing new. That may sound a bit strange, but this section will simply apply and expand upon what you have already learned. We will combine what you have learned about series and parallel circuits in a systematic approach to solving voltages and currents in overall series circuits (series-parallel circuits).

Finding Voltage Drops

To begin, we examine the simplest of all overall series circuits as shown in Example 7-4. In the analysis of this circuit, we calculate the voltage across each resistor and the current through each resistor. As a starting point, we calculate total resistance, then find total current ($I_t = E_s/R_t$). Notice that the total current in the circuit must flow through R_1. If the total current is first calculated, the voltage across R_1 can be calculated using Ohm's Voltage Law ($V_{R1} = I_t \cdot R_1$). Once we calculate V_{R1} we can subtract it from the source voltage E_s to determine the voltage dropped across the parallel section ($R_2 \parallel R_3$) (Kirchhoff's Voltage Law). It is important for you to remember here that the voltage is the same across parallel branches. In other words, $V_{R2} = V_{R3}$. The parallel resistors act as one to produce this voltage drop. Therefore,

$$V_{R2} = V_{R3} = E_s - V_{R1}$$

Notice that V_{R2} and V_{R3} were determined in an indirect way. First V_{R1} was calculated, then it was subtracted from E_s to find V_{R2} and V_{R3}. There is another way to find V_{R2} and V_{R3} more directly. The voltage across R_2 and R_3 can be found by multiplying the total current by the equivalent resistance of ($R_2 \parallel R_3$). Remember, parallel sections can be replaced with an equivalent resistance (R_{eq}). Therefore, $V_{R2} = V_{R3} = I_t \cdot R_{eq}$. Let's summarize our methods for finding voltage drops.

EXAMPLE 7-4

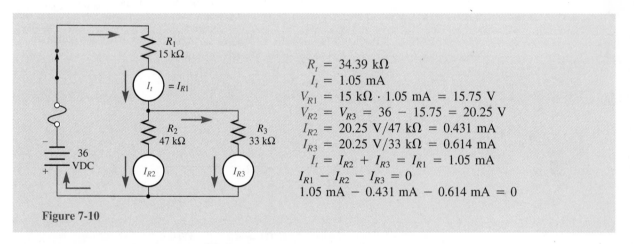

$R_t = 34.39\ k\Omega$
$I_t = 1.05\ mA$
$V_{R1} = 15\ k\Omega \cdot 1.05\ mA = 15.75\ V$
$V_{R2} = V_{R3} = 36 - 15.75 = 20.25\ V$
$I_{R2} = 20.25\ V/47\ k\Omega = 0.431\ mA$
$I_{R3} = 20.25\ V/33\ k\Omega = 0.614\ mA$
$I_t = I_{R2} + I_{R3} = I_{R1} = 1.05\ mA$
$I_{R1} - I_{R2} - I_{R3} = 0$
$1.05\ mA - 0.431\ mA - 0.614\ mA = 0$

Figure 7-10

Calculating Voltage Drops in Overall Series Circuits

1. Voltage drops across single series resistors in an overall series circuit are found by multiplying each individual resistor by the total current. ($V_R = I_t \cdot R$)

2. In a simple overall series circuit, the voltage drop across the parallel section can be found by subtracting the individual series resistor's voltage drop from the source voltage. ($V_{R2 \parallel R3} = E_s - V_{R1}$)

3. The voltage drop across any parallel section of any overall series circuit can be found by multiplying the total current by the parallel section's equivalent resistance. ($V_{R2 \parallel R3} = V_{R2} = V_{R3} = I_t \cdot R_{eq}$, where $R_{eq} = R_2 \parallel R_3$)

Solving for Currents

Solving for total current is no real problem. You already know how to do that from many previous examples. You simply divide the total resistance into the source voltage ($I_t = $

E_s/R_t). The big question here is: How do we determine how this total current is divided in the parallel section? All of the current goes through R_1 but is divided between R_2 and R_3, as is seen in Example 7-5. Again, we use Ohm's Law to solve for current in R_2 and R_3. In order to solve for current, you must know resistance and voltage ($I = E/R$). The resistances of R_2 and R_3 are given. If we can find the voltage across $R_2 \parallel R_3$, we can use this voltage to find the individual currents through R_2 and R_3. Therefore, $I_{R2} = V_{R2}/R_2$ and $I_{R3} = V_{R3}/R_3$ (where $V_{R2} = V_{R3}$). As long as you can find the voltage across a parallel section in any complex circuit, you can solve for individual branch currents. Also, you can use the current divider formula from Section 6-3 to determine branch currents ($I_B = I_t \cdot R_{eq}/R_B$) and the voltage divider formula to solve for parallel-section voltage drops ($V_{parallel\ section} = E_s \cdot R_{eq}/R_t$). Let's summarize this procedure in a step-by-step fashion.

Calculating Parallel Section Branch Currents

1. First solve for total resistance and current ($I_t = E_s/R_t$).

2. Next, find the voltage drop across the parallel section by using Ohm's Voltage Law ($V_{parallel\ section} = I_t \cdot R_{eq}$) or the voltage divider formula ($V_{parallel\ section} = E_s \cdot R_{eq}/R_t$).

3. Now, use Ohm's Current Law ($I_B = V_{parallel\ section}/R_B$) or the current divider formula ($I_B = I_t \cdot R_{eq}/R_B$) to solve for branch currents.

EXAMPLE   7-5

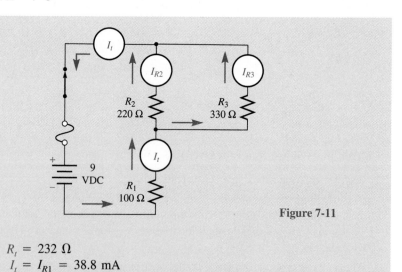

Figure 7-11

$R_t = 232\ \Omega$

$I_t = I_{R1} = 38.8\ \text{mA}$

$V_{R1} = 100\ \Omega \cdot 38.8\ \text{mA} = 3.88\ \text{V}$

$V_{R2} = V_{R3} = 9\ \text{V} - 3.88\ \text{V} = 5.12\ \text{V}$

$I_{R2} = 5.12\ \text{V}/220\ \Omega = 23.3\ \text{mA}$

$I_{R3} = 5.12\ \text{V}/330\ \Omega = 15.5\ \text{mA}$

$I_t = I_{R2} + I_{R3} = 23.3\ \text{mA} + 15.5\ \text{mA} = 38.8\ \text{mA}$

Take time to study Example 7-6 to reinforce what we have discussed about finding voltage drops and currents in an overall series circuit.

EXAMPLE   7-6

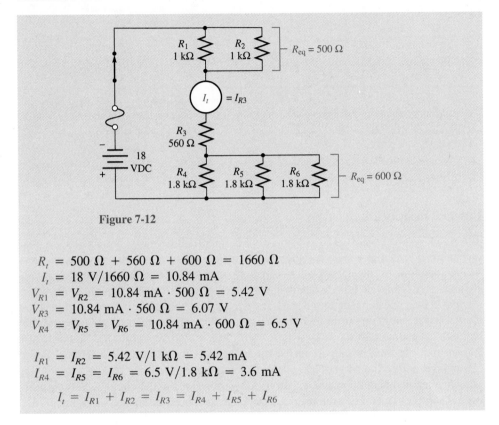

Figure 7-12

$R_t = 500\ \Omega + 560\ \Omega + 600\ \Omega = 1660\ \Omega$

$I_t = 18\ \text{V}/1660\ \Omega = 10.84\ \text{mA}$

$V_{R1} = V_{R2} = 10.84\ \text{mA} \cdot 500\ \Omega = 5.42\ \text{V}$

$V_{R3} = 10.84\ \text{mA} \cdot 560\ \Omega = 6.07\ \text{V}$

$V_{R4} = V_{R5} = V_{R6} = 10.84\ \text{mA} \cdot 600\ \Omega = 6.5\ \text{V}$

$I_{R1} = I_{R2} = 5.42\ \text{V}/1\ \text{k}\Omega = 5.42\ \text{mA}$

$I_{R4} = I_{R5} = I_{R6} = 6.5\ \text{V}/1.8\ \text{k}\Omega = 3.6\ \text{mA}$

$I_t = I_{R1} + I_{R2} = I_{R3} = I_{R4} + I_{R5} + I_{R6}$

Now, let's make this a little more challenging. In Example 7-7, the only values given are E_s, R_1, R_2, and V_{R1}. In this example, we want to find I_t, I_{R2}, I_{R3}, V_{R2}, V_{R3}, and R_3. We must start with the information that is given and work our way systematically through the problem. Follow Example 7-7 through very carefully to see how this is done.

It is always wise, and often essential, to start with a component or part of the circuit that you know two things about.

In this circuit, R_1 and V_{R1} are given. This allows you to calculate I_{R1}, which is the total current. To find the current through R_2, you must know the voltage across R_2. In this case, the voltage across R_2 is the voltage left over from the R_1 voltage drop. In other words, $V_{R2} = V_{R3} = E_s - V_{R1}$. With the voltage across R_2, you can calculate the current through R_2 ($I_{R2} = V_{R2}/R_2$). Now you are able to solve for I_{R3} because you know that $I_{R2} + I_{R3} = I_t$ (Kirchhoff's Current Law). Therefore, $I_{R3} = I_t - I_{R2}$. Finally, the value of R_3 can be solved because you know V_{R3} and I_{R3}. Therefore, $R_3 = V_{R3}/I_{R3}$. Problems like these are puzzles that must be fit together piece by piece.

EXAMPLE   7-7

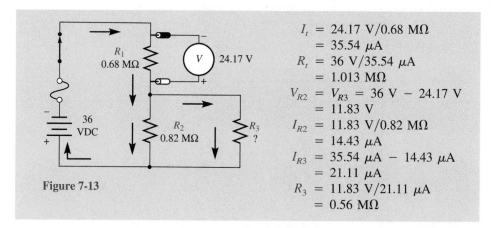

$$I_t = 24.17 \text{ V}/0.68 \text{ M}\Omega$$
$$= 35.54 \ \mu\text{A}$$
$$R_t = 36 \text{ V}/35.54 \ \mu\text{A}$$
$$= 1.013 \text{ M}\Omega$$
$$V_{R2} = V_{R3} = 36 \text{ V} - 24.17 \text{ V}$$
$$= 11.83 \text{ V}$$
$$I_{R2} = 11.83 \text{ V}/0.82 \text{ M}\Omega$$
$$= 14.43 \ \mu\text{A}$$
$$I_{R3} = 35.54 \ \mu\text{A} - 14.43 \ \mu\text{A}$$
$$= 21.11 \ \mu\text{A}$$
$$R_3 = 11.83 \text{ V}/21.11 \ \mu\text{A}$$
$$= 0.56 \text{ M}\Omega$$

Figure 7-13

Power Dissipation

The actual power dissipated by a component is important because components are man-ufactured and rated to handle certain maximum amounts of power. In previous chapters, you learned how to calculate power dissipation for resistors in series and parallel circuits. By now, you should be fully aware that, regardless of the circuit configuration, the power dissipated by a resistor is determined by any two of three things: resistance, voltage across the resistance, and current through the resistance. Any of the three power formulas may be used, depending on what is known ($P_R = I_R \cdot V_R = V_R^2/R = I_R^2 \cdot R$). You have also learned that the total power dissipated by a circuit is the sum of all individual power dissipations in the circuit regardless of circuit configuration (series or parallel). Everything you have learned about power calculations for series and parallel circuits is also true for overall series circuits. For example, let's calculate the power dissipated in Figure 7-13.

$$
\begin{aligned}
P_{R1} &= I_{R1} \cdot V_{R1} & = 35.54 \ \mu\text{A} \cdot 24.17 \text{ V} &= 859 \ \mu\text{W} \\
P_{R2} &= I_{R2} \cdot V_{R2} & = 14.43 \ \mu\text{A} \cdot 11.83 \text{ V} &= 170.7 \ \mu\text{W} \\
+ \ P_{R3} &= I_{R3} \cdot V_{R3} & = 21.11 \ \mu\text{A} \cdot 11.83 \text{ V} &= \underline{249.7 \ \mu\text{W}} \\
P_t &= \qquad P_{R1} + P_{R2} + P_{R3} & &= 1{,}279 \ \mu\text{W} = 1.279 \text{ mW}
\end{aligned}
$$

Also,

$$P_t = I_t \cdot E_s = 35.54 \ \mu\text{A} \cdot 36 \text{ V} = 1{,}279 \ \mu\text{W} = 1.279 \text{ mW}$$

Now, take a few moments to test your skills on Self-Check 7-2.

7-3 Overall Parallel Circuits

See color photo insert
page B6. Photograph 35.

In this section, we shall examine the overall parallel type of circuit, shown in Figure 7-16. Notice that there are two bus wires, which will carry current to three separate branches. No single resistor, junction, or wire within the circuit will carry total current. Only ter-

1.

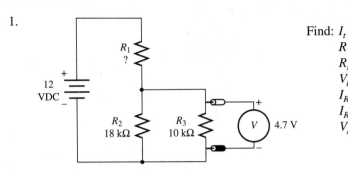

Find: I_t =
R_1 =
R_t = P_t =
V_{R1} = P_{R1} =
I_{R2} = P_{R2} =
I_{R3} = P_{R3} =
V_{R2} =

Figure 7-14

2.

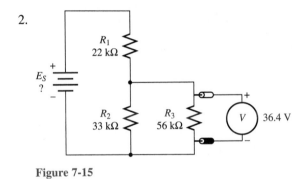

Find: I_t =
E_s =
R_t = P_t =
I_{R1} = P_{R1} =
I_{R2} = P_{R2} =
I_{R3} = P_{R3} =
V_{R1} =

Figure 7-15

minals A and B, outside the circuit, carry total current. As you can probably see, the overall parallel circuit is identified as follows:

Identifying Characteristics of Overall Parallel Circuits

1. Bus wires carry and distribute current to the circuit. (One bus wire is usually a common ground such as copper foil or a metal chassis.)

2. There are separate and identifiable parallel branches for current flow.

3. There is no single component, junction, or wire within the circuit through which total current will flow.

4. The circuit can be reduced to a simple parallel circuit.

Solving for Total Resistance

Figure 7-16 may be reduced to a simple parallel circuit by first replacing each branch with a single equivalent resistance. The first branch contains R_1, R_2, and R_3. The equivalent resistance for this branch is simply the sum of the three series resistors. Therefore, branch one equals $R_1 + R_2 + R_3$. Branch two is simply R_4. Branch three is $R_5 + R_6$.

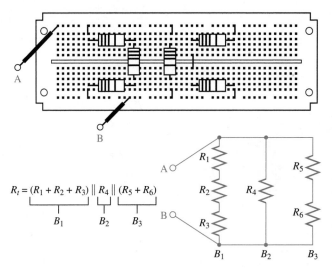

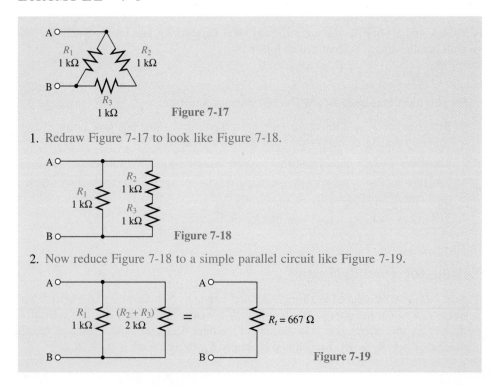

A
B
$R_t = (R_1 + R_2 + R_3) \,||\, R_4 \,||\, (R_5 + R_6)$
B_1 B_2 B_3
A
B
R_1
R_2
R_3
R_4
R_5
R_6
B_1 B_2 B_3

Figure 7-16 Resistors arranged to form an overall parallel circuit.

Now that the circuit is in its simplest form, we can solve for total resistance using an appropriate parallel-resistance formula such as the reciprocal formula (conductance formula).

As you will see, it is often very helpful to redraw and simplify the circuit in stages as you solve for total resistance. Consider Example 7-8.

EXAMPLE 7-8

A
B
R_1 1 kΩ
R_2 1 kΩ
R_3 1 kΩ

Figure 7-17

1. Redraw Figure 7-17 to look like Figure 7-18.

A
B
R_1 1 kΩ
R_2 1 kΩ
R_3 1 kΩ

Figure 7-18

2. Now reduce Figure 7-18 to a simple parallel circuit like Figure 7-19.

A
B
R_1 1 kΩ
$(R_2 + R_3)$ 2 kΩ
=
A
B
$R_t = 667\ \Omega$

Figure 7-19

3. Finally, solve for R_t;

$$R_t = \frac{R_1 \times (R_2 + R_3)}{R_1 + (R_2 + R_3)} = \frac{1\ k\Omega \cdot 2\ k\Omega}{1\ k\Omega + 2\ k\Omega} = 667\ \Omega$$

Let's examine the step-by-step procedure before considering another example.

Calculating R_t for Overall Parallel Circuits

1. Redraw if necessary.
2. Simplify by combining series resistors in each branch.
3. Solve for R_t using the appropriate parallel formula.

Now, study Example 7-9 and Example 7-10.

EXAMPLE   7-9

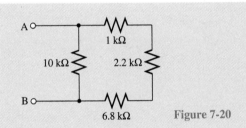

Figure 7-20

1. Redraw Figure 7-20 to look like Figure 7-21.

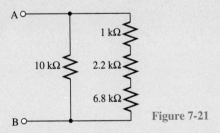

Figure 7-21

2. Simplify Figure 7-21 to look like Figure 7-22.

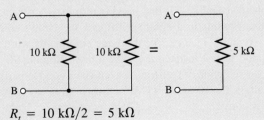

Figure 7-22

$R_t = 10\ k\Omega/2 = 5\ k\Omega$

Suppose we move terminal B, in Example 7-10, to the left of the 68 Ω resistor. Would the total resistance between A and B still be about 13 Ω? Let's find out. Study Example 7-11.

EXAMPLE   7-10

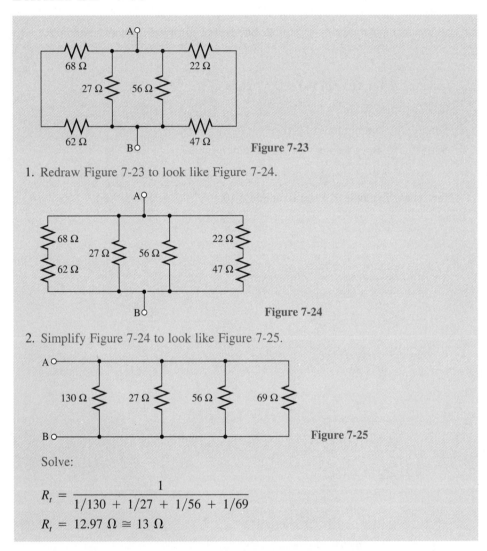

Figure 7-23

1. Redraw Figure 7-23 to look like Figure 7-24.

Figure 7-24

2. Simplify Figure 7-24 to look like Figure 7-25.

Figure 7-25

Solve:

$$R_t = \frac{1}{1/130 + 1/27 + 1/56 + 1/69}$$

$$R_t = 12.97 \ \Omega \cong 13 \ \Omega$$

We can conclude, from Examples 7-10 and 7-11, that moving terminals, or even test leads, from your ohmmeter to different points in a circuit will make the circuit entirely different! Even though Examples 7-10 and 7-11 look like the same circuits, they are not. Where you connect your power supply, or test leads, does make a big difference! If you have access to an ohmmeter and a good selection of resistors, it would be a good experience for you to construct and verify Examples 7-10 and 7-11. It is always good to prove to yourself that what you are learning is correct.

EXAMPLE 7-11

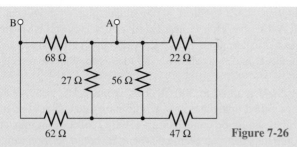

Figure 7-26

1. Redraw Figure 7-26 to look like Figure 7-27.

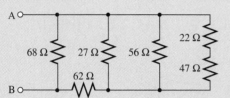

Figure 7-27

2. Simplify Figure 7-27 to look like Figure 7-28.

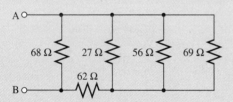

Figure 7-28

3. Simplify Figure 7-28 to look like Figure 7-29.

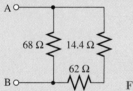

Figure 7-29

4. Simplify Figure 7-29 to look like Figure 7-30.

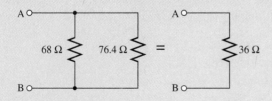

Figure 7-30

Solve:

$$R_t = \frac{68\ \Omega \times 76.4\ \Omega}{68\ \Omega + 76.4\ \Omega} = 36\ \Omega.$$

A Practical Example

Example 7-12 is an interesting one. It is a simple TV/FM signal splitter. As you can see, it is an overall parallel circuit. A TV antenna is connected to terminals A and B via ribbon cable. One television is connected to terminals B and C, while another television or FM receiver is connected to terminals C and A as shown. The idea is to allow two receivers to share the same antenna.

The standard TV ribbon cable is rated at 300 Ω. This is not a DC resistance but, rather, an AC type of resistance called *impedance*. (You will learn much about this later.) For best results, this ribbon cable should be connected to a 300 Ω load, or an impedance close to that value. Therefore, the impedance at terminals A and B should be approximately 300 Ω.

EXAMPLE 7-12 A SIMPLE SIGNAL SPLITTER

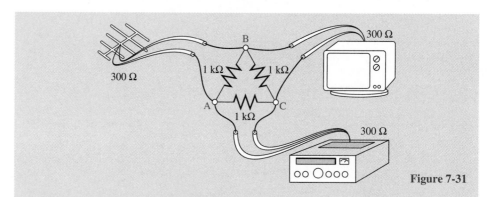

Figure 7-31

Problem: Find total resistance at terminals A and B (with the antenna ribbon cable removed). Redraw Figure 7-31 to look like Figure 7-32.

Figure 7-32

Solve for parallel resistances and simplify to look like Figure 7-33.

$$R = \frac{1{,}000 \ \Omega \times 300 \ \Omega}{1{,}000 \ \Omega + 300 \ \Omega} = \frac{300{,}000 \ \Omega^2}{1{,}300 \ \Omega} = \underline{230.8 \ \Omega}$$

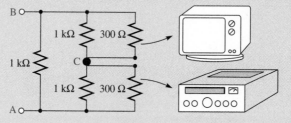

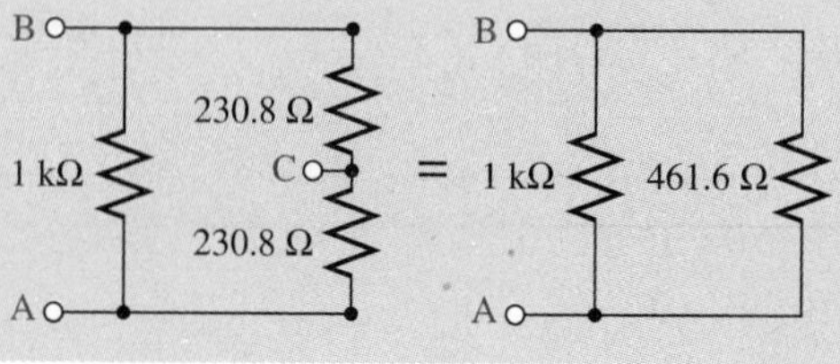

Figure 7-33

Now solve for total resistance from A to B.

$$R_t = \frac{1,000\ \Omega \times 461.6\ \Omega}{1,000\ \Omega + 461.6\ \Omega} = \frac{461,600\ \Omega^2}{1,461.6\ \Omega} = 316\ \Omega$$

Note: Our calculated value of 316 ohms is sufficiently close to the ideal 300 Ω for this circuit to work well.

Example 7-12 shows you how to calculate the impedance at terminals A and B to see if, in fact, it is close to the ideal 300 Ω value. If you want to construct a circuit to verify your calculations, you must use 300 Ω resistors in place of the TV and FM radio. Once again, this is because the 300 Ω receiver connections are not a DC resistance and therefore are not measurable with an ohmmeter.

Handling a Seemingly Complex Circuit

Solving for total resistance in a seemingly complex overall parallel circuit involves the use of basic skills in a step-by-step process. It involves: (1) combining parallel resistors, (2) combining series resistors, (3) redrawing the circuit, and (4) combining again until the circuit is reduced to its simplest form. Example 7-13 illustrates this process.

EXAMPLE   7-13

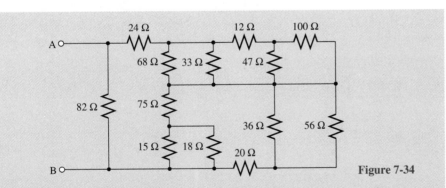

Figure 7-34

Step #1: Identify parallel resistors and replace them with equivalent resistances.

Figure 7-34 has the following parallel pairs:

$(68 \parallel 33)$, $(47 \parallel 100)$, $(36 \parallel 56)$, $(15 \parallel 18)$

$\downarrow$ $\downarrow$ $\downarrow$ $\downarrow$

22.2 Ω 32 Ω 22 Ω 8.2 Ω

(Continues next page)

EXAMPLE 7-13 (Continued)

Step #2: Redraw the circuit with equivalent resistances as shown in Figure 7-35.

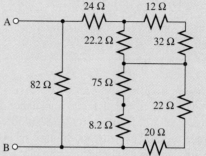

Figure 7-35

Step #3: Combine series resistors. Any two or more series resistors that do not have a branch in between them may be added together. Step #2 has three such cases:

$$75\ \Omega + 8.2\ \Omega = 83.2\ \Omega \qquad\qquad 12\ \Omega + 32\ \Omega = 44\ \Omega$$

$$22\ \Omega + 20\ \Omega = 42\ \Omega$$

Step #4: Redraw the circuit in reduced form again like Figure 7-36.

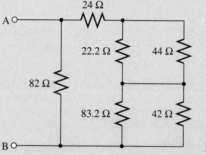

Figure 7-36

Step #5: Replace parallel resistors with equivalent resistances.

$$(22.2\ \Omega \parallel 44\ \Omega) \quad \text{and} \quad (83.2\ \Omega \parallel 42\ \Omega)$$
$$\downarrow \qquad\qquad\qquad\qquad \downarrow$$
$$14.8\ \Omega \qquad\qquad\qquad 28\ \Omega$$

Step #6: Redraw and combine series resistors as in Figure 7-37.

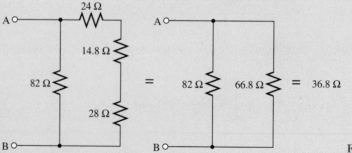

Figure 7-37

Now, apply what you have learned in Example 7-13 to Self-Check 7-3.

Solve for total resistance between terminals A and B for Figures 7-38 and 7-39.

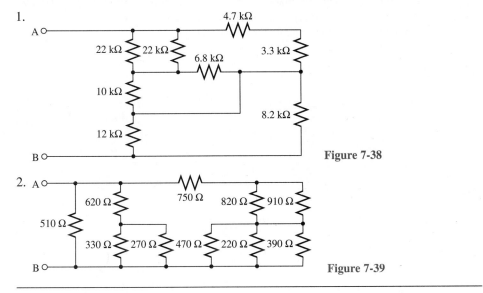

Figure 7-38

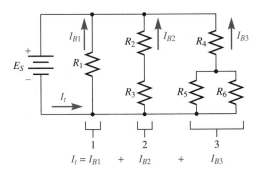

Figure 7-39

7-4 Voltage, Current, and Power in Overall Parallel Circuits

We calculate voltages, currents, and power in overall parallel circuits much as we do in overall series circuits. Think of the overall parallel circuit as two or more separate overall series circuits all connected to the same voltage source. Each separate overall series circuit is simply another branch of the overall parallel circuit. Your knowledge of overall series circuits and simple parallel circuits will enable you to quickly understand this section. If you are still not too comfortable with overall series circuit analysis, you should go back and review the previous sections of this chapter before continuing here.

Figure 7-40 illustrates a typical overall parallel circuit. Notice that the source voltage is available to several branches and each branch is either a single resistor, a simple series circuit, or an overall series circuit. We can solve for the total resistance and current of

Figure 7-40 This circuit demonstrates overall parallel circuit theory.

each branch, considered as a separate problem. As a final step, we can simply add these branch currents together to solve for the total current in the overall parallel circuit. Power dissipation is calculated using any of the power formulas for each resistor. Total power is found by adding all individual dissipations or by applying total current, total resistance, and source voltage to any of the power formulas. Now let's take a look at Example 7-14 and solve for voltage drops, currents, and total power.

EXAMPLE 7-14

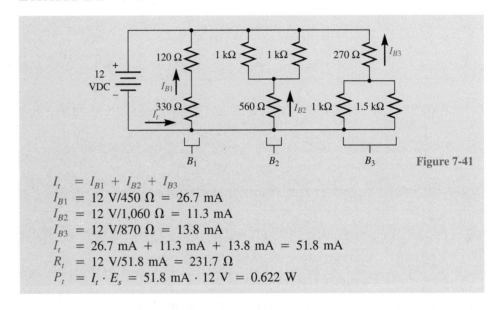

$$I_t = I_{B1} + I_{B2} + I_{B3}$$
$$I_{B1} = 12 \text{ V}/450 \ \Omega = 26.7 \text{ mA}$$
$$I_{B2} = 12 \text{ V}/1{,}060 \ \Omega = 11.3 \text{ mA}$$
$$I_{B3} = 12 \text{ V}/870 \ \Omega = 13.8 \text{ mA}$$
$$I_t = 26.7 \text{ mA} + 11.3 \text{ mA} + 13.8 \text{ mA} = 51.8 \text{ mA}$$
$$R_t = 12 \text{ V}/51.8 \text{ mA} = 231.7 \ \Omega$$
$$P_t = I_t \cdot E_s = 51.8 \text{ mA} \cdot 12 \text{ V} = 0.622 \text{ W}$$

Figure 7-41

Let's review our overall parallel circuit-analysis steps before we continue to the next example.

Calculating I_t and Voltages for Overall Parallel Circuits

1. Solve for total resistance, total current, and voltage drops in each branch, treating each branch as though it were a separate problem.

2. Add all branch currents together to solve for the total current.

Consider Example 7-15.

EXAMPLE   7-15

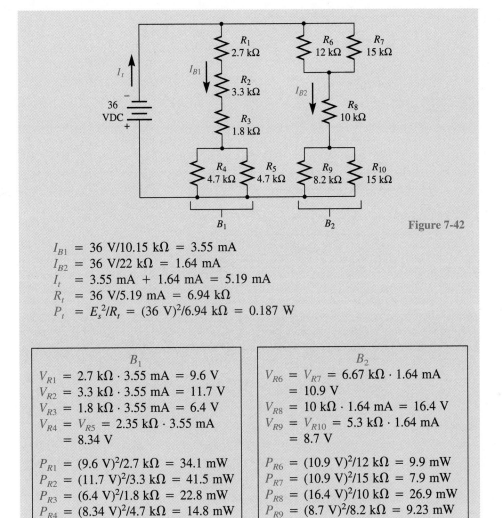

Figure 7-42

I_{B1} = 36 V/10.15 kΩ = 3.55 mA
I_{B2} = 36 V/22 kΩ = 1.64 mA
I_t = 3.55 mA + 1.64 mA = 5.19 mA
R_t = 36 V/5.19 mA = 6.94 kΩ
P_t = E_s^2/R_t = (36 V)^2/6.94 kΩ = 0.187 W

B_1
V_{R1} = 2.7 kΩ · 3.55 mA = 9.6 V
V_{R2} = 3.3 kΩ · 3.55 mA = 11.7 V
V_{R3} = 1.8 kΩ · 3.55 mA = 6.4 V
V_{R4} = V_{R5} = 2.35 kΩ · 3.55 mA
$\qquad$ = 8.34 V
P_{R1} = (9.6 V)^2/2.7 kΩ = 34.1 mW
P_{R2} = (11.7 V)^2/3.3 kΩ = 41.5 mW
P_{R3} = (6.4 V)^2/1.8 kΩ = 22.8 mW
P_{R4} = (8.34 V)^2/4.7 kΩ = 14.8 mW
P_{R5} = (8.34 V)^2/4.7 kΩ = 14.8 mW

B_2
V_{R6} = V_{R7} = 6.67 kΩ · 1.64 mA
$\qquad$ = 10.9 V
V_{R8} = 10 kΩ · 1.64 mA = 16.4 V
V_{R9} = V_{R10} = 5.3 kΩ · 1.64 mA
$\qquad$ = 8.7 V
P_{R6} = (10.9 V)^2/12 kΩ = 9.9 mW
P_{R7} = (10.9 V)^2/15 kΩ = 7.9 mW
P_{R8} = (16.4 V)^2/10 kΩ = 26.9 mW
P_{R9} = (8.7 V)^2/8.2 kΩ = 9.23 mW
P_{R10} = (8.7 V)^2/15 kΩ = 5.05 mW

Now it's your turn. Try your skills on Self-Check 7 - 4.

SELF-CHECK 7-4

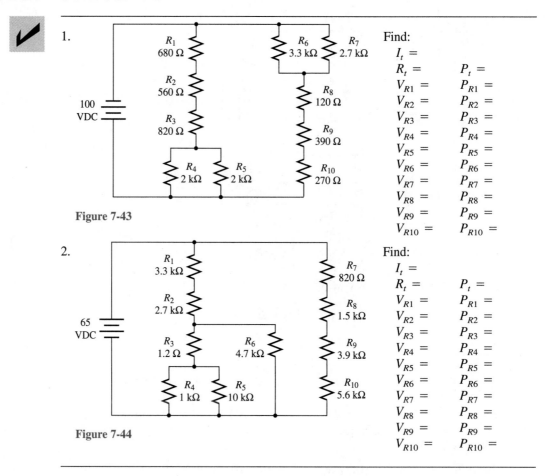

1.

Find:

$I_t =$

$R_t =$ $P_t =$

$V_{R1} =$ $P_{R1} =$

$V_{R2} =$ $P_{R2} =$

$V_{R3} =$ $P_{R3} =$

$V_{R4} =$ $P_{R4} =$

$V_{R5} =$ $P_{R5} =$

$V_{R6} =$ $P_{R6} =$

$V_{R7} =$ $P_{R7} =$

$V_{R8} =$ $P_{R8} =$

$V_{R9} =$ $P_{R9} =$

$V_{R10} =$ $P_{R10} =$

Figure 7-43

2.

Find:

$I_t =$

$R_t =$ $P_t =$

$V_{R1} =$ $P_{R1} =$

$V_{R2} =$ $P_{R2} =$

$V_{R3} =$ $P_{R3} =$

$V_{R4} =$ $P_{R4} =$

$V_{R5} =$ $P_{R5} =$

$V_{R6} =$ $P_{R6} =$

$V_{R7} =$ $P_{R7} =$

$V_{R8} =$ $P_{R8} =$

$V_{R9} =$ $P_{R9} =$

$V_{R10} =$ $P_{R10} =$

Figure 7-44

7-5 The Wheatstone Bridge

As shown in Figure 7-45, the **Wheatstone bridge** is basically two simple voltage dividers placed side by side with a meter bridged between them. The purpose of the circuit, when it was originally conceived, was to determine the value of an unknown resistor. Even today, sophisticated versions of the Wheatstone bridge are used to measure very high or very low resistances. However, this simple circuit has many other applications in industrial control circuits. Before we discuss applications, let's see how the circuit functions and what it basically does.

How It Works

Notice, in Figure 7-45, one resistor, labeled R_X, is connected to special terminals (usually threaded terminals with thumb or wing nuts). In the conventional Wheatstone bridge, R_X is the unknown resistor. Connected in series with R_X is R_V. R_V is a Variable resistor connected to a very carefully scaled and calibrated knob. R_V is adjusted to determine the

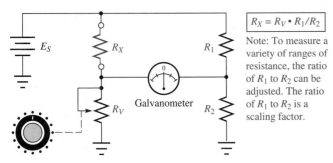

$$R_X = R_V \cdot R_1/R_2$$

Note: To measure a variety of ranges of resistance, the ratio of R_1 to R_2 can be adjusted. The ratio of R_1 to R_2 is a scaling factor.

Figure 7-45   The galvanometer forms the bridge in a Wheatstone bridge circuit.

unknown R_X. Notice the meter that bridges the gap between the two voltage dividers. This meter is a special meter called a **galvanometer**. The galvanometer is a meter whose needle is normally at rest in the center of the scale. This allows the needle to move in either direction, depending on the polarity of voltage applied to its terminals. R_V is adjusted for a **balanced bridge**, as indicated by the center scale needle position of the galvanometer. At center scale, the voltage across the meter is **nulled** to 0 volts. In other words, there is no difference of potential across the meter. Once the bridge is successfully balanced by adjusting R_V the value of R_X can be read from the calibrated scale on the knob of R_X (more on this later). To summarize, the R_X is placed on the test terminals, R_V is adjusted for a **null** on the galvanometer, and the value of R_X is read from the calibrated scale of R_V.

What Makes It Work

But what is actually happening as the galvanometer is being nulled? First, you must understand that a null occurs when the voltage across R_V is equal to the voltage across R_2. In other words, if the voltage across R_V is 6 V and the voltage across R_2 is 6 V, there is no difference of potential. The two potentials are both 6 V, and the difference between them is 0 volts. Thus, the galvanometer will be in its normal center position due to lack of voltage, and therefore lack of current.

Second, it is important for you to understand that the resistance of R_V does not have to equal the resistance of R_2, nor does the current through R_V have to equal the current through R_2. It is only important that the voltages across R_V and R_2 match for a null to

---

Sir Charles Wheatstone (1802–1875)

Charles Wheatstone was a British physicist and well-known inventor. He was an early pioneer in the field of electricity and electronics with many scientific contributions to his credit. Among his inventions were a concertina, a polar clock, and a cryptographic machine. Wheatstone and W. F. Cook patented an electrical telegraph system in 1837 at about the same time that Samuel Morse developed the telegraph system in the United States. Interestingly, Wheatstone did not invent the bridge that carries his name. However, he did improve it. (Photo courtesy of The Bettmann Archive.)

occur. Also, you should see that if $V_{RV} = V_{R2}$ then V_{RX} must equal V_{R1} and V_{RX}/V_{RV} must equal V_{R1}/V_{R2}.

From this information, the formula used to solve for R_X, under balanced bridge conditions, can be derived as follows:

$$\frac{V_{RX}}{V_{RV}} = \frac{I_{RX} \cdot R_X}{I_{RX} \cdot R_V} = \frac{V_{R1}}{V_{R2}} = \frac{I_{R1} \cdot R_1}{I_{R1} \cdot R_2}$$

therefore,

$$\frac{I_{RX} \cdot R_X}{I_{RX} \cdot R_V} = \frac{I_{R1} \cdot R_1}{I_{R1} \cdot R_2} = \frac{R_X}{R_V} = \frac{R_1}{R_2}$$

So,

$$R_X = R_V \cdot R_1/R_2 \tag{7.1}$$

EXAMPLE SET   7-16

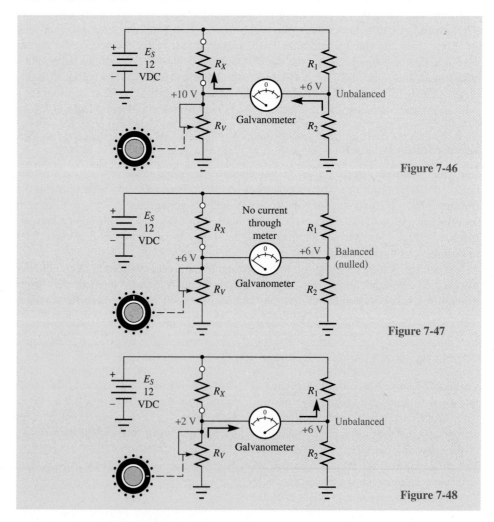

Figure 7-46

Figure 7-47

Figure 7-48

Formula 7.1 can be used to solve for an unknown resistance R_X in any Wheatstone bridge circuit when the other three resistors are known. Notice that the ratio of R_1 to R_2 definitely affects the value of R_X, since $R_X = R_V \cdot R_1/R_2$. For this reason, the ratio of R_1 to R_2 is called a **scaling factor**, which must be multiplied by the reading taken from the calibrated knob.

Now study Example Set 7-16 very carefully. Notice the position of the needle on the galvanometer for each of the three examples shown.

Modern Applications

The Wheatstone bridge is a very useful circuit that has found its way into many modern applications in industry. It is primarily used as a position-sensing circuit that can be implemented in many different ways. One very common implementation of the Wheatstone bridge is in motor-positioning circuits. Figure 7-49 illustrates this particular implementation. The purpose of the circuit is to be able to control the actual position of the motor shaft by remotely adjusting a potentiometer.

As you can see, a second potentiometer is mechanically connected to the motor shaft via gears, a belt, or a chain. When the setting of the remote potentiometer is changed, the bridge between the potentiometers is unbalanced. A power amplifier, which drives the motor, amplifies the resulting difference of potential and causes the motor to turn in a particular direction. The relative polarity of the unbalanced bridge (as discussed earlier) determines the direction of shaft rotation just as it did the needle direction on the galvanometer. As the shaft rotates, the second potentiometer is adjusted until the bridge is once again balanced (nulled). The motor shaft will of course do more than simply null the bridge. In addition to the nulling potentiometer, the shaft may be connected to components in a large broadcast transmitter to automatically and remotely tune the transmitter, or to the arm or end effector of a robot, or to a dial to indicate the position of a device, or to any number of other mechanical systems. You will undoubtedly see the Wheatstone bridge used in many different applications throughout your career.

To make sure you understand the mathematics and theory behind the Wheatstone bridge, take time to answer the questions of Self-Check 7-5.

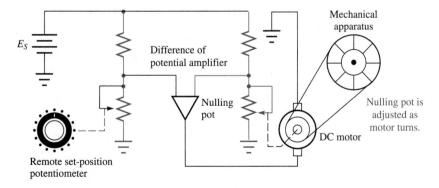

Figure 7-49   The bridge circuit is used in motor shaft position control systems.

SELF-CHECK 7-5

1. What was the purpose for the original Wheatstone bridge?

2. What does it mean to null the meter?

3. If V_{RV} is $+8.5$ V and V_{R2} is set at $+6$ V, what is the difference of potential? Which voltage would you consider to be relatively negative?

4. If $R_1 = 10$ kΩ, $R_2 = 22$ kΩ and $R_V = 150$ kΩ, what value should R_X be in order for the bridge to be balanced?

5. If the value of R_X in Question 4 is 470 kΩ and the source voltage is $+10$ VDC, what is the difference of potential across the bridge? Describe the direction of current flow.

7-6 The Loaded Voltage Divider

One Important Fact

The loaded voltage divider is simply a voltage divider to which a practical load has been connected (such as a resistor, DC motor, light bulb, transistor circuit, etc.). It sounds simple enough, doesn't it? Well it is, once you understand one very important fact: *The load device will always cause the output voltage of the voltage divider to decrease, or drop down, when it is connected.* Therefore, if you design the voltage divider to provide a 6 V output and then connect the load, the voltage will drop down. It may drop a lot, say down to 3 or 4 volts, or it may drop very little, say down to 5.8 volts or so. In this section, we will cover the hows and whys of this very important fact.

The Loading Effect

So, what determines if the voltage drops a lot or a little? If the resistance of the load device is greater than ten times the resistance of the output resistor of the voltage divider, the **loading effect** will not be very significant. What does that mean? First of all, loading effect means the effect that the load has on the output voltage of the voltage divider. If the load resistance is larger, the loading effect will be minimal and if the load resistance is small, the loading effect will be significant. The term **load** actually refers to the amount of current that is drawn by a device. A smaller resistance will actually be a larger load and vice versa. If the load is large (small load resistance and large current demand), the loading effect is great, and if the load is small (large load resistance and small current demand), the loading effect is small.

The Typical Loaded Voltage Divider

Now that you have some basic ideas about loaded voltage dividers, let's take a moment to analyze a very simple, and typical, voltage divider by examining Figure 7-50. Notice that the circuit is identified in two parts: the voltage divider and the load. First, an analysis of only the voltage divider is done to discover the unloaded output voltage. This is a simple series circuit with the output taken across R_2. Next, the switch is closed to connect the

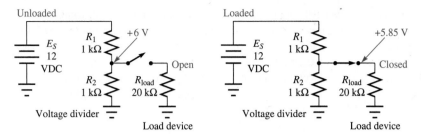

Figure 7-50   The loaded voltage divider is a series-parallel circuit.

load resistor. Now the circuit becomes an overall series, or series-parallel, type of circuit. The output voltage is actually the voltage that appears across the parallel combination of R_2 and R_{load}. Therefore, the loaded output voltage can be found in one of two ways: (1) $V_{out} = I_t \cdot (R_2 \parallel R_{load})$, or (2) $V_{out} = E_s - V_{R1}$. Example Set 7-17 is designed to illustrate the change in output voltage that occurs when different loads are connected to the voltage divider. Make sure you see the relationship between size of load and amount of loading effect in the output.

EXAMPLE SET 7-17

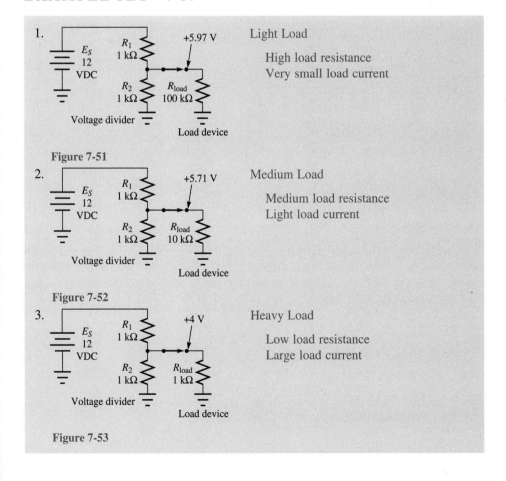

1.   Light Load

High load resistance
Very small load current

Figure 7-51

2.   Medium Load

Medium load resistance
Light load current

Figure 7-52

3.   Heavy Load

Low load resistance
Large load current

Figure 7-53

When the voltage divider is loaded down and the output voltage drops, what happens to the "lost" voltage? When the load is connected to the voltage divider, another path for current is provided. Due to the load, this additional current must go through R_1, the series dropping resistor. The additional current causes the voltage across R_1 to increase ($V_{R1} = I_t \cdot R_1$). Therefore, there is less voltage left over from the source voltage for R_2 and the load ($V_{out} = V_{R2} = V_{load} = E_s - V_{R1}$). So, the lost voltage, due to loading, is added to the voltage across R_1, the series dropping resistor.

Multiple-Output Loaded Voltage Dividers

The simple two-resistor voltage divider with a single load is by far the most common and practical loaded voltage divider that you will encounter. That does not mean that you will not encounter more complex loaded-voltage-divider circuits. Example 7-18 shows a three-resistor voltage divider with two loads. The main thing you will discover as we examine these multiple-output loaded voltage dividers is that the two loads do have an effect on each other's voltage drop. You will discover that changing a load resistance will not only affect the voltage across that load resistance, it will also change the voltage across the other load resistance (or other load resistances, in more complex loaded voltage dividers).

E X A M P L E   7 - 1 8

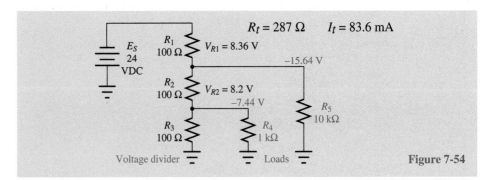

Figure 7-54

Analysis Method

In Example 7-18, the voltages across the load resistors were calculated just as you would calculate voltage drops in any overall series circuit. The method you might follow in solving voltage drops in this circuit (a method that can be applied to other loaded voltage dividers) is summarized as follows:

Analysis Steps for Example 7-18

1. Solve for total resistance.
2. Solve for total current.
3. Find the voltage drop across R_1. ($V_{R1} = I_t \cdot R_1$)

4. Find the voltage drop across load resistor R_5. ($V_{R5} = E_s - V_{R1}$)
5. Find the current through R_2. ($I_{R2} = I_t - I_{R5}$)
6. Find the voltage drop across R_2. ($V_{R2} = I_{R2} \cdot R_2$)
7. Find voltage drop across R_3 and load resistor R_4. ($V_{R4} = V_{R5} - V_{R2}$) or [$V_{R4} = I_{R2} \cdot (R_3 \| R_4)$]
8. If needed, find the current through R_3 and R_4. ($I_{R3} = V_{R3}/R_3$) and ($I_{R4} = V_{R3}/R_4$)

Notice that this method involves a systematic approach in solving for output voltages of a loaded voltage divider. It is a method that can be applied to any size loaded voltage divider from the simplest to the most complex.

The Proof Is in the Calculations

Earlier, it was stated that changing a load will not only affect the voltage across that load, it will also change the voltage across any other load(s). Let's prove that now. Example 7-19 is the same circuit as Example 7-18 except load resistor R_5 has been changed. Notice what has happened to all other voltage drops in the circuit. Did the voltage across load resistor R_4 remain the same as it was in Example 7-18? No, It did not! Why? Because changing a load affects total current which, in turn, affects the voltage drop across the series dropping resistor, in this case R_1. Naturally, this causes all other voltage drops to change.

EXAMPLE   7-19

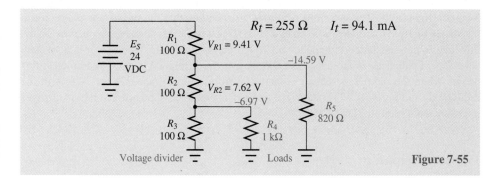

$R_t = 255\ \Omega$ $I_t = 94.1$ mA

E_S 24 VDC

R_1 100 Ω $V_{R1} = 9.41$ V

−14.59 V

R_2 100 Ω $V_{R2} = 7.62$ V

−6.97 V

R_5 820 Ω

R_3 100 Ω

R_4 1 kΩ

Voltage divider   Loads   **Figure 7-55**

One thing you should be discovering here is that the load devices of a loaded voltage divider should not be permitted to change in resistance. In other words, all loads should be of a fixed resistance in order to maintain all voltages at a constant predetermined design value. The loaded voltage divider is a fairly simple circuit that has many uses. But it does have practical limitations. The following example, Example 7-20, is offered to illustrate circumstances under which a loaded voltage divider will not work well.

EXAMPLE   7-20

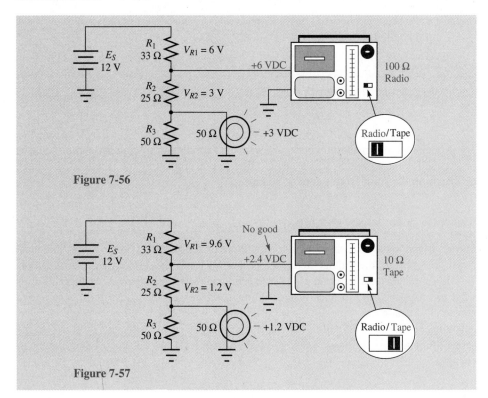

Figure 7-56

Figure 7-57

As you can see, one of the loads in Example 7-20 is a radio/tape player and the other load is a light bulb used as a power-on indicator. Notice that the radio/tape player has a resistance of 100 Ω with the radio on and only 10 Ω when the tape player is on. Obviously, this load is not constant and depends on which part of the radio/tape player is activated. Two sets of calculations are needed to show the great difference in circuit parameters that occurs when the radio/tape player is switched from radio to tape. Figure 7-56 is in the radio mode and Figure 7-57 is in the tape mode. Also, note how the power-on light changes intensity from radio to tape.

Power Dissipation

Power dissipation in the loaded-voltage-divider network (circuit) is a very important consideration. The series dropping resistor, or resistors, in a loaded voltage divider will, in many cases, be dissipating a rather significant amount of power. Naturally, these resistors must be conservatively rated to insure reliability. You will recall that you must know at least two parameters pertaining to a particular resistor in order to calculate its power dissipation (I_R and V_R, or I_R and R, or V_R and R). As in all other DC circuits, the total power dissipation in a loaded voltage divider, including loads and voltage divider, can be found using E_s, I_t, and/or R_t, or by adding all individual power dissipations together.

Self-Check 7-6 will give you a chance to test your skills and check your own work.

1. What does the phrase ''loading effect'' mean?

2. Which would be a ''heavier'' load, a 10 kΩ resistor or a 100 Ω resistor?

3. What does the term *load* actually refer to?

4. What is the difference between a loaded and an unloaded voltage divider?

5. Calculate all resistor voltages for Figure 7-58.

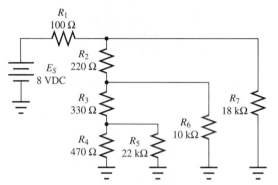

Figure 7-58

6. Calculate the power dissipation for each resistor in Figure 7-58.

7-7 Troubleshooting the Loaded Voltage Divider

Once again we turn our attention to the important subject of troubleshooting. Thus far, we have discussed the theory and calculations involved in loaded voltage dividers. In this section, we will examine the use of the multimeter in locating a trouble such as a short or an open. To enrich your experience in electricity and electronics, you will want to build some of the example circuits and duplicate the troubles. (To duplicate an open, simply disconnect one lead of the component. To duplicate a short, place a jumper wire across the component.)

Seeking Out Shorts

To begin, it is important to recall the basic characteristics of a short and what readings might be expected on an ammeter, a voltmeter and an ohmmeter when seeking one out. You will remember that a short is a very low- or no-resistance path for current around or through a component or components in a circuit. An abnormally low, or zero-ohm, resistance reading will be indicated on an ohmmeter. A short will cause total circuit current to increase, since some of the circuit resistance is eliminated by the short. Thus, an ammeter properly placed to read total current will show an abnormally high reading. Also, if the current, due to the short, is greater than the fuse or circuit breaker rating, the fuse or circuit breaker will open. Finally, a very obvious characteristic of the short is the lack of voltage drop across it. A shorted resistor, for example, will not have its normal voltage drop. Instead, the voltage across a short is always zero volts (if the short is in fact zero ohms). Now, it's time to practice applying what you know about the short.

EXAMPLE   7-21

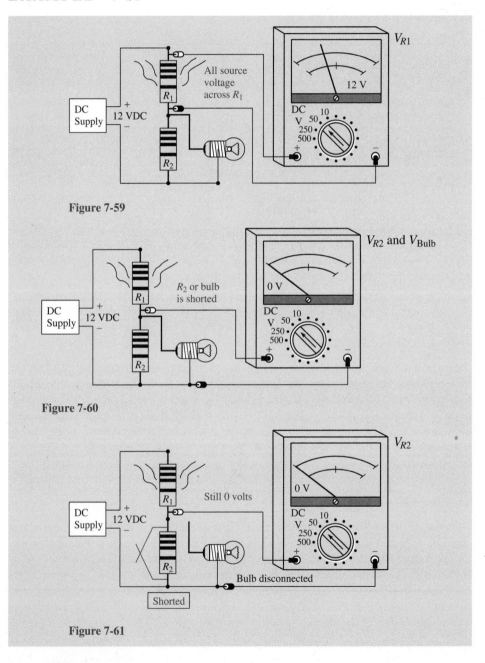

Figure 7-59

Figure 7-60

Figure 7-61

DC Circuits and Measurements

Measuring the Total Resistance of a Simple Series Circuit

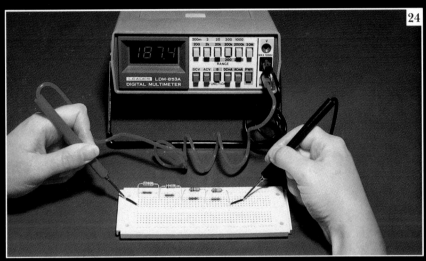

24) Even though lethal voltage is not present, you should not touch the metal ends of the test probes. If both hands touch the metal tips of the probes, your body will be in parallel with the circuit being measured. If the circuit is a high resistance circuit, such as the circuit shown, your body will have a significant effect on the resistance reading. The reading indicates the total resistance to be 187.4 kΩ. Parallel body loading could drop the reading down to 170 kΩ or less. (Section 5-1) (test instruments courtesy of Leader Instruments Corp., Hauppauge, NY)

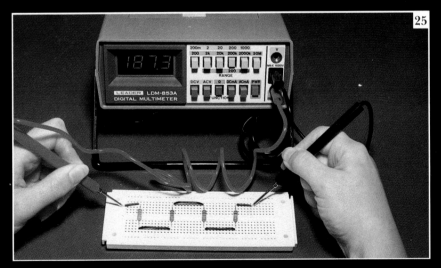

25) This photo shows the same resistors arranged in series in a different physical layout. Notice the black jumpers used to "daisy chain" the resistors. Also, notice the meter reading is now 187.3 kΩ instead of 187.4 kΩ. This is due to the biggest limiting factor of digital instruments in general. That is, the accuracy of the instrument is always specified to include ± 1 or ± 2 counts of the least significant digit. For this reason, a range should be selected that makes use of all available display digits whenever possible. (Section 5-1) (test instruments courtesy of Leader Instruments Corp., Hauppauge, NY)

Measuring a Series Voltage Drop:

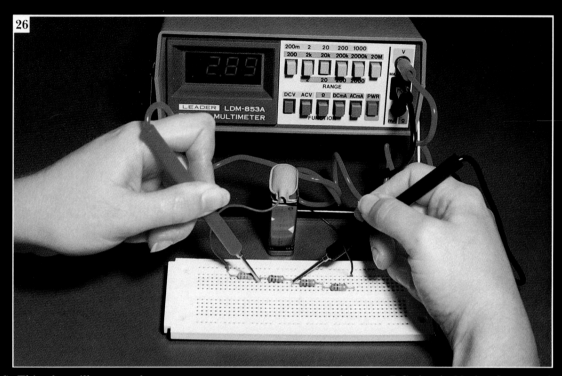

26) This photo illustrates the proper way to measure a voltage drop in a DC circuit. Notice the meter is set for the DCV function and the 20 V range. Proper polarity is being observed with the red probe (+) on the side of the resistor closest to the positive side of the DC source and the black probe (−) on the side closest to the negative side of the DC source. A reading of 2.89 V is indicated. If the test probes were reversed, the reading would be −2.89 V with no damage to the meter (damage could occur to an analog meter). (Sections 5-2 to 5-5 and 5-9) (test instruments courtesy of Leader Instruments Corp., Hauppauge, NY)

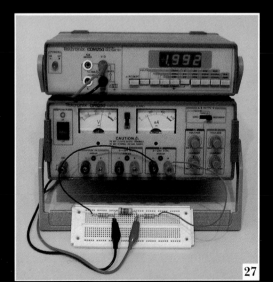

27) Here, 15 V is being applied to a series circuit consisting of a 51 Ω resistor, a 15 Ω resistor, and another 51 Ω resistor. The voltage across the 15 Ω resistor is measured to be 1.992 V. $R_T = 51\ \Omega + 15\ \Omega + 51\ \Omega = 117\ \Omega$, $I_T = 15\ V/117\ \Omega = 128\ mA$, $V15\ \Omega = 128\ mA \cdot 15\ \Omega = 1.92\ V$. The difference between measured and calculated voltage is due to tolerances of the resistors. Notice the polarity of the DMM test leads with respect to the polarity of the power supply leads. (Chapter 5) (equipment courtesy of Tektronix, Inc.)

Common Ground

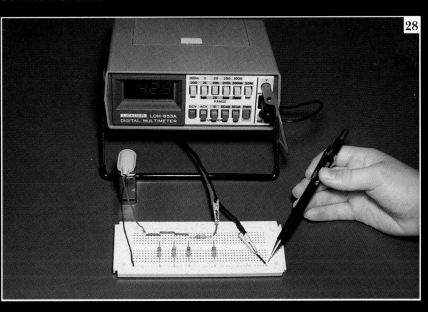

28) This photo demonstrates the use of the lower bus on the proto-board as a common ground. In this case, the common ground is negative since it is connected to the negative side of the DC source. Notice the negative meter lead is also connected to the common ground for voltage measurements. The circuit shown is overall parallel (parallel-series). (Section 5-6) (test instruments courtesy of Leader Instruments Corp., Hauppauge, NY)

A Dual Polarity DC Supply

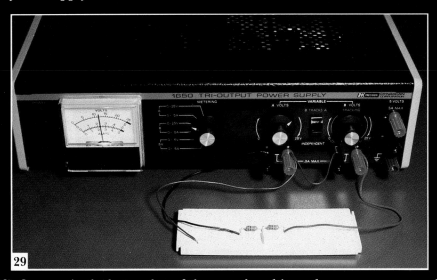

29) Once again the bottom bus of the proto-board is used as a common ground. In this case, the bus serves as common ground for both a plus and minus DC supply. Notice the green wire used to interconnect the positive terminal of one DC source to the negative terminal of the second DC source then to the common bus on the proto-board. The resistor on the left has a negative potential applied to it and the resistor on the right has a positive potential applied to it. This bipolar DC power supply permits independent adjustment of the two DC sources or source B can track source A. The supply is presently in the tracking mode. Both sources, A & B, are adjusted at the same time by adjusting the A volts knob. (ref. Section 5-6) (test instruments courtesy of B&K PRECISION/Dynascan Corporation)

Measuring Total Resistance of a Simple Parallel Circuit

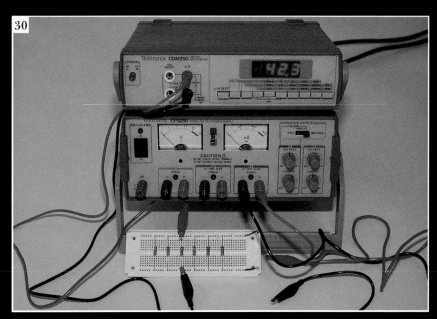

30) The total resistance of this parallel circuit is measured to be 42.3 Ω. Notice all resistors are in parallel using the top and bottom bus strips of the protoboard. The test leads could be placed across any of the resistors to measure total resistance. (Chapter 6) (equipment courtesy of Tektronix, Inc.)

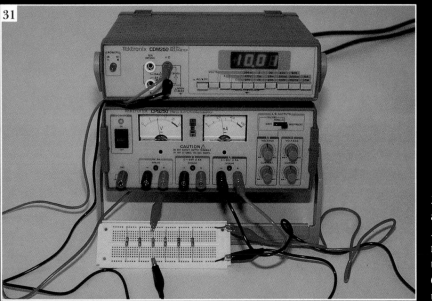

31) In this photograph, 10 V is applied to the parallel circuit. All parallel branches (resistors) receive the same 10 V via the upper and lower bus. (Chapter 6) (equipment courtesy of Tektronix, Inc.)

Measuring Branch Current in a Parallel Circuit

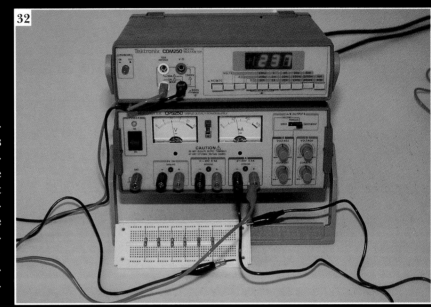

32) Here, total current supplied to the parallel circuit is measured by placing the ammeter (DMM) in series with the DC supply as shown. Note that the red (+) lead of the ammeter (DMM) is connected to the positive post of the power supply. Ohm's Law: $I = E/R = 10 V/42.3 \Omega = 236$ mA. (Chapter 6) (equipment courtesy of Tektronix, Inc.)

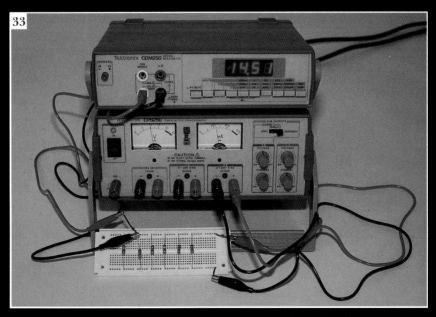

33) In this photograph, the current through a single branch resistor is measured and found to be 14.5 mA. Notice the ammeter (DMM) is placed in series with the second resistor from the left (680 Ω). Ohm/s Law: $I = E/R = 10 V/680 \Omega = 14.7$ mA. The difference is due to resistor tolerance. (Chapter 6) (equipment courtesy

An Overall Series Arrangement of Resistors

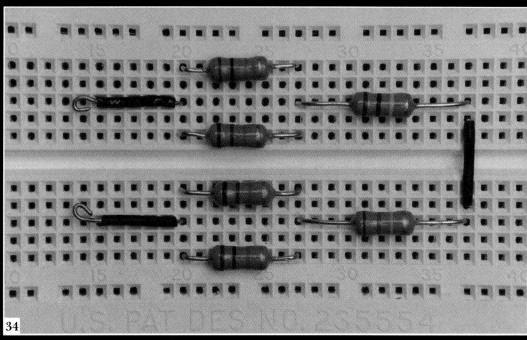

34

34) Notice that this simple circuit contains two parallel sections in an overall series arrangement (a series-parallel circuit). (Section 7-1)

An Overall Parallel Arrangement of Resistors

35

35) This is a five branch parallel resistor circuit. Notice that branches four and five are simple two resistor series circuits. Thus, the entire circuit is overall parallel (parallel-series circuit). (Section 7-3)

Measuring Total Current

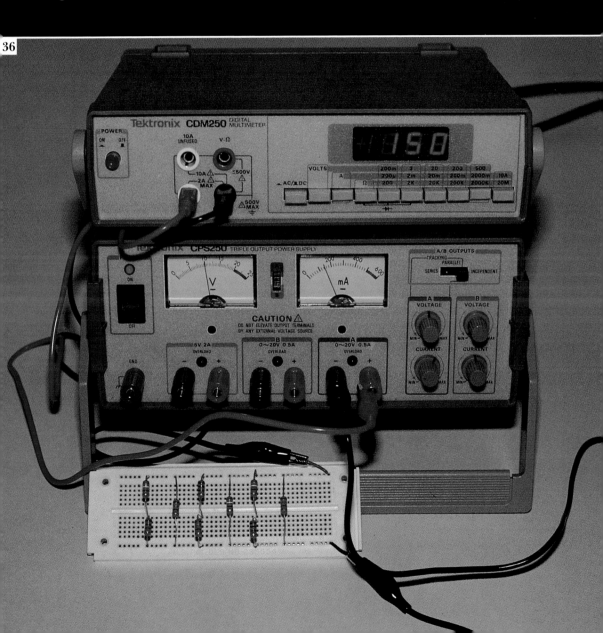

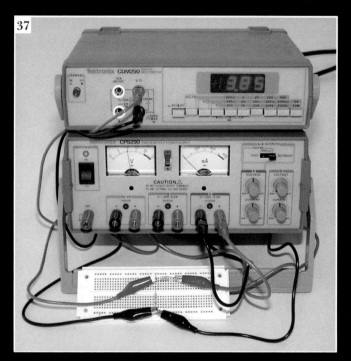

## 36

37) This photograph illustrates how a voltmeter can load down the circuit when used to measure voltage. Two 4.7-MΩ resistors form a simple voltage divider with 10 V applied. The voltage across each resistor is really 5 V when the voltmeter is disconnected. (Chapter 7) (equipment courtesy of Tektronix, Inc.)

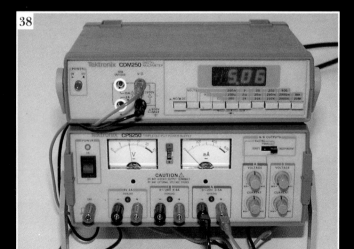

38) Here, two 1-kΩ resistors are used to form the voltage divider and the voltage across the bottom resistor is again measured. This

We start our troubleshooting discussion by examining the simplest and most common type of loaded voltage divider, the single-load voltage divider shown in Example 7-21. Let's assume that we are alerted to circuit trouble because R_1 has become very warm and the light bulb, our load device, is no longer on. You should recognize these two symptoms as an indication of the presence of a short. The series dropping resistor R_1 is unusually warm because of excessive current flowing through it. From this, we might assume that either the light bulb or R_2 is shorted. We will need our multimeter to confirm our assumptions.

The voltmeter proves most useful as we investigate this circuit. First, we can measure the voltage drop across R_1. In doing so, we discover that all of the applied voltage appears across R_1. This explains why R_1 is much warmer than normal ($P_{R1} = V_{R1}{}^2/R_1$). To reinforce theory, we can measure the voltage across R_2 and the light bulb as shown in Figure 7-60. Naturally, our voltmeter indicates 0 volts. You always measure 0 V across a dead short. So, either the light bulb or the resistor R_2 is shorted. Now, how do you determine which it is? By the process of elimination. Disconnect the light bulb, by removing one of its wires, and measure the voltage across R_2 as shown in Figure 7-61 (turn power off until your work is done then reapply power). Is the voltage across R_2 still 0 volts? Let's say that it is. Then, R_2 is in some way externally shorted. (Resistors are rarely shorted internally.) Make a visual inspection of the resistor's terminals with the surrounding wiring and chassis. But, what if the voltage across R_2 jumped up after the light bulb was removed and power was turned back on? That would indicate that the light bulb was shorted. However, light bulbs don't normally short. They open. Still, the light-bulb socket could be shorted and it's worth investigating.

So, R_2 is found to be shorted (externally). All the symptoms of a short were there to point us toward R_2: the overheated series dropping resistor, the total source voltage across the series dropping resistor, and the absence of voltage across R_2 and the light bulb.

Now, let's look at our light bulb circuit one more time with a completely different set of symptoms and measurements. Example 7-22 illustrates the following symptoms: the light bulb is much brighter than normal, the supply voltage is measured and is found to be normal, the total supply voltage is measured across the light bulb and R_2 (Figure 7-62), and there is no voltage across the series dropping resistor, R_1 (Figure 7-63). At this point, you have made observations and measurements that indicate R_1 is shorted (externally).

EXAMPLE   7-22

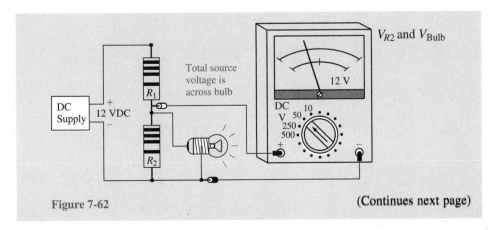

Figure 7-62

(Continues next page)

EXAMPLE 7-22 (Continued)

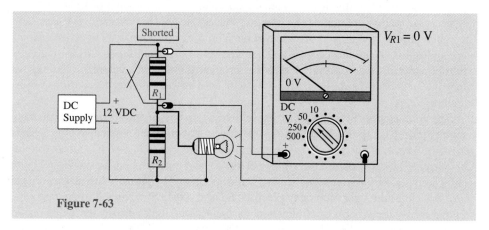

Figure 7-63

Closing in on an Open

Before we discuss how to close in on an open, we should once again take time for a little review. Recall that an open is a break in current. The normal path for current has been disrupted because of a cracked foil, a broken conductor, or an overstressed and burned component. The overall result is that the total circuit current is lower than normal. Also, the voltage across the open is higher than normal, or in some cases equal to the source voltage (depending upon the location of the component in the circuit). Recall that, when an open is checked with the ohmmeter, an infinite ($\infty\ \Omega$) reading is indicated.

EXAMPLE 7-23

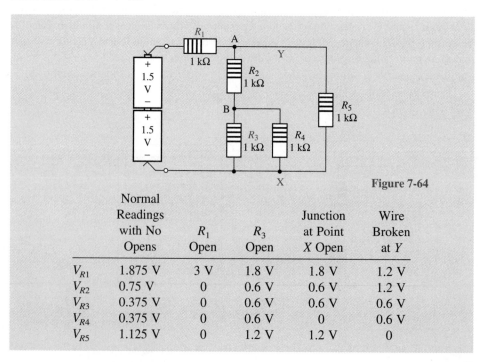

Figure 7-64

	Normal Readings with No Opens	R_1 Open	R_3 Open	Junction at Point X Open	Wire Broken at Y
V_{R1}	1.875 V	3 V	1.8 V	1.8 V	1.2 V
V_{R2}	0.75 V	0	0.6 V	0.6 V	1.2 V
V_{R3}	0.375 V	0	0.6 V	0.6 V	0.6 V
V_{R4}	0.375 V	0	0.6 V	0	0.6 V
V_{R5}	1.125 V	0	1.2 V	1.2 V	0

Figure 7-64 in Example 7-23 contains an open. But where is the open? Is it an open resistor, a broken conductor, or a loose connection? All of these are likely and real possibilities. Again, the voltmeter is a valuable tool to close in on the open. In this example, we will assume four different possibilities and see how meter readings are analyzed to identify each. The possible troubles to explore are: R_1 open, R_3 open, a loose connection at point X, and a broken conductor at point Y. Notice that a chart listing the normal circuit voltages and the actual voltages obtained from measurements is provided in Example 7-23. It is always good to make a list of symptoms and measurements as you troubleshoot a circuit. This will help you pinpoint the trouble with a minimum of confusion.

In our first case, R_1 is open. Notice that the voltage across R_1 is the same as the source voltage. This alone is a strong indication that R_1 is open. All other voltages are 0 volts because R_1 has effectively disconnected the positive side of the supply from the rest of the circuit. (We are discussing opens here, but you should realize that the same measurements would be obtained if R_5 were shorted. Think about it!) To be sure that R_1 is open, you should remove it and measure it with an ohmmeter.

Our second case is R_3 open. If R_3 is open, all circuit voltages, except the source voltage, will be abnormal. After listing the measured voltages, you would compare them to normal voltages and begin your analysis. First of all, the voltage across R_1 is lower than normal. This is a strong indication of an open somewhere other than R_1. The lower voltage drop across R_1 indicates less total current than normal, which indicates an open somewhere. We can close in on the open a little closer by applying more circuit theory. If R_2 was open, the voltage across R_3 and R_4 would be 0 volts because R_3 and R_4 would effectively be disconnected from the circuit. So, R_2 must be good. The possibilities have been narrowed to R_3, R_4, and R_5. At this point, you have a choice of several different ways to proceed. You could narrow the possibilities a little further with some mathematical circuit analysis. In other words, you could calculate voltage drops with R_3 open, then R_4 open, and, finally, R_5 open. This process would reveal that the problem is either R_3 or R_4. You still would not know which one of the two is open. In a large circuit, a more practical approach would be to measure currents in various branches. This identifies the open resistor without any doubt. In this case, there would be a reasonable amount of current flowing through R_5 and R_4 but no current flowing through R_3. Once again, the suspected resistor should be measured with an ohmmeter for verification.

Many opens in circuits are loose connections caused by poor soldering and mechanical vibration. In this case, let's consider R_4 as having a loose connection at point X. Again, we take our voltage measurements and discover that the voltage across R_4 is 0 volts. At the same time, R_3 has a voltage drop of 0.6 volts. If R_4 is open, you measure the 0.6 volts across R_4, which is actually the voltage across R_3. If R_4 is shorted, there is no voltage across R_3. All indications point to a loose connection at either point B or point X. You can place your voltmeter across each junction to find out which one is open. If the junction is a good, solid connection, no voltage will be measured. You must have resistance to have a voltage drop. A good connection has 0 ohms of resistance and no voltage drop. Many times a careful visual inspection will reveal loose connections.

Moving parts or poorly mounted PC boards are common causes of broken conductors or cracked foils that interrupt current. These types of opens are sometimes difficult to find. Broken conductors hide inside plastic insulation, and the cracks in copper foils on PC boards are very fine and difficult to see. Often, this kind of an open is overlooked by the technician because he or she assumes the trouble to be a component. Once again, the voltmeter can be used to locate the problem. In Example 7-23, let's consider an open in the conductor at point Y. Notice the voltage measurements that have been listed for this particular case. We see that there is no voltage across R_5 but there is voltage across R_2, R_3, and R_4. If R_5 were open, we would definitely read a voltage across it. If R_5 were shorted, there would be no voltage across R_2, R_3, or R_4. So, the problem must be either a loose connection or a broken conductor. To identify exactly where the open is, you would simply place your voltmeter across the full length of each conductor that connects R_5 to the rest of the circuit. Remember, a voltage is indicated across an open.

Voltmeter Loading Effects

As you know, the voltmeter is a very valuable tool when it comes to troubleshooting a circuit. However, there is something else you need to know. The voltmeter has internal resistance, and in some cases, placing the voltmeter across a component to measure its voltage will have the same effect as adding a resistor in parallel. It may cause the currents and voltages in the circuit to change measurably. In other words, the voltmeter becomes a load itself. As you know, adding a load to a voltage divider causes the voltage to drop. Therefore, as shown in Figure 7-65, the voltmeter will actually give you false readings. So, what to do?

See color photo insert page B8.

If the meter resistance is greater than ten times the resistance of the component whose voltage is being measured, the drop or error in voltage will not be overly significant. Naturally, the higher the resistance of the voltmeter, the more accurate the readings will be, due to less loading effect. So, you ask, how much resistance does a voltmeter have? That depends on the type of voltmeter you have and its **sensitivity**. The low-priced, analog-type multimeter may have a volts function rating, or sensitivity, of 20,000 Ω/V (ohms per volt) or, in some cases, 30,000 Ω/V, or even 50,000 Ω/V. This means that for every volt of a particular range the voltmeter has a specified amount of resistance. If a 20 kΩ/V voltmeter is set on the 10 V range, the total voltmeter resistance will be 200 kΩ (20 kΩ/V · 10 V = 200 kΩ). If the voltmeter is set on the 100 V range, the total voltmeter resistance will be 2 MΩ (20 kΩ/V · 100 V = 2 MΩ). As you can see, this kind of voltmeter would have a significant loading effect on a low-voltage circuit made up of high-resistance components. In other words, a 200 kΩ voltmeter would have a significant loading effect across a 100 kΩ resistor. As a rule, this type of analog voltmeter would only be

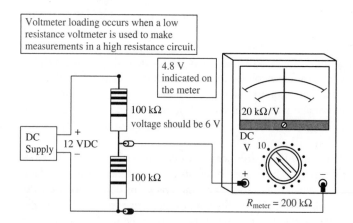

Figure 7-65 A voltmeter will load down a circuit as voltage is being measured.

Figure 7-66   Here's a handy handheld DMM probe. (Courtesy of A. W. Sperry Instruments, Inc.)

used on low-resistance circuits, say 10 kΩ or less. As you can see, the higher the voltmeter sensitivity, the more accurate your measurements will be.

Not all analog voltmeters have such a relatively low resistance. Some analog volt-meters have an amplifier with a very high input resistance that will not significantly load down even high-resistance circuits. These high-resistance amplifiers are called **Field Effect Transistors (FETs)**. Voltmeters that have an FET amplifier normally have a 10 MΩ re-sistance regardless of which range the meter is set on. This 10 MΩ meter resistance allows accurate measurements across almost any circuit.

The most common multimeters used by technicians and engineers today are digital multimeters. Most digital multimeters have a 10 MΩ voltmeter resistance, and usually will not significantly load down a circuit when measurements are made. Figure 7-66 demon-strates a very useful hand-held digital multimeter probe.

Take time now to reinforce what you have learned on this section by answering the questions of Self-Check 7-7.

S E L F - C H E C K   7 - 7

1. The presence of a short in a circuit will cause the total current to (increase/decrease).

2. When the test leads of a voltmeter are placed across a short, the meter will indicate (zero volts/maximum volts).

3. If the current in a circuit is less than normal, it is because of a (short/open).

4. If one of three parallel resistors is shorted, will the voltage across the other two resistors be higher or lower than normal?

5. If a DC voltmeter is rated at 10 kΩ/V and the range selector is set on the 50 V range, what is the resistance of the meter?

6. Explain why the voltmeter described above would not give an accurate reading if used to measure the voltage across a 680 kΩ resistor.

Summary

FORMULAS

(7.1) $R_X = R_V \cdot R_1/R_2$

where R_X is the unknown resistor value in a Wheatstone bridge.

CONCEPTS

- An overall series circuit can be reduced to a simple series circuit.
- An overall parallel circuit can be reduced to a simple parallel circuit.
- Power dissipation for any resistor in any simple or complex circuit is calculated using the voltage across the resistor, the current through the resistor, and/or the resistor's value.
- Total power in a DC circuit is always the sum of all individual power dissipations regardless of circuit type or complexity. Also, total power can be calculated using total resistance, total current, and/or the source voltage.
- The Wheatstone bridge is used to measure very large or very small resistances by adjusting a variable resistor that balances two voltage dividers that are bridged by a galvanometer.
- Connecting a load to a voltage divider causes its output voltage to decrease, or drop down.
- A load connected to a voltage divider is actually current demanded by some device.
- The voltage across the series dropping resistor in a voltage divider increases if the load is increased.
- Loaded voltage dividers are very inefficient and impractical for varying loads. They are used mainly for fixed loads.
- A low-resistance voltmeter acts as a load when measuring voltage in high-resistance circuits. This affects the accuracy of measurements.

PROCEDURES

Calculating R_t for Overall Series Circuits

1. Reduce any parallel sections to their equivalent resistance.
2. Add all series resistors and equivalent resistances.

Calculating Voltage Drops in Overall Series Circuits

1. Each individual resistor can be multiplied by the total current.
2. In a simple overall series circuit, the voltage drop across the parallel section can be found by subtracting the series resistor's voltage drop from the source voltage.
3. The voltage drop across any parallel section can be found by multiplying the total current by the equivalent resistance of that parallel section.

Calculating Parallel-Section Branch Currents

1. First solve for total resistance and total current.
2. Calculate the voltage across the parallel section.
3. Use Ohm's Law to solve for current in each branch of the parallel section. Also, the current divider formula can be used to find branch current from total current.

Calculating R_t for Overall Parallel Circuits

1. Redraw the circuit if necessary.
2. Simplify by combining series resistors in each branch.
3. Calculate R_t for the simplified circuit.

Calculating I_t and Voltages for Overall Parallel Circuits

1. Consider each branch as a separate problem. Solve for total resistance, total current, and voltage drops in each branch.
2. Add all branch currents to find the total current.

SPECIAL TERMS

- Overall series (series-parallel)
- Overall parallel (parallel-series)
- Wheatstone bridge
- Galvanometer
- Balanced bridge, nulled, null
- Load, loading effect
- Voltmeter loading effect
- Sensitivity

Need to Know Solution

If R_1 were open, the total source voltage would be measured across it. If R_6 were shorted, no voltage could be measured.

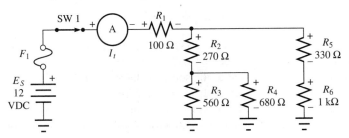

$V_{R1} = 2.39$ V	$I_{R1} = 23.9$ mA
$V_{R2} = 4.5$ V	$I_{R2} = 16.66$ mA
$V_{R3} = 5.1$ V	$I_{R3} = 9.1$ mA
$V_{R4} = 5.1$ V	$I_{R4} = 7.5$ mA
$V_{R5} = 2.4$ V	$I_{R5} = 7.23$ mA
$V_{R6} = 7.2$ V	$I_{R6} = 7.23$ mA

Questions

7-1 Overall Series Circuits

1. What is one characteristic of an overall series circuit?
2. What is the first step in calculating total resistance for an overall series circuit?

7-2 Voltage, Current, and Power in Overall Series Circuits

3. Explain how to calculate the voltage drop across a parallel section in an overall series circuit.

7-3 Overall Parallel Circuits

4. What is one characteristic of an overall parallel circuit?
5. What is a beginning step you might take in solving for total resistance in an overall parallel circuit?

7-4 Voltage, Current, and Power in Overall Parallel Circuits

6. Explain how to solve for total current in an overall parallel circuit.

7-5 The Wheatstone Bridge

7. What was the original purpose for the Wheatstone bridge?
8. What is a galvanometer?
9. How is a balanced bridge indicated on the galvanometer?
10. What does it mean to null the meter of a Wheatstone bridge?
11. Explain how a positive voltage can look relatively negative compared to another positive voltage.
12. Describe a modern application for a Wheatstone bridge.

7-6 The Loaded Voltage Divider

13. What always happens when a load is connected to a voltage divider?
14. If the load resistance is very high compared to the voltage divider resistors, will the loading effect be (a) great or (b) small?
15. What does the term *load* actually refer to?

7-7 Troubleshooting the Loaded Voltage Divider

16. If the total current supplied to a circuit is much lower than normal, does this indicate a possible short exists or that a possible open exists?
17. The voltage across a short is always equal to the source voltage. (True or False?)
18. The voltage across an open is always greater than normal. (True or False?)
19. Can a voltmeter load down a circuit when making voltage measurements? Explain.
20. Which is better, a voltmeter sensitivity of 30 kΩ/V or a voltmeter sensitivity of 50 kΩ/V? Why?

Problems

7-1 Overall Series Circuits

1. Calculate total resistance between terminals A and B of Figure 7-67.

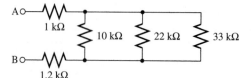

Figure 7-67

2. Calculate total resistance between terminals A and B of Figure 7-68.

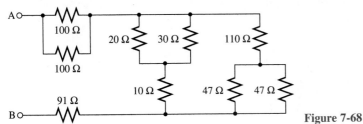

Figure 7-68

3. Calculate total resistance between terminals A and B of Figure 7-69.

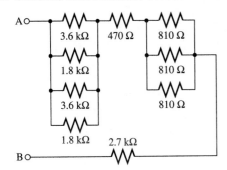

Figure 7-69

7-2 Voltage, Current, and Power in Overall Series Circuits

4. Calculate the listed unknowns for Figure 7-70.

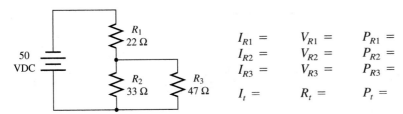

$I_{R1} =$ $V_{R1} =$ $P_{R1} =$

$I_{R2} =$ $V_{R2} =$ $P_{R2} =$

$I_{R3} =$ $V_{R3} =$ $P_{R3} =$

$I_t =$ $R_t =$ $P_t =$

Figure 7-70

5. Calculate the listed unknowns for Figure 7-71.

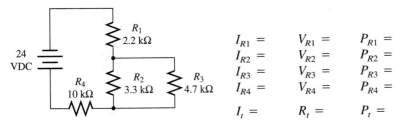

$I_{R1} =$ $V_{R1} =$ $P_{R1} =$

$I_{R2} =$ $V_{R2} =$ $P_{R2} =$

$I_{R3} =$ $V_{R3} =$ $P_{R3} =$

$I_{R4} =$ $V_{R4} =$ $P_{R4} =$

$I_t =$ $R_t =$ $P_t =$

Figure 7-71

6. Calculate the following unknowns for Figure 7-72. Remember to start with a component or section that you know two things about.

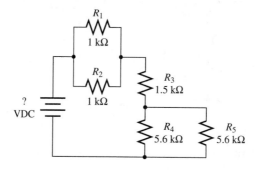

$I_{R1} =$ $V_{R1} =$ $P_{R1} =$
$I_{R2} =$ $V_{R2} =$ $P_{R2} =$
$I_{R3} =$ $V_{R3} = 7.5$ V $P_{R3} =$
$I_{R4} =$ $V_{R4} =$ $P_{R4} =$
$I_{R5} =$ $V_{R5} =$ $P_{R5} =$

$I_t =$ $E_s =$ $R_t =$ $P_t =$

Figure 7-72

7. Calculate all listed parameters for Figure 7-73.

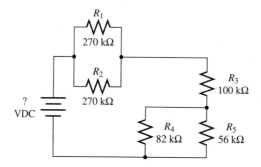

$I_{R1} =$ $V_{R1} =$ $P_{R1} =$
$I_{R2} =$ $V_{R2} =$ $P_{R2} =$
$I_{R3} =$ $V_{R3} =$ $P_{R3} =$
$I_{R4} =$ $V_{R4} =$ $P_{R4} =$
$I_{R5} =$ $V_{R5} = 6.82$ V $P_{R5} =$

$I_t =$ $E_s =$ $R_t =$ $P_t =$

Figure 7-73

7-3 Overall Parallel Circuits

8. Calculate total resistance between terminals A and B in Figure 7-74.

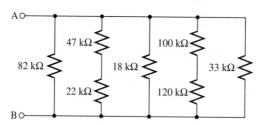

Figure 7-74

9. Calculate total resistance between terminals A and B in Figure 7-75.

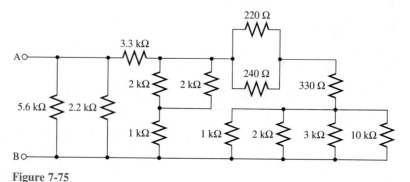

Figure 7-75

7-4 Voltage, Current, and Power in Overall Parallel Circuits

10. Solve for currents and voltage drops in Figure 7-76.

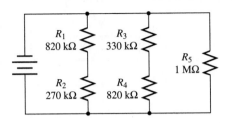

$I_{R1} =$ $V_{R1} =$
$I_{R2} =$ $V_{R2} =$
$I_{R3} =$ $V_{R3} =$
$I_{R4} =$ $V_{R4} =$
$I_{R5} = 20 \; \mu A$ $V_{R5} =$

$I_t =$ $E_s =$

Figure 7-76

11. Solve for currents and voltage drops in Figure 7-77.

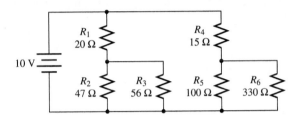

$I_{R1} =$ $V_{R1} =$
$I_{R2} =$ $V_{R2} =$
$I_{R3} =$ $V_{R3} =$
$I_{R4} =$ $V_{R4} =$
$I_{R5} =$ $V_{R5} =$
$I_{R6} =$ $V_{R6} =$

$I_t =$ $R_t =$

Figure 7-77

12. Solve for currents, voltage drops, and power in Figure 7-78.

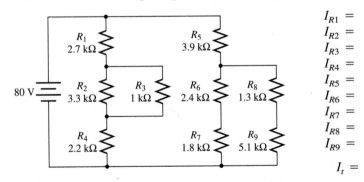

$I_{R1} =$ $V_{R1} =$
$I_{R2} =$ $V_{R2} =$
$I_{R3} =$ $V_{R3} =$
$I_{R4} =$ $V_{R4} =$
$I_{R5} =$ $V_{R5} =$
$I_{R6} =$ $V_{R6} =$
$I_{R7} =$ $V_{R7} =$
$I_{R8} =$ $V_{R8} =$
$I_{R9} =$ $V_{R9} =$

$I_t =$ $P_t =$

Figure 7-78

7-5 The Wheatstone Bridge

13. In a balanced Wheatstone bridge, what is the value of the unknown resistor (R_X) if $R_V = 3 \; k\Omega$, $R_1 = 10 \; k\Omega$, and $R_2 = 56 \; k\Omega$?
14. In a Wheatstone bridge, what value would R_X have to be in order for the bridge to be balanced if $R_1 = 47 \; k\Omega$, $R_2 = 33 \; k\Omega$, and $R_V = 15 \; k\Omega$?
15. If the value of R_X, in Problem 14, were 10 kΩ and if the source voltage were 20 V, what would the difference of potential be across the meter?
16. Describe the direction of current through the galvanometer in Problem 13.

7-6 The Loaded Voltage Divider

17. Calculate the parameters listed in the chart for Figure 7-79 with SW_1 open, then closed.

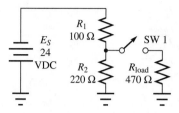

Figure 7-79

	SW1	
	OPEN	CLOSED
V_{R1}		
V_{R2}		
P_{R1}		
P_{R2}		

18. Calculate the parameters listed in the chart for Figure 7-80 with SW_1 in position A, then position B.

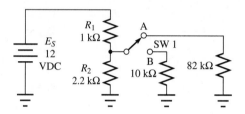

Figure 7-80

	SW1	
	A	B
V_{R1}		
V_{R2}		
P_{R1}		
P_{R2}		

19. Calculate the parameters listed in the chart for Figure 7-81.

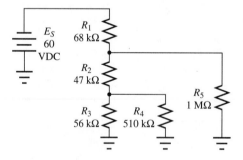

Figure 7-81

V_{R1}	
V_{R2}	
V_{R3}	
V_{R4}	
V_{R5}	
P_{R1}	
P_{R2}	
P_{R3}	

20. Calculate the parameters listed in the chart for Figure 7-82.

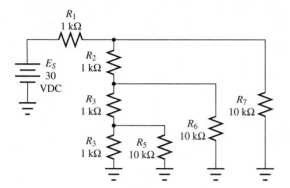

V_{R1}	
V_{R2}	
V_{R3}	
V_{R4}	
V_{R5}	
V_{R6}	
V_{R7}	

Figure 7-82

7-7 Troubleshooting the Loaded Voltage Divider

21. Identify the trouble in Figure 7-83.

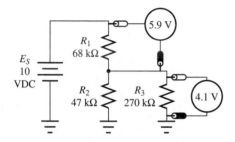

Figure 7-83

22. Identify the trouble in Figure 7-84.

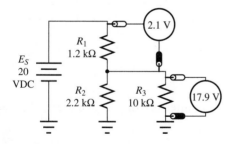

Figure 7-84

23. Identify the trouble in Figure 7-85.

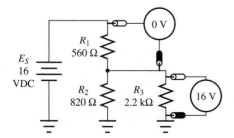

Figure 7-85

Answers to Self-Checks

Self-Check 7-1

1. $R_t = 78.7 \, \Omega$
2. $R_t = 7.03 \, k\Omega$
3. $R_t = 73 \, k\Omega$

Self-Check 7-2

1. $I_t = 0.731$ mA; $R_1 = 10 \, k\Omega$; $R_t = 16.4 \, k\Omega$; $I_{R1} = 0.731$ mA; $I_{R2} = 0.261$ mA; $I_{R3} = 0.47$ mA; $V_{R1} = 7.3$ V; $P_{R1} = 5.34$ mW; $P_{R2} = 1.23$ mW; $P_{R3} = 2.21$ mW; $P_t = 8.77$ mW
2. $I_t = 1.75$ mA; $E_s = 75$ V; $R_t = 42.8 \, k\Omega$; $I_{R1} = 1.75$ mA; $I_{R2} = 1.1$ mA; $I_{R3} = 0.65$ mA; $V_{R1} = 38.6$ V; $P_{R1} = 67.4$ mW; $P_{R2} = 40.2$ mW; $P_{R3} = 23.7$ mW; $P_t = 131$ mW

Self-Check 7-3

1. $R_t = 10.1 \, k\Omega$
2. $R_t = 248 \, \Omega$

Self-Check 7-4

1. $I_t = 76.8$ mA; $R_t = 1{,}302 \, \Omega$; $V_{R1} = 22.24$ V; $V_{R2} = 18.3$ V; $V_{R3} = 26.8$ V; $V_{R4} = 32.7$ V; $V_{R5} = 32.7$ V; $V_{R6} = 65.6$ V; $V_{R7} = 65.6$ V; $V_{R8} = 5.28$ V; $V_{R9} = 17.2$ V; $V_{R10} = 11.9$ V; $P_t = 7.68$ W; $P_{R1} = 727$ mW; $P_{R2} = 598$ mW; $P_{R3} = 876$ mW; $P_{R4} = 535$ mW; $P_{R5} = 535$ mW; $P_{R6} = 1.30$ W; $P_{R7} = 1.59$ W; $P_{R8} = 232$ mW; $P_{R9} = 759$ mW; $P_{R10} = 524$ mW
2. $I_t = 14.2$ mA; $R_t = 4{,}572 \, \Omega$; $V_{R1} = 28.78$ V; $V_{R2} = 23.54$ V; $V_{R3} = 7.22$ V; $V_{R4} = 5.48$ V; $V_{R5} = 5.48$ V; $V_{R6} = 12.7$ V; $V_{R7} = 4.51$ V; $V_{R8} = 8.25$ V; $V_{R9} = 21.45$ V; $V_{R10} = 30.8$ V; $P_t = 923$ mW; $P_{R1} = 251$ mW; $P_{R2} = 205$ mW; $P_{R3} = 43.4$ mW; $P_{R4} = 30$ mW; $P_{R5} = 3$ mW; $P_{R6} = 34.3$ mW; $P_{R7} = 24.8$ mW; $P_{R8} = 45.4$ mW; $P_{R9} = 118$ mW; $P_{R10} = 169$ mW

Self-Check 7-5

1. To measure unknown resistances
2. Make circuit adjustments that will reduce the difference of potential across the meter to zero volts.
3. The difference of potential is 2.5 volts, with the $+6$ V looking relatively negative.
4. 68.2 kΩ
5. 4.5 V; current would flow from R_V through the meter to R_1

Self-Check 7-6

1. *Loading effect* refers to the effect that a load has when it is connected to a voltage divider. The load will cause the voltage of the voltage divider to drop somewhat, depending on how heavy a load it is (how much current is drawn).
2. The 100 Ω resistor is the heavier load because it will draw the most current.
3. The term *load* actually refers to the amount of current a device draws.

4. An unloaded voltage divider is a simple series circuit and a loaded voltage divider is an overall series, or series-parallel, circuit.

5. $V_{R1} = 0.8$ V; $V_{R2} = 1.67$ V; $V_{R3} = 2.32$ V; $V_{R4} = 3.21$ V; $V_{R5} = 3.21$ V; $V_{R6} = 5.53$ V; $V_{R7} = 7.2$ V

6. $P_{R1} = 6.4$ mW; $P_{R2} = 12.7$ mW; $P_{R3} = 16.3$ mW; $P_{R4} = 22$ mW; $P_{R5} = 0.47$ mW; $P_{R6} = 3.1$ mW; $P_{R7} = 2.9$ mW

Self-Check 7-7

1. Increase
2. Zero volts
3. Open
4. Lower (zero volts)
5. 500 kΩ
6. The 500 kΩ meter would severely load down the voltage across the 680 kΩ resistor. The meter resistance of a voltmeter must be much greater than the resistance of the device whose voltage is being measured.

SUGGESTED PROJECTS

1. Add the formulas and important concepts from this chapter to your Electricity and Electronics Notebook.

2. Build a Wheatstone bridge. Use a potentiometer for one of the resistors (R_V). You probably don't have a galvanometer, but a digital multimeter will do just as well. The display of the multimeter will read ± some value when the bridge is not nulled. Experiment with scaling factors (R_1/R_2) while trying different resistor values for R_X. Use a 9 V transistor radio battery with all resistor values greater than 1 kΩ (for slow battery drain).

3. Build a few of the loaded-voltage-divider circuits shown in this chapter and verify your calculations with a multimeter.

EDISON EMBOSSING TRANSLATING
TELEGRAPH

Edison's Embossing Translating Telegraph was said by Edison to be the father of the phonograph. A pattern perforated on the rotating disk was traced by a needle and transmitted by wire to another such machine where the pattern was reproduced on another rotating disk. Simultaneous two-way transmission was possible. It appears that this could be considered a forerunner of the fax machine.

Chapter 8

Circuit Analysis Tools (Theorems and Procedures)

OBJECTIVES

After studying this chapter, you will understand and be able to use:

• the Superposition theorem

• Thevenin's theorem

• Norton's theorem

• Millman's theorem

• the maximum power transfer theorem

• nodal analysis

• mesh analysis

• delta and wye conversions

INTRODUCTION

Have you ever had a job to do or a task to perform and found that you were unable to accomplish it simply because you did not have the correct tool or tools? I am sure some of you can relate to this kind of situation. There are times when a screwdriver and pliers are simply not adequate for the task at hand. There is no getting around the basic fact that without the proper tools some jobs are just about impossible.

And so it is in electronics. Naturally, specialized tools are needed for the assembly or construction of electrical and electronics projects. These are hardware tools that you will accumulate and learn to use as time goes on. But there are other tools that you will use with just as much skill. These are analysis tools that you will use to analyze circuits of greater complexity than those that have been covered thus far. As with hardware tools, there are special circuit-analysis tools that are needed for special analysis tasks. Ohm's Law, by itself, is not always adequate to thoroughly analyze a circuit.

In this chapter we will begin to place a variety of special analysis tools in your circuit-analysis tool kit. You will find that each tool has a special approach in solving complex circuit problems. You will need to study them carefully to understand their differences and applications.

NEED TO KNOW

This Need to Know will introduce you to the type of circuit we will discuss in this chapter. Throughout this chapter, you will be introduced to a variety of analytical tools needed to solve for currents and voltage drops in circuits such as this:

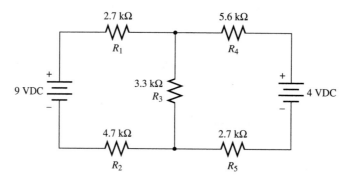

After careful examination of the circuit shown above, can you answer the following questions?

1. How much current is flowing in each resistor?
2. What is the direction of current in each resistor?
3. What are the resistor voltage drops?
4. What is the polarity of each voltage drop?
5. Are both voltage sources supplying current, or is one source receiving current?

At this point, most people would not be able to answer the above questions. Perhaps you are one of those people. No big deal! This merely establishes your need to know. You will soon find yourself busily solving this type of circuit as you complete this chapter.

8-1 The Superposition Theorem

The first handy tool in your circuit-analysis tool kit is called the **superposition theorem**. The idea behind the superposition theorem is that each power source in a circuit containing multiple sources creates its own set of currents and voltage drops, which are superimposed on others in the same circuit. The overall effect of the separate voltage sources is simply the algebraic sum (the sum of signed quantities) of the separate currents and voltage drops that each source has produced. The superposition theorem can be stated as follows:

Superposition Theorem

The actual currents and voltage drops in a circuit containing multiple sources are the algebraic sums of currents and voltage drops produced by each individual source.

The superposition theorem is often referred to as a **network theorem**. That is, a theory that is applied to a network. In this case, we are talking about an electrical or electronic network. It is an arrangement of components, conductors, and electrical sources. A network is not a new or special type of circuit. In fact, any circuit that we have discussed thus far may be referred to as an electrical network.

Using Superposition

What do Ohm's Law and superposition have in common? The superposition theorem simply establishes a procedure for using Ohm's Law to solve for currents and voltage drops in multiple-source circuits. By following this procedure, you will be able to use common series and parallel circuit-analysis techniques that employ Ohm's Law. Super-position is a very simple and straightforward procedure that centers on three basic steps:

Step #1

Analyze the circuit using only one source voltage at a time. All other source voltages are removed and replaced with their internal resistances. Usually the internal resistance of a source is low enough compared to circuit resistance that the source may be temporarily replaced with a short. Use Ohm's Law and series and parallel circuit-analysis techniques to solve for currents and voltage drops in all branches. Label all current directions and magnitudes and voltage polarities in the circuit.

Step #2

Repeat step #1 for each voltage source. Keep track of the currents and voltage drops for each source. Colored pencils are helpful here. You may want to use a different color pencil for each voltage source or you may prefer to redraw the circuit for each voltage source.

Step #3

Take the algebraic sum of all currents in each branch. Use the resultant branch currents to find voltage drops or simply add voltage drops, taking polarity into consideration. Place all voltage sources back in the circuit.

Now let's apply the superposition theorem in Example 8-1. Study it carefully as you apply the steps of superposition.

EXAMPLE   8-1

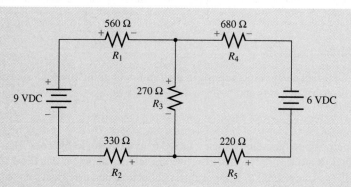

Figure 8-1

STEPS #1 & 2: Analyze using one voltage source at a time. Replace the unused
source with a short.

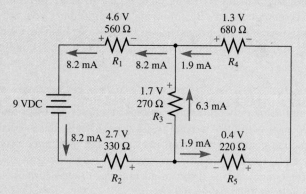

Figure 8-2

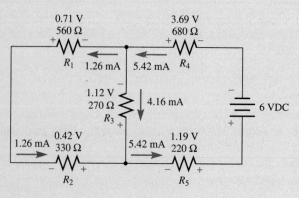

Figure 8-3

STEP #3: Take the algebraic sum of all branch currents and calculate voltage drops using Ohm's Law or simply add voltage drops created by each source.

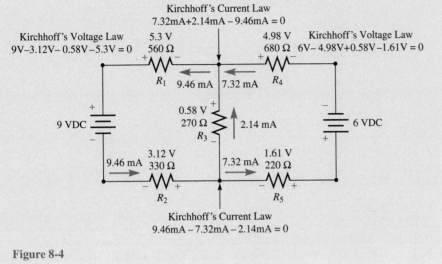

Figure 8-4

Now consider Example 8-2. It's not as complicated as it might seem.

EXAMPLE 8-2

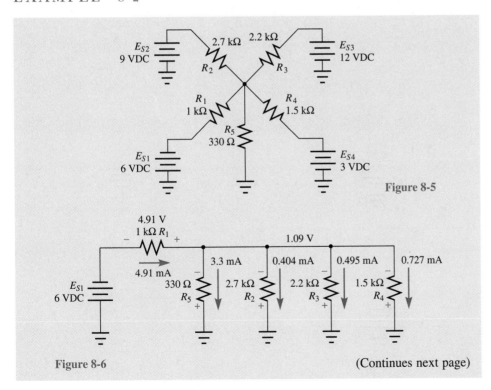

Figure 8-5

Figure 8-6   (Continues next page)

EXAMPLE 8-2 (Continued)

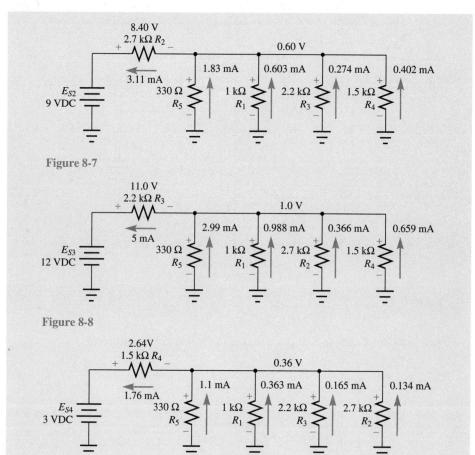

Figure 8-7

Figure 8-8

Figure 8-9

Now take the algebraic sum of all currents or voltage drops for each resistor. The result is shown in Figure 8-10.

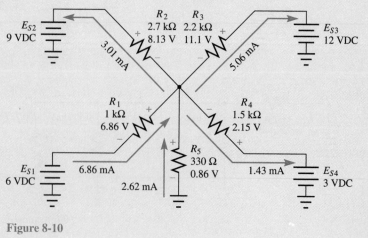

Figure 8-10

$V_{R1} = 6.86$ V	You can use Kirchhoff's Voltage Law
$V_{R2} = 8.13$ V	to demonstrate that these voltage
$V_{R3} = 11.13$ V	drops are correct. The algebraic sum
$V_{R4} = 2.15$ V	of all voltage drops in any loop must
$V_{R5} = 0.86$ V	equal the source voltage for that loop.

S E L F - C H E C K   8 - 1

1. Use the superposition theorem to solve for current magnitudes and directions, along with voltage drops and polarities, in Figure 8-11. The solution is shown at the end of the chapter.

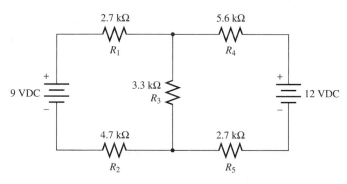

Figure 8-11

8-2 Thevenin's Theorem

One of the most useful tools for your circuit-analysis tool kit is **Thevenin's theorem**. This theorem is a tool that allows you to reduce any simple or complex circuit to a single voltage source and a single series resistor to which a load may be connected. This is very useful, since the effects of connecting different loads to the complex circuit can be quickly and easily evaluated. As an example, suppose you have designed a loaded voltage divider and you want to determine the output voltage that would be supplied across load resistances of various values. In previous chapters, you learned how to analyze loaded voltage dividers. The process involved many mathematical steps that finally led to solving for the voltage drop across the load. If the load is changed, all of the calculations have to be repeated to re-solve for the voltage drop across the new load. Needless to say, this can be very time-consuming. Thevenin's theorem saves you much time and effort for this and other circuit-analysis tasks. We begin our discussion of Thevenin's theorem by first examining a definition or statement of the theorem. Then, we will investigate the Thevenized circuit as a practical voltage source. Finally, we will use Thevenin's theorem to solve loaded output voltages of simple and more complex circuits.

A Bit of History

Léon Charles Thevenin was born in Meaux, France. He graduated from the Ecole Polytechnique in 1876. In 1883, Thevenin published his Equivalent Generator Theorem, which today is known as Thevenin's theorem. However, many historians give credit to Hermann von Helmholtz, since he published a version of the theorem in 1853. Helmholtz's version was related to animal electricity and Thevenin's version to telecommunication systems.

Understanding Thevenin's Theorem

Thevenin's Theorem

Any linear circuit, consisting of resistances and one or more sources of voltage and having two output terminals, can be replaced by a single voltage source and a single series resistance for the purpose of circuit analysis.

Even though Thevenin's theorem is a new concept to you, it does not have to be a difficult one. The statement of the theorem given above starts with the phrase, "Any linear circuit." The word **linear** simply means that in order for Thevenin's theorem to apply to a particular circuit, the circuit must consist of components whose current will either increase or decrease in direct proportion to an increase or decrease in voltage. Naturally, resistors fall into this category, along with other components such as capacitors and inductors. (You will learn much about capacitors and inductors in future chapters.)

The output terminals, mentioned in the statement of the theorem, are simply two circuit terminals to which you might connect a load or various loads, one at a time. Then the entire circuit, connected to the output terminals and load, can be considered as a **practical voltage source**, which includes a series resistance. The entire circuit, up to the output terminals, is replaced with a single voltage source, called the **Thevenin voltage** and designated as V_{Thevenin} or V_{Th}, and a single series resistance, called the **Thevenin resistance** and designated as R_{Thevenin} or R_{Th}. Figure 8-12 illustrates Thevenin's theorem.

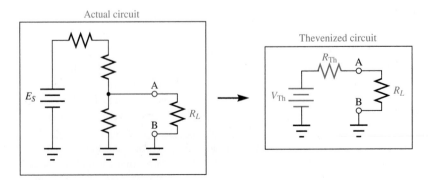

Figure 8-12   Thevenin's theorem.

Finding V_{Th} and R_{Th}

Example 8-3 demonstrates how to find V_{Th} and R_{Th} from a typical voltage divider circuit. The process of converting a circuit to its Thevenin equivalent is called **Thevenizing**. There are only three basic steps needed to Thevenize a circuit:

Thevenizing

1. Find V_{Th} by removing the load from the output terminals and solve for the unloaded output voltage, using Ohm's Law and/or some circuit-analysis technique.

2. Find R_{Th} by replacing the voltage source, or sources, with a resistance representing the internal resistance of each voltage source (usually a short), then find the total resistance of the circuit from the open terminals.

3. Draw the Thevenized circuit with the load connected. Calculate the loaded output voltage using Ohm's Law or the voltage divider proportion formula.

E X A M P L E   8 - 3

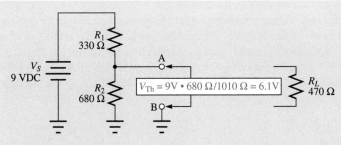

Figure 8-13

STEP #1: Remove the load and calculate the terminal voltage designated as V_{Th}.

$V_{Th} = 9V \cdot 680\,\Omega/1010\,\Omega = 6.1\,V$

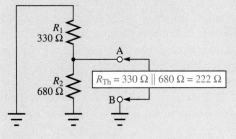

Figure 8-14

STEP #2: Replace the source voltage with a short and find the total circuit resistance, R_{Th}, from the output terminals.

$R_{Th} = 330\,\Omega \parallel 680\,\Omega = 222\,\Omega$

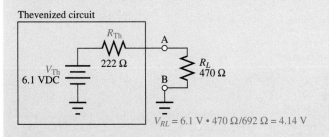

Figure 8-15

STEP #3: Draw the Thevenized circuit with the load connected. Calculate the output voltage.

$V_{RL} = 6.1\,V \cdot 470\,\Omega/692\,\Omega = 4.14\,V$

performed. The most peculiar part of the entire procedure is the method used to determine R_{Th}. Notice that the source voltage is removed and replaced with a wire (short). The total circuit resistance, designated R_{Th}, is then found from the output terminals. Normally, you would find total resistance from the input of the circuit, where the source voltage is connected. In this case, you want to see how the entire circuit looks from the load. This will help you see how current and voltage, supplied to a load resistance by any circuit, is supplied through an equivalent series resistance. Thus, any circuit, be it simple or complex, can be reduced to what amounts to a practical voltage source containing an internal series resistance.

The Thevenized Circuit as a Practical Voltage Source

In the final analysis, the actual voltage supplied to any load will be $V_{Th} - V_{RTh}$. In other words, R_{Th} of the circuit drops some of the voltage depending upon the size of the load (the amount of load current). R_{Th} is the internal resistance of the entire circuit that is supplying voltage to a load through a set of terminals. Thus, the entire circuit is acting like a practical voltage source that has internal resistance, which in fact drops some of the voltage across itself. A practical voltage source, therefore, can be an entire circuit that is supplying a particular load or it may simply be a real-world DC source such as a battery or an AC to DC converter.

The phrase real-world is used here because any real DC source does have some internal resistance, similar to R_{Th}. A flashlight battery may have approximately 1 Ω of internal resistance. This internal resistance is mainly due to the inefficiency of the chemical production of electrons. The internal resistance of the battery increases as the battery discharges. Thus, the voltage available at its output terminals decreases due to the increased apparent internal voltage drop.

Can Source Resistance Be Ignored?

In spite of the real-world fact that voltage sources have internal resistance, we do not always concern ourselves with it. In schematics, we show the voltage source as being merely a voltage. There is usually no indication that the voltage source has an internal resistance. The reason for this is a practical one. Most of the time, the internal resistance of the voltage source is very small compared to the external circuit, so its effect is insignificant and can be ignored. Thus, many practical voltage sources can be treated as though they were **ideal voltage sources** (ones that have no internal resistance). Shown in Figure 8-16 is a voltage source that has no internal resistance at all. It is a voltage source that can

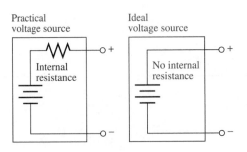

Figure 8-16   Practical and ideal voltage sources.

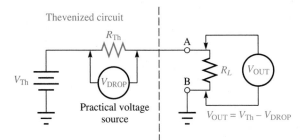

Figure 8-17   The Thevenized circuit as a practical voltage source.

supply a fixed amount of voltage to any load with no internal voltage drop (since there is no internal resistance to drop voltage across).

The practical DC source normally has a relatively low internal resistance, which can be ignored. It is for this reason that we are able to simply replace the voltage source or sources with a short when a circuit is Thevenized to find R_{Th}. This is why we seemed to ignore the part of the Thevenizing procedure that says to replace all voltage sources with resistors that represent their internal resistances. In practice, it is simply not always necessary to consider this internal resistance.

As shown in Figure 8-17, the resistance of a circuit that has been Thevenized must be considered. In most cases R_{Th} will drop a significant amount of voltage. This voltage drop must be considered when determining the output voltage supplied to the load. And in fact, that is the whole purpose for Thevenizing a circuit: to transform the complex circuit into a practical voltage source with a significant internal resistance R_{Th}. In this way, the complex circuit is reduced to a voltage V_{Th}, a series dropping resistor R_{Th}, and a load R_L.

Using Thevenin's Theorem

Thevenizing the Loaded Voltage Divider

Example 8-4 illustrates a typical loaded voltage divider. Notice the three steps involved in the Thevenizing process: determine V_{Th}, determine R_{Th}, and draw the Thevenized circuit with load connected. After Thevenizing, you can quickly determine what effect a change in load resistance has on output voltage. Notice also that in step #2, R_3 seems to disappear. R_3 was deliberately placed in the circuit to show that any components that are directly in parallel with a source voltage are eliminated when the source voltage is replaced with a short. In effect, R_3 is eliminated because it is totally bypassed with a short.

EXAMPLE   8-4

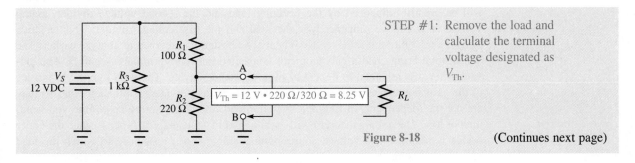

STEP #1: Remove the load and calculate the terminal voltage designated as V_{Th}.

$V_{Th} = 12 \text{ V} \cdot 220\ \Omega / 320\ \Omega = 8.25 \text{ V}$

Figure 8-18

(Continues next page)

EXAMPLE   8-4 (Continued)

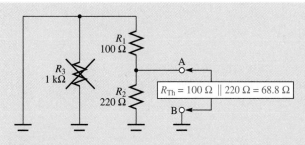

STEP #2: Replace the source voltage with a short and find the total circuit resistance, R_{Th}, from the output terminals.

$R_{Th} = 100\ \Omega\ \|\ 220\ \Omega = 68.8\ \Omega$

Figure 8-19

Thevenized circuit

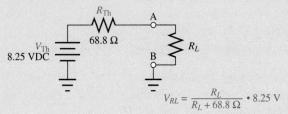

STEP #3: Draw the Thevenized circuit with the load connected. Calculate the output voltage.

$$V_{RL} = \frac{R_L}{R_L + 68.8\ \Omega} \cdot 8.25\ V$$

Figure 8-20

Now the output voltage can quickly be determined for any value of load resistance.

R_L	Calculations	V_{RL}
10 kΩ	(10,000/10,068.8) · 8.25	= 8.19 V
1 kΩ	(1,000/1,068.8) · 8.25	= 7.72 V
100 Ω	(100/168.8) · 8.25	= 4.89 V
10 Ω	(10/78.8) · 8.25	= 1.05 V

Thevenizing the Resistive Bridge

Example 8-5 shows how a bridge circuit can be solved by Thevenizing. It is important for you to understand that a simple circuit analysis using Ohm's Law alone is not possible in a circuit of this type. The load resistance, connected between terminals A and B, will cause the two voltage dividers to be interdependent. In other words, the voltage from A to ground will be greatly influenced by the bridging load and the second voltage divider, as the voltage at B will be influenced by the load and the first voltage divider. This resistive bridge is not quite the same as the Wheatstone bridge you learned to analyze earlier. The Wheatstone bridge had a galvanometer connected between terminals A and B. The galvanometer was assumed to be a very high-resistance device. The high internal resistance of the galvanometer effectively isolates the two voltage dividers from each other so they are not interdependent. In Example 8-5, the load resistance can be any value you want, high or low. Thevenin's theorem will greatly simplify the analysis process since the entire bridge is reduced to a voltage source with a single series resistance to which the load resistance is connected.

EXAMPLE 8-5

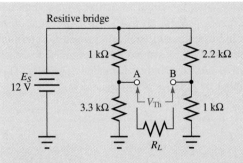

Resitive bridge

E_S 12 V

1 kΩ 2.2 kΩ

A B

3.3 kΩ 1 kΩ

V_{Th}

R_L

STEP #1: Remove the load and calculate the terminal voltage (V_{Th}).

V_{Th} is the difference of potential between terminals A and B in Figure 8-21.

Figure 8-21

Find the voltage at A and the voltage at B in Figure 8-21.

$$V_A = (3.3 \text{ k}\Omega/4.3 \text{ k}\Omega) \cdot 12 \text{ V} = +9.21 \text{ V}$$
$$V_B = (1 \text{ k}\Omega/3.2 \text{ k}\Omega) \cdot 12 \text{ V} = +3.75 \text{ V}$$
$$V_{Th} = +9.21 \text{ V} - 3.75 \text{ V} = \underline{5.46 \text{ V}}$$

STEP #2: Replace the voltage source with a short and find R_{Th} for each terminal as shown in Figure 8-22.

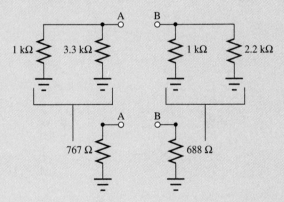

A B

1 kΩ 3.3 kΩ 1 kΩ 2.2 kΩ

A B

767 Ω 688 Ω

Figure 8-22

STEP #3: Draw the Thevenized circuit and connect the load. Calculate the output voltage as shown in Figure 8-23.

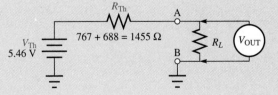

R_{Th}

A

V_{Th} 5.46 V

767 + 688 = 1455 Ω

R_L V_{OUT}

B

Figure 8-23

Now the output voltage across terminals A and B can quickly be determined for any load!

For example: If $R_L = 1$ kΩ then the output voltage is as follows:

$$V_{out} = (1 \text{ k}\Omega/2{,}455 \text{ } \Omega) \cdot 5.46 \text{ V} = 2.22 \text{ V}$$

Thevenizing Multiple-Voltage-Source Circuits

Example 8-6 shows how to Thevenize a more complex circuit containing more than one voltage source. As you can see, the Thevenizing process always follows the same three basic steps regardless of the type of circuit. In this case, superposition is used to determine V_{Th}. This requires that the voltage from terminal A to ground be determined for each voltage source. Then, all terminal A voltages are superimposed (added together). This procedure is followed regardless of the number of separate voltage sources.

EXAMPLE 8-6

STEP #1: Remove the load and calculate the terminal voltage (V_{Th}).

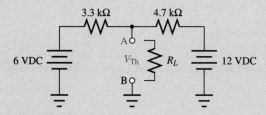

Figure 8-24

Use the superposition theorem to find the voltage from terminal A to ground as shown in Figure 8-25.

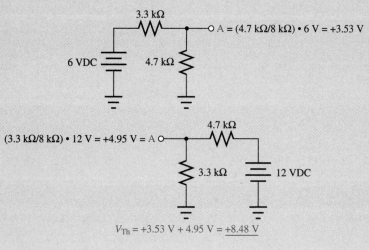

$$A = (4.7\ k\Omega/8\ k\Omega) \cdot 6\ V = +3.53\ V$$

$$(3.3\ k\Omega/8\ k\Omega) \cdot 12\ V = +4.95\ V = A$$

$$V_{Th} = +3.53\ V + 4.95\ V = \underline{+8.48\ V}$$

Figure 8-25

STEP #2: Replace the voltage sources with shorts and find R_{Th} as shown in Figure 8-26.

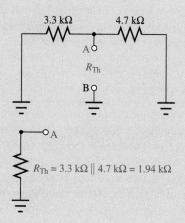

Figure 8-26

STEP #3: Draw the Thevenized circuit and connect the load. Calculate the output voltage as shown in Figure 8-27.

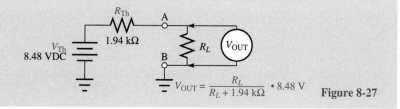

$$V_{OUT} = \frac{R_L}{R_L + 1.94 \text{ k}\Omega} \cdot 8.48 \text{ V}$$

Figure 8-27

SELF-CHECK 8-2

Thevenize the following circuits. Show all work and solve for the output voltage across the load.

1.

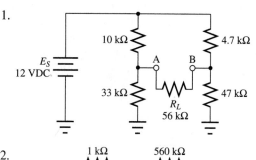

Figure 8-28

2.

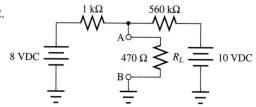

Figure 8-29

Once the circuit has been completely Thevenized, the terminal voltage for any load resistance can be quickly determined. As you can see, the tremendous advantage of using Thevenin's theorem is that you do not have to repeat the superposition process over and over again every time the load resistance is changed. The complex multiple-voltage-source circuit is reduced to a very simple single-source series circuit. I am sure you agree that Thevenin's theorem is a powerful circuit-analysis tool. Learn to use it well.

8-3 Norton's Theorem

Norton's theorem is another unique circuit-analysis tool that can be very useful in analyzing somewhat complex circuits. In this section, you will see that Norton's theorem is actually a direct complement to Thevenin's theorem. It was formulated by Edward L. Norton of AT&T in 1926. Many circuits can be analyzed using either or both of these theorems. As with Thevenin's theorem, Norton's theorem is used to reduce a complex circuit to a much simpler equivalent form. Then the effects of connecting different loads can quickly be analyzed. We will investigate the theorem and associated concepts, then discuss several practical applications.

Understanding Norton's Theorem

Norton's Theorem

Any linear circuit, consisting of resistances and one or more voltage sources and having two output terminals, can be replaced with a single constant-current source I_N and a single parallel resistance R_N for the purpose of circuit analysis.

Notice the similarity and differences between Norton's theorem and Thevenin's theorem. The similarity lies in the manner in which the theorems are stated. The important difference between the two lies in the fact that Thevenin's theorem deals with a voltage source and a series resistor, while Norton's theorem deals with a constant-current source and a parallel resistor. So, what is a constant-current source? A **constant-current source** is an electrical device, or circuit, that can supply a constant amount of current regardless of the load resistance. It is symbolized with a circle containing an arrow that indicates the direction of current. Figure 8-30 illustrates the constant-current source combined with a parallel resistance, forming a Nortonized circuit. The Nortonized circuit represents the more complex circuit and simplifies circuit analysis for different loads. As with Thevenin's

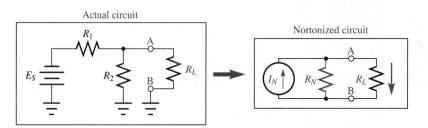

Figure 8-30   Norton's theorem.

theorem, the complex circuit is simplified to an electrical source and a resistance to which a load is connected. The current from the constant-current source I_N is divided between the parallel resistor R_N and the load R_L, just as voltage is divided between a series resistor R_{Th} and the load in a Thevenized circuit. An **ideal current source** would be a constant-current source with an infinite parallel resistance R_N, so all current would go to the load resistance R_L regardless of its size. An ideal current source is actually a constant-current source.

Finding I_N and R_N

Example 8-7 illustrates how to convert a voltage divider circuit to its Norton equivalent for the purpose of simplifying further circuit analysis. As you might guess, this process is referred to as **Nortonizing**. And, as in Thevenizing, there are three basic steps.

Nortonizing

1. Find I_N by shorting the output terminals and calculating the amount of current that will flow through the short.

2. Find R_N by replacing the voltage source or sources with a resistance representing the internal resistance of each voltage source, then find the total resistance of the circuit from the open terminals.

3. Draw the Nortonized circuit with the load connected. The direction of current, indicated by the constant-current source, should be consistent with the polarity of the original voltage source. Calculate the amount of current supplied to the load resistance using the current division formula.

EXAMPLE 8-7

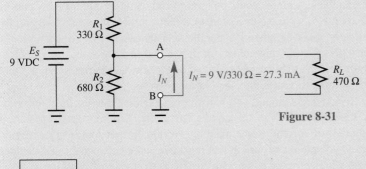

STEP #1: Find I_N by shorting the output terminals and calculating the amount of current that will flow through the short as shown in Figure 8-31.

Figure 8-31

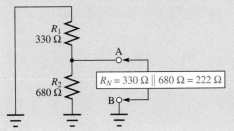

STEP #2: Replace the source voltage with a short and find the total circuit resistance, R_N, from the output terminals as shown in Figure 8-32.

Figure 8-32

(Continues next page)

EXAMPLE   8-7 (Continued)

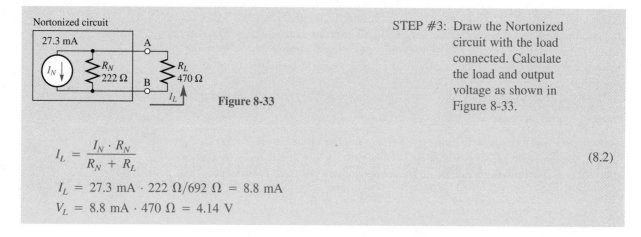

STEP #3: Draw the Nortonized circuit with the load connected. Calculate the load and output voltage as shown in Figure 8-33.

Figure 8-33

$$I_L = \frac{I_N \cdot R_N}{R_N + R_L} \tag{8.2}$$

$I_L = 27.3 \text{ mA} \cdot 222 \ \Omega/692 \ \Omega = 8.8 \text{ mA}$

$V_L = 8.8 \text{ mA} \cdot 470 \ \Omega = 4.14 \text{ V}$

You have probably noticed that the procedure for finding R_N is the same as that used to find R_{Th}—the voltage source is replaced with a short and total resistance is calculated from the unloaded output terminals. Again, when we replace a voltage source with a short, we are assuming the voltage source to have a very low internal resistance as compared to the actual circuit. Thus, for all practical purposes, the voltage source is considered to be ideal.

The Nortonized Circuit as a Practical Current Source

When a load resistance is connected to the Nortonized circuit model, does all of the Norton current (I_N) flow through the load? No, it does not. Why not? Because the Nortonized circuit is not a perfect, or ideal, constant-current source. The Nortonized circuit has a parallel resistance that demands a share of the current provided by the constant-current source. Ideally, this parallel resistance would be infinite and therefore not take any current at all. Hence, an ideal current source is a constant-current source with an infinite parallel resistance (∞). Practically speaking, there is a parallel resistance and it is quite a bit lower than infinite since it represents the original circuit. It is a real resistance that, in fact, does demand a certain amount of the total current supplied by the constant-current source. Thus, the actual load is the Norton current minus the Norton resistor current: $I_L = I_N - I_{RN}$.

Once you have Nortonized a circuit, you will want to examine the results obtained by connecting various load resistances to the circuit. Naturally, different load resistances demand different amounts of current. Remember, the Norton current source is a constant-current source. It is a fixed amount of current that must be shared between R_N and the load resistance. So, how do you determine how much current goes to each? Recall Formula 6.7 from your study of current dividers:

$$I_B = I_t \cdot R_t/R_B \tag{6.7}$$

Applying this formula to Norton's theorem, the variables become:

$$I_L = \frac{I_N \cdot (R_N \parallel R_L)}{R_L} \tag{8.1}$$

As you can see, Formula 8.1 requires that you first calculate the total resistance of R_L in parallel with R_N. We may simplify this formula somewhat as follows:

$$(R_N \parallel R_L)/R_L = \frac{(R_N \cdot R_L)/(R_N + R_L)}{R_L} = \frac{R_N \cdot R_L}{(R_N + R_L) \cdot R_L} = \frac{R_N}{(R_N + R_L)}$$

Now, by substitution, Formula 8.1 becomes:

$$I_L = I_N \cdot \frac{R_N}{R_N + R_L} = \frac{I_N \cdot R_N}{R_N + R_L} \tag{8.2}$$

Formula 8.2 allows us to determine, very quickly, the amount of current that a particular load resistance will receive. Example 8-8 makes use of Formula 8.2 in a practical Nortonized circuit with many different load resistances. As you can see, once a complex circuit has been Nortonized, it is a fairly simple task to analyze the results of connecting various loads to its output terminals. As with Thevenizing, Nortonizing is a powerful tool that can eliminate much repetitive work.

EXAMPLE 8-8

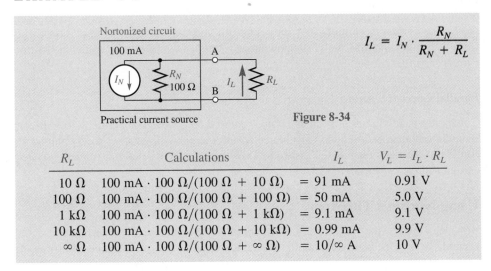

$$I_L = I_N \cdot \frac{R_N}{R_N + R_L}$$

Figure 8-34

R_L	Calculations	I_L	$V_L = I_L \cdot R_L$
10 Ω	100 mA · 100 Ω/(100 Ω + 10 Ω)	= 91 mA	0.91 V
100 Ω	100 mA · 100 Ω/(100 Ω + 100 Ω)	= 50 mA	5.0 V
1 kΩ	100 mA · 100 Ω/(100 Ω + 1 kΩ)	= 9.1 mA	9.1 V
10 kΩ	100 mA · 100 Ω/(100 Ω + 10 kΩ)	= 0.99 mA	9.9 V
∞ Ω	100 mA · 100 Ω/(100 Ω + ∞ Ω)	= 10/∞ A	10 V

Notice also in Example 8-8 that the load voltage, or loaded terminal voltage, increases as the load resistance is made larger. This is exactly what you would expect from a resistive circuit containing a series dropping resistor, such as a Thevenized circuit. It is also interesting to note that the highest possible terminal voltage is 10 volts, which of course equals the 100 mA Norton current times the 100 Ω Norton resistance. You may recall that the highest output voltage from a Thevenized circuit is the Thevenin voltage itself. In this case, the highest output voltage is determined by the Norton current. The Thevenized model would have a 10 V source and a 100 Ω series resistor, while the Norton model has a 100 mA current source and a 100 Ω parallel resistor. Whichever model is used, the load voltage and current for any load resistance will be the same.

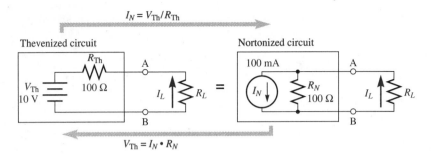

Figure 8-35   The Thevenin-Norton interchange.

The Thevenin/Norton Interchange

It is important to realize that the Norton and Thevenin circuit models are interchangeable. There are cases when it is necessary, or much simpler, to convert from one model to the other in the solution of a complex problem. Later, we will examine one or two such cases. For now, we will discuss the methods by which you will be able to quickly convert from one model to the other. Figure 8-35 illustrates the two models and shows the conversion from one to the other. As you can see, the same resistor is used for each model. (You may recall that R_N and R_{Th} are found in the same manner.) To convert from Norton to Thevenin, the Norton current is multiplied by the resistor to find the Thevenin voltage. To convert from Thevenin to Norton, the Thevenin voltage is divided by the resistor to find the Norton current. It is that simple!

Parallel Current Sources

Combining parallel current sources is much more direct. Consider Figure 8-36. All constant-current sources are added algebraically and the parallel resistors are combined according to the rules for parallel resistors. This immediately results in a single-equivalent current source, or Norton model.

Using Norton's Theorem

Now, let's apply what we have discussed about Norton's theorem and current sources in general. To do this, we shall examine the same examples that were used to illustrate the

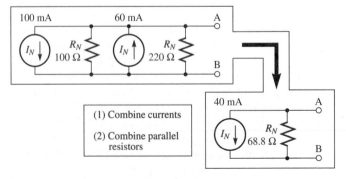

Figure 8-36   Combining parallel current sources.

use of Thevenin's theorem. This will allow you to see that Norton's theorem is indeed interchangeable with Thevenin's theorem.

EXAMPLE 8-9

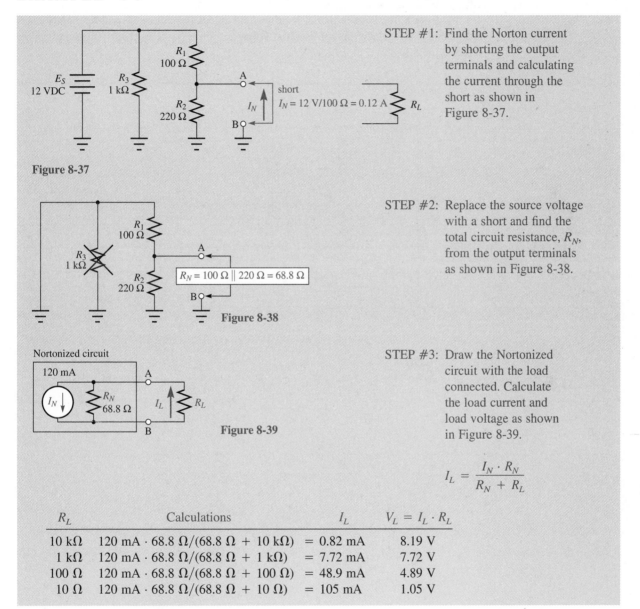

STEP #1: Find the Norton current by shorting the output terminals and calculating the current through the short as shown in Figure 8-37.

Figure 8-37

STEP #2: Replace the source voltage with a short and find the total circuit resistance, R_N, from the output terminals as shown in Figure 8-38.

$R_N = 100\ \Omega\ \|\ 220\ \Omega = 68.8\ \Omega$

Figure 8-38

Nortonized circuit

STEP #3: Draw the Nortonized circuit with the load connected. Calculate the load current and load voltage as shown in Figure 8-39.

Figure 8-39

$$I_L = \frac{I_N \cdot R_N}{R_N + R_L}$$

R_L	Calculations	I_L	$V_L = I_L \cdot R_L$
10 kΩ	120 mA · 68.8 Ω/(68.8 Ω + 10 kΩ) = 0.82 mA		8.19 V
1 kΩ	120 mA · 68.8 Ω/(68.8 Ω + 1 kΩ) = 7.72 mA		7.72 V
100 Ω	120 mA · 68.8 Ω/(68.8 Ω + 100 Ω) = 48.9 mA		4.89 V
10 Ω	120 mA · 68.8 Ω/(68.8 Ω + 10 Ω) = 105 mA		1.05 V

Nortonizing the Loaded Voltage Divider

Example 8-9 illustrates the Norton solution to the loaded voltage divider originally shown in Example 8-4. It is important for you to notice, in step #1, that the Norton current is determined by R_1 alone, since R_1 is the only resistor in series with the short that has replaced the source voltage. The current through R_3 has nothing to do with output terminals A and B. Also, R_2 is not considered, since it is eliminated by the short across the output terminals.

In step #2, R_N is found the same way that R_{Th} was found earlier. Notice that R_3 is still of no concern since it is eliminated by the shorted voltage source.

In step #3, the Norton model is drawn and calculations for different load resistances are made. The resulting output terminal voltages are the same as those calculated earlier in Example 8-4.

EXAMPLE   8-10

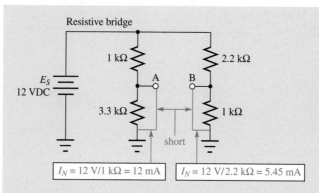

STEP #1: Short each terminal to ground and calculate Norton current for each branch as shown in Figure 8-40.

$I_N = 12\ \text{V}/1\ \text{k}\Omega = 12\ \text{mA}$ $I_N = 12\ \text{V}/2.2\ \text{k}\Omega = 5.45\ \text{mA}$

Figure 8-40

STEP #2: Replace the voltage source with a short and find R_N for each terminal as shown in Figure 8-41.

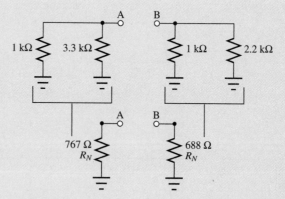

Figure 8-41

STEP #3: Draw the two current sources for the circuit as shown in Figure 8-42.

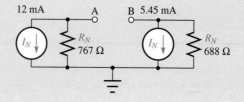

Figure 8-42

STEP #4: Convert current sources to Thevenin equivalents then combine voltage
sources and resistances as shown in Figure 8-43.

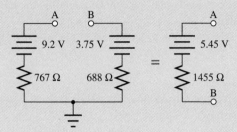

Figure 8-43

*Convert back to the Norton model if desired, as shown in Figure 8-44.

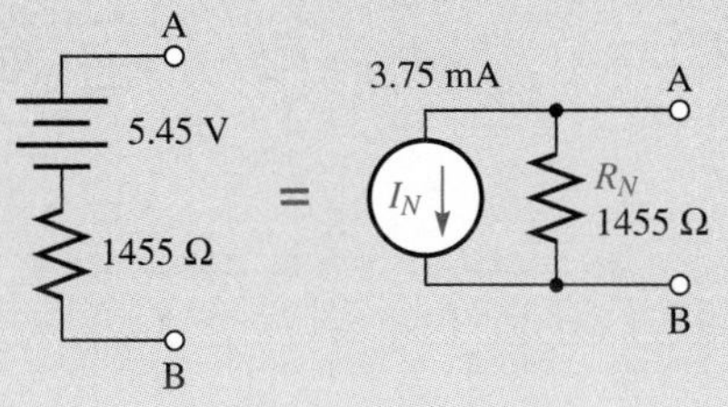

Figure 8-44

Nortonizing the Resistive Bridge

Example 8-10 uses the same circuit as that of Example 8-5. In this case, Norton's theorem
is used along with Thevenin's theorem to acquire a final solution to the problem. This is
a good example, since it illustrates many of the things we have discussed, such as Norton
to Thevenin conversion and series current sources.

Recall from Example 8-5 that the difference of potential between terminals A and B
was found by subtracting the voltage across B to ground from the voltage across A to
ground. The voltages at A and B are two separate voltages from what are effectively two
separate circuits (except for the common source voltage) that are caused to interact through
a common load resistance connected between A and B. Thus, just as it was necessary to
Thevenize each terminal separately, it is necessary to Nortonize each terminal separately.
This results in two Norton models, or current sources, connected in series through the
common ground.

The series current sources must be converted to voltage sources to easily combine
them into one voltage source. Notice that the two voltage sources are opposite in polarity,
as were the current sources. The resulting Thevenin voltage is the difference between the
two opposing voltage sources. The Thevenin resistance is the sum of the two series resistors. This final Thevenin model could be converted back to the Norton model if desired.
However, it is not necessary to do so.

Nortonizing Multiple-Voltage-Source Circuits

Example 8-11 illustrates how multiple-voltage-source circuits can be Nortonized. Again,
we are using a circuit that was Thevenized earlier, in Example 8-6. One Norton model is
created for each voltage source connected across terminals A and B. As you can see, the

current sources are in parallel and therefore can be added together. The Norton resistors are in parallel and therefore can be combined. The Norton resistors are in parallel and must be combined using parallel rules. We quickly arrive at a single Norton model for the entire circuit. For analysis, various loads may be connected to this model or it can be converted to the Thevenin equivalent.

E X A M P L E   8-11

STEP #1: Short the output terminals and find the Norton current for each voltage source as shown in Figure 8-45.

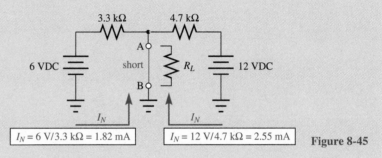

$$I_N = 6 \text{ V}/3.3 \text{ k}\Omega = 1.82 \text{ mA}$$ $$I_N = 12 \text{ V}/4.7 \text{ k}\Omega = 2.55 \text{ mA}$$

Figure 8-45

STEP #2: Find the Norton resistance for each current source as shown in Figure 8-46.

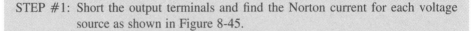

Figure 8-46

STEP #3: Draw the Norton models and combine them into one as shown in Figure 8-47.

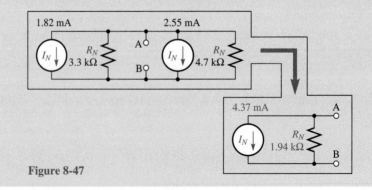
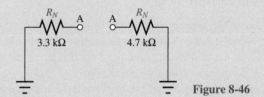

Figure 8-47

Take time now to test your skill at using this new circuit-analysis tool by completing Self-Check 8-3.

Go back and Nortonize the circuits of Self-Check 8-2. Write out all of your work. As always, a solution is provided at the end of this chapter.

8-4 Millman's Theorem

Millman's theorem is another powerful tool. Yet it is very simple to use. It combines what you have learned about voltage sources and current sources into one convenient formula. Millman's formula allows you to quickly reduce a complicated circuit composed of parallel voltage sources to a single voltage source, or Thevenin model. In some cases, this eliminates the need to use superposition. You are going to like this one!

Understanding Millman's Theorem

> **Millman's Theorem**
>
> Any circuit composed of any number of parallel voltage sources can be reduced to a single equivalent voltage source by: (1) converting each voltage source to a current source, (2) combining all parallel current sources and resistors into one current source, and (3) converting back to a single voltage source from the single current source.

The three steps stated in the theorem above are combined in Millman's formula:

$$V_{AB} = \frac{(V_1/R_1) + (V_2/R_2) + (V_3/R_3) + \cdots + (V_N/R_N)}{(1/R_1) + (1/R_2) + (1/R_3) + \cdots + (1/R_N)} \tag{8.3}$$

Figure 8-48 illustrates a typical circuit to which Millman's theorem can be applied. As you can see, the voltage sources are obviously in parallel and share common output terminals. In some circuit diagrams it is not immediately obvious that the voltage sources are in parallel. You may have to redraw the circuit and/or Thevenize portions of it before you are able to apply Millman's theorem. Before we discuss the application of Millman's theorem, it is important to understand the intermediate steps that lead from the multiple parallel voltage sources to the single voltage source. As mentioned earlier, Millman's

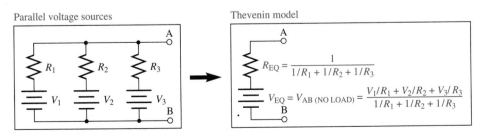

Figure 8-48   Millman's theorem.

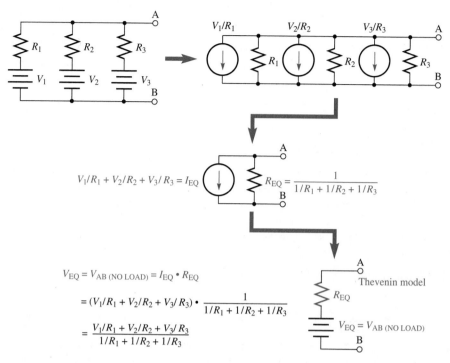

Figure 8-49   Deriving Millman's formula.

formula contains all three steps stated in the theorem. As you examine Figure 8-49, you will gain a much deeper understanding of Millman's theorem by seeing how the formula is derived.

Deriving the Formula

The first step in Millman's theorem is to convert each voltage source to a current source. To do this, each voltage source is divided by its series resistance. The second step is to combine the parallel current sources into one. This is accomplished by adding all current sources algebraically and by combining all parallel resistors according to the reciprocal formula. Notice that all current sources are added together in the numerator of Millman's formula, since each V/R term represents a constant-current source. Therefore, the entire numerator represents the overall equivalent constant-current source. You should recall that in order to convert back to a voltage source, which is what we finally want to do, the constant current is multiplied by the parallel resistance. The equivalent parallel resistance, found by combining the parallel resistors of the current sources, is expressed by the reciprocal formula. Thus, the output voltage (V_{AB}) is the equivalent constant current (numerator terms) times the equivalent parallel resistance (the reciprocal formula). The resulting formula is given as Millman's formula.

Using Millman's Theorem

To illustrate the power and convenience of Millman's theorem, we will reexamine the circuit of Example 8-2. This is shown in Example 8-12. You will readily see and appreciate the usefulness of Millman's theorem. Study the following examples carefully to see exactly how and why Millman's theorem applies. You will want to use Millman's theorem wherever possible in any circuit analysis that you do in the future.

EXAMPLE 8-12

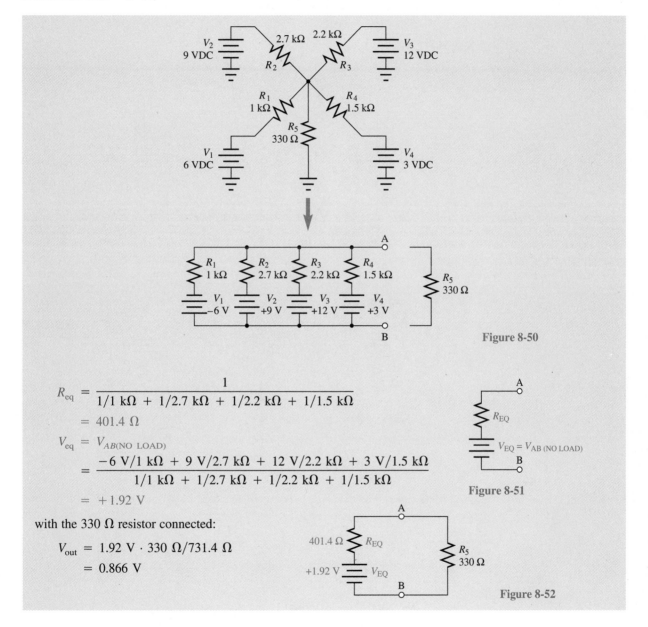

Figure 8-50

$$R_{eq} = \frac{1}{1/1 \text{ k}\Omega + 1/2.7 \text{ k}\Omega + 1/2.2 \text{ k}\Omega + 1/1.5 \text{ k}\Omega}$$

$$= 401.4 \ \Omega$$

$$V_{eq} = V_{AB(NO\ LOAD)}$$

$$= \frac{-6 \text{ V}/1 \text{ k}\Omega + 9 \text{ V}/2.7 \text{ k}\Omega + 12 \text{ V}/2.2 \text{ k}\Omega + 3 \text{ V}/1.5 \text{ k}\Omega}{1/1 \text{ k}\Omega + 1/2.7 \text{ k}\Omega + 1/2.2 \text{ k}\Omega + 1/1.5 \text{ k}\Omega}$$

$$= +1.92 \text{ V}$$

Figure 8-51

with the 330 Ω resistor connected:

$$V_{out} = 1.92 \text{ V} \cdot 330 \ \Omega/731.4 \ \Omega$$

$$= 0.866 \text{ V}$$

Figure 8-52

The first example, Example 8-12, is the same circuit as that used in Example 8-2 where superposition was used. R_5 is the load resistance and is temporarily removed. The four voltage sources are in parallel since they are all connected between ground and a common junction. Millman's theorem is used to find the value of the single voltage source

(V_{eq}) and the reciprocal formula is used to determine the equivalent series resistance (R_{eq}). Once the equivalent voltage source, or Thevenin model, is determined, the load may be reconnected and the actual loaded output voltage can be found. The voltage drop across each of the four series resistors may be determined using Kirchhoff's Voltage Law (i.e., $V_{R1} = 6\,\text{V} - V_{R5}$). As you can see, in this example, Millman's theorem is much more convenient than the superposition theorem.

EXAMPLE   8-13

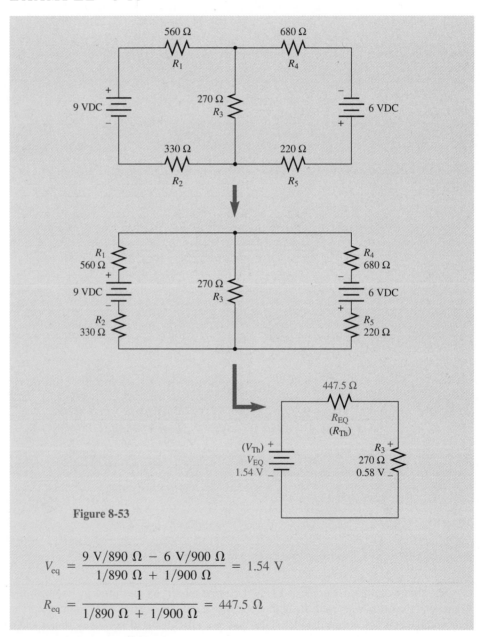

Figure 8-53

$$V_{eq} = \frac{9\,\text{V}/890\,\Omega - 6\,\text{V}/900\,\Omega}{1/890\,\Omega + 1/900\,\Omega} = 1.54\,\text{V}$$

$$R_{eq} = \frac{1}{1/890\,\Omega + 1/900\,\Omega} = 447.5\,\Omega$$

$$V_{R3} = 1.54 \text{ V} \cdot 270 \text{ }\Omega/717.5 \text{ }\Omega = 0.58 \text{ V}$$
$$V_{(R1+R2)} = 9 \text{ V} - 0.58 \text{ V} = 8.42 \text{ V}$$
$$V_{R1} = 8.42 \text{ V} \cdot 560 \text{ }\Omega/890 \text{ }\Omega = 5.3 \text{ V}$$
$$V_{R2} = 8.42 \text{ V} \cdot 330 \text{ }\Omega/890 \text{ }\Omega = 3.12 \text{ V}$$
$$V_{(R4+R5)} = 6 \text{ V} + 0.58 \text{ V} = 6.58 \text{ V}$$
$$V_{R4} = 6.58 \text{ V} \cdot 680 \text{ }\Omega/900 \text{ }\Omega = 4.97 \text{ V}$$
$$V_{R5} = 6.58 \text{ V} \cdot 220 \text{ }\Omega/900 \text{ }\Omega = 1.61 \text{ V}$$

In Example 8-13, notice that we must combine R_1 and R_2, and R_4 and R_5, in the use of Millman's formula. In other words, each voltage source must fit the Thevenin model of having a single series resistance. R_3 is considered the load resistance and is temporarily removed. Once the parallel voltage sources have been reduced to a single voltage source or Thevenin model, the load can be connected and the loaded output voltage determined. The individual voltage drops across R_1, R_2, R_4, and R_5 are determined using Kirchhoff's Voltage Law and the proportion formula for series resistors.

Take time now to use this new tool to solve for output voltage in each of the circuits of Self-Check 8-4.

SELF-CHECK 8-4

1. Use Millman's theorem to find the voltage across R_3 in Figure 8-54.

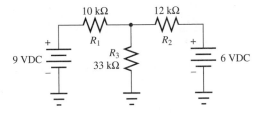

Figure 8-54

2. Use Millman's theorem to find the voltage across R_3 in Figure 8-55.

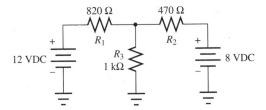

Figure 8-55

8-5 Combined Current and Voltage Sources

When a circuit-analysis problem involves a current source and a voltage source or sources, it is necessary to convert all sources to either current or voltage sources to obtain an equivalent circuit model. The final model may be either the Thevenin or the Norton model. Consider the following examples.

EXAMPLE   8-14

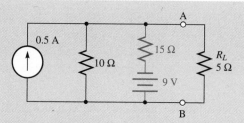

Figure 8-56

Convert the voltage source in Figure 8-56 to a current source as shown in Figure 8-57. Recall: $R_N = R_{Th}$ and $I_N = \dfrac{V_{Th}}{R_{Th}} = \dfrac{9\text{ V}}{15\text{ }\Omega} = 0.6\text{ A}$

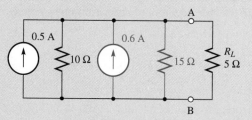

Figure 8-57

Combine the current sources of Figure 8-57. Recall that the sources are added algebraically and the resistors are combined as parallel resistors.

$0.5\text{ A} + 0.6\text{ A} = 1.1\text{ A}$

$10\text{ }\Omega \parallel 15\text{ }\Omega = 6\text{ }\Omega$

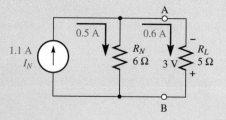

Figure 8-58

$I_N = 1.1\text{ A} \qquad R_N = 6\text{ }\Omega$

$I_L = I_N \cdot \dfrac{R_N}{R_N + R_L} = 1.1\text{ A} \cdot \dfrac{6\text{ }\Omega}{6\text{ }\Omega + 5\text{ }\Omega} = 0.6\text{ A}$

$V_L = 0.6\text{ A} \cdot 5\text{ }\Omega = 3\text{ V}$

Figure 8-59 is the Thevenin equivalent model.

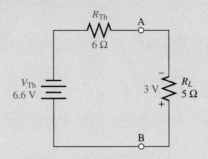

Figure 8-59

$V_{Th} = I_N \cdot R_N = 1.1 \text{ A} \cdot 6 \text{ } \Omega = 6.6 \text{ V}$

$R_{Th} = R_N = 6 \text{ } \Omega$

$V_L = 6.6 \text{ V} \cdot \dfrac{5 \text{ } \Omega}{5 \text{ } \Omega + 6 \text{ } \Omega} = 3 \text{ V}$

EXAMPLE   8-15

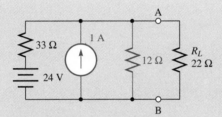

Figure 8-60

Convert the current source in Figure 8-60 to a voltage source and redraw the circuit as shown in Figure 8-61.

Recall: $R_{Th} = R_N$, $V_{Th} = I_N \cdot R_N = 1 \text{ A} \cdot 12 \text{ } \Omega = 12 \text{ V}$

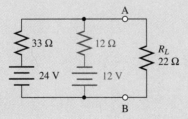

Figure 8-61

Now use Millman's Theorem to combine the two voltage sources as shown in Figure 8-62.

(Continued next page)

EXAMPLE   8-15 (Continued)

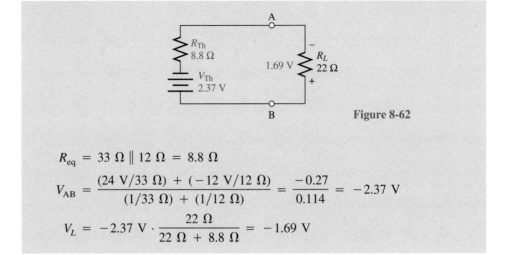

Figure 8-62

$$R_{eq} = 33\ \Omega \parallel 12\ \Omega = 8.8\ \Omega$$

$$V_{AB} = \frac{(24\ \text{V}/33\ \Omega) + (-12\ \text{V}/12\ \Omega)}{(1/33\ \Omega) + (1/12\ \Omega)} = \frac{-0.27}{0.114} = -2.37\ \text{V}$$

$$V_L = -2.37\ \text{V} \cdot \frac{22\ \Omega}{22\ \Omega + 8.8\ \Omega} = -1.69\ \text{V}$$

EXAMPLE   8-16

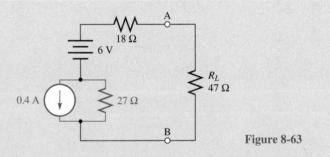

Figure 8-63

First, convert the current source in Figure 8-63 to an equivalent voltage source as shown in Figure 8-64.

$$R_{Th} = R_N = 27\ \Omega,\ V_{Th} = I_N \cdot R_N = 0.4\ \text{A} \cdot 27\ \Omega = 10.8\ \text{V}$$

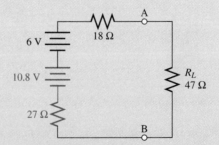

Figure 8-64

Now, simply add the series-aiding voltage sources and resistors as shown in Figure 8-65.

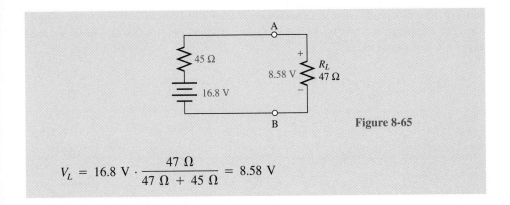

$$V_L = 16.8 \text{ V} \cdot \frac{47 \ \Omega}{47 \ \Omega \ + \ 45 \ \Omega} = 8.58 \text{ V}$$

Now try your skills on the problems of Self-Check 8-5.

1. Solve for load voltage in Figure 8-66.

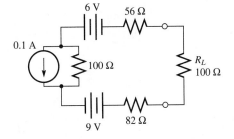

Figure 8-66

2. Solve for load voltage in Figure 8-67.

Figure 8-67

8-6 The Maximum Power Transfer Theorem

Thus far in this chapter, we have discussed voltage sources and current sources as they relate to various circuit-analysis tools. But what about power sources? As you know, power is the product of current and voltage. Any voltage source will supply current and any current source will supply voltage. Therefore, voltage and current sources supply power. So, how much power can a voltage or current source supply? First of all, in a very practical way, the amount of power a source can supply depends upon the design of the source. If the power source is an AC to DC converter and it is designed with low-power components, naturally the amount of power the converter can handle and deliver is limited accordingly. Thus the maximum amount of power a source can supply is limited, in a very practical way, by component ratings.

EXAMPLE   8-17

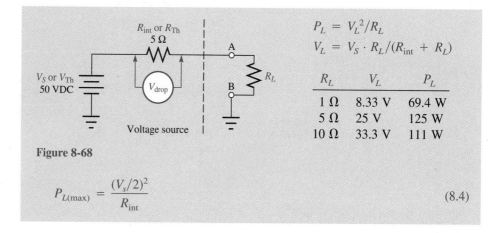

$$P_L = V_L^2/R_L$$
$$V_L = V_S \cdot R_L/(R_{int} + R_L)$$

R_L	V_L	P_L
1 Ω	8.33 V	69.4 W
5 Ω	25 V	125 W
10 Ω	33.3 V	111 W

Figure 8-68

$$P_{L(max)} = \frac{(V_s/2)^2}{R_{int}} \qquad (8.4)$$

But what about batteries, photovoltaic cells, and other DC sources? What limits the maximum amount of power that these sources can supply? The maximum amount of power is limited by the internal resistance of the source and follows the principles of the **maximum power transfer theorem**, stated as follows:

The Maximum Power Transfer Theorem

The amount of power any source can supply to a load is limited by the internal resistance of the source and will be at a maximum value when the load resistance is equal to the internal resistance of the source.

This theorem simply says that a maximum amount of power will be delivered to a load resistance that is equal to the internal resistance of the source. As shown in Example 8-17, if a 50 V source has an internal resistance of 5 Ω and a 5 Ω load is connected to its output terminals, a maximum power of 125 W is delivered to the load. Since the load and the internal resistance are in series, 25 V is dropped across the internal resistance, leaving 25 V for the load. Therefore, the maximum power dissipation by the load is $(25 \text{ V})^2/5 \text{ }\Omega$ or 125 W. Less power would be delivered to any load resistance that is less than or greater than 5 Ω. Notice that the power P_L delivered to the load is only 69.4 W with a 1 Ω load resistance and 111 W with a 10 Ω load

$$P_{L(max)} = \frac{(V_s/2)^2}{R_{int}} \qquad (8.4)$$

resistance. You can try other values of load resistances yourself to see if the output power is ever greater than 125 W. You will find that no other load resistance will draw as much

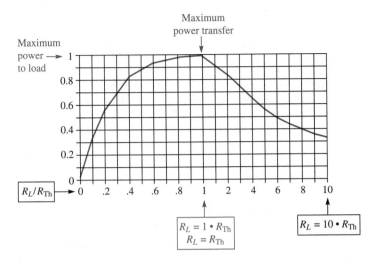

Figure 8-69   The maximum power transfer curve clearly shows that maximum power is transferred when $R_L = R_{Th}$.

power as the 5 Ω resistance. The only way you would be able to get more power out of the source is to decrease its internal resistance. We can conclude that the maximum power than can be delivered by a voltage source is equal to one half of the source voltage, squared, divided by the internal resistance of the source, since maximum output power occurs when the load resistance is equal to the internal resistance of the source (see Formula 8.4).

Figure 8-69 graphically illustrates the theorem of maximum power transfer. The scale along the bottom of the graph represents various ratios of load resistance to source resistance (or the Thevenin equivalent R_{Th}). The scale along the left of the graph represents various coefficients of power delivered to the load. The 1 indicates maximum power, and naturally all lesser coefficients represent load-power dissipations less than maximum. As you can see, maximum power is transfered to the load only when the load resistance is equal to the internal source resistance (Thevenin equivalent resistance R_{Th}). All values of load resistance greater than or less than the internal source resistance will receive and dissipate less power.

EXAMPLE   8-18

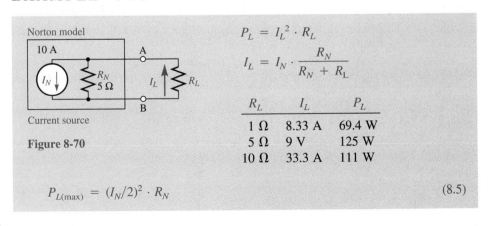

$$P_L = I_L^2 \cdot R_L$$

$$I_L = I_N \cdot \frac{R_N}{R_N + R_L}$$

R_L	I_L	P_L
1 Ω	8.33 A	69.4 W
5 Ω	9 V	125 W
10 Ω	33.3 A	111 W

Figure 8-70

$$P_{L(max)} = (I_N/2)^2 \cdot R_N \tag{8.5}$$

The maximum power transfer theorem applies to current sources as well. Example 8-18 (page 281) shows the Norton equivalent of the voltage source of Example 8-17. Notice that, once again, the maximum output power occurs when the load resistance is equal to the internal resistance of the source. Also, when the load resistance matches the internal resistance of the current source, the source current is divided in half (half for the load and half for the internal parallel resistance). Therefore, maximum output power is equal to one half of the constant current, squared, times the internal resistance of the current source (Formula 8.5). As you can see, less power will be delivered to any load resistance lower or higher than 5 Ω.

$$P_{L(\text{max})} = (I_N/2)^2 \cdot R_N \tag{8.5}$$

In this section, we have seen that the maximum amount of power that a voltage or current source can deliver to a load can be predicted if the internal resistance and voltage or current of the source are known. Formulas 8.4 and 8.5 are used for this purpose. We have also shown that in order for maximum power to be delivered to a load, the load resistance must be equal to the internal resistance of the source. Take a few minutes now to test your skills on Self-Check 8-6.

SELF-CHECK 8-6

1. Given a Thevenin model with the parameters $V_{\text{Th}} = 7$ V, $R_{\text{Th}} = 15$ Ω
   a. What should the load resistance be for maximum output power?
   b. What is the maximum possible output power?

2. Given a Norton model with the parameters $I_N = 200$ mA, $R_N = 40$ Ω
   a. What should the load resistance be for maximum output power?
   b. What is the maximum possible output power?

8-7 Nodal Analysis

Even though Kirchhoff's Voltage and Current Laws may seem very simple and of limited value, they are actually very powerful circuit-analysis tools that can be used to solve for currents and voltages in very complex circuits. In this section, you will learn how to apply Kirchhoff's Laws using *nodal* circuit analysis.

As you know, some circuits do not easily lend themselves to simple analysis using Ohm's Law. Figure 8-71 is a case in point. Notice that this circuit has two voltage sources. Can you determine the proper direction of current in each resistor? How about the voltage drops and polarities? Not sure? You can be sure by following the method called **nodal analysis**.

But what does *nodal* mean? **Nodal** means, "that which applies to a node." Fine, you say! What is a node? A **node** is simply a connecting point, or junction, for two or more components. Figure 8-71 has six nodes—marked A, B, C, D, E, and F. The nodes marked C and D are called *major, primary,* or *principal* nodes. These are nodes that are formed

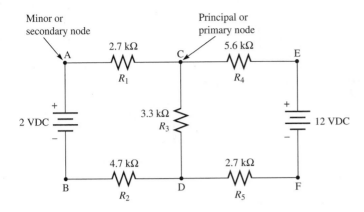

Figure 8-71

at the junction of three or more components. The focus of our attention will be on these **principal nodes**.

Our objective for analyzing the circuit of Figure 8-71 is to determine all currents and voltage drops in the circuit. Using **nodal analysis**, you will first be concerned with solving for the voltage that exists between two primary nodes, then node current and remaining voltage drops are easily found.

EXAMPLE   8-19

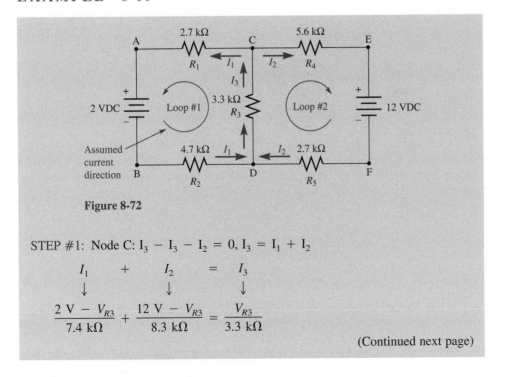

Figure 8-72

STEP #1: Node C: $I_3 - I_3 - I_2 = 0$, $I_3 = I_1 + I_2$

$$I_1 \quad + \quad I_2 \quad = \quad I_3$$

$$\downarrow \qquad\qquad \downarrow \qquad\qquad \downarrow$$

$$\frac{2\,V - V_{R3}}{7.4\ k\Omega} + \frac{12\,V - V_{R3}}{8.3\ k\Omega} = \frac{V_{R3}}{3.3\ k\Omega}$$

(Continued next page)

EXAMPLE   8-19   (Continued)

STEP #2

$$\frac{8.3 \text{ k}\Omega \,(2 \text{ V} - V_{R3}) + 7.4 \text{ k}\Omega \,(12 \text{ V} - V_{R3})}{61.42 \cdot 10^6} = \frac{V_{R3}}{3.3 \text{ k}\Omega}$$

$$\frac{16{,}600 - 8{,}300 \, V_{R3} + 88{,}800 - 7{,}400 \, V_{R3}}{61.42 \cdot 10^6} = \frac{V_{R3}}{3.3 \text{ k}\Omega}$$

$$\frac{105{,}400 - 15{,}700 \, V_{R3}}{61.42 \cdot 10^6} = \frac{V_{R3}}{3.3 \text{ k}\Omega}$$

$$105{,}400 - 15{,}700 \, V_{R3} = \frac{61.42 \cdot 10^6 \cdot V_{R3}}{3{,}300}$$

$$105{,}400 - 15{,}700 \, V_{R3} = 18{,}612 \, V_{R3}$$

$$105{,}400 = 34{,}312 \, V_{R3} \qquad \boxed{V_{R3} = 3.07 \text{ V}}$$

STEP #3

$$I_1 = \frac{2 \text{ V} - 3.07 \text{ V}}{7{,}400 \, \Omega} = \frac{-1.07 \text{ V}}{7{,}400 \, \Omega} = -0.145 \text{ mA}$$

$$I_2 = \frac{12 \text{ V} - 3.07 \text{ V}}{8{,}300 \, \Omega} = \frac{8.93 \text{ V}}{8{,}300 \, \Omega} = 1.08 \text{ mA}$$

Current is flowing opposite to what was assumed.

$$I_3 = \frac{3.07 \text{ V}}{3{,}300 \, \Omega} = 0.930 \text{ mA}$$

STEP #4

$$V_{R1} = 0.145 \text{ mA} \cdot 2{,}700 \, \Omega = 0.392 \text{ V}$$

$$V_{R2} = 0.145 \text{ mA} \cdot 4{,}700 \, \Omega = 0.682 \text{ V}$$

$$V_{R4} = 1.08 \text{ mA} \cdot 5{,}600 \, \Omega = 6.05 \text{ V}$$

$$V_{R5} = 1.08 \text{ mA} \cdot 2{,}700 \, \Omega = 2.92 \text{ V}$$

In Example 8-19, we use nodal analysis to solve for the voltage drop that exists between nodes C and D and is shared by both loops. Then node currents and remaining voltage drops are found. To begin, we *assume* that loop currents are in the directions as indicated in Figure 8-72.

The basic approach in nodal analysis is to define a nodal current equation in terms of voltage and resistance.

Now, let's begin a step-by-step analysis of our example.

Step #1

We begin with a current equation for nodes C and D and substitute voltage/resistance equivalents for each current term. Notice, in the example, that each current term is expressed in terms of voltage divided by resistance. It is very important for you to understand these terms. The object is to define all voltage drops in terms of the voltage drop between the two primary nodes C and D using D as a reference node for voltage polarities. There-fore, each current term has a voltage expression in the numerator that involves the voltage drop across R_3. The I_3 term is naturally expressed as $V_{R3}/3.3$ kΩ. But, notice how the I_1 term is expressed. R_1 and R_2 are combined as 7.4 kΩ and the sum of the voltage drops across them is equal to the source voltage minus V_{R3}. In other words, the combined voltage drops of R_1, R_2, and R_3 must equal the source voltage of loop #1 ($V_{R1} + V_{R2} + V_{R3} =$ 2 V). Therefore, $V_{R1} + V_{R2}$ must equal 2 V $- V_{R3}$. (For now, it is important to combine R_1 and R_2 as though they were a single resistor. Once we have determined the value of V_{R3}, we can solve for I_1 and determine V_{R1} and V_{R2} using Ohm's Law.) In similar fashion, the I_2 term is expressed as 12 V $- V_{R3}$ divided by 8.3 kΩ, the combined resistance of R_4 and R_5. Note that in this example both voltage sources are positive with respect to node D.

Step #2

The next step involves the algebraic manipulation of the equation in order to solve for V_{R3}. If your skills in algebra are weak, you will want to follow the solution through very carefully. You may also want to review Appendix C, "Algebraic Operations."

Step #3

Substitute the value of V_{R3} into each term of the nodal current equation to solve for I_1, I_2, and I_3. Notice that I_1 is found to be negative. The negative sign indicates that this current is flowing in the direction opposite to what you had assumed. You are alerted to this by the negative sign. The loop #1 current direction can now be corrected.

Step #4

Use node currents to determine unknown voltage drops in each loop. I_1 is used to deter-mine V_{R1} and V_{R2}, and I_2 is used to determine V_{R4} and V_{R5}. Figure 8-73 shows the correct current directions and voltage drops.

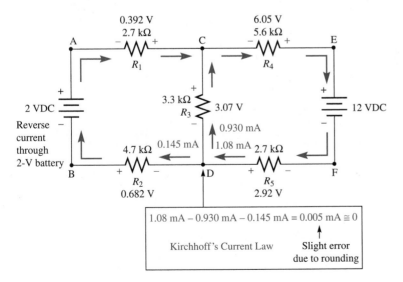

Figure 8-73

Now let's make it a little more challenging by examining a circuit in which more than one primary node must be considered. In Example 8-20 we will use nodal analysis to develop two equations that pertain to the same circuit from the vantage points of two primary nodes.

EXAMPLE 8-20

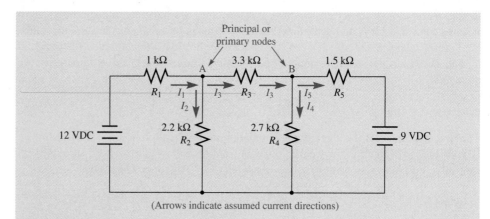

Figure 8-74

• Node A:

$$I_1 - I_2 - I_3 = 0$$

$$\frac{-12 \text{ V} - (-V_{R2})}{1 \text{ k}\Omega} - \frac{-V_{R2}}{2.2 \text{ k}\Omega} - \frac{-V_{R2} - (-V_{R4})}{3.3 \text{ k}\Omega} = 0$$

$$\frac{-12 \text{ V}}{1 \text{ k}\Omega} + \frac{V_{R2}}{1 \text{ k}\Omega} + \frac{V_{R2}}{2.2 \text{ k}\Omega} + \frac{V_{R2}}{3.3 \text{ k}\Omega} + \frac{-V_{R4}}{3.3 \text{ k}\Omega} = 0$$

$$-0.012 + 0.001V_{R2} + 0.000455V_{R2} + 0.000303V_{R2} - 0.000303V_{R4} = 0$$

$$-0.012 + 0.00176V_{R2} - 0.000303V_{R4} = 0$$

$$0.00176V_{R2} - 0.000303V_{R4} = 0.012$$

• Node B:

$$I_3 - I_4 - I_5 = 0$$

$$\frac{-V_{R2} - (-V_{R4})}{3.3 \text{ k}\Omega} - \frac{-V_{R4}}{2.7 \text{ k}\Omega} - \frac{-V_{R4} - (+9 \text{ V})}{1.5 \text{ k}\Omega} = 0$$

Note: $V_{R5} = 9 \text{ V} + V_{R4}$ if the assumed current direction for I_4 is correct.

$$\frac{-V_{R2}}{3.3 \text{ k}\Omega} + \frac{V_{R4}}{3.3 \text{ k}\Omega} + \frac{V_{R4}}{2.7 \text{ k}\Omega} + \frac{V_{R4}}{1.5 \text{ k}\Omega} + \frac{9 \text{ V}}{1.5 \text{ k}\Omega} = 0$$

$$-0.000303 \, V_{R2} + 0.000303 \, V_{R4} + 0.00037 \, V_{R4}$$
$$+ \, 0.006 + 0.000667 \, V_{R4} = 0$$

$$-0.000303 \, V_{R2} + 0.00134 \, V_{R4} + 0.006 = 0$$

$$0.000303 \, V_{R2} - 0.00134 \, V_{R4} = 0.006$$

- Now solve for V_{R2} using subtraction. We must eliminate the V_{R4} terms in the two simultaneous equations. To do this, we must make the terms equal. So, we will multiply all terms in the Node A equation by 4.422.

$$\begin{array}{rrr} 0.00176\ V_{R2} & -\ 0.000303\ V_{R4} = & 0.012 \\ \times 4.422 & \times 4.422 & \times 4.422 \\ \hline 0.00778\ V_{R2} & -\ 0.00134\ V_{R4} = & 0.0531 \end{array}$$

Next we must subtract the Node A equation from the Node B equation.

$$\begin{array}{rrr} 0.000303\ V_{R2} & -\ 0.00134\ V_{R4} = & 0.006 \\ -(0.00778\ \ \ V_{R2} & -\ 0.00134\ V_{R4} = & 0.0531) \\ \hline -0.00748\ \ \ V_{R2} & +\quad 0\quad\ V_{R4} = & -0.0471 \end{array}$$

Solve for V_{R2}.

$$V_{R2} = \frac{-0.0471}{-0.00748} = 6.30\ \text{V}$$

- Use the Node B equation to solve for V_{R4}.

$$(0.000303 \cdot 6.30\ \text{V}) - 0.00134\ V_{R4} = 0.006$$

$$0.00191 - 0.00134\ V_{R4} = 0.006$$

$$-0.00134\ V_{R4} = 0.006 - 0.00191$$

$$-0.00134\ V_{R4} = 0.00409$$

$$V_{R4} = \frac{0.00409}{-0.00134} = -3.05\ \text{V}$$

⌐Indicates I_4 is in a
direction opposite to
what was assumed

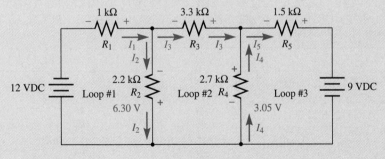

Figure 8-75

- Now solve for all other parameters. (absolute value of parameters)

$$V_{R1} = 12\ \text{V} - 6.30\ \text{V} = 5.70\ \text{V}$$

$$I_1 = I_{R1} = 5.70\ \text{V}/1\ \text{k}\Omega = 5.70\ \text{mA}$$

$$I_2 = 6.30\ \text{V}/2.2\ \text{k}\Omega = 2.86\ \text{mA}$$

(Continued next page)

EXAMPLE 8-20 (Continued)

$$I_3 = I_1 - I_2 = 5.70 \text{ mA} - 2.86 \text{ mA} = 2.84 \text{ mA}$$
$$V_{R3} = I_3 \cdot R_3 = 2.84 \text{ mA} \cdot 3.3 \text{ k}\Omega = 9.37 \text{ V}$$
$$I_4 = 3.05 \text{ V}/2.7 \text{ k}\Omega = 1.13 \text{ mA}$$
$$I_5 = I_3 + I_4 = 2.84 \text{ mA} + 1.13 \text{ mA} = 3.97 \text{ mA}$$
$$V_{R5} = I_5 \cdot R_5 = 3.97 \text{ mA} \cdot 1.5 \text{ k}\Omega = 5.96 \text{ V}$$

• Use Kirchhoff's Voltage Law to prove the voltage drops in each of the three sectional loops.

#1: $-5.70 \text{ V} - 6.30 \text{ V} + 12 \text{ V} = 0$

#2: $-9.37 \text{ V} + 3.05 \text{ V} + 6.30 \text{ V} = -0.02 \cong 0$

#3: $-3.05 \text{ V} - 5.96 \text{ V} + 9 \text{ V} = -0.01 \cong 0$

Design Note 8-1 can be used to solve for variables in two simultaneous equations.

DESIGN NOTE 8-1: Simultaneous Equations

THE SIMULTANEOUS EQUATIONS SHOULD BE EXPRESSED IN THE FOLLOWING FORM AND RELATION TO EACH OTHER

#1	$A \cdot$ (VARIABLE #1) + $B \cdot$ (VARIABLE #2) = C
#2	$D \cdot$ (VARIABLE #1) + $E \cdot$ (VARIABLE #2) = F

WHERE LETTERS A, B, D, and E ARE COEFFICIENTS OF THE UNKNOWN VARIABLES. THE VARIABLES WILL USUALLY BE LOOP CURRENTS. C and F ARE KNOWN QUANTITIES SUCH AS THE SOURCE VOLTAGE IN A LOOP

EXPLANATION	"BASIC" PROGRAM SOLUTION
THE VARIABLES OF THE EQUATIONS ARE SOLVED IN THIS BASIC PROGRAM BY THE METHOD OF SUBTRACTION AND SUBSTITUTION. EQUATION #2 IS SUBTRACTED FROM EQUATION #1 IN ORDER TO ELIMINATE VARIABLE #2 IN BOTH EQUATIONS. THEN VARIABLE #1 IS SOLVED AND SUBSTITUTED BACK INTO EQUATION #1 TO SOLVE FOR VARIABLE #2. PROGRAM LINES 20 TO 70 PROVIDE "ON SCREEN" INSTRUCTIONS. LINE 90 ASKS FOR THE UNITS BEING USED, SUCH AS AMPERES OR VOLTS.	10 CLS 20 PRINT"SIMULTANEOUS EQUATIONS" 30 PRINT"" 40 PRINT"THE TWO SIMULTANEOUS EQUATIONS MUST TAKE THE FOLLOWING FORM:" 50 PRINT"" 60 PRINT"A * (VARIABLE #1) + B * (VARIABLE #2) = C" 70 PRINT"D * (VARIABLE #1) + E * (VARIABLE #2) = F" 80 PRINT"":PRINT"" 90 INPUT"TYPE THE UNITS OF THE VARIABLES. (i.e., VOLTS, AMPS, etc.)";U$ 100 INPUT"ENTER COEFFICIENT A-";A 110 INPUT"ENTER COEFFICIENT B-";B

EXPLANATION	"BASIC" PROGRAM SOLUTION
LINES 100 THROUGH 150 ALLOW YOU TO IN-PUT THE VALUES OF THE COEFFICIENTS AND QUANTITY EACH EQUATION IS SET EQUAL TO. THE *X* FACTOR, IN LINE 160, IS USED TO MAKE THE *B* COEFFICIENT IN EQUATION #1 EQUAL TO COEFFICIENT *E* IN EQUATION #2. LINES 170 AND 180 ARE DERIVED AS FOL-LOWS:	120 INPUT"ENTER QUANTITY C-";C 130 INPUT"ENTER COEFFICIENT D-";D 140 INPUT"ENTER COEFFICIENT E-";E 150 INPUT"ENTER QUANTITY F-";F 160 X = E/B 170 V1 = (X ∗ C − F)/(X ∗ A − D) 180 V2 = (C − A ∗ V1)/B 190 PRINT"" 200 PRINT"VARIABLE #1 IS ";V1;" ";U$;"." 210 PRINT"VARIABLE #2 IS ";V2;" ";U$;"." 220 PRINT"" 230 INPUT"ANOTHER PROBLEM? (Y/N)";A$ 240 IF A$ = "Y" THEN CLEAR:GOTO 50 250 CLS:END

$A \cdot (\text{VARIABLE \#1}) + B \cdot (\text{VARIABLE \#2}) = C$
$D \cdot (\text{VARIABLE \#1}) + E \cdot (\text{VARIABLE \#2}) = F$

FORMULA DERIVATION (LINE 170)

$$V_1 = (X \cdot C - F)/(X \cdot A - D)$$

EQUATION #1 IS MULTIPLIED BY *X* SO VARIABLE #2 MAY BE ELIMINATED. RECALL THAT $X = E/B$.

$$X \cdot A \cdot (\text{VAR. \#1}) + X \cdot B \cdot (\text{VAR. \#2}) = X \cdot C$$

NOW, EQUATION #2 IS SUBTRACTED FROM EQUATION #1.

$$X \cdot A \cdot (\text{VAR. \#1}) + X \cdot B \cdot (\text{VAR. \#2}) = X \cdot C$$
$$\underline{- D \cdot (\text{VAR. \#1}) - E \cdot (\text{VAR. \#2}) = -F}$$
$$((X \cdot A) - D) \cdot (\text{VAR. \#1}) = (X \cdot C) - F$$

NEXT, A SOLUTION IS FOUND FOR VARIABLE #1

$$(\text{VAR. \#1}) = V_1 = \frac{(X \cdot C) - F}{(X \cdot A) - D} = (X \cdot C - F)/(X \cdot A - D)$$

> MULTIPLICATION HAS PRIORITY
> OVER SUBTRACTION

FORMULA DERIVATION (LINE 180)

$$V_2 = (C - A \cdot V_1)/B$$

THE VALUE OF VARIABLE #1 IS SUBSTITUTED INTO EQUATION #1 IN ORDER TO SOLVE FOR VARIABLE #2.

$$A \cdot V_1 + B \cdot V_2 = C$$
$$B \cdot V_2 = C - (A \cdot V_1)$$
$$V_2 = \frac{C - (A \cdot V_1)}{B} = (C - A \cdot V_1)/B$$

SELF-CHECK 8-7

1. Solve for all currents and voltage drops in Figure 8-76 using nodal analysis.

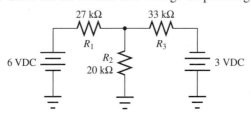

Figure 8-76

8-8   Mesh Analysis

Thus far, you have added superposition, Thevenin, Norton, Millman, and nodal analysis to your circuit-analysis tool kit. All of these tools are great, but they are not the only choices available to you for solving currents and voltages in a complex circuit. So, before you put away your tool kit, here is one more handy circuit-analysis tool that you may decide you really like, or may even prefer over nodal analysis. We call this tool **mesh analysis**.

The word **mesh** refers to an interlaced or interconnected structure such as a screen or net. Walls or sides that make up a mesh are shared with adjoining sections. The circuits that we have been discussing in this chapter can be viewed as electrical **meshworks** with a resistor in common with adjoining sections. Each section of the meshwork has its own current, which interacts in the common mesh resistor with the current of the adjoining section, either aiding or opposing that current. Using mesh analysis, you will pretend that the mesh currents remain separated in the common mesh resistor as you establish loop equations for each mesh section. You will find that mesh analysis is a fairly simple step-by-step process. Refer to Example 8-21 as each step is discussed.

EXAMPLE   8-21

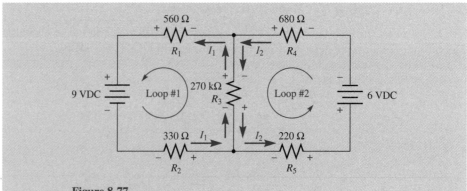

Figure 8-77

STEP #1: Assign a loop current in each section of the meshwork as shown in Figure 8-77.

STEP #2: Form mesh equations for each loop.

(LOOP #1) $1{,}160\,I_1 - 270\,I_2 = +9\text{ V}$

(LOOP #2) $1{,}170\,I_2 - 270\,I_1 = +6\text{V}$

STEP #3: Solve for the two unknown currents in the two simultaneous equations.

$$4.3333 \cdot \left.\begin{array}{r} [1{,}160\,I_1 - 270\,I_2 = +9\text{ V}] \\ -270\,I_1 + 1{,}170\,I_2 = +6\text{ V} \end{array}\right] =$$

$$\left.\begin{array}{r} 5{,}027\,I_1 - 1{,}170\,I_2 = +39\text{ V} \\ -270\,I_1 + 1{,}170\,I_2 = +6\text{ V} \end{array}\right]$$

$$4{,}757\,I_1 + \quad 0\,I_2 = +45\text{ V}$$

$$4{,}757\,I_1 = +45\text{ V}$$

$$I_1 = +9.46\text{ mA}$$

NOW WE CAN SOLVE FOR I_2.

$1{,}160\,(9.46\text{ mA}) - 270\,I_2 = +9\text{ V}$

$10.97\text{ V} - 270\,I_2 = +9\text{ V}$

$10.97\text{ V} - 9\text{ V} = 270\,I_2$

$1.97\text{ V}/270 = +7.3\text{ mA}$

Figure 8-78 shows the actual current directions. The directions originally assumed are correct.

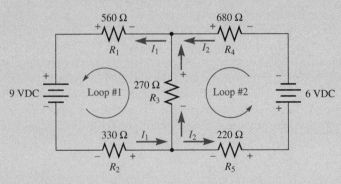

Figure 8-78

NOTE: If I_1 or I_2 were calculated as negative quantities, that would indicate the current is actually flowing in the opposite direction to what you originally thought.

STEP #4: Solve for all voltage drops and mark polarities as shown in Figure 8-78.

$V_{R1} = 560\ \Omega \cdot 9.46\text{ mA} \quad = \quad 5.3\text{ V}$

$V_{R2} = 330\ \Omega \cdot 9.46\text{ mA} \quad = \quad 3.12\text{ V}$ LOOP #1 = 9 V

$V_{R3} = \left[\begin{array}{ll} 270\ \Omega \cdot 9.46\text{ mA} & = \quad 2.55\text{ V} \\ 270\ \Omega \cdot -7.3\text{ mA} & = \ -1.97\text{ V} \end{array} \right.$

$\qquad\qquad\qquad\qquad\quad V_{R3} = \quad 0.58\text{ V}$

$9\text{ V} - 3.12\text{ V} - 0.58\text{ V} - 5.3\text{ V} = 0$

(Continued next page)

EXAMPLE 8-21 (Continued)

$$V_{R4} = 680\ \Omega \cdot 7.3\ \text{mA} \quad = \quad 4.96\ \text{V}$$
$$V_{R5} = 220\ \Omega \cdot 7.3\ \text{mA} \quad = \quad 1.61\ \text{V}$$

LOOP #2 = 6 V

$$V_{R3} = \begin{bmatrix} 270\ \Omega \cdot -9.46\ \text{mA} = -2.55\ \text{V} \\ 270\ \Omega \cdot \quad 7.3\ \text{mA} = \quad 1.97\ \text{V} \end{bmatrix}$$
$$V_{R3} = -0.58\ \text{V}$$

6 V − 4.96 V + 0.58 V −
1.61 V = 0.01 ≅ 0

Indicates opposite polarity in loop #2

Step #1

Assign all loop currents. One loop current is indicated for each section of the meshwork, as is shown in Example 8-21.

Step #2

The next step is to form simultaneous **mesh equations** that satisfy each loop or section of the meshwork. Each voltage in the left side of the mesh equation is expressed as a loop current times resistance. Each mesh equation will always contain a positive term composed of the loop's current times the total resistance of the loop. For example, notice that mesh equation #1 has the following positive term: $1{,}160\ I_1$. The 1,160 is the sum of R_1, R_2, and R_3. The next term in mesh equation #1 is negative $(-270\ I_2)$. This indicates that there is an opposing voltage drop across R_3 that is caused by the opposing current of loop #2. If the currents were aiding in R_3, the I_2 term would be positive. Finally, these two terms are set equal to the source voltage, whose polarity is always positive. This procedure must be followed exactly in order to obtain correct results. Read this step again and study the example to make sure you understand how to form these mesh equations.

Step #3

The next step is to solve for the individual loop currents of the two simultaneous mesh equations. This can be done by the method of addition, subtraction, and substitution, or by using determinants (or on your computer using Design Note 8-1). We are using the method of addition and substitution in this example. (For a discussion on the use of determinants, see Appendix E, "Simultaneous Equations and Determinants.") As you can see in Example 8-21, the I_2 terms in both mesh equations are eliminated by first multiplying each term of the loop #1 mesh equation by 4.3333. This causes the I_2 terms of each mesh equation to have the same coefficient (1,170). When the two mesh equations are added together, the I_2 terms drop out, since they are opposite in sign. This leaves an equation with only one unknown variable (I_1). Once I_1 is found, it can be substituted into one of the original mesh equations to find I_2. If I_1 or I_2 were calculated as negative quantities, this would indicate that the assumed direction for current in the loop was incorrect. As it turns out, the actual direction of current in each loop is as originally assumed.

Step #4

The final step in the mesh analysis procedure is to correct the loop current directions (if necessary), mark voltage drop polarities, and calculate the individual voltage drops. Ohm's Law is used for this final step. It is important for you to see in this example that the actual loop currents oppose each other in R_3. Therefore, the currents must be subtracted from one another. I_1 is the greater current. Thus, the polarity of the voltage drop across R_3 is determined by the direction of I_1.

Other Cases

What do you do differently if a circuit has more than two mesh sections? As far as the procedure goes, there is no difference. Each step must be followed exactly as was done for Example 8-21. The only real difference is the fact that there are more loop currents and mesh equations. There is one loop current and one mesh equation for each section of the meshwork. Naturally, the solution of three or more unknown loop currents in three or more mesh questions involves much more work. However, even though your algebra skills are challenged a bit more, the overall mesh-analysis procedure remains the same.

Suppose one or more of the meshwork sections does not have a source voltage included in its loop? How would the mesh equation be written? If a particular section of a meshwork does not have a source voltage, the section's mesh equation will simply be set equal to zero.

Example 8-22 illustrates the cases we have just discussed. Take time to study this example by retracing the steps of mesh analysis. When you feel you are ready, try your new skills on Self-Check 8-8.

EXAMPLE   8-22

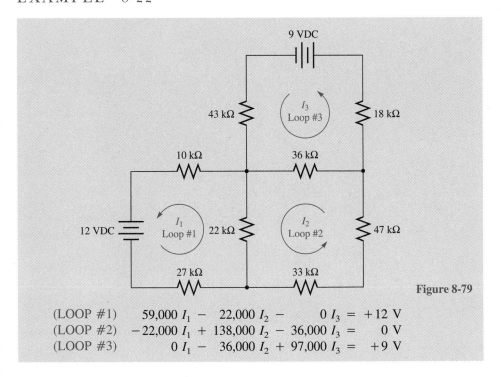

Figure 8-79

(LOOP #1) $59{,}000\, I_1 - 22{,}000\, I_2 - 0\, I_3 = +12$ V
(LOOP #2) $-22{,}000\, I_1 + 138{,}000\, I_2 - 36{,}000\, I_3 = 0$ V
(LOOP #3) $0\, I_1 - 36{,}000\, I_2 + 97{,}000\, I_3 = +9$ V

SELF-CHECK 8-8

1. Solve Example 8-19 using mesh analysis. Your answers should, of course, match those already found using nodal analysis.

8-9 Let Delta (Δ) = Wye (Y)

You are probably wondering what "Let Delta (Δ) = Wye (Y)" means. What do the symbols, Δ and Y, have to do with circuit analysis? Since too much suspense is not good for anyone, I'll tell you right away! The Δ depicts a triangular arrangement of components and the Y depicts a wye arrangement. In many of the circuits that you analyze, you will recognize portions of the circuit arranged in a Δ or Y configuration. There are cases in which a solution to a complicated circuit would be far simpler if a portion of the circuit could be converted to the opposite configuration, Δ → Y or Y → Δ. Fortunately, this can be done. A portion of a circuit that is in the Δ configuration can be converted to the Y configuration, and vice versa, for the purpose of circuit analysis. What we have here is one more interesting tool to add to your circuit-analysis tool kit.

Figure 8-80 illustrates the two component configurations and the set of formulas needed to convert from one to the other. In each case, three formulas are needed to complete a conversion. As you can see, each of these configurations has three terminals by which it may be connected to a larger circuit. It is important for you to understand that the resistance or voltage drop between any two of the terminals is the same for either configuration in any circuit. Substituting one configuration for the other in a circuit will not change overall circuit parameters, such as total resistance and total current. However,

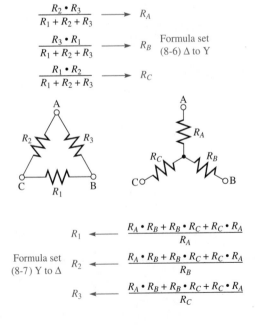

Figure 8-80   Delta Δ/Wye Y conversion.

the overall circuit will function differently, as can be seen by the obvious change in the circuit layout. You will understand this more clearly in later examples. For now, we will focus on the two configurations apart from any specific application. Example Set 8-23 illustrates two examples of Δ ⇌ Y conversion, one in each direction. If an ohmmeter were connected between terminals A and B of each configuration, no difference in resistance would be seen from one configuration to the other (likewise for terminals B and C, and C and A). The measured, or calculated, resistance between any two corresponding terminals of equivalent delta and wye resistor configurations is always the same.

EXAMPLE SET   8-23

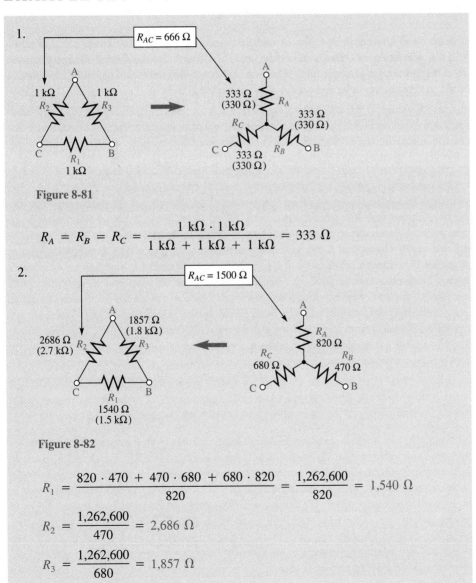

1.

$R_{AC} = 666\ \Omega$

Figure 8-81

$$R_A = R_B = R_C = \frac{1\ \text{k}\Omega \cdot 1\ \text{k}\Omega}{1\ \text{k}\Omega + 1\ \text{k}\Omega + 1\ \text{k}\Omega} = 333\ \Omega$$

2.

$R_{AC} = 1500\ \Omega$

Figure 8-82

$$R_1 = \frac{820 \cdot 470 + 470 \cdot 680 + 680 \cdot 820}{820} = \frac{1{,}262{,}600}{820} = 1{,}540\ \Omega$$

$$R_2 = \frac{1{,}262{,}600}{470} = 2{,}686\ \Omega$$

$$R_3 = \frac{1{,}262{,}600}{680} = 1{,}857\ \Omega$$

Now that you see how easy the conversion from one configuration to the other is, we will consider a classic example of its use. Example 8-24 shows the bridge circuit once again. This time we are able to convert the bridge to a simple overall series circuit. This enables us to find the bridge voltage using simple circuit-analysis techniques. Notice that the bridge circuit of Figure 8-83 has two Δ configurations: (R_1, R_3, and R_L) and (R_2, R_4, and R_L). It also has two Y configurations: (R_1, R_L, and R_2) and (R_3, R_L, and R_4). The circuit may be analyzed by Δ to Y conversion or Y to Δ conversion.

The most direct solution is obtained by converting either the top or bottom delta configuration to a wye. As shown in Figure 8-84, the upper delta has arbitrarily been chosen for the conversion. Note how quickly this simplifies the circuit. Once the conversion is made, simple circuit-analysis techniques are used to find the difference of potential between terminals B and C. The overall circuit parameters of total current and resistance are not affected in the conversion process. However, it is very apparent that the circuit is no longer the original bridge. This simplified circuit is merely a model for analysis purposes. It is a model of the original bridge taking into account the 6.8 kΩ bridging resistor (R_L). If the bridging resistor, or load, is changed to a different value, a new model must be derived using the delta to wye conversion formulas.

As you can see, the $\Delta \rightarrow$ Y conversion technique can be a very handy tool that will allow you to simplify otherwise difficult circuits. As with any circuit-analysis tool, practice will improve your skill. Take time to study Design Note 8-2, then test your skills on Self-Check 8-9.

EXAMPLE 8-24

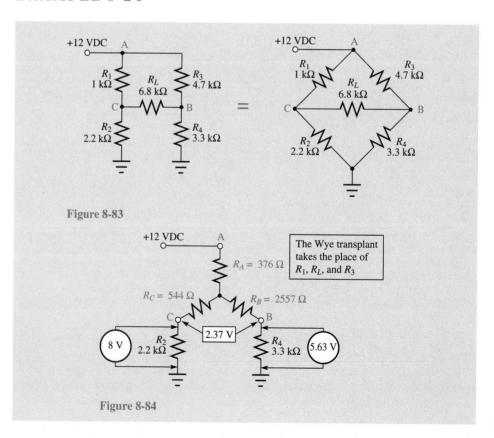

Figure 8-83

Figure 8-84

$$R_A = \frac{R_1 \cdot R_3}{R_1 + R_3 + R_L} = \frac{1 \text{ k}\Omega \cdot 4.7 \text{ k}\Omega}{12.5 \text{ k}\Omega} = 376 \ \Omega$$

$$R_B = \frac{R_3 \cdot R_L}{R_1 + R_3 + R_L} = \frac{4.7 \text{ k}\Omega \cdot 6.8 \text{ k}\Omega}{12.5 \text{ k}\Omega} = 2{,}557 \ \Omega$$

$$R_C = \frac{R_L \cdot R_1}{R_1 + R_3 + R_L} = \frac{6.8 \text{ k}\Omega \cdot 1 \text{ k}\Omega}{12.5 \text{ k}\Omega} = 544 \ \Omega$$

SOLVE FOR THE VOLTAGE FROM *B* TO GROUND AND THE VOLTAGE FROM *C* TO GROUND. THEN, TAKE THE DIFFERENCE OF POTENTIAL BETWEEN THE TWO (V_{CB}).

$$R_t = 376 \ \Omega + (544 \ \Omega + 2.2 \text{ k}\Omega \parallel 2{,}557 \ \Omega + 3.3 \text{ k}\Omega)$$
$$= 376 \ \Omega + 1{,}869 \ \Omega$$
$$= 2{,}245 \ \Omega$$
$$V_{R_A} = 12 \text{ V} \cdot 376 \ \Omega/2{,}245 \ \Omega = 2 \text{ V}$$
$$V_{(C \text{ to GND})} = 10 \text{ V} \cdot 2.2 \text{ k}\Omega/2{,}744 \ \Omega = 8 \text{ V}$$
$$V_{(B \text{ to GND})} = 10 \text{ V} \cdot 3.3 \text{ k}\Omega/5{,}857 \ \Omega = 5.63 \text{ V}$$
$$V_{CB} = 8 \text{ V} - 5.63 \text{ V} = \boxed{2.37 \text{ V}}$$

DESIGN NOTE 8-2:   Delta/Wye Conversion

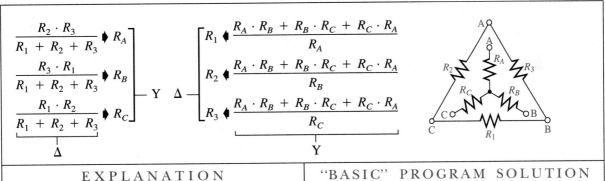

$$\frac{R_2 \cdot R_3}{R_1 + R_2 + R_3} \blacktriangleright R_A$$
$$\frac{R_3 \cdot R_1}{R_1 + R_2 + R_3} \blacktriangleright R_B \quad \bigg] \ \text{Y} \quad \Delta \ \bigg[\quad R_1 \blacktriangleleft \frac{R_A \cdot R_B + R_B \cdot R_C + R_C \cdot R_A}{R_A}$$
$$\frac{R_1 \cdot R_2}{R_1 + R_2 + R_3} \blacktriangleright R_C \qquad\qquad R_2 \blacktriangleleft \frac{R_A \cdot R_B + R_B \cdot R_C + R_C \cdot R_A}{R_B}$$
$$R_3 \blacktriangleleft \frac{R_A \cdot R_B + R_B \cdot R_C + R_C \cdot R_A}{R_C}$$

$$\Delta \qquad\qquad\qquad\qquad \text{Y}$$

EXPLANATION	"BASIC" PROGRAM SOLUTION
CLEAR THE SCREEN AND PROGRAM TITLE	10 CLS 20 PRINT"DELTA/WYE CONVERSION PROGRAM" 30 PRINT"":PRINT""
SELECT THE DIRECTION IN WHICH A CONVERSION IS TO BE MADE. JUMP TO LINE 200 FOR WYE TO DELTA CONVERSION.	40 PRINT"(1) DELTA TO WYE CONVERSION" 50 PRINT"(2) WYE TO DELTA CONVERSION" 60 PRINT"":INPUT"WHICH CONVERSION? (1 OR 2)";C 70 PRINT"" 80 IF C = 2 THEN 200
DELTA TO WYE CONVERSION SECTION. ENTER THE VALUES OF THE THREE DELTA RESISTORS IN CONSECUTIVE ORDER. USE E NOTATION IF DESIRED.	90 INPUT"ENTER THE VALUES OF R1, R2, AND R3. (R1, R2, R3) – ";R1,R2,R3 100 S = R1 + R2 + R3 110 RA = R2 * R3/S 120 RB = R3 * R1/S (Continued next page)

DESIGN NOTE 8-2:   (Continued)

EXPLANATION	"BASIC" PROGRAM SOLUTION
THE VALUES OF THE WYE RESISTORS ARE THEN PRINTED ON THE SCREEN. JUMP TO LINE 320 TO END.	130 RC = R1 * R2/S 140 PRINT"" 150 PRINT"RA = ";RA;" OHMS, RB = ";RB;" OHMS, RC = ";RC;" OHMS." 155 PRINT"" 160 GOTO 300
WYE TO DELTA CONVERSION SECTION. ENTER THE WYE RESISTORS IN CONSECUTIVE ORDER USING E NOTATION IF DESIRED. THE VALUES OF THE DELTA RESISTORS ARE THEN PRINTED ON THE SCREEN. LINE 320 ENDS THE PROGRAM.	200 INPUT"ENTER THE VALUES OF RA, RB, AND RC. (RA,RB,RC) – ";RA,RB,RC 210 S = RA * RB + RB * RC + RC * RA 220 R1 = S/RA 230 R2 = S/RB 240 R3 = S/RC 250 PRINT"" 260 PRINT"R1 = ";R1;" OHMS, R2 = ";R2;" OHMS, R3 = ";R3;" OHMS." 270 PRINT"" 300 INPUT"ANOTHER PROBLEM? (Y/N)";A$ 310 IF A$ = "Y" THEN CLEAR:GOTO 30 320 CLS:END

SELF-CHECK 8-9

1. Find the voltage across R_L in Figure 8-85. Use Δ to Y conversion.

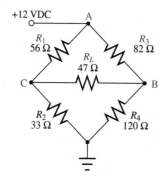

Figure 8-85

Summary

FORMULAS

$$(8.1)\ I_L = \frac{I_N \cdot (R_N \parallel R_L)}{R_L}$$

(current divider formula)

(8.2) $I_L = I_N \cdot \dfrac{R_N}{R_N + R_L} = \dfrac{I_N \cdot R_N}{R_N + R_L}$

(current divider formula for the Norton model)

(8.3) $V_{AB} = \dfrac{(V_1/R_1) + (V_2/R_2) + (V_3/R_3) + \cdots + (V_N/R_N)}{(1/R_1) + (1/R_2) + (1/R_3) + \cdots + (1/R_N)}$

(Millman's formula used to find the output terminal voltage for parallel voltage sources)

(8.4) $P_{L(\text{max})} = \dfrac{(V_s/2)^2}{R_{\text{int}}}$

(to determine the maximum output power from a voltage source)

(8.5) $P_{L(\text{max})} = (I_N/2)^2 \cdot R_N$

(to determine the maximum output power from a current source)

(8.6) $R_A = \dfrac{R_2 \cdot R_3}{R_1 + R_2 + R_3}$

$R_B = \dfrac{R_3 \cdot R_1}{R_1 + R_2 + R_3}$

$R_C = \dfrac{R_1 \cdot R_2}{R_1 + R_2 + R_3}$

(formula set for Δ to Y conversion)

(8.7) $R_1 = \dfrac{R_A \cdot R_B + R_B \cdot R_C + R_C \cdot R_A}{R_A}$

$R_2 = \dfrac{R_A \cdot R_B + R_B \cdot R_C + R_C \cdot R_A}{R_B}$

$R_3 = \dfrac{R_A \cdot R_B + R_B \cdot R_C + R_C \cdot R_A}{R_C}$

(formula set for Y to Δ conversion)

CONCEPTS

- The Superposition Theorem: The actual currents and voltage drops, in a circuit containing multiple source voltages, are the algebraic sum of currents and voltage drops produced by each individual source voltage.
- Thevenin's theorem states that any linear complex circuit can be reduced to a single source voltage with a series resistor for the purpose of circuit analysis.
- A practical voltage source will have some internal resistance represented as a series resistor.
- An ideal voltage source has *no* internal resistance.
- Norton's theorem states that any linear complex circuit can be reduced to a constant-current source and a parallel resistance for the purpose of circuit analysis.
- A practical current source has an internal parallel resistance.
- An ideal current source is a constant-current source with an infinite parallel resistance.
- Millman's theorem states that any number of parallel voltage sources can be reduced to a single practical voltage source by converting all parallel voltage sources to practical

current sources, then combining them into a single practical current source and converting it back to a single practical voltage source.

- The maximum power transfer theorem states that the amount of power any source can supply to a load is limited by the internal resistance of the source and will be at a maximum value when the load resistance equals the internal resistance of the source.
- A node is a junction or connecting point in a circuit.
- A mesh, or meshwork, is a complex circuit that is composed of interconnecting sections having some components in common between sections.
- If the calculated current in a mesh equation is negative in sign, it is because the actual direction of current is opposite to that which was originally assumed.
- $\Delta \rightleftharpoons Y$ conversion can be used to greatly simplify some circuits for analysis.

PROCEDURES

Superposition

1. Completely analyze the circuit using only one source voltage at a time. All other source voltages are temporarily shorted (replaced with a wire).

2. Repeat step #1 for each source voltage. Label current directions and voltage drops for each resistor for each voltage source.

3. Take the algebraic sum of all currents or voltage drops for each component.

Thevenizing Procedure

1. Remove the load and solve for V_{Th} at the unloaded output terminals.

2. Replace all voltage sources with a short and calculate R_{Th} from the unloaded output terminals.

3. Draw the Thevenin model with the load connected. Calculate the loaded output voltage.

Nortonizing Procedure

1. Find I_N by replacing the load with a short and calculating the current that would flow through the short.

2. Replace all voltage sources with a short and calculate R_N from the unloaded output terminals.

3. Draw the Norton model with the load connected and calculate the amount of current supplied as the load.

Norton to Thevenin Conversion

The Norton current is multiplied by the parallel resistance to find the Thevenin voltage, $R_{Th} = R_N$.

Thevenin to Norton Conversion

The Thevenin voltage is divided by the series resistance to find the Norton current. $R_{Th} = R_N$.

Nodal Analysis

1. Form a primary nodal current equation and express each current in terms of voltage and resistance. All voltages must be expressed in terms of the voltage drop between the two primary nodes.

2. Solve the nodal current equation for the primary nodes' voltage drop.

3. Substitute the voltage drop back into each term of the nodal current equation and solve for each nodal current.

4. Use the node currents to solve for all unknown voltage drops in the circuit.

Mesh Analysis

1. Assign all loop currents.

2. Form mesh equations by expressing each voltage drop as an *IR* term and setting the sum of all drops equal to the source voltage of each section.

3. Solve for the unknown currents using a simultaneous-equation technique.

4. Calculate all voltage drops and correct current directions and voltage drop polarities as necessary.

SPECIAL TERMS

- Superposition theorem, Thevenin voltage and resistance
- Network theorem, network
- Norton current and resistance
- Closed loop, loop equation
- Nodal analysis, nodal, node, principal node
- Mesh analysis, meshwork, mesh, mesh equation
- Delta (Δ) configuration, wye (Y) configuration

Need to Know Solution

ACTUAL CURRENT DIRECTIONS ARE SHOWN BY LOOP ARROWS

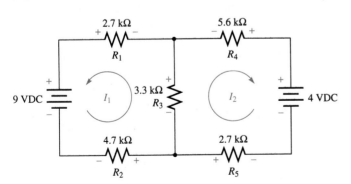

$$10{,}700 \cdot I_1 + \ 3{,}300 \cdot I_2 = +9 \qquad \text{SIMULTANEOUS}$$
$$+3{,}300 \cdot I_1 + 11{,}600 \cdot I_2 = +4 \qquad \text{MESH EQUATIONS}$$

$I_1 = +0.805$ mA

INDICATES CURRENT IS
FLOWING AS ASSUMED.

$I_1 = 0.805$ mA	$I_2 = 0.116$ mA
$V_{R1} = 2.174$ V	$V_{R4} = 0.65$ V
$V_{R2} = 3.784$ V	$V_{R5} = 0.313$ V
$+V_{R3} = 3.04$ V	$+V_{R3} = 3.04$ V
$E_{S1} = 8.998$ V	$E_{S2} = 4.003$ V

Questions

8-1 The Superposition Theorem

1. Write the superposition theorem in your own words.
2. What do superposition and Ohm's Law have in common?
3. List the three main steps of the superposition method.

8-2 Thevenin's Theorem

4. What does it mean to Thevenize a circuit?
5. What is a linear circuit?
6. Why is the Thevenin model so useful for load analysis?
7. What is the difference between a practical and an ideal voltage source?

8-3 Norton's Theorem

8. How is Norton's theorem different from Thevenin's theorem?
9. What is a constant-current source?
10. An ideal current source has a very low internal parallel resistance. (True or False)
11. Why is the Norton model considered a practical current source?

8-4 Millman's Theorem

12. Describe Millman's theorem in your own words.
13. When parallel voltage sources are combined, they are first converted to parallel _____ _____ .

8-5 Combined Current and Voltage Sources

14. Describe how you would analyze a circuit that contains a voltage source and a current source.
15. How is a current source converted to a voltage source?

8-6 The Maximum Power Transfer Theorem

16. Describe the maximum power transfer theorem in your own words.
17. What two basic factors limit the amount of power a source can deliver to a load?
18. If a voltage source has an internal series resistance of 10 Ω, what value load resistor will draw the most power from the source?
19. A practical current source is delivering maximum power to a load when all of the current from the constant-current source is sent to the load, with no current going through the internal parallel resistance of the current source. (True or False)

8-7 Nodal Analysis

20. What is a principal node?
21. Summarize the nodal analysis procedure.
22. What does it mean when simultaneous loop equations are solved for an unknown current in a circuit, and it is found that the current is a negative quantity?

8-8 Mesh Analysis

23. What does *mesh* mean as it pertains to an electrical circuit?
24. Any mesh equation is composed of three terms. What are they?
25. What determines the number of mesh equations that you must form in order to solve for unknown currents and voltage drops in a complex circuit?
26. How would you write a mesh equation for a section of an electrical meshwork that does not have a source voltage?

8-9 Let Delta (Δ) = Wye (Y)

27. What will happen to the total resistance and total current of a circuit if a delta section within the circuit is replaced by an equivalent wye section?

Problems

8-1 The Superposition Theorem

1. Use the superposition theorem to solve for currents and voltage drops in Figure 8-86.

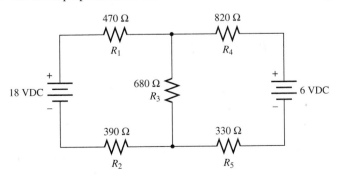

Figure 8-86

2. Use the superposition theorem to solve for currents and voltage drops in Figure 8-87.

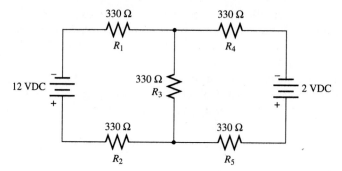

Figure 8-87

8-2 Thevenin's Theorem

3. Thevenize Figure 8-88.

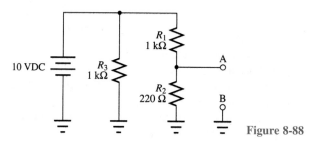

Figure 8-88

4. Thevenize Figure 8-89.

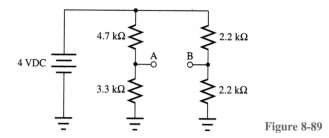

Figure 8-89

5. Thevenize Figure 8-90.

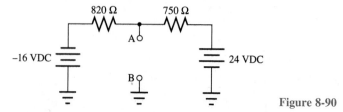

Figure 8-90

6. What will the voltage be across a 10 kΩ resistor connected to the output of Figure 8-89?

7. What will the voltage be across a 470 Ω resistor connected to the output of Figure 8-90?

8-3 Norton's Theorem

8. Convert the Thevenin model of Problem 3 into a Norton model.
9. Convert the Thevenin model of Problem 5 into a Norton model.
10. Combine the current sources of Figure 8-91 into one.

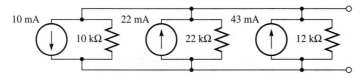

Figure 8-91

11. Nortonize Figure 8-92 and calculate load voltage.

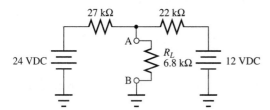

Figure 8-92

12. Nortonize Figure 8-93 and calculate the voltage across the load.

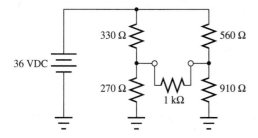

Figure 8-93

13. Convert the Norton model of Problem 11 into the Thevenin equivalent.

8-4 Millman's Theorem

14. Reduce Figure 8-94 to a single practical voltage source using Millman's formula.

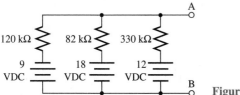

Figure 8-94

15. Solve for the voltage across the load in Figure 8-95.

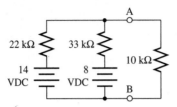

Figure 8-95

8-5 Combined Current and Voltage Sources

16. Convert the current source in Figure 8-96 to a voltage source and reduce the circuit to a Thevenin model.

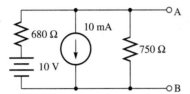

Figure 8-96

17. Convert the voltage source in Figure 8-97 to a current source and reduce the circuit to a Norton model.

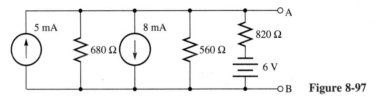

Figure 8-97

18. Determine load voltage and current if a 1 kΩ load resistor is connected to terminals A and B of Figure 8-96.

19. Determine load voltage and current if a 100 Ω resistor is connected to terminals A and B of Figure 8-97.

8-6 The Maximum Power Transfer Theorem

20. Add a resistive load to Figure 8-98 that will draw the maximum amount of power. Then calculate how much power that is.

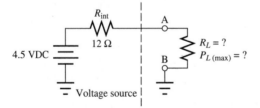

Figure 8-98

21. Add a resistive load to Figure 8-99 that will draw the maximum amount of power. Then calculate how much power that is.

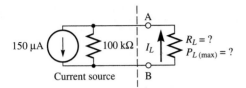

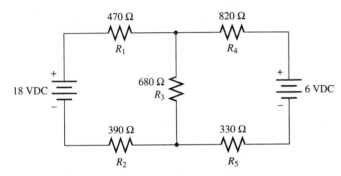

Figure 8-99

8-7 Nodal Analysis

22. Use the nodal analysis procedure to solve for currents and voltage drops in Figure 8-86.

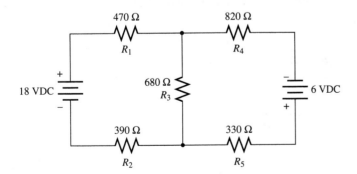

Figure 8-86

23. Use the nodal analysis procedure to solve for currents and voltage drops in Figure 8-100. (Note that this is the same circuit as that shown in Figure 8-86 except the 6 V supply is reversed in polarity.)

Figure 8-100

24. Use the nodal analysis procedure to solve for currents and voltage drops in Figure 8-101. Show all of your work.

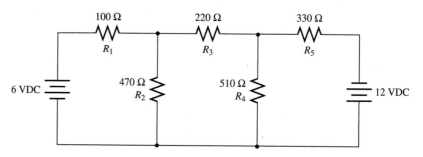

Figure 8-101

8-8 Mesh Analysis

25. Form the mesh equations that would be necessary to solve for unknown currents in Figure 8-102.

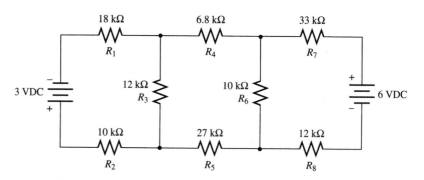

Figure 8-102

26. Use mesh analysis to solve for currents and voltage drops in Figure 8-87. Show all work.

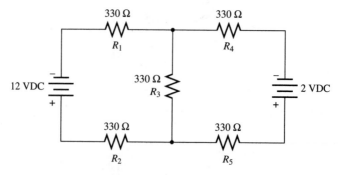

Figure 8-87

8-9 Let Delta (Δ) = Wye (Y)

27. Convert Figure 8-103 into an equivalent wye configuration.

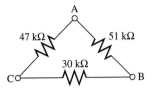

Figure 8-103

28. Convert Figure 8-104 into an equivalent delta configuration.

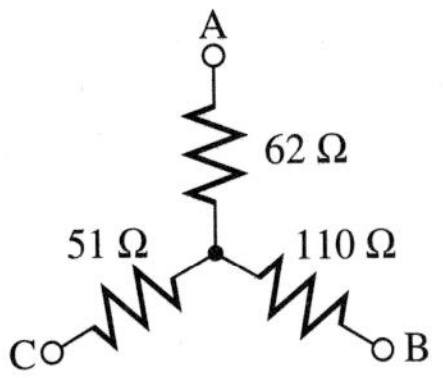

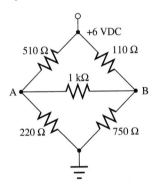

Figure 8-104

29. Solve for the output voltage between terminals A and B in Figure 8-105. Use delta to wye conversion and show all your work.

Figure 8-105

Answers to Self-Checks

Self-Check 8-1

Steps
1.

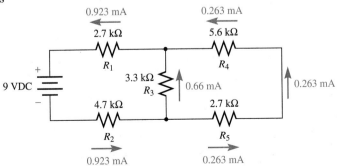

Figure 8-106

2.

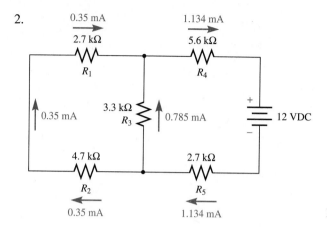

Figure 8-107

3.

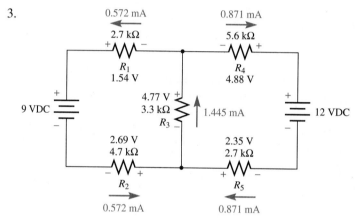

Figure 8-108

Self-Check 8-2

1. $V_{Th} = 1.7$ V; $R_{Th} = 11.94$ kΩ; $V_L = 1.4$ V
   Terminal B is more positive than terminal A.
2. $V_{Th} = +9.28$ V; $R_{Th} = 359$ Ω; $V_L = +5.26$ V

Self-Check 8-3

1. $I_N = 0.142$ mA; $R_N = 11.94$ kΩ; $I_L = 25$ μA
2. $I_N = 25.8$ mA; $R_N = 359$ Ω; $I_L = 11.2$ mA

Self-Check 8-4

1. $+6.65$ V
2. $+7.28$ V

Self-Check 8-5

1. $V_{Th} = 8.6$ V, $R_{Th} = 68.8$ Ω
   $V_L = 4.27$ V
2. $V_{Th(total)} = 5$ V, $R_{Th(total)} = 238$ Ω
   $V_L = 1.48$ V

Self-Check 8-6

1. a. The load resistance should be 15 Ω for maximum output power.
   b. The maximum output power will be 0.82 W.
2. a. The load resistance should be 40 Ω for maximum output power.
   b. The maximum output power will be 0.4 W.

Self-Check 8-7

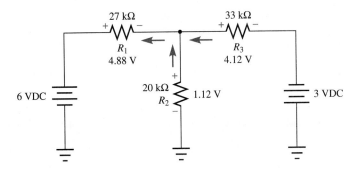

Figure 8-109

1. According to Kirchhoff's Current Law:

$$I_{R1} = I_{R2} + I_{R3}$$

Use voltage and resistance to express each current:

$$\frac{6\text{ V} - V_{R2}}{27\text{ k}\Omega} = \frac{V_{R2}}{20\text{ k}\Omega} + \frac{V_{R2} - (-3\text{ V})}{33\text{ k}\Omega}$$

Solve for V_{R2}: $V_{R2} = 1.12$ V

Solve for V_{R1}: $V_{R1} = 6$ V $- 1.12$ V $= 4.88$ V

Solve for V_{R3}: $V_{R3} = V_{R2} - (-3$ V$) = 1.12$ V $+ 3$ V $= 4.12$ V

Self-Check 8-8

1. First form the mesh equations:

(Mesh #1) $10,700\ I_1 + 3,300\ I_2 = +2$ V

(Mesh #2) $3,300\ I_1 + 11,600\ I_2 = +12$ V

Solve the simultaneous mesh equations for the unknown currents:

$I_1 = 0.145$ mA; $I_2 = 1.08$ mA

Solve for I_{R3}: $I_{R3} = I_2 - I_1 = 1.08$ mA $- 0.145$ mA

$= 0.935$ mA

Solve all voltage drops:

$$V_{R1} = 0.145 \text{ mA} \cdot 2{,}700 \ \Omega = 0.392 \text{ V}$$
$$V_{R2} = 0.145 \text{ mA} \cdot 4{,}700 \ \Omega = 0.682 \text{ V}$$
$$V_{R3} = \ \ 0.935 \text{ mA} \cdot 3{,}300 \ \Omega = 3.09 \text{ V}$$
$$V_{R4} = \ \ 1.08 \text{ mA} \ \cdot 5{,}600 \ \Omega = 6.05 \text{ V}$$
$$V_{R5} = \ \ 1.08 \text{ mA} \ \cdot 2{,}700 \ \Omega = 2.92 \text{ V}$$

Self-Check 8-9

Δ to Y conversion $= (R_A = 24.8 \ \Omega; R_B = 20.8 \ \Omega; R_C = 14.2 \ \Omega)$
$V_C = +4.93 \text{ V}; V_B = +6.01 \text{ V}; V_{CB} = V_{RL} = 1.08 \text{ V}$

SUGGESTED PROJECTS

1. Add the analytical tools from this chapter to your Electricity and Electronics Notebook.

2. Build one or two of the complex DC circuits shown in this chapter and verify your calculations with a multimeter.

3. Think of a way to determine the internal resistance of a voltage source experimentally. Enough information was given in this chapter for you to be able to do this. (*Hint:* It has something to do with maximum power transfer.) Now, have a friend create a voltage source (Thevenin model) by connecting a resistor of unknown value in series with a battery. Then experimentally determine the resistor's value using the experimental procedure you devised. (Using the ohmmeter is illegal here, of course.)

EDISON SLOW-SPEED MOTOR
1889

As early as 1890 an Edison battery-operated fan became a necessity in our homes and places of business

Part III

Magnetism and Magnetic Devices

PART OUTLINE

In the same year that Edison invented the first practical electric light bulb, 1879, he made significant design improvements to the dynamo, a DC electric generator. Building on the work of Englishman Michael Faraday, and American and German inventers Wallace and Siemens, Edison developed generators that were approximately 90% efficient compared to the old 40% to 50% efficiency. The generators on these two pages are examples of his work. You will learn more about Edison's generators, motors, and meters in the openings of the following Part III chapters. (Photographs by Mark E. Hazen, courtesy of Edison-Ford Winter Estates, Fort Myers, FL 33901)

Edison's Generators, Motors, and Meters

(1) These "Long Waisted Mary Anns", as Edison called them, were installed then activated on July 4, 1883 to light the homes and streets of Sunbury, Pennsylvania. What a marvelous sight it must have been to see the night illuminated with Edison's lights and generating system. (2) Shown in this photograph is Edison's chemical meter used to determine electrical usage. Approximately 1% of the electrical current used by the customer passed through the glass jar containing a zinc sulfate solution and two zinc electrodes. The DC current caused zinc from one electrode to be removed and deposited on the second. Each month, the jars were collected and the zinc electrodes weighed to determine electrical usage. These meters were found to be very accurate. The thermostatically controlled light bulb shown was used to warm the box in the winter.

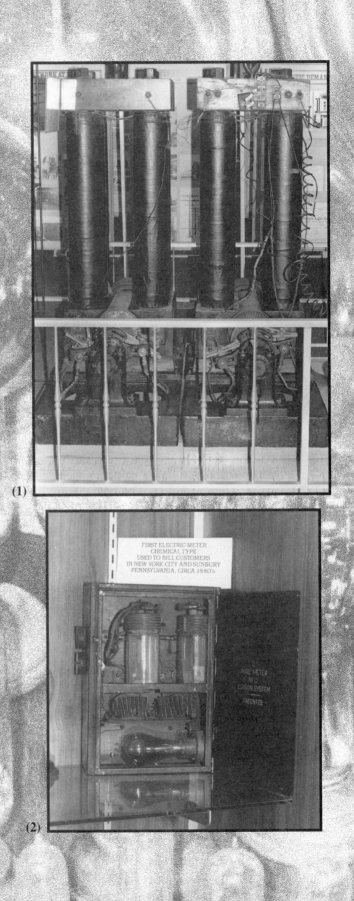

(1)

(2)

Chapter 9

Magnetism and Electromagnetic Induction

OBJECTIVES

After studying this chapter, you will be able to
- explain the characteristics of magnetic polarity, magnetic attraction and repulsion, and the magnetic field.
- calculate magnetic quantities such as total flux, magnetomotive force, field intensity, flux density, and permeability.
- describe the characteristics of magnetic and nonmagnetic materials.
- explain how some materials can be magnetized and demagnetized.
- determine the magnetic polarity of a solenoid by observing the direction of current and the manner in which the coil is wound.

- calculate electromagnetic quantites of magnetomotive force, field intensity, and flux density when the amount of current, number of turns of the coil, core permeability, and solenoid length are known.
- understand and explain magnetic hysteresis.
- understand and explain the Hall effect.
- list and explain the three basic requirements for electromagnetic induction.
- demonstrate and use the left-hand rule for induction.
- understand and use Lenz's Law to determine the magnetic polarity of an induction coil.
- explain Faraday's Law.

- use Faraday's Law to calculate induced voltage in a coil.
- determine the polarity of induced voltage and current in a coil.
- describe how alternating current (AC) is generated.

- describe the basic parts of an alternator.
- explain how direct current (DC) is generated.
- describe the basic parts of a DC generator.
- explain the basic principles of magnetohydrodynamic (MHD) power generation.

INTRODUCTION

Magnetism is something that has fascinated all of us since we were small children. At one time or another, we have all amused ourselves by dangling tacks or paper clips from toy magnets, or perhaps by chasing one magnet with another across a tabletop. This fascination with magnetism is not unique to modern man. Ancient writings of the Chinese and Greeks told of a mysterious black stone that had the power to attract other black stones and iron metals. These mysterious stones became known as lodestones, later to be identified as a mineral called magnetite.

Mankind's understanding of magnetism really began in the mid-thirteenth century with a French experimenter named Petrus de Maricourt. It is he who discovered and traced magnetic lines of force and identified magnetic poles. But scientific progress was very slow in the Middle Ages (fifth to fifteenth centuries), perhaps due to primitive methods of communication, limited scientific writings, and much superstition. It wasn't until about 1600 that the earth's magnetic field was discovered, by an English doctor named William Gilbert. For the first time, it was possible to formulate a scientific explanation for the strange actions of the lodestone compass used by early explorers. With little more progress, another 200 years would go by before man would begin to see the relationship between electricity and magnetism.

In the early 1800s, men such as Hans Christian Oersted of Denmark, Michael Faraday and James Maxwell of Britain, and André Ampère of France led the way with discoveries tying magnetism and electricity closely together. By the late 1800s, mankind was using the telegraph, telephones, centrally generated and distributed electricity, and even electric cars (battery-powered cars were popular between about 1880 and 1920), all making use of magnetic and electromagnetic principles. Today mass-transit vehicles race along magnetic tracks at hundreds of km/h suspended in midair and propelled by powerful magnetic fields. We have learned well how to use this mysterious force. In this chapter, you will learn in a very short time what it took mankind hundreds of years to discover.

NEED TO KNOW

Let's suppose you are living on an island in the South Pacific and you have recently purchased your own gasoline-powered generator to provide electricity for your home (hut). One night, not long after putting the generator into service, you turn on nearly every light in the house and proceed to cook your evening meal. The small stove you are using is electric. You notice, when you turn on the stove, that the lights get dim and the gasoline engine driving the generator slows down and starts coughing deeply. You turn off the electric stove, the lights brighten, and the gasoline engine stops coughing. How can that generator possibly know when you turn on the electric stove? You soon discover that you have a choice of starving in the light or cooking in the dark.

Why did the engine slow down and begin laboring (coughing) when all the lights and the electric stove were on? Is the engine defective? Is the generator defective? How did the generator know when the electric stove was on or off? Why did the lights dim? There must be a scientific explanation! The explanation is up to you. You are supposed to be the expert. The lights dim, the engine slows and coughs, and you can't cook your meal. What's happening? You need to know! Read on.

9-1 Magnetism and the Magnetic Field

Magnetic Polarity

All magnets, be they natural or man-made, have a **magnetic polarity**. In other words, a magnet will have two poles which are opposites. Just as an electrical source has a + and − pole representing electrical polarity, a magnetic source will have a **north** (N) and **south** (S) **pole** representing magnetic polarity. Magnetic lines of force are depicted as extending from the N end of the magnet to the S end, shown by arrowed lines in Figure 9-1.

See color photo insert page C1, photograph 39.

The polarity of the magnet is not arbitrarily determined. A bar magnet that is suspended at its center from a string, for example, will align itself with the true magnetic north of the earth (magnetic north is about 15° off from geographical north). The needle of a compass is actually a carefully balanced magnet whose north seeking end is labeled north. Thus, the north seeking end of any magnet is considered the north end of that magnet.

Magnetic Field

The **magnetic field** is an invisible force field that extends outward in loops from any magnetic source, such as a current-carrying conductor, a magnetized iron or iron-alloy object, or a natural iron-oxide mineral called magnetite. As depicted in Figure 9-1, the

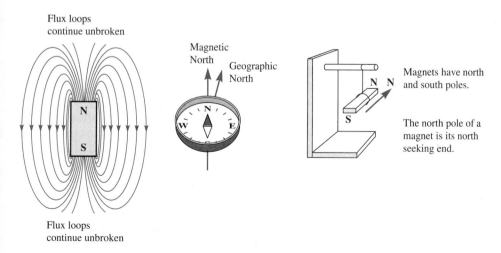

Figure 9-1   The north end of a magnet is the north-seeking end.

James Clerk Maxwell (1831 – 1879)

James Maxwell was a British scientist considered to be one of the greatest mathematicians and physicists of his time. He is well known for his contributions to science in the areas of electricity, magnetism, and kinetic theory of gases. In 1864, he proposed that energy traveled through space as electromagnetic waves. Maxwell concluded that light waves are electromagnetic in nature. In 1873, Maxwell's *Treatise on Electricity and Magnetism* was published. It is recognized as the foundation for modern electromagnetic theory. (Photo courtesy of The Bettmann Archive.)

force field is organized into actual lines of force that run parallel to each other, extending from the north end of the magnet, or magnetized material, to the south end. These magnetic lines of force are called **magnetic flux**, **flux lines**, or simply **flux**. These flux lines form continuous loops that may be traced from the north end of the magnet through the air, or surrounding material, to the south end. Each flux loop is continued within the magnet from the south end to the north end. The flux lines are evenly spaced within the magnet and symmetrically arranged in the outer loops, extending from north to south, as long as the magnetic material and surrounding medium are homogeneous (uniform in structure and composition).

Field Strength

Flux (Φ) and the Weber (Wb)

The strength of the magnetic field, and therefore the magnetic source, is determined by the number and concentration of flux lines. The number of flux lines is expressed in units of **webers (Wb)**, a standard SI unit. A very strong half-pound permanent magnet may have between 2,000 and 3,000 lines of flux (Φ). The weber, named after the German physicist Wilhelm Weber (1804 – 1890), is a very large unit, equal to 100 million flux lines (1 Wb = $1 \cdot 10^8$ Φ, and 1 Φ = $1 \cdot 10^{-8}$ Wb = 0.01 μWb). Since the weber is such an enormous unit, it is usually prefixed with micro or milli (μWb or mWb).

EXAMPLE   9-1   2,000 Φ is equal to 20 μWb (2,000 $\Phi \cdot 1 \cdot 10^{-8}$ Wb/Φ = $2 \cdot 10^{-5}$ Wb = 20 μWb).

Flux Density (B)

The concentration of flux is referred to as **flux density**, symbolized with an uppercase B. The flux density is simply a ratio of the number of flux lines that pass through a given cross-sectional area. In other words, B is equal to magnetic flux per unit area expressed as follows:

$$B = \Phi/A \qquad \text{(flux per unit area)} \tag{9.1}$$

Flux density (B) is expressed in **tesla**. One tesla (1 T) is equal to 100,000,000 flux lines passing through a unit area of one square meter (1 T = 100 MΦ/m^2). The flux density of the earth's magnetic field is approximately 50 μT along the earth's surface. The flux

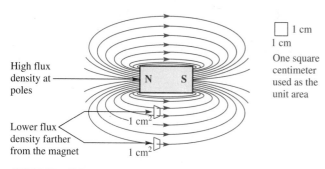

High flux density at poles

Lower flux density farther from the magnet

1 cm²

1 cm²

1 cm
1 cm
One square centimeter used as the unit area

(NOTE: Not all flux lines are shown in this illustration.)

Figure 9-2   Flux density (*B*) is the number of flux passing through a unit area.

density of a toy horseshoe magnet may be in the neighborhood of 2 to 20 mT at its poles, and a very powerful permanent magnet may have a flux density of 0.1 T at its poles.

Figure 9-2 illustrates the flux surrounding a permanent magnet. In this example, a unit area of 1 cm² is used because the magnet is small. Note that the flux density is greater at the poles of the magnet than at some location away from the poles. Note also that flux density decreases with increasing distance from the magnet. The change in flux density is the inverse square of the change in distance. If the distance is doubled, the flux density is ¼ of what it was ($\frac{1}{2}^2 = \frac{1}{4}$). If the distance is tripled, the flux density will be ⅑ of what it was ($\frac{1}{3}^2 = \frac{1}{9}$).

If the flux density is 1 mT at a distance of 5 cm from a magnet, at a distance of 10 cm the flux density is 0.25 mT ($B = (5 \text{ cm}/10 \text{ cm})^2 \cdot 1 \text{ mT} = 0.25 \text{ mT}$).

EXAMPLE   9-2

As you can see, the magnetic field strength decreases rapidly with increasing distance. We will discuss flux density again later.

Attraction and Repulsion

Magnetism is used in a wide variety of instruments, devices, and machines. The widespread usefulness of magnetism is based on its characteristics of attraction and repulsion. Hi-Fi speakers, electric motors, and linear induction motors are just a few of the many useful

See color photo insert pages C2 through C4.

Nikola Tesla (1856–1943)

Nikola Tesla is an American inventor who was born in Smiljan, Austria-Hungary (now Yugoslavia). He moved to the United States in 1884, bringing with him a wealth of knowledge and creativity. He is very well known for his work with electromagnetic devices—such as AC induction motors, high-voltage apparatus (Tesla coil), and a wireless transmission system. Tesla held many patents on his inventions—including the induction motor, which he sold to George Westinghouse. (Photo courtesy of The Bettmann Archive.)

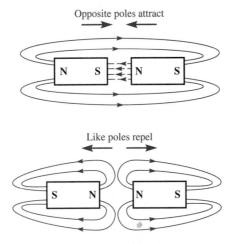

Figure 9-3 Like poles repel while opposite poles attract.

devices that employ the principles of magnetic attraction and repulsion. Two or more magnetic fields will either join, and develop a force of attraction, or separate, developing a force of repulsion. Whether the force developed is that of attraction or repulsion depends upon the relative polarity of the magnetic sources involved. The basic physical law is stated as follows:

Opposite poles attract, while like poles repel.

The north pole of one magnet will repel the north pole of another. Likewise, the south pole of one magnet will repel the south pole of another. However, if a north pole of one magnet is brought close to the south pole of another, a force of attraction will result. Figure 9-3 shows the magnetic flux pattern for each case, attraction and repulsion. Notice that the flux lines join between poles of opposite polarity and repel between poles of like polarity.

In this section, you were introduced to basic magnetic theory, calculations, and principles. Take time to review this section before answering the questions of Self-Check 9-1.

SELF-CHECK 9-1

1. How can you determine the north end of an unmarked bar magnet?

2. If a magnet generates 5,000 flux lines, what is its magnetic flux expressed in webers?

3. If 200 flux lines pass through a cross-sectional area of 4 cm², what is the flux density measured in tesla?

4. If the flux density at a distance of 4 cm from a magnetic source is 8 mT, what is the flux density at a distance of 16 cm from the source?

5. If a magnetic vehicle is to float above a magnetic surface and the magnetic surface is of the north polarity, what polarity should the bottom of the magnetic vehicle be? Why?

9-2 Magnetic and Antimagnetic Materials

Magnetic Material Characteristics

Permeability (μ)

The **permeability** of a material is a measure of the ease with which flux can be established, or set up, in and throughout the material. The symbol used for permeability is the Greek letter μ (mu). All materials or substances may be classified magnetically, using permeability as the comparative figure of merit (value used in classification). If materials are to be classified, they must be compared to some standard. In this case, the standard material by which all others are compared is air (space or vacuum). Air is assigned a permeability value of one (1), and all other materials are rated accordingly. This is known as **relative permeability** (μ_r), since the permeability of all other materials is determined relative to air. The relative permeability of various iron alloys may range as high as 10,000 (100,000 for permalloy). Metals with very high permeabilities ($\geq 1,000$) are referred to as high-mu metals. One very soft iron alloy is actually named **mumetal** (the ''mu'' standing for ''μ'' as in permeability). Mumetal is commonly used for magnetic shielding, since the magnetic flux is collected and trapped within the walls of the shield. Because the permeability of mumetal is so high, very little flux escapes into the air on the other side. Thus, instead of the magnetic field being repelled by the shield, it is readily accepted and contained. Figure 9-4 demonstrates the shielding effect of high-mu metals.

Reluctance ($\mathcal{R}$ or R_m)

Reluctance is another figure of merit by which materials may be magnetically categorized. It is the opposite of permeability. **Reluctance** is the measure of opposition to the establishment of flux in a material. Some materials actually resist magnetic lines of force, just as resistance in a circuit will oppose or resist current (this analogy will be developed further later). If an object made of a material that has a high reluctance, such as copper, is placed in a magnetic field, most of the flux lines will go around the object instead of through it. The air around the copper object has a higher permeability and a lower reluctance than the object. Thus, as shown in Figure 9-5, the flux will tend to follow the paths of least reluctance $\mathcal{R}$ (magnetic resistance R_m).

Retentivity

When a material is brought within a magnetic field, the magnetic field attempts to induce a magnetic polarity in the material. Naturally the success of this will depend upon the permeability and reluctance of the material. When the material is removed from the mag-

Figure 9-4   High-μ metal provides exceptional magnetic shielding.

Figure 9-5 illustration:
Less reluctance in the air

Flux

N

HIGH RELUCTANCE METAL

S

Flux

Most flux will follow path of least reluctance

Figure 9-5 Flux would rather flow through air than through high-reluctance materials.

netic field, a certain amount of magnetic polarity, or magnetism, may remain in the material. Figure 9-6 demonstrates this. The magnetism remaining in the material is referred to as *residual magnetism*. The units of measure for residual magnetism are gauss (cgs) or tesla (SI). The ability of the material to retain a quantity of residual magnetism is known as **retentivity**.

Magnetic Material Categories

Ferromagnetic Materials

The prefix "ferro-" comes from *ferrum*, the Latin word for iron. Thus, **ferromagnetic** materials are those materials that have magnetic properties similar to iron. In other words, they are materials that have high permeability, low reluctance, and are easily magnetized. Some of the most common ferromagnetic materials are iron, nickel, cobalt, and various iron alloys such as alnico and permalloy.

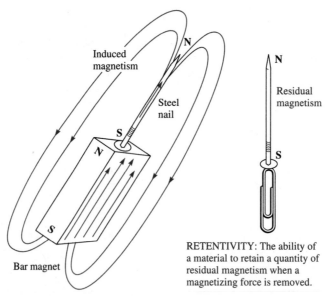

Induced magnetism

N

Steel nail

S

N

Bar magnet

S

N

Residual magnetism

S

RETENTIVITY: The ability of a material to retain a quantity of residual magnetism when a magnetizing force is removed.

Figure 9-6 Materials with retentivity have residual magnetism when the magnetizing force is removed.

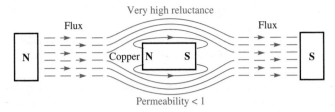

Figure 9-7   Diamagnetic materials such as copper actually produce opposing magnetic fields when placed in a magnetic field.

Paramagnetic Materials

The prefix "para-" means "besides" or "other than." **Paramagnetic**, therefore, is a term that means "other than magnetic." Materials such as air, chromium, platinum, and aluminum are considered to be paramagnetic. They have relative permeabilities of one (1), or slightly greater, and become only slightly magnetized under the influence of a very strong magnetic field. The polarity of this slight magnetization is the same as that of the magnetizing field. When the magnetizing field is removed, there is no residual magnetism remaining in the paramagnetic material.

Diamagnetic Materials

This time the prefix is "dia-" or "di-" meaning "of two." Diamagnetic materials are those materials that will become slightly magnetized in the presence of a very strong magnetic field. However, as shown in Figure 9-7, the polarity of magnetism is opposite to that of the magnetizing field. Thus, there are two magnetic fields, or magnetic polarities, in opposition to each other, such as two north poles or two south poles. When the magnetizing field is removed, there is no residual magnetism left in the diamagnetic material. Some of the more common diamagnetic materials are copper, bismuth, silver, gold, and mercury. The permeability of these materials is less than one (1) and they exhibit a very high reluctance.

Take time to briefly review the terms explained in this section. Then test your knowledge by answering the following Self-Check questions.

SELF-CHECK 9-2

1. What is the term for the measure of opposition to the establishment of flux in a material?

2. What is the term for the ability of a material to retain a quantity of residual magnetism?

3. Define ferromagnetic materials in your own words.

4. Copper is classified as what kind of magnetic material?

9-3 Permanent Magnets

Permanent magnets in natural form were discovered long ago in ancient times and recorded in early manuscripts by the Chinese, Greeks, and other early civilizations. This natural magnetic material is commonly known as magnetite, a ferro–ferric oxide mineral (Fe_3O_4). Thirteenth-century Europeans discovered that an elongated piece of this marvelous mineral suspended from a string would align itself north and south. Thus, the magic of this stone led many explorers to new conquests. It readily became known as a "leading stone," for obvious reasons, and later became known as a "lodestone."

The Magnetic Structure

It is believed that each atom in ferromagnetic materials has an overall magnetic polarity. For this reason, the atoms of ferromagnetic materials are often referred to as **atomic dipoles**, meaning they have two magnetic poles.

Figure 9-8 shows how neighboring atomic dipoles form small magnetically organized clusters called **domains**. Under normal conditions, these domains do not organize any further, due to thermal agitation from ambient temperatures. However, under the influence of an external magnetic field, thermal effects are overcome and the domains begin to align with each other and the magnetic field. A strong external magnetic field will cause more domains to align than a weak field. If the external magnetizing force is steadily increased, a point will be reached at which all domains in the material are aligned. At this point, the ferromagnetic material is said to be **saturated**. Those materials whose domains remain significantly aligned after the external magnetic field is removed are termed **permanent magnets**.

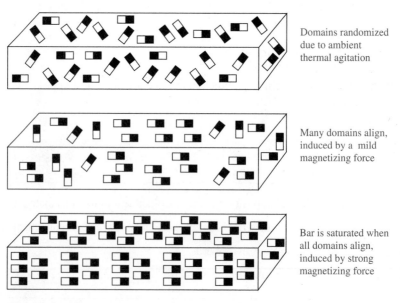

Domains randomized due to ambient thermal agitation

Many domains align, induced by a mild magnetizing force

Bar is saturated when all domains align, induced by strong magnetizing force

Figure 9-8   Magnetic domains in ferromagnetic materials.

Magnetism and Electromagnetism

Bar Magnet

39) This photo illustrates the magnetic lines of force surrounding a bar magnet. Note the flux lines exposed by the powdered iron. Also note that the south end of the compass needle is attracted to the north end of the magnet. (Sections 9-1 to 9-3)

Electromagnet

40) Use the left hand rule for electron flow or the right hand rule for conventional flow to verify the magnetic polarity of this electromagnet. Once again, the magnetic flux is exposed by the powdered iron. (Section 9-4)

Magnetism and Electromagnetism at Work

41) Many highly skilled technicians are needed in the medical
field to install and maintain sophisticated equipment such as this
nuclear magnetic resonance im-
aging system (NMR). (Hank
Morgan/Science Source/Photo Researchers)

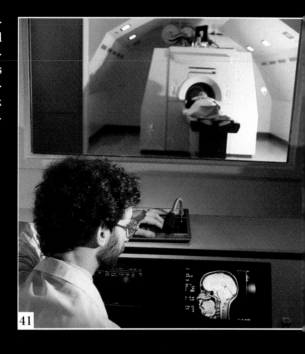

42) Electromagnetic principles are applied to
high speed mass transportation systems known as
Maglev (magnetic levitation). (Courtesy of Kaku
Kurita/Gamma-Liason)

43) The Maglev vehicle is suspended by a frictionless magnetic field above the track. Propulsion is
accomplished by a huge linear induction motor
formed between the length of the vehicle's chassis
and the track. (Graig Davis/SYGMA)

Magnetism and Electromagnetism at Work

44) Electromagnetism has been at work in electric vehicles since the turn of the century. Electric wires can be seen charging the batteries in the trunk of this early automobile. (Courtesy of Arizona Public Service Co.)

45) The modern "electrek" fiberglass electric vehicle from Unique Mobility is able to cruise effortlessly and silently at highway speeds. (Courtesy of Arizona Public Service Co.)

46) Electromagnetic waves provide communication and broadcast links vital to the world community. This microwave dish is a state-of-the-art link with satellite television broadcasting. (Courtesy of M/A COM, Inc.)

47) This Syncom IV-3 satellite is virtually hand launched by astronaut James D. Van Hoften. (Courtesy of NASA)

Magnetism and Electromagnetism at Work

48)

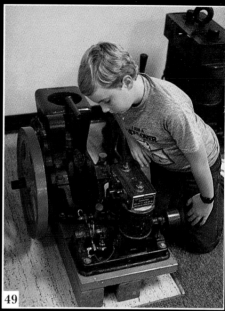

49)

48) This 180-ton superconducting electromagnet was built by Argonne scientists for a cooperative research program in magnetohydrodynamics (MHD). The magnet produces an intense magnetic field used to induce electrical current in highly accelerated ionized hot coal gases. (Courtesy Argonne National Laboratory)

49) This future technician examines one of the first portable steam-powered electric generators designed and built by Thomas Edison. (Courtesy of Edison Winter Home, Fort Myers, FL)

50)

51)

50) Experimental vertical axis wind turbines can also be seen at work in the Cochella Valley.

51) Thousands of wind turbine generators cover the constantly windblown floor of the Cochella Valley, North Palm Springs, California. Millions of watts of

Man-Made Permanent Magnets

ALNICO and Permalloy Magnets

A great variety of man-made permanent magnets are in use today, and engineers continue to develop stronger and more compact permanent magnets for modern applications. Among the most common permanent magnets manufactured today are ALNICO magnets. **ALNICO** magnets are an alloy of **AL**uminum, **NI**ckel, and **CO**balt, with some iron. They are fairly strong magnets with a high flux density and high retentivity. One of the most common uses for ALNICO magnets is in PM loudspeakers (PM = permanent magnet). Small ALNICO magnets (from a couple of hundred grams to a couple of kilograms) will have a total flux ranging from a couple of hundred to tens of thousands. Close cousins to ALNICO magnets are permalloy magnets, which have very high flux densities and very high retentivity. Permalloy magnets are made of iron and nickel.

Ceramic Magnets

One very special category of man-made permanent magnets is that of ceramic magnets, often referred to as *ferrites*. Ceramic magnets are modeled from a paste of barium carbonate and ferric oxide and hardened in a firing process, yielding a wide variety of shapes and sizes. The application of a strong magnetizing field during the manufacturing process establishes the magnetic domains forming the permanent magnet. These magnets are commonly found in small DC motors, analog meter movements, and a myriad of other general applications. Ferrites have very high internal resistances due to the isolation of ferric oxide particles in the ceramic compound. As a result, ceramic magnets are insulators instead of conductors. This is very useful in reducing the electrical losses due to large internal currents (and subsequent power losses) that are associated with ferromagnetic materials used in power and radio-frequency applications.

Permanent vs Temporary Magnets

What determines whether a ferromagnetic material will be a permanent magnet or just a temporary magnet? Before we answer that question, let's discuss the definition of a temporary magnet. A **temporary magnet** is a piece of ferromagnetic material that is magnetized only under the influence of some magnetizing force while the force is applied. When the external magnetizing force (magnetic field) is removed, very little, if any, residual magnetism remains in the material. Ambient thermal energy quickly scrambles the magnetic domains. Thus, the material is magnetized only temporarily, in the presence of a magnetizing force.

Ferromagnetic materials can be further classified as magnetically hard or magnetically soft. Materials that act as temporary magnets are said to be magnetically soft. A **magnetically soft** material has a very low reluctance, which means the domains are easily arranged and then disarranged. In the presence of a magnetizing force, the domains quickly align, and when the force is removed, thermal energy sets the domains in disarray. Therefore, in magnetically soft material, the domains are easily influenced. **Magnetically hard** materials have a relatively high reluctance. A very strong magnetizing force must be applied to a magnetically hard material in order to align domains. However, once the domains have been aligned, the magnetically hard material becomes a permanent magnet, with ambient temperatures having little or no disrupting effect.

Demagnetization of Permanent Magnets

Temperature Effects

In the discussion of temporary magnets, it was stated that ambient thermal energy was sufficient to scramble domains in magnetically soft material when the magnetizing force was removed. Heat is one of the biggest enemies of all types of magnets. Permanent magnets will also lose their magnetism if the ambient temperature is high enough. The temperature at which a particular permanent magnet loses all of its magnetism is known as the **Curie temperature**. At that temperature, the permeability of the material drops to unity (1). The magnetism does not automatically return when the material cools. It must be remagnetized.

Mechanical Vibration

As well as heat, sudden shock and vibration can have a damaging effect on a magnet. Permanent magnets that are relatively weak and made of material that is not very hard, magnetically, can be weakened further by the sudden shock of being dropped or jarred. This causes some of the domains to shift out of alignment, thus weakening the field.

It has also been found that mechanical vibration, applied to a ferromagnetic material placed in a magnetizing field, will aid the alignment of domains in the material.

Alternating Current

Perhaps the most common method of demagnetizing an object is the use of a rapidly reversing magnetic field created by an alternating current (AC). Alternating current is current that continually changes direction at a particular rate. If this alternating current is applied to a coil of wire, an alternating magnetic field is created about the coil. This alternating magnetic field can then be used to scramble the magnetic-domain alignment of a material. Direct current applied to the coil will magnetize ferromagnetic materials, while alternating current applied to the coil will demagnetize them. Figure 9-9 shows some applications where alternating current is used to demagnetize (degauss) a color TV picture tube and a magnetic record/playback head in a tape machine.

The Air Gap

Air Gap?

Magnets are shaped according to purpose or desired application. A **bar magnet** may serve as the needle on a compass or as a means of attaching an object to a ferromagnetic surface. The bar magnet is relatively weak, however, because of the linear separation of the north and south poles. Flux lines have to extend through the air from one end of the magnet to the other. The distance between the north and south ends of a magnet is called the **air gap**. If the air gap is large, as in the case of a bar magnet, the flux lines will have great opportunity to spread out, thus reducing field strength or flux density (B) at all locations between poles. If the bar magnet could be bent around to bring the north end closer to the south end, the air gap would be shortened and the flux density between poles would increase. Such is the case with horseshoe magnets, as shown in Figure 9-10.

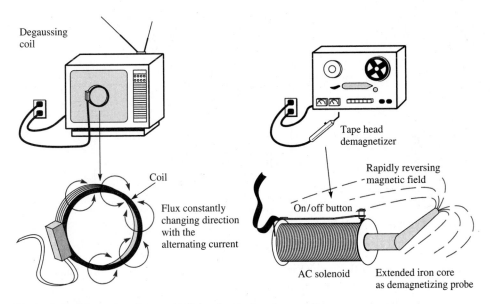

Figure 9-9   Alternating Current (AC) can be used to demagnetize.

Horseshoe magnets are designed to concentrate magnetic flux between poles. Flux density in the air gap is much greater than it would be in a linear bar-magnet arrangement. Though frequently thought of as toys, horseshoe magnets are commonly used in analog meter movements, microwave amplifiers and oscillators, and other fairly sophisticated pieces of equipment. In fact, the record/playback head of any tape recorder is actually a horseshoe-shaped electromagnet with a very narrow air gap.

Closing the Gap

What would happen if the air gap of, say, a horseshoe magnet were bridged (closed) with a piece of ferromagnetic material? If the ferromagnetic material has a high permeability, as most do compared to air, the magnetic flux will be highly concentrated and contained in the bridging material. Very few, if any, flux lines will remain in the air between the poles, since the ferromagnetic material offers a very low-reluctance (low-resistance) path to the flux. Any flux that may be in the air in spite of the ferromagnetic bridge is referred to as **leakage flux** and is described as **flux leakage**. Since the bridge offers a low-reluctance path to the flux, the total flux of the magnet becomes self-contained, with direct and

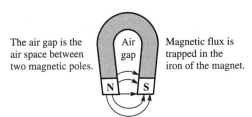

The air gap is the air space between two magnetic poles.

Air gap

Magnetic flux is trapped in the iron of the magnet.

Leakage flux is concentrated in the air gap.

Figure 9-10   The space between the north and south poles is called the air gap.

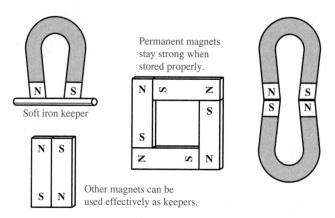

Permanent magnets stay strong when stored properly.

Soft iron keeper

Other magnets can be used effectively as keepers.

Figure 9-11 Closing the air gap will help preserve the magnet.

continuous loops formed through the horseshoe magnet and bridge. Just as magnetic flux cannot escape, external flux, from another magnetic source, cannot enter. Thus, the bridge will keep internal flux in and external flux out. For this reason, the bridge itself is called a **keeper**. Ideally, all magnets should be stored with a keeper, to maintain their strength in the presence of an invading magnetic field. Two horseshoe magnets may be joined in opposite polarity to form a continuous magnetic loop, or two bar magnets may be stored side by side in opposite polarity. Each magnet will act as a keeper for the other. See Figure 9-11.

Take time to review the terms and concepts in this section. Then, test your understanding by answering the questions of Self-Check 9-3.

SELF-CHECK 9-3

1. What is a magnetic domain?

2. What does it mean if a material is magnetically hard?

3. List three ways to demagnetize.

4. What is a keeper?

Hans Christian Oersted (1777–1851)

Hans Oersted was born in Rudkobing, Denmark. He was known for his work in physics and chemistry. He is most noted for his discovery that current-carrying conductors produce magnetic fields. His observation of a compass needle being influenced by a current-carrying conductor was the beginning of the age of electromagnetism. It is also thought that Oersted was the first to produce aluminum. (Photo courtesy of The Bettmann Archive.)

9-4 Electromagnetism

The Solenoid

The Straight Conductor

In 1820, Hans Christian Oersted discovered that the needle of a compass would align itself broadside and perpendicular to a conductor carrying a direct current. This was the first evidence that electrical current creates a magnetic field. Figure 9-12 illustrates this fact. Notice that the magnetic flux is established in a plane perpendicular to and concentric with the conductor. The needle of the compass aligns itself in such a way that the circular flux lines from the conductor flow through the needle from south to north, since the internal magnetic flux of the magnetized needle flows from south to north.

See color photo insert page C1, photograph 40.

It was also discovered that the direction of current in the conductor determines the direction of magnetic flux surrounding the conductor. One very useful method used to determine the direction of magnetic flux surrounding a current-carrying conductor is the **left-hand rule for conductors**, demonstrated in Figure 9-12. The magnetic flux can be

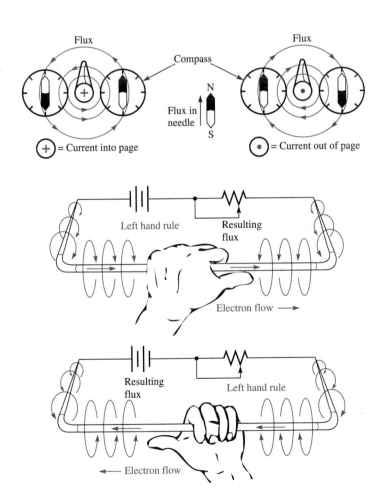

Figure 9-12   Magnetic flux surrounds current-carrying conductors.

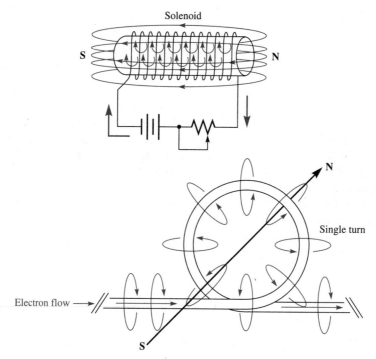

Figure 9-13   Windings of wire produce a solenoid having a concentrated magnetic field.

visualized by mentally wrapping the fingers of your left hand around the conductor with your thumb pointing in the direction of electron flow. The direction of the flux will be the same as the direction in which your fingers are pointing.

The Coiled Conductor

It was soon discovered that the strength of the magnetic field about a current-carrying conductor could be increased by forming one or more loops with the conductor. Such a configuration is commonly known as a *coil*, or **solenoid**. As shown in Figure 9-13, magnetic flux is concentrated in the center of each winding of the solenoid, thereby increasing the flux density within the coil. All flux lines within the windings are aiding one another, since they are heading in the same direction, and can be compared to the flux lines running south to north within a permanent magnet. Each individual winding contributes to the overall magnetic field of the solenoid.

Just as each individual winding has a magnetic polarity, the entire coil, or solenoid, has the same polarity. This polarity may be determined by applying the left-hand rule for conductors to a single loop or by applying the **left-hand rule for solenoids** as shown in Figure 9-14. Notice that the fingers of the left hand surround the solenoid in the direction of current flow while the thumb points to the north pole. The magnetic polarity of the solenoid is determined not only by the direction of current in the conductor but also by the direction in which the coil is actually wound.

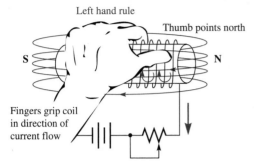

Left hand rule

Thumb points north

S   N

Fingers grip coil
in direction of
current flow

Figure 9-14   Determining the magnetic
polarity of a solenoid.

The Toroid

As shown in Figure 9-15, a **toroid** is a donut-shaped core upon which a coil, or solenoid, may be wound. Any flux produced by the coil is totally contained in the toroidal core. Just as flux lines form continuous loops, the toroid is a continuous loop of high-permeability material. Since the reluctance of the core material is very, very low, the flux lines will naturally flow in the loop of the toroid instead of escaping into the air. For this reason, toroids are said to be *self-shielding*. Toroidal cores are made of ferrite materials for power, audio, and radio frequency applications.

Magnetomotive Force (MMF)

Ampere-Turns (At)

The total flux of the magnetic field produced by a solenoid is directly proportional to the amount of current I flowing in the coil and the number of turns N. Increasing current increases magnetic flux, and increasing the number of turns on the coil also increases magnetic flux. The product of current applied to the solenoid and the number of turns of the solenoid is considered to be the force that actually produces the total magnetic flux inside and surrounding the coil. This force is called the **magnetomotive force (MMF)**, comparable to electromotive force. The MMF is the magnetic difference of potential, or pressure, that determines the number of flux lines created by the electromagnet (solenoid). The SI unit of measure for MMF is the **ampere-turn (At)**. One ampere of current flowing through one turn of a coil produces one At of MMF ($1 A \cdot 1 t = 1 At$). Each turn of the

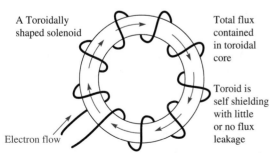

A Toroidally
shaped solenoid

Total flux
contained
in toroidal
core

Toroid is
self shielding
with little
or no flux
leakage

Electron flow

Figure 9-15   The toroid.

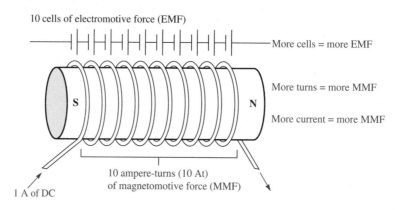

Figure 9-16   Magnetomotive force MMF.

coil represents one magnetic cell containing a unit quantity of MMF. A series of consecutive turns is like a series arrangement of voltaic cells (batteries), each cell adding to the total EMF. Likewise, each consecutive turn of a coil will add to the total MMF. Thus, as shown in Figure 9-16, if 1 A of current flows through 10 turns of a coil, the MMF is 10 At (1 A · 10 t = 10 At). It is important to realize that the same amount of MMF may be obtained by 0.1 A of current flowing through 100 t of a coil (0.1 A · 100 t = 10 At). Thus, a low-current solenoid can be as strong as a high-current solenoid if it has many more turns.

$$MMF = N \cdot I = NI \qquad \text{[measured in ampere-turns (At)]} \qquad (9.2)$$

where N = number of turns and I = current.

Opposition to MMF

Just as electromotive force E must work against resistance R to produce current I, MMF must work against reluctance ($\mathcal{R}$) to produce total flux (Φ). Therefore, the total flux developed by a solenoid is determined as follows: (magnetic Ohm's Law formula)

$$\Phi = MMF/\mathcal{R} = \frac{At}{At/Wb} \qquad (9.3)$$

where Φ is the number of flux lines in webers, MMF is the magnetomotive force in ampere-turns, and $\mathcal{R}$ is the reluctance in ampere-turns per weber. Hence

$$\mathcal{R} = MMF/\Phi = At/Wb$$

Since reluctance may be expressed as MMF/Φ, the unit for reluctance is simply the At/Wb. One ampere-turn per weber is often referred to as one Rel (1 At/Wb = 1 Rel: **Rel** = **Rel**uctance).

EXAMPLE   9-3   A solenoid has an MMF of 1,000 At and its core material has reluctance of $4 \cdot 10^6$ Rels. The total flux is: $\Phi = 1,000$ At$/4 \cdot 10^6$ At/Wb = 250 μWb.

Field Intensity (H)

Length and Strength

The MMF, or number of ampere-turns, is not the only factor that determines the strength of a magnetic field developed by a solenoid. The spacing of the turns and therefore the overall length of the coil are also very important. If the turns are widely spaced, the flux lines will not be able to aid each other with the same intensity as they would if the turns were tightly wound. Increasing the spacing between turns increases the reluctance between turns and thereby decreases the overall intensity of the field. Closely winding the coil reduces reluctance and allows the individual windings to reinforce each other's magnetic field. This increases the intensity or strength of the flux lines. While MMF is the force that produces total flux, the intensity of the total flux produced is properly called **field intensity** (H). As you will see, the field intensity (H) is responsible for determining the flux density within a solenoid core. If the field intensity is high, the flux density is high. Field intensity is determined from the following mathematical relationship:

$$H = \text{MMF}/l = NI/l = \text{At/m} \tag{9.4}$$

where H is the field intensity measured in At/m, N is the total number of turns, I is the current in amperes, and l is the length in meters.

The standard SI unit of measure for field intensity is **ampere-turns per meter (At/m)**. At/m is actually a quantity of MMF spread out over a unit length. Figure 9-17 illustrates the relationship between length and field strength. The actual length of an air-core coil is simply measured from end to end of the windings or magnetic pole to pole. If a solenoid has an iron core that extends beyond the ends of the windings, the length of the coil is considered to be equal to the length of the iron core. If the core is a toroid, the length (l) is the mean, or average, circumference of the core found by using the mean diameter [D_m = (outer diameter + inner diameter)/2]. Thus, $l = \pi D_m$ and $H = NI/\pi D_m$ for a toroidal solenoid.

Another Look at Permeability

Permeability (μ) and Flux Density (B)

A magnetic field that has more flux per unit area than another is the stronger of the two fields. In other words, if the flux density of one field is greater than the flux density of another field, the higher-density field is the stronger of the two. So, what determines the amount of flux density? Two main factors determine flux density for a solenoid: (1) the field intensity of the coil (H), and (2) the absolute permeability (μ) of the core of the solenoid. If both factors are known, the flux density for a solenoid can be calculated as follows:

$$B = \mu \cdot H = \mu \cdot NI/l \tag{9.5}$$

As you can see, the absolute permeability of the core is just as important as field intensity in determining flux density. Increasing the absolute permeability (by changing core material), and/or increasing field intensity (by increasing current or number of turns), increases flux density ($B = \mu NI/l$).

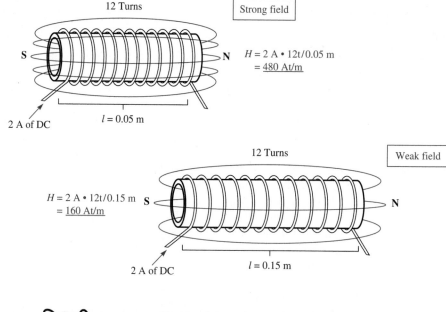

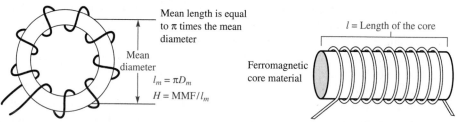

Figure 9-17   The field intensity (H) depends on the amount of current, the number of turns, and the length of the coil (solenoid).

Absolute Permeability (μ)

What is absolute permeability? **Absolute permeability** can be defined in terms of flux density and field intensity. If Formula 9.5 is rearranged, the absolute permeability can be expressed as follows:

$$\mu = B/H \tag{9.6}$$

In the SI system, the absolute permeability of air is *not* one (1). Air has an absolute permeability of 1.26 μT/At/m and is symbolized as μ_0.

As you know, the relative permeability μ_r of a material is the measure of the material's permeability compared to air (or vacuum, or space). If the relative permeability μ_r of any material is known, the absolute permeability μ for that material can be found by multiplying its relative permeability by the absolute permeability of air, μ_0. In SI units, the absolute permeability of air is 1.26 μT/At/m. Thus, the following formula may be used to find the absolute permeability of any material whose relative permeability is known:

$$\mu = \mu_r \cdot \mu_0 = \mu_r \cdot 1.26 \; \mu\text{T/At/m} \tag{9.7}$$

where μ is the absolute permeability of a material in units of T/At/m.

If the absolute permeability of a material is found experimentally by dividing a calculated value of field intensity H into a measured flux density B, ($\mu = B/H$), the relative permeability for that material can be readily found as follows:

$$\mu_r = \mu/\mu_0 \qquad (9.8)$$

The following examples will help clarify the relationship between absolute and relative permeability:

An iron alloy core has a relative permeability of 650. What is the absolute permeability of this core in T/At/m?   E X A M P L E   9-4
Ans. $\mu = 650 \cdot 1.26 \ \mu T/At/m = 819 \ \mu T/At/m$

If a special core material has an absolute permeability of 1.26 mT/At/m, what is its relative permeability?   E X A M P L E   9-5

Ans. $\mu_r = \dfrac{1.26 \ mT/At/m}{1.26 \ \mu T/At/m} = 1000$ (units cancel)

Calculating Flux Density ($B = \mu \cdot H$)

Once the absolute permeability μ of a core material and the field intensity H of the coil are known, the flux density B can be calculated using Formula 9.5.
Study the following examples:

A solenoid has a core whose relative permeability is 400. The coil develops a field intensity of 100 At/m. What is the flux density in the core?   E X A M P L E   9-6
Ans. $B = 400 \cdot 1.26 \ \mu T/At/m \cdot 100 \ At/m = 0.0504$ T

A toroidal solenoid has a mean diameter of 4 cm, has a 200-turn coil with 0.25 A of current, and its core has a relative permeability of 500. Find the flux density in its core.   E X A M P L E   9-7
Ans. $B = \mu \cdot H$

First find μ

$\mu = \mu_r \cdot \mu_0 = 500 \cdot 1.26 \ \mu T/At/m = 630 \ \mu T/At/m$

Now find H

$H = NI/\pi D_m = (200 \ t \cdot 0.25 \ A)/(\pi \cdot 0.04 \ m) = 398 \ At/m$

Now find B

$B = 630 \ \mu T/At/m \cdot 398 \ At/m = 0.25$ T

Table 9-1 is a summary of the SI units discussed in this section. The table is also helpful in converting between the SI and older cgs system of units. Take time to study Design Note 9-1. Then, test your knowledge by answering the questions of Self-Check 9-4.

TABLE 9-1 Electromagnetic Units

Magnetic Quantity	cgs Units	Conversion	SI Units
Flux (Φ)	Maxwell (Mx) 1 Mx = 1 Φ	# Mx/100 M Φ = # Wb # Mx = 100 M Φ · # Wb	Weber (Wb) 1 Wb = 100 MΦ
Magnetomotive force MMF = NI	Gilbert (Gb)	# Gb · 0.794 At/Gb = # At # Gb = 1.26 Gb/At · # At	Ampere-turn (At)
Field Intensity (H) $H = NI/l$	Oersted (Oe) 1 Oe = 1 Gb/cm = 100 Gb/m	# Oe · 79.4 At/m/Oe = # At/m # Oe = 0.0126 Oe/At/m · # At/m	Ampere-turns per meter (At/m)
Flux Density (B) $B = \mu \cdot H$	Gauss (G) 1 G = 1 Mx/cm²	# G · 100 μT/G = # T # G = 10 kG/T · # T	Tesla (T) 1 T = 1 Wb/m²
Absolute Permeability (μ) $\mu = \mu_r \cdot \mu_0{}^\dagger$	μ = # Gb/Oe $\mu_0{}^\dagger$ = 1 Gb/Oe	# G/Oe · 1.26 μT/At/m/G/Oe = # T/At/m # G/Oe = 794 kG/Oe/T/At/m · # T/At/m	Teslas per Ampere-turns per meter (T/At/m)
Reluctance ($\mathcal{R}$) $\mathcal{R}$ = MMF/Φ	Gilberts per maxwell (Gb/Mx)	# Gb/Mx · 79.4 μAt/Wb/Gb/ Mx = # At/Wb # Gb/Mx = 0.0126 μGb/Mx/At/Wb · # At/Wb	Ampere-turns per weber (At/Wb)

$^\dagger\mu_0$ = absolute permeability of air, vacuum, or space = 1.26 μT/At/m = 1 Gb/Oe

SELF-CHECK 9-4

1. Which end of the solenoid in Figure 9-18 is the north end?

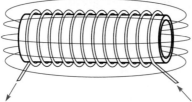

Electron flow   **Figure 9-18**

2. Using SI units, calculate the MMF of a 500-turn coil that is drawing 100 mA.

3. Given a coil with a fixed number of turns, what will happen to the field intensity produced by the coil if the coil is stretched out, or lengthened?

4. What factors, or values, must you know in order to determine the field intensity of a solenoid?

5. Using SI units, find the flux density in the core of a solenoid having the following parameters: current = 200 mA, 400 turns, 10 cm long, iron core with a relative permeability of 300.

DESIGN NOTE 9-1:   Flux Density (B)

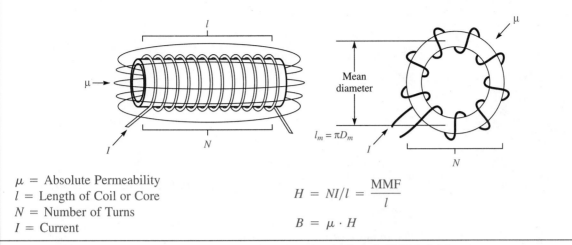

μ = Absolute Permeability
l = Length of Coil or Core
N = Number of Turns
I = Current

$$H = NI/l = \frac{MMF}{l}$$

$$B = \mu \cdot H$$

"BASIC" PROGRAM SOLUTION

```
10   CLS
20   PRINT"FLUX DENSITY (B)"
30   PRINT"":PRINT""
40   PRINT"THIS PROGRAM WILL CALCULATE THE
     FLUX DENSITY FOR STRAIGHT OR TOROIDAL IN-
     DUCTORS"
50   PRINT""
55   PRINT"SELECT INDUCTOR TYPE"
60   INPUT"(1) STRAIGHT - (2) TOROIDAL ";C
70   IF C = 2 THEN GOTO 200
80   IF C<>1 AND C<>2 THEN GOTO 50
90   PRINT""
100  INPUT"ENTER THE LENGTH OF THE COIL OR
     CORE IN METERS-";L
110  INPUT"ENTER THE AMOUNT OF COIL
     CURRENT-";I
120  INPUT"ENTER THE NUMBER OF COIL TURNS-";N
130  INPUT"ENTER THE RELATIVE PERMEABILITY OF
     THE CORE MATERIAL-";UR

140  U = UR * 1.26E-06
150  GOTO 300
200  PRINT""
210  INPUT"ENTER THE MEAN DIAMETER OF THE
     TOROIDAL CORE-";D
220  L = 3.1416 * D
230  GOTO 110
300  PRINT""
310  H = N * I/L
320  B = U * H
330  PRINT"THE FLUX DENSITY IS ";B;" TESLAS."
340  G = B * 10000
350  PRINT"THE FLUX DENSITY IN GAUSS IS ";G;" G."
360  PRINT""
370  INPUT"CALCULATE ANOTHER? (Y/N)";A$
380  IF A$ = "Y" THEN CLEAR:GOTO 30
400  CLS:END
```

9-5 Magnetic Hysteresis

The $B-H$ Magnetization Curve

You should now be aware of the fact that the amount of flux density (B) in the core of a solenoid is determined by the field intensity (H) of the coil and the absolute permeability (μ) of the core material. If the current applied to a solenoid is increased, the MMF will increase, causing the field intensity to increase, which causes the flux density to increase. One important question that should come to mind here is: If the current to the coil is continually increasing, will the flux density in the core continue increasing without limit? The answer is no. The reason is a very practical one. A ferromagnetic core, for example,

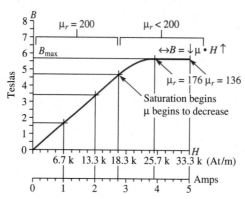

NOTE: This *B–H* Curve is for a 400-turn coil
that is 0.06m in length. The core has a relative
permeability of 200.

Figure 9-19 The B–H magnetization curve.

has a certain number of magnetic domains that can be induced to align with the flux generated by the MMF. As current to the coil is increased, more and more of the domains are induced to align with the magnetizing field. Finally, a limit is reached at which nearly all domains have aligned with the magnetizing field. Any further increase in current, or MMF, will not produce a significant increase in flux density in the core. This is called **saturation**. The core has been saturated with a practical upper limit in flux density. According to the formula ($B = \mu \cdot H$), if H continues to increase and B does not, it must mean that the absolute permeability μ is effectively decreasing. Recall that permeability is the measure of the ease with which flux is established in a core. If the core becomes saturated, it is not easy for further flux to become established. Thus, the effective permeability decreases as the core becomes more and more saturated.

Figure 9-19 graphically illustrates the relationship between field intensity and flux density for a certain core material in a characteristic **B – H magnetization curve**. As you can see, a certain level of field intensity is reached after which flux density no longer increases proportionately. As soon as the graph starts to curve, saturation has been reached. Note from the graph shown in Figure 9-19 that the effective relative permeability of the core starts to drop below 200 after saturation has begun. Further increases in field intensity force the core deeper into saturation until the curve finally levels off, indicating an absolute maximum level of flux density. After this maximum level of flux density is reached, the effective permeability drops at a rate that balances out the further increase in field intensity, thus maintaining a constant maximum level of flux density.

The Hysteresis Loop

Residual Flux

What happens to the flux density in a core when the field intensity is reduced to zero from some high value? The $B – H$ curve of Figure 9-19 seems to indicate that if H is reduced to zero (0), flux density will return to zero (0). However, the $B – H$ curve in Figure 9-19 only illustrates the $B – H$ relationship when flux is first established with an increasing DC to the coil of a solenoid that has an initially unmagnetized core. If the current is removed, thereby removing the magnetizing force, some residual flux remains. Thus, when H drops to zero (0), there will be a remaining value of flux density. In order to reduce the flux density to zero (0), a magnetizing force must be applied in opposite polarity to the residual magnetism. This is called the **coercive force**; it is created by reversing the direction of current in the coil.

Lagging Flux

What does all of this mean? It means the change in flux density B in the core is lagging behind the change in field intensity H. When the H drops back to zero (0), B is still at some residual value and when H is increased in the opposite direction, B finally drops to zero (0). This phenomenon is known as **hysteresis**, meaning ''to lag behind.'' The hysteresis is caused by the reluctance of magnetic domains to quickly rearrange themselves in line with a changing magnetizing force.

AC and Hysteresis

When an alternating current (AC) is applied to a solenoid, the hysteresis effect is continual. That is, as the alternating current applied to the coil increases and decreases with alternating changes in direction, the corresponding change in flux density lags behind. The graph of flux density vs alternating current, or alternating field intensity, is called the *hysteresis loop*.

Understanding the Loop

Figure 9-20 demonstrates the relationship between 1¼ cycles of AC and the lagging flux density. Notice that there are two quantities of residual magnetism ($+B_r$ and $-B_r$) when H is zero (0) and two quantities of coercive force ($-H_c$ and $+H_c$) needed to bring residual B back to zero (0). As shown in the graph, magnetic flux is first established in the core between points 0 and 1. At point 1 the AC has reached a maximum positive peak with a corresponding maximum flux density ($+B_{max}$). From point 1 to point 2, the AC decreases in positive amplitude (positive level) until it momentarily drops to zero at point 2. Even though the current is zero at point 2, there is still some residual flux in the core ($+B_r$). The AC continues in the negative, or opposite, direction from point 2 to point 3. At the point marked $-H_c$, the reversed field intensity, caused by reversed current, is strong enough to coerce the residual flux density back to zero. At point 3, the flux density again

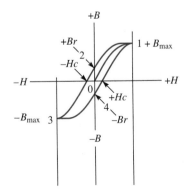

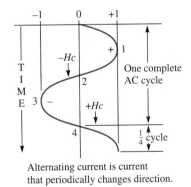

Alternating current is current
that periodically changes direction.

Figure 9-20   The hysteresis loop.

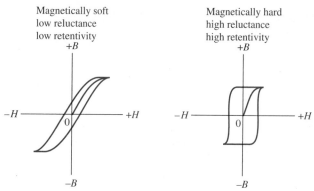

Figure 9-21   Characteristic hysteresis loops.

reaches a maximum value ($-B_{max}$), this time in opposite polarity. From point 3 to point 4, the magnetic domains once again resist change, leaving a residual magnetism at $-B_r$, which corresponds to zero current at point 4. The hysteresis loop is completed from point 4 back to point 1. Again, a coercive force ($+H_c$) is needed to overcome the residual flux density ($-B_r$) in the core. The loop is repeated over and over again with every cycle of alternating current. Figure 9-21 illustrates two extreme hysteresis loops. The narrow loop is typical for transformer cores, while the wide loop is representative of ferrite materials and materials used for permanent magnets.

Take time now to review this section by answering the questions of Self-Check 9-5.

SELF-CHECK 9-5

1. What happens to the effective permeability of a core when the magnetizing field intensity is increased to the region of saturation?

2. When the field intensity produced by a coil drops back to zero from some high value, will the flux density of the core material drop back to zero? Explain.

3. What is the coercive force, as pertaining to the hysteresis loop?

9-6 The Hall Effect

Throughout this chapter, you may have been wondering how the flux density, and therefore the strength, of a magnet or solenoid, might actually be measured. It would be very useful to be able to measure the density of a magnetic field, since other quantities, such as absolute permeability of a core material ($\mu = B/H$), could then be calculated. But what kind of special apparatus will it take to measure invisible magnetic fields? It will require some kind of meter that can be calibrated in gauss (or tesla) and some type of probe or sensor that will produce a voltage or current in proportion to the density of the magnetic field in which the probe is placed. Over the years, such an apparatus has been developed. It is referred to as a *Magnetometer* or a *Gaussmeter*. Figure 9-22 is an example of a commercially available digital Gaussmeter. The object mounted on top of the meter is a special probe that is sensitive to magnetic fields.

Figure 9-22 A portable gaussmeter. (Courtesy of Walker Scientific, Inc.)

But how do the probe and Gaussmeter work? The probe is able to sense and monitor a magnetic field by use of a phenomenon called the **Hall effect**. The effect was first discovered in 1897 by the American physicist Edwin H. Hall. Hall discovered that if a conductor carrying a DC current is placed in a magnetic field perpendicular to the flux lines, a small voltage is actually developed across the width of the conductor perpendicular to the flux lines and the flow of current. The Hall effect is illustrated in Figure 9-23. The reason the small potential, known as the Hall voltage (V_n), exists is because electrons flowing through the probe are also flowing through a perpendicular magnetic field. The moving electrons create their own magnetic fields, which react with the stationary field that is being measured. As a result, the electrons do not flow in a straight line down the probe. The electrons tend to flow more to one side of the probe, thereby creating a difference of potential across the probe's width. This difference of potential is proportionately larger in a stronger magnetic field. In other words, the relationship between gauss and voltage is a linear one. The amount of difference of potential is also proportionate to the amount of current flowing in the probe. This current is held constant at various specified

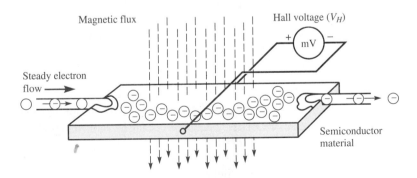

Figure 9-23 The Hall effect.

levels for each of several ranges of measurement. A larger probe current increases the magnetic-field sensitivity of the probe. Thus, the Hall voltage (V_H) is directly proportional to probe current (I_P) and flux density (B).

Hall probes are made of semiconductor materials instead of conductors, since it has been found that the Hall voltage is much greater for semiconductor materials. Hall voltages may range from μV to 100 mV (using semiconductor probes made of materials such as indium arsenide). The probe is placed in a magnetic field perpendicular to the flux lines and carefully rotated from side to side to obtain the highest reading possible in that part of the field. The Hall voltage produced by the probe is then amplified and fed to the analog or digital meter, which is calibrated in gauss.

SELF-CHECK 9-6

 As a review to this section, take a blank sheet of paper and draw a diagram illustrating the Hall effect. Try to do this completely from memory. Compare your results with Figure 9-23.

9-7 The Principles of Induction

Electromagnetic Induction

See color photo insert
page C4.

Electromagnetic induction is a means by which current is forced to flow in a conductor under the influence of an external magnetic field. In other words, a magnetic field can cause current to flow in a neighboring conductor. You are aware from the preceding sections that a current flowing in a conductor produces a magnetic field surrounding the conductor. If a current-carrying conductor can generate a magnetic field, is it not also possible that an established magnetic field can generate a current in a conductor? This is the question that occurred to many experimenters in the early 1800s, following Oersted's discovery of electromagnetism in 1820.

Requirements for Electromagnetic Induction

How does a magnetic field induce a current in a conductor? This is the question the early scientists had to answer. They soon discovered that there are three basic requirements for electromagnetic induction to occur: (1) there must be a conductor, (2) there must be a magnetic field, and (3) there must be a relative motion between the conductor and the magnetic field.

1. A Conductor

The first requirement for electromagnetic induction is: There must be a conductor in which current can be induced. This may seem obvious to you but there may be some things about this requirement that are not so obvious. In order for current to flow in this conductor, a continuous current loop must be formed. In other words, the ends of the conductor must be connected to some kind of a load to form a complete circuit. Then, as current is induced in the conductor, it flows through the load and performs work. If the load is not present, there is no complete path for current, and therefore no current can flow or be induced.

2. Magnetic Field

It is the magnetic field that influences the current to flow. This magnetic field may be provided by an electromagnet or a permanent magnet. In any case, the conductor must be placed in the vicinity of the air gap of the magnet, since that is where free flux, or leakage flux, is present. The flux must cut through the conductor at a right angle. In other words, the conductor must be placed perpendicular to the flux lines, which are flowing north to south in the air gap. If the air gap is small, yet sufficient to accommodate the presence of the conductor, the flux density in the air gap will be great. This is important, since a stronger magnetic field will induce a greater current in the conductor. If an electromagnet is used to provide the magnetic field, an increase in current to the electromagnet will result in an increase in induced voltage and current supplied to a load. This comes about because an increase in current to the electromagnet causes an increase in the strength of its magnetic field, thus inducing more voltage across the conductor.

3. Relative Motion

Having a conductor and a magnetic field is not enough to induce a current. Placing a conductor properly within a strong magnetic field is not enough either. There must be relative motion between the conductor and the magnetic field. This means the magnetic field can remain stationary while the conductor is moved perpendicularly through the flux, or the conductor can be held stationary while the magnet is moved, or both the conductor and the magnet can be in motion, as long as they are moved in different directions.

Why the relative motion? If a magnetic field is stationary, because neither the conductor nor the magnet are in motion, free electrons and holes (positive ions) in the conductor will not be disturbed. Any interaction between the external magnetic field and the magnetic field of each electron will be very localized, causing each electron to remain fairly stationary with respect to holes in the atomic structure. However, if a series of external flux lines sweep through, or cut through, the conductor, the electrons will react and begin to move, and continue to move with each passing flux. Thus, the relative motion of conductor and flux generates an organized electron flow in one direction and a relative hole flow in the other direction (conventional flow). The relative motion of conductor and flux with resulting current is known as **generator action**.

Figure 9-24 demonstrates, in a very simplified way, what we have discussed. In this illustration, the conductor is held stationary and the magnet is being moved down, thus

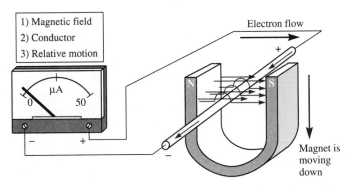

Figure 9-24   Generator action involves relative motion of a conductor in a magnetic field.

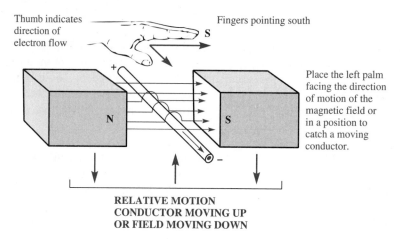

Figure 9-25 Determining the direction of induced current in a conductor.

providing relative motion. The same results could be obtained by holding the magnet stationary and moving the conductor up. The ammeter completes the circuit allowing current to flow. The question here is: How can we determine the direction in which current will flow in the conductor?

Using the Left-Hand Rule for Induction

The direction of flow is determined by two factors: (1) the direction of the magnetic field, or orientation of magnetic poles, and (2) the motion of the flux cutting across and through the conductor. As shown in Figure 9-25, the **left-hand rule for induction** applies these two factors, yielding a quick determination of direction of current. When using this left-hand rule, place all four fingers in the direction of flux (fingers pointing toward the south pole). The palm of the hand should move in the direction the magnetic field is moving as it cuts through the conductor. If the conductor is moving instead of the magnetic field, the palm should be placed in such a manner as to catch the conductor. In either case, the extended thumb indicates the direction of electron flow. Study Figure 9-25 until you fully understand the use of the left-hand rule for induction.

Now, take time to reinforce your knowledge by answering the questions of Self-Check 9-7.

Heinrich F. E. Lenz (1804–1865)

Heinrich Lenz was a German-born scientist who spent many years of his life working in Russia. Much of his work was in the areas of physics and electricity. He discovered the basic principle of countermagnetomotive force, which occurs when current is induced in a conductor. He discovered that induced current produces its own magnetic field, which opposes the motion that is generating it. Lenz demonstrated that work must be done to generate electricity. Thus, he was the first to offer an explanation for the conversion of mechanical to electrical energy. He is also credited with discovering the effects of temperature on material resistance.

1. What is electromagnetic induction?

2. List the three basic requirements for induction.

3. How must the conductor be placed with respect to the magnetic field?

4. What does *relative motion* mean?

5. Determine the direction of electron flow in the conductor of Figure 9-26.

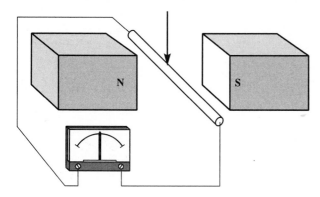

Figure 9-26

9-8 Lenz's Law

Lenz's Law is actually another physical principle of induction that can be used to determine the direction of induced current in a conductor. It is basically stated as follows:

> **Lenz's Law**
>
> The induced current is in such a direction that the resulting magnetic field it produces opposes the external magnetic field that generated it.

This law applies to straight conductors, coiled conductors, and all electrical generators. Recall that a current-carrying conductor develops a magnetic field about itself. When induction takes place and current is made to flow, a magnetic field is created around the conductor. This magnetic field is a **countermagnetomotive force (CMMF)** that opposes the external magnetic field that induced the current in the first place. Therefore, there is opposition to the relative motion. This is why a generator requires a proportionate amount of mechanical power to generate electrical power (theoretically 746 W per 1 Hp). If more current is demanded from the generator by a load, the magnetic field surrounding the conductor in the generator becomes stronger, thus increasing the opposition to generator shaft rotation. Therefore, an increase in load (current demand) causes an increase in demand for mechanical energy.

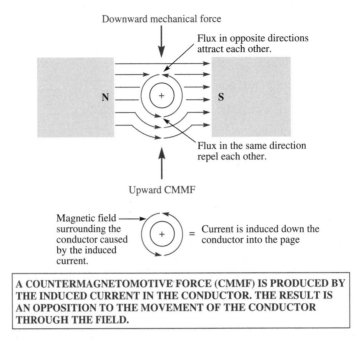

Figure 9-27 Lenz's Law for a straight conductor.

Figure 9-27 illustrates Lenz's Law for a straight conductor. Here, the external magnetic field is held stationary and the conductor is moved down through the field. The left-hand rule for induction indicates that current is flowing into the page as shown (fingers point to south pole, palm is positioned to catch the conductor, thumb indicates the direction of current). As a result of the direction of current in the conductor, the magnetic field generated around the conductor is in a counterclockwise direction. Notice that the lines of magnetic flux above the conductor are in opposite directions. This causes a force of attraction (opposites attract) above the conductor. Below the conductor the magnetic flux lines are in the same direction, causing a force of repulsion. Thus, a countermagnetomotive force (CMMF) is produced that opposes the motion of the conductor through the magnetic field.

Figure 9-28 illustrates Lenz's Law for coils. A bar magnet is moved in and out of the center region of a coil. In what direction will current flow when the magnet is moved into the coil? In what direction will current flow when the magnet is moved out of or away from the coil? Fortunately, Lenz's Law simplifies these determinations. Simply keep in mind that the coil will develop a magnetic field that opposes any motion of the magnet. When the south end of the magnet is moved into the coil, the coil must develop a south pole to oppose it (because like poles repel). If this is true, and it is, the current induced in the coil must be flowing in a direction that can produce a south pole on that end of the coil. The left-hand rule for solenoids is used to determine the direction of current. The thumb of the left hand is pointed to where the north pole should be and the fingers indicate the direction of current around the coil.

If the south end of the magnet is then pulled out of the coil, the coil must respond by producing a north pole that opposes the outward motion. This means that current in the coil must flow in the opposite direction. Thus, when the magnet is moved into the coil, induced current flows one way and when the magnet is pulled out of the coil, induced

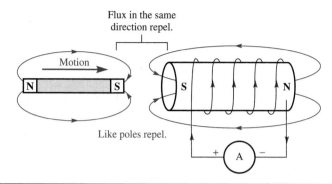

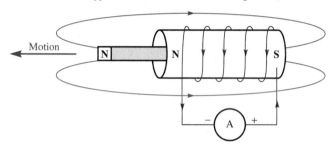

Figure 9-28   Lenz's Law for coils (solenoids)

current reverses direction. This is the basic principle involved in alternating current (AC) generators. (AC generators will be discussed in a later section.) Make sure you understand the application of Lenz's Law before you continue to the next section. Self-Check 9-8 will help reinforce what has been discussed here.

S E L F - C H E C K   9 - 8

1. Write Lenz's Law in your own words.

2. Why is mechanical power needed to drive a generator?

3. Determine the magnetic polarity of, and the direction of current in, the coil in Figure 9-29.

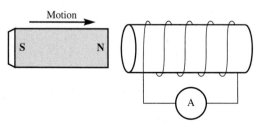

Figure 9-29

9-9 Faraday's Law

The Law

In the previous sections, we discussed generator action or electromagnetic induction in terms of current. We said that the relative motion of a conductor in a magnetic field induces a current in a conductor. But what about voltage? The generator action produces a difference of potential across the conductor just as it produces a current in the conductor, assuming a load is provided for a complete circuit. As you know, if a load is not connected to the conductor, no current will flow. However, a difference of potential, or voltage, is still developed at the output terminals of the conductor.

How is this difference of potential created? The concept is illustrated in Figure 9-30. When the relative motion first begins, free electrons are forced to move to one end or terminal of the conductor. One terminal of the conductor has a surplus of electrons, the other has a deficiency (an abundance of holes). Thus, a difference of potential, or voltage, exists between terminals. In Figure 9-31 the generator coil is represented schematically as a voltage source with a small internal coil resistance (R_g). If a load is connected to the generator terminals, the voltage will be present at the generator terminals and across the load (minus a small voltage drop across the internal resistance of the coil). The amount of current is determined by the amount of generator terminal voltage and the amount of load resistance ($I = V_G/R_L$).

So, how can the amount of voltage a generator will produce be determined? The answer to that is found in Faraday's Law, which is stated as follows:

> **Faraday's Law**
>
> The amount of voltage induced across a coiled conductor is dependent upon three main factors: (1) the amount of flux (Φ) cutting across the coil, (2) the number of turns of the coil (N), and (3) the rate at which the flux cuts across the turns of the coil ($\Delta\Phi/\Delta t$).

Amount of Induced Voltage

1. Voltage and Flux (Φ)

The amount of flux is related to the strength of the external magnetic field. A strong magnet will have more flux between its air gap than a weaker one. The stronger the magnetic field is, the greater is the induced generator voltage.

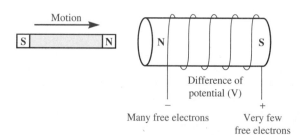

Motion

Difference of potential (V)

Many free electrons   Very few free electrons

Figure 9-30 A difference of potential is produced through generator action.

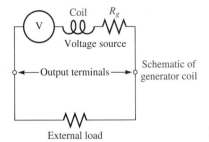

Coil R_g

Voltage source

Output terminals

Schematic of generator coil

External load

Figure 9-31 The schematic of a generator.

2. Voltage and Number of Turns (N)

More turns means more voltage. Each turn of the coil has a voltage induced across it. Since all turns of the coil are in series with each other, the total generator voltage is equal to the sum of all separately induced turn voltages.

3. Voltage and the Rate of Cutting ($\Delta\Phi/\Delta t$)

Increasing the rate at which the flux cuts across the conductor, or coil, also increases the voltage developed across it. Using the basic generator model shown in Figure 9-32, we can visualize the concept of a change in flux per unit of time ($\Delta\Phi/\Delta t$). The delta symbol (Δ) stands for "a change in." Therefore, a change in flux is represented as $\Delta\Phi$ and a change in time is represented as Δt. As the conductor in Figure 9-32 is moved down through the flux, it cuts through a certain number of flux lines per unit of time. The unit of time may be microseconds, milliseconds, or seconds. As each unit of time passes, during which the conductor is moving down through the field, more and more flux lines will have been cut by the conductor. If the conductor moves at a steady rate, the number of flux lines cut per unit of time is constant. This constant rate generates a steady current and voltage. If the conductor is moved at a faster rate, more flux lines will be cut per unit of time. More flux lines cut per unit of time produces a higher current and voltage.

Faraday's Formula

If a coil is passed through a magnetic field, the effect of $\Delta\Phi/\Delta t$ is multiplied by the number of turns in the coil. Faraday's formula provides a means of calculating the voltage induced

Michael Faraday (1791–1867)

Michael Faraday is considered to have been one of the greatest chemists and physicists of his time. Born near London, Faraday grew to become a renowned lecturer and scientist. His greatest discovery was that of electromagnetic induction—the very principle by which all electrical generators operate. It is interesting to note that Joseph Henry, an American physicist, actually discov-ered electromagnetic induction before Faraday but failed to publish his findings. Faraday's notability gained him immediate recognition for the discovery. (Photo courtesy of The Bettmann Archive.)

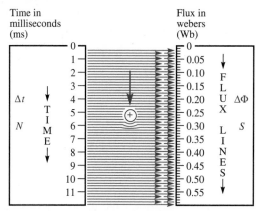

Flux lines are being cut at the rate of 0.05 Wb/ms

From $t = 1$ ms to $t = 2$ ms THE CHANGE IN TIME IS 1 ms.
$\Delta t = 1$ ms

From $t = 1$ ms to $t = 2$ ms THE CHANGE IN FLUX IS 0.05 Wb.
$\Delta\Phi = 0.05$ Wb

$\Delta\Phi/\Delta t = 0.05$ Wb/1 ms = 50 Wb/s

Figure 9-32   The rate of cutting flux determines the amount of induced voltage.

across a coil when the rate of cutting ($\Delta\Phi/\Delta t$) and number of turns (N) are known. The formula is for a coil with an iron core and is given as follows:

$$V_{ind} = N \cdot \Delta\Phi/\Delta t \qquad (9.9)$$

where V_{ind} is the induced voltage, N is the number of turns, and $\Delta\Phi/\Delta t$ is the change in flux per unit of time measured in Webers per second, Wb/s. As you can see, the $\Delta\Phi$ is expressed in Webers and the unit of time is the second.

EXAMPLE   9-8   If a 200-turn coil cuts through a magnetic field at the rate of 5 Wb/s, the induced voltage will be 1,000 volts (200 · 5 Wb/s = 1,000 V).

If the number of turns is increased, the induced voltage will increase. If the flux density of the field is increased, more flux is cut every second, thus increasing the induced voltage. If the coil is moved more quickly, taking less time to move through the field, the rate of cutting is increased and the induced voltage increases. The increased cutting rate is the result of time being decreased.

EXAMPLE   9-9   If it takes a quarter of a second (0.25 s) instead of one second to cut through 5 Wb of flux, the induced voltage increases to 4,000 volts (200 · 5 Wb/0.25 s = 200 · 20 Wb/s = 4,000 V).

Thus, all three factors given in Faraday's voltage law are shown mathematically in his formula. Any or all of these factors may be adjusted to set the generator output to the desired voltage.

The Polarity of the Induced Voltage

How can the polarity of the induced voltage be determined? The polarity of the generator voltage is determined the same way the direction of current is determined—by using Lenz's Law and the left-hand rule for solenoids. Always remember that a coil entering a magnetic field must develop a magnetic field of its own that will oppose the external magnetic field. From this, the magnetic polarity of the coil can be determined, along with the direction of current in the coil. Ultimately, the polarity of voltage at the generator's output terminals is determined by the direction of motion, the polarity of the magnetic field, and the method in which the coil is wound.

Take time now to answer the questions of Self-Check 9-9.

S E L F - C H E C K   9 - 9

1. List the three factors that determine the amount of voltage induced across a coiled conductor.

2. What does $\Delta\Phi/\Delta t$ stand for?

3. How much voltage is induced across a 120-turn coil that is cutting through flux at the rate of 1.5 Wb/s?

4. List three ways of increasing the amount of voltage induced across the coil of Question 3.

9-10   AC Generation

Alternating current (AC) is current that alternates in direction. It is current that is periodically reversing direction in the circuit. For this reason, generators that are used to produce AC are often called alternators. In this section, we will concentrate on the devices that produce AC.

Basic Alternator Construction

Figure 9-33 illustrates the basic part of an **AC generator**, or **alternator**. The magnetic field needed for generation is provided by either a permanent-magnet arrangement or electromagnets. When electromagnets are employed, they are placed on opposite sides of the generator in such a way as to provide a north and south pole. The coil windings of the electromagnets are fixed, or **stat**ionary, and therefore are called **stators** or **stator windings**.

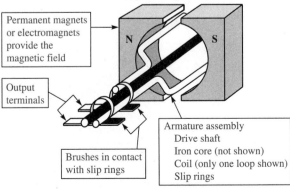

Permanent magnets or electromagnets provide the magnetic field

Output terminals

Brushes in contact with slip rings

Armature assembly
Drive shaft
Iron core (not shown)
Coil (only one loop shown)
Slip rings

An EMF is developed as the armature is rotated

Figure 9-33   The AC generator (alternator).

Since these stator windings provide the magnetic field for the generator, they are often referred to as **field coils** or **field windings**.

The movable part of the generator that rotates on the generator drive shaft is called the **armature**. By definition, an armature is the movable part of an electromagnetic device. In this case, the armature consists of a drive shaft, an iron core, armature coil windings, and slip rings. The drive shaft is connected to an external source of mechanical energy and is rotated to create relative motion between the armature winding and the magnetic field.

An iron core is provided for the armature coil to increase permeability, lower total reluctance between the north and south magnetic poles (by reducing the air gap to a very small space on all sides of the armature), and thereby increase the flux density through which the armature is rotated. The iron core is not shown in the illustration for sake of simplicity.

Although only one loop of the armature coil is shown, the coil is actually formed by many turns of enamel-coated wire. As you know from Faraday's Voltage Law, the number of turns the armature coil has will affect the amplitude of the output voltage (more turns = more voltage). However, it is not always practical to depend on a large number of turns for a high voltage output. It may be necessary for the armature coil to have relatively few turns of very large-gauge wire or even copper strapping. Such is the case for high-power generators that must supply very large amounts of current. The windings must be thick to handle the current demand, or load. The larger-size conductor leaves room for fewer turns in a given space. Therefore, a high voltage output must be obtained by a stronger magnetic field and/or a higher rate of cutting flux (faster armature rotation).

The **slip rings** are necessary to allow the armature to rotate freely. Special low-resistance solid carbon (graphite) brushes ride on the slip rings to provide a path for current to a practical load device. There are two slip rings, one for each end of the armature coil. Alternating current is passed back and forth through the slip rings and brushes to the load. Naturally, friction will wear down the brushes and slip rings over long-term operation.

Alternator Operation

Polarity Reversal

How is alternating current produced? Figure 9-34 illustrates how the current, induced in the armature coil, is caused to alternate in direction. Notice that armature rotation is in the

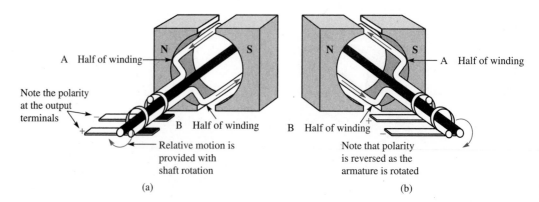

(a)   (b)

Figure 9-34 Current reverses direction and voltage changes polarity with every 180° of shaft rotation.

clockwise direction. In Figure 9-34a, the A half of the winding is moving up through the magnetic field that extends from north to south. Using the left-hand rule for induction, we see that current is flowing forward toward the slip ring as shown. However, as the armature continues to rotate, the current in the A half of the winding is forced to reverse direction, as shown in Figure 9-34b. Again, the left-hand rule for induction may be used to visualize this. At the same time, the current in the B half of the winding is also reversing direction, to aid and complement the A half. Notice that the polarity of voltage at the output terminals of the generator continues to change or alternate as the armature is rotated.

Output Voltage vs Cutting Angle

The output voltage at an alternator terminal does not quickly change from some high positive value to a high negative value and back to a high positive value again. Instead of an immediate jump in voltage, the terminal voltage rises in amplitude (amount) according to the angle at which the conductor is cutting the flux lines. Figure 9-35 illustrates this important concept. The figure shows three coil positions as the armature is rotated through 90° in a counterclockwise direction. In position #1, the rotary motion of the coil is parallel to the flux lines. Therefore, no flux lines are being cut by the conductor and no current or voltage is induced. In position #2, the coil is cutting the flux at an angle of 45°. The current induced in the coil at this location is equal to the trigonometric sine of 45° times the maximum amount of current produced when the coil cuts the flux at a right angle. If the maximum current induced at position #3 is 10 A, the current induced at position #2 is 7.07 A (10 A · sine 45° = 7.07 A). The graph in Figure 9-35 illustrates the relationship between the angle of cutting flux and the amount of induced voltage or current. As you

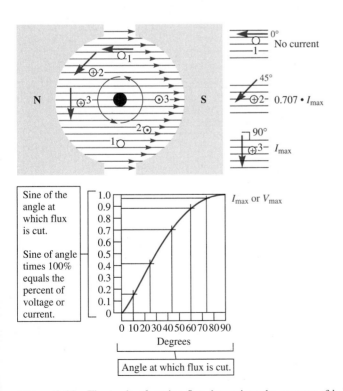

Figure 9-35   The angle of cutting flux determines the amount of induced voltage and current.

can see, the voltage or current at a particular point in time (instantaneous voltage or current) is a function of the sine of the angle of cutting flux. The percentage of maximum voltage or current at any angle of cutting flux is found by multiplying the sine of the angle by 100%. As an example, if the angle of cutting is 60°, the sine of 60° is 0.866 and the instantaneous amplitude of the voltage or current at that angle will be 86.6% of whatever the maximum value is.

The AC Sine Wave

As the armature of the alternator is rotated through one complete revolution, a **sine wave** is produced at its output terminals. This sine wave is the waveform for both the voltage and current. In the preceding discussion, you learned that the generator output, voltage or current, is a function of the sine of the angle of cutting flux. As the armature rotates, the conductor will cut the flux at every possible angle from 0° to 360°. If the sine of each degree of rotation were graphically plotted, the resulting graph would naturally be that of the sine wave. Figure 9-36 demonstrates this fact. The chart shows the amplitude and

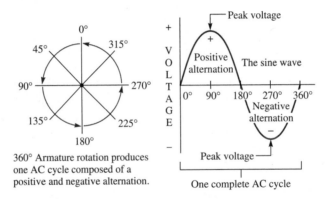

360° Armature rotation produces one AC cycle composed of a positive and negative alternation.

> **THE AMPLITUDE OF THE VOLTAGE AT THE OUTPUT TERMINALS OF THE GENERATOR IS A FUNCTION OF THE SINE OF THE ANGLE OF THE COIL AS IT ROTATES AND CUTS THROUGH THE FLUX.**

Maximum peak voltage = 100 V

Angle	Sine ∠	Output
0°	0	0
45°	0.707	70.7 V
90°	1.0	100 V
135°	0.707	70.7 V
180°	0	0
225°	−0.707	−70.7 V
270°	−1.0	−100 V
315°	−0.707	−70.7 V
360°	0	0

Schematic symbol for the AC Generator indicates a sine wave output.

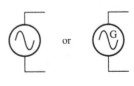

Figure 9-36   A sine wave is produced as the armature of an alternator rotates through its magnetic field.

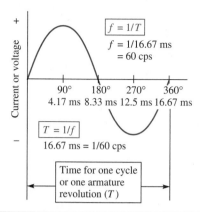

$f = 1/T$
$f = 1/16.67$ ms
$= 60$ cps

90°   180°   270°   360°
4.17 ms 8.33 ms 12.5 ms 16.67 ms

$T = 1/f$
16.67 ms $= 1/60$ cps

Time for one cycle
or one armature
revolution (T)

IN THIS ILLUSTRATION, THE ARMATURE IS ROTATING AT
60 REVOLUTIONS PER SECOND (RPS) EQUAL TO 3600
REVOLUTIONS PER MINUTE (RPM).
3600 RPM YIELDS 60 RPS WHICH DEVELOPS 60 CPS = 60 Hz

Figure 9-37   The relationship between revolutions, frequency, and cycle time.

polarity of output voltage taken in 45° steps through 360° of armature rotation. Maximum voltage, or peak voltage, is given as 100 V_p. The output reaches the peak value of 100 V at 90° and 270° of rotation. Of course the polarity of the peak voltage alternates between positive and negative.

If the alternator has a magnetic field consisting of one polar pair (one north pole and one south pole), the waveform at the output terminals will consist of one positive alternation and one negative alternation. An **alternation** is that part of the waveform that is either all positive or all negative. If the voltage is being measured at one of the output terminals with respect to the other, that terminal will alternate between positive and negative potentials. Two such alternations (a positive and a negative) will constitute one cycle. An AC **cycle** may be defined as (1) two consecutive alternations, or (2) the waveform that exists between two points having the same amplitude, polarity, and direction. Notice, in Figure 9-36, that the voltage at the 360° point is the same as the voltage at the 0° point (0 V). Also notice that the voltage at the 360° point will start to rise in the positive direction just as the voltage at the 0° point begins to rise in the positive direction. Thus, the waveform begins to repeat itself. When the waveform begins to repeat itself, one cycle has ended and a new cycle has begun. The number of armature revolutions per second determines the frequency of the AC voltage or current. Thus, **frequency** is measured in **cycles per second (cps)** or the more modern unit of **hertz (1 cps = 1 Hz)**. See Figure 9-37.

Multiphase AC Generators

Multiphase Alternators

Figure 9-38 illustrates a typical multiphase alternator. In this case, the alternator has three phases. What is a phase? A **phase** is a single output from an alternator whose waveform is shifted in time, or electrical degrees, as compared to other outputs, or phases, from the same alternator. In other words, the voltage available at the three separate outputs do not

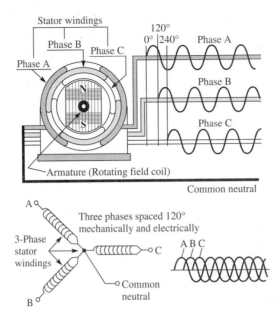

Figure 9-38 A three-phase AC generator.

all reach a positive peak at the same time. Neither do they reach a negative peak at the same time. In fact, because of the mechanical construction of the three-phase generator, the output waveforms will never be in phase with each other. The word "phase" refers to the relative position of the armature in terms of degrees of rotation. Since armature rotation produces AC cycles, the mechanical degrees are translated into electrical degrees. Thus, the word "phase" applies to electrical degrees of the AC cycle. Each cycle contains 360 electrical degrees. Therefore, any number of AC waveforms can be compared to each other in terms of electrical degrees, or phase. Also, each electrical degree represents a small fraction of time. For example, a 60-cps sine wave has a period of 16.67 ms. The 16.67 ms is the time for the 360 electrical degrees contained in one cycle. Therefore, each electrical degree has a duration of 46.3 μs per degree (16.67 ms/360° = 46.3 μs/°).

As you see, in Figure 9-38, the three-phase alternator is designed with three sets of stator windings that are mechanically arranged 120° apart. As you know, a stator is a stationary coil. In many alternator designs, the stators provide the output current and voltage instead of the armature. In such cases, the armature is actually a rotating electromagnet providing both the magnetic field and relative motion. It is often called a **rotating field coil** for this reason. Current is supplied to the armature through an arrangement of brushes and slip rings. Since the stators are placed 120° apart, the voltages induced by the rotating field coil will reach maximum peaks 120° electrically separated from each other. The phase relationship of the three AC outputs is shown schematically as three coils drawn with 120° spacing and graphically as three waveforms separated by 120 electrical degrees.

Three-phase alternators are far more common than you might expect. Power companies use three-phase alternators to supply very high voltages throughout large systems. Factories and other concerns with high electrical usage use three-phase power for most of their equipment, since the equipment will operate at a higher efficiency (lower cost). Another common use for three-phase alternators is in the charging system of nearly every motor vehicle. The three stator outputs of the automotive alternator are connected to rectifiers that convert the AC into DC. The DC is then applied to the battery to maintain its charge. The voltage regulator keeps the charge voltage at the appropriate level by controlling the amount of current supplied to the rotating field coil. This, of course, varies with the speed of the engine.

Take time now to go back and review the terms and concepts from this section. It would be helpful for you to study the illustrations once again. Then test your understanding by answering the questions of Self-Check 9-10.

1. What is a stator winding?

2. The movable part of the generator that rotates on a shaft is called what?

3. In your own words, explain how the amplitude of the generator output voltage is related to the angle of cutting flux.

4. Why is the waveform produced by an alternator called a sine wave?

5. Explain how alternator revolution and waveform frequency are related.

9-11 DC Generation

Generating DC is very similar to generating AC: a magnetic field is needed, a conductor is needed, and relative motion is still needed. So, if everything is the same, what's the difference? The difference is in the manner in which the generator voltage is supplied to the output terminals.

The Difference Is the Commutator

What Is a Commutator?

An AC generator becomes a DC generator by the manner in which the voltage, or current, is supplied to the generator output terminals. Look at Figure 9-39. This shows the basic DC generator construction. Notice that there is a magnetic field, an armature, and two brushes. But what happened to the slip rings? This is the basic difference between an alternator and a DC generator. The alternator has two slip rings and the DC generator has a single segmented slip ring called a **commutator**. The word *commutator* comes from the words *commute* and *commutate*, which mean "to change." The commutator is changing or switching the output terminals at the right time to maintain a constant polarity. As the armature starts to develop a negative alternation, the commutator switches the polarity of the output terminals via the brushes. This keeps all positive alternations on one terminal and all negative alternations on the other.

The Commutator at Work

You are still not sure how the commutator works, right? Perhaps Figure 9-40 will help. Let's start with Figure 9-40a and rotate the armature counterclockwise through one complete revolution. When the armature is positioned vertically, no current or voltage can be induced. The direction of the conductor is parallel with the flux. Thus, in diagram (a),

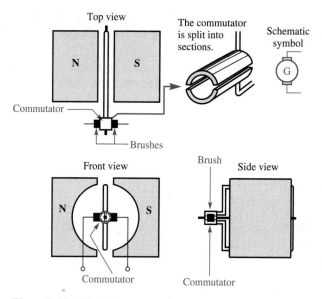

Figure 9-39   The DC generator has a commutator which is a segmented slip ring.

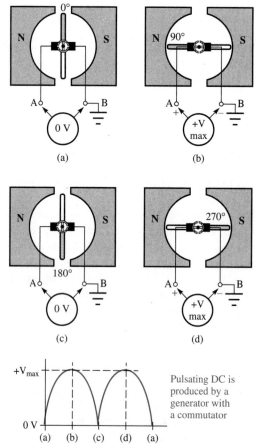

Pulsating DC is produced by a generator with a commutator

Figure 9-40   This simple DC generator produces pulsating DC.

there is no output at the terminals. Notice that terminal B is being used as a common reference or ground. We will concentrate on the waveform that appears at terminal A using terminal B as the reference.

In Figure 9-40b, the armature has rotated through 90°. The armature is now in a maximum output position. Here, we see that terminal A is at a maximum positive voltage. The graph in Figure 9-40 illustrates the increase in voltage as the armature has rotated through the first 90° [from (a) to (b)]. As the armature continues through the next 90°, the positive voltage at terminal A decreases to zero output. Notice, in diagram (c), that the brushes are bridging across the gap, or spacing, between commutator halves. The brushes have momentarily shorted across the armature. However, there is no problem here. When the armature is in the vertical position, there is no induced current to short out. As you see, it is very important to design the armature so the slits between commutator halves will be in contact with the brushes at zero crossing. In other words, the brushes should be across the commutator gaps, or slits, at the point in rotation where no current is induced in the armature coil.

As the armature continues to rotate from the 180° position to the 270° position (Figure 9-40d), the commutator switches the armature coil with the brushes. Again, the A terminal will receive a positive rising DC pulse as shown in the graph. The commutator will make sure that the A terminal is always connected to the left half of the armature coil that is rotating through 0° to 180°, and the B terminal will always be connected to the half of the armature coil that is rotating through 180° to 360°. The resulting generator output is a **pulsating DC** as shown in the graph. Naturally, this pulsating DC is not smooth and steady like the DC of a battery.

Improving the DC Generator Output

DC Ripple

Most practical DC generators do not have severely pulsating DC available at their output terminals. Instead of deep DC pulsations, the output is a shallow ripple. Figure 9-41 illustrates a DC generator that produces a **DC ripple** voltage instead of deep pulsations. Notice that the +DC is never allowed to drop all the way down to zero crossing. This is accomplished by using a two-coil armature and a commutator that has four segments. The two armature coils are mechanically arranged 90° apart. This causes the induced current in each coil to be separated by 90 electrical degrees. As you can see, coil #1 reaches maximum positive voltage 90° before the induced voltage of coil #2 reaches maximum. Coil #2 therefore fills in for coil #1 as the armature is rotated. As a result, the voltage at terminal A is always at a high level. The commutator makes sure that positive peak voltage is always being passed to terminal A. At the same time, terminal B is receiving negative peak voltages. The series of peaks created by the two coils and four-segment commutator is known as a DC **ripple voltage**.

Improving DC Ripple

Can you think of a way to further improve the ripple voltage at the output of a DC generator? In other words, how can the generator output be made to look more like a smooth DC instead of a rippled DC? You are right! The output can be smoothed out even

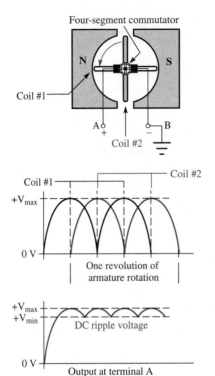

Four-segment commutator

N   S

Coil #1

A ○
   +

○   B
     −

Coil #2

Coil #1 ─┐   ┌─ Coil #2

$+V_{max}$

0 V

One revolution of
armature rotation

$+V_{max}$
$+V_{min}$
DC ripple voltage

0 V
Output at terminal A

Figure 9-41   The four-segment commutator
produces a smoother DC known as DC ripple.

further by adding more armature coils and more commutator segments. The more coils
and segments, the smoother the output voltage and current. Some DC generators have 8,
10, or even more, commutator segments as shown in Figure 9-42. This reduces the ripple
to a very small amount and makes it very easy to filter the generator output to a smooth,
pure, DC.

Figure 9-42   The commutator and brushes of an early DC generator designed by Thomas
Edison. (Courtesy of Edison Winter Home, Fort Myers, FL)

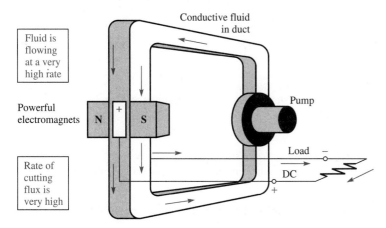

Figure 9-43   Magnetohydrodynamic power generation.

A DC Generator with No Ripple?

Is it possible to generate a DC current, or voltage, with no ripple at all? Yes, it is. And, it is being done by many power companies across the United States, and perhaps other countries as well. So, how is the ripple eliminated? In order to eliminate the ripple, we must design a generator that will allow the conductor to continually cut the flux at a 90° angle (angle for maximum output). This eliminates the rotating shaft. We must employ some means by which the conductor will continue to cut through the flux at a right angle. In other words, our design must be more like that shown in Figure 9-43.

This method of DC generation is known as **magnetohydrodynamic (MHD) power generation**. It means ''magnetic fluid dynamics'' and is sometimes called **hydromagnetic power generation**. The principle is very simple. As shown in Figure 9-43, a conductive fluid—a liquid metal or an ionized gas (plasma)—is pumped or compressed through a large closed-loop duct, or conduit. A strong magnetic field is applied through a section of the conduit. Thus, the conductive fluid flows through the flux of the magnetic field. The flowing fluid provides the relative motion and a current is induced perpendicular to the direction of fluid flow and perpendicular to the magnetic flux. Large metal contacts are positioned to deliver this DC current to a load.

You will want to take a few moments to review this section before answering the questions of Self-Check 9-11.

SELF-CHECK   9-11

1. What part of the DC generator makes it notably different from an AC generator?

2. Briefly explain how pulsating DC is generated.

3. What is ripple voltage?

4. What effect will increasing the number of armature coils and commutator segments have on the ripple voltage?

5. Describe an MHD power generation system.

Summary

FORMULAS

(9.1) $B = \Phi/A$

where B is the flux density measured in teslas, Φ is number of flux lines in webers, and A is the area in square meters

(9.2) $MMF = NI$

where MMF is the magnetomotive force measured in ampere-turns (At), N is the number of turns in the coil, and I is the amount of current flowing through the coil

(9.3) $\Phi = MMF/\mathcal{R}$

where the total number of flux lines is determined by the amount of MMF and $\mathcal{R}$ (reluctance)—this is similar to Ohm's Law

(9.4) $H = MMF/l$

where H is the field intensity measured in ampere-turns per meter (At/m), MMF is measured in ampere-turns (At), and l is the length of the coil measured in meters

(9.5) $B = \mu \cdot H$

where B is the flux density measured in teslas (T), μ is the absolute permeability measured in teslas per ampere-turns per meter (T/At/m), and H is the field intensity measured in ampere-turns per meter (At/m)

(9.6) $\mu = B/H$

(9.7) $\mu = \mu_r \cdot \mu_0$

where μ is the absolute permeability measured in teslas per ampere-turns per meter (T/At/m), μ_r is the relative permeability, having no units, and μ_0 is the absolute permeability of air, space, or vacuum given as 1.26 μT/At/m

(9.8) $\mu_r = \mu/\mu_0$

(9.9) $V_{\text{ind}} = N \cdot \Delta\Phi/\Delta t$

where V_{ind} is the induced voltage, N is the number of turns, and $\Delta\Phi/\Delta t$ is the change in flux per unit of time measured in webers per second (Wb/s)

CONCEPTS

- A magnetic field is made up of invisible lines of force called flux lines, or flux.
- Flux density (B) is the number of flux lines passing through a unit square area.
- Opposite poles attract; like poles repel.
- Permeability is the ease with which flux can be established in and throughout a material.
- Reluctance ($\mathcal{R}$) is the measure of opposition to the establishment of flux in and throughout a material.
- Retentivity is the ability of a material to retain some residual magnetism when a magnetizing force is removed.
- Paramagnetic materials become only slightly magnetized in the presence of a magnetizing force with no residual magnetism when the force is removed.
- Diamagnetic materials actually develop a weak magnetic field that opposes the magnetizing field.

- Ferromagnetic materials have atomic dipoles that form magnetic domains.
- Magnetic domains can become disarranged by heat, mechanical vibration, or the application of an alternating magnetic field created by alternating current in a coil.
- Current-carrying conductors are surrounded by magnetic flux.
- Magnetomotive force (MMF) is a magnetic pressure created by current flowing through turns of a coil.
- Field intensity (H) is the measure of concentration of MMF for a coil of a given length.
- The hysteresis effect in a core is caused by residual magnetism in the core, which must be overcome by a reversed field intensity called a coercive force.
- The Hall effect: If a current-carrying conductor or semiconductor is placed in a magnetic field perpendicular to the flux lines, a small voltage will be developed across the width of the conductor perpendicular to the flux lines and perpendicular to the flow of current.
- Current can be induced to flow in a conductor if the following requirements are met:

  1. A magnetic field must exist.
  2. The conductor must form a complete circuit.
  3. There must be relative motion between the conductor and the magnetic flux. The conductor must cut the flux at a right angle for maximum induction.

- Lenz's Law states that the induced current is in such a direction that the magnetic field produced by the current will oppose the external magnetic field that generated the current.
- The opposing magnetic field created by the induced current is called a *countermagnetomotive force* (CMMF).
- Faraday's Law states that the amount of voltage induced across a coiled conductor depends upon three main factors: (1) the amount of flux (Φ) cutting across the coil, (2) the number of turns of the coil, and (3) the rate at which the flux cuts across the turns of the coil ($\Delta\Phi/\Delta t$).
- Coils that are used to create a magnetic field for induction are called *field coils*.
- Generator windings that are stationary, or fixed in placed, are called *stators* or *stator windings*.
- The moving part of an electromagnetic device is normally called the *armature*.
- In an alternator, slip rings and brushes are used to connect the rotating armature to the outside world.
- The output waveform of an alternator is a sine wave, since the output voltage follows the trigonometric sine of the angle of rotation of the armature.
- Every cycle of a sine wave contains a positive and a negative alternation.
- Peak voltage is generated as the armature coil cuts the flux at a right angle (90°).
- Multiphase alternators have a rotating field coil that induces current in more than one set of stator windings.
- Replacing the slip rings of an alternator with a commutator will turn the alternator into a DC generator.
- A DC generator with one armature coil and only two commutator segments produces a rough DC output called *pulsating DC*.
- A DC generator that has more than one armature coil and more than two commutator segments produces a DC output known as *DC ripple voltage*.
- Magnetohydrodynamic (MHD) power generation is a means by which pure DC power is generated using conductive fluid motion. The conductive fluid is pumped at a very high rate through a duct that intersects a magnetic field.

PROCEDURES

Using the Left-Hand Rule for Induction

1. Place all four fingers of the left hand in the direction of the south magnetic pole.

2. If the magnetic field is in motion, the palm of your left hand should move in the direction the field is moving.
   If the conductor is moving, the palm of your left hand should be in such a position as to catch the conductor.

3. The thumb will indicate the direction of electron flow.

Using Lenz's Law to Determine Direction of Induced Current in a Coil

1. Determine what magnetic polarity of the coil is necessary to oppose the motion of the external magnetic field.

2. Use the left-hand rule for solenoids to determine the direction of induced current.

SPECIAL TERMS

- Magnetic polarity, north pole, south pole
- Magnetic field, magnetic flux, flux lines, flux
- Total flux (Φ), weber (Wb)
- Flux density (B), tesla (T)
- Permeability, relative permeability (μ_r)
- Absolute permeability (μ), mumetal
- Reluctance ($\mathcal{R}$), retentivity
- Electromagnetic induction
- Generator action
- Left-hand rule for induction
- Lenz's Law, countermagnetomotive force (CMMF)
- Faraday's Law, rate of cutting flux ($\Delta\Phi)/\Delta t$)
- Alternator, AC generator
- Stator windings, stators, field coil
- Armature, slip rings, brushes
- Sine wave, cycle, alternation
- Phase, rotating field coil
- DC generator, commutator
- Pulsating DC, ripple voltage, DC ripple
- Magnetohydrodynamic (MHD) power generation

Need to Know Solution

By now you realize the gasoline-powered generator did not have a mind of its own. When the electric stove was turned on, much more current was being demanded from the generator. This increased load caused the countermagnetomotive force of the generator to increase. The CMMF works against the magnetic field of the generator and opposes the motion of the armature. With the extra load and increased CMMF, the gasoline engine had to work harder. Apparently, the engine was not able to produce enough horsepower to generate the needed electric power. Thus, the lights dimmed and the engine began laboring heavily. There is nothing wrong with the generator unit itself. You simply chose an underpowered unit for your hut. Looks like a dim future to me.

HOW TO CHOOSE. You must know your worst-case load requirement. Add up all of the individual power requirements for every light and appliance that must be on at the same time. Add an additional 25% as an extra power margin. Order the generator unit that will meet or exceed your calculated requirements and look forward to a bright future.

Questions

9-1 Magnetism and the Magnetic Field

1. How can the polarity of a magnet be determined?
2. What is flux?
3. Which direction do lines of force run within the magnet?
4. How many flux lines are there in one weber?
5. Define *flux density*.
6. If the north pole of one magnet is placed close to the north pole of another magnet, will the force developed be that of attraction or repulsion?

9-2 Magnetic and Antimagnetic Materials

7. What is relative permeability?
8. If it is very difficult to establish flux in a material, the material is said to have a high (a) retentivity, (b) permeability, (c) reluctance.
9. The ability of a material to retain a quantity of residual magnetism when a magnetizing force is removed is known as (a) retentivity, (b) permeability, (c) reluctance.
10. What is a ferromagnetic material? Give three examples.
11. What is a paramagnetic material? Give three examples.
12. What is a diamagnetic material? Give three examples.

9-3 Permanent Magnets

13. For what are ALNICO permanent magnets commonly used?
14. What are ferrite magnets?
15. What is the difference between a permanent magnet and a temporary magnet?
16. What is the Curie temperature?
17. Explain how alternating current is used to demagnetize.
18. What is a magnetic air gap?
19. What is leakage flux?

9-4 Electromagnetism

20. What is a solenoid?
21. What is the purpose of forming a solenoid?

22. Use the left-hand rule to determine the magnetic polarity of the solenoids in Figure 9-44.

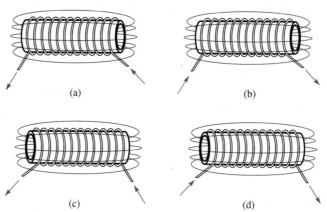

(a)   (b)

(c)   (d)   **Figure 9-44**

23. What advantage does a toroidal solenoid have over a straight solenoid?
24. What two factors determine the amount of MMF?
25. How is MMF similar to EMF?
26. What two factors determine the amount of total flux generated by a solenoid?
27. How is magnetic field intensity affected by coil length?

9-5 Magnetic Hysteresis

28. What does a *B–H* magnetization curve demonstrate?
29. What is saturation?
30. What is residual flux?
31. Explain coercive force.

9-8 The Hall Effect

32. What is the name of an instrument that can measure the flux density of a magnetic field?
33. Explain the Hall effect.

9-7 The Principles of Induction

34. List the three basic requirements for electromagnetic induction.
35. Describe the manner in which the conductor must cut the magnetic flux.
36. What is meant by *generator action*?
37. Determine the direction of electron flow induced in the conductors of Figures 9-45a and 9-45b.

9-8 Lenz's Law

38. Write Lenz's Law in your own words.
39. Explain CMMF.
40. Explain why an increase in generator load causes an increased demand for mechanical energy.

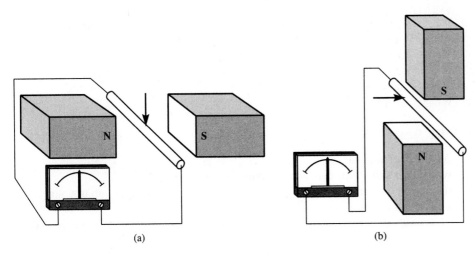

Figure 9-45

41. Determine the direction of induced current in the coils of Figures 9-46a and 9-46b using Lenz's Law and the left-hand rule for solenoids.

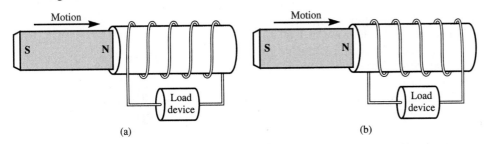

Figure 9-46

9-9 Faraday's Law

42. Describe how the amount of flux is related to the amount of induced voltage across a coil.
43. Describe how the number of turns of a coil is related to the amount of induced voltage.
44. Describe how the rate of cutting flux is related to the amount of induced voltage across a coil.
45. Ultimately, the polarity of voltage at the generator's output terminals will be determined by what three factors?

9-10 AC Generation

46. What is an alternator?
47. What is a stator?
48. What is an armature?
49. What does the speed of armature rotation have to do with the amount of generator output voltage?

50. What does the number of coil turns have to do with the amount of output voltage?
51. What is a slip ring and what does it do?
52. Why is the alternator output voltage called a sine wave?
53. At what angle of cutting flux is maximum current and voltage induced?
54. What is an alternation?
55. What is an AC cycle?

9-11 DC Generation

56. What is a commutator?
57. What is the difference between pulsating DC and DC ripple?
58. How can a conventional DC generator be designed to produce an output voltage with a minimum of DC ripple?
59. How much DC ripple does an MHD generator produce compared to a conventional DC generator?

Problems

9-1 Magnetism and the Magnetic Field

1. 25,500 Φ equals how many webers?
2. 35 μWb equals how many flux lines?
3. How many flux lines are in 5 Wb?
4. How many webers are in 1 T?
5. 3,000 Φ/cm^2 equals how many teslas?
6. How many Φ/m^2 is 0.45 tesla?
7. If the flux density is measured as 5,000 μT at a distance of 8 cm from the magnetic source, what will the flux density be at 16 cm from the magnetic source?

9-4 Electromagnetism

8. If 0.3 A of current flows through a 350-turn coil, what is the MMF in ampere-turns?
9. How much current would an 800-turn coil have to have to equal the MMF of the coil in Problem 8?
10. 10 At is equal to how many gilberts? (See Table 9-1.)
11. If a coil has 900 turns and 0.5 A of current, and is 0.075 m long, what is the field intensity?
12. If a coil has 900 turns and 0.5 A of current, and is 0.0375 m long, what is the field intensity?
13. If a coil has 900 turns and 1 A of current, and is 0.0375 m long, what is the field intensity?
14. If a coil has 450 turns and 0.5 A of current, and is 0.075 m long, what is the field intensity?
15. If a toroidal solenoid has an inner diameter of 3 cm and an outer diameter of 5 cm, has 150 turns of wire, and has 1.5 A of current flowing through its windings, what is the field intensity developed?
16. Convert 250 At/m to oersteds. (See Table 9-1.)

17. Convert 35 Oe to ampere-turns per meter. (See Table 9-1.)
18. If an iron-alloy core has a relative permeability of 150, what is its absolute permeability?
19. If a solenoid has 600 turns, 0.25 A of current, is 0.02 m long, and has a core with a relative permeability of 300, what is the flux density developed in its core?
20. If a solenoid has 100 turns and 0.5 A of current, is 0.08 m long, and has a core with a relative permeability of 700, what is the flux density developed in its core?
21. If a toroidal solenoid has an inner diameter of 4 cm and an outer diameter of 5 cm, has 300 turns of wire, has 0.5 A of current flowing through its windings, and has a core with a relative permeability of 350, what is the flux density developed in its core?

9-9 Faraday's Law

22. Find the amount of voltage induced across a 1,000-turn coil that is cutting flux at the rate of 0.03 Wb/s.
23. Find the amount of voltage induced across a 500-turn coil that is cutting flux at the rate of 0.5 Wb/s.
24. Find the amount of voltage induced across a 100-turn coil that is cutting flux at the rate of 0.75 Wb/s.

Answers to Self-Checks

Self-Check 9-1

1. Balance it from a string. The north end will seek magnetic north. Also, a marked bar magnet may be used, in which case the north end of the unmarked magnet will seek the south end of the marked magnet.
2. 50 μWb
3. 5 mT
4. 0.5 mT
5. North, because like poles repel

Self-Check 9-2

1. Reluctance
2. Retentivity
3. Materials that have magnetic characteristics similar to iron. They usually have high permeability and low reluctance.
4. Diamagnetic

Self-Check 9-3

1. A minute formation or alignment of atomic dipoles
2. It is a ferromagnetic material with a fairly high reluctance used for permanent magnets.
3. Apply heat, vibration or shock, or use an alternating magnetic field.
4. A keeper is a soft iron bar that is used to close the air gap of a magnet for storage. It helps prevent demagnetization.

Self-Check 9-4

1.

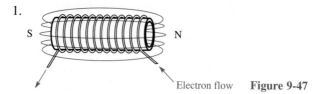

S   N

Electron flow   **Figure 9-47**

2. 50 At
3. Field intensity decreases.
4. Number of turns, amount of current, and length of the coil or core
5. 0.3024 T

Self-Check 9-5

1. The permeability of the core decreases.
2. No. Normally there is some residual magnetism in a ferromagnetic core.
3. The coercive force is the reversed field intensity needed to return the flux density in the core to zero (0).

Self-Check 9-6

See Figure 9-23.

Self-Check 9-7

1. Electromagnetic induction is a method by which electrons are forced to flow in a conductor under the influence of an external magnetic field.
2. A conductor, a magnetic field, relative motion
3. The conductor must be placed at a right angle with respect to the magnetic flux that is being cut.
4. The conductor is moving while the magnetic field is stationary, the magnetic field is moving while the conductor is stationary, or both are moving in different directions.
5.

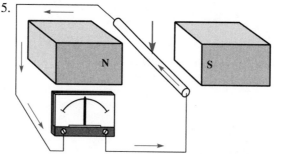

N   S

Figure 9-48

Self-Check 9-8

1. See Lenz's Law as stated in the beginning of Section 9-8.
2. The induced current will develop a magnetic field of its own that will oppose the magnetic field of the generator. This opposition is felt by the device or system that is

driving the generator. Thus, mechanical energy is used to turn the generator and create electrical energy.

3.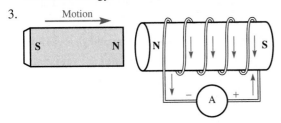

Figure 9-49

Self-Check 9-9

1. Number of coil turns, amount of flux cutting across the coil (or being cut by the coil), and the rate at which flux is being cut
2. The change in flux ($\Delta\Phi$) per change in time (Δt)
3. 180 V
4. Increase number of turns, increase flux density of magnetic field, and increase the rate of cutting flux

Self-Check 9-10

1. A stationary winding
2. The armature
3. The amplitude is related to the sine of the angle of cutting flux. Maximum amplitude is obtained as the conductor cuts the flux at a right angle (90°).
4. The AC waveform is called a sine wave because the amplitude of the waveform (voltage or current) varies according to the sine of the angle of armature rotation or angle of cutting flux.
5. The frequency measured in cps is determined by alternator armature revolutions measured in rps. A two-pole alternator will produce 1 cps for every 1 rps.

Self-Check 9-11

1. The commutator
2. The commutator switches the armature with the output terminals at the point of waveform zero crossing, which has the effect of keeping both alternations either positive or negative. The polarity of voltage and current at the output terminals will therefore remain the same even though the voltage and current is varying greatly between zero and some peak value.
3. Ripple voltage is a rapidly varying DC that rises periodically to a peak then falls to a lower value. Unlike pulsating DC, ripple does not drop clear down to zero volts and return to a peak.
4. The ripple voltage will decrease, or become shallower. Thus, the DC will be smoother, with less variation in amplitude.
5. See Figure 9-43.

SUGGESTED PROJECTS

1. Add some of the main concepts, formulas, and procedures from this chapter to your Electricity and Electronics Notebook.

2. Obtain an old automobile alternator from a junkyard. Clean it up and disassemble it. Examine it carefully to see how the armature and stators are actually wound.

3. Can a small DC motor be used as a DC generator? Obtain a small permanent-magnet DC motor from a discarded toy. Connect a DC voltmeter to the two terminals on the motor. Spin the armature of the motor as fast as you can with your fingers. Will the meter indicate a voltage is being generated? Try it and find out. You may want to carefully disassemble the motor to see how it is similar to a DC generator.

(1) Edison greatly improved upon electric motors beginning with early experiments in 1879. This slow-speed motor was developed in 1889 and was later manufactured at the Edison General Electric plant in Schenectady, New York. Edison is credited with inventing the ''Universal Motor'' in 1907 which would operate from AC or DC. This motor is used in power tools of all kinds today. (2) Notice Edison's trade mark signature on the upper right of this electric fan. Edison's battery-operated fans were a welcome addition to many homes and businesses as early as 1890.

(1)

EDISON SLOW-SPEED MOTOR
1889

(2)

As early as 1890 an Edison battery-operated fan became a necessity in our homes and places of business.

Chapter 10

Electromagnetism at Work

OBJECTIVES

After studying this chapter, you will be able to

- list and explain the requirements for motor action.
- understand and use the right-hand rule for motor action.
- explain how the direction of motor action is reversed.
- explain the importance of counterelectromotive force (CEMF) in motor action.
- describe basic permanent-magnet (PM) DC motor construction and operation.
- explain practical methods of DC motor speed and direction control.
- explain basic electromagnetic (EM) DC motor construction and operation.

- identify the basic types of EM DC motors and understand how to control their speed and direction of rotation.
- explain the basic construction and theory of operation of AC induction motors.
- describe the construction and theory of operation of AC synchronous motors.
- describe the construction and theory of operation of electromagnetic control devices such as solenoids, contactors, and control relays.
- describe the construction and theory of operation of electromagnetic transducers such as dynamic microphones, PM loudspeakers, and tape heads.

INTRODUCTION

In this chapter, you will explore electromagnetism at work. You will begin this exploration with an investigation into the theory and principles of electromagnetic motor action. Once those basic principles are understood, you will be ready to apply them to DC motors, AC motors, electromagnetic control devices, and electromagnetic transducers. This chapter answers many of the how and why questions that naturally come to mind as you ponder the inner workings of these everyday electromagnetic and electromechanical devices. A basic understanding of the subjects covered in this chapter is essential for you as a technician working in a world filled with automation.

NEED TO KNOW

How much do you already know about electromagnetic devices such as DC motors, AC motors, relays, loudspeakers, tape heads, etc.? Not sure? OK, it might be interesting to take a short quiz to find out. Without looking ahead, see if you can answer the following questions (correctly, of course). When you are finished, turn to the end of the chapter, the "Need to Know Solution" section, to find the answers. Then, keep these questions in mind as you read this chapter (especially the ones you answer incorrectly). This quiz will help establish your need to know.

1. Three basic requirements must be met for motor action to occur: (1) a magnetic field, (2) a conductor placed perpendicular to and within the magnetic field, and (3) _____ .
   a. relative motion between the conductor and field
   b. copper-wire conductors
   c. an electrical current flowing through the conductor
   d. a source of mechanical energy

2. When voltage is first applied to a motor, the current demanded by the motor is very large. The reason there is this initial surge in current is that _____ .
   a. the motor is empty of electrons and needs them badly
   b. there is no CEMF to oppose the applied EMF
   c. the motor branches are temporarily shorted
   d. the electrons are all excited and anxious to move out

3. If the shaft of an operating motor is overloaded, the motor will eventually burn out (assuming no thermal overload protection). Why?
   a. The heavy load slows the motor, reducing the CEMF and causing an increase in armature current.
   b. The heavy load slows the motor, reducing the applied EMF and causing an increase in armature current.
   c. The heavy load slows the motor, reducing the applied EMF and causing a decrease in armature current.
   d. The motor gets tired like anything else.

4. A practical DC motor will have a minimum of how many commutator segments?
   a. one
   b. two
   c. three
   d. four

5. The direction of DC motor armature rotation can be reversed by doing what?
   a. Turn the motor around.
   b. Reduce the applied EMF.
   c. Reverse the polarity of EMF applied to the armature.
   d. Reverse the magnetic field polarity and the applied EMF polarity.

6. With reference to AC induction motors, slip is what?
   a. The freedom of armature rotation.
   b. The smoothness of the brushes rubbing on the armature.
   c. The size of the air gap between the rotor and the magnets.
   d. The difference in speed between the rotor speed and the synchronous speed.

7. AC synchronous motors are used for _____ .
   a. water pumps
   b. electric vehicles
   c. fans and blowers
   d. phonograph turntables

10-1 Motor Action

Requirements for Motor Action

In this section, you will discover how closely related generator action is to another electromagnetic characteristic, motor action. You will learn how motor action and generator action actually occur at the same time.

See color photo insert pages C2 through C4.

Motor action is a *result* or an *effect*. Just as generator action is the electrical result of cutting flux with a conductor, **motor action** is the mechanical result of current flowing through a conductor that is placed in a magnetic field. In order for motor action to occur, the three basic requirements listed in Figure 10-1 must be met. Notice the similarity between the requirements for generator action and those for motor action. The only difference is in requirement #3. To achieve motor action, current is required, and to achieve generator action, relative motion is required. This is a case in which cause and effect are interchangeable. In other words, if an electromotive force (EMF), acting as the cause, is applied to a conductor that is placed in a magnetic field, the effect is motion. Conversely, if mechanical motion, acting as the cause, is applied to a conductor that is placed in a magnetic field, the effect will be an induced EMF.

How can a current flowing through a conductor create motor action? That is a very important and good question. Interestingly, you already know the answer. Consider what

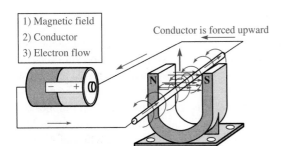

1) Magnetic field
2) Conductor
3) Electron flow

Conductor is forced upward

Figure 10-1 Motor action results when a current-carrying conductor is placed in a magnetic field.

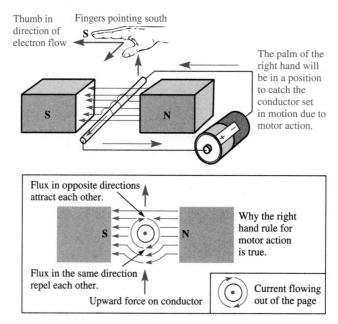

Figure 10-2   Determining the direction of motor action.

you have already learned about magnetism and electromagnetism. Recall that a magnetic field is developed around a current-carrying conductor (left-hand rule for conductors). Remember also that magnetic fields will either repel or attract each other, depending upon their polarities. Motor action is actually the interaction of the magnetic field of the conductor with the stationary magnetic field. Along one side of the conductor, flux of the conductor and flux of the stationary field are of opposite polarity, causing a force of attraction. Along the opposite side of the conductor, flux of the conductor and flux of the stationary field are of the same polarity, causing a force of repulsion. (See the inset of Figure 10-2). One side of the conductor is being pulled while the opposite side is being pushed by magnetic forces. We can summarize Figure 10-1 and motor action as follows: A conductor is placed in a stationary magnetic field, an EMF is applied to the conductor, the EMF produces a current through the conductor, the current through the conductor is an MMF that produces flux around the conductor, and the flux surrounding the conductor interacts with the stationary flux, resulting in mechanical motion.

The Right-Hand Rule for Motor Action

The actual direction in which a current-carrying conductor will move in a magnetic field can be predicted, or determined, in one of two ways: (1) use the left-hand rule for conductors, then determine which side of the conductor is repelled and which side is attracted by the stationary flux, or (2) use the **right-hand rule for motor action**. Figure 10-2 illustrates the use of this rule. Many people prefer this method over method number one simply because it is quicker to use. The fingers of your right hand must point toward the south stationary magnetic pole. Your right thumb must point in the direction of electron flow through the conductor. With the right fingers and thumb properly aligned, the palm will be in a position to catch the conductor.

The inset of Figure 10-2 demonstrates why the right-hand rule is true. It is true because the direction of flux above the conductor is opposite to the flux of the stationary field and the direction of flux below the conductor is the same as the flux of the stationary field. Thus, above the conductor is a force of attraction and below the conductor is a force of repulsion. If the battery is reversed, will the right-hand rule still be true? Yes! However, you will have to properly place your right hand so your thumb will indicate the new direction of current. With the battery reversed, you will have to place your right hand below the conductor, palm facing up, to properly indicate the direction of current. The upward facing palm then indicates the conductor is moving downward. Remember, your right palm is positioned to catch the conductor.

Reversing Direction of Motor Action

Reversing the direction of motor action can be accomplished in one of two ways: (1) Reverse the direction of current in the conductor, or (2) reverse the magnetic field polarity. Figure 10-3 illustrates this. Notice, doing both will *not* cause a reversal in motor action. Apply the right-hand rule for motor action to each of the diagrams of Figure 10-3 to see how current and magnetic polarity affect the direction of motor action. Later in this chapter, we will apply these basic principles to practical, real motors.

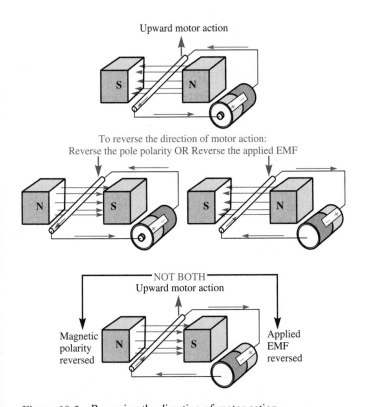

Figure 10-3 Reversing the direction of motor action.

Counterelectromotive Force (CEMF)

Do you remember the countermagnetomotive force (CMMF) that was discussed in the previous chapter? It is an induced magnetomotive force, which creates flux that opposes the stationary flux of the generator as its armature is turned. The CMMF creates a motor action that opposes the mechanical rotation of the armature. Naturally, the mechanical force used to rotate the armature is always a little stronger than the opposition created by the CMMF. Just as generator action produces CMMF, motor action produces counterelectromotive force (CEMF).

Counterelectromotive force (CEMF) is a voltage that is induced across the windings of the motor armature as it rotates. In other words, while the motor is acting like a motor it is also acting as a generator. The polarity of this induced voltage (CEMF) is opposite to that of the external electromotive force that causes motor action in the first place. Thus, the name *counter*electromotive force (CEMF). Figure 10-4 illustrates the concept of CEMF.

The inset of Figure 10-4 illustrates the effects of CEMF on the applied EMF. Since the induced voltage is in opposition to the external applied voltage, the external difference of potential is effectively reduced. Also, the amount of current supplied to the motor is reduced, due to the countercurrent induced in the armature as it rotates. This fact is very

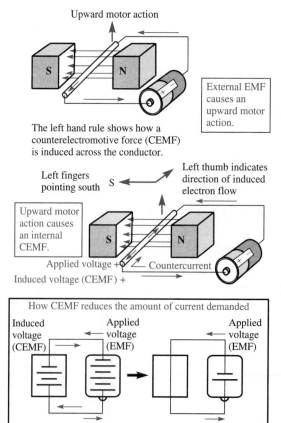

Figure 10-4   Counterelectromotive force CEMF.

significant! It is the reason motors do not burn up due to large amounts of current from the external source of EMF. The windings on the armature of a real motor are usually made of low-resistance copper wire or strapping. When an EMF is first applied to a motor, only the resistance of the copper limits the amount of current. Thus, motor starting current is very high. This small amount of resistance is not enough to prevent the motor from burning up. As the motor builds up speed, more CEMF is induced across the armature. Very quickly, a balance is reached between the EMF, the CEMF, and the mechanical load. The difference between the EMF and CEMF will be the difference of potential needed to turn the armature working against a particular mechanical load. If the mechanical load is very light, the difference in potential between the EMF and the CEMF will be very small (e.g., EMF = 12 V and CEMF = 11.75 V). Thus, the actual amount of current in the armature will be very small. Conversely, if the mechanical load on a motor is large, or heavy, the difference in potential between the EMF and the CEMF will be much greater (e.g., EMF = 12 V and CEMF = 6 V). If the mechanical load on the motor is great, the difference between the EMF and the CEMF is great, and therefore the amount of current supplied to the motor is great. Under a heavy load the motor could burn up, since the CEMF is so small and the effective applied voltage is so much greater. Under normal load conditions, a motor will run warm, or even very warm, with no particular concern as long as ventilation, or cooling, is adequate.

Take time now to reinforce these new concepts by answering the questions of Self-Check 10-1.

SELF-CHECK 10-1

1. List the three basic requirements for motor action.
2. Describe one method for determining the direction of motor action.
3. What is counterelectromotive force?
4. Briefly explain why a free-running motor will not burn up even though the resistance of the armature windings is very low.

10-2 DC Motors

Basic PM DC Motor Construction and Theory

The basic **permanent-magnet (PM) DC motor** is very similar to a permanent-magnet DC generator. It has a pair of magnetic poles, a wirewound armature, and a commutator with graphite brushes. Figure 10-5 illustrates the basic components of the DC motor and its schematic symbol.

See color photo insert page C3.

Radically Varying Torque and CEMF

Figure 10-6 illustrates an impractical single-armature-coil DC motor. Note the radically varying amount of torque created by motor action as the shaft rotates through one revolution. It is no coincidence that the graph looks the same as the pulsating DC generated by a similarly designed basic generator. In fact, the graph also represents the CEMF that

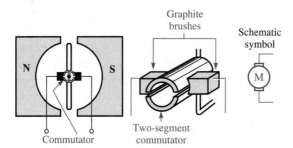

Figure 10-5   The basic DC motor.

is being generated as motor action occurs. The rotational torque and induced voltage (CEMF) are a function of the sine of the angular position of the armature winding in the stationary magnetic field. At 180° (Figure 10-6c) there is no rotational torque or CEMF, since the sine of 180° is zero (0). At the 270° position (Figure 10-6d) the torque and CEMF are once again maximum, since the sine of 270° is one (1).

Naturally, this motor would not qualify as smooth-running. The armature rotation is a throbbing rotary motion. Under medium load conditions, the armature will want to stall at the 0° and 180° positions. What is worse, the brushes and commutator segments will wear quickly, due to repeated shorting as the armature turns. This design must definitely be improved.

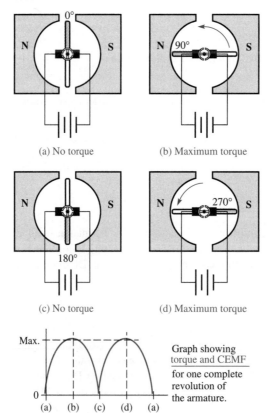

Figure 10-6   Pulsating torque and CEMF are produced by this simple DC motor.

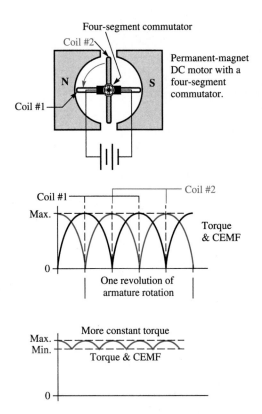

Permanent-magnet
DC motor with a
four-segment
commutator.

Figure 10-7   Torque is greatly improved
with this DC motor.

Improving the DC Motor Design

The DC motor can be improved the same way the DC generator was improved, by adding more commutator segments and armature coils. Figure 10-7 illustrates the very desirable results. The two armature coils, mechanically placed 90° apart, will insure a more consistent rotary torque on the shaft. When one coil is in the no-torque position (0°/180°), the other coil will be in the maximum-torque position (90°/270°). The motor will not stall, the brushes are never shorted by commutator segments, and, unlike Figure 10-6, the motor is self-starting (no fingers needed).

Minimum Practical Armature Requirements

Figure 10-8 shows a small permanent-magnet (PM) DC motor that meets minimum practical design requirements. The armature has three coils spaced 120° apart and three commutator segments. At any given time, at least two of the coils will be in a position of rotational torque. This allows the motor to be self-starting whatever the initial armature position. The three commutator segments ensure that the brushes cannot become shorted through a single segment. Current supplied by the brushes is always applied to the armature coils for uninterrupted torque. DC motors of this design are normally used for light-duty applications such as electric trains, electric model racing cars, and various other toys. Where a more continuous and stronger amount of torque is required, armature designs employ 8, 10, or more commutator segments and associated coils.

Figure 10-8   A small permanent-magnet DC motor.

Permanent-Magnet DC Motor Control

Reversing the Direction of Armature Rotation

Part of the fun of operating an electric train is to be able to back the engine onto a siding and hook onto a waiting string of box cars, then switch the controls to forward and come roaring out of the siding onto the main line with your load. How and why does the engine's DC motor reverse direction with the flick of a switch? Figure 10-9 shows how, and the right-hand rule for motor action tells why. Recall that the direction of motor action can be reversed by reversing the direction of current in the conductor or reversing the magnetic

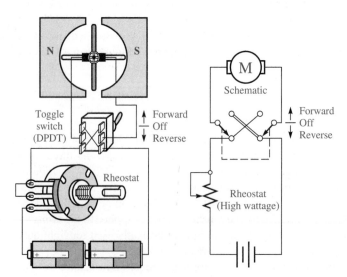

Figure 10-9   PM DC motor speed control.

poles, but not both at the same time. In Figure 10-9, a DPDT center-off switch is used to change the polarity of applied EMF and turn the motor off (in the center position). A separate SPST switch could have been used along with a two-position DPDT switch instead of the single DPDT center-off switch shown. The DPDT switch is a very common means of obtaining forward/reverse control. The direction of rotation can also be controlled by switching the magnetic poles back and forth. However, that, of course, is not usually practical with PM DC motors.

PM DC Motor Speed Control

Figure 10-9 also demonstrates a means of controlling the speed of a permanent-magnet motor by controlling the amount of current supplied to the armature. This can be done in many different ways, the simplest of which is a high-wattage rheostat in series with the applied EMF. The rheostat controls the amount of current supplied to the armature and therefore dissipates a lot of heat itself. More efficient transistor switching circuits can be used to turn the current to the armature rapidly on and off. The current on time and off time are electronically varied to control motor speed. If the pulses are on longer than they are off, the motor runs fast and vice versa.

Electromagnetic DC Motors

Basic Construction

Figure 10-10 illustrates the **electromagnetic (EM) DC motor** and its schematic representation. As you can see, the only real difference between the EM and the PM DC motor is the way the magnetic field is generated. In the EM DC motor, a field coil is used to create a magnetic field. This design adds greater operating flexibility, since the strength of the magnetic field can be varied by controlling the amount of current supplied to the field windings. Figure 10-11 shows a typical EM DC motor with one end housing, or bell housing, removed. Field-coil windings can be seen on the left and right. The armature is wound with 10 overlapping coils and 10 commutator segments. Spring-loaded graphite brushes are contained in the white plastic casing on each side of the commutator.

The EM DC motor shown in Figure 10-12 is called a *separately excited* motor. Its field coil and armature can be controlled, or excited, separately. In the following discussion, this motor will be used as a model for understanding the theory of EM DC motor operation and control.

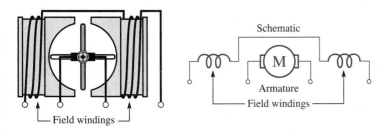

Figure 10-10   An electromagnetic (EM) DC motor.

Figure 10-11   This automobile fan motor is a typical EM DC motor.

Controlling Speed and Torque

The speed of an EM DC motor can be varied by controlling the amount of current supplied to its armature or the amount of current supplied to its field coil from a variable source of EMF (or a rheostat in series with a source of EMF). As the EMF applied to the armature is increased, the magnetic field of the armature will increase in strength, resulting in an increased torque and armature speed. As the armature increases in speed, the CEMF induced across the armature windings will increase. If the applied EMF is held constant, a balance will be reached between EMF, CEMF, and the amount of energy required to accommodate a mechanical load.

The speed of the armature may also be controlled by the amount of EMF applied to the field coil, but not in the same way. If the amount of EMF applied to the field coil is increased, the resulting increase in MMF will cause the stationary magnetic field to become stronger (an increase in total flux and flux density). Naturally, this will increase the amount of torque. However, the speed of the motor actually decreases.

Why will the motor decrease in speed with an increase in magnetic field strength? The answer to this question is found in the answer to a second question. When the magnetic field strength is increased, what happens to the amount of CEMF being induced across the armature coil? That's right! The induced CEMF will increase. For an instant of time, the

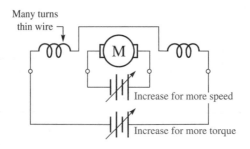

Many turns
thin wire

Increase for more speed

Increase for more torque

Figure 10-12   Separately excited EM DC motor.

CEMF may be equal to or greater than the external EMF. This means the source of EMF will not be able to supply any current to the armature. Therefore, there is no armature magnetic field to interact with the stationary magnetic field. Remember, this occurs for only a very short period of time. During this short period of time, the armature has no choice but to slow down. As the armature decreases in speed, the amount of induced CEMF is decreased. The decrease in CEMF allows the external EMF to once again supply current to the armature. Then, a new balance (between EMF, CEMF, and energy required to accommodate the mechanical load) is reached at a lower rpm. If we wish to return to the original speed at this higher level of torque, the EMF supplying the armature must be increased.

Types of Electromagnetic DC Motors

Most practical EM DC motors operate from a single source of EMF. Single-supply EM DC motors are typed according to the manner in which the field coil or coils are wound and electrically connected to the supply and armature. There are basically three main types: (1) the series EM DC motor, (2) the parallel, or shunt, EM DC motor, and (3) the compound EM DC motor. Figure 10-13 schematically illustrates these basic types and provides a table for comparison.

Series Motor

Because of the series armature and field-coil arrangement, the **series motor** is ideal for applications that require large amounts of starting torque. These motors are commonly used in electric vehicles, since high starting torque is very important. They are also found in many high-torque electric tools that operate from alternating current. That's right. Many series DC motors will also operate from a comparable AC source. An electric hand drill, for example, will operate from 115 VAC or 115 VDC. Recall that a DC motor will not reverse direction if the polarity of both the armature and magnetic field are reversed. When AC is applied, the current and voltage polarity reversals are applied to both the armature and the field coil at the same time. This allows the armature to continue rotation in the same direction. Because of the AC/DC flexibility, these motors are often called **universal motors**. Other types of EM DC motors will also run on AC but at very poor efficiencies, for reasons beyond the scope of this discussion.

Parallel (Shunt) Motor

The **parallel** or **shunt motor** has a field coil made of a relatively large number of turns of a smaller-gauge wire. The amount of current delivered to the shunt field coil is independent of the amount of current supplied to the armature coil. Since the amount of current flowing through the field coil is determined by the amount of applied EMF and the resistance of the field-coil windings, the MMF and total flux is constant. This type of EM DC motor is very similar in characteristics to the PM DC motor, since the field flux is constant. As a result, the shunt motor has a relatively low starting torque but very good speed regulation. For this reason, the shunt motor is used in applications where it is necessary for the motor to run at a constant speed under varying load conditions. Some typical applications might be certain machine tools, fans, and blower motors.

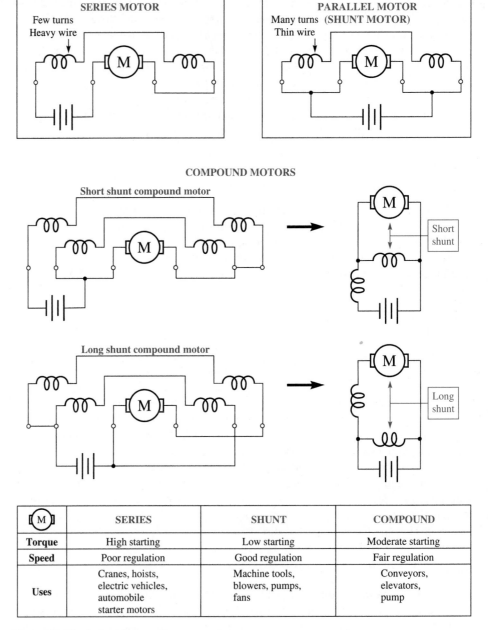

Figure 10-13   Series, shunt, and compound motors.

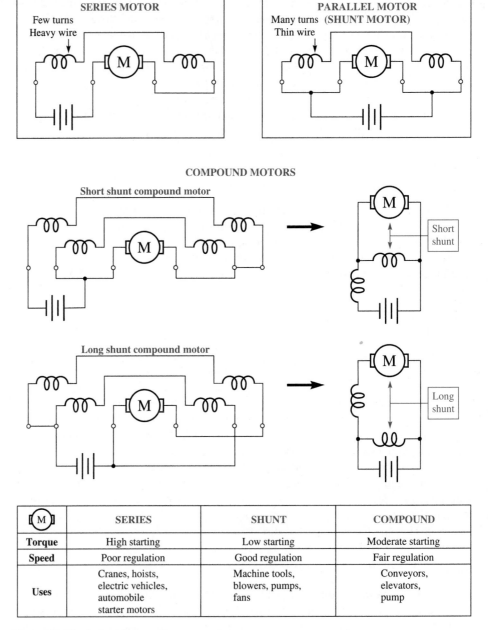	SERIES	SHUNT	COMPOUND
Torque	High starting	Low starting	Moderate starting
Speed	Poor regulation	Good regulation	Fair regulation
Uses	Cranes, hoists, electric vehicles, automobile starter motors	Machine tools, blowers, pumps, fans	Conveyors, elevators, pump

Compound Motors

The largest category of the three types of motors is that of the compound motors. Many variations of this type exist. Here, we will only mention some of the varieties and discuss compound motors in general.

The compound motor has two field coils. One field coil is connected in series with the armature and the other is connected in parallel. As in the series motor, the series coil is made of relatively few turns of heavy-gauge wire and, as in the shunt motor, the parallel coil is made of a large number of turns of small-gauge wire. As you have probably guessed, the reason for using both field coils is to gain the advantages of the series and shunt motor all in one motor. The series coil insures a strong starting torque while the parallel coil insures good speed regulation.

Take time now to review this section. Then, test your understanding of the construction, operation, and control of DC motors by answering the questions of Self-Check 10-2.

SELF-CHECK 10-2

1. How should a DC motor armature be designed in order to obtain a more continuous rotational torque?

2. What prevents the armature of a free-running motor from burning up? Explain.

3. Increasing field-coil current in an EM DC motor will have what effect on torque? Explain.

4. Decreasing the amount of current supplied to the armature will have what effect on speed? Explain.

5. What is the main advantage of the series motor?

6. Of the three main types of single-supply EM DC motors, which has the best speed regulation?

10-3 AC Motors

AC Induction Motors

AC induction motors are much different in design and operation from the DC motors previously discussed. In fact, AC induction motors are often preferred over DC motors for many applications for the following reasons: simple construction, low maintenance, high efficiency, good constant speed regulation, and low cost. In this section, we will examine the construction and theory of operation of AC induction motors along with the various types in common use. The basic principles of AC induction motor action are the same electromagnetic principles discussed earlier. However, the application of these principles will be new to you. Let's begin our discussion by examining basic AC motor construction.

See color photo insert page C2.

Basic Construction

AC induction motors are divided into two basic categories: (1) **squirrel-cage rotor**, and (2) **wirewound rotor**. The most common category is that of the squirrel-cage rotor AC induction motor. This kind of AC motor has an armature (squirrel-cage rotor) that requires no brushes, slip rings, or commutator for connection to the outside world. The rotor is completely self-contained. The wirewound rotor, however, does require slip rings and brushes. This permits the wirewound rotor to be connected to an external rheostat that is

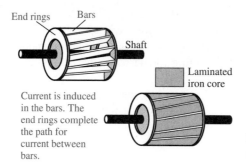

End rings   Bars

Shaft

Laminated
iron core

Current is induced
in the bars. The
end rings complete
the path for
current between
bars.

Figure 10-14   The squirrel-cage rotor—the
armature of an AC induction motor.

used to control the amount of current induced in the rotor. In this way, the torque of the wirewound rotor is controlled. Because squirrel-cage rotor AC induction motors are by far the most common, our discussion of induction motors will be limited to them.

Figure 10-14 illustrates the typical squirrel-cage rotor found in most AC induction motors. As you can see, it is very different from the armatures found in DC motors. Notice the solid crossbars running parallel and slightly askew to the shaft (the crossbars are skewed to reduce electromagnetically generated noise and vibration). These crossbars are heavy conductors (usually aluminum or copper) in which currents are induced. These currents are able to circulate through the conductive end rings, completing a current loop through another crossbar. Naturally, the circulating currents produce magnetic fields that interact with the external field that induced the current in the first place. The interaction of the fields produces motor action. The magnetic field of the rotor is strengthened with the addition of a **laminated iron core**.

Figure 10-15 shows a disassembled AC induction motor. As can be seen, contained within the motor housing and surrounding the squirrel-cage rotor are electromagnetic coils. As in other electromagnetic devices, these coils are called *stators*, since they are stationary. The overall efficiency of the AC induction motor will depend heavily upon the number of

Figure 10-15   A small AC induction motor.

stator coils and the manner in which AC is applied to them. This leads us to the theory of operation of AC induction motors.

Theory of Operation

The theory of operation can be stated in a few sentences. Alternating current is applied to the stator, or stators, in such a manner as to create a rotating magnetic field. The rotating magnetic field induces current in the crossbars of the squirrel-cage rotor. The induced currents produce magnetic fields that oppose the magnetic fields that created them (Lenz's Law). This opposition produces motor action (rotary action of the rotor). Thus, the rotor will attempt to follow the rotating magnetic field of the stator or stators.

To help clarify the theory of operation, let's examine the illustrations of Figure 10-16. The induction rotor used in Figure 10-16 is a very simple one consisting of two

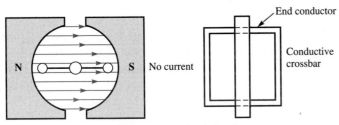

(a) No relative motion and no induced current

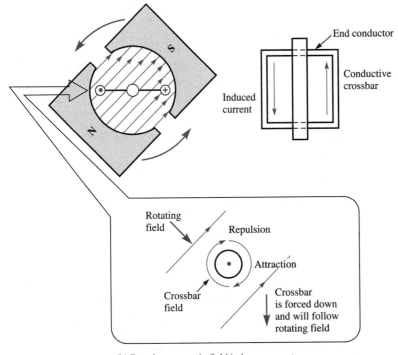

(b) Rotating magnetic field induces current

Figure 10-16   Basic induction motor operation.

crossbars and two end connectors forming one complete loop. Permanent magnets are used in the illustrations to simplify the discussion. In practice, the magnetic field would be created by stator coils.

In Figure 10-16a, the rotor is motionless in the midst of a stationary magnetic field. As a result, no current is induced in the rotor crossbars, since there is no relative motion. However, suppose we find some way to cause the magnetic field to rotate as shown in Figure 10-16b? The result of rotating the magnetic field is an induced current in the single-loop rotor. The induced current in the rotor develops a magnetic field, surrounding the conductive crossbars, that opposes the rotating external magnetic field (Lenz's Law). Thus, the counterclockwise rotation of the magnetic field will result in a counterclockwise rotation of the single-loop rotor. Carefully examine the magnified inset of Figure 10-16b.

Slip Is Important

It is important for you to understand that the current is induced in the rotor because of the relative motion between the rotating magnetic field and the rotor itself. If the magnetic field continues to rotate and the rotor continues to follow the rotating magnetic field *at the same speed*, there will be no relative motion, no induced current, and no motor action. If there is no motor action, the rotor must slow down. The rotating magnetic field must be rotating at a higher speed than the rotor to maintain relative motion, induced current, and motor action. The necessary difference in speed between the rotating magnetic field and the rotor is called **slip**.

All AC induction motors have slip. Slip is essential for the operation of an induction motor. The amount of slip depends upon the amount of mechanical load connected to the rotor shaft. As mechanical load is increased, rotor speed decreases, the amount of slip increases, and as a result, the amount of relative motion increases. The increased relative motion induces more current in the rotor, which increases the amount of torque on the rotor. A balance is quickly established as load is added or taken away. Usually, the rotor speed is in the neighborhood of 95% of the speed of the rotating magnetic field. Rotor speed varies only a couple percent above or below the 95% value over a wide range of mechanical loads. Good speed regulation is one of the advantages of the AC induction motor.

Types of AC Induction Motors

Split-Phase Induction Motor

The **split-phase induction motor** operates from a single-phase AC source as shown in Figure 10-17. Since these motors are designed to operate from a single AC phase, they are normally designed for relatively low-power applications, usually in the neighborhood of one horsepower (1 Hp) or less. They are commonly found in domestic appliances such as automatic clothes washers and dryers.

Split-phase induction motors have two field windings, or coils: a starter winding and a main winding. The starter winding is usually made of few turns of relatively large-gauge wire, and the main winding is made of many turns of relatively small-gauge wire. In many split-phase designs, the starter winding is used only to increase starting torque and efficiency. As the rotor reaches 60 to 70% of the maximum running speed, a centrifugal switch is often used to automatically remove power from the starter winding. Some split-

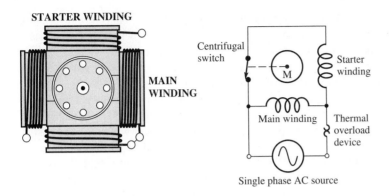

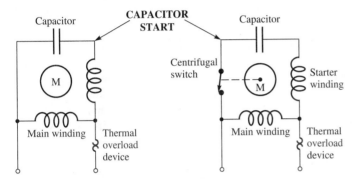

Applied AC is split into two phases ideally 90° apart

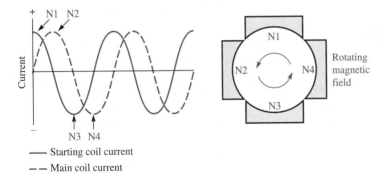

—— Starting coil current

– – Main coil current

Figure 10-17   The split-phase AC induction motor.

phase induction motors may not have this centrifugal switch, in which case the starter winding remains energized all the time. The main winding is always energized and is often called the *running winding*. As shown in Figure 10-17, a thermal overload device is often placed in series with both windings. This device is a temperature-sensitive normally closed switch, mounted inside the motor housing, that will open if the stator windings reach a certain temperature.

The name *split-phase* is given to these motors because the single-phase AC source is divided into two phases. Normally this is accomplished by placing a high-voltage capacitor, of a specific capacitive value, in series with the starter winding. This capacitor causes the current in the starter winding to lead the current in the main winding. For greatest efficiency, it is desirable for the starter winding current to lead the main winding current by 90 electrical degrees. If the starter capacitor is not used, the phase shift will be much less than 90°, resulting in very poor starting torque and efficiency.

The graph of Figure 10-17 shows the ideal 90° phase relationship between the starter winding current and the main winding current. Notice, starting on the left of the graph, that the starter winding current is at a positive peak while the main winding current is at zero crossing (0 A). As time continues to the right, the main winding current increases as the starter winding current decreases. When the main winding current reaches a positive peak and the starter winding current reaches zero crossing, 90 electrical degrees have passed, or one quarter of a cycle. As time continues, the 90° phase shift between the two currents remains constant. It is this 90° phase shift between the two currents that creates the rotating magnetic field needed for induction to take place.

How do these two currents and field coils create a rotating magnetic field? Notice the magnetic-pole diagram next to the currents graph. As a starting point, let's assume that the upper electromagnetic pole will be a north magnetic polarity when the starter winding current is at a positive peak (N1). As the starter winding current decreases, the upper magnetic pole decreases in strength. At the same time, the main winding current increases and the left electromagnetic pole increases as a north magnetic polarity (N2). Already, the north magnetic polarity has rotated from the upper pole to the left pole. This pole rotation continues as the starter winding current reaches a negative peak and the lower electro-magnetic pole becomes a north polarity (N3). As N3 increases in strength, N2 decreases in strength. The right electromagnetic pole becomes north in polarity as the main winding current reaches a negative peak at N4, during which time N3 decreases in strength. Finally, the starter winding current returns to a positive peak, with the upper electromagnetic pole once again becoming a strong north magnetic polarity. Thus, the north magnetic polarity has rotated a full 360°, or one revolution, with one complete AC cycle.

The speed of pole rotation is called the **synchronous speed**, since it is synchronized with the applied alternating current. The synchronous speed therefore depends upon the frequency of the applied AC. If the frequency of the applied AC is 60 cycles per second (cps), the synchronous speed will be 60 revolutions per second (rps), or 3,600 revolutions per minute (rpm). As you know, the rotor cannot keep up with the rotating magnetic field. In order for induction to occur, there must be slip. Since a typical value of slip is about 5%, the actual rotor speed will be about 3,400 rpm.

Shaded-Pole Induction Motor

Shaded-pole induction motors are basically designed for low-power applications such as fans and inexpensive record players (usually less than 0.1 Hp). They are very simple in construction and very low in cost. As shown in Figure 10-18, a **shaded-pole induction motor** has one main field winding, which is connected to a single-phase AC source. This field winding is made of many turns of relatively small-gauge wire, even though only a few turns are shown in the illustration. Each electromagnetic pole has a single loop formed from a copper strap, or heavy-gauge copper wire. This single loop is called a **shading strap**. The shading strap will have a current induced in it by the expanding and collapsing main field. As a result, the shading strap produces a magnetic field that opposes any change in flux in the main field (Lenz's Law). This opposition from the shading strap causes the

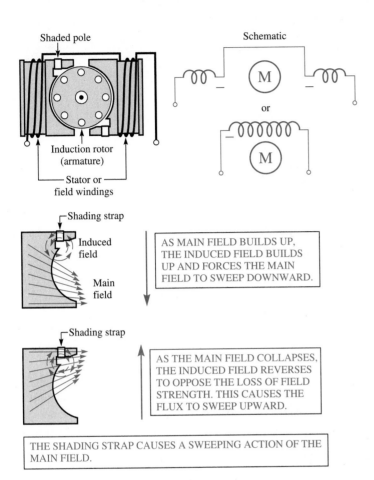

Figure 10-18   The shaded-pole AC induction motor.

main magnetic field to become warped, or offset. Thus the name **shaded-pole** was given to this motor design.

So, how does a shaded-pole induction motor create a rotating magnetic field? In this case, the magnetic field does not rotate all the way around the rotor. Instead, relative motion of the main field and rotor is provided by a sweeping action of the main field. While this is less efficient than a fully rotating magnetic field, it is fine for low-power applications. As shown in Figure 10-18, the shading strap causes the magnetic flux to sweep down and up across the face of the pole. As the main field is building up, the induced and opposing magnetic field of the shading strap forces the main field to sweep downward. When the main field starts to collapse, due to alternating current cycling, the field of the shading strap actually reverses in opposition to the decrease in flux. Thus, the main field is strengthened and attracted upward toward the shading strap. This sweeping action is repeated with every cycle of alternating current. As one pole sweeps in one direction, the opposite pole is sweeping in the other direction, producing the relative motion needed to induce current in the rotor. The induced current produces a magnetic field that opposes the main field, and motor action results.

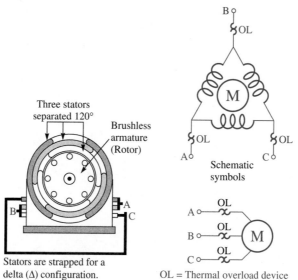

Three stators separated 120°

Brushless armature (Rotor)

Stators are strapped for a delta (Δ) configuration.

OL = Thermal overload device

Figure 10-19   A three-phase AC induction motor.

Three-Phase Induction Motor

The three-phase induction motor is designed for high-power industrial applications (usually greater than 1 Hp). It is far more efficient than single-phase induction motors, since the rotor torque is developed by three phases of alternating current, which produce a smooth rotating magnetic field. As shown in Figure 10-19, the three-phase motor has three sets of stator windings that work together to develop the rotating magnetic field needed for induction. These windings may be wired in a delta (Δ) configuration, as shown, or in a wye (Y) configuration.

AC Synchronous Motors

Figure 10-20 illustrates a typical three-phase **AC synchronous motor**. The reason this type of motor is termed *synchronous* is that the rotor will rotate at the same speed as the

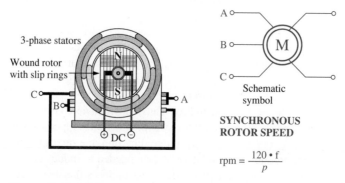

3-phase stators

Wound rotor with slip rings

Schematic symbol

SYNCHRONOUS ROTOR SPEED

$$\text{rpm} = \frac{120 \cdot f}{p}$$

Figure 10-20   A three-phase synchronous AC motor.

rotating magnetic field (synchronous speed). The rotor of the synchronous motor is either a permanent magnet or an electromagnet. The rotor shown in Figure 10-20 is wirewound and externally energized from a DC source via brushes and slip rings. As a result, the electromagnetic rotor stays in step (remains synchronized) with the rotating magnetic field. No slip occurs, nor is slip necessary for motor action. In an induction motor, slip is absolutely necessary to create a relative motion that will induce current in the rotor. In a synchronous motor, the current is provided externally and induction is not needed. However, many high-power synchronous motors have an electromagnetic rotor that also has conductive crossbars. These crossbars act like squirrel-cage rotor crossbars and help increase the starting torque of the synchronous motor. Once the rotor is up to synchronous speed, the cross bars do nothing. (Remember, there is no slip or relative motion at synchronous speed.) Like the three-phase AC induction motor, the direction of rotation of the three-phase synchronous motor may be reversed by switching any two of the three phases.

Synchronous motors are used in applications where fixed, or constant, speed is extremely important, such as audio and video tape machines, record turntables, clocks, and industrial applications where speed and timing are important. It is interesting to note that the synchronous motor may also be used as an AC generator by using a mechanical source of energy to rotate the wound rotor while a DC excitation voltage is applied to it.

Take a few minutes to review this section. Then test your understanding by answering the questions of Self-Check 10-3.

1. How does a squirrel-cage rotor differ from the armature of a DC motor?

2. What is slip and why is it necessary?

3. List two advantages of an AC induction motor over a DC motor.

4. What is synchronous speed?

5. What is the basic physical difference between AC induction motors and AC synchronous motors.

10-4 Electromagnetic Control Devices

The Control Solenoid

Many machines, apparatuses, or pieces of electromechanical equipment require electrically energized devices that create linear mechanical motion. The linear motion may be needed to reposition a tool or an electronic sensor, to engage the starter motor with the flywheel of an automobile engine, to control the fuel-to-air mixture in a carburetor, or to open and close a valve or damper by remote control. In most cases, this electrically controlled linear mechanical motion can be obtained through the use of a control solenoid.

Figure 10-21 illustrates the construction and operation of a typical solenoid. As you know, the term *solenoid* refers to a coil of wire that produces a magnetic field when a

Solenoid

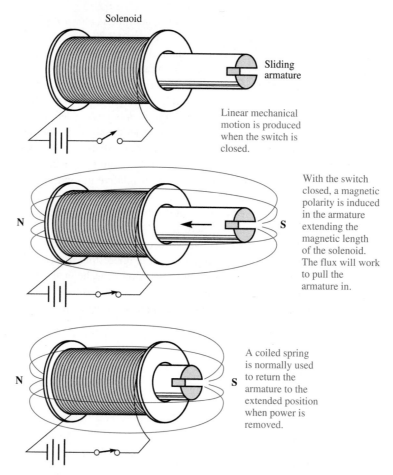

Sliding armature

Linear mechanical motion is produced when the switch is closed.

N   S

With the switch closed, a magnetic polarity is induced in the armature extending the magnetic length of the solenoid. The flux will work to pull the armature in.

N   S

A coiled spring is normally used to return the armature to the extended position when power is removed.

Figure 10-21   How the electromechanical solenoid works.

current is applied to it. However, the term *solenoid* is often used to describe control devices that include more than just a coil. **Control solenoids** have soft iron cores that are pulled in when electrical current is applied to the coil. The sliding iron core is actually called an *armature*. It may not look like the armature of a motor or generator, nevertheless an armature is defined as the moving part of an electromagnetic device.

As shown in the figure, an applied EMF will create an MMF that will generate flux through and around the solenoid. Because of the very low reluctance and high permeability of the armature, flux will extend from the center of the coil through the armature. The force exerted by the flux will pull the armature into the center of the coil. This force can be very strong when large amounts of current are applied to the coil. When the EMF is removed, the armature will either remain in the center of the coil or, if it is spring loaded, return to the extended position. In some applications, two coils share the same armature in a pull-pull arrangement, so one of two mechanical positions may be chosen depending on which coil is energized. Many solenoids operate equally well from DC or AC. However, most are designed specifically for either AC or DC operation.

Heavy-Duty Contactors

The **contactor** is a close cousin to the control solenoid. It uses a solenoid/armature arrangement to pull heavy-duty electrical contacts together. Contactors are used to control the large amounts of current and high voltages required by heavy-duty industrial equipment, such as three-phase motors and large capacity power supplies. Figure 10-22 illus-

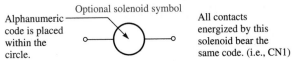

Figure 10-22 The heavy-duty contactor for high-voltage high-current use.

trates a typical contactor. There are many variations of the basic design shown. The armature is typically either a simple I bar or a T bar. Although not shown, the armatures are spring loaded to return them to a normally extended position. As you can see, a phenolic insulator containing conductive crossbars is bonded to the armature. The crossbars have heavy-duty copper contacts at each end, which are mated to stationary contacts when the solenoid is energized. The crossbar double-contact design is used to effectively double the air space between contacts, to reduce the amount of high-voltage arcing that often occurs when contacts separate. The iron core and armature of AC energized contactors are laminated to block induced currents in them that would convert a large amount of energy to heat. More on this in the chapter on transformers.

Figure 10-22 also includes a schematic illustrating a typical contactor application. Note the symbols used to represent the solenoid and contacts. Often, a circle symbol is used to represent the solenoid in a schematic that is very complicated (involving many contactors). Each circle is labeled with an alphanumeric code designating a specific contactor. The contacts associated with that particular contactor bear the same code. This makes it easy to associate a set of contacts drawn on one end of a schematic with the proper solenoid drawn on the other end of the diagram.

Control Relays

Another important electromagnetic control device is the control relay. The **control relay** is in the same family as the solenoid and contactor. It employs an AC or DC solenoid to pull an armature in to mate one or more sets of contacts. Control relays are designed for electronic circuit control and low-power electrical control. They are manufactured in almost countless variations of shape, size, and ratings. Figure 10-23 displays a modest fraction of the variations available. Some are made for PC board mounting, some are made for socket mounting, and some are made for chassis mounting with solder lugs. In spite of the vast number of varieties available, most control relays follow the same general design as that shown in Figure 10-24. As you can see, the basic relay consists of a solenoid with an iron core, an iron frame, an iron armature, a return spring, and electrical contacts with associated terminals.

The operation of the relay is very simple. The solenoid has an iron core, an iron frame, and an iron armature to conduct magnetic flux when energized. An air gap exists between

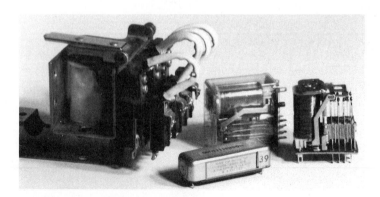

Figure 10-23 A variety of typical control relays.

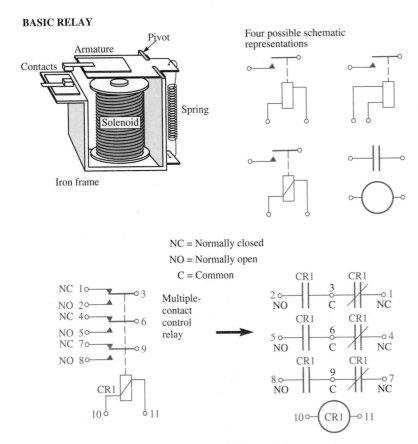

BASIC RELAY

Four possible schematic representations

NC = Normally closed
NO = Normally open
C = Common

Multiple-contact control relay

Figure 10-24 The control relay and schematic symbols.

the armature and one end of the solenoid. When an EMF is applied to the solenoid, magnetic flux will bridge the air gap and pull the armature in. When the EMF is removed, a spring will return the armature to its normally out resting position.

Figure 10-24 also illustrates some of the schematic symbols used to represent relays. It is sometimes difficult to differentiate between a control relay and a contactor in a schematic. Most electronic circuit schematics use the switch-contact symbols with the dashed line to the solenoid. This is commonly seen in schematics of tape recorders, radio communications transceivers, burglar alarm systems, etc. However, electrical control schematics, as in industrial control applications, normally use the coded circle symbol for the solenoid and the parallel bars for contacts. Figure 10-25 shows a typical application in which the relay is wired to be self-holding.

Contactor and Relay Specifications

Some of the more important relay and contactor specifications are operating voltage, pickup voltage, dropout voltage, and contact current and voltage ratings. The **operating voltage** is the coil voltage, specified by the manufacturer, that will insure quick and solid operation of the relay or contactor. If the specified operating voltage is used, the armature

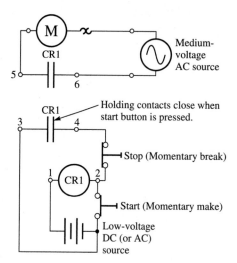

Figure 10-25   A self-holding control relay circuit.

will be held firmly in place. The **pickup voltage** is a voltage lower than the specified operating voltage. It is the minimum solenoid voltage at which armature motion will occur. The **dropout voltage** is a voltage lower than the pickup voltage. It is the solenoid voltage at which an energized relay or contactor will release. Contacts are rated according to the amount of current and voltage they can handle without excessive arcing, flashover (ionization of the air between contacts), or welding of the contacts. The **contact current and voltage ratings** are specified for resistive loads and are normally derated by 20 – 30% if the contacts are used to control power supplied to inductive loads (motors, transformers, etc.). (Inductive loads create excessive voltages when contacts are opened, due to collapsing magnetic flux surrounding the windings. Often, the induced voltage will arc between the contacts as they are opened, causing pitting on the contact faces.)

Before continuing to the next section, take a moment to answer the questions of Self-Check 10-4.

SELF-CHECK 10-4

1. How might you electrically control the flow of a fluid in a pipe?

2. For what are contactors used?

3. What is the difference between pickup voltage and dropout voltage as they pertain to relays and contactors?

10-5 Electromagnetic Transducers

In this section, we will explore a variety of electromagnetic transducers. Let's begin by defining what a transducer is. A **transducer** is a device that is able to convert one form of energy to another. An **electromagnetic transducer** is any device using electromagnetic principles that converts another form of energy to electricity or converts electricity to another form of energy. Here, we will discuss three common household electromagnetic transducers: the dynamic microphone, the permanent-magnet loudspeaker, and the record/reproduce tape head. These devices all share common electromagnetic principles.

The Dynamic Microphone

The **dynamic microphone** has been a mainstay of the broadcast and recording industry for as long as the industry has existed. It is fairly simple in construction, rugged, reliable, and offers a very good frequency response to voice and music [on an average of 30 Hz (cps) to 15 kHz (kcps)]. Figure 10-26 illustrates the basic construction of a typical dynamic microphone. At the heart of the microphone is a coil of very fine copper wire. This coil, known as the **voice coil**, is made to move very rapidly in the air gap of a permanent-magnet arrangement. This rapid motion is caused by sound waves impinging on a flexible **diaphragm** (usually a plastic membrane) to which the voice coil is mounted. The resulting in and out displacement of the voice coil is very slight, yet sufficient to induce a small, rapidly varying, voltage across the coil (generator action). In other words, the sound vibrations provide relative motion to the coil within the magnetic field, which results in an induced AC that follows the pattern of the sound vibrations.

The PM Loudspeaker

Permanent-magnet (PM) loudspeakers are perhaps the most common of all electromagnetic transducers. They are employed in everything from headphones, to portable radios and tape players, to expensive home stereo systems, and the list goes on. We will not attempt to cover all there is to know about these devices. We will, however, consider their construction and theory of operation as electromagnetic transducers.

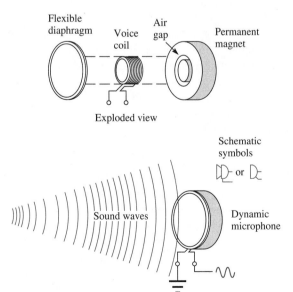

Figure 10-26   The dynamic microphone.

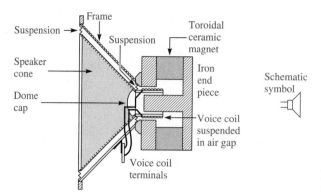

Figure 10-27   The permanent-magnet (PM) loudspeaker.

Figure 10-27 shows a cross-sectional view of a typical PM speaker. One thing you will soon realize is that the PM speaker is very similar to the dynamic microphone, only on a larger scale. While the dynamic microphone is designed to convert acoustical energy (sound energy) to electrical energy, the loudspeaker is designed to do the reverse (convert electrical energy to acoustical energy). Notice that in the figure a voice coil is suspended in the air gap of a permanent-magnet arrangement. While some inexpensive speakers have small ALNICO magnets, most have donut-shaped ceramic (ferrite) magnets. The voice coil is part of a heavy paper cone that is suspended in a metal frame. Flexible braided wires extend from the cone to a terminal strip mounted to the metal frame.

The theory of operation is very simple. A high-level AC signal, following the pattern of voice, music, and so on, is applied to the speaker terminals. This electrical energy is transferred to the voice coil via the flexible braided wires. The AC acting in the turns of the voice coil provides the MMF needed to generate flux. The flux will vary in polarity and intensity with the AC as it interacts with the permanent magnetic field. These flux variations will cause very rapid in and out movement (motor action) of the voice coil and cone. The vibrating cone will cause molecular air disturbances that are perceived as sound.

While a loudspeaker is designed to handle a great range of power (fractions of a watt to hundreds of watts) and to convert electrical energy to acoustical energy, it may also be used as a microphone. Many intercom systems use a single speaker for both the talk and listen functions. The performance of a speaker acting as a microphone is surprisingly good, considering that it is designed for high-power operation as a speaker. It is also interesting to note that a dynamic microphone may be used as an extremely low-power speaker, as in headphones. The dual capability of these devices is known as **reciprocity**. Each, however, will perform best when being used for that which it was designed.

The Record/Reproduce Tape Head

Another very common and important electromagnetic transducer is the **record/reproduce tape head**. Figure 10-28 illustrates a half-track monophonic record/reproduce head. Notice the very basic construction. A coil, wrapped with many turns of very fine wire, is wound around a toroidal core made of very hard ferromagnetic material. Although exaggerated in the illustration, a very fine air gap is left in the toroidal core. This air gap is very thin— 0.0001 to 0.0002 inches wide. The toroidal assembly is placed in a ferromagnetic housing that has mounting tabs.

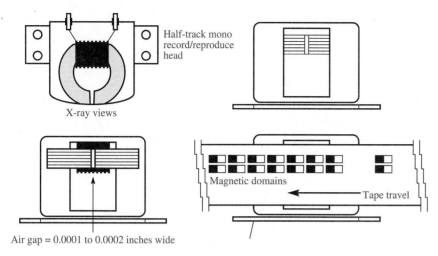

Half-track mono record/reproduce head

X-ray views

Magnetic domains

Tape travel

Air gap = 0.0001 to 0.0002 inches wide

Figure 10-28 A record/reproduce tape head.

A record/reproduce tape head converts electrical energy to magnetic energy and vice versa. Like the loudspeaker and dynamic microphone, the tape head is reciprocal in operation. In the record mode, an AC signal containing audio information is applied to the tape head terminals. The audio AC generates flux in the toroidal core. Leakage flux is then produced at the air gap. The recording tape passes across the air gap at a constant speed and is influenced by the leakage flux. The tape is coated with a thin layer of ferromagnetic material that has a high retentivity. As the tape passes by, the leakage flux will induce an organized pattern of magnetic domains in the tape coating. The pattern of the domains will follow the amplitude and frequency variations of the audio AC.

In the reproduce mode, magnetic energy is converted back to electrical energy. As the tape moves across the air gap, the prerecorded magnetic domains in the tape induce a rapidly varying magnetic flux in the toroidal core. The rapidly varying flux induces a small voltage in the coil. This small voltage (low millivolts) is then amplified and sent to a loudspeaker or headphones.

As a review of this section, take a moment to answer the questions of Self-Check 10-5.

SELF-CHECK 10-5

1. Briefly describe the operation of a dynamic microphone.

2. How is a PM loudspeaker similar to a dynamic microphone?

3. Is the operation of a tape head reciprocal? Explain.

Summary

CONCEPTS

- Motor action requires:

    1. a magnetic field.

    2. a conductor placed in the magnetic field perpendicular to the flux lines.

    3. a source of EMF supplying current to the conductor.

- Motor action is reversed by:

    1. reversing the polarity of applied EMF, or

    2. reversing the polarity of the magnetic field.

    Note: If #1 and #2 are done at the same time, the motor action will not reverse.

- Counterelectromotive force (CEMF) is developed due to generator action as a result of the relative motion created by motor action. It opposes the applied EMF, thereby controlling the amount of current supplied to the motor armature.
- PM DC motor speed control is accomplished by controlling the amount of current supplied to the armature.
- Increasing the current supplied to the field coil of an EM DC motor will increase motor torque and decrease speed.
- The series EM DC motor has poor speed regulation and very high starting torque. It is also known as a *universal motor*, since it will operate on AC or DC.
- The shunt EM DC motor has very good speed regulation and low starting torque.
- The compound EM DC motor has fair speed regulation and moderate starting torque.
- The direction of rotation of an EM DC motor is reversed by reversing the polarity of applied EMF to either the field coil or the armature, but not both.
- The squirrel-cage rotor is used in most AC induction motors. It has no slip rings or need for brushes.
- AC applied to stator windings creates a rotating magnetic field in an induction motor.
- The induced current in an induction rotor will produce a magnetic field that opposes the rotating magnetic field (Lenz's Law). This opposition provides motor action and armature rotation.
- Slip is absolutely essential for the operation of AC induction motors.
- The split-phase AC induction motor operates from single-phase AC.
- The shaded-pole AC induction motor is a single-phase AC motor that uses a shading strap on each pole face to cause magnetic flux to sweep up and down.
- Three-phase AC induction motors operate at a much higher efficiency than single-phase motors.
- The direction of rotation for a three-phase AC induction motor, or synchronous motor, may be reversed by switching any two of the three phases.
- An AC synchronous motor has a permanent-magnet or electromagnetic rotor. Rotor speed is the same as the synchronous speed of the rotating magnetic field created by the stators.
- Control solenoids provide linear mechanical motion when an EMF is applied to the coil.
- Contactors are heavy-duty, solenoid-operated devices with double contacts, used to control large amounts of power.

- Control relays are used for low-power electronic circuit control.
- The dynamic microphone and the PM loudspeaker are reciprocal devices, within limits.
- The record/reproduce tape head is a reciprocal electromagnetic transducer that converts electrical energy to magnetic energy and vice versa.

PROCEDURES

Using the Right-Hand Rule for Motor Action

1. Point the fingers of the right hand in the direction of the south magnetic pole.

2. The thumb must point in the direction of electron flow in the conductor.

3. The right palm will be in a position to catch the conductor. Motor action will force the conductor toward the palm.

SPECIAL TERMS

- Motor action
- Right-hand rule for motor action
- Counterelectromotive force (CEMF)
- PM DC motor
- EM DC motor
- Series motor (universal motor), shunt motor
- Compound motor
- AC induction motor, squirrel-cage rotor, wirewound rotor
- Laminated iron core, slip
- Split-phase induction motor, synchronous speed
- Shaded-pole induction motor, shading strap, shaded pole
- Three-phase induction motor
- AC synchronous motor
- Control solenoid, contactor, control relay
- Operating voltage, pickup voltage, dropout voltage
- Contact current and voltage ratings
- Electromagnetic transducer, dynamic microphone
- PM loudspeaker, reciprocity, record/reproduce tape head

Need to Know Solution

Answers to the Need to Know Quiz

1. c

2. b

3. a

4. c

5. c

6. d

7. d

Questions

10-1 Motor Action

1. List the three basic requirements for motor action.
2. Use the right-hand rule to determine the direction of motor action in Figures 10-29a, b, c, and d.

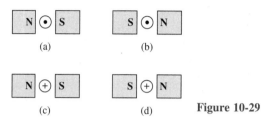

(a)   (b)

(c)   (d)   **Figure 10-29**

3. Describe two ways to reverse the direction of motor action.
4. How is CEMF produced?
5. What does CEMF have to do with the amount of current delivered to the armature from the applied EMF?
6. Why will a motor burn up if the shaft becomes jammed and cannot turn while an EMF is applied?

10-2 DC Motors

7. How are a DC motor and DC generator similar in mechanical construction?
8. Why will the addition of more armature coils and more than two commutator segments improve the performance of a DC motor?
9. What is the minimum number of commutator segments for a practical DC motor?
10. Draw a diagram of a DC motor circuit that includes a means of motor-speed control and direction control.
11. What is the basic difference between a PM DC motor and an EM DC motor?
12. If the EMF applied to the armature of an EM DC motor is increased, what will happen to the armature speed? What will happen to motor torque?
13. Why do shunt EM DC motors have good speed regulation?
14. What is a compound EM DC motor?
15. Which type of EM DC motor is called a *universal motor* and why?
16. List two applications for a series motor.
17. List two applications for a shunt motor.
18. Describe how you might control the direction of any EM DC motor.

10-3 AC Motors

19. What are some obvious differences between the wirewound rotor and the squirrel-cage rotor for AC induction motors?
20. What does Lenz's Law have to do with the operation of a squirrel-cage rotor?
21. What is the purpose of the rotating magnetic field in an AC induction motor?
22. What is slip?
23. How much slip (in percent) does an induction motor normally have?
24. How many field windings (coils or stators) does a split-phase induction motor have?

25. Why do some split-phase motors have a centrifugal switch?
26. The rotating magnetic field of an induction motor rotates at synchronous speed. What does that mean?
27. What is the purpose of the shading strap in each pole of a shaded-pole motor?
28. What are shaded-pole induction motors typically used for?
29. Why is the three-phase induction motor more efficient than a single-phase induction motor?
30. How can the speed and direction of a three-phase induction or synchronous motor be controlled?
31. What is the greatest physical difference between the induction motor and the synchronous motor?
32. What is the greatest operational difference between the induction and synchronous motors?
33. Where would a synchronous motor be used?

10-4 Electromagnetic Control Devices

34. Explain how a control solenoid works.
35. What are heavy-duty contactors used for?
36. How do control relays differ from contactors?
37. What is the difference between operating voltage and pickup voltage in regard to relays and contactors?
38. Why should contact voltage and current ratings be derated when the relay or contactor is controlling power to an inductive load?

10-5 Electromagnetic Transducers

39. What is an electromagnetic transducer?
40. Briefly explain how the dynamic microphone operates.
41. What is the meaning of the term *reciprocity* in regard to dynamic microphones and PM loudspeakers?
42. What is the purpose of the thin air gap in a record/reproduce tape head?
43. Briefly explain the operation of the tape head as it is recording information on a magnetic tape.

Answers to Self-Checks

Self-Check 10-1

1. A magnetic field, a conductor placed in the magnetic field perpendicular to the flux, and a source of EMF applied to the conductor
2. Use the right-hand rule for motor action. With the fingers pointing toward the south pole and the thumb in the direction of electron flow, the right palm will be positioned to catch the conductor.
3. It is a voltage induced across the armature coil as the armature rotates from motor action. It opposes the applied EMF, thus reducing current in the armature.
4. The speed of a free-running motor is very high; therefore, the induced CEMF is very high. The CEMF opposes the applied EMF, acting like a large resistance.

Self-Check 10-2

1. The armature should have more than one armature coil and more than two commutator segments.
2. The induced CEMF will control the amount of running current. The induced CEMF is maximum when the motor is free-running.
3. Increasing field-coil current will increase the magnetic field strength, which increases torque.
4. Decreasing current supplied to the armature will decrease armature speed because torque will be reduced. The difference between the applied current and the induced countercurrent will be insufficient to create enough torque to maintain the original speed. As the armature slows, CEMF is reduced and a new balance between EMF, CEMF, torque, and speed is established.
5. Very high starting torque
6. The shunt motor

Self-Check 10-3

1. The squirrel-cage rotor has no commutator or need for brushes. It is a self-contained unit.
2. Slip is the difference in speed between the rotor and the rotating magnetic field. It is necessary because slip is the relative motion needed for inducing current in the rotor crossbars.
3. Simple construction, low maintenance, high efficiency, good speed regulation, low cost
4. Synchronous speed is the speed of the rotating magnetic field as it is synchronized with the applied AC.
5. The armatures are different. The synchronous motor has a permanent-magnet armature or an electromagnetic armature with slip rings and brushes.

Self-Check 10-4

1. Use a solenoid-operated control value.
2. Contactors are used to control the large amounts of current and high voltages applied to high-power equipment.
3. The pickup voltage is the minimum voltage needed to cause the armature to move. The dropout voltage is the voltage at which the relay or contactor releases.

Self-Check 10-5

1. Sound waves cause the diaphragm to vibrate. The voice coil will move in and out very rapidly with the vibrations. Since the voice coil is mounted in the air gap of a magnet arrangement, the vibrations will cause a rapidly varying voltage and current to be induced.
2. The loudspeaker has the same basic parts: diaphragm (cone), voice coil, permanent magnet.
3. Yes. The tape head will convert electrical energy to magnetic energy and magnetic energy to electrical.

1. Add some of the main concepts, formulas, and procedures from this chapter to your personal Electricity and Electronics Notebook.

2. Obtain a small PM DC motor from a discarded toy or from an electronics or hobby store. Carefully disassemble it and study its construction. You may want to try your hand at rewinding the motor to operate at a lower or higher voltage. For operation at a higher voltage, the copper windings should be replaced with many more turns of a smaller-gauge wire. For operation at a lower voltage, the copper windings should be replaced with fewer turns of a larger-gauge wire. Each coil must be wound tightly and with the same number of turns for electrical and physical balance. The method of winding should be exactly the same as for the windings you removed. Be careful to note how the original armature was wound.

3. Use a small PM DC motor to experiment with motor speed and direction control. You will need a DPDT switch and a 50 Ω, 10 to 25 W wirewound rheostat.

(1)

(1) Meters are related to motors and generators. It's not surprising that Edison invented some meters of his own. This particular Edison meter is a laboratory galvanometer. The curved scale is perched atop a post in the center of the apparatus and the needle (gone) would move left or right over the scale depending on the direction of current through the coils.
(2) This early Edison galvanometer used an electromagnet to pull an iron plunger that was countered by a spring and balance weight.

Edison experimental model
galvanometer 214wh

(2)

Chapter 11

Electromagnetic Instruments and Circuits

OBJECTIVES

After studying this chapter, you will be able to

- describe the construction, operation, and characteristics of various D'Arsonval/Weston meter movements and the electrodynamometer movement.
- design multirange ammeter circuits using basic single shunts or the Ayrton shunt.
- explain the cause and effects of ammeter circuit loading.
- design multirange voltmeter circuits using single multipliers or a series multiplier arrangement.

- explain the cause and effects of voltmeter circuit loading.
- explain the need for zeroing the ohmmeter before use and the effects of battery drain on ohmmeter accuracy.
- design a multirange ohmmeter circuit.
- describe how various analog meter circuits are combined to make use of a single meter movement as a multimeter.
- describe the differences between, and advantages and disadvantages of, analog and digital multimeters.

INTRODUCTION

In 1820, Hans Christian Oersted discovered the relationship between electricity and magnetism; in 1830, Michael Faraday and Joseph Henry demonstrated electromagnetic induction; by 1880, many generator and electric motor designs were available in Europe and the United States. In light of these facts, it is amazing that it was not until 1881 that a French scientist, Jacques Arsène d'Arsonval, designed and built the first laboratory moving-coil meter movement (the galvanometer). His galvanometer worked on the principles of electromagnetic motor action, principles that had been known for decades.

That first laboratory galvanometer was a very fragile and cumbersome instrument. It was constructed in two separate units: the moving-coil assembly and the scale assembly. The moving-coil assembly consisted of an iron-core coil suspended vertically in the air gap of a permanent magnetic field by spring-loaded wires. The upper and lower suspension wires also served to deliver current to the coil. On top of the suspended moving coil, a small mirror was mounted. Placed across the room, opposite the moving-coil assembly, was the scale assembly. The scale was marked with a zero (0) midscale, with increasing numbers left and right. Mounted under the scale was a light source directed at the small mirror mounted on the moving coil. With no current applied to the coil, the light would reflect from the mirror and illuminate the zero (0) at midscale (most of the time). With current applied, the coil would move in one direction or the other, depending on the direction of current, and rotate the small mirror, thus reflecting light to a different spot on the scale. Naturally, the meter was most useful in dim light, was not portable, and was difficult to calibrate and maintain accuracy.

Fortunately, an enterprising young inventor and businessman from England, Edward Weston, picked up on D'Arsonval's idea and made significant improvements to the moving-coil meter design. The most significant improvement was that of making the galvanometer portable and self-contained as one unit. Weston began manufacturing his meters in 1882 and established the Weston Electric Company in the United States in 1888. To this day, Weston meters are famous the world over.

Because of these significant contributions by D'Arsonval and Weston, modern DC moving-coil meter movements are named in their honor. The basic moving-coil meter movement is often referred to as a D'Arsonval movement or as a D'Arsonval/Weston movement. In this chapter, you will become familiar with the D'Arsonval/Weston movement and various DC instrument circuits in which this movement is used. You will also be introduced to other meter-movement designs. Finally, you will explore multimeter circuits and examine the comparative merits of the analog and the modern digital multimeter.

NEED TO KNOW

Let's suppose your boss comes to you one day on the job with three meter movements that he found collecting dust in the stockroom. He wants you to choose one of the meter movements and make a four-range voltmeter with a range-select switch. The four ranges he wants are: 1 VDC, 10 VDC, 100 VDC, and 1,000 VDC. You take a close look at each of the meter movements and discover the following: the first meter is a jewel D'Arsonval/Weston movement with a coil resistance of 1,000 Ω and a sensitivity of 50 μA; the second meter is a jewel D'Arsonval/Weston movement with a meter resistance of 400 Ω and a sensitivity of 100 μA; and the third meter is a taut-band D'Arsonval/Weston movement with a meter resistance of 750 Ω and a sensitivity of 25 μA.

Which of the three meter movements would you choose and how would you design the multirange voltmeter circuit around it? Remember, you are a "Supertech" and you can't disappoint your boss. You should have it designed and built in about two hours. You might even have time to read this chapter to learn how to do it! Your boss has established your need to know for you. Go for it!

11-1 Meter Movements

The analog meter movement is one of the most common devices used for electrical testing and measurement. It is found at the heart of portable multimeters, wattmeters, ammeters, voltmeters, ohmmeters, temperature meters, galvanometers, gas and radiation detectors, and so on. Even in the modern age of digital dominance, the analog meter movement is alive and well. These meters are still found in modern broadcast transmitters, various test instruments, and many technicians' and engineers' tool cases. In this section you will examine some of the more common types of analog meter movements, comparing them in physical construction, operation, and use.

The D'Arsonval/Weston Movement

Basic Construction and Operation

Figure 11-1 illustrates the basic construction of the D'Arsonval/Weston movement. It is actually a PM DC motor whose armature is limited to about 90° of rotation. A moving coil wound on a core of iron or aluminum will rotate due to motor action when a DC current is applied. A needle or pointer attached to the armature assembly is made to deflect across a scale in direct proportion to the amount of applied current. If half of the full scale current is applied, the needle will deflect halfway. If one-fourth of the full scale current is applied, the needle will deflect one-fourth of the way, etc. The resulting scale is said to be *linear*, due to this direct relationship between current and deflection. The amount of current needed for **full scale deflection (FSD)** is known as the meter's **sensitivity**. Figure 11-2 shows two panel meters, each with linear scales. The scales may be labeled as Amperes, mA, μA, Volts, mV, μV, Degrees Celsius, Pounds Pressure, or any other quantity. Regardless of the quantity marked on the scale, it is current that makes the pointer deflect.

Visible in the lower center of each of the meter covers is a small screwhead used for **meter zeroing**. When the screwhead is turned, a small camlike arm will cause the pointer to move into alignment with the zero line marked on the left end of the scale. In this way, the meter can be mechanically zeroed or calibrated for accuracy.

The metal bobbin or core of the D'Arsonval/Weston movement is used to create a damping effect on needle deflection. As the armature rotates, its motion will actually induce currents in the core that create magnetic fields in opposition to the stationary magnetic field. This slight magnetic opposition will ensure that the needle does not overshoot, bounce, or hunt upon arrival at its proper indicating position. Without this damping effect, it would take quite some time for the pointer to finally come to rest.

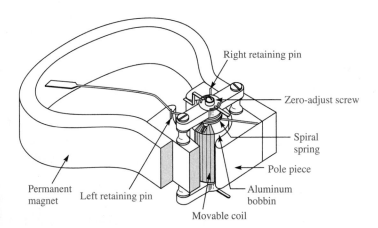

Figure 11-1   The basic D'Arsonval/Weston analog meter movement.

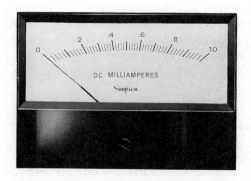

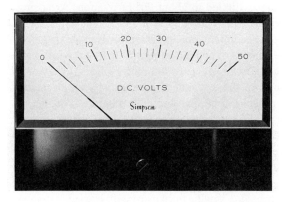

Figure 11-2   Typical linear meter scales. (Courtesy of Simpson Electric Co.)

The accuracy of these meter movements is typically in the neighborhood of 1 to 3% of FSD current and is rated by the manufacturer. If, for example, a 1 mA meter movement is rated at 2% accuracy, the indicated current at any point on the scale could be off by as much as 0.02 mA (1 mA · 0.02 = 0.02 mA). This might be a significant amount for readings taken on the low end of the scale. For this reason, it is always best to operate the meter so that readings fall in the upper half of the scale. A range switch usually makes this possible.

D'Arsonval/Weston meter movements are polarized. That is, the measured current must be applied to the meter in proper polarity. If the measured current is applied in reverse polarity, the meter will attempt to deflect backwards. For this reason, meter terminals are marked with a + and − sign to facilitate proper connection. An exception to what has just been stated would be the galvanometer, whose pointer is normally at rest midscale and can deflect left or right depending on the direction in which the monitored current is flowing.

The Jewel Movement

Perhaps the most common of the D'Arsonval/Weston type of movements is the **jewel movement** (shown in Figure 11-3). Found in most panel meters and portable multimeters, it is among the least expensive of all meter movements. In this type of movement, the armature assembly pivots on a pointed axle arrangement. The end points are mated into conical end caps forming what are known as *jewel bearings*. The end point and conical end cap must be made of very hard materials to insure long life and minimize friction. It

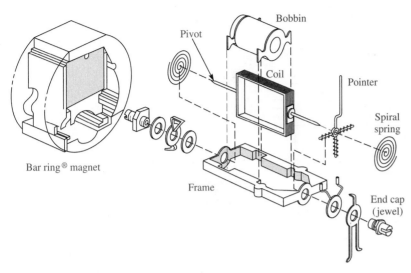

Figure 11-3 The jewel movement. (Courtesy of the Triplett Corporation)

has been said that the name *jewel bearing* really has nothing to do with precious stones. In fact, however, it does. Often the jewel bearing is constructed from either man-made or natural jewels, such as low-grade sapphires. Current is applied to the armature coil via spiral end springs (sometimes called *hair springs*). These springs also provide counter torque to the armature assembly and return the pointer to the zero, or resting, position when current is removed. Typical jewel-movement meter sensitivities range from 1 mA down to about 50 μA for FSD.

The Taut-Band Movement

The taut-band movement is a more expensive and more sensitive version of the D'Arsonval/Weston meter movement. As shown in Figure 11-4, the entire armature assembly is suspended between two spring-loaded bands. The bands are held taut by two leaf springs that protect the bands from damage due to severe mechanical shock. This method of armature mounting is practically frictionless, allowing accurate and repeatable measurements. The torque or twisting of the bands will return the pointer to the zero position when current is removed. Meter sensitivities for this type of movement range down as low as a couple μA.

The Electrodynamometer Movement

The electrodynamometer, shown in Figure 11-5, is another meter movement designed for a special application. It is designed primarily to measure AC or DC power. As you can see, the electrodynamometer has a stationary field coil and a moving coil. The field coil is normally connected in parallel with a voltage source, while the moving coil is connected in series with the load that is demanding current. The magnetic strength of the field coil is proportional to the amount of source voltage and the magnetic strength of the moving coil is proportional to the amount of load current. The amount of pointer deflection depends upon the field strength of both coils. As a result, the electrodynamometer acts as an analog

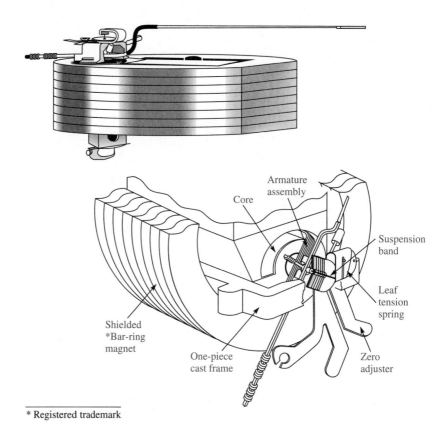

* Registered trademark

Figure 11-4   The taut-band movement. (Courtesy of the Triplett Corporation)

computer—multiplying the load current and source voltage and yielding the power dissipation of the load ($P = I \cdot E$). The scale of the meter is, of course, labeled in watts.

In the following sections, you will investigate various meter circuits that are designed around the D'Arsonval/Weston movement. Before you continue, take time to review this section by answering the questions of Self-Check 11-1.

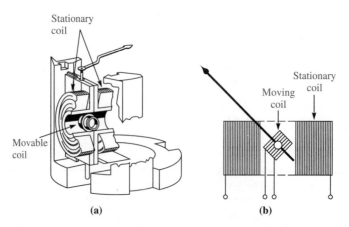

Figure 11-5   The electrodynamometer. (Courtesy of the Triplett Corporation)

1. How does the jewel movement differ from the taut-band movement?

2. Which type of meter movement is used to measure power?

3. What is meter damping and why is it needed?

11-2 Ammeter Circuits

In this section, you will take an inside look at practical ammeter circuits. After studying this section, you will be able to design your own ammeter circuits when meter sensitivity and meter resistance are known. This should be an interesting and enjoyable experience for you, since it will activate your creative juices.

Before we begin, let's take a few moments to recall what you already know about ammeters. First, you know the ammeter is used to measure current in a circuit. As such, it must be placed in series with the circuit whose current you wish to monitor. Since the ammeter is placed in series in the circuit, it should have a very low internal resistance (ideally 0 Ω) in order not to add significantly to the resistance already in the circuit. Also, you know the ammeter is a polarized device and proper polarity must be observed when it is placed in the circuit. Finally, because the ammeter has a very low internal resistance, it is *never* placed in parallel with a voltage source. If this is accidentally done, the ammeter circuit and movement will almost surely be destroyed.

The Basic Shunt

Scaling

The basic ammeter shunt circuit is shown in Figure 11-6. The purpose of the meter shunt is to bypass most of the circuit current around the meter movement itself. Suppose you have a D'Arsonval/Weston meter movement that has a sensitivity of 50 μA but you want to measure currents as high as 100 mA. Obviously, a meter movement with a 50 μA sensitivity cannot be used to measure 100 mA of current unless a shunt is used. The **shunt** allows most of the current to bypass the meter movement. The current that flows through the meter movement is proportionate to, and representative of, the total measured current.

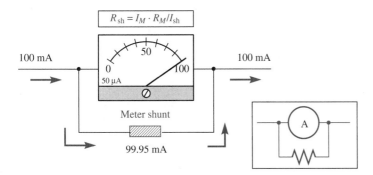

Figure 11-6   The basic meter shunt.

Thus, the scale on the meter face can be calibrated from 0 to 100 mA even though the actual maximum current through the movement will only be 50 μA. This is called **scaling** or **ranging** and is accomplished by choosing a shunt resistor of the proper value.

Calculating the Shunt

So, how can we determine the proper value of a shunt resistor for a particular ammeter scale or range? To begin with, you must know the sensitivity and amount of internal resistance of the meter movement you intend to use. Then the value of the shunt resistor is determined from the following rationale and formula. Since the voltage across the meter (E_M) is the same as the voltage across the shunt (E_{sh}), the meter current (I_M) times the meter resistance (R_M) will equal the shunt current (I_{sh}) times the shunt resistance (R_{sh}). In other words: $E_{sh} = E_M$, therefore $I_{sh} \cdot R_{sh} = I_M \cdot R_M$. Rearranging the equation to solve for the shunt resistance yields:

$$R_{sh} = \frac{I_M \cdot R_M}{I_{sh}} \tag{11.1}$$

Now, let's apply Formula 11.1 to our example. We have a 50 μA meter movement and we want to measure up to 100 mA with it. Before we can use the formula, we must know one more thing: the internal meter resistance (the resistance of the moving-coil windings). Let's assume the manufacturer has specified this movement to have a meter resistance of 500 Ω. Now we have all the information we need to calculate the proper shunt resistance. The meter current is 50 μA when the shunt current is 99.95 mA. Remember, the total measured current is divided between the meter and the shunt (100 mA − 50 μA = 99.95 mA). Applying these values of I_{sh}, I_M, and R_M to Formula 11.1, we find R_{sh} to be 0.25 Ω (R_{sh} = 50 μA $\cdot$ 500 Ω/99.95 mA = 0.25 Ω). The shunt resistor will be a precision wirewound or metal film resistor.

Multiranging Using Basic Shunts

Figure 11-7 illustrates a multirange ammeter circuit. A high-quality rotary switch is used to select ranges. Notice that the rotary switch shown has a "make before break" sliding contact. This eliminates extreme needle bounce and unnecessary breaking of the monitored

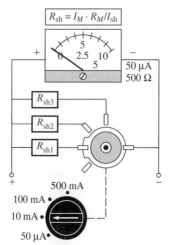

Figure 11-7 A multirange ammeter circuit using multiple shunts.

circuit as different ranges are selected. Each shunt, for each range, is calculated using Formula 11.1. Again, you must know the meter resistance and its sensitivity before the shunts can be calculated. The actual ranges are up to you. Notice, however, that the meter face must be scaled with numbers that match the various ranges. For example, a scale of 0 to 10 could be used for a 10 mA, 100 mA, 1 A, or 10 A range while a scale of 0 to 5 could be used for a 5 mA, 50 mA, 500 mA, or 5 A range. Naturally, the lowest range is determined by the sensitivity of the chosen meter movement.

The Ayrton (Universal) Shunt

Advantages Over the Basic Shunt

The **Ayrton** or **universal shunt** is another very common method of scaling used in multirange ammeter circuits. Example 11-1 illustrates a three-range ammeter circuit employing the Ayrton shunt. As before, a make before break rotary switch is used for range selection

EXAMPLE   11-1

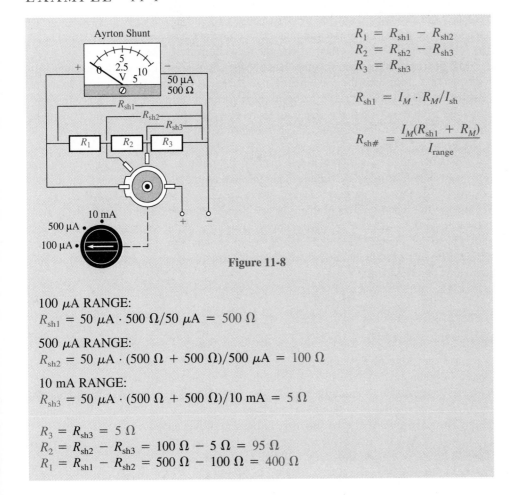

$$R_1 = R_{sh1} - R_{sh2}$$
$$R_2 = R_{sh2} - R_{sh3}$$
$$R_3 = R_{sh3}$$

$$R_{sh1} = I_M \cdot R_M / I_{sh}$$

$$R_{sh\#} = \frac{I_M(R_{sh1} + R_M)}{I_{range}}$$

Figure 11-8

100 μA RANGE:
$R_{sh1} = 50\ \mu A \cdot 500\ \Omega/50\ \mu A = 500\ \Omega$

500 μA RANGE:
$R_{sh2} = 50\ \mu A \cdot (500\ \Omega + 500\ \Omega)/500\ \mu A = 100\ \Omega$

10 mA RANGE:
$R_{sh3} = 50\ \mu A \cdot (500\ \Omega + 500\ \Omega)/10\ mA = 5\ \Omega$

$R_3 = R_{sh3} = 5\ \Omega$
$R_2 = R_{sh2} - R_{sh3} = 100\ \Omega - 5\ \Omega = 95\ \Omega$
$R_1 = R_{sh1} - R_{sh2} = 500\ \Omega - 100\ \Omega = 400\ \Omega$

to avoid an interruption in circuit current when the switch is rotated. Notice that, as higher current ranges are selected, shunt resistance decreases and resistance in series with the meter increases. When the range switch is rotated to the 500 μA position, R_1 will be in series with the meter and $R_2 + R_3$ will serve as a shunt. When the 10 mA range is selected, $R_1 + R_2$ will be in series with the meter and R_3 will be the shunt. As far as the meter movement itself is concerned, $R_1 + R_2 + R_3$ is always across its terminals regardless of the range selected. This is an improvement over the basic shunt because the meter movement always has a shunt resistance connected to its terminals and that shunt resistance is always the same value. This is an advantage because the combination of $R_1 + R_2 + R_3$ serves as a constant load for the meter. This contributes to the damping action of the meter. When the moving coil is in motion, a CEMF is induced, which creates a magnetic field that opposes the motion. The CEMF and damping characteristics are therefore partially controlled by the meter shunting. This may seem like a small point but it is one many engineers consider.

Ayrton Shunt Design

The design of the Ayrton shunt is a little more involved than that of the basic shunt, although not difficult. We will use Example 11-1 to demonstrate the design process step by step. First, you must know the meter sensitivity I_M and the meter resistance R_M of the movement you wish to use, and you must decide on the current ranges you wish to have. The lowest range should be at least twice the meter sensitivity ($2I_M$).

The next step is to determine the value of the shunt needed for the lowest range. In Example 11-1, the shunt for the lowest range is R_{sh1} (the sum of R_1, R_2, and R_3). The value of R_{sh1} is calculated using the basic shunt formula (Formula 11.1). Since the lowest range is 100 μA, and the meter is to receive 50 μA, the shunt current will be 50 μA. As shown in the example, R_{sh1} is found to be 500 Ω. Keep in mind that $R_{sh1} = R_1 + R_2 + R_3 = 500\ \Omega$.

When the range switch is rotated to the 500 μA position, $R_2 + R_3$ will form a shunt (R_{sh2}) and R_1 will be in series with the meter movement. If 500 μA is applied to the circuit, 50 μA will go through the meter and R_1 while 450 μA goes through R_3 and R_2 (R_{sh2}). We must now find R_{sh2}. The problem is, we do not know the value of R_1. If R_1 were known, R_{sh2} could be found by subtracting R_1 from R_{sh1}. What we need is a formula that allows us to determine R_{sh2} even though R_1 is not known. The following formula is used to find the value of all remaining shunts (because of lack of space, the derivation of this formula is not provided):

$$R_{sh\#} = \frac{I_M \cdot (R_{sh1} + R_M)}{I_{range}} \tag{11.2}$$

Since the value of R_{sh1} does not change regardless of the number of resistors and ranges, Formula 11.2 may be used to solve for any shunt values (in this case, R_{sh2} and R_{sh3}). As you can see in Example 11-1, $R_{sh2} = 100\ \Omega$ and $R_{sh3} = 5\ \Omega$. R_1, R_2, and R_3 are found by taking the difference between consecutive shunt values. See the Ayrton shunt design summary included in the summary at the end of this chapter.

Design Note 11-1 is provided as an aid to designing and understanding the Ayrton shunt. Take a few moments to study it before you continue.

DESIGN NOTE 11-1: Ayrton Shunt

$R_1 = R_{sh1} - R_{sh2}$
$R_2 = R_{sh2} - R_{sh3}$
$R_3 = R_{sh3}$

$(11.1)\ R_{sh1} = I_M \cdot R_M / I_{sh}$

$(11.2)\ R_{sh\#} = \dfrac{I_M(R_{sh1} + R_M)}{I_{range}}$

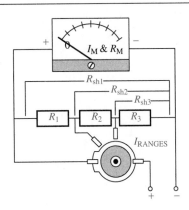

DESIGN SUMMARY

1. METER MOVEMENT SENSITIVITY AND RESISTANCE MUST BE KNOWN.

2. DECIDE ON RANGES. THE LOWEST RANGE SHOULD BE AT LEAST TWICE THE METER SENSITIVITY.

3. DRAW A SCHEMATIC OF YOUR CIRCUIT.

4. DETERMINE THE VALUE OF THE FIRST SHUNT—FOR THE LOWEST RANGE USING FORMULA (11.1).

5. DETERMINE ALL REMAINING SHUNT VALUES USING FORMULA (11.2).

6. SOLVE FOR INDIVIDUAL RESISTORS BY TAKING THE DIFFERENCE BETWEN CONSECUTIVE SHUNT VALUES.

7. CONSTRUCT YOUR CIRCUIT USING PRECISION WIREWOUND OR METAL FILM RESISTORS.

"BASIC" PROGRAM SOLUTION

```
  5 DIM IR(10), RSH(10), R(10)
 10 CLS
 20 PRINT"AYRTON SHUNT DESIGN PROGRAM"
 30 PRINT"":PRINT""
 40 INPUT"ENTER THE METER SENSITIVITY IN
    AMPS-";IM
 50 INPUT"ENTER THE METER'S INTERNAL RESIS-
    TANCE-";RM
 60 INPUT"ENTER THE NUMBER OF RANGES-";N
 70 PRINT""
 80 PRINT"ENTER THE RANGE CURRENTS ONE AT A
    TIME STARTING WITH THE LOWEST RANGE
    FIRST"
 85 PRINT"*** THE LOWEST RANGE MUST BE
    GREATER THAN THE METER SENSITIVITY ***"
 90 PRINT""
100 FOR X = 1 TO N
110 PRINT"RANGE #";X;" = ";:INPUT IR(X)
115 IF IR(X) < = IM THEN GOTO 85
120 NEXT X
130 REM
140 REM
150 REM SOLVE FOR FIRST SHUNT – RSH1
160 ISH = IR(1) – IM
170 RSH(1) = IM * RM/ISH
180 REM
190 REM
200 REM SOLVE FOR ALL OTHER SHUNTS
210 FOR X = 2 TO N
220 RSH(X) = IM * (RSH(1) + RM)/IR(X)
230 NEXT X
240 REM
```

Continued next page

D E S I G N   N O T E   1 1 - 1:   (Continued)

E X P L A N A T I O N	"B A S I C" P R O G R A M S O L U T I O N
	250 REM
	260 REM SOLVE FOR INDIVIDUAL RESISTORS
	270 FOR X = 1 TO N
	280 IF X = N THEN R(X) = RSH(X): GOTO 300
	290 R(X) = RSH(X) − RSH(X + 1)
	300 NEXT X
	310 REM
	320 REM
	330 REM DISPLAY RESULTS
	340 PRINT"":PRINT""
	350 FOR X = 1 TO N
	360 PRINT"R";X;" = ";R(X)
	370 NEXT X
	400 END

Ammeter Loading Effects

Earlier it was stated that an ammeter or ammeter circuit should have a very low internal resistance (ideally 0 Ω) so that it will not significantly add to the resistance of the circuit being measured. This, of course, is true. But what about the 50 μA meter movement discussed earlier? It has a meter resistance of 500 Ω. Isn't that a significant amount of resistance? The answer is: It's relative. If the 50 μA meter movement is used without a shunt to measure a current below 50 μA, it would most likely be placed in series with a circuit of very high resistance (perhaps 2 MΩ or more). Compared to 2 million ohms, 500 Ω is relatively insignificant, and the decrease in circuit current due to the additional 500 Ω is extremely insignificant.

However, if the 500 Ω meter is placed in series with a 2 kΩ circuit, the additional meter resistance would be significant and the measured current would be quite a bit less than the actual circuit current without the meter. This kind of severe ammeter circuit loading is rare, since most low-resistance circuits conduct larger currents, which must be measured with higher-current, lower-resistance meter movements or ammeter circuits that have shunts. The shunt greatly reduces the overall resistance of the ammeter circuit. In Example 11-1, a 5 Ω shunt is used for the 10 mA range. Thus, the overall meter circuit resistance is slightly less than 5 Ω. Needless to say, a 5 Ω ammeter circuit placed in series with a 2 kΩ circuit will have no loading effect at all. Even though the ammeter loading effect is usually very small, it is good for you to be aware of it. If a measured current is lower than expected, the ammeter could be at fault by having a higher internal resistance than you thought. It is therefore often useful to know the internal resistance of your ammeter or ammeter circuit.

Before you continue, take time to review this section by answering the questions of Self-Check 11-2.

1. Given a meter movement whose resistance is 100 Ω and whose sensitivity is 1 mA, determine the value of a basic shunt needed to scale the meter to a 100 mA range.

2. Use the meter movement of Problem 1 above to design a two-range Ayrton shunt ammeter circuit. The two ranges shall be 10 mA and 100 mA.

11-3 Voltmeter Circuits

The voltmeter is probably the most often used of all test and measurement instruments. It is an instrument that is connected to a circuit while the circuit is functioning, or malfunctioning. Supply voltages, bias voltages, and control voltages can be monitored without interrupting or adversely affecting the circuit. In this section, you will take an inside look at practical voltmeter circuits to see how they are designed and better understand how they are used.

As we begin, let's take a moment to recall what you already know about voltmeters. First, and very obviously, you know the voltmeter is used to measure voltage in or across a circuit. As such, it must be placed in parallel with the circuit or component whose voltage you wish to monitor. Since the voltmeter is placed in parallel with a component, it must have a very high internal resistance (ideally ∞ Ω) in order not to significantly load down the voltage being measured. Also, you know the DC voltmeter is a polarized device and proper polarity must be observed when it is placed in the circuit. Finally, because the voltmeter has a very high internal resistance, it is never placed in series with a circuit and voltage source. If this is done, the voltmeter will prevent the circuit being tested from functioning properly and most of the supply voltage will appear across the voltmeter instead of the circuit.

Multiplier Resistors (Multipliers)

Scaling

A basic voltmeter circuit is shown in Figure 11-9. The circuit consists of a DC meter movement (D'Arsonval/Weston), a series resistance called a **multiplier** (R_X), and polar-

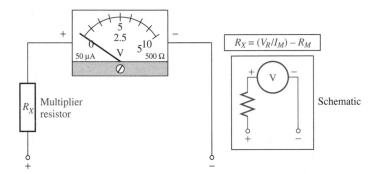

Figure 11-9   Voltmeter multiplier resistors used for scaling.

ized input terminals. The meter movement itself is a sensitive ammeter. The multiplier must limit the meter current to a value less than or equal to the movement's sensitivity. Thus for higher voltage ranges, the multiplier must be a larger value. The meter face is marked to match the voltage range (V_R) for which the multiplier is chosen. The scale on the meter face is linear (evenly spaced divisions) because the meter movement and the multiplier are linear devices. Current through the multiplier is directly proportional to the amount of voltage being measured and the resulting deflection of the meter pointer is directly proportional to the amount of current.

Calculating the Multiplier

As in the design of ammeter circuits, you must first know the sensitivity and internal resistance of the meter movement you intend to use for your voltmeter circuit. The value of the multiplier is basically found by using Ohm's Law ($R = E/I$). Total meter-circuit resistance can be found by dividing the range voltage by the meter-movement sensitivity ($R_t = V_R/I_M$). This total circuit resistance is the sum of the multiplier resistance and the meter resistance ($R_t = R_X + R_M$). Therefore, the actual value of the multiplier is found by subtracting meter resistance from the total resistance. This results in the following formula for R_X:

$$R_X = (V_R/I_M) - R_M \qquad (11.3)$$

Let's suppose we want to design a voltmeter with a 10 VDC range using the movement of Figure 11-9. We see that the sensitivity of the movement is 50 μA and it has an internal resistance of 500 Ω. Applying this information to Formula 11.3, we find R_X to be 200,000 Ω − 500 Ω, or 199,500 Ω ($R_X = (10 \text{ V}/50 \ \mu\text{A}) - 500 \ \Omega = 199,500 \ \Omega$).

The Significance of Meter Resistance

Where do we get a 199,500 Ω resistor? Why not use a 200,000 Ω resistor instead? When a single precision resistor is not available in a value that matches the calculated R_X value, two or more precision resistors may be connected in series or parallel to obtain the needed value. However, in our example here we might wonder if there is really a practical difference between 200,000 Ω and 199,500 Ω. In other words, how much error would there be if we used a 200,000 Ω multiplier instead of the calculated value of 199,500 Ω? Using a 200 kΩ multiplier, the total circuit resistance is 500 Ω more than needed (200,500 Ω). As a result, the percentage of error is (500 Ω/200.5 kΩ) · 100% = 0.25%. The accuracy of the meter movement itself is usually in the range of 2 to 3%. For all practical purposes, an additional 0.25% can be considered insignificant and a 200 kΩ multiplier resistor may be used.

However, meter resistance should not be totally ignored. There are cases in which the meter resistance is significant. For example, suppose we want a voltage range of 100 mV, using the same meter movement as our previous example. The value of R_X will only be 1,500 Ω ($R_X = (100 \text{ mV}/50 \ \mu\text{A}) - 500 \ \Omega = 1,500 \ \Omega$). If we ignore the 500 Ω meter resistance and fail to subtract it from the 2,000 Ω total circuit resistance, the multiplier will be 2,000 Ω. Using a multiplier of 2,000 Ω, the actual total circuit resistance will be 2,500 Ω. Obviously, the 500 Ω meter resistance cannot be ignored unless you are willing to accept an error of 20% [(500 Ω/2,500 Ω) · 100% = 20%]. A 1,500 Ω precision resistor is readily available and should definitely be used.

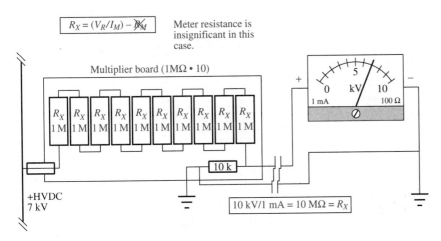

Figure 11-10   An array of high-voltage multiplier resistors forming a high-voltage probe.

High-Voltage Voltmeter Circuit

High-voltage voltmeter circuits are needed to monitor high voltages in electronic systems such as high-power broadcast transmitters. Figure 11-10 illustrates a case in point. It is very common to see a 1 mA meter movement used in high-voltage circuits. This permits the use of 1 MΩ multipliers for every 1,000 V of range needed (1,000 V/1 mA = 1 MΩ). The internal resistance of the meter movement is usually about 100 Ω and can be considered very insignificant compared to the 1 MΩ multiplier or multipliers. The meter circuit illustrated in Figure 11-10 is designed to measure high voltages (up to 10 kV). Therefore, ten 1 MΩ multipliers are connected in series to provide the necessary 10 MΩ of multiplier resistance (R_X = 10,000 V/1 mA = 10 MΩ). The high voltage is distributed among the 10 resistors to reduce the possibility of high-voltage arcing. Special high-voltage resistors are used. Characteristically, they are about 2 to 3 inches long (5 to 8 cm) and are rated at 5 or 10 W. The 10 kΩ resistor, also mounted on the multiplier board, is a safety feature. If the meter should open, the 10 kΩ resistor prevents high voltage from appearing at the meter terminals. The voltage is limited to a maximum of 10 V [(10 kΩ/10.01 MΩ) · 10 kV = 9.99 V = 10 V].

Multiranging Using Multipliers

When a voltmeter is used as a general purpose test and measurement instrument, it is desirable for the meter to have **multirange** capability (to have more than one range). Figure 11-11 illustrates two methods of obtaining multirange capability. Individual multipliers (one for each voltage range), or a series multiplier arrangement may be used. When individual multipliers are used for each range, each multiplier is calculated using Formula 11.3. When a series multiplier arrangement is used, each multiplier will make use of resistance provided by other multipliers used for lower ranges. For example, if the highest range is chosen, R_{X3} will make use of the resistance contributed by R_{X2}, R_{X1}, and, of course, R_M. Therefore, R_{X1} must be calculated first, then R_{X2}, and finally R_{X3}. To accommodate this multiplier sequence, Formula 11.3 is modified as follows:

$$R_{X\#} = (V_R/I_M) - (R_M + \text{all previous } R_X \text{ values}) \qquad (11.4)$$

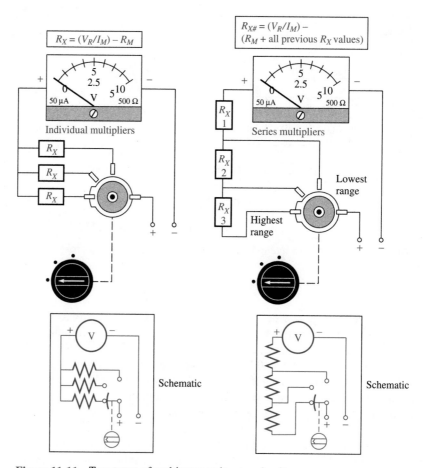

Figure 11-11 Two types of multirange voltmeter circuits.

Example 11-2 illustrates the use of Formula 11.4 in designing a series-multiplier, multirange voltmeter. The meter sensitivity and the meter resistance must be known. Once the desired ranges are chosen, multiplier resistors are calculated starting with the lowest voltage range. Notice the meter resistance is considered significant since ignoring it would create an error of almost 5% on the 0.5 V range [(500/10,500) · 100% = 4.7%]. The multirange voltmeter design is summarized at the end of the chapter.

Design Note 11-2 is provided as an aid to the design and understanding of series-multiplier, multirange voltmeter circuits. You may want to take time to study it before continuing.

Voltmeter Sensitivity

The **voltmeter sensitivity** is directly related to the sensitivity of the meter movement itself. Yet, it is expressed in ohms per volt (Ω/V). For example, many garden-variety analog DC voltmeters have a voltmeter sensitivity of 20 kΩ/V. What does that indicate? It is an indication of two important facts: (1) the voltmeter circuit has 20,000 Ω of resistance for every volt of any range the meter is set to, and (2) the sensitivity of the meter movement

EXAMPLE   11-2

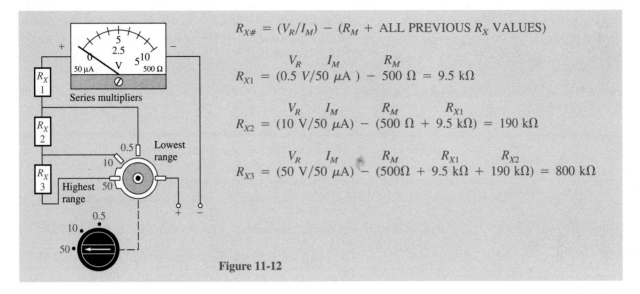

$$R_{X\#} = (V_R/I_M) - (R_M + \text{ALL PREVIOUS } R_X \text{ VALUES})$$

$$R_{X1} = \overset{V_R}{(0.5 \text{ V}}/\overset{I_M}{50 \text{ }\mu\text{A})} - \overset{R_M}{500 \text{ }\Omega} = 9.5 \text{ k}\Omega$$

$$R_{X2} = \overset{V_R}{(10 \text{ V}}/\overset{I_M}{50 \text{ }\mu\text{A})} - (\overset{R_M}{500 \text{ }\Omega} + \overset{R_{X1}}{9.5 \text{ k}\Omega}) = 190 \text{ k}\Omega$$

$$R_{X3} = \overset{V_R}{(50 \text{ V}}/\overset{I_M}{50 \text{ }\mu\text{A})} - (\overset{R_M}{500\Omega} + \overset{R_{X1}}{9.5 \text{ k}\Omega} + \overset{R_{X2}}{190 \text{ k}\Omega}) = 800 \text{ k}\Omega$$

Figure 11-12

itself is 50 μA (1 V/20,000 Ω = 50 μA). If, for example, the range switch is set on a 100 V range, the total meter-circuit resistance is 2 MΩ (100 V $\cdot$ 20 kΩ/V = 2MΩ). However, on a 1 V range, the total meter circuit resistance is only 20 kΩ (1 V $\cdot$ 20 kΩ/V = 20 kΩ).

Example 11-3 uses the circuit of Example 11-2 to illustrate how meter resistance depends upon the voltmeter range setting and voltmeter sensitivity. As you can see, the

EXAMPLE   11-3

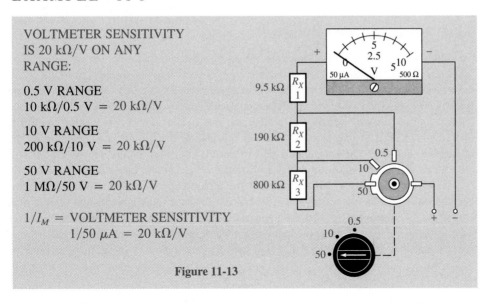

VOLTMETER SENSITIVITY
IS 20 kΩ/V ON ANY
RANGE:

0.5 V RANGE
10 kΩ/0.5 V = 20 kΩ/V

10 V RANGE
200 kΩ/10 V = 20 kΩ/V

50 V RANGE
1 MΩ/50 V = 20 kΩ/V

$1/I_M$ = VOLTMETER SENSITIVITY
    1/50 μA = 20 kΩ/V

Figure 11-13

voltmeter-circuit resistance increases as the voltage range is increased. However, voltmeter sensitivity does not change regardless of which range the meter is set to. The voltmeter sensitivity is a function of meter-movement current sensitivity: the lower the meter-movement current sensitivity, the higher the voltmeter sensitivity (more ohms per volt).

Voltmeter Loading Effects

Earlier it was stated that a voltmeter, or voltmeter circuit, should have a very high resistance (ideally ∞ Ω) so that it will not significantly load down the voltage of the component being measured. The higher the voltmeter sensitivity, the higher the voltmeter circuit resistance, and the less loading effect there is. The accuracy of any voltage measurement depends upon the range the voltmeter is on, the voltmeter sensitivity, and the resistance of the component whose voltage is being measured. (See also "Voltmeter Loading Effects" in Chapter 7.)

Take time now to test your understanding of voltmeter circuits by answering the questions of Self-Check 11-3.

S E L F - C H E C K   1 1 - 3

1. Given a meter movement whose resistance is 100 Ω and whose sensitivity is 1 mA, determine the value of a multiplier needed to scale the meter to a 500 V range.

2. Use the meter movement of Problem 1 to design a three-range series multiplier voltmeter circuit. The three ranges shall be 10 V, 100 V, and 1,000 V.

11-4 Ohmmeter Circuits

The ohmmeter is another very practical and useful test and measurement instrument. As you know, the ohmmeter is used to measure unknown values of resistance, to check for out-of-tolerance resistors, to check for opens or shorts, and to check continuity—looking for broken wires, bad switches, etc. The ohmmeter can also be used to check capacitors, inductors, transformers, transistors, diodes, and many other devices. It is a very versatile instrument indeed. In this section, you will investigate the design and operation of the ohmmeter. This will not only enable you to design your own ohmmeter circuits, it will also clarify the use and operation of the ohmmeter as a test and measurement instrument.

First, let's begin with a review of what you already know about ohmmeters. Remember, the ohmmeter is a test instrument that is only used on components and circuits that are turned off, disconnected, or have their power source removed. The ohmmeter is never used with circuit power on. You know ohmmeters have their own internal DC source. Also, recall that any component that is tested or measured with the ohmmeter must be

DESIGN NOTE 11-2:   Series Multipliers

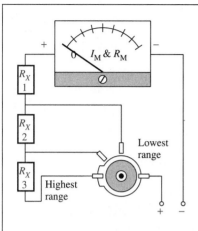

Lowest range

Highest range

$$(11.4) \quad R_{X\#} = (V_R/I_M) - (R_M + \text{ALL PREVIOUS } R_X \text{ VALUES})$$

$$R_{X1} = (V_R/I_M) - R_M$$

$$R_{X2} = (V_R/I_M) - (R_M + R_{X1})$$

$$R_{X3} = (V_R/I_M) - (R_M + R_{X1} + R_{X2})$$

DESIGN SUMMARY

1. METER MOVEMENT SENSITIVITY AND RESISTANCE MUST BE KNOWN.

2. DECIDE ON RANGES AND DRAW A SCHEMATIC OF YOUR CIRCUIT.

3. BEGIN CALCULATIONS ON THE LOWEST RANGE USING FORMULA (11.4)

4. DO NOT IGNORE THE METER RESISTANCE IN YOUR CALCULATIONS. IT IS MOST SIGNIFICANT ON THE LOWER RANGES.

5. CONSTRUCT YOUR CIRCUIT USING PRECISION WIREWOUND OR METAL FILM RESISTORS.

"BASIC" PROGRAM SOLUTION

```
10 CLS
20 DIM VR(20)
30 PRINT"SERIES MULTIPLIERS"
40 PRINT"":PRINT""
50 PRINT"THIS PROGRAM WILL CALCULATE THE
   SERIES MULTIPLIER RESISTORS NEEDED FOR"
60 PRINT"A MULTIRANGE VOLTMETER CIRCUIT."
70 PRINT""
80 INPUT"ENTER THE METER SENSITIVITY. ";IM
90 INPUT"ENTER THE METER RESISTANCE. ";RM
100 SUM = RM
110 INPUT"ENTER THE NUMBER OF VOLTAGE
    RANGES. ";N
120 PRINT""
130 PRINT"ENTER THE VOLTAGE RANGES ONE AT A
    TIME STARTING WITH THE LOWEST RANGE."
140 FOR X = 1 TO N
150 PRINT"RANGE #";X;" = ";:INPUT VR(X)
160 NEXT X
170 PRINT""
180 FOR X = 1 TO N
190 SUM = SUM + R
200 R = VR(X)/IM − SUM
210 PRINT "R";X;" = ";R
260 NEXT X
300 PRINT""
310 INPUT"ANOTHER PROBLEM?   (Y/N)";A$
320 IF A$ = "Y" THEN GOTO 70
330 CLS:END
```

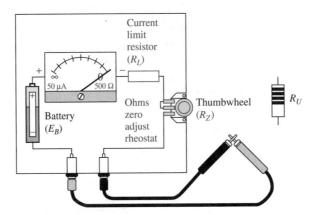

Figure 11-14 A basic ohmmeter circuit.

isolated first. In other words, if the component being tested is in a circuit, at least one of its leads must be disconnected from the circuit before it is measured. Now, let's continue from here to see how ohmmeters are designed.

The Basic Ohmmeter

Basic Components

A basic ohmmeter circuit is shown in Figure 11-14. As you can see, the ohmmeter circuit consists of a sensitive DC meter movement, a series current-limiting resistor, an **ohms-zero-adjust** rheostat, and a DC source (battery). Soon, you will see how important a sensitive meter movement is in the design of a practical ohmmeter circuit. The ohms-zero-adjust rheostat permits frequent calibration of the ohmmeter circuit, and the battery serves as the voltage source needed to deflect the meter pointer.

Theory of Operation

Figure 11-15 illustrates the design of a basic ohmmeter circuit. Before you can actually design an ohmmeter circuit, you must understand the theory behind its operation. The meter movement requires current for pointer deflection; the series combination of the meter resistance (R_M), half of the ohms-zero-adjust rheostat ($0.5\,R_Z$—the rheostat is initially set at half range to allow for pointer adjustment above or below the zero line), and the current-limiting resistor (R_L) will provide the correct amount of resistance ($R_t = R_M + 0.5\,R_Z + R_L$) to limit current to the amount needed for full scale deflection. The battery provides the EMF needed for current ($I_M = E_B/R_t$). Before the ohmmeter is used, the test leads are deliberately shorted together to cause the pointer to deflect full scale (usually from left to right). The scale on the meter face starts at infinite ohms ($\infty\Omega$) on the left end of the scale, ranging down to zero ohms ($0\,\Omega$) on the right. The shorted test leads represent a short ($0\,\Omega$) and the pointer moves to the extreme right end of the scale to indicate zero ohms. However, the pointer may not come to rest exactly on the zero line. If it does not, the ohms-zero-adjust rheostat can be adjusted to align the pointer with the zero line. This is referred to as **zeroing the ohmmeter**. Thus, when the leads are shorted together, full scale current flows through the meter movement, causing the pointer to indicate a short ($0\,\Omega$).

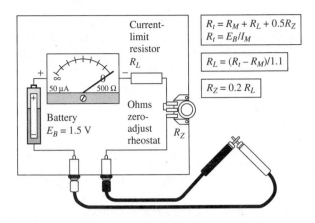

$$R_t = R_M + R_L + 0.5 R_Z$$
$$R_t = E_B / I_M$$

$$R_L = (R_t - R_M)/1.1$$

$$R_Z = 0.2\,R_L$$

$R_t = 1.5\ \text{V}/50\ \mu\text{A} = 30\ \text{k}\Omega$
$R_L = (30\ \text{k}\Omega - 500\ \Omega)/1.1 = 26.82\ \text{k}\Omega$ (Use a 27 kΩ Res.)
$R_Z = 0.2 \cdot 26.82\ \text{k}\Omega = 5.36\ \text{k}\Omega$ (Use a 5 kΩ Pot.)

Figure 11-15 Designing a basic ohmmeter circuit.

When an unknown resistance (R_U) is measured with the ohmmeter, the test leads are placed across the resistor, including it in the circuit. The amount of meter current and pointer deflection is then determined by the sum of the ohmmeter circuit resistance and the unknown [$I_M = E_B/(R_t + R_U)$]. The larger the unknown resistor, the less the meter current and pointer deflection.

Basic Ohmmeter Design

Now that you have a fairly thorough understanding of the ohmmeter theory of operation, you can proceed with the actual design. As in previous meter-circuit designs, you must know the meter-movement sensitivity and meter resistance. You must also decide the battery voltage you wish to use (usually 1.5 or 3 V). Next, you must calculate the total ohmmeter circuit resistance needed to limit current to the amount needed for full scale deflection (I_M) when the test leads are shorted together.

$$R_t = E_B/I_M \tag{11.5}$$

Once the total circuit resistance is calculated, using Ohm's Law, the values of the current-limiting resistor and the ohms-zero-adjust rheostat must be found. The meter resistance is a known value. Therefore, the current-limiting resistor and the zero-adjust rheostat can be mathematically represented in terms of R_t and R_M as follows:

$$R_t = R_M + 0.5\,R_Z + R_L$$

We can rearrange the equation to show

$$R_t - R_M = 0.5\,R_Z + R_L$$

or

$$R_L + 0.5\,R_Z = R_t - R_M$$

In order to determine the actual value of R_L or R_Z, we must establish a design criterion that defines their relative values. The limit resistor is much larger than the ohms-zero-adjust rheostat, since the rheostat is meant for fine adjustment. An adequate criterion, therefore, would define half of the rheostat as equal to $\frac{1}{10}$ of the limit resistor. In other words, $0.5\ R_Z = 0.1\ R_L$ or $R_Z = 0.2\ R_L$. With this criterion in mind, we can substitute $0.1\ R_L$ for $0.5\ R_Z$ in the formula above, to solve for R_L as follows:

$$R_L + 0.1\ R_L = R_t - R_M$$

therefore

$$1.1\ R_L = R_t - R_M$$

and

$$R_L = (R_t - R_M)/1.1 \tag{11.6}$$

The value of the zero-adjust rheostat can now be determined as follows:

$$R_Z = 0.2\ R_L \tag{11.7}$$

Notice in Figure 11-15 that the meter parameters are 50 μA and 500 Ω. The battery is 1.5 V and the total circuit resistance is found to be 30 kΩ using Formula 11.5. The current-limiting resistor is found to be 26.82 kΩ using Formula 11.6, and the value of the zero-adjust rheostat is found to be 5.36 kΩ. Practical choices for R_L and R_Z would be a 27 kΩ fixed resistor and a 5 kΩ potentiometer wired as a rheostat.

Designing the Scale

Not only must you design the ohmmeter circuit, you must design the scale on the meter face as well. Figure 11-16 illustrates the scale that is needed for a 50 μA meter movement and a 1.5 V battery. As you can see, the scale is definitely not linear. The right half of the

R_U	240 kΩ	120 kΩ	60 kΩ	30 kΩ	10 kΩ	5 kΩ	1 kΩ
I_M (μA)	5.6	10	16.7	25	37.5	43	48.4
Current ratio	$\frac{5.6}{50}$	$\frac{10}{50}$	$\frac{16.7}{50}$	$\frac{25}{50}$	$\frac{37.5}{50}$	$\frac{43}{50}$	$\frac{48.4}{50}$
Scale factor	0.11	0.2	0.33	0.5	0.75	0.86	0.97
% Scale	11%	20%	33%	50%	75%	86%	97%

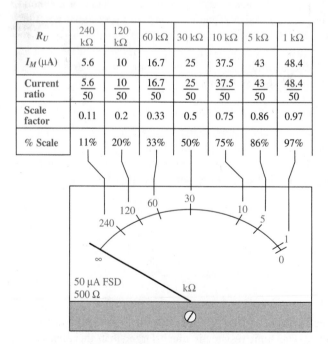

Figure 11-16   The ohmmeter scale is nonlinear and reversed.

scale indicates values of resistance less than 30 kΩ, and the left half of the scale indicates resistances from 30 kΩ to infinity (∞ Ω). The overcrowding on the left side of the scale makes readings very difficult and greatly reduces accuracy.

Why is the scale nonlinear, and why is the center of the scale marked 30 kΩ? Let's start with the second question. The center of the scale is marked as 30 kΩ because, if a 30 kΩ resistor (R_U) is being measured, the meter current will be cut in half, to 25 μA, which causes the needle to deflect by half or 50% of the scale:

$$I_M = 1.5 \text{ V}/(R_U + R_t) = 1.5 \text{ V}/(30 \text{ k}\Omega + 30 \text{ k}\Omega) = 25 \text{ } \mu\text{A}$$

As a rule, in a basic ohmmeter circuit such as this, midscale is always marked according to the value of R_t. What about the remaining scale markings? The table in Figure 11-16 shows how different values of measured resistance (R_U) result in different values of meter current, which in turn are certain percentages of the maximum meter current needed for FSD.

Accuracy

There are basically three factors that will contribute to the accuracy, or inaccuracy, of your ohmmeter readings: (1) the accuracy of the meter movement itself, (2) the care you take in reading the pointer position on the scale, and (3) the voltage level of the battery. The first factor can be minimized by using a good-quality meter movement, and the second can be minimized with practice. The battery voltage, however, turns out to be a bigger problem. The meter face of Figure 11-16 is designed assuming a constant 1.5-volt source. Will the battery always be 1.5 volts? Of course not. As the battery ages and is used, its voltage drops. The ohms-zero-adjust rheostat is used to rezero the meter at full scale and compensate for battery aging.

However, when you rezero the meter, does this guarantee the accuracy of the rest of the scale? Let's use a little math to find out. Suppose the battery has dropped down to 1.4 volts. The total resistance needed for FSD meter current is now 28 kΩ ($R_t = 1.4$ V/50 μA $= 28$ kΩ). This means the pointer will indicate midscale when a 28 kΩ resistor is measured, since the 28 kΩ resistor plus the 28 kΩ R_t will double resistance, cutting meter current in half. But wait. The meter is marked as 30 kΩ midscale. Obviously we have an error here of 2 kΩ or 7% (2 kΩ/28 kΩ $= 0.07$ or 7%). The weaker the battery gets, the worse the error becomes. If extreme accuracy is absolutely necessary, a low-voltage regulator circuit should be used to supply a constant voltage that is unaffected by aging or use.

The Multirange Ohmmeter

The multirange ohmmeter gives you the capability of measuring a wide range of resistances with the same instrument. Figure 11-17 illustrates a typical multirange ohmmeter circuit. At first glance, it seems that this circuit is very difficult to design. However, as you will quickly see, it is not difficult to design at all. Notice that one ohmmeter scale is being used for all three ranges. Each range is actually a scale multiplier. If the selector switch is set at $\times 1$ and the pointer is straight up, indicating 10, the measured resistance is 10 $\times$ 1 or 10 Ω. If the selector switch is set at $\times 10$ and the pointer is straight up, indicating 10, the measured resistance is 10 $\times$ 10 or 100 Ω, and so on.

Notice also that when the selector is in the $\times 1$ position, a 10 Ω range resistor is placed in parallel across the meter circuit (Figure 11-17a). If a 10 Ω resistor is being

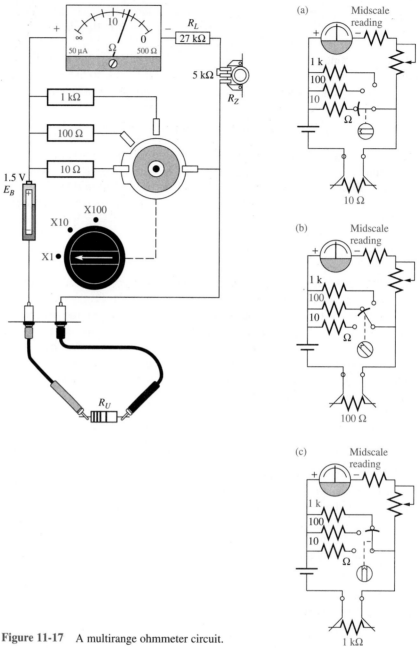

Figure 11-17   A multirange ohmmeter circuit.

measured, it will appear in series with the 10 Ω range resistor and the battery. The 30 kΩ meter circuit will have very little loading effect on the 10 Ω range resistor. As a result, the voltage supplied to the meter circuit is half of the battery voltage, or 0.75 V. Naturally, this voltage will produce only half of the FSD meter current in the meter circuit, resulting in half scale deflection. Thus the 10 Ω range resistor determines that the midscale point will be equal to 10 Ω. In reality, you would purchase a meter movement with a standard ohmmeter scale already marked on it and the midscale value would determine the size of range resistor for the ×1 position. In this case, let's assume we have purchased a meter movement with a marked ohmmeter scale that indicates 10 midscale. This dictates that the ×1 range resistor must be 10 Ω.

The ×10 range is shown in Figure 11-17b. As before, when the selector switch is set to the ×10 position, a 100 Ω range resistor is placed in parallel with the meter circuit. In this case, the pointer will deflect to midscale when a 100 Ω resistor is measured. The pointer indicates 10 and the range is ×10, so the measured resistance is 10 × 10 Ω or 100 Ω. The 30 kΩ meter circuit still has little loading effect on the range resistor, since 30 kΩ is much greater than 100 Ω.

The ×100 range is shown in Figure 11-17c. When the selector switch is set to the ×100 position, a 1,000 Ω range resistor is placed in parallel with the meter circuit. In this case, the pointer will deflect to midscale when a 1,000 Ω resistor is measured. The pointer indicates 10 and the range is ×100, so the measured resistance is 100 × 10 Ω or 1,000 Ω. The 30 kΩ meter circuit still has little loading effect on the range resistor, since 30 kΩ is much greater than 1,000 Ω. However, this is about the highest practical range for this circuit without introducing too much error, due to a meter-circuit loading effect. In other words, the next range resistor would be 10 kΩ and a 30 kΩ meter-circuit resistance would definitely load it down. Higher ranges could be obtained by using a higher battery voltage (i.e., 3 V). This would require that the meter-circuit resistances (R_L and R_Z) be higher to limit the current to the maximum full scale current, in this case 50 μA. If the meter circuit resistance is higher, a higher range resistance can be used.

A Case in Point

Figure 11-18 illustrates a small, general-purpose multimeter. Notice the ohms scale (the uppermost scale). As you can see, the midscale position is 35. This dictates that the first range resistor, in the R × 1 position, must be 35 Ω. In the R × 100 position, the range resistor is 3,500 Ω. Notice also that the VDC meter sensitivity is marked as 20 kΩ/VDC on the left lower corner of the meter face. This indicates that the meter movement is a 50 μA movement (1 V/20 kΩ = 50 μA). If the internal battery voltage were known, the

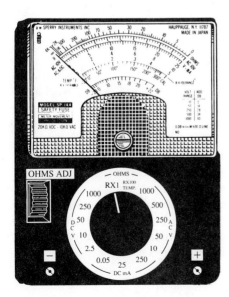

Figure 11-18   A handy pocket-sized analog multimeter. (Courtesy of A.W. Sperry Instruments, Inc.)

entire ohmmeter circuit could be calculated and drawn without opening the meter case or looking at the manufacturer's schematic of the circuit. A design summary is provided at the end of this chapter.

Take time now to test your understanding of ohmmeter circuits by answering the questions of Self-Check 11-4.

SELF-CHECK 11-4

1. Given a 25 μA meter movement with a meter resistance of 2 kΩ and a 3 V battery, find the values of R_L and R_Z needed to make a basic ohmmeter circuit.

2. In the ohmmeter circuit of #1, what should the midscale point be marked as? (What number marking?)

3. If you purchase a meter movement whose face is already scaled as an ohmmeter, and the midscale point is marked as a 20, what should the value of the first range resistor (for the $\times$1 range) be in a multirange ohmmeter circuit?

11-5 Multimeters

Without any doubt, the multimeter is the single most useful piece of test equipment available to the technician or engineer. It is lightweight and compact, and it contains all of the basic meter circuits discussed in previous sections in one package. In this section, we will explore the two general categories of multimeters, analog and digital, comparing their relative strengths and weaknesses.

The Analog Multimeter (MM or VOM)

The Basic Analog VOM

The analog **MultiMeter (MM)** is often referred to as a **Volt-Ohm-Milliammeter (VOM)**. It is a single instrument containing three switch-selected meter circuits and one D'Arsonval/Weston meter movement. The accuracy of the instrument depends on the quality of the meter movement and other circuit elements. An analog MM using a taut-band meter movement is, as a rule, more accurate and more sensitive than a less expensive jewel movement. Precision film resistors and/or wirewound resistors are used to insure accuracy and resist value changes due to aging. Also, high-quality switches, connectors, and jacks are needed to insure very low-resistance connections and to resist corrosion in high-humidity environments (silver- or gold-plated contacts). In some cases, the multimeter housing is waterproofed to guarantee environmental immunity.

Figure 11-19 is a schematic of a VOM combining the individual circuits designed in previous sections. Notice that a three-section rotary switch is used to select functions (ammeter, ohmmeter, voltmeter) and ranges at the same time. Each function, or meter circuit, is designed using the procedures outlined in the previous sections. The single meter movement is passed from one circuit to another via SW1c.

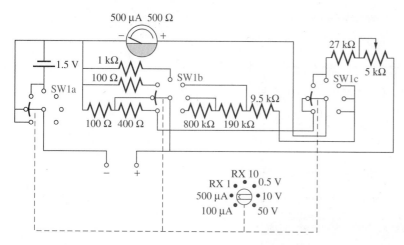

Figure 11-19   A typical analog volt-ohm-milliammeter circuit (VOM).

Additional Functions

Some portable multimeters include functions beyond the standard VOM. Figure 11-20 is an example of a multimeter that includes a transistor tester circuit used to determine the current gain in various transistors. The scale corresponding to this function is labeled hFE, a symbol used for DC transistor current gain (the higher the reading, the higher the gain).

This multimeter also includes a piezoelectric transducer that generates an audible tone when the musical note function is selected and the test leads are shorted together. An audible tone can be a very useful troubleshooting aid when checking the continuity of long cables or identifying cable pairs in large unmarked bundles. Note also that this multimeter boasts of being drop-proof, within reason (lower left corner of meter face). This is due to the spring suspension holding the taut bands and the armature in place.

Figure 11-20   A taut-band analog multimeter. (Courtesy of A.W. Sperry Instruments, Inc.)

Figure 11-21   An FET VOM. (Courtesy of A.W. Sperry Instruments, Inc.)

FET Multimeter

As shown in Figure 11-21, some analog multimeters use a **field effect transistor (FET)** amplifier to increase the DC voltmeter circuit resistance. The field effect transistor has a very high input impedance that enables DC voltmeter circuits to have a very high, fixed input impedance of, usually, 10 MΩ. Regardless of which DCV range the meter is set to, the meter resistance between test probes is a fixed 10 MΩ. The advantage of the FET multimeter over a standard VOM is that of little or no loading effect when voltage measurements are made.

Multifunction Recorder

Figure 11-22 shows a Simpson Multicorder, a multifunction chart, or strip, recorder. An instrument such as this can be very useful in monitoring and recording AC line-voltage variations or DC power-supply variations over a period of time. Some troubleshooting situations require the use of an instrument such as this to demonstrate, or prove, that an irregularity exists or an intermittent trouble. We need not discuss the obvious advantages of letting an instrument such as this monitor a circuit instead of a human eternally perched on a stool with a multimeter in one hand and test probes in the other (dread the thought that he/she should blink as the glitch occurs).

A Specialized Multimeter

The multimeter shown in Figure 11-23 is specialized, designed for electricians to test cable and wiring insulation and the resistance of high-voltage circuits with respect to earth ground. When using the MΩ range, for insulation testing, the ohmmeter circuit voltage is

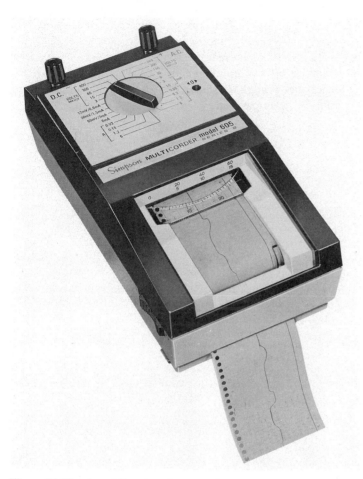

Figure 11-22   A multifunction chart recorder that records variations in voltage and current over time.

Figure 11-23   A voltmeter and megohm insulation tester. (Courtesy of A.W. Sperry Instruments, Inc.)

very high, in this case 500 VDC. Naturally, caution must be exercised when using the MΩ range, in order to maintain your professional composure. This high voltage is necessary to produce enough current through very high resistances (hundreds of megohms) to cause some pointer deflection. These instruments are often called "meggers," since they are capable of measuring resistances in the high MΩs and GΩs. Notice this multimeter also has normal R × 1 and R × 100 Ohms ranges and a 600 VAC range.

The Digital Multimeter (DMM or DVOM)

DMMs

Digital multimeters have many advantages over their analog cousins. One obvious advantage is the digital readout itself. You do not have to read the meter head-on to avoid a reading error. Basically, the reading is exactly what the digital display says it is. That is not to say DMMs are perfect. Most are rated at some percent accuracy, usually within less than 1% and ±1 count of the least significant digit on the display. However, even the least expensive of digital meters are more accurate than the higher-priced analog meters. Almost all DMMs have 10 MΩ input impedances for measuring voltage, which means there is little or no loading effect on a circuit while measurements are being taken.

Special Features

Most DMMs today include practical features such as auto ranging, data hold, and low-battery alert. DMMs that have auto-ranging capability conspicuously lack a range-selector switch. Only the function switch is needed. The DMM automatically selects the range needed as a measurement is being made. The data hold is a nice feature to have. A small button is provided that allows you to freeze a reading on the digital display. This could come in handy when you are required to take a reading in a confined area and you can't quite read the meter and safely handle the test probes while balancing on the ball of one foot. Or maybe you simply want to share your readings with someone in the next room. The low-battery alert has its obvious advantages. Other special functions that may be included in the DMM are an analog bar-graph display (a segmented LCD bar graph that is not intended for accuracy), an auxiliary analog meter movement, a continuity beeper, a diode test, conductance, a transistor hFE test, capacitance, an LED test, a digital logic test, and a digital frequency counter. A popular DMM with many of these features is shown in Figure 11-24.

Analog vs Digital

With all of the accuracy, ease of use, and flexibility of the DMM, why are analog MMs still in use? Analog meters are still needed for nulling, peaking, and trend observation, and in some cases they are less susceptible to reading errors caused by strong radio-frequency electromagnetic fields. Most technicians prefer using an analog meter when tuning a radio circuit, since it is easy to watch the meter pointer reach a peak, or null, with signal level. It is relatively difficult to use a DMM for this same application. The DMM will have a short delay (usually 0.1 to 0.5 sec.) in updating the display, which makes it difficult to find an exact peak or null when tuning a circuit. For the same reason, the DMM is difficult to use for trend observation. If a voltage source is varying slowly up and down, an analog meter will clearly show the up and down trend. The DMM shows a confusing

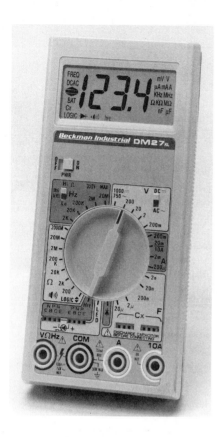

Figure 11-24 The DM27XL by Beckman Industrial is a nearly complete handheld test bench and measures AC/DC voltage, AC/DC current, resistance, capacitance, transistor gain (h_{FE}), frequency, and diode and continuity checks. (Courtesy of Beckman Industrial, Inc.)

sequence of display changes that may seem difficult to follow. As a general rule, analog meters are best used for measuring slowly changing voltages in cases where maximums, minimums, or trends are to be observed, and DMMs are best used for accurately measuring fixed AC or DC voltages. The DMM is more accurate than the analog MM in all functions as long as the parameter being measured is not slowly varying or being influenced by an electromagnetic field. Table 11-1 gives a brief comparative overview of analog and digital multimeters.

Before continuing, take time for a brief review by answering the questions of Self-Check 11-5.

SELF-CHECK 11-5

1. What is the advantage of an FET analog DVM over a general-purpose 20 kΩ/V DVM?

2. What is a multifunction chart recorder?

3. Which type of meter, (a) analog or (b) digital, is easier to use for peaking or nulling?

4. List four advantages of the DMM over the analog MM.

TABLE 11-1   Analog vs Digital

ANALOG	DIGITAL
Meter error in the range of ±1 to 3% of actual.	Meter error less than ±1% of actual + one or more counts of the least significant digit.
Reading errors can occur due to parallax.	Reading errors are not likely, since a digital display is used.
Must "zero" the ohmmeter each time before use on each range.	Digital circuitry does not require ohms zeroing at all.
Accuracy of meter readings is higher on the right half of the scale for all functions.	Accuracy of reading is not as severely affected by range setting and parameter magnitude.
General purpose* analog voltmeters can have a severe loading effect on a circuit when measurements are taken.	DMMs have very high input impedances, usually 10 MΩ. Circuit loading is still a concern for high impedance circuits.
Analog meters are particularly suited for "nulling," "peaking," and trend observation as in slowly changing voltage levels.	Digital meters are better suited for measuring steady state voltages and currents with extreme accuracy.
Most general-purpose* analog meters work well in the presence of electromagnetic fields.	Many digital meters are adversely affected by electromagnetic fields.

*General purpose means: meter/resistor circuits not having amplifiers, such as field effect transistors

Summary

FORMULAS

(11.1) $R_{sh} = \dfrac{I_M \cdot R_M}{I_{sh}}$ (ammeter basic shunt)

where R_{sh} is the shunt resistance, I_M is the meter current sensitivity for FSD, R_M is the meter coil resistance, and I_{sh} is the shunt current

(11.2) $R_{sh\#} = \dfrac{I_M \cdot (R_{sh1} + R_M)}{I_{range}}$ (Ayrton shunt ammeter)

(11.3) $R_X = (V_R/I_M) - R_M$ (single range voltmeter multiplier)

where R_X is the multiplier resistor, V_R is the range voltage, I_M is the FSD meter current, and R_M is the meter resistance

(11.4) $R_{X\#} = (V_R/I_M) - (R_M + \text{all previous } R_X \text{ values})$

(multirange voltmeter shunt) (see Example 11-2)

(11.5) $R_t = E_B/I_M$ (ohmmeter circuit resistance)

where R_t is the total ohmmeter-circuit resistance, E_B is the internal battery voltage, and I_M is meter-current sensitivity for FSD

(11.6) $R_L = (R_t - R_M)/1.1$ (ohmmeter current-limiting resistor)

(11.7) $R_Z = 0.2 R_L$ (ohms-zero-adjust rheostat)

CONCEPTS

- D'Arsonval/Weston meter movements are moving-coil instruments that function according to the principles of motor action and have linear scales.
- A meter's current sensitivity is the amount of current needed to cause full scale pointer deflection (FSD).
- Meter readings are most accurately taken from the right half of the meter scale.
- Taut-band meter movements are generally more rugged and more accurate, and are available in lower FSD meter currents than jewel-type movements.
- The electrodynamometer is a moving-coil meter using an electromagnetic field developed around a stationary coil.
- The electrodynamometer is an AC or DC instrument.
- Meter-coil resistance is very important and must be considered when determining multiplier values for voltmeter circuits, especially on the low ranges.
- Voltmeter sensitivity is related to meter-movement current sensitivity in the sense that voltmeter sensitivity is the reciprocal of meter-movement current sensitivity ($1/I_M$). Voltmeter sensitivity is measured in ohms per volt (Ω/V).
- Battery aging in an analog ohmmeter will severely decrease meter accuracy.
- Analog multimeters are best used for peaking, nulling, or trend observation.
- Digital multimeters are best used for steady AC or DC voltage and current measurements.
- DMMs provide much greater accuracy than analog MMs.

PROCEDURES

Ayrton Shunt Design Procedure

1. Meter-movement sensitivity and resistance must be known.
2. Decide on ranges. The lowest range should be at least twice the value of meter sensitivity.
3. Draw a schematic of your multirange ammeter circuit.
4. Determine the value of the first shunt (for the lowest range) using Formula 11.1.
5. Determine all remaining shunt values using Formula 11.2.
6. Solve for individual resistors by taking the difference of consecutive shunt values.
7. Construct your circuit using precision wirewound or metal film resistors.

Voltmeter Circuit Design Procedure

1. The meter movement sensitivity and resistance must be known.
2. Determine desired ranges and draw the schematic.
3. If you are using the individual multiplier method, calculate each multiplier using Formula 11.3 for each range.
4. If you are using the series multiplier method, begin multiplier calculations on the lowest voltage range using Formula 11.4.
5. Do not ignore the meter resistance in your calculations. The greatest percentage of error will occur in the lowest voltage range if meter resistance is ignored.

Multirange Ohmmeter Design Procedure

1. Obtain a DC meter movement marked with a standard ohmmeter scale. You will need to know its FSD sensitivity and internal resistance.

2. Decide on a battery voltage (usually 1.5 or 3 V).

3. Determine the values of the current limiting resistor and the ohms-zero-adjust rheostat using the basic ohmmeter circuit formulas (11.5, 11.6, 11.7).

4. The lowest range resistor ($\times 1$) will be the same value as indicated midscale on the meter face. The next range resistor will be 10 times the midscale value, the next 100 times the midscale value, and so on. As a rule of thumb, the meter-circuit resistance should be greater than 10 times the highest-range resistor to ensure a reasonable amount of accuracy on the highest range. A meter movement with the lowest possible sensitivity should be used, since this will ensure a high meter-circuit resistance.

5. Draw a diagram of your circuit and label all components.

SPECIAL TERMS

- D'Arsonval/Weston meter movement
- Full scale deflection (FSD), sensitivity
- Meter zeroing
- Jewel movement, taut-band movement
- Electrodynamometer movement
- Linear scale
- Shunt, scaling, ranging, Ayrton (universal) shunt
- Multiplier, multiranging, voltmeter sensitivity
- Ohms zero adjust
- Nonlinear scale
- Multimeter (MM), volt-ohm-milliammeter (VOM)
- Field-effect transistor (FET) VOM
- Multifunction chart recorder
- Megger
- Digital multimeter (DMM)
- Digital volt-ohm-milliammeter (DVOM)

Need to Know Solution

The taut-band D'Arsonval/Weston movement is the best choice of the three, since taut-band movements are generally the most accurate and this particular movement has the highest sensitivity (25 μA). A series multiplier design should be used following the voltmeter circuit-design procedure and using Formula 11.4. The following values should be calculated: $R_1 = 39.25$ kΩ; $R_2 = 360$ kΩ; $R_3 = 3.6$ MΩ; $R_4 = 36$ MΩ.

Questions

11-1 Meter Movements

1. Who were the two men credited with the invention of the moving-coil meter movement?

2. Define meter-movement current sensitivity.
3. What is mechanical meter zeroing in regard to analog meter movements?
4. Meter readings should be taken from what portion of an analog meter scale? Why?
5. How does a taut-band movement differ from a jewel movement?
6. What are electrodynamometers normally used for?
7. Explain how the electrodynamometer is an analog computer.

11-2 Ammeter Circuits

8. Briefly explain how an ammeter should be placed in a circuit for measurements and why it must be placed that way.
9. What is a meter shunt?
10. What is meant by *scaling* or *ranging*?
11. How is the Ayrton shunt an improvement over the basic meter-shunt design?

11-3 Voltmeter Circuits

12. Describe how the voltmeter should be placed when measuring a voltage in a circuit.
13. What is a multiplier?
14. How is the voltmeter range on a multirange analog voltmeter related to the voltmeter loading effect?
15. Is it necessary to observe polarity when using an analog voltmeter in a circuit? Explain.

11-4 Ohmmeter Circuits

16. List the components of a basic ohmmeter circuit.
17. The zero (0) is on which end of the ohmmeter scale?
18. Explain how to electrically zero an analog ohmmeter.
19. What three factors affect the accuracy of your analog ohmmeter readings?

11-5 Multimeters

20. What does VOM stand for?
21. What is the main advantage of an FET multimeter over a general-purpose multimeter?
22. What is a "megger" and what is it normally used for?
23. List three advantages of the digital multimeter over the analog multimeter.
24. Analog multimeters are best used for what purpose?

Problems

11-2 Ammeter Circuits

1. Calculate the value of the shunt needed for a 500 mA ammeter circuit using a 100 Ω, 1 mA meter movement.
2. Calculate the value of the shunt needed for a 1 A ammeter circuit using a 500 Ω, 50 μA meter movement.
3. Design an Ayrton shunt ammeter circuit using a 600 Ω, 25 μA meter movement. The meter ranges shall be 50 μA, 1 mA, and 10 mA. Draw and label the schematic.
4. Design an Ayrton shunt ammeter circuit using a 1 kΩ, 10 μA meter movement. The meter ranges shall be 50 μA, 500 μA, and 1 mA. Draw and label the schematic.

11-3 Voltmeter Circuits

5. If a 700 Ω, 100 μA meter movement is used for a 500 mV voltmeter circuit, will the meter resistance be significant, or can the meter resistance simply be ignored? Why?

6. How many high-voltage 1 MΩ multiplier resistors would you need for a 20 kV voltmeter circuit using a 1 mA meter movement?

7. Design a series-multiplier, multirange voltmeter circuit using a 500 Ω, 50 μA meter movement. The voltage ranges shall be 2 V, 50 V, 200 V, and 500 V. Draw and label the schematic.

8. Design a series-multiplier, multirange voltmeter circuit using a 100 Ω, 1 mA meter movement. The voltage ranges shall be 1 V, 5 V, 10 V, and 50 V. Draw and label the schematic.

9. Calculate the voltmeter sensitivity for a 100 V voltmeter circuit that has a 25 μA meter movement.

10. Calculate the voltmeter sensitivity for a 5 V voltmeter circuit that has a 50 μA meter movement.

11. A DC voltmeter has a voltmeter sensitivity of 50 kΩ/V. What is the total voltmeter circuit resistance if the voltage range of the voltmeter is 100 V? What is the current sensitivity of the meter movement itself?

11-4 Ohmmeter Circuits

12. Design a basic ohmmeter circuit using a 300 Ω, 250 μA meter movement and a 1.5 V battery. Solve for R_L and R_Z.

13. Design a basic ohmmeter circuit using a 600 Ω, 50 μA meter movement and a 3 V battery. Solve for R_L and R_Z.

14. What value should the midscale marking be for the ohmmeter of Problem 13? Why?

15. Design a multirange ohmmeter circuit using a 750 Ω, 25 μA meter movement and a 3 V battery. The ohmmeter scale is already marked on the face of the movement. The midscale marking is 15. Solve for R_L, R_Z, and the range resistors needed for the R $\times$ 1, R $\times$ 10, R $\times$ 100, and R $\times$ 1,000 ranges. Draw a schematic and label the values of all components.

Answers to Self-Checks

Self-Check 11-1

1. The jewel movement uses an axle pivot/end cap arrangement for coil mounting and suspension, whereas the taut-band movement uses spring-loaded bands. Taut-band movements are more sensitive and more accurate.

2. The electrodynamometer is used to measure power.

3. Meter damping is a means of stabilizing the pointer as it deflects to an indicating position. Damping greatly reduces pointer overshoot and pointer bounce.

Self-Check 11-2

1. Shunt = 1.01 Ω
2. R_1 = 10 Ω; R_2 = 1.11 Ω

Self-Check 11-3

1. 499,900 Ω or 500 kΩ would be fine.
2. $R_1 = 9{,}900 \ \Omega$; $R_2 = 90$ kΩ; $R_3 = 900$ kΩ

Self-Check 11-4

1. $R_L = 107{,}273 \ \Omega$ or 110 kΩ; $R_Z = 21{,}455 \ \Omega$ or 20 kΩ
2. 120 with the scale labeled kΩ
3. 20 Ω

Self-Check 11-5

1. The FET DC voltmeter has a very high input impedance (usually 10 MΩ) and therefore will not load down a circuit as much as a 20 kΩ/V voltmeter would.
2. The multifunction chart recorder is an instrument that can be used to monitor and make a permanent record of AC or DC voltage or current variations. It is an analog device that produces an ink trace on a continuously fed strip of paper. The paper is calibrated for current and voltage measurements.
3. Analog
4. (1) The DMM is more accurate. (2) Reading errors are not likely, since digital display is used. (3) DMMs have very high input impedances, usually 10 MΩ. (4) The ohmmeter function does not require zeroing.

SUGGESTED PROJECTS

1. Add some of the main concepts, formulas, and procedures from this chapter to your personal Electricity and Electronics Notebook.

2. Obtain a D'Arsonval/Weston meter movement and design and build a multirange ammeter circuit, a multirange voltmeter circuit, and a basic ohmmeter circuit.

Part IV

Inductors, Capacitors, and Pulse Response

During the years of 1887 and 1888, among other projects, Edison sought a way to project moving pictures. At the same time, an American by the name of Hannibal W. Goodwin developed a transparent film that could be light sensitized with chemicals. George Eastman manufactured the film and motion picture cameras, leading the way for Edison to develop the motion picture projector known as the kinetoscope. On October 6, 1889, Edison gave a demonstration of his invention at his West Orange laboratory. The movie was accompanied by synchronized sound from one of his phonographs. He applied for a patent on August 24, 1891. On April 14, 1894, he opened the first "peephole" parlor on Broadway in New York. The first commercial showing of motion pictures took place on April 23, 1896, at Koster & Bial's Music Hall in New York using an Edison Vitascope. (Photograph from The Bettmann Archive.)

In the openings of the Part IV chapters to follow, you will learn more about his amazing motion picture inventions.

Thomas Edison and the Kinetoscope

Edison continued to improve the kinetoscope from his first patent in 1891. The kinetoscope shown here has a large square box on top to contain the film and feed it down behind the lens. The large box in the rear contained the powerful electric arc lamp. The later model in the background of the photo has open feed and takeup reels.

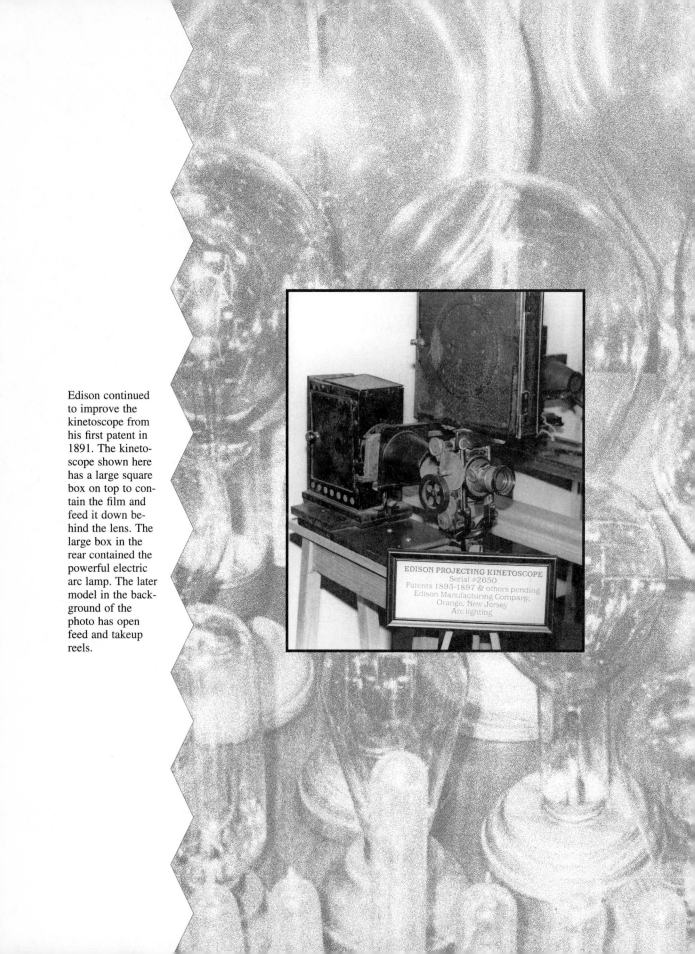

EDISON PROJECTING KINETOSCOPE
Serial #2650
Patents 1893-1897 & others pending
Edison Manufacturing Company,
Orange, New Jersey
Arc lighting

Chapter 12

Inductance

OBJECTIVES

After studying this chapter, you will be able to
- explain inductance, or self-inductance.
- calculate inductance when the number of coil turns is known and the change in flux per change in current is known.
- calculate the amount of induced voltage across a coil when the inductance of the coil and the rate of change in current are known.
- explain how energy is stored in an inductor and calculate the quantity of energy stored.
- list the physical factors that affect inductance and explain the effect of each.
- calculate inductance given the number of turns, absolute permeability of the core, cross-sectional area of the core, and length of the core.

- describe basic types of inductors and their schematic symbols.
- explain the purpose for the laminations in laminated iron-core inductors.
- calculate total inductance for series and parallel inductors.
- explain stray inductance.
- troubleshoot an inductor using an ohmmeter, a DC voltmeter, and a DC ammeter.

INTRODUCTION

In this chapter, you will discover how the principles of electromagnetic induction apply to individual coils. You will learn that coils have the ability to induce a voltage across themselves as a result of any change in current. Because of this characteristic of self-induction, the coil is appropriately named an *inductor*. Also, you will discover that induc-

455

tors not only induce voltage across themselves, in so doing they oppose any change in circuit current. Thus, the inductor acts as a conductor offering no opposition to steady DC, but it is very reactive to AC, a constantly changing current.

The inductor's ability to react to changes in current and store energy in its magnetic field make it very useful in a great number of AC circuits. In some applications, inductors are used as filters to block AC and pass DC. In many applications, inductors are used for tuning, frequency selection, and oscillation in radio, television, and countless related circuits. This chapter should be considered as an introduction to inductance and inductors, laying a foundation for future study.

NEED TO KNOW

In previous chapters, we referred to the coil as a solenoid in a variety of electromechanical devices, such as relays and electric valves. We labeled coils as stator windings and armature windings in generators and motors. In each of these applications, you were able to see how the magnetic field of a coil was used to perform mechanical work or produce an electrical voltage and current. In spite of all you have learned about coils, there is still more to learn.

Some coils, used in electrical and electronic circuits, are referred to by still another name. They are called *inductors*. Inductors are very much the same as the coils to which you have already been introduced. They are generally coils of wire, wound on a cylindrical form, with an air, ferrite, or iron core. It is the applications to which these coils are put that makes them unique and worthy of a special name, for in these applications certain coil characteristics become very important.

But what are these special coil characteristics? They make up what may be called "the nature of inductance" and pertain to inductors used as components in electrical and electronic circuits. Perhaps you are already somewhat familiar with inductors and their characteristics. See if you can finish the following sentences that describe the nature of inductors. These are important characteristics that are significant to the topic of inductors in many of the remaining chapters of this textbook.

1. It is the nature of an inductor to _____ a voltage across _____ .

2. It is the nature of an inductor to _____ any change in _____ .

3. It is the nature of an inductor to store _____ in its _____ .

Joseph Henry (1797 – 1878)

Joseph Henry was born in Albany, NY. He was a student, and later a mathematics and physics teacher, at Albany Academy. Henry was a professor of philosophy at Princeton University from 1832 to 1846. He later became the first secretary and director of the Smithsonian Institute. Although not initially recognized for his work in physics and electromagnetism, Henry was be-

stowed the honor of having his name used as the standard unit of measure for inductance. (Photo courtesy of The Bettmann Archive.)

12-1 The Nature of Inductance

Inductance and Inductors

Inductance (*L*)

What is **inductance**? Inductance is related to induction. Recall that induction is the process of generating a voltage across a coil, and a current in a coil, as a result of relative motion between magnetic flux and the coil. Inductance is related to induction in that inductance (sometimes referred to as **self-inductance**) is the measure of a coil's ability to induce a voltage *across itself* as a result of a change in current (Δi) in its windings. All coils are able to induce voltage across themselves. Consequently, coils, used in a wide range of electrical and electronic circuits, are commonly referred to as inductors. The mathematical symbol, or variable, for inductance is *L* and the unit of measure for inductance is the **henry (H)**.

Inductance Explained

But how is it possible for an inductor to induce voltage across itself? In Chapter 9, we discussed the following three requirements for induction: a magnetic field, a conductor (coil), and relative motion between the two. In our discussion, you learned that current is caused to flow through the coil and a difference of potential is established across the coil as a result of the coil cutting through the magnetic flux. You also learned that the amount of induced voltage was determined by the rate of cutting flux ($\Delta\Phi/\Delta t$) and the number of turns of the coil (Faraday's formula: $V_{ind} = N \cdot \Delta\Phi/\Delta t$). The same theory and principles hold true for self-induction. Let's see how.

Induced CEMF

Consider Figure 12-1. When the switch is set to position B, current flows through the inductor and a magnetic field is produced. From Chapter 9, you learned that a magnetic field exists about the turns of a current-carrying coil (solenoid). What we did not discuss was the fact that the current in the coil and the magnetic field surrounding the coil do not reach maximum levels instantly when the switch is closed. In fact, it takes a certain period of time, though it is usually very short, for the current and magnetic flux to reach a maximum level. The reason for this is the action of self-inductance, or inductance. As current is applied to the coil, magnetic flux is created and expands through the coil windings. The expansion of magnetic flux provides a relative motion, which in turn induces a voltage across the coil windings. This induced voltage is the opposite polarity of the external voltage applied to the coil. Thus, the induced voltage is a counterelectromotive force (CEMF) that opposes the applied EMF. Naturally, a countercurrent is also induced along with the CEMF. This does not mean the circuit has two currents flowing in opposite directions at the same time. It does mean, however, that the current being supplied to the coil is being opposed and cannot reach a maximum level instantaneously. Since the current applied to the inductor is not able to reach a maximum level instantaneously, it is said to be a **transient current** (in the process of changing level or amplitude).

After the transient period has passed, the magnetic field and current of the coil will have reached maximum levels. Since the source voltage is a DC source, the current in the coil and magnetic field surrounding the coil will remain constant until the switch is opened. Therefore, since the magnetic field is no longer expanding, there is no relative motion to

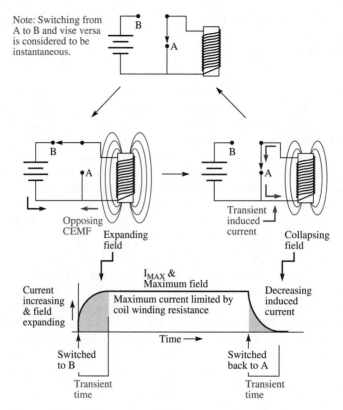

Figure 12-1   Inductance—self-inductance.

induce a CEMF. The only time a CEMF is induced is during times when the applied current and resulting magnetic flux are changing (increasing or decreasing during a transient period).

Induced EMF

When the switch is set back to position A, the stationary magnetic field will once again be set in motion as it collapses around the inductor. Note that the field does not instantly disappear. Another transient period occurs as the field collapses. During this transient time, the relative motion of the collapsing field again induces a voltage across the inductor. This time, since the direction of relative motion has changed (collapsing instead of expanding), the polarity of induced voltage and direction of induced current will change. In fact, this transient induced current will attempt to sustain the original applied current, opposing the sudden drop caused by the opened switch.

Inductors Oppose Change in Current

From the preceding discussion you should note that a major characteristic of inductors is the opposition they offer to any change in current. As current, under pressure from an applied EMF, increases in the inductor, a CEMF is generated across the inductor which opposes the increase. Then, as current applied to the inductor decreases, the coil opposes the change by attempting to sustain the current as it was.

Inductance (L) and Self-Induced Voltage (v_L)

Inductance Related to a Change in Flux

Earlier, it was stated that inductance is the measure of a coil's ability to induce a voltage across itself as a result of a change in current (Δi) in the coil. The change in current produces a change in magnetic flux ($\Delta \Phi$). It is the change in magnetic flux that induces the CEMF, or EMF, across the coil. If one conductor is able to induce more voltage across itself than a second inductor, we say that the first inductor has a *greater inductance* than the second. The greater inductance of the first inductor comes from its ability to produce a greater change in flux for a certain change in current as compared to the second inductor. This can be shown mathematically in the following formula:

$$L = N(\Delta \Phi / \Delta i) \qquad\qquad (12.1)$$

where L is inductance measured in henries (H), N is the number of turns of the inductor, $\Delta \Phi$ is the change in flux measured in webers (Wb) (recall, 1 Wb $= 1 \cdot 10^8$ flux lines), and Δi is the change in current measured in amperes (A).

Formula 12.1 clearly shows the relationship between the change in flux and the amount of inductance of the coil.

A 300-turn coil produces a change in flux of 0.003 Wb as a result of a change in current of 1 A. The inductance of the coil is 0.9 H (henries) ($L = 300$ turns $\cdot$ 0.003 Wb/1 A $= 0.9$ H).

EXAMPLE   12-1

Inductance Related to Induced Voltage

What does the amount of inductance (L, measured in henries) have to do with **self-induced voltage**? They are directly related. In fact, by taking a closer look at Faraday's formula for induced voltage, we will discover a mathematical relationship between them. Consider Faraday's formula and the following legitimate expansion of the formula:

$$V_{ind} = N \cdot (\Delta \Phi / \Delta t) \qquad \text{(Faraday's formula)}$$

where V_{ind} = self-induced voltage.

Expand the formula as follows:

$$\frac{\Delta \Phi}{\Delta t} = \frac{\Delta \Phi}{\Delta i} \cdot \frac{\Delta i}{\Delta t}$$

therefore,

$$N \cdot \frac{\Delta \Phi}{\Delta t} = N \cdot \frac{\Delta \Phi}{\Delta i} \cdot \frac{\Delta i}{\Delta t} = V_{ind}$$

Notice that the expansion above has revealed the presence of the inductance formula (12-1) contained in Faraday's formula for induced voltage. By substituting L for $N \cdot (\Delta \Phi / \Delta i)$, the following formula for induced voltage can be obtained:

$$V_{ind} = v_L = L \cdot (\Delta i / \Delta t) \qquad\qquad (12.2)$$

where v_L is the self-induced voltage, L is the inductance of the coil measured in henries (H), Δi is a change in current measured in amperes (A), and Δt is a change in time measured in seconds (s). Therefore, the amount of self-induced voltage depends on the amount of inductance of the coil (number of henries) and the rate of change of current through the coil.

E X A M P L E   1 2 - 2   If the current in a 0.5 H inductor is changing at the rate of 100 mA/s, the self-induced voltage is 0.05 V (v_L = 0.5 H · 100 mA/s = 0.05 V = 50 mV).

E X A M P L E   1 2 - 3   If the current in a 2 H inductor changes by 300 mA in a period of 400 ms, the self-induced voltage is 1.5 V (v_L = 2 H · 300 mA/400 ms = 1.5 V). The rate of change in current is 300 mA/400 ms = 0.75 A/s.

The Henry (H)

At this point in our discussion, it is possible to offer a definition for the henry (H). By rearranging Formula 12.2, the following relationship is revealed:

$$L = \frac{v_L}{\Delta i / \Delta t} \tag{12.3}$$

From Formula 12.3, we may define the henry (H) as the amount of inductance required to self-induce a CEMF of 1 V with a 1 A/s rate of change of current. In other words, one henry equals one induced volt per one ampere per second change in a current (1 H = 1 V/1 A/s).

Inductance and Stored Energy

Inductors are able to store energy in the form of a magnetic field. An ideal inductor has no conductor resistance and therefore does not use power in the same sense that resistors do. Resistance in a conductor or component converts electrical energy to heat energy. Power is dissipated as heat in the resistance of the component according to the I^2R power formula. Inductors usually have very little resistance. Therefore, very little electrical power is dissipated in the form of heat. However, inductors do appear to use large amounts of power because the electrical energy is converted to a magnetic field. This magnetic field stores the energy and returns it to the circuit as the field collapses.

The amount of energy a coil stores depends upon two primary factors: the size of the coil (the number of henries) and the maximum current in the coil. Recall that electrical energy is expressed in joules (J) and 1 W of power is one joule of energy converted every second. Stored inductor energy, in units of joules, is found from the following simple formula:

$$E = L \cdot I^2 / 2 \tag{12.4}$$

where E is the energy in joules, L is the inductance in henries (H), and I is the maximum inductor current.

E X A M P L E   1 2 - 4   Consider a 10 H inductor with a maximum current of 2 A. The amount of energy stored in its magnetic field is 20 J (E = 10 · 2^2/2 = 10 · 4/2 = 40/2 = 20 J).

The Nature of Inductors Summarized

It is the nature of an inductor to:

• induce a voltage across itself (CEMF) with any change in current in its windings.
• oppose any change in current in its windings.
• oppose AC and pass DC.
• store energy in the form of a magnetic field.

Take time to review this section by answering the questions of Self-Check 12-1.

1. Define inductance.

2. What is a major characteristic of inductors?

3. What is the amplitude of the self-induced voltage across a 3 H coil whose current is changing at the rate of 250 mA per second?

4. Define the henry in terms of self-induced voltage and rate of current change.

5. How many joules of energy will a 5 H coil store in its magnetic field if the maximum coil current is 2 A?

12-2 Physical Factors Affecting Inductance

Physical Factors and a Formula

The amount of flux an inductor produces and the amount of voltage the inductor induces across itself is ultimately determined by the physical construction of the inductor. There are four primary physical factors that contribute to the inductance of a coil: (1) the number of turns of wire (N), (2) the absolute permeability of the core material (μ), (3) the cross-sectional area of the coil (A), and (4) the end-to-end length of the coil (l). Figure 12-2 illustrates these four factors. The mathematical relationship between these factors is given as follows (*Note:* There are many formulas used to calculate inductance depending on the type of inductor. This formula is a close approximation for coils whose length is at least 10 times the diameter.):

$$L = \frac{N^2\, \mu A}{l} \tag{12.5}$$

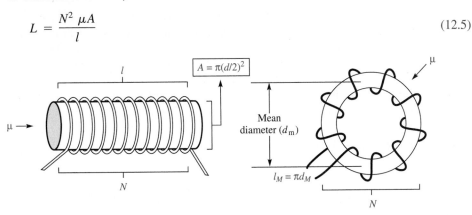

l = The length of the coil or core if core is longer than coil
N = The total number of turns
μ = The absolute permeability of the core
A = The cross-sectional area of the core

Figure 12-2   Physical factors of inductance.

where L is the inductance measured in henries (H), N is the number of turns, μ is the absolute permeability measured in teslas per ampere turns per meter (T/At/m), A is the cross-sectional area measured in square meters (m^2), and l is the coil length measured in meters (m).

Number of Turns (N)

As you can see in Formula 12.5, the number of turns of wire composing the coil has a significant effect on coil inductance, since the N factor is squared. Whether the coil is a single layer or multilayer coil, the total number of turns is used in the formula. Naturally, the greater the number of turns, the greater the inductance will be.

Absolute Permeability (μ)

The absolute permeability is also very significant. Recall that the **absolute permeability (μ)** of a core material is found by multiplying the material's **relative permeability (μ_r)** by the absolute permeability of air ($\mu_0 = 1.26\ \mu$T/At/m). As a review, if the relative permeability of a certain core material is known to be 200, its absolute permeability is 252 μT/At/m ($\mu = \mu_r \cdot \mu_0 = 200 \cdot 1.26\ \mu$T/At/m). The higher the permeability of the core material, the greater the inductance of the coil.

Area (A)

Inductors having large cross-sectional areas have greater inductance than coils with smaller cross-sectional areas. A larger cross-sectional area means the core is larger and there is more room for magnetic flux. More flux means more self-induced voltage (greater inductance).

Length (l)

The length of the coil is the only factor that affects coil inductance inversely. All else being equal, a long coil will have less inductance than a shorter coil. The compactness of the shorter coil provides shorter distances over which magnetic flux must span. As a result, the magnetic field is stronger or more intense. Since all windings are encompassed in a concentrated field, the inductance is increased.

Take a few minutes now to carefully consider the examples illustrated in Example Set 12-5 and Design Note 12-1.

Changing Inductance

Adjustable Cores

Some inductors have adjustable ferrite cores (**slugs**) that are threaded into the center of the coil form. The inductance is varied by changing the position of the ferrite slug inside the coil. If the ferrite slug is not all the way in the coil, some of the coil windings will have air as a core. This reduces the inductance of the coil, since the permeability of air is much less than that of ferrite. Thus, the overall permeability of the core within the center of the coil is varied by the position of the ferrite slug. This is known as **permeability tuning**.

EXAMPLE SET   12-5

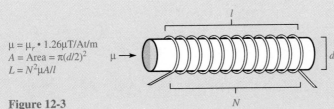

$\mu = \mu_r \cdot 1.26\mu T/At/m$
$A = \text{Area} = \pi(d/2)^2$
$L = N^2\mu A/l$

Figure 12-3

1. For Figure 12-3, $l = 0.1$ m, $d = 0.01$ m, $N = 50$ turns, and $\mu_r = 1$
   Find: L
   First find the absolute permeability and the cross-sectional area.
   $\mu = 1 \cdot 1.26 \ \mu T/At/m$
   $A = \pi \cdot (0.005 \ m)^2 = \pi \cdot 25 \ \mu m^2 = 78.5 \ \mu m^2$
   Now solve for L.

   $L = 50^2 \cdot 1.26 \ \mu T/At/m \cdot 78.5 \ \mu m^2/0.1 \ m$
   $\quad = 2,500 \cdot 1.26 \ \mu T/At/m \cdot 78.5 \ \mu m^2/0.1 \ m$
   $\quad = \underline{2.47 \ \mu H}$

2. For Figure 12-3, $l = 0.2$ m, $d = 0.15$ m, $N = 100$ turns, and $\mu_r = 150$
   Find: L
   First find the absolute permeability and the cross-sectional area.
   $\mu = 150 \cdot 1.26 \ \mu T/At/m = 189 \ \mu T/At/m$
   $A = \pi \cdot (0.075 \ m)^2 = \pi \cdot 5.625 \ mm^2 = 17.7 \ mm^2$
   Now solve for L.

   $L = 100^2 \cdot 189 \ \mu T/At/m \cdot 17.7 \ mm^2/0.2 \ m$
   $\quad = 10,000 \cdot 189 \ \mu T/At/m \cdot 17.7 \ mm^2/0.2 \ m$
   $\quad = \underline{0.167 \ H}$

Close Wound vs Space Wound

Another method of changing inductance is related to the length of the coil. If a coil's windings are wound tightly together (usually enamel-coated copper wire), the coil is said to be **close wound**. By **space winding** the coil, the inductance is reduced. This, of course, lengthens the coil as it is stretched out leaving air spaces between turns. The inductance of heavy-gauge, solid-wire (or tubing) air-core coils may be changed using this method.

Layer Wound

Layer winding is a method used to increase the inductance of a coil. By winding many layers of wire on the same coil form, the cross-sectional area and number of turns is increased without increasing coil length. Since the inductance increases directly with area and the number of turns squared, the increase in inductance using layer wnding is very significant.
   Time for a short review.

DESIGN NOTE 12-1: Inductance

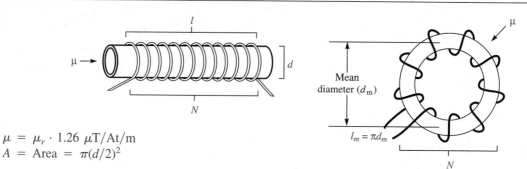

$\mu = \mu_r \cdot 1.26\ \mu T/At/m$

$A = \text{Area} = \pi(d/2)^2$

$L = N^2\mu A/l$ (where the ratio of $l/d > 10$)

$l = $ the length of the coil or core if core is longer than coil

"BASIC" PROGRAM SOLUTION

```
10 CLS
20 PRINT"INDUCTANCE"
30 PRINT"":PRINT""
40 PRINT"THIS PROGRAM WILL CALCULATE THE INDUCTANCE OF A STRAIGHT OR TOROIDAL INDUCTOR"
50 PRINT""
55 PRINT"SELECT INDUCTOR TYPE"
60 INPUT"(1) STRAIGHT  - (2) TOROIDAL ";C
70 IF C = 2 THEN GOTO 200
80 IF C<>1 AND C<>2 THEN GOTO 50
90 PRINT""
100 INPUT"ENTER THE LENGTH OF THE COIL OR CORE IN METERS-";L
110 INPUT"ENTER THE DIAMETER OF THE CORE-";D
120 INPUT"ENTER THE NUMBER OF COIL TURNS-";N
130 INPUT"ENTER THE RELATIVE PERMEABILITY OF THE CORE MATERIAL-";UR
140 U = UR * 1.26E-06
150 GOTO 300
200 PRINT""
210 INPUT"ENTER THE INNER DIAMETER OF THE TOROIDAL CORE-";ID
215 INPUT"ENTER THE OUTER DIAMETER OF THE TOROIDAL CORE-";OD
216 D = (OD - ID)
217 X = ID + (OD - ID)/2
220 L = 3.1416 * X
230 GOTO 120
300 PRINT""
310 A = 3.1416 * (D/2)^2
320 IND = N^2 * U * A/L
330 PRINT"THE INDUCTANCE OF THE COIL IS ";IND;" H."
360 PRINT""
370 INPUT"CALCULATE ANOTHER?   (Y/N)";A$
380 IF A$ = "Y" THEN CLEAR:GOTO 30
400 CLS:END
```

1. Given an inductor with a length of 0.2 m, a diameter of 0.02 m, 200 turns, and an iron core with a relative permeability of 180, find the inductance.

2. How many turns of the coil in #1 should you remove to reduce the coil's inductance to half? (Assume all other factors remain unchanged.)

12-3 Types of Inductors

Air-Core Inductors

Air-core inductors represent one large general category of inductors. They are inductors that, of course, have only air in the core of the coil ($\mu_r = 1$). For the most part, these inductors have inductances measured in the microhenries (μH). They may be fixed or variable. Air-core coils are usually used in radio and television circuits to help tune the particular circuit to a specific frequency. As you will see in a later chapter, changing inductance in a tuned circuit will change the frequency at which the circuit is sensitive and responsive.

Air-core inductors are manufactured in all shapes and sizes for many different applications. They may be wound on a resin, plastic, paper, or ceramic coil form. Others are made of thick wire or tubing and need no form at all. Many variable air inductors are mounted inside a special insulating frame. Figure 12-4 shows such an inductor. This is a heavy-duty air-core variable inductor used in radio-frequency transmitters. The close-up, Figure 12-4b, shows the rotating wiper arm inside the coil. As the shaft is rotated, more or fewer turns of the coil are used. The inductor shown is used to tune the final amplifier in a 1 kW or 2 kW transmitter. However, variable inductors of this type become man-size and larger in very-high-power international broadcasting transmitters ranging from 500 kW to several MW of output power.

Ferrite Core Inductors

Ferrite Cores

Ferrite-core inductors are also manufactured in a variety of shapes and sizes for many different applications. The inductance of ferrite-core coils ranges from microhenries (μH) to henries (H). Figure 12-5 shows three common types of ferrite cores. From left to right they are cup cores, beads, and toroids. The **cup cores** are designed so that the coil, wound on a plastic spool, is completely enclosed in the hollow middle of two cup halves. The result is total flux containment and shielding. This prevents the magnetic field of the coil from interfering with a neighboring coil and vice versa. Cup-core coils generally have inductances in the millihenries and are often used in audio filter circuits. **Toroid cores** are also self-shielding, with all flux contained inside the ferrite core. Toroidal coils are now being used for nearly all frequency ranges. Their inductances range from microhenries to henries. Ferrite **beads** are used for very small inductances in the nano and picohenry ranges (nH $= 10^{-9}$ H and pH $= 10^{-12}$ H).

(a)

(b)

Figure 12-4 (a) A variable air-core inductor. (b) Closeup showing the wiper arm for tuning.

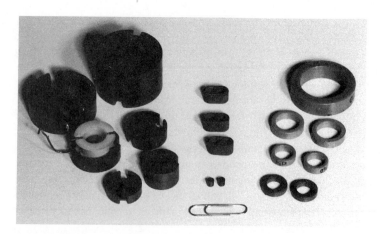

Figure 12-5 A variety of ferrite cores—cup, beads, toroidal.

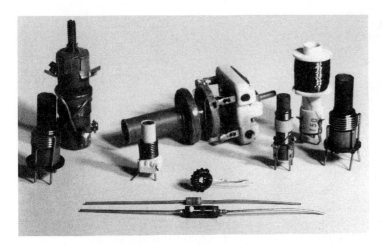

Figure 12-6   A variety of ferrite-core inductors for radio applications.

A Variety of Ferrite Inductors

Figure 12-6 shows a small variety of ferrite inductors, both fixed and variable. Many variable ferrite inductors have a tapped coil form into which the ferrite core (slug) is threaded. The position of the slug with respect to the coil determines the amount of inductance. Maximum inductance is obtained by positioning the slug completely within the length of the windings. Minimum inductance results when the slug is backed all the way out. Inductors shown in the photo range from μH to mH.

Laminated Iron-Core Inductors

Laminated iron-core inductors have a core made of many layers of thin iron plates (**laminations**). Each plate is electrically insulated with a baked-on enamel coating. The purpose of the laminations is to greatly reduce power losses in the core caused by eddy currents. **Eddy currents** are induced in the core by the expanding and collapsing magnetic field of the coil. The laminations interrupt the paths eddy currents normally take, thus eliminating unwanted power losses. Laminated iron-core inductors have inductances ranging in the millihenries to henries and are normally used for audio and power frequency filtering (50 and 60 Hz).

Chokes

Figure 12-7 shows a variety of **chokes**. As you can see, chokes are inductors. These inductors are referred to as *chokes* because they are designed for a special purpose: that of blocking AC while passing DC. The choke will choke off any AC signal that may be riding on a DC line. As an example, a laminated iron-core choke can be used in an automobile to block alternator noise (AC) from getting into a car radio via the $+12$ VDC wire. The DC passes and the alternator whine is blocked. Chokes are used in audio, radio, and television equipment for a similar purpose; to block unwanted AC noise and pass pure DC.

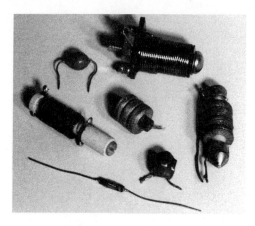

Figure 12-7   A variety of chokes that block AC and pass DC.

Answer the following questions before continuing to the next section.

SELF-CHECK 12-3

1. What is the general range of inductance for air-core inductors?
2. What is the advantage of ferrite-cup cores and toroid cores?
3. What is the purpose for the laminations in laminated iron-core inductors?
4. What is the purpose for a choke?

12-4 Total Inductance

Inductors in Series

Finding the total inductance of a series circuit, composed totally of inductors, is as simple as finding the total resistance of a series resistor circuit. You simply add all of the individual inductances. This is assuming there is no magnetic interaction between the inductors—i.e., they are shielded, self-shielding such as cup cores and toroids, or are separated sufficiently from one another. Consider the following formula and study Example Set 12-6.

$$L_{t(\text{series})} = L_1 + L_2 + L_3 + \cdots \tag{12.6}$$

where all inductances are expressed in henries (H).

EXAMPLE SET 12-6

1.

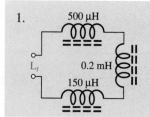

500 μH

L_t

0.2 mH

150 μH

Figure 12-8

$$L_t = L_1 + L_2 + L_3 + \cdots$$

$$L_t = 500\ \mu\text{H} + 0.2\ \text{mH} + 150\ \mu\text{H}$$
$$= \underline{850\ \mu\text{H}} = \underline{0.85\ \text{mH}}$$

2.
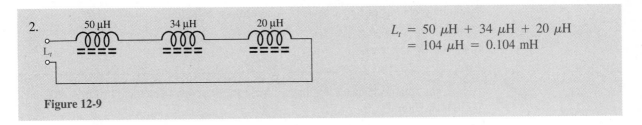

$$L_t = 50 \ \mu H + 34 \ \mu H + 20 \ \mu H$$
$$= 104 \ \mu H = 0.104 \ mH$$

Figure 12-9

Inductors placed electrically in parallel are also treated like resistors in parallel. In other words, you have your choice of three formulas for solving total parallel inductance: the same values formula, the product over the sum formula, and the reciprocal formula. Once again, we are assuming there is no magnetic interaction among the inductors. As before, consider the following formulas and study Example Set 12-7

$$L_{t(parallel)} = L/N \tag{12.7}$$

where L is the value of each and every inductance and N is the number of inductors having the same value.

$$L_{t(parallel)} = \frac{L_1 \cdot L_2}{L_1 + L_2} \tag{12.8}$$

where there are only two inductors in parallel.

$$L_{t(parallel)} = \frac{1}{1/L_1 + 1/L_2 + 1/L_3 + \cdots} \tag{12.9}$$

where there are two or more inductors in parallel.

EXAMPLE SET   12-7

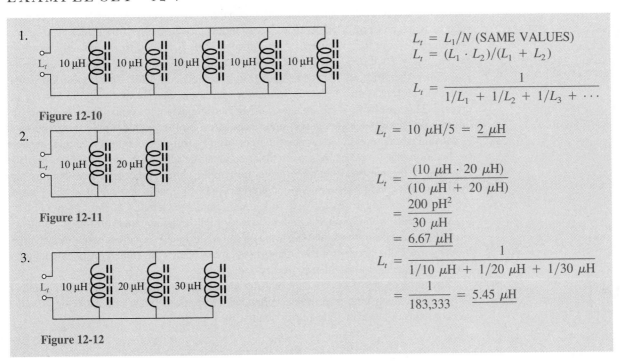

1.

Figure 12-10

2.

Figure 12-11

3.

Figure 12-12

$$L_t = L_1/N \ \text{(SAME VALUES)}$$
$$L_t = (L_1 \cdot L_2)/(L_1 + L_2)$$

$$L_t = \frac{1}{1/L_1 + 1/L_2 + 1/L_3 + \cdots}$$

$$L_t = 10 \ \mu H/5 = \underline{2 \ \mu H}$$

$$L_t = \frac{(10 \ \mu H \cdot 20 \ \mu H)}{(10 \ \mu H + 20 \ \mu H)}$$
$$= \frac{200 \ pH^2}{30 \ \mu H}$$
$$= \underline{6.67 \ \mu H}$$

$$L_t = \frac{1}{1/10 \ \mu H + 1/20 \ \mu H + 1/30 \ \mu H}$$
$$= \frac{1}{183,333} = \underline{5.45 \ \mu H}$$

Work the problems of Self-Check 12-4 before you continue.

SELF-CHECK 12-4

1. Find the total inductance for the following four series inductors: 100 mH, 200 mH, 50 mH, 700 mH

2. Find the total inductance for the following two parallel inductors: 3 H, 2.5 H

3. Find the total inductance for the following three parallel inductors: 900 mH, 1,500 mH, 500 mH

12-5 Stray Inductance

Inductance is not confined to components that are called inductors. All electrical and electronic components and conductors have some inductance. It is not necessary for a conductor to be looped into the shape of a coil for it to have inductance. A perfectly straight piece of wire has some inductance. These small inductances, which are not officially recognized or labeled in schematics, are classified as **stray inductances**. Bus wires, hookup wires, interconnect cables, wire leads on components, and copper foil on printed circuit boards all have some stray inductance.

Is that good or bad? That depends! Usually the effect stray inductance has on the operation of a circuit depends on the AC frequency that the circuit is handling. Stray inductance is usually very small (measured in nanohenries—nH). At very low frequencies, such as voice and power frequencies, stray inductance has very little effect. The reason you already know. Recall that the amount of CEMF is related to the rate of change in current: $v_L = L \cdot (\Delta i/\Delta t)$. If the AC is a low frequency, the rate of change in current is low. Since the stray inductance is also low, the induced CEMF and opposition to the change in current is small. As the frequency increases, the rate of change in current increases. This results in more induced CEMF and greater opposition. Thus, at some frequency, the stray inductance has a significant effect on the operation of the circuit, such as causing a reduction in signal amplitude or even causing the circuit to become unstable (a condition that would cause the circuit to oscillate, producing an unwanted AC signal).

That is the reason design engineers are careful about component lead lengths and miscellaneous conductor lengths—including copper foil paths on printed circuit boards. As a general rule, it is always wise to keep all conductor lengths as short as possible to avoid problems with stray inductances. In very-high-frequency and ultra-high-frequency circuits (frequencies above 30 MHz), the circuit is designed to include stray inductances. In many cases, the copper foils on the printed circuit boards are etched to specific lengths, widths, and shapes to provide the needed inductance.

Try to be aware of the concept of stray inductance as you assemble lab experiments and projects. Keep your wiring neat, organized, and short. Now, a quick review before you continue. Answer the questions of Self-Check 12-5.

1. Describe stray inductance in your own words.

2. At what frequencies are the effects of stray inductance most severe, very low or very high? Why?

12-6   Troubleshooting Inductors

Using the Ohmmeter to Test Windings

Low Resistance Is Probably Good—Open Is Bad

The ohmmeter is one of the most useful instruments used in troubleshooting inductors. With the ohmmeter, a good or bad determination can readily be established. Basically, the ohmmeter is being used as a continuity checker. In other words, the pointer on the meter face will deflect up-scale if ohmmeter current is able to flow through the inductor from one test lead to the other. In many cases, it may be assumed the inductor is probably good if there is continuity indicated by a very low resistance reading on the ohmmeter (analog or digital). If the ohmmeter indicates a very high, or infinite, resistance, the inductor is open, definitely bad. Figure 12-13 illustrates this basic test. As in testing resistors, the inductor must be isolated from its circuit by disconnecting at least one of its leads. Otherwise, an alias path may exist through neighboring components.

See color photo insert page D2, photograph 53.

Is It Really Good?

Does an indication of low resistance and continuity always mean the inductor is good? The answer is no. While it is true the inductor is merely wire, and therefore should have a very low resistance, the ohmmeter has no way of testing the inductor for correct value. The inductor may actually be out of tolerance and far from its marked or supposed value. This could be caused by shorted windings or a cracked ferrite core. An ohmmeter is not capable of testing for correct value. To test for correct value, a special test circuit can be

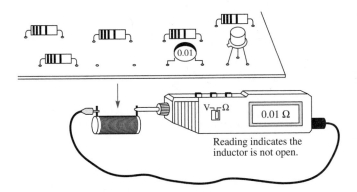

Figure 12-13   A simple ohmmeter test.

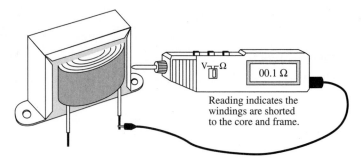

Figure 12-14   Windings-to-core short test—should read open.

assembled and a bench signal generator and oscilloscope may be used. Inductance meters and bridge circuits are available commercially that are capable of quickly determining the actual value of the inductor. However, the subjects of special circuits and special equipment are beyond the scope of this present discussion and must be held for a later chapter.

One More Ohmmeter Test

Figure 12-14 illustrates one more very important inductor ohmmeter test. Laminated iron-core inductors are often chassis mounted. This means the metal frame of the inductor will be at ground potential (grounded). The windings of the inductor should not be shorted to the laminated core or metal frame. The simple test illustrated in Figure 12-14 will reveal whether the windings are shorted to the frame. An open indication is desirable and should be observed on the ohmmeter.

Test your understanding of inductor troubleshooting by answering the question in Self-Check 12-6.

S E L F - C H E C K   1 2 - 6

1. Can an ohmmeter be used to determine if an inductor is really good? Explain.

Summary

FORMULAS

(12.1) $L = N (\Delta\Phi/\Delta i)$

> where L is inductance measured in henries (H), N is the number of turns of the inductor, $\Delta\Phi$ is the change in flux measured in webers (Wb), and Δi is the change in current measured in amperes (A).

(12.2) $V_{\text{ind}} = v_L = L \cdot (\Delta i/\Delta t)$

> where v_L is the self-induced voltage, L is the inductance of the coil measured in henries (H), Δi is a change in current measured in amperes (A), and Δt is a change in time measured in seconds (s).

(12.3) $L = \dfrac{v_L}{\Delta i / \Delta t}$

(12.4) $E = L \cdot I^2 / 2$

where E is the energy in joules, L is the inductance in henries (H), and I is the maximum inductor current.

(12.5) $L = \dfrac{N^2 \mu A}{l}$

where L is the inductance measured in henries (H), N is the number of turns, μ is the absolute permeability measured in teslas per ampere turns per meter (T/At/m), A is the cross-sectional area measured in square meters (m^2), and l is the coil length measured in meters (m).

(12.6) $L_{t(\text{series})} = L_1 + L_2 + L_3 + \cdots$

(12.7) $L_{t(\text{parallel})} = L/N$

where L is the value of each and every inductance and N is the number of inductors having the same value.

(12.8) $L_{t(\text{parallel})} = \dfrac{L_1 \cdot L_2}{L_1 + L_2}$

where there are only two inductors in parallel.

(12.9) $L_{t(\text{parallel})} = \dfrac{1}{1/L_1 + 1/L_2 + 1/L_3 + \cdots}$

where there are two or more inductors in parallel.

CONCEPTS

- Inductance, or self-inductance, is the measure of a coil's ability to induce a voltage across itself as a result of a change in current in its windings.
- A transient current is a current that is in the process of changing level, or amplitude.
- The amount of voltage self-induced across a coil depends upon the inductance of the coil (L) and the rate of change in current in the coil ($\Delta i / \Delta t$).
- One henry (1 H) is the amount of inductance required to self-induce a CEMF of 1 V with a 1 A/s change of current.
- It is the nature of an inductor to induce a voltage across itself, to oppose any change in current, and to store energy in the form of a magnetic field.
- The inductance of a coil varies directly with the square of the number of turns, directly with the permeability of its core material, directly with the cross-sectional area of its core, and inversely with the length of the coil, or length of the core if the core is longer than the coil.
- Close winding and layer winding increases inductance as compared to space winding.
- Chokes are used to block AC and pass DC.
- The inductance of wiring, component leads, and copper foil is known as stray inductance.
- An ohmmeter can be used to show that an inductor is definitely open, and therefore bad. However the ohmmeter cannot show that the inductor is the correct value.

SPECIAL TERMS

- Self-inductance, inductance (L), henry (H)
- Transient current (i)
- Self-induced voltage (v_L)
- Absolute permeability (μ), relative permeability (μ_r)
- Permeability tuning, slug
- Close winding, space winding, layer winding
- Cup core, toroid core, beads
- Laminated iron core, laminations, eddy currents
- Chokes
- Stray inductance

Need to Know Solution

1. It is the nature of an inductor to **induce** a voltage across **itself**.
2. It is the nature of an inductor to **oppose** any change in **current**.
3. It is the nature of an inductor to store **energy** in its **magnetic field**.

Questions

12-1 The Nature of Inductance

1. Define *inductance*.
2. Does the current in a coil reach a maximum level instantly when an EMF is applied? Explain.
3. What is a transient current?
4. Will an inductor oppose any current or any *change* in current? Explain the difference.
5. Define the henry (H) in terms of induced voltage and rate of change in current.
6. How does an inductor store energy?
7. It is the nature of an inductor to _____ a voltage across itself, to oppose any _____ in current, and to store energy in the form of a _____ _____ .

12-2 Physical Factors Affecting Inductance

8. List the four physical factors that affect the inductance of a coil.
9. Will increasing only the length of a coil (a) increase or (b) decrease the inductance of the coil?
10. Will increasing the cross-sectional area of a coil (a) increase or (b) decrease the inductance of the coil?
11. What is permeability tuning?

12-3 Types of Inductors

12. In what kind of circuits are air-core coils usually used?
13. What is a cup core?
14. What do toroid cores have in common with cup cores?
15. Describe the maximum inductance position of the threaded ferrite core in a variable ferrite inductor.
16. What are laminations in a laminated iron core? What is their purpose?

12-5 Stray Inductance

17. Define and give three examples of stray inductance.
18. Explain why stray inductance has a greater effect as frequency is increased.
19. What precautions should you take in regard to stray inductance as you assemble lab experiments or projects?

Problems

12-1 The Nature of Inductance

1. Find the inductance of a 50 turn coil that produces a 0.2 μWb change in flux for a 1 A change in current.
2. Find the inductance of a 100 turn coil that produces a 0.5 μWb change in flux for a 100 mA change in current.
3. Find the voltage induced across a 5 mH inductor when the inductor's current changes at the rate of 250 mA/s.
4. Find the voltage induced across a 0.5 H inductor when the inductor's current changes at the rate of 0.3 A/s.
5. What is the inductance of a coil whose self-induced voltage is 2 V as a result of a 0.5 A/s change rate of current?
6. What is the inductance of a coil whose self-induced voltage is 50 mV as a result of a 200 mA/s change rate of current?
7. If the maximum current through a 500 mH inductor is found to be 250 mA, how much energy will the inductor store (in joules)?
8. If the maximum current through a 1 H inductor is found to be 2 A, how much energy will the inductor store (in joules)?

12-2 Physical Factors Affecting Inductance

9. A certain coil is 10 cm long, has an air core, has 40 turns, and has a cross-sectional diameter of 1 cm. What is the inductance of the coil?
10. A certain coil is 20 cm long, has a ferrite core whose relative permeability is 150, has 200 turns, and a cross-sectional diameter of 1.5 cm. What is its inductance?
11. If a long 100 μH inductor is cut exactly in half, what will the inductance of each half be?
12. If a long 25 μH inductor is cut into four inductors of equal length, what will the inductance of each quarter be?

12-4 Total Inductance

13. Find the total inductance of the following inductances placed in *series*: 45 mH, 100 mH, 500 μH, 10 mH
14. Find the total inductance for the following inductances placed in *series*: 2 H, 1.5 H, 750 mH
15. Find the total inductance for the following inductances placed in *parallel*: 45 mH, 100 mh, 500 μH, 10 mH
16. Find the total inductance for the following inductances placed in *parallel*: 2 H, 1.5 H, 750 mH
17. Find the total inductance for the following inductances placed in *parallel*: 10 H, 10 H

18. Find the total inductance for the following inductances placed in *parallel*: 5 H, 8 H, 5 H, 8 H
19. Find the total inductance for the following inductances placed in *parallel*: 100 mH, 100 mH, 25 mH, 50 mH
20. Find the total inductance for the following inductances placed in *parallel*: 1 H, 2 H, 3 H, 4 H, 5 H

Answers to Self-Checks

Self-Check 12-1

1. Inductance—the measure of a coil's ability to induce a voltage across itself as a result of a change in current in its windings.
2. Inductors oppose any change in current.
3. 0.75 V
4. One henry is the amount of inductance required to self-induce a CEMF of 1 V with a 1 A/s rate change in current.
5. 10 J

Self-Check 12-2

1. 14.25 mH
2. Remove 58.6 turns and leave 141.4 turns. The only factor that is changed is the number of turns. Therefore, the remaining 141.4 turns are stretched out to the original length.

Self-Check 12-3

1. In the microhenries (μH)
2. Total flux containment and shielding
3. To reduce losses due to eddy currents
4. To block AC and pass DC

Self-Check 12-4

1. 1,050 mH = 1.05 H
2. 1.364 H
3. 264.7 mH

Self-Check 12-5

1. Small inductances contributed to the circuit by lead lengths, copper foil paths, and wiring.
2. Very high—because the rate of change in current is higher at the higher frequencies, which causes the self-induced CEMF to be greater (as demonstrated in Faraday's formula)

Self-Check 12-6

1. No. It is not always possible to determine if some of the windings are shorted together. Also, the ohmmeter cannot determine if the inductor is the correct value (i.e., the core may be cracked).

SUGGESTED PROJECTS

1. Add some of the main concepts and formulas from this chapter to your Electricity and Electronics Notebook.

2. Remove the cover from a radio and see how many inductors you can identify. If the radio is AC-powered, be sure to unplug the radio before removing the cover.

3. Obtain a random selection of inductors and examine their construction. Also, measure the resistance of their windings with your ohmmeter.

Edison is seen here working in his optics laboratory. This amazing man's knowledge covered major topics including electricity, chemistry, light, optics, photography, motors, generators, mechanical mechanisms, botany, and physics. The invention of the kinetoscope was a natural step for him combining his knowledge of electric lighting, electric motors, optics, and mechanics.

Chapter 13

Capacitance

CHAPTER OUTLINE

OBJECTIVES

After studying this chapter, you will be able to

- describe the electrical property called *capacitance.*
- explain how an electrical charge and energy is stored in a capacitor.
- explain the relationship between flux density, electric field intensity, and dielectric permittivity.
- calculate the quantity of charge, capacitance, charge voltage, or stored energy of a capacitor.
- calculate capacitance when physical factors are known.
- list major capacitor types and describe their uses.
- calculate the total capacitance of series and parallel capacitor circuits.
- explain stray capacitance.
- troubleshoot capacitors with an ohmmeter and a capacitance meter.
- list many of the similarities and differences between inductors and capacitors.

INTRODUCTION

In this chapter, you will discover an electrical property that is opposite in nature to inductance. That property is capacitance. You will learn what capacitance is and how it is characterized.

Capacitors are very important components in both AC and DC circuits. They are every bit as important as resistors and inductors. Their uniqueness makes them indispensable in virtually every piece of electronic equipment, from transistor radios to satellites.

NEED TO KNOW

In the introduction to this chapter, it is stated that capacitors are very important components in AC and DC circuits. You might be wondering why they are so important and how they are used. That is a little difficult to discuss if you are not already somewhat familiar with basic electronic circuits such as amplifiers and power supplies. Without capacitors, electronic circuits such as these would be of little use. Consider the following list of capacitor applications.

- Filter capacitors in DC power supplies
- AC coupling capacitors in amplifier circuits
- Audio- and radio-frequency filter capacitors
- Bypass capacitors in amplifiers
- Timing capacitors in timer circuits

It is probably safe to say that you are not familiar with any of the listed capacitor applications. At this point, it is only important that you realize that capacitors do play a significant role in many areas of electronics. As you study this chapter, keep the following questions in mind and seek answers to them: What is a capacitor? How are capacitors made? How do they work? How are they used?

13-1 The Nature of Capacitance

Capacitance and Capacitors

Capacitance (C)

Capacitance (C) is the measure of a capacitor's ability (capacity) to store an electric charge on its plates. This simple definition requires a significant amount of explanation. Let's begin with a basic description of capacitor construction and theory. Figure 13-1 illustrates simple capacitor construction. As shown, a capacitor is an electrical component that consists of two conductors (**plates**) separated by an insulator (the **dielectric**).

Capacitance and the Electric Field

When a capacitor is connected to a DC voltage source, as shown in Figure 13-2, free and flowing electrons are conducted to one of the plates. As electrons accumulate on one of

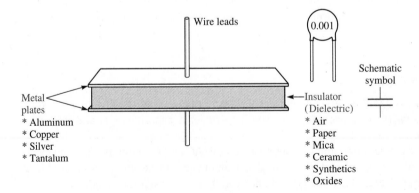

Figure 13-1   A capacitor is formed when two conductors are separated by an insulator.

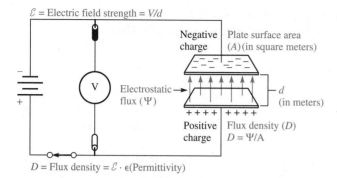

$\mathscr{E}$ = Electric field strength = V/d

Negative charge

Plate surface area (A) (in square meters)

Electrostatic flux (Ψ)

d (in meters)

Positive charge

Flux density (D)
$D = \Psi/A$

V

D = Flux density = $\mathscr{E} \cdot \epsilon$ (Permittivity)

Figure 13-2   Capacitance and the electric field.

the plates, the opposite plate becomes positively charged by **electrostatic induction** (the presence of the negative charge on one plate repels negative charges on the neighboring plate). Thus, the opposite plate receives an equal and opposite positive charge of holes.

Two things occur at the same time when the switch is closed. First, a large quantity of current flows in the circuit as though the circuit were completed with a solid, uninterrupted conductor (a short). The induced charge on the positive plate, caused by repulsion of electrons away from the plate, gives the *appearance* of current through the insulator (dielectric). Second, as electrons accumulate on one plate and a positive charge is induced on the other, a difference of potential (voltage) is developed on the capacitor across its plates. This capacitor voltage increases as positive and negative plate charges continue to accumulate. As the capacitor charges, the rate of charge steadily decreases to zero, at which time the capacitor is fully charged. Charge current ceases because the capacitor voltage matches the source voltage. The capacitor acts as a voltage source itself, countering the applied source voltage.

The difference of potential between the two parallel plates of the capacitor produces an electric field. The **electric field** is similar to a magnetic field in that flux lines containing energy extend between the plates. These electrostatic flux lines are conventionally thought of as extending from the positive charge to the negative charge (positively charged plate to the negatively charged plate). The **electrostatic flux** is symbolized with the Greek letter psi (Ψ). Since the plates are uniform and parallel, the flux lines are evenly distributed with a constant **flux density (D)**. Flux density is expressed as the number of flux lines per unit area. In this case, the standard unit for area is the square meter. Therefore, flux density is expressed as flux per square meter ($D = \Psi/m^2$).

The density of the flux in a capacitor will depend upon two principal factors: the field intensity ($\mathscr{E}$) and the permittivity (ϵ) of the dielectric. **Electrostatic field intensity** is similar to magnetic field intensity. It is a pressure that is producing the flux in the presence of opposition from the medium through which the flux must pass. While magnetic field intensity is measured in At/m, electrostatic field intensity is measured in V/m (volts per meter). Electrostatic field intensity depends on the amount of capacitor charge voltage and the distance between the two plates (where $\mathscr{E} = V/d$). The opposing medium is an insulator (vacuum, air, ceramic, glass, etc.). If one medium, or dielectric, will permit more flux to pass than another, it is said that it has a higher **permittivity** (similar to permeability of magnetic-core materials). Therefore, a higher field intensity and/or a higher permittivity will result in a higher flux density. The following formula for electrostatic flux density shows the mathematical relationship between field intensity and permittivity:

$$D = \mathscr{E} \cdot \epsilon \qquad\qquad (13.1)$$

where D is the flux density in flux per square meter, $\mathscr{E}$ is the field intensity in V/m, and ϵ is the permittivity of the dielectric material in farads per meter (F/m).

Capacitor Charge and Discharge

Notice in Figure 13-3 how the capacitor voltage increases over a short period of time once the switch is set to position A. The voltage across the capacitor does not jump instantly to a maximum value (equal to the applied source voltage). But why doesn't it? When the capacitor is fully charged, there is a large quantity (Q) of electrons on one plate. These electrons arrive at the plate through circuit resistance (wire resistance, source resistance, resistors, etc.), and as you know, resistance limits the amount of current in the circuit. Thus, the total capacitor charge arrives a little at a time over a period of time. Once more, as the charge increases, the capacitor voltage increases in a polarity that opposes the applied

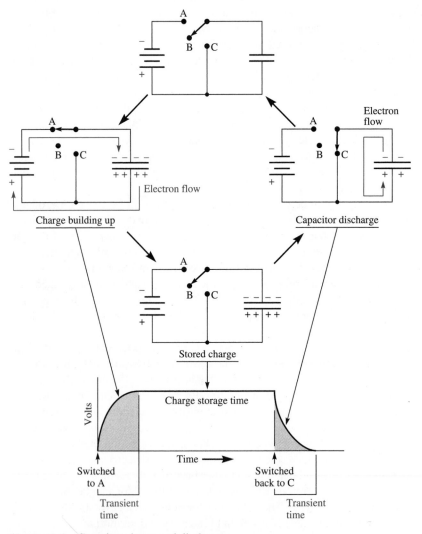

Figure 13-3 Capacitor charge and discharge.

source voltage. As a result, the charge current is delivered at an ever-decreasing rate through any circuit resistance. The amount of current at any instant of time is the rate of charge at that instant of time. In other words, the change in quantity of electrons for a very short change in time is the current, or rate of charge, for that time ($i = \Delta Q/\Delta t$). By the time the capacitor is fully charged, the charge current will have dropped to zero. Therefore, *current only flows during the time the capacitor is charging or discharging.*

Once charged, the capacitor remains charged until a conductive path is provided that permits the capacitor plates to neutralize. With the switch in Figure 13-3 set to position B, the charge will remain on the plates of the capacitor for an indefinite length of time (assuming a perfect insulator between the capacitor's plates).

When the switch is set to position C, a conductive path is provided for electrons to flow to the positive side of the capacitor. Remember, the positive plate is positive because it is greatly lacking in electrons. The quantity of electrons needed to bring the positive plate back to a neutral potential is exactly the same as the surplus of electrons on the negatively charged plate ($\# - e = \# + $ holes). The rate of discharge is determined by circuit resistance and capacitor voltage. The discharge rate constantly decreases, since the capacitor voltage is decreasing during discharge. When this transient period is past, the capacitor voltage will be zero and, of course, the circuit current will be zero.

The Farad (F)

C, Q, and V

The **farad (F)** is the standard unit of measure for capacitance. But, what is a farad? The farad is a very large unit of capacitance defined as the amount of capacitance needed by a capacitor to store one coulomb (1 C) of electrons under the influence of one volt. In other words, if a capacitor is able to store one coulomb of charge for every volt of charge voltage, its capacitance is said to be one farad (1 F). The following formula is useful in determining capacitance when the quantity of charge and voltage are known:

$$C = Q/V \tag{13.2}$$

where C is the capacitance measured in farads (F), Q is the **quantity of charge** measured in coulombs (C), and V is the voltage of the capacitor. Consider Example Set 13-1.

EXAMPLE SET   13-1

1. A capacitor charges to a potential of 10 V while storing 5 C of electrons on one plate. The capacitance is 0.5 F ($C = Q/V = 5$ C/10 V).

2. A capacitor charges to a potential of 0.3 V while storing 1 C of electrons on one plate. The capacitance is 3.33 F ($C = Q/V = 1$ C/0.3 V).

Formula 13.2 may also be rearranged to solve for quantity of charge (Q) or voltage as follows:

$$Q = C \cdot V \tag{13.3}$$

$$V = Q/C \tag{13.4}$$

Consider Example Set 13-2.

EXAMPLE SET   13-2

1. A 0.03 F capacitor is charged to 5 V. The quantity of stored charge is 0.15 C ($Q = CV = 0.03$ F $\cdot$ 5 V).
2. 0.002 C of charge is stored on the plates of a 0.04 F capacitor. The voltage across the capacitor is 0.05 V or 50 mV ($V = Q/C = 0.002$ C$/0.04$ F).

Smaller Units of Capacitance

The farad (F) is a very large unit of capacitance. While some manufacturers do make 1 F capacitors, most practical capacitors are measured in units of microfarads (μF) or picofarads (pF). For example, a DC power supply might use a relatively large amount of capacitance of 2,000 to 10,000 μF to filter the DC (remove any voltage and current variations (ripple) that normally exist in the conversion from AC to DC). Computer power supplies may use filter capacitors ranging from 10,000 to 100,000 μF. On the other extreme, radio and television circuits use capacitors valued in the range of a few picofarads (pF) for very-high-frequency tuned circuits. Some of the most common general-purpose values you will work with are 10 μF, 0.1 μF, and 0.01 μF. These common values are used to filter, bypass, and couple AC in countless circuits.

Capacitance and Stored Energy

We have already mentioned that capacitors store electrical energy. This energy can be viewed as being stored in the flux between the plates, since it is the electrostatic field that prevents any accumulation of illegitimate charge on the plates, thereby maintaining the established difference of potential (i.e., electrons cannot arbitrarily return to the positively charged plate, since the negative charge on the opposite plate will continually repel electrons from the positively charged plate).

While the electrostatic flux are important in maintaining the charge and difference of potential, it is important to remember that they are the result of the presence of the accumulated charge. Thus, it is the accumulated charge and subsequent difference of potential that is used to calculate the amount of stored energy. As you know, the unit used for electrical energy is the joule (J). Recall from Chapter 4 that one joule is equal to one volt-coulomb. In other words, the number of joules of electrical energy is determined by multiplying voltage by the quantity of charge ($E = V \cdot Q$). In the case of a capacitor, the stored energy is equal to the amount of work required to charge it. As the capacitor charges, the amount of work increases. This is because the capacitor develops a voltage that opposes the charging voltage. As the capacitor charge increases, so does the opposition it offers to further accumulation of charge. When the capacitor first begins to charge, there is no opposition (no capacitor voltage). When the capacitor is fully charged, there is maximum opposition and no further charge current flows. Thus, the average opposition against which the charge must work is one half the total charge voltage ($V/2$). Using this average voltage, the total stored energy may be calculated as follows:

$$E = Q \cdot V/2 \tag{13.5}$$

where E is the stored energy measured in joules (J), Q is the quantity of charge measured in coulombs (C), and V is the capacitor voltage. Consider Example Set 13.3.

EXAMPLE SET   13-3

1. A capacitor stores 0.03 C of charge on a plate and the difference of potential is 5 V. The amount of stored energy is 0.075 J ($E = Q \cdot V/2 = 0.03$ C $\cdot$ 5 V/2).
2. A capacitor stores 0.001 C of charge on a plate and the difference of potential is 100 mV. The amount of stored energy is 50 μJ ($E = Q \cdot V/2 = 0.001$ C $\cdot$ 100 V/2 = 50 μJ).

Formula 13.3 may be used to transform Formula 13.5 into a more practical and useful formula. Since $Q = C \cdot V$, $C \cdot V$ may be substituted for Q in Formula 13.5. This results in $E = C \cdot V \cdot V/2$ or

$$E = C \cdot V^2/2 \qquad\qquad (13.6)$$

Consider Example Set 13-4.

EXAMPLE SET   13-4

1. A 50 μF capacitor is charged to 10 V. The amount of stored energy is 2.5 mJ ($E = C \cdot V^2/2 = 50$ μF $\cdot$ (10 V)^2/2 = 2.5 mJ).
2. A 5,000 μF capacitor is charged to 40 V. The amount of stored energy is 4 J ($E = C \cdot V^2/2 = 5,000$ μF $\cdot$ (40 V)^2/2 = 4 J).

Summary

It is the nature of a capacitor to

- store a quantity of charge on its plates.
- adjust the amount of charge on its plates in accordance with any change in applied source voltage.
- block DC and pass AC, since DC provides a fixed charge and AC is continually changing, causing the plates to continuously adjust their charge.
- store energy in the electrostatic flux between its plates (the amount of flux depending on the quantity of charge).

Review this section. Then test your understanding by answering the questions of Self-Check 13-1.

SELF-CHECK   13-1

1. Define *capacitance*.

2. Define the farad.

3. If 0.01 C of electrons are stored on a plate of a capacitor whose voltage is measured to be 12 V, what is the capacitance of the capacitor?

(Continued next page)

4. Is 10,000 μF a large amount of capacitance or a small amount?

5. If the voltage across a 1,000 μF capacitor is 24 V, what is the quantity of charge measured in coulombs?

6. Convert 22,000 pF to μF.

7. How much energy (in joules) is stored in a 100 μF capacitor that is charged to 40 V?

13-2 Physical Factors Affecting Capacitance

Physical Factors and a Formula

There are three physical factors that contribute to the capacitance of the capacitor: (1) the area of its plates, (2) the absolute permittivity of its dielectric, and (3) the spacing, or distance, between the two plates. Figure 13-4 illustrates these three factors; the mathematical relationship between them is given as follows:

$$C = \epsilon \cdot A/d \qquad\qquad (13.7)$$

where C is the capacitance measured in farads (F, μF, pF), ϵ is the absolute permittivity of the dielectric measured in farads per meter (F/m), A is the area of one plate in square meters, and d is the distance between the plates measured in meters (m).

Area (A)

The surface area of the plates is directly related to the amount of capacitance. The larger the surface area, the greater the capacitance. The area of one plate is used in the formula. Plate area must be expressed in square meters.

Absolute Permittivity (ϵ)

The higher the absolute permittivity, the greater the capacitance. But what is absolute permittivity? The concept of permittivity is very similar to permeability. Recall that a given core material has a relative permeability (permeability relative to air or vacuum) and an absolute permeability. In like manner, all insulators have relative and absolute permittivities. The **relative permittivity**, often referred to as the **dielectric constant**, is related to the permittivity of a vacuum. The **absolute permittivity** of a vacuum is the

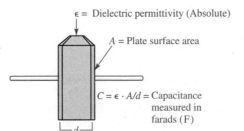

ϵ = Dielectric permittivity (Absolute)

A = Plate surface area

$C = \epsilon \cdot A/d$ = Capacitance measured in farads (F)

Figure 13-4 Physical factors that determine the amount of capacitance.

T A B L E   1 3 - 1   Dielectric Materials and Permittivity

Material	Relative Permittivity— (Dielectric Constant—K)	Absolute Permittivity—ϵ $\epsilon = \epsilon_r \cdot \epsilon_0$[†]
		pF/m
Vacuum	1	8.85
Air	1.0006	8.855
Teflon	2.1	18.59
Polyethylene	2.3	20.35
Polystyrene	2.6	23.01
Rubber	3 – 5	26.55 – 44.25
Paper	3 – 4	26.55 – 35.40
Mylar	3	26.55
Polyester	4	35.40
Glass (Pyrex, Window)	5 – 8	44.25 – 70.8
Mica	5 – 8	44.25 – 70.8
Bakelite	6 – 7	53.1 – 61.95
Porcelain	5 – 7	44.25 – 61.95
Aluminum oxide	7	61.95
Tantalum oxide	25	221.3
Water (distilled)	80	708.0
Ceramic		
(Titanium dioxide)	10 – 110	88.5 – 974.0
(Barium – Strontium Titanate)	to 7500	to 66,380

[†]$\epsilon_0 = 8.85$ pF/m

constant 8.85 pF/m (picofarads per meter). Therefore, the absolute permittivity of all other insulators (dielectrics) is based on the absolute permittivity of a vacuum. If the relative permittivity (dielectric constant) of a dielectric is known, the following formula is used to determine its absolute permittivity:

$$\epsilon = \epsilon_r \cdot \epsilon_0 = \epsilon_r \cdot 8.85 \text{ pF/m} \tag{13.8}$$

where ϵ is the absolute permittivity of the dielectric measured in F/m, ϵ_r is the relative permittivity of the dielectric (having no units), and ϵ_0 is the absolute permittivity of a vacuum, equal to 8.85 pF/m.

Table 13-1 is a listing of common insulators used as capacitor dielectrics. The dielectric constants shown in the table are the relative permittivities for the various materials. The absolute permittivities are also given.

Distance (d)

The distance between the capacitor's plates is inversely related to the capacitance. The greater the spacing, the lower the capacitance. The distance must be expressed in meters (m) in Formula 13.7.

Study the examples of Example Set 13-5 to see how capacitance is calculated when the physical factors are known. Then, study Design Note 13-1.

EXAMPLE SET 13-5

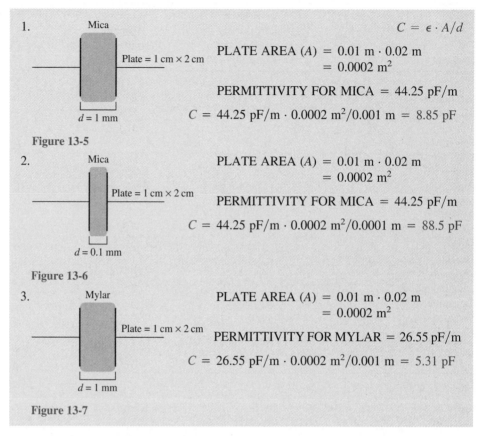

1.

Mica

Plate = 1 cm × 2 cm

$d = 1$ mm

Figure 13-5

$C = \epsilon \cdot A/d$

PLATE AREA (A) = 0.01 m · 0.02 m
= 0.0002 m²

PERMITTIVITY FOR MICA = 44.25 pF/m

C = 44.25 pF/m · 0.0002 m²/0.001 m = 8.85 pF

2.

Mica

Plate = 1 cm × 2 cm

$d = 0.1$ mm

Figure 13-6

PLATE AREA (A) = 0.01 m · 0.02 m
= 0.0002 m²

PERMITTIVITY FOR MICA = 44.25 pF/m

C = 44.25 pF/m · 0.0002 m²/0.0001 m = 88.5 pF

3.

Mylar

Plate = 1 cm × 2 cm

$d = 1$ mm

Figure 13-7

PLATE AREA (A) = 0.01 m · 0.02 m
= 0.0002 m²

PERMITTIVITY FOR MYLAR = 26.55 pF/m

C = 26.55 pF/m · 0.0002 m²/0.001 m = 5.31 pF

Changing Capacitance

In order to change the capacitance of a capacitor, you naturally must be able to change one or more of the three physical factors. In the next section, we will discuss various capacitor types. Many capacitors are fixed in value as determined by the manufacturer. These capacitors are sealed and there is no convenient way to modify their capacitance. However, variable capacitors offer some means by which a physical factor may be adjusted. Usually it is the spacing between the plates that is adjusted, or the overlapping areas of two plates (one plate is fixed while the other plate is movable or rotatable).

Time for a short review. Take a moment to answer the questions of Self-Check 13-2.

SELF-CHECK 13-2

1. List the three physical factors that affect capacitance.

2. If a dielectric has a relative permittivity of 55, what is its absolute permittivity?

3. A capacitor's plate dimensions are 0.01 m by 0.02 m, the plate spacing is 0.0001 m, and the relative permittivity of the dielectric is 20. Find the capacitance.

4. What effect does increasing plate spacing have on capacitance?

DESIGN NOTE 13-1: Capacitance

ϵ = Dielectric Permittivity = $\epsilon_r \cdot \epsilon_0$

ϵ_r = RELATIVE PERMITTIVITY = DIELECTRIC CONSTANT = K

ϵ_0 = ABSOLUTE PERMITTIVITY OF VACUUM = 8.85 pF/m

A = Plate Surface Area
(in square meters)

d (in meters)

$C = \epsilon \cdot A/d$ = CAPACITANCE MEASURED IN FARADS (F)

"BASIC" PROGRAM SOLUTION

```
10  CLS
20  PRINT"CAPACITANCE"
30  PRINT"":PRINT""
40  PRINT"THIS PROGRAM WILL CALCULATE ANY OF THESE PARAMETERS:"
50  PRINT"CAPACITANCE - PLATE AREA - PLATE SPACING - DIELECTRIC CONSTANT (K)"
55  PRINT""
60  PRINT"********** ENTER A VALUE FOR ALL KNOWN PARAMETERS ************"
70  PRINT"*** SIMPLY PRESS RETURN (ENTER) FOR THE UNKNOWN PARAMETER ***"
80  PRINT""
90  INPUT"ENTER THE DESIRED CAPACITANCE IN FARADS. USE E NOTATION. ";C
100 INPUT"ENTER THE AREA OF ONE PLATE IN SQUARE METERS. ";A
110 INPUT"ENTER THE DISTANCE BETWEEN THE PLATES. ";D
120 INPUT"ENTER THE DIELECTRIC CONSTANT (K). ";K
130 E = 8.85E-12 * K
140 PRINT""
150 IF C = 0 THEN C = E * A/D:PRINT"CAPACITANCE = ";C;" FARADS."
160 IF A = 0 THEN A = C * D/E:PRINT"PLATE AREA = ";A;" SQUARE METERS."
170 IF D = 0 THEN D = E * A/C:PRINT"PLATE SPACING = ";D;" METERS."
180 IF K = 0 THEN E = C * D/A:K = E/8.85E-12:PRINT"DIELECTRIC CONSTANT (K) = ";K
190 PRINT""
200 INPUT"ANOTHER PROBLEM? (Y/N)";A$
210 IF A$ = "Y" THEN CLEAR: GOTO 55
220 CLS:END
```

13-3 Types of Capacitors

Fixed-Value Capacitors

Mica Capacitors

Mica capacitors are fabricated in a **stacked-sandwich** type of construction as illustrated in Figure 13-9. **Mica** is a mineral that can be separated into thin sheets. These thin sheets are an excellent dielectric that has a very high dielectric resistance (>1 GΩ) and very little dimensional change with changes in temperature. The mica sheets are placed between metal foil plates, leads are attached, and the entire stack is **hermetically sealed** (total airtight encapsulation). The capacitance of a capacitor constructed in this manner is

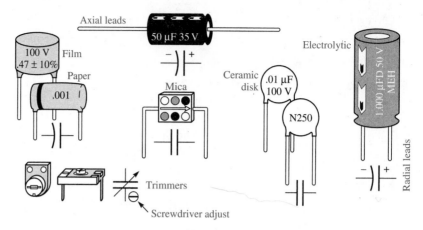

Figure 13-8

determined by the number of capacitor sections, the surface area of a plate, the thickness of the dielectric, and the dielectric constant (permittivity). The overall plate area is the surface area of a single plate times the number of dielectric layers (in this case, mica layers). Values of capacitance range from a few picofarads (pF) to thousands of picofarads. Voltage ratings range from hundreds to thousands of volts. Mica capacitors are very temperature-stable and are used in circuits that require this stability, such as critically tuned circuits found in radio receivers and transmitters.

Paper Capacitors

Paper capacitors use a very thin tissue paper as the dielectric and thin foil for the plates. The paper is thoroughly impregnated with paraffin to prevent infiltration of moisture. These capacitors are fabricated in a **rolled-sandwich** construction as illustrated in Figure 13-10. Two long strips of metal foil, separated by the waxed paper, are tightly rolled and encased

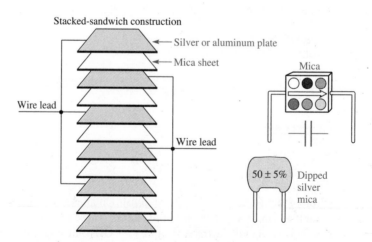

Figure 13-9 The stacked-sandwich construction of mica capacitors.

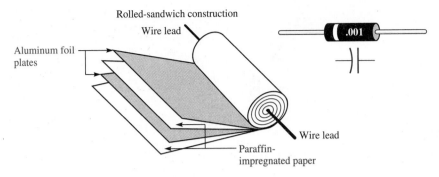

Figure 13-10   The rolled-sandwich construction of paper capacitors.

in a sealed tube or hermetically sealed by being dipped in a resin compound. Of course, wire leads are connected to the foils for circuit connection. The body of capacitors fabricated in this fashion is often marked to indicate which wire lead is connected to the outer foil. Paper capacitors range in value from hundreds of picofarads to about one microfarad, with voltage ratings in the hundreds of volts. Paper capacitors are not as temperature-stable as mica capacitors. Thus, they are not used in applications where extreme stability is required. However, they are used for signal coupling and bypassing.

Ceramic Capacitors

Ceramic capacitors use an oven-fired ceramic dielectric. The ceramic compound can be mixed to tailor the relative permittivity (ϵ_r = dielectric constant = K) to any value from below 100 to thousands. This means a large range of capacitor values is possible in standard fixed-dimension packages. For any given value of capacitance and voltage rating, the ceramic capacitor is smaller than paper, mica, and many other capacitor types. The type of ceramic compound used in the capacitor is very important in terms of temperature stability and capacitance value. As a general rule, ceramic compounds of very high permittivity are not temperature-stable. In general, ceramic capacitors range in value from less than 1 pF to 1 μF, with voltage ratings ranging to tens of thousands of volts. Dielectric resistance for most ceramics exceeds 1 GΩ.

Ceramic capacitors are a rather large family of capacitors available in a variety of styles. Figures 13-11 and 13-12 illustrate the most common styles of ceramic capacitors. The **disc** style is a very common general-purpose capacitor, available in a wide range of capacitance, voltage, and temperature stability. Its construction is very straightforward. A ceramic disc is silver-plated on each side. Wire leads are attached on each side and the entire unit is encapsulated in a resin compound. **Tubular** ceramic capacitors are similar to the discs, except the ceramic compound is formed into a tubular shape. A silver coating is placed on the inner wall and outer surface of the tube, thus forming the two plates. Tubular ceramic capacitors are usually of higher voltage, lower value (<1 pF to hundreds of pF), and tighter tolerance than the disc style. **Monolithic (chip-style)** ceramic capacitors are fabricated using stacked-sandwich construction. Layers of ceramic dielectric are alternated with metal plating, and the entire sandwich is fired at high temperatures to form a single monolithic component. This style of fabrication offers a very high amount of capacitance in a given volume of space. As a result, chip capacitors have become very

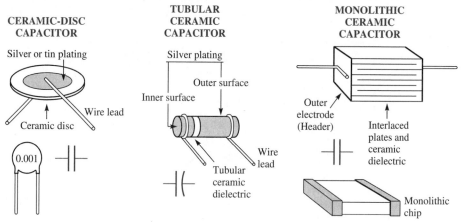

CERAMIC-DISC CAPACITOR

Silver or tin plating

Wire lead

Ceramic disc

0.001

TUBULAR CERAMIC CAPACITOR

Silver plating

Outer surface

Inner surface

Wire lead

Tubular ceramic dielectric

MONOLITHIC CERAMIC CAPACITOR

Outer electrode (Header)

Interlaced plates and ceramic dielectric

Monolithic chip

Figure 13-11 Different types of ceramic capacitors.

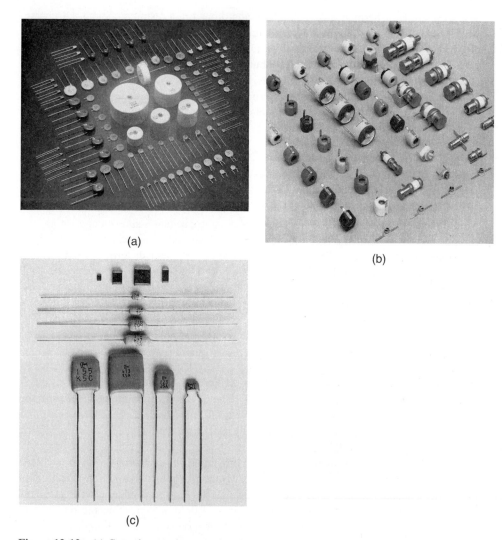

(a)

(b)

(c)

Figure 13-12   (a) Ceramic capacitors ranging from 12 V to 30 kV, (b) ceramic trimmer capacitors, (c) monolithic ceramic capacitors. (Courtesy of Murata Erie, Inc., and KSA Associates, Inc., Smyrna, Georgia 30080.)

popular as single radial- or axial-lead components, multipackaged arrays in ''single in-line'' and ''dual in-line'' packages (SIPs and DIPs), and surface-mount-technology (SMT) components.

Ceramic capacitors are found in nearly all applications. They are used for signal coupling and bypassing, for filtering, and in tuned circuits. Since ceramic capacitors are manufactured in such a wide range of dielectric constants and temperature stability, it is very important to match the proper ceramic with each application. Many common ceramic capacitors change capacitance widely with temperature variations and therefore cannot be used in tuned circuits used for frequency selection, filtering, or oscillation. However, these same capacitors would be fine for signal coupling and bypassing in most applications.

Film Capacitors

Many film capacitors are very similar in construction to paper capacitors. A very thin insulating film is rolled between two metal foils. Wire leads are connected radially or axially to the foils and the entire unit is encapsulated in a resin compound. The outer foil lead is often marked on the capacitor body and, as with paper capacitors, should be connected to ground, or the lowest potential, in a circuit. Some film capacitors are fabricated using the stacked-sandwich construction. The thin-film dielectric is a type of synthetic such as polyester, polyethylene, polystyrene, or Mylar®. These dielectric materials are very stable with temperature variations. Film capacitors range in capacitance from a few pF to about 1 μF. Voltage ratings are in hundreds of volts. Figure 13-13 shows a variety of film capacitors.

Mylar® (polyester) capacitors are often used in medium- and low-frequency circuits for filtering, coupling, and bypassing. They are also temperature-stable enough to be used in timer circuits. Polystyrene capacitors have very good temperature stability and are suitable for radio-frequency circuits such as filters, frequency-selective circuits, and oscillators.

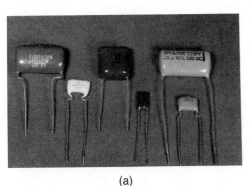

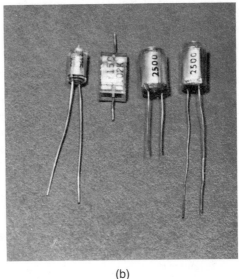

(a)   (b)

Figure 13-13 (a) A variety of film capacitors used for filtering, coupling AC signals, and bypassing resistors, (b) polystyrene film capacitors used for radio-frequency circuits.

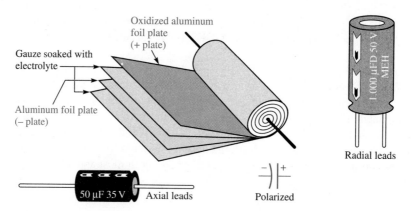

Figure 13-14   The rolled-sandwich construction of aluminum electrolytic capacitors.

Aluminum Electrolytic Capacitors

Electrolytic capacitors are designed to provide large amounts of capacitance in a single component. Electrolytics range from about 1 μF to a full farad. Such large amounts of capacitance are possible due to two primary factors: very large plate-surface areas, and a very thin dielectric (very close plate spacing). As shown in Figure 13-14, the **aluminum electrolytic capacitor** is fabricated in a rolled-sandwich construction. The surface area of the rolled plates is very large. One of the aluminum plates is heavily oxidized on its inner surface (aluminum oxide coating). The **aluminum oxide** on this plate is an insulator and serves as the dielectric. Naturally, this is a very thin dielectric, allowing for very large amounts of capacitance (close plate spacing). But, how can a close uniform contact between the oxidized plate and the other plate be insured? This is accomplished with a wet, pasty conductive substance called an **electrolyte**. The electrolyte is absorbed in a thin paper or gauze layer between the two plates. The fluidity of the electrolyte then ensures perfect contact between the nonoxidized plate and the aluminum oxide layer of the other plate.

The chemistry of the electrolytic capacitor requires that it be **polarized**. The oxidized plate is positive (anode) and the nonoxidized plate is negative (cathode). The plates are wrapped so that the negative plate is on the outside of the roll and is connected to the aluminum can in which the capacitor is encased. When the capacitor is connected in a circuit in the proper DC polarity, electrolysis takes place inside the capacitor, which maintains the aluminum oxide layer, thus insuring long life for the capacitor. If the polarity of DC voltage applied to the electrolytic is incorrect, the aluminum oxide quickly dissolves and the capacitor becomes shorted. Electrolytics often explode in such cases.

Electrolytic Capacitor Characteristics

- low dielectric resistance (500 kΩ to 10 MΩ) and a relatively high leakage current
- relatively short shelf life (electrolyte tends to dry out and the dielectric disintegrates)
- relatively low voltage ratings, due to very thin oxide dielectric, as compared to mica, ceramic, etc.
- high amount of inductance due to the size of the plates and the rolled construction
- polarized (except for nonpolarized electrolytics)

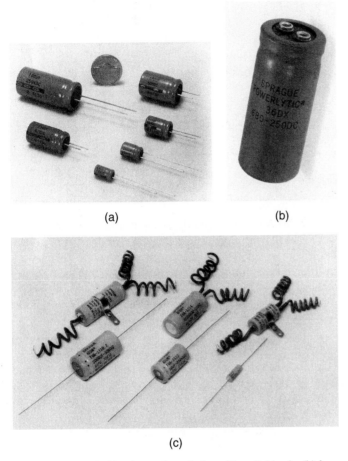

(a)　　　　　　　　(b)

(c)

Figure 13-15 (a) Aluminum electrolytics with radial leads, (b) large electrolytic with aluminum screw inserts, (c) pigtail and axial-lead electrolytic capacitors. (Courtesy of Sprague Electric Company.)

Electrolytics, such as those shown in Figure 13-15, are used as low-frequency filters (usually below 20 kHz), as low-frequency coupling and bypass capacitors, and as power-supply filter capacitors. They are found in audio amplifiers and DC power supplies.

Nonpolarized Electrolytic Capacitors

Nonpolarized electrolytics are essentially the same as polarized electrolytics, except an oxide layer is provided on the inside surface of each plate. Thus, regardless of the applied DC polarity, the capacitor will not short. Nonpolarized electrolytics are used to couple pure AC low-frequency signals. For example, nonpolarized electrolytics are used to couple AC to a stator of a split-phase motor. Nonpolarized electrolytics should be used in any application in which the voltage applied to the capacitor actually reverses polarity.

Tantalum Electrolytic Capacitors

Tantalum electrolytic capacitors are polarized and operate on the same principle as aluminum electrolytics; they are composed of two plates, an oxide layer, and an electrolyte.

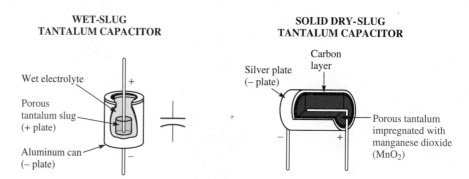

Figure 13-16   Tantalum electrolytic capacitors.

In this case, the anode (+ electrode) is tantalum and the oxide is a tantalum oxide. The cathode is aluminum, tin, copper, or silver. Some tantalums are fabricated in a **rolled-sandwich type** of construction with a wet electrolyte, just like aluminum electrolytics. Others are fabricated much differently. Solid or **slug-type tantalums** contain a formed lump, or slug, of porous tantalum (see Figure 13-16). The porous slug is made by packing powdered tantalum into a form (cylindrical or block) and inserting a tantalum wire to be used as a connecting terminal. The form is then sintered (fired) at high temperature. This creates a hard, yet very porous, tantalum slug. **Wet-slug tantalums** are packaged with a liquid electrolyte that is absorbed into the porous tantalum. A forming voltage is then used to form the tantalum oxide on the inside surfaces of each of the countless tiny chambers existing in the porous tantalum. In some cases, an acid is used to form the tantalum oxide layer before capacitor assembly. **Dry-slug tantalums** are formed by introducing the porous slug to manganese nitrate, which creates manganese dioxide throughout the porous tantalum. The manganese dioxide serves as the negative electrode. See Figure 13-17.

Tantalum Capacitor Characteristics

- very high capacitance in a small volume
- tantalum oxide has a higher permittivity (dielectric constant) than aluminum oxide (7 for aluminum oxide and 25 for tantalum oxide)
- longer shelf life and a much higher dielectric resistance (less leakage current) than aluminum electrolytics
- more costly and less forgiving to voltage surges and polarity reversals than aluminum electrolytics

Variable Capacitors

Vacuum Variables

Vacuum variable capacitors contain one fixed and one movable plate in an evacuated chamber. The chamber is usually glass or ceramic. The amount of capacitance is adjusted by changing the spacing between plates or changing the amount of plate-surface overlap. They normally have low capacitance ranges (i.e., 5 to 20 pF or 10 to 100 pF, etc.) and very high voltage ratings (up to 50 kV and more). These are very-high-quality capacitors used in tuned radio and television circuits (usually in transmitters).

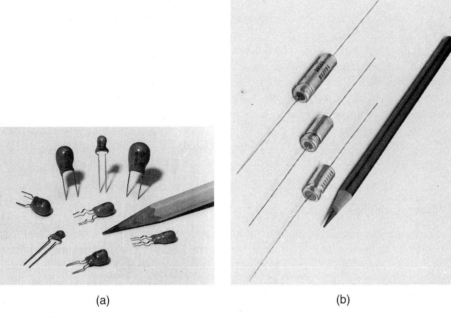

(a)   (b)

Figure 13-17 (a) Dipped radial-lead tantalum capacitors, (b) axial-lead tantalum capacitors. (Courtesy of Sprague Electric Company.)

Air Variables

Air variable capacitors have their plates exposed to the air and have no other dielectric than the air. The plates are usually two sets of metal vanes that can be intermeshed without touching. One set is held stationary (stator vanes) and is insulator-mounted to a metal frame that is usually mounted to chassis ground. The movable set of plates (rotor vanes) is mounted on the main shaft, which is electrically connected to the metal frame and chassis ground. The rotor vanes are intermeshed with the stator vanes as the shaft is rotated. The degree to which the two sets of vanes are intermeshed determines the amount of capacitance. The capacitance ranges for this type of variable capacitor may be as low as 1 to 10 pF or as high as 100 to 1,000 pF. Voltage ratings range up to 1 or 2 kV. Air variables are used extensively in radio receivers and low-power radio transmitters.

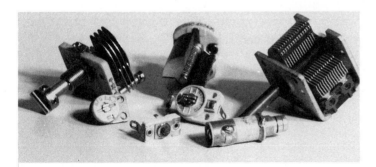

Figure 13-18 Variable capacitors—air, ceramic, and mica.

Mica Variables

Mica variables are usually small **trimmer capacitors** of the **compression type**. They are very low in value, with ranges below 100 pF. One plate is stationary, while the other plate is like a flat leaf spring that is forced down on the mica dielectric sheet separating the two plates. A screw is usually the mechanical means by which the spring plate is compressed onto the mica dielectric. If the screw is not very tight, air will become part of the dielectric along with the mica sheet. Ultimately, the mica sheet prevents the two plates from shorting when they are compressed. Typical capacitance ranges for mica compression trimmers are very low, such as 5 to 30 pF or 10 to 40 pF. Voltage ratings are in the hundreds of volts. Mica variables are used in radio and television circuits to trim a tuned circuit or oscillator to the correct frequency. They allow fine tuning or calibration of radio circuits.

Ceramic Variables

Ceramic variable capacitors are usually low-value, small-size trimmer capacitors. Again, one plate is stationary and the second is movable. **Rotary ceramic trimmer capacitors** have two ceramic discs that are plated only on half of a surface. Maximum capacitance is obtained when the plated half of the movable disc is perfectly matched with the plated half of the stationary disc. Another type of ceramic variable is a **variable tubular** type. A ceramic tube or cylinder is plated on the outside, and a threaded conductor is inserted inside. The amount of overlap between the threaded conductor and the outer plating will determine the amount of capacitance. As with mica variables, the capacitance ranges for ceramic trimmers is in the low picofarads and voltage ratings are in the hundreds of volts. Ceramic variable capacitors are used for the same purposes as mica variables.

Teflon® Variables

Teflon® variable capacitors are very similar to air variable capacitors, having two sets of metal vanes as plates. The difference is that thin Teflon® sheets are used instead of air to separate the plates. As a result, the variable capacitor can be much smaller for a given range of capacitance. Teflon® variables are used for frequency selection and oscillator adjustment in small portable radios.

SELF-CHECK 13-3

1. What applications are fixed mica capacitors suited for?

2. List three fabrication styles for fixed ceramic capacitors.

3. Explain why electrolytics can have very high capacitances.

4. Which type of electrolytic has less leakage current, (a) aluminum or (b) tantalum?

5. For what are air variable capacitors normally used?

13-4 Capacitor Ratings

Capacitance and Tolerance

Common Labeling

Many capacitors of all types are labeled in a literal fashion. For example, a ceramic, paper, or polyester capacitor may be labeled as follows: .01 10%. This indicates the capacitor is 0.01 μF and has a 10% tolerance. However, the capacitor body may or may not be marked to indicate microfarads or picofarads. If the unit of capacitance is indicated and is in microfarads, it may be signified as follows: μF, mfd, MF, or MFD. If the capacitance is in picofarads, it may be signified as: pF, PF, pfd, PFD, $\mu\mu$F, or MMF.

As a basic rule of thumb, if a capacitor, other than an electrolytic or tantalum, is marked with a whole number, such as 5, 100, 500, etc., the unit of capacitance is picofarads (pF). Many low-value ceramic, mica, and polystyrene capacitors are labeled in this way. If the capacitor is labeled with a decimal number less than 1, such as .002, .01, or .047, the unit of capacitance is microfarads (μF).

Many capacitors, less than 1 μF, are labeled with a number code. For example, a Mylar® capacitor might be labeled as follows: 202K100. This code indicates the capacitor's capacitance and voltage rating. The 202K is the capacitance and the 100 is the voltage rating. But, what does the 202K mean? 202K is equal to 2,000 pF which is equal to 0.002 μF. The first two digits are taken literally as the first two significant digits, 2<u>0</u>2K. The next digit is the multiplier digit, indicating the exponent of the power of ten needed (indicates how many zeros to add to the first two digits); 20<u>2</u>K. Thus, 202 = 20 $\cdot$ 10^2 = 2,000 pF. The K stands for ±10% tolerance. See Table 13-2. Consider Example Set 13-6.

EXAMPLE SET   13-6

1. 473K100 = 47,000 pF = 0.047 μF and a 100 V rating, ±10% tol.
2. 222J400 = 2,200 pF = 0.0022 μF and a 400 V rating, ±5% tol.
3. 104G200 = 100,000 pF = 0.1 μF and a 200 V rating, ±2% tol.

TABLE   13-2   Tolerance Designators

Designator	Tolerance
B	±0.10%
C	±0.25%
D	±0.50%
F	±1.0%
G	±2.0%
J	±5.0%
K	±10.0%

Capacitor Color Code

Some capacitors, such as some micas, ceramics, and tubular capacitors, are labeled with a color code, as illustrated in Figure 13-22. Color-coded capacitor values are expressed in picofarads. After examining Figure 13-22, study the color code examples of Example Set 13-7.

EXAMPLE SET   13-7

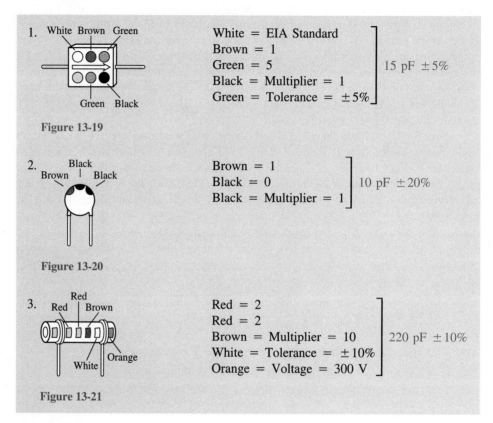

1.

White = EIA Standard
Brown = 1
Green = 5   ⎤
Black = Multiplier = 1   ⎬ 15 pF ±5%
Green = Tolerance = ±5%   ⎦

Figure 13-19

2.

Brown = 1
Black = 0   ⎤ 10 pF ±20%
Black = Multiplier = 1   ⎦

Figure 13-20

3.

Red = 2
Red = 2
Brown = Multiplier = 10   ⎤
White = Tolerance = ±10%   ⎬ 220 pF ±10%
Orange = Voltage = 300 V   ⎦

Figure 13-21

Voltage Rating

Working Voltage DC (WVDC)

The **voltage rating** (often labeled as **WVDC** and **Working Voltage DC**) of a capacitor is the maximum DC voltage that can safely be applied across the capacitor without arcing, or voltage breakdown, occurring in its dielectric. The difference of potential on the capacitor plates creates a stress on the atomic structure of the dielectric. If the difference of potential is too high, electrons will be torn from their orbits in the dielectric material, producing an electrical arc. A small carbon path known as a **puncture** is created. The puncture is permanent, except in the case of self-healing dielectric materials such as air, oil, and some metal films. As you might expect, a punctured capacitor is no longer usable, since the puncture acts as a resistive path through which the capacitor is continually discharged. In effect, the capacitor becomes a resistor, but is considered to be shorted.

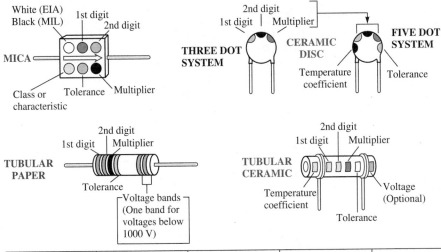

Color	Significant figures	Multiplier	Tolerance	Voltage rating
Black	0	1	± 20%	
Brown	1	10	± 1%	100
Red	2	100	± 2%	200
Orange	3	1,000	± 3%	300
Yellow	4	10,000		400
Green	5	100,000	± 5% (EIA)	500
Blue	6	1,000,000	± 6%	600
Violet	7			700
Gray	8		± 30%	800
White	9		± 10%	900
Gold		0.1	± 5% (MIL)	1,000
Silver		0.01	± 10%	2,000
No band			± 20%	500

EIA - Electronic Industries Association   MIL - Military Standard

Figure 13-22   The capacitor color code.

AC Rating Considerations

If AC is applied to the capacitor, the peak AC voltage must not exceed the capacitor's voltage rating. Also, if a DC voltage is present with the AC voltage, the sum of the peak AC and DC voltages must not exceed the capacitor's voltage rating. In most cases, it is wise to operate the capacitor well below its rated voltage. An exception to this is electrolytic capacitors. Electrolytics depend on the capacitor voltage to maintain the oxide dielectric. Many manufacturers recommend that electrolytic capacitor voltage be maintained just below their WVDC rating (10 to 20% below the rated voltage).

Dielectric Strength

There are basically two factors that determine voltage rating: the thickness of the dielectric (plate spacing) and the dielectric strength. **Dielectric strength** is the measure of a

TABLE   13-3   Dielectric Strength

Material	Dielectric Strength V/mil	Dielectric Strength V/mm
Vacuum	20	800
Air	20	800
Teflon®	1,500	60,000
Polyethylene	400	16,000
Polystyrene	400	16,000
Rubber	700	28,000
Paper	1,300	50,000
Mylar®	400	16,000
Polyester	400	16,000
Glass (Pyrex, Window)	700 – 2,000	28k – 80k
Mica	600 – 2,000	24k – 80k
Bakelite	400	16,000
Porcelain	200	8,000
Ceramic	500 – 1,300	20k – 50k

dielectric's ability to withstand a difference of potential without breakdown, or arcing. The dielectric strength is specified in volts per mil ($\frac{1}{1000}$ of an inch thickness). The thicker the dielectric and the higher its dielectric strength, the higher the capacitor's voltage rating. Table 13-3 includes a list of common dielectric materials with their dielectric strengths in V/mil and V/mm. The values shown are general, ballpark values and vary depending on the actual composition of the dielectric material.

Temperature Coefficient

N###, P###, and NP0

Most capacitors are affected by temperature. If a capacitor's capacitance increases with temperature the capacitor is said to have a **positive temperature coefficient**. If a capacitor's capacitance decreases as the temperature increases, the capacitor is said to have a **negative temperature coefficient**. Many capacitors are rated to stay within a certain percentage of change of capacitance over a specified range of temperature. Others are marked according to the amount and direction of capacitance variation for every degree centigrade change in temperature. Ceramic capacitors are often marked this way. For example, if a capacitor is marked with N200, it means the capacitor has a negative temperature coefficient (N) and its capacitance will change 200 parts per million (ppm) for every degree centigrade change in temperature above or below a reference temperature (usually 25°C). If the ambient temperature surrounding this capacitor is 40°C, its capacitance will be 3,000 ppm lower than the marked value. If this capacitor is 1 μF at 25°C, its capacitance at 40°C will be 3,000 pF less than 1 μF (since there are 1 million pF per μF). If the capacitor is marked with a P instead of an N (i.e., P150), it indicates the capacitor has a positive temperature coefficient. A marking of NP0 (*N*egative-*P*ositive-Zero) indicates the capacitor is stable and its capacitance should not change with temperature.

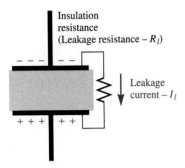

Insulation
resistance
(Leakage resistance – R_l)

Leakage
current – I_l

+ + +   + +

Figure 13-23   Dielectric resistance and leakage current.

Insulation Resistance and Leakage Current

Leakage Current

Ideally, an insulator (dielectric) should have an infinite resistance, meaning it will conduct absolutely no current. In practice, dielectrics do exhibit some very low level of conductivity. In other words, a very low **leakage current** does flow. A charged capacitor eventually discharges itself due to this leakage current. There are actually three leakage paths through which the capacitor might discharge: through the dielectric, through impurities on the surface of the dielectric or capacitor body, and through the air surrounding the capacitor. In most cases, the sum total of all of these is insignificant.

Dielectric Resistance

The **dielectric resistance**, sometimes called the *insulation resistance* or *leakage resistance* (R_l), is normally very, very high (see Figure 13-23). It is usually measured in hundreds of megohms or in gigohms ($1 \cdot 10^9$ Ω). The exception to this is electrolytics, which characteristically have fairly high leakage currents and low dielectric resistances (500 kΩ to 10 MΩ). However, the electrolytic leakage current is usually small compared to the quantity of stored charge on the capacitor's plates.

Before continuing, test your understanding by answering the questions of Self-Check 13-4.

S E L F - C H E C K   1 3 - 4

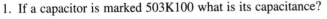

1. If a capacitor is marked 503K100 what is its capacitance?

2. What two factors determine the voltage rating of a capacitor?

3. What does the capacitor marking N100 indicate?

4. What is capacitor leakage current?

13-5 Total Capacitance

Capacitors in Series

Placing capacitors in series with one another will not increase capacitance. In fact, *the total capacitance of series-connected capacitors is less than the lowest-value capacitor.* This is very different from series resistances and series inductances. Therefore, you must keep this in mind when calculating total capacitance for series-connected capacitors. To calculate total capacitance (C_t) for series capacitors, you will actually use the parallel resistor (product over the sum, same values, and reciprocal) formulas. Why? Placing capacitors in series has the same effect as increasing the distance, or spacing, between the plates of a single capacitor. As you know, increasing plate separation decreases capacitance. In a functioning circuit, each of the series capacitors receives a charge through the others. Therefore, the charge on any one of the capacitors is limited by the charge capability of the lowest-value capacitor. This, coupled with the increased plate spacing, is why the total series capacitance is less than the lowest-value capacitor. The following three formulas reveal this fact:

$$C_t = C/N \tag{13.9}$$

(Same values formula, where all capacitors are the same value and total capacitance is found by dividing one of the capacitors by the number of series capacitors)

$$C_t = \frac{C_1 \cdot C_2}{C_1 + C_2} \tag{13.10}$$

(Product over the sum formula, where there are only two capacitors in series having different values)

$$C_t = \frac{1}{1/C_1 + 1/C_2 + 1/C_3 + \cdots} \tag{13.11}$$

(Reciprocal formula, where there are more than two series capacitors all of different values)
  Example Set 13-8 clarifies the use of these three formulas. Always remember:

Total series capacitance is calculated in the same manner as total parallel resistance and inductance.

EXAMPLE SET   13-8

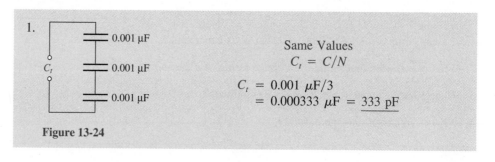

1.

0.001 μF
0.001 μF
0.001 μF
C_t

Same Values
$C_t = C/N$

$C_t = 0.001 \ \mu F/3$
   $= 0.000333 \ \mu F = \underline{333 \ pF}$

Figure 13-24

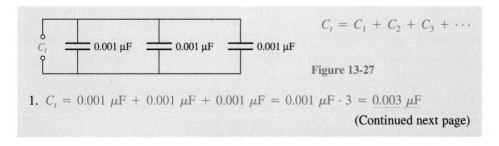

2.

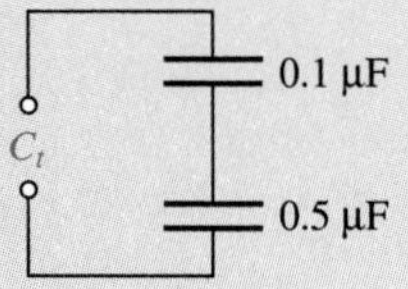

0.1 μF

C_t

0.5 μF

Two Different Values

$$C_t = \frac{C_1 \cdot C_2}{C_1 + C_2}$$

Figure 13-25

$$C_t = \frac{0.1 \ \mu F \cdot 0.5 \ \mu F}{0.1 \ \mu F + 0.5 \ \mu F}$$

$$= \frac{0.05 \ pF^2}{0.6 \ \mu F}$$

$$= \underline{0.083 \ \mu F}$$

3.

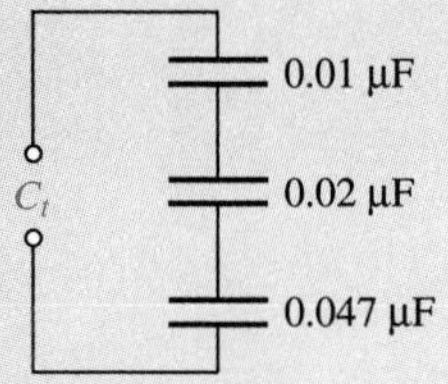

0.01 μF

C_t

0.02 μF

0.047 μF

More Than Two Different Values

$$C_t = \frac{1}{1/C_1 + 1/C_2 + 1/C_3 + \cdots}$$

Figure 13-26

$$C_t = \frac{1}{\dfrac{1}{0.01 \ \mu F} + \dfrac{1}{0.02 \ \mu F} + \dfrac{1}{0.047 \ \mu F}} = 0.0058 \ \mu F$$

Capacitors in Parallel

Total capacitance is increased by placing capacitors in parallel. The effect here is that of increasing plate surface area. In other words, each individual capacitor contributes its plate area to the plate area of other paralleled capacitors. A large plate area yields a greater capacitance. Finding total capacitance for capacitors placed in parallel is very straightforward. You simply add the values of all individual paralleled capacitors.

$$C_t = C_1 + C_2 + C_3 + \cdots \tag{13.12}$$

(Parallel capacitance formula, where the total capacitance for any number of parallel capacitors is the sum of all individual capacitances.)

Study Example Set 13-9.

EXAMPLE SET   13-9

C_t

0.001 μF   0.001 μF   0.001 μF

$$C_t = C_1 + C_2 + C_3 + \cdots$$

Figure 13-27

1. $C_t = 0.001 \ \mu F + 0.001 \ \mu F + 0.001 \ \mu F = 0.001 \ \mu F \cdot 3 = \underline{0.003 \ \mu F}$

(Continued next page)

EXAMPLE SET   13-9   (Continued)

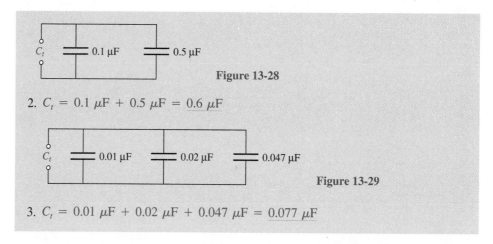

Figure 13-28

2. $C_t = 0.1\ \mu F + 0.5\ \mu F = \underline{0.6\ \mu F}$

Figure 13-29

3. $C_t = 0.01\ \mu F + 0.02\ \mu F + 0.047\ \mu F = \underline{0.077\ \mu F}$

Now, test your skills by answering the questions of Self-Check 13-5.

SELF-CHECK   13-5

1. Given three *series* capacitors of 10 μF, 5 μF, and 20 μF, find C_t.

2. Given four *parallel* capacitors of 100 pF, 200 pF, 150 pF, and 300 pF, find C_t.

3. Given two *series* capacitors of 30 pF and 50 pF, find C_t.

4. Given five *series* capacitors of 35 pF each, find C_t.

13-6 Capacitors in DC Circuits

Charging Current and Quantity of Charge

The amount of capacitor charge current depends on the amount of applied voltage, the circuit resistance, and the amount of charge voltage currently on the capacitor. When the DC source voltage is first applied to a capacitor, the capacitor acts like a short and demands a great amount of current. At this first instant of time, circuit resistance is the only current-limiting factor. As the capacitor charge increases, the current decreases. However, during the charge period the capacitor will have collected a total charge measured in coulombs. This total quantity of charge for series and parallel capacitor circuits is illustrated in Figures 13-30 and 13-31.

Series Capacitors

When capacitors are connected in series, the total quantity of charge supplied by the voltage source is the same amount as that which is received by each individual series capacitor.

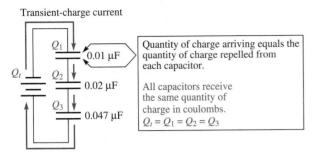

Transient-charge current

Quantity of charge arriving equals the quantity of charge repelled from each capacitor.

All capacitors receive the same quantity of charge in coulombs.
$Q_t = Q_1 = Q_2 = Q_3$

Figure 13-30   Each series capacitor receives the same quantity of charge (Q) measured in coulombs (C).

This is because there is only one path for current. As shown in Figure 13-30, a charge that is repelled from one plate of one capacitor is picked up by a plate of the next capacitor, and so on. Since the quantity of repelled charge is always equal to the number of accumulated charge, every capacitor receives the same quantity of charge at the same time. In equation form:

$$Q_t = Q_1 = Q_2 = Q_3 = \cdots \tag{13.13}$$

(Total charge for series capacitors, where the quantity of charge Q_t provided by the voltage source is equal to the quantity of charge arriving at and leaving each series capacitor—Q_1, Q_2, Q_3, etc.)

Parallel Capacitors

When capacitors are placed in parallel, each capacitor receives its own quantity of charge directly from the voltage source, as is shown in Figure 13-31. Since the voltage source is supplying a charge to each parallel capacitor, the total supplied charge is equal to the sum of all individual capacitor charges. This is expressed mathematically as follows:

$$Q_t = Q_1 + Q_2 + Q_3 + \cdots \tag{13.14}$$

(Total charge for parallel capacitors, where the quantity of charge Q_t provided by the voltage source is equal to the sum of all individual capacitor charges)

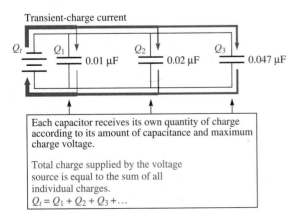

Transient-charge current

Each capacitor receives its own quantity of charge according to its amount of capacitance and maximum charge voltage.

Total charge supplied by the voltage source is equal to the sum of all individual charges.
$Q_t = Q_1 + Q_2 + Q_3 + \ldots$

Figure 13-31   Each parallel capacitor receives its own quantity of charge.

Total Capacitance and Total Charge

Recall that the total quantity of charge Q is determined by the capacitance C multiplied by the capacitor's acquired voltage V (Formula 13.3). Since $Q = C \cdot V$, the total quantity of charge for capacitors in a DC circuit can be determined as follows:

$$Q_t = C_t \cdot E_s \tag{13.15}$$

where Q_t is the total supplied charge for a DC capacitor circuit, C_t is the total circuit capacitance, and E is the applied DC source voltage. Consider Examples 13-10 and 13-11.

EXAMPLE   13-10

12 VDC is applied to a *series* circuit containing three capacitors whose values are 100 μF, 100 μF, and 50 μF. The total circuit capacitance is 25 μF [$C_t = 1/(1/100$ μF + 1/100 μF + 1/50 μF)]. The total quantity of charge, which is equal to the charge on each capacitor, is found using Formula 13.14 as follows: $Q_t = 25\ \mu$F $\cdot$ 12 V = 300 μC.

EXAMPLE   13-11

50 VDC is applied to a *parallel* circuit containing three capacitors whose values are 1,000 μF, 2,000 μF, and 500 μF. The total circuit capacitance is 3,500 μF ($C_t = 1,000\ \mu$F + 2,000 μF + 500 μF). The total quantity of charge is $Q_t = 3,500\ \mu$F $\cdot$ 50 V = 0.175 C.

Capacitor Charge Voltages

Recall that Kirchoff's Voltage Law states that the algebraic sum of all voltages in a series circuit must equal the applied voltage. This law is also true for series-connected capacitors. The source voltage applied to a series arrangement of capacitors is divided along the capacitors according to their capacitance. The series capacitor having the least capacitance receives the largest charge voltage and the capacitor having the greatest capacitance receives the least charge voltage. The proof of this is found in Formula 13.4. The capacitor voltage is equal to the quantity of charge divided by the capacitor's capacitance ($V = Q/C$). If the quantity of charge is held constant, the amount of voltage on a capacitor will depend entirely on its capacitance. Consider Example Set 13-12.

EXAMPLE SET   13-12

1.

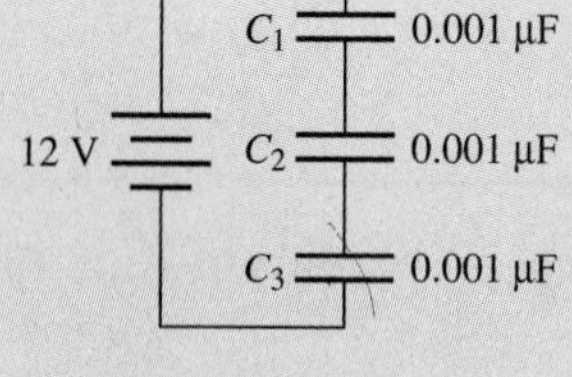

Figure 13-32

$C_t = 0.001\ \mu$F / 3
$\quad = 0.000333\ \mu$F = 333 pF

$Q_t = 12$ V $\cdot$ 333 pF
$\quad = 4$ nC = 0.004 μC

$V_{C1} = 0.004\ \mu$C / 0.001 μF
$\quad = 4$ V = $V_{C2} = V_{C3}$

$Q_t = E_s \cdot C_t$
$V_x = Q_t/C_x$

2.

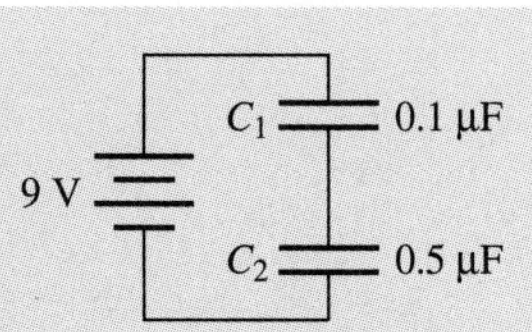

Figure 13-33

$$C_t = \frac{0.1\ \mu F \cdot 0.5\ \mu F}{0.1\ \mu F + 0.5\ \mu F} = \underline{0.083\ \mu F}$$

$Q_t = 9\ V \cdot 0.083\ \mu F = 0.75\ \mu C$

$V_{C1} = 0.75\ \mu C\ /\ 0.1\ \mu F = 7.5\ V$
$V_{C2} = 0.75\ \mu C\ /\ 0.5\ \mu F = 1.5\ V$

3.

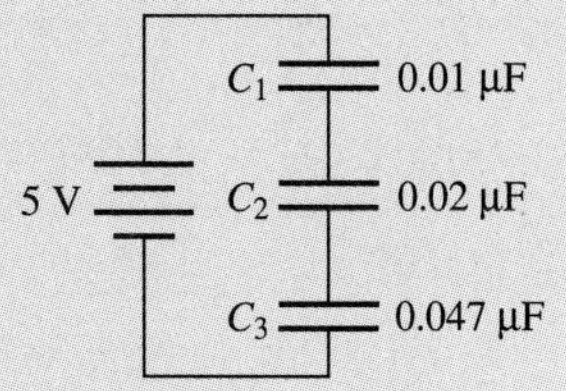

Figure 13-34

$C_t = 1\ /\ (1/0.01\ \mu F + 1/0.02\ \mu F + 1/0.047\ \mu F) = 0.0058\ \mu F$

$Q_t = 5\ V \cdot 0.0058\ \mu F = 0.029\ \mu C$

$V_{C1} = 0.029\ \mu C\ /\ 0.01\ \mu F = \underline{2.9\ V}$
$V_{C2} = 0.029\ \mu C\ /\ 0.02\ \mu F = \underline{1.45\ V}$
$V_{C3} = 0.029\ \mu C\ /\ 0.047\ \mu F = \underline{0.62\ V}$

Capacitor Voltage Divider Rule

As you can see, the amount of voltage on a given capacitor is inversely related to the capacitor's capacitance. It is an inverse relationship, or inverse proportion. As with resistor voltage drops, it is possible to derive a proportion formula (voltage divider rule) to determine a capacitor's charge voltage according to its relative capacitance (relative or in proportion to total capacitance). Let's designate any one of the series capacitors as being capacitor x (C_x). The charge on that capacitor is Q_x and is equal to the total supplied charge (Q_t) ($Q_x = Q_t$). Now, follow this derivation:

Since $Q_x = Q_t$ in a series circuit, and $Q_x = C_x \cdot V_x$ and $Q_t = C_t \cdot E_s$, $C_x \cdot V_x = C_t \cdot E_s$. Solving for V_x: $V_x = C_t \cdot E_s / C_x = E_s \cdot C_t / C_x$, so

$$V_x = E_s \cdot C_t / C_x \qquad (13.16)$$

(proportion formula for series capacitors, where V_x is the charge voltage on a specified series capacitor, E_s is the applied source voltage, C_t is the total series circuit capacitance, and C_x is the capacitance of the specified series capacitor).

Example Set 13-13 illustrates how the charge voltages of Example Set 13-12 are calculated using Formula 13.15.

EXAMPLE SET 13-13

1.

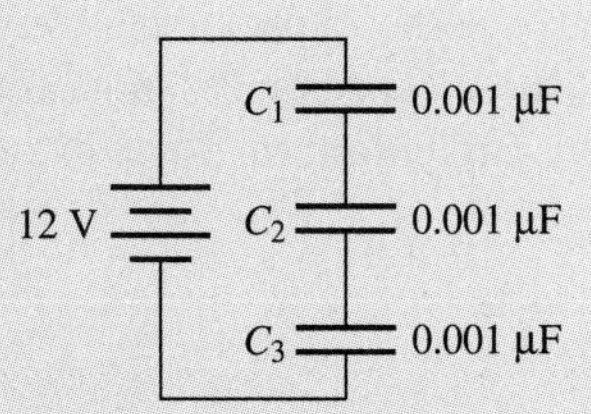

Figure 13-35

$C_t = 333\ pF$ $\qquad\qquad V_x = E_s \cdot C_t / C_x$

$V_{C1} = 12\ V \cdot 333\ pF\ /\ 0.001\ \mu F$

$\quad = \underline{4\ V} = V_{C2} = V_{C3}$

(Continued next page)

EXAMPLE SET 13-13 (Continued)

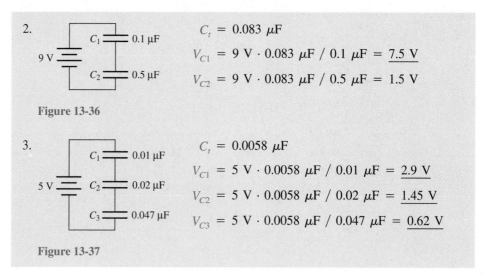

2.

$C_t = 0.083 \ \mu F$

$V_{C1} = 9 \text{ V} \cdot 0.083 \ \mu F \ / \ 0.1 \ \mu F = \underline{7.5 \text{ V}}$

$V_{C2} = 9 \text{ V} \cdot 0.083 \ \mu F \ / \ 0.5 \ \mu F = 1.5 \text{ V}$

Figure 13-36

3.

$C_t = 0.0058 \ \mu F$

$V_{C1} = 5 \text{ V} \cdot 0.0058 \ \mu F \ / \ 0.01 \ \mu F = \underline{2.9 \text{ V}}$

$V_{C2} = 5 \text{ V} \cdot 0.0058 \ \mu F \ / \ 0.02 \ \mu F = \underline{1.45 \text{ V}}$

$V_{C3} = 5 \text{ V} \cdot 0.0058 \ \mu F \ / \ 0.047 \ \mu F = \underline{0.62 \text{ V}}$

Figure 13-37

Now, test your skills and knowledge by solving the problems of Self-Check 13-6.

SELF-CHECK 13-6

1. The total quantity of charge supplied by a DC source to a *series* capacitor circuit is equal to what?

2. The total quantity of charge supplied by a DC source to a *parallel* capacitor circuit is equal to what?

3. If the total capacitance of a certain capacitor circuit is 20 μF and the source voltage is 12 V, what is the total quantity of charge?

4. Given a series capacitor circuit having a source voltage of 24 VDC, with capacitors of 0.1 μF, 0.05 μF, 0.01 μF, and 0.2 μF, find the voltage drop across each capacitor.

13-7 Stray Capacitance

Like inductance, capacitance is not confined only to components called capacitors. All electrical and electronic components and conductors have some capacitance. Any two conductors separated by an insulator exhibit capacitance. These capacitances, not officially recognized or labeled in schematics, are classified as **stray capacitances**. Some examples of stray capacitance are shown in Figure 13-38.

Instances of Stray Capacitance

There are many instances of stray capacitance in the typical circuit. Wires used to connect one circuit to another have stray capacitance between themselves and to a neighboring

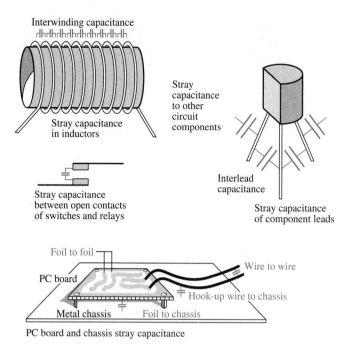

Interwinding capacitance

Stray capacitance in inductors

Stray capacitance to other circuit components

Interlead capacitance

Stray capacitance between open contacts of switches and relays

Stray capacitance of component leads

Foil to foil

Wire to wire

PC board

Hook-up wire to chassis

Metal chassis   Foil to chassis

PC board and chassis stray capacitance

Figure 13-38   Stray capacitance is everywhere.

ground or circuit. Component leads exhibit stray capacitance between themselves and the surrounding circuit. Open switches and relay contacts have a certain amount of capacitance between contacts. While the open switch blocks DC, it may not block a very-high-frequency AC signal (acting as a small capacitor, the contacts will block DC and pass AC). The copper foil on a printed circuit board exhibits stray capacitance to nearby conductors, chassis ground, etc. Inductors have stray capacitance from the coil to ground, coil to neighboring components, and even between windings from one turn to the next. For this reason many chokes are stagger wound and separated into sections. The stagger winding helps cancel interwinding capacitance and dividing the choke into sections effectively places the undesired interwinding capacitance of each section in series with the others (recall that capacitance is reduced when capacitors are in series). Plug-in prototyping boards have a fairly significant amount of stray capacitance between conductive rows and columns. This reduces their usefulness for prototyping circuits that operate at frequencies above about 1 MHz (1,000,000 Hz).

Effects of Stray Capacitance

The effects of stray capacitance are usually that of AC signal attenuation (reduction or loss in amplitude) and circuit instability. The circuit may tend to become unstable causing an oscillation to occur at an undesired AC signal frequency. Many circuits are designed to be amplifiers raising a signal voltage, or power, to a higher level. Stray capacitance can cause such a circuit to turn into a signal generator instead of a signal amplifier.

Stray capacitance is usually very low, a few pF or so. As with any capacitor, the amount of stray capacitance is determined by surface area of the conductors, spacing between conductors, and the permittivity of the insulating material (dielectric constant).

Since stray capacitance is very small, it has no effect on DC and very little effect on audio frequencies (below 20 kHz). The effect of stray capacitance does increase with frequency. The following self-check will help reinforce your understanding of stray capacitance.

SELF-CHECK 13-7

1. Describe stray capacitance.

2. Does stray capacitance have a greater effect at (a) high or (b) low frequencies?

13-8 Troubleshooting Capacitors

Shorted Capacitors

See color photo insert page D2, photograph 54.

Shorted capacitors are capacitors that have developed a low-resistance direct current path between their plates. The plates may actually be touching, or a low-resistance impurity path (such as a carbon or metal oxide path) may have developed, bridging the plates. Naturally, shorted capacitors are not able to store a charge. In practice, shorted capacitors are usually easy to identify. Physically, the capacitor may be very warm to the touch while the circuit is in operation or may be discolored from heat. Electrically, DC flowing through the capacitor can be measured (as by a series ammeter) and the DC voltage across the capacitor will be very low or nil (recall zero volts is measured across a dead short).

Figure 13-39 illustrates the manner in which a capacitor can be checked with the ohmmeter to determine if it is shorted. Remember, the ohmmeter circuit has its own internal voltage source. A good capacitor will receive a charge from the ohmmeter's internal battery. As shown in Figure 13-39, first remove the capacitor from its circuit and make sure it is completely discharged by shorting its leads together. Then, connect the ohmmeter's test leads across the capacitor. For large-value capacitors (>1 μF), if the capacitor is not

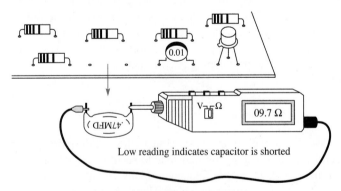

Low reading indicates capacitor is shorted

> Remove suspected capacitor from its circuit.
> Discharge the capacitor by shorting its leads.
> Test with the ohmmeter.

Figure 13-39 Using an ohmmeter to see if a capacitor is shorted.

shorted, the ohmmeter will indicate a steadily increasing resistance from some low initial value (increasing from or near 0 Ω). This indicates the capacitor is building up a charge. Low-value capacitors (<1 μF) charge too quickly to be seen on the ohmmeter. If the ohmmeter indicates a fixed low resistance, the capacitor is considered to be shorted (regardless of capacitor value).

Leaky Capacitors

Leaky capacitors are capacitors that have developed a high-resistance direct current path between their plates. This may have been caused by an internal arc through the dielectric or across an edge of the dielectric. The arc forms an impurity path for charge flow between the plates. It is not common to find leaky ceramic, mica, or polyester capacitors. However, it is fairly common to find electrolytics or tantalums that are leaky. Electrolytics age and have a certain shelf life. During the aging process, the dielectric oxide layer dissolves in the electrolyte paste. If this continues for too long, the electrolytic will be unusable. The electrolytic will last much longer in a circuit under the influence of a DC charge. The charge reforms and maintains the oxide layer. If the oxide layer is damaged due to aging or internal arcing, the electrolytic becomes shorted or leaky.

Once again, the ohmmeter may be used to check for leaky capacitors. The same procedure used to check for shorted capacitors is used to check for leaky ones. If the capacitor is leaky, the ohmmeter will indicate a fixed lower-than-normal resistance after a short charge period. Above 0.5 MΩ is normal for electrolytics and tantalums. The resistance of solid dielectric capacitors should be in the hundreds of megohms. (If manufacturer's data sheets are available, the normal dielectric resistance will be specified.)

Open Capacitors

An open capacitor is a capacitor that will not pass DC or AC. This is usually because one of the leads has broken free from the plate connection in the body of the capacitor. Capacitors that normally show a charge current on an ohmmeter are easily identified as being open if there is no change in ohmmeter reading when the meter is connected to the capacitor. If the capacitor is of a very low value, the ohmmeter will not be very useful in determining the open condition, since the charge occurs more quickly than the ohmmeter can register. Good low-value capacitors will be indicated as open with most ohmmeters. So, how can we determine if the capacitor is open? In later chapters, you will learn how capacitors act in AC circuits. This knowledge will enable you to test the capacitor in its circuit under operating conditions. You will be able to predict the amount of AC voltage drop the capacitor should have if it is good. However, there are other pieces of test equipment that can be used to determine an open capacitor. One such piece of equipment is the capacitance meter.

The Capacitance Meter

A capacitor will often appear to be good when checked with the ohmmeter, when in fact it is not good. A low-value capacitor is shown as an open using the ohmmeter, which is what you would expect. However, the capacitor may be far out of tolerance (off its value) or may actually be open. In such cases, a capacitance meter can provide a definite conclusion regarding the condition of the capacitor. Figure 13-40 shows two popular capacitance

See color photo page D2, photograph 55, and page D3.

Figure 13-40   Model 3000 capacitance meter. (Courtesy of Global Specialties, an Interplex Electronics Company, New Haven, CT.)

meters. Both are portable and easy to use. The capacitor in question is placed between the test terminals, and the actual capacitance reading is digitally displayed. Accuracy of these instruments is as good as within $\pm 0.2\%$, depending on the particular range. Capacitance meters are also useful for determining the value of unmarked or unreadable capacitors.

S E L F - C H E C K   1 3 - 8

1. How will an ohmmeter indicate a shorted capacitor?

2. How will an ohmmeter indicate a leaky capacitor?

3. Explain why the ohmmeter cannot determine whether a low-value capacitor is really open or not.

4. What are two uses for the capacitance meter?

13-9   Comparing Inductors and Capacitors

Inductors and capacitors are complementary components. That is, they are opposites in many ways, yet they work together to perform a specific task in many electrical and electronic circuits. In later chapters, you will see how these two very different components are used in many practical circuits. For now, it is sufficient to consider some of the facts about inductors and capacitors that make them different, yet in some ways similar and perfect complements of one another. This type of comparison is best done with a parallel table. Take time to study Table 13-4.

TABLE 13-4   Comparing Inductors and Capacitors

L ⟿ (H)	C ⊣⊢ (F)
1. Unit of measure is the Henry.	1. Unit of measure is the farad.
2. Pass DC.	2. Blocks DC.
3. Opposition to AC increases as the AC frequency increases.	3. Opposition to AC decreases as the AC frequency increases.
4. Energy is stored in a magnetic field.	4. Energy is stored in an electric field.
5. Stray capacitance between windings.	5. Stray inductance in plates and wire leads.
6. Some resistance in the windings.	6. Some resistance in the wire leads and plates.
7. Acts like an OPEN the first instant a potential is applied.	7. Acts like a SHORT the first instant a potential is applied.
8. Inductor current increases over a short transient period when a potential is applied.	8. Capacitor charge current decreases over a short transient period when a potential is applied.
9. Magnetic flux is concentrated in the core material.	9. Electric flux is concentrated in the dielectric.
10. Total series and parallel inductance is calculated the same as total resistance.	10. Total series and parallel capacitance is calculated opposite to the manner in which total resistance is calculated.

SELF-CHECK 13-9

1. List three opposite facts comparing inductors and capacitors.

Summary

FORMULAS

(13.1) $D = \mathscr{E} \cdot \epsilon$

where D is the flux density measured in flux per square meter, $\mathscr{E}$ is the electric field intensity measured in V/m, and ϵ is the absolute permittivity of the dielectric material measured in farads per meter (F/m).

(13.2) $C = Q/V$

where C is the capacitance measured in farads (F), Q is the quantity of charge measured in coulombs (C), and V is the capacitor charge voltage.

(13.3) $Q = C \cdot V$

(13.4) $V = Q/C$

(13.5) $E = Q \cdot V/2$

where E is the stored energy measured in joules (J).

(13.6) $E = C \cdot V^2/2$

where E is the stored energy measured in joules (J).

(13.7) $C = \epsilon \cdot A/d$

where C is the capacitance measured in farads, ϵ is the absolute permittivity of the dielectric material measured in F/m, A is the surface area of one plate, and d is the distance between plates.

(13.8) $\epsilon = \epsilon_r \cdot \epsilon_0 = \epsilon_r \cdot 8.85 \text{ pF/m}$

where ϵ is the absolute permittivity of the dielectric measured in F/m, ϵ_r is the relative permittivity or dielectric constant (K) having no units, and ϵ_0 is the absolute permittivity of a vacuum—equal to 8.85 pF/m.

(13.9) $C_t = C/N$

same values formula, where C_t is the total capacitance for a series capacitor circuit in which N capacitors are all the same value, and C is the value of one of the capacitors.

(13.10) $C_t = \dfrac{C_1 \cdot C_2}{C_1 + C_2}$

product over the sum formula, where C_t is the total capacitance for two series capacitors C_1 and C_2.

(13.11) $C_t = \dfrac{1}{1/C_1 + 1/C_2 + 1/C_3 + \cdots}$

reciprocal formula, where C_t is the total capacitance for more than two series capacitors.

(13.12) $C_t = C_1 + C_2 + C_3 + \cdots$ (total capacitance for parallel capacitors)

(13.13) $Q_t = Q_1 = Q_2 = Q_3 = \cdots$

where Q_t is the total quantity of charge supplied to *series* capacitors.

(13.14) $Q_t = Q_1 + Q_2 + Q_3 + \cdots$

where Q_t is the total quantity of charge supplied to *parallel* capacitors.

(13.15) $Q_t = C_t \cdot E_s$

where Q_t is the quantity of charge, C_t is total circuit capacitance, and E_s is the source voltage.

(13.16) $V_x = E_s \cdot C_t/C_x$

where V_x is a capacitor voltage drop in a series circuit, and C_x is the capacitance of the capacitor.

CONCEPTS

- A quantity of electrical charge is stored on the plates of a capacitor and energy is stored in the electric flux between the plates.
- Capacitors do not charge instantaneously when a DC voltage is applied, nor do they discharge instantaneously when a discharge path is provided.
- It is the nature of a capacitor to store energy in the electrostatic flux between its plates, adjust the amount of charge on its plates in accorance with any change in applied source voltage, block DC, and pass AC.

- The capacitance of a capacitor is determined by the dielectric permittivity, the surface area of a plate, and the distance between the plates.
- The voltage rating of a capacitor is determined by plate spacing and dielectric strength.
- Total capacitance for series capacitors is calculated using parallel resistor formulas.
- Total capacitance for parallel capacitors is the sum of the capacitances of all parallel capacitors.
- Series capacitors all receive the same quantity of charge when a voltage source is applied.
- In a series capacitor circuit, the capacitor with the least capacitance receives the largest charge voltage.
- Stray capacitance is present between component leads, open switch and relay contacts, copper foils, coil windings, hookup wires, and so on.
- A capacitor should have a very high dielectric resistance (>100 MΩ for most, 500 kΩ to 10 MΩ for electrolytics). If the resistance is less than it should be, the capacitor is either leaky or shorted (very low resistance).
- Inductors pass DC and capacitors block DC.
- Inductors store energy in magnetic flux and capacitors store energy in electrostatic flux.
- As the frequency of an applied AC signal increases, inductor opposition increases while capacitor opposition decreases.

PROCEDURES

Testing a Capacitor with the Ohmmeter

1. Remove the capacitor from its circuit.

2. Short the capacitor's leads to make sure it is fully discharged.

3. Set the ohmmeter to the highest range and connect its test leads across the capacitor. Remember, capacitors whose values are less than 1 μF may not show any charge current on the ohmmeter.

SPECIAL TERMS

- Capacitance (C), plates, dielectric
- Electrostatic induction, electric field
- Electrostatic flux (Ψ), flux density (D)
- Electrostatic field intensity ($\mathscr{E}$)
- Absolute permittivity (ϵ), relative permittivity (ϵ_r)
- Dielectric constant (K)
- Quantity of charge (Q), farad (F)
- Polarized capacitor, nonpolarized electrolytic
- Voltage rating, shelf life
- Wet- and dry-slug tantalums
- Working voltage DC (WVDC), dielectric strength
- Positive and negative temperature coefficient
- Dielectric resistance, leakage current
- Stray capacitance
- Leaky capacitors

Need to Know Solution

- What is a capacitor?
  A capacitor is an electrical component that can store an electrical charge on its plates and energy in an electric field.
- How are capacitors made?
  Capacitors are formed by placing an insulator between two metal plates. Wire leads or terminals are affixed to the plates for electrical connection.
- How do they work?
  When a difference of potential is applied to the capacitor's plates, electrons flow to one plate and repel electrons from the other plate. No current actually flows through the dielectric between the plates. Any change in the externally applied EMF causes a change in the amount of charge on the capacitor's plates. The resulting change in quantity of charge on the plates causes a current in the leads of the capacitor. Thus, capacitors pass AC and block DC.
- How are they used?
  Capacitors are used to pass AC signals while blocking any DC. Thus, they are used as coupling capacitors, bypass capacitors, and tuning components in frequency-selective circuits.

Questions

13-1 The Nature of Capacitance

1. Define *capacitance*.
2. Describe basic capacitor construction.
3. What is electrostatic induction and how does it apply to capacitors?
4. What two factors determine the flux density between the plates of a capacitor?
5. What is the permittivity of a dielectric?
6. What is electric field intensity and what is its unit of measure?
7. Briefly explain why a capacitor will not charge instantly when a voltage is applied.
8. If a voltage source connected across a capacitor is increased in voltage, what happens to the quantity of charge on the capacitor's plates?

13-2 Physical Factors Affecting Capacitance

9. List the three physical factors that affect capacitance.
10. A vacuum has a permittivity of 1. Is that its absolute permittivity or its relative permittivity?
11. What is the relative permittivity of paper?
12. Will increasing plate spacing increase or decrease capacitance?
13. Will increasing plate area increase or decrease capacitance?

13-3 Types of Capacitors

14. What does *hermetically sealed* mean?
15. List three applications for mica capacitors.
16. Describe the construction of a typical paper capacitor.
17. List three types of ceramic capacitors.
18. List four general uses for ceramic capacitors.

19. What is the general range of capacitance for film capacitors?
20. Which film capacitor is very suitable for frequency-selective circuits and oscillators?
21. What is the actual dielectric in an aluminum electrolytic capacitor?
22. Why is the capacitance so high in electrolytic capacitors as compared to film or paper capacitors?
23. Why must proper polarity be observed when connecting an electrolytic capacitor in a circuit?
24. Explain why an aluminum electrolytic has more stray inductance than a mica capacitor.
25. What is a nonpolarized electrolytic capacitor?
26. Explain how slug tantalums can have relatively large amounts of capacitance for their size.
27. For what are vacuum variable capacitors used?
28. For what are air variable capacitors normally used?

13-4 Capacitor Ratings

29. What actually determines a capacitor's voltage rating?
30. What is the dielectric strength of rubber in V/mil?
31. What does it mean for a capacitor to have a negative temperature coefficient?
32. What does a marking of NP0 mean?
33. What is capacitor leakage current?

13-5 Total Capacitance

34. How does finding total capacitance for series capacitors differ from finding total resistance for series resistors?

13-6 Capacitors in DC Circuits

35. The quantity of charge on a capacitor is measured in what units?
36. If two capacitors of different capacitance are connected in series with each other and a DC source, which capacitor will receive the largest voltage charge, the one with more capacitance or less capacitance? Why?

13-7 Stray Capacitance

37. Define *stray capacitance* in your own words.
38. List at least five sources or examples of stray capacitance.
39. Does the effect of stray capacitance increase or decrease as frequency increases?

13-8 Troubleshooting Capacitors

40. If a mica capacitor measures infinite with an ohmmeter, is the capacitor definitely good? Explain.
41. Should you expect a 0.001-μF capacitor to show any charge when an ohmmeter is connected? Why?
42. What should the dielectric resistance of a good aluminum electrolytic capacitor be when measured with the ohmmeter? (give a general range)
43. What piece of test equipment can you use to definitely determine whether a capacitor is open, out of tolerance, or good?

13-9 Comparing Inductors and Capacitors

44. How does the way a capacitor stores energy differ from the way an inductor stores energy?
45. At the first instant a voltage source is applied to an inductor, it acts like a(n) (open/short). At the first instant a voltage source is applied to a capacitor, it acts like a(n) (open/short).

Problems

13-1 The Nature of Capacitance

1. If a capacitor is charged to 10 V and has a plate spacing of 2 mm, what is the electric field intensity?
2. If a capacitor is charged to 20 V, has a plate spacing of 1 mm, and has a dielectric with an absolute permittivity of 44.25 pF/m, what is the flux density between the plates?
3. If a capacitor is charged to 12 V and has a charge of 120 μC, what is its capacitance?
4. If a capacitor is charged to 15 V and has a charge of 1 mC, what is its capacitance?
5. If a 1,000 μF capacitor is charged to 40 V, what is the quantity of charge on its plates?
6. If a 0.022 μF capacitor is charged to 100 V, what is the quantity of charge on its plates?
7. What is the voltage across a 5 μF capacitor that has a 12 mC charge?
8. What is the voltage across a 450 pF capacitor that has a 200 μC charge?
9. Calculate the amount of energy stored in a 20 μF capacitor that has been charged to 30 V.
10. Calculate the amount of energy stored in a 10,000 μF capacitor that has been charged to 5 V.

13-2 Physical Factors Affecting Capacitance

11. What is the absolute permittivity of a dielectric material that has a dielectric constant of 7?
12. What is the absolute permittivity of a dielectric material that has a dielectric constant of 100?
13. Calculate the capacitance of a capacitor that has the following physical factors: plate dimensions are 2 cm $\times$ 10 cm, plate spacing is 0.1 mm, dielectric is air.
14. Calculate the capacitance of a capacitor that has the following physical factors: plate dimensions are 2 cm $\times$ 2 cm, plate spacing is 0.4 mm, dielectric is polyethylene.

13-4 Capacitor Ratings

15. What is the capacitance of a capacitor that is labeled 100MFD?
16. If a mica capacitor is labeled simply as 200, what is its capacitance?
17. What is the capacitance of a Mylar® capacitor that is labeled 472K100? What is its voltage rating?
18. What is the capacitance of a Mylar® capacitor that is labeled 103K400? What is its voltage rating?

19. A mica capacitor has the following color code: 1st digit = orange, 2nd digit = black, multiplier = red, tolerance = green. What is the capacitance?
20. A ceramic disc capacitor has the following color code: 1st digit = green; 2nd digit = black; multiplier = black; tolerance = silver. What is the capacitance?
21. A ceramic-disc capacitor is marked with a P750. What does the P750 indicate?

13-5 Total Capacitance

22. Calculate the total capacitance for three *series* capacitors having the following values: 20 μF, 20 μF, and 10 μF.
23. Calculate the total capacitance for five *series* capacitors having the following values: 1 μF, 0.5 μF, 3 μF, 1 μF, and 2 μF.
24. Calculate the total capacitance for two *series* capacitors having the following values: 4,000 μF, and 8,000 μF.
25. Calculate the total capacitance for three *parallel* capacitors having the following values: 20 μF, 20 μF, and 20 μF.
26. Calculate the total capacitance for three *parallel* capacitors having the following values: 100 pF, 20 pF, and 50 pF.

13-6 Capacitors in DC Circuits

27. If a DC voltage source supplies a total of 3 mC of charge to three *series*-connected 1,000 μF capacitors, how much charge does each capacitor receive?
28. If a DC voltage source supplies a total of 3 mC of charge to three *parallel*-connected 1,000 μF capacitors, how much charge does each capacitor receive?
29. Calculate the total charge delivered by a source voltage of 15 V to the following three *parallel*-connected capacitors: 470 μF, 1,000 μF, 200 μF.
30. Calculate the total charge delivered by a source value of 15 V to the following three *series*-connected capacitors: 470 μF, 1,000 μF, 200 μF.
31. 100 V is applied to a *series* circuit containing three capacitors whose values are 100 pF, 50 pF, and 200 pF. What is the total quantity of charge delivered to the circuit?
32. 20 V is applied to a *series* circuit containing three capacitors whose values are 50 μF, 50 μF, and 200 μF. What is the total quantity of charge delivered to the circuit?
33. 10 V is applied to a *parallel* circuit containing three capacitors whose values are 0.01 μF, 0.02 μF, and 0.047 μF. What is the total quantity of charge delivered to the circuit?
34. Solve for the charge voltage across each capacitor in Problem 27.
35. Solve for the charge voltage across each capacitor in Problem 30.
36. Solve for the charge voltage across each capacitor in Problem 31.

Answers to Self-Checks

Self-Check 13-1

1. Capacitance is the measure of a capacitor's ability to store an electric charge on its plates.
2. One farad is the amount of capacitance needed by a capacitor to store one coulomb of electrons under the influence of an applied voltage of one volt.
3. 833 μF

4. Large
5. 0.024 C
6. 0.022 μF
7. 0.08 J

Self-Check 13-2

1. Plate area, plate spacing, dielectric permittivity
2. 487 pF/m
3. 354 pF
4. Decreases capacitance

Self-Check 13-3

1. Frequency-selective circuits and oscillators
2. Disc, tubular, monolithic chip
3. Large plate areas and very thin dielectrics
4. Tantalum
5. Tuned circuits in low-power transmitters and radio receivers

Self-Check 13-4

1. 50,000 pF or 0.05 μF
2. Plate spacing and dielectric strength
3. A negative temperature coefficient of 100 ppm
4. Current that actually flows through the dielectric

Self-Check 13-5

1. 2.86 μF
2. 750 pF
3. 18.75 pF
4. 7 pF

Self-Check 13-6

1. Equal to the charge on each individual capacitor
2. Equal to the sum of the charges on all parallel capacitors
3. 240 μC
4. 1.778 V, 3.556 V, 17.778 V, 0.889 V

Self-Check 13-7

1. Stray capacitance is random capacitance between any two conductors separated by an insulator—such as interlead capacitance, coil interwinding capacitance, capacitance between hookup wires, etc.
2. High

Self-Check 13-8

1. Very low resistance reading
2. Lower reading than what is normal

3. Capacitors lower than 1 μF show no charge on the ohmmeter and look open when they may actually be normal.
4. To test for open or out-of-tolerance capacitors

Self-Check 13-9

1. Inductors pass DC, capacitors block DC; inductors store energy in a magnetic field, capacitors store energy in an electric field; inductors act like an open the first instant a voltage is applied; capacitors act like a short the first instant a voltage is applied.

SUGGESTED PROJECTS

1. Add some of the main concepts, formulas, and procedures from this chapter to your Electricity and Electronics Notebook.

2. Visit a local electronics parts store and examine the variety of available fixed and variable capacitors. This will give you a stronger mental image of the various capacitor types we have discussed in this chapter.

3. If you have access to a capacitance meter, try to obtain a variety of junk-box capacitors and test them for out of tolerance or open. Use an ohmmeter to check for shorted, open, or leaky capacitors.

Edison will always be known as one of the fathers of modern moving pictures. In 1913, he introduced the Kinetophone for talking motion pictures. Very little significant improvement has been made to motion picture projectors since his time. Some of Edison's larger, more advanced kinetophones are shown here.

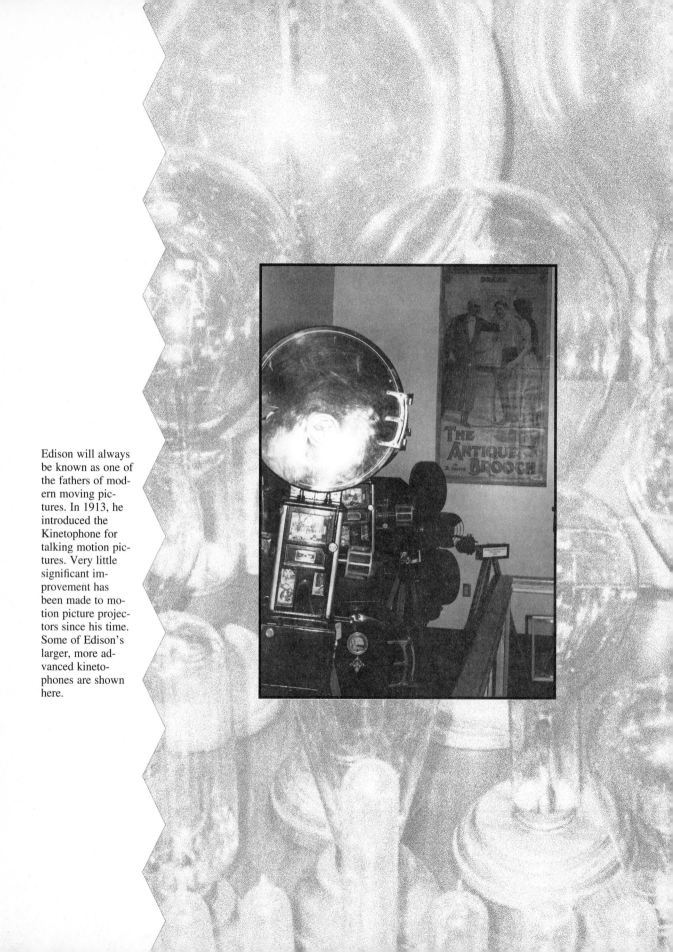

Chapter 14

RC and *RL* Transient Response

OBJECTIVES

After studying this chapter, you will be able to

- calculate the length of time required for a capacitor to fully charge or discharge in any DC-powered *RC* transient circuit.
- calculate the charge, or discharge, voltage and current for an *RC* transient circuit at any instant in time from any initial value of capacitor voltage.
- calculate the length of time required for an inductor to fully build up or decay in any DC-powered *RL* transient circuit.
- calculate the buildup or decay, current, and voltage for an *RL* transient circuit at any instant in time from any initial value of inductor current.
- calculate the length of time required to arrive at a specific value of current or voltage in an *RC* or *RL* transient circuit.
- explain the difference between pure DC and pulsed DC.
- explain how an *RC* or *RL* circuit responds to an applied pulsed DC.
- explain what integrated and differentiated waveforms are and how they are produced in a transient circuit.
- predict short-, medium-, and long-time-constant transient waveforms for transient circuits.
- explain the basic uses for *RC* and *RL* transient circuits.

INTRODUCTION

In this chapter, we complete the bridge between the world of DC and AC circuits. Here, you will examine circuits that produce voltages and currents that are **transient** (changing) over a determinable length of time. Our discussion will include the transient characteristics

525

of *RC* and *RL* circuits as powered by DC and pulsed-DC voltage sources. You will discover how the capacitor opposes changes in DC voltage and how the inductor opposes changes in DC current. Finally, you will be made aware of the importance of *RC* and *RL* transient characteristics in many significant electrical and electronic circuits. A basic understanding of the material contained in this chapter is essential to understand the electronic circuits and systems to which you will be introduced.

NEED TO KNOW

How is it possible to control the speed of a DC motor with an applied DC voltage and current that is rapidly turned on and off (pulsed)? What does a resistor and a capacitor have to do with the length of time a traffic light stays red, yellow, and green? How can resistors and capacitors be used to convert a positive DC source to a negative DC source? How can a resistor and an inductor be used to produce very narrow negative pulses from an applied positive pulsed DC? Why do inductors generate extremely high voltages when a DC source is removed?

Well? Come up with any answers? No? That's OK. That's why you are ready to study this chapter—right? Right! I guess you might say that once again you have a need to know.

14-1 The *RC* Time Constant

The *RC* Time Constant and Charge Time

Charge Time Factors

How long does it take for a capacitor to fully charge? As you know from the previous chapter, the capacitor must charge through any circuit resistances such as actual resistors, lead resistance, plate resistance, etc. The circuit resistance limits the amount of current, which is actually the rate of charge. In addition to this, a large capacitor obtains a greater quantity of charge than a smaller capacitor over its charge period. Thus, a large-value capacitor takes longer to charge than a low-value capacitor through a given circuit resistance. The point is, the length of time required for the capacitor to fully charge depends on two factors: the value of the capacitor and the amount of resistance through which the capacitor must charge.

The *RC* Time Constant

How is the charge time actually calculated? It was just stated that the capacitor charge time is determined by the amount of resistance and capacitance. In fact, the product of the circuit resistance and capacitance is a time period during which the capacitor will charge to 63% of full charge. This time period is known as a **time constant (TC)** and is often symbolized with the greek letter tau (τ). Figure 14-1 illustrates how a time constant is related to capacitor charge time. Notice the capacitor requires approximately five time constants to fully charge, and

$$1 \text{ TC} = 1 \tau = R \cdot C \tag{14.1}$$

where $1\ \tau$ is the time (in seconds) required for the capacitor to charge to 63% of its full charge voltage, or 63% of the remaining charge voltage, R is the series resistance through which the capacitor must charge, and C is the capacitance of the capacitor.

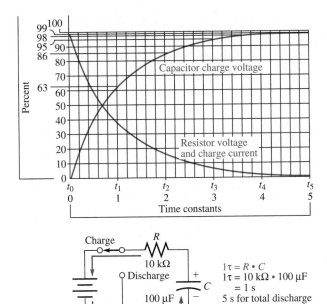

Figure 14-1   It takes five time constants for a capacitor to fully charge.

$1\tau = R \cdot C$
$1\tau = 10 \ k\Omega \cdot 100 \ \mu F$
$\quad = 1 \ s$
5 s for total discharge

The Charging Process

A capacitor can be considered to be fully charged after a period of five time constants, where $5 \ \tau = 5 \cdot R \cdot C$. During each of the five time constants, the capacitor charges to an additional 63% of the remaining voltage. The following example and Figure 14-1 will help you visualize this process. Let's suppose we have a 12 V battery, a switch, a 10 kΩ resistor, and a 100 μF electrolytic capacitor. Using these components, a circuit is assembled as shown in Figure 14-1. We assume we are starting with a fully discharged (neutralized) capacitor and the switch is open.

Time Zero to Time One $(t_0 \rightarrow t_1)$ The capacitor begins to charge at time zero when the switch is first set to the charge position. The first time period is equal to 1 TC (1 τ) and extends from time zero to time one $(t_0 \rightarrow t_1)$. This first time period is equal to $R \cdot C$, which in this case is 10 k$\Omega \cdot 100 \ \mu$F or 1 s. During this first one-second period, the capacitor charges to 63% of the applied voltage. Since the battery voltage is 12 V, the capacitor will have charged to 7.56 V at the end of the first time constant (7.56 = 0.63 $\cdot$ 12 V).

Time One to Time Two $(t_1 \rightarrow t_2)$ The second time constant is a one-second time period extending from time one to time two $(t_1 \rightarrow t_2)$. During this time period, the capacitor charges an additional 63% of the remaining voltage. The remaining voltage is the difference between the present capacitor voltage and the applied source voltage. In this case, the remaining voltage is 4.44 V (4.44 V = 12 V − 7.56 V). Thus, at the end of the second time period, or time constant, the capacitor voltage has increased an additional 2.8 V (2.8 V = 0.63 $\cdot$ 4.44 V). At the end of the second TC, the capacitor has reached a voltage of 10.36 V (10.36 V = 7.56 V + 2.8 V), or 86.3% of the applied source voltage.

Time Two to Time Three $(t_2 \rightarrow t_3)$ The third time constant is a one-second time period extending from time two to time three $(t_2 \rightarrow t_3)$. During this time period, the capacitor charges an additional 63% of the remaining voltage. In this case, the remaining voltage is 1.64 V (1.64 V = 12 V − 10.36 V). Thus, at the end of the third time period, or time constant, the capacitor voltage has increased an additional 1.03 V (1.03 V = 0.63 · 1.64 V). At the end of the third TC, the capacitor has reached a voltage of 11.39 V (11.39 V = 10.36 V + 1.03 V), or 94.93% of the applied source voltage.

Time Three to Time Four ($t_3 \rightarrow t_4$)　The fourth time constant is a one-second time period extending from time three to time four ($t_3 \rightarrow t_4$). During this time period, the capacitor charges an additional 63% of the remaining voltage. In this case, the remaining voltage is 0.61 V (0.61 V = 12 V − 11.39 V). Thus, at the end of the fourth time period, or time constant, the capacitor voltage has increased an additional 0.384 V (0.384 = 0.63 · 0.61 V). At the end of the fourth TC, the capacitor has reached a voltage of 11.774 V (11.774 V = 11.39 V + 0.384 V), or 98.12% of the applied source voltage.

Time Four to Time Five ($t_4 \rightarrow t_5$)　The fifth time constant is a one-second time period extending from time four to time five ($t_4 \rightarrow t_5$). The remaining voltage is now 0.226 V (0.226 V = 12 V − 11.774 V). Thus, at the end of the fifth time period, or time constant, the capacitor voltage increases an additional 0.142 V (0.142 V = 0.63 · 0.226 V). At the end of the fifth TC, the capacitor has reached a voltage of 11.916 V (11.916 V = 11.774 V + 0.142 V), or 99.31% of the applied source voltage. For all practical purposes, the capacitor is considered to be fully charged.

Resistor Voltage and Current During Charge

As the capacitor charges, the series resistor is placed between two opposing voltage sources: the applied source voltage and the capacitor charge voltage. The difference of potential across the resistor, resulting from the difference between capacitor charge voltage and source voltage, determines the amount of current (charge rate) in the circuit at any point in time during the transient period. At time zero (t_0), when the switch is first closed, the capacitor acts as a short and has no charge voltage. It is at this first instant of time that the circuit current is maximum and the resistor has maximum voltage across itself. In our example, the resistor voltage is 12 V at the very first instant of time. It is at this first instant of time that charge current is maximum, as demonstrated by Ohm's Law ($I = E/R =$ 12 V/10 kΩ = 1.2 mA). At the end of the first time constant (at t_1), the capacitor voltage increases by 63% to 7.56 V, leaving 4.44 V across the 10 kΩ resistor (12 V − 7.56 V = 4.44 V, Kirchhoff's Voltage Law). The charge current at that point in time has decreased 63% to 444 μA (444 μA = 4.44 V/10 kΩ). As the capacitor continues to charge with each time constant, the charge current and resistor voltage continue to decrease. Finally, at t_5, resistor voltage and charge current have dropped to zero.

The *RC* Time Constant and Discharge Time

Discharge Time Factors

The same factors affect discharge time as charge time: circuit resistance and capacitance. If the capacitor is discharged through the same resistance through which it was charged, the discharge time will be equal to the charge time, which is equal to 5 τ, or 5 · R · C.

The Discharge Process

Figure 14-2 illustrates the relationship between capacitor discharge and the five time constants. As before, the capacitor is 100 μF and the resistor is 10 kΩ. Therefore, 1 τ is equal

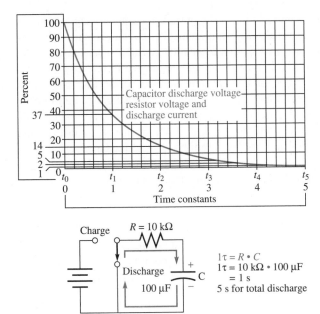

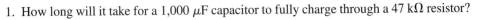

$$1\tau = R \cdot C$$
$$1\tau = 10 \text{ k}\Omega \cdot 100 \text{ }\mu\text{F}$$
$$= 1 \text{ s}$$
5 s for total discharge

Figure 14-2   It takes five time constants for a capacitor to fully discharge.

to 1 s. Notice that for every second of time (every time constant), the capacitor voltage drops an additional 63% of remaining charge voltage. The discharge curve shown is the inverse of the charge curve. Also, note that the resistor voltage is the capacitor voltage. Thus, the resistor voltage and resistor current decrease over the 5 τ period along with the capacitor voltage.

SELF-CHECK 14-1

1. How long will it take for a 1,000 μF capacitor to fully charge through a 47 kΩ resistor?

2. Does resistor voltage increase or decrease as the capacitor charges? Why?

3. If a resistor and a capacitor are connected in series with a 15 V source, what will the capacitor charge voltage be after two time constants?

14-2 The *L/R* Time Constant

The *L/R* Time Constant and Current Buildup

Buildup Time Factors

You should recall that inductor current and magnetic flux build up in and around a coil over a short transient period when a voltage source is applied. The current through the inductor and the flux surrounding the inductor do not instantly appear at their maximum values the moment an EMF is applied. Recall that a CEMF, which opposes the increasing current and expanding magnetic field, is generated by the inductor. If the inductor is large in value, the induced CEMF and resulting opposition to buildup will be significant. It is for this reason that a larger inductor has a longer buildup time than a smaller inductor.

However, the amount of inductance (*L*) is not the only factor contributing to buildup time. The amount of circuit resistance is also very important. The circuit resistance consists of coil-winding resistance and any actual resistors. If the total circuit resistance is low, the maximum buildup current will be high, since the circuit resistance is the only current-limiting factor in the circuit at the end of the transient buildup period (no longer any CEMF). The high maximum current requires that the inductor take longer to build up. A large circuit resistance would limit the maximum buildup current to a lower value. This lower maximum buildup current is reached more quickly by a given value inductor than a high maximum current would be. Thus, the buildup time is short for large series resistances and long for low series resistances. The overall buildup time, therefore, is determined by two factors: the amount of inductance and the inverse of circuit resistance.

The *L/R* Time Constant

How is buildup time actually calculated? As in *RC* transient circuits, charge or buildup is accomplished in a five-time-constant period. In *RL* transient circuits, the length of a time constant is determined by the amount of inductance and the inverse of circuit resistance, as shown in the following mathematical relationship:

$$1 \text{ TC} = 1 \ \tau = L/R \tag{14.2}$$

where $1 \ \tau$ is equal to the inductance *L* divided by the circuit resistance *R*

The Buildup Process

The inductor current and magnetic flux will build up to maximum values over a period of five time constants, where $5 \ \tau = 5 \cdot L/R$. As illustrated in Figure 14-3, the inductor current

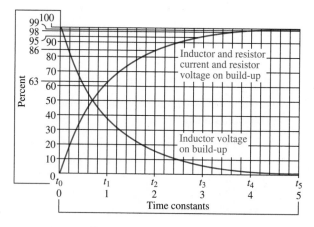

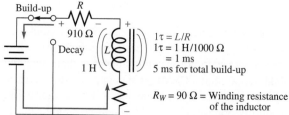

Figure 14-3   It takes five time constants for an inductor to fully build up.

increases by 63% of the remaining buildup current with each time constant. The following discussion describing the current buildup process is very similar to the capacitor charge process. In this case, current is building up instead of voltage. For the purpose of our discussion, our circuit consists of a 12 V battery, a switch, a 1 H inductor that has 90 Ω of winding resistance, and a 910 Ω resistor. Since the inductance is 1 H and the total circuit resistance is 1 kΩ (90 Ω + 910 Ω), each time-constant period will be 1 ms (L/R = 1 H/1 kΩ = 1 ms).

Time Zero to Time One ($t_0 \rightarrow t_1$)   During this first time constant period, the inductor current increases from 0 A to 63% of the maximum buildup current. At the first instant of time t_0, when the switch is first closed, the inductor current is zero and the inductor acts like an open. From t_0 to t_1, the inductor current increases facing opposition from induced CEMF. In order to determine the value of inductor current at t_1, the maximum buildup current must be determined. The maximum buildup current after 5 τ is limited only by the circuit resistance, since the magnetic flux is fully expanded and there is no more CEMF. Thus, the maximum current is found using Ohm's Law ($I = E/R$ = 12 V/1 kΩ = 12 mA). At the end of the first time constant, the inductor current will be 63% of 12 mA, or 7.56 mA (7.56 mA = 0.63 · 12 mA).

Time One to Time Two ($t_1 \rightarrow t_2$)   The second time constant is a 1 ms time period extending from time one to time two ($t_1 \rightarrow t_2$). During this period, the inductor current increases an additional 63% of the remaining possible current (from 7.56 mA to the maximum 12 mA). Thus, at the end of the second time constant t_2, the current has increased by: 2.8 mA [2.8 mA = 0.63 · (12 mA − 7.56 mA)]. The inductor current at t_2 is therefore 10.36 mA (10.36 mA = 7.56 mA + 2.8 mA).

Time Two to Time Three ($t_2 \rightarrow t_3$)   During this 1 ms time period, the inductor current increases an additional 63% of the remaining 1.64 mA (1.64 mA = 12 mA − 10.36 mA). Thus, at t_3, 1.03 mA has been added to the 10.36 mA for a total of 11.39 mA [11.39 mA = 10.36 mA + (0.63 · 1.64 mA)].

Time Three to Time Four ($t_3 \rightarrow t_4$)   During this 1 ms time period, the inductor current increases an additional 63% of the remaining 0.61 mA (0.61 mA = 12 mA − 11.39 mA). Thus, at t_4, 0.38 mA has been added to the 11.39 mA for a total of 11.77 mA [11.77 mA = 11.39 mA + (0.63 + 0.61 mA)].

Time Four to Time Five ($t_4 \rightarrow t_5$)   During this 1 ms time period, the inductor current increases an additional 63% of the remaining 0.23 mA (0.23 mA = 12 mA − 11.77 mA). Thus, at t_5 0.14 mA has been added to the 11.77 mA for a total of 11.91 mA [11.91 mA = 11.77 mA + (0.63 · 0.23 mA)]. For all practical purposes, the inductor current is considered to have reached its maximum level and buildup is complete.

Resistor Voltage and Current, and Inductor Voltage During Buildup

Since the *RL* circuit is a series circuit, the resistor current is the same as the inductor current. As the inductor current builds up, the current through the resistor increases. Thus, resistor current starts at 0 A at time zero and increases to a maximum level at time five. The resistor voltage follows the buildup current and also reaches a maximum level at time five. Thus, the curve for inductor buildup in Figure 14-3 also represents resistor current and resistor voltage.

While inductor current, resistor current, and resistor voltage are increasing, the inductor voltage is decreasing over the buildup transient period. At time zero, the inductor current is 0 A and the inductor looks like an open for the first instant of time (at t_0). It is at this first instant of time that the inductor voltage is maximum (equal to the source voltage), since the source voltage is always measured across an open in a simple series circuit. At the end of the transient period (t_5), the inductor current has reached a maximum level with no further magnetic-field expansion. In the example used for this discussion, the inductor now looks like a 90 Ω resistor with a 1.08 V voltage drop (1.08 V = 12 mA · 90 Ω). The maximum resistor voltage is 10.92 V (10.92 V = 12 mA · 910 Ω). In most cases, the inductor winding resistance is very small compared to other circuit resistances. Therefore, winding resistance is often ignored.

The *L/R* Time Constant and Decay Time

Decay Time Factors

The decay time is affected by the same factors that affect buildup time: the amount of inductance and the amount of resistance. As in buildup, the *RL* circuit requires five time constants to fully decay, where $5\,\tau = 5 \cdot L/R$.

The Decay Process

Figure 14-4 illustrates the inductor-current and magnetic-field decay process. Again, the circuit is composed of a 910 Ω resistor and a 1 H inductor that has a 90 Ω winding resistance. As before, one time constant is 1 ms and the inductor will require 5 ms to fully decay. It is assumed that the switch is *instantly* moved to the decay position. At time zero

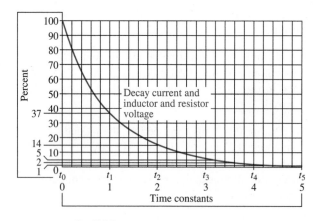

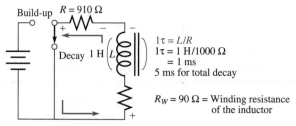

Figure 14-4 It takes five time constants for an inductor to fully decay.

the magnetic flux begins to collapse and induces a maximum voltage across the inductor. For a short five-time-constant transient period, the inductor acts like a generator supplying current to the 910 Ω resistor. It is important to note that the induced inductor voltage is the opposite polarity to the original source voltage used on buildup. You will recall that an inductor opposes any change in current. In this case, the inductor self-induces a current that is in the same direction as the buildup current. Thus, the inductor current replaces the buildup current when the switch is instantly changed to the decay position. In order for the induced decay current to be in the same direction as the buildup current, the self-induced inductor voltage must be of the opposite polarity.

The self-induced decay current starts at the maximum value of 12 mA and decays to 0 A in a period of 5 τ. In a manner similar to buildup, the decay current will decrease by 63% during the first TC period and decrease an additional 63% of the current existing at the beginning of each TC period until the magnetic field is completely collapsed. Naturally, the resistor voltage starts at maximum with the inductor and drops to zero over the 5 τ period as the inductor decays.

In the previous discussions, note that the amount of applied source voltage has nothing to do with inductor buildup and decay time. These times are determined solely by the size of the inductor and resistor value.

Test your understanding by answering the questions of Self-Check 14-2.

1. How is one time constant of an *RL* transient circuit determined?

2. How many time constants does an *RL* transient circuit require for buildup or decay?

3. Calculate the transient time required for a 100 mH coil to fully build up through a 4.7 kΩ resistor.

4. What happens to inductor voltage during the buildup transient period?

5. Does the inductor current change direction from buildup to decay? Explain.

14-3   The Universal Time-Constant Chart

The Universal Chart

In the previous two sections, you discovered it takes five time constants (5 τ) for a capacitor to charge and discharge and an inductor to build up and decay. You also discovered that the transient voltages and currents rise and fall at a nonlinear rate. The transient graphs are rising and falling curves instead of straight lines. It is important to realize that all *RC* and *RL* transient circuits follow the same nonlinear curves and require a five-time-constant transient period. Every *RC* or *RL* combination will charge, or build up, to: 63% at 1 τ, 86% at 2 τ, 95% at 3 τ, 98% at 4 τ, and 99% at 5 τ. Also, every transient circuit will discharge, or decay, to: 37% at 1 τ, 14% at 2 τ, 5% at 3 τ, 2% at 4 τ, and 1% at 5 τ. In other words, the transient *RC* and *RL* characteristics are **universal** for all combinations of *RC* and *RL*.

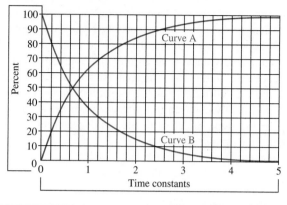

Curve A		Curve B	
RC	*RL*	*RC*	*RL*
v_C on charge	v_R and i_R and i_L on build-up	v_C on discharge v_R and i_R and i_C on charge and discharge	v_L on build-up and decay v_R and i_R and i_L on decay

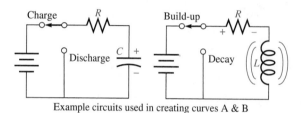

Example circuits used in creating curves A & B

Figure 14-5   The universal time constant chart.

Since the transient characteristics for all combinations of *RC* and *RL* are the same, a chart can be made and used as a tool to quickly determine the percentage of charge and discharge, or buildup and decay, for any transient circuit. Figure 14-5 illustrates such a chart, appropriately named the **universal time-constant chart**. Take time to carefully study the chart and try to visualize which transient currents and voltages are represented by curve A and curve B. You will discover that this chart is actually a review of the previous two sections covering *RC* and *RL* time constants.

Using the Chart

Finding Values of Transient Voltage and Current

The universal time-constant chart can be used to find values of voltage or current at any instant in time during the transient period. This requires two simple steps:

1. Convert the specified length of time into an equivalent number of time constants.
2. Read the percentage of current or voltage from the appropriate curve and convert the percentage to actual voltage or current.

Let's examine a few examples to see how this is done.

Given: an *RC* transient circuit as shown in Figure 14-5, *R* is 10 kΩ, *C* is 1 μF, and the applied DC is 24 V. What is the charge on the capacitor 25 ms after the switch is set to the charge position?

E X A M P L E   1 4 - 1

Solution: First convert the 25 ms to an equivalent number of time constants. To do this, divide the time for one time constant into the specified length of time. In this case, 1 TC = 10 kΩ · 1 μF = 0.01 s = 10 ms. The number of time constants is 25 ms/10 ms = 2.5 τ. Next, locate 2.5 τ on the base line of the universal chart and follow it vertically to intersect with curve A (curve A is used since we want to find capacitor voltage on charge). From the point of intersection, move to the left to the percent scale and read the percent of charge. In this example, the percent of charge is approximately 92%. The charge on the capacitor after 25 ms is 92% of 24 V, or 22.1 V (22.1 V = 0.92 · 24 V).

Given: an *RL* transient circuit as shown in Figure 14-5, *R* is 47 kΩ, *L* is 5 mH, and the applied DC is 12 V. The resistance of the coil windings is negligible. What is the buildup current in the inductor 0.212 μs after the switch is set to the buildup position?

E X A M P L E   1 4 - 2

Solution: First, the specified length of time must be converted to an equivalent number of time constants. In this case, 1 τ = *L/R* = 5 mH/47 kΩ = 0.106 μs. Thus, the number of time constants is 0.212 μs/0.106 μs, or 2 τ. Using the A curve, we move up from the 2, intersect with the curve, then move left to read 86% buildup. At 0.212 μs, the inductor and resistor current has built up to 86% of the maximum current. The maximum current will be 255 μA. Therefore, at 0.212 μs the inductor and resistor current is 219 μA (219 μA = 0.86 · 255 μA).

Given: an *RL* transient circuit as shown in Figure 14-5, *R* is 1 kΩ, *L* is 0.2 H, and the applied DC is 6 V. The resistance of the coil windings is negligible. What is the resistor voltage 0.4 ms after the switch is set to the decay position? (Assume the inductor current and field were fully built up before the switch is set to the decay position.)

E X A M P L E   1 4 - 3

Solution: As before, start by converting the specified time to the equivalent number of time constants. In this case, 1 τ = *L/R* = 0.2 H/1 kΩ = 0.2 ms. Thus, the number of time constants is 0.4 ms/0.2 ms = 2 τ. In this example, the B curve is used because we are finding the resistor voltage on decay. Reading up and across to the percent scale we find that 14% of the maximum voltage is across the resistor 0.4 ms into decay. Thus, the resistor voltage, also equal to the inductor voltage, is 0.84 V (0.84 V = 0.14 · 6 V).

Finding the Length of Time Required for Specific Values of Voltage or Current

The universal time-constant chart can also be used to determine the length of time required to reach a specific voltage or current. There are three basic steps involved in this procedure:

1. Convert the specific voltage or current to a percentage of the maximum value.

2. Use the appropriate curve on the time-constant chart to translate the percentage to the corresponding number of time constants.

3. Convert the number of time constants to actual time by multiplying the number of time constants by the time for one time constant.

Let's consider a couple of examples.

E X A M P L E   14-4   **Given:** an *RC* transient circuit with a DC source voltage of 10 V, $R = 2.7$ kΩ and $C = 4.7$ μF. How long will it take for the capacitor charge voltage to reach 3 V? (Assume the capacitor started to charge from 0 V.)
Solution: First, the 3 V charge is converted to a percentage of the 10 V source. 3 V is 30% of 10 V (30% = 3 V/10 V · 100%). Second, use the TC chart to determine the number of time constants corresponding to the 3 V charge (curve A). In this case, it is approximately 0.33 τ. Finally, convert 0.33 τ to actual time. The actual time is 0.33 τ = 0.33 · 2.7 kΩ · 4.7 μF = 0.0042 s = 4.2 ms. Thus, it takes 4.2 ms for the capacitor voltage to reach 3 V.

E X A M P L E   14-5   **Given:** an *RL* transient circuit with a DC source voltage of 15 V, $R = 47$ kΩ and $L = 100$ mH (negligible winding resistance), how long will it take for the inductor buildup current to reach 0.2 mA? (Assume the buildup started from 0 A.)
Solution: First, convert the 0.2 mA to a percentage of full buildup current. The full buildup current is found using Ohm's Law: $I = E/R = 15$ V/47 kΩ = 0.32 mA. The specified current is found to be 62.5% of the full buildup current (62.5% = 0.2 mA/0.32 mA · 100%). Second, use the TC chart to convert the percentage to the corresponding number of time constants (curve A). In this case, 62.5% is approximately 1 τ. Finally, convert 1 τ to the actual time. The actual buildup time is $L/R = 100$ mH/47 kΩ = 2.13 μs. Thus, the inductor current will build up to 0.2 mA in 2.13 μs.

Exponential Curves and Equations

Curves A and B of the universal time-constant chart are referred to as **exponential curves**. That is, the curves are a function of an exponential equation. In fact, if the exponential equations are known and an engineering calculator is available, the values of voltage, current, and time can be determined without using the TC chart. The following discussion presents these equations with examples of their use.

General Equations for Voltage and Current

The following two equations are general formulas that can be used to determine instantaneous values of voltage or current for curve A or B (*RC* and *RL* transient circuits):

$$v = V_F + (V_I - V_F) \cdot e^{-t/\tau} \tag{14.3}$$

$$i = I_F + (I_I - I_F) \cdot e^{-t/\tau} \tag{14.4}$$

where v and i are instantaneous values of voltage and current at a particular instant in time (t), V_F and I_F are **F**inal values of voltage or current at the end of a 5 τ period, V_I and I_I are **I**nitial or starting values of voltage or current, e is the base of natural logarithms and is equal to 2.718, and τ is one time constant for *RC* or *RL* transient circuits ($\tau = R \cdot C$ or L/R).

As you can see, Formulas 14.3 and 14.4 use **natural logarithms**—to the base 2.718. This should not pose any real problem for you. Most scientific and engineering calculators have a natural log key labeled **e^x**. The use of this key will make solving these equations almost as simple as Ohm's Law. The first step in using these formulas is to clearly identify the value of each variable in the needed formula. This is accomplished by examining the parameters given for a particular transient circuit. Let's look at some examples to see how these formulas and the e^x key are used.

Given: an *RC* transient circuit with a 20 VDC source, $R = 1 \text{ k}\Omega$ and $C = 10 \ \mu\text{F}$. What   E X A M P L E   14-6
is the charge voltage after 15 ms of charging? The capacitor charges from 0 V.
Solution: First, identify the value of all variables on the right side of the equation, in
this case Formula 14.3. In this example, the initial voltage is 0 V, the final voltage at
the end of 5 τ is 20 V, the specified length of time t is 15 ms, and 1 τ = R · C =
1 kΩ · 10 μF = 10 ms. With these values, Formula 14.3 becomes:

$$v = 20 \text{ V} + (0 \text{ V} - 20 \text{ V}) \cdot e^{-15\text{ms}/10\text{ms}}$$

The calculator solution is best approached by starting on the right end of the equation.
Enter $-15 \text{ E}-3 \div 10 \text{ E}-3$, then press the equals key [=]. Then press the e^x key (this
may require that a shift key be pressed first). Next, multiply the e^x key result by the
evaluated quantity in parentheses (-20). Then add 20. For many calculators the keys
should be pressed in the following order: (The final answer should round off to 15.5 V.)

[1] [5] [±] [EXP] [3] [±] [÷] [1] [0] [EXP] [3] [±] [=]

[SHIFT] [e^x] [X] [2] [0] [±] [+] [2] [0] [=] 15.5 V

Given: an *RL* transient circuit with a 12 VDC source, $R = 470 \ \Omega$ and $L = 10$ mH   E X A M P L E   14-7
(winding resistance is negligible). What is the decay current after 50 μs of decay? The
inductor decays from full buildup current.
Solution: First, identify the value of all variables on the right side of the equation, in
this case Formula 14.4. In this example, the initial current is 25.5 mA, the final current
at the end of 5 τ is 0 A, the specific length of time t is 50 μs, and 1 τ = L/R =
10 mH/470 Ω = 21.3 μs. With these values, Formula 14.4 becomes:

$$i = 0 \text{ A} + (25.5 \text{ mA} - 0 \text{ A}) \cdot e^{-50\mu\text{s}/21.3\mu\text{s}}$$

As before, start on the right end of the equation. Enter $-50 \text{ E}-6 \div 21.3 \text{ E}-6$, then
press the equals key (=). Then press the e^x key (this may require that a shift key be
depressed first). Next, multiply the e^x key result by the evaluated quantity in parentheses
(25.5 mA). The calculator keys should be pressed in the following order: (The final answer
should round off to 2.44 mA.)

[5] [0] [±] [EXP] [6] [±] [÷] [2] [1] [.] [3] [EXP] [6] [±] [=]

[SHIFT] [e^x] [X] [2] [5] [.] [5] [EXP] [3] [±] [=] 2.44 mA

Take a few minutes to study Design Note 14-1 before answering the questions of
Self-Check 14-3. Note that Design Note 14-1 has a program for calculating instantaneous
values of transient voltage and current and a program for calculating transient time. These
programs represent a software version of the universal time-constant chart.

S E L F - C H E C K   14-3

1. Use the universal TC chart to determine the percentage of voltage on a capacitor after
   2 τ of charging.

2. Use the universal TC chart to determine the percentage of inductor current 1.5 τ into
   decay.

(Continued next page)

3. Without looking, list the transient voltages and currents represented by the B curve.

4. Given: an *RC* transient circuit with a 10 VDC source, $R = 8.2$ kΩ and $C = 10,000$ μF. What is the remaining resistor voltage after 100 s of discharge? Assume the capacitor to have been fully charged before the switch is placed in the discharge position.

5. At the first instant of a charge period, is capacitor current maximum or minimum?

6. At the first instant of a decay period, is inductor current minimum or maximum?

14-4 *RC* Transient Response and Waveforms
The Pulsed-DC Square Wave

Describing a Pulsed-DC Square Wave

In our discussion of *RC* transient response and waveforms, a pulsed DC is applied to the *RC* circuit. Pulsed DC is often referred to as a *square wave*, or *rectangular* wave, since the waveform looks like a square-cornered box when viewed on an oscilloscope. Figure 14-6 illustrates the pulsed-DC waveform and offers a circuit analogy demonstrating how the waveform might be produced mechanically. As you see, an SPDT switch is connected to a positive DC source. When the switch is set to the ON position, the common terminal of the switch instantly jumps to + 10 VDC. When the switch is toggled to the OFF position, the common terminal drops instantly to 0 V (shorted to ground). A positive pulse occurs every time the switch is set to the ON position. The amplitude of the pulse is referred to as the **peak pulse amplitude** and is measured from the 0 VDC reference to the positive, or negative, **crown**. The length of time from the instant the switch is switched on to the time it is switched off is called the **pulse duration (PD)** or **pulse width (PW)**. One *on* time plus one *off* time equals one **pulse cycle**. The length of time for a pulse cycle is the **cycle time (T)**. The percentage of *on* time (pulse duration) as compared to the cycle time is known as the pulse **duty cycle**. In our discussions, we will use pulsed-DC square waves that have a 50% duty cycle, meaning the pulse duration is half the cycle time, which also means the *on* and *off* times are equal.

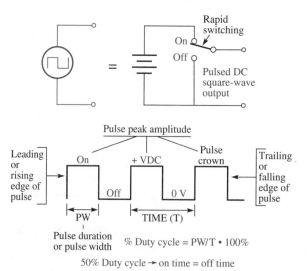

Figure 14-6   Pulsed DC and duty cycle.

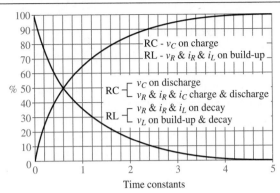

Time constants

$$v = V_F + (V_I - V_F) \cdot e^{-t/\tau} \qquad\qquad i = I_F + (I_I - I_F) \cdot e^{-t/\tau}$$

TRANSIENT VOLTAGE OR CURRENT CALCULATOR PROGRAM	TRANSIENT TIME CALCULATOR PROGRAM

```
10  CLS
20  PRINT"TRANSIENT V OR I"
30  PRINT""
40  PRINT"THIS PROGRAM WILL CALCULATE THE
    INSTANTANEOUS VOLTAGE OR CURRENT OF
    AN RC OR RL"
50  PRINT"CIRCUIT AT A SPECIFIED TIME (t)."
70  PRINT""
80  INPUT"IS THE TRANSIENT CIRCUIT RC OR RL?
    (ENTER RC OR RL) ";A$
90  IF A$ = "RL" THEN GOTO 120
100 INPUT"ENTER THE CAPACITANCE IN FARADS.
    USE E NOTATION. ";C
110 GOTO 130
120 INPUT"ENTER THE INDUCTANCE IN HENRIES. USE
    E NOTATION IF DESIRED. ";L
130 INPUT"ENTER THE RESISTANCE IN OHMS. ";R
140 INPUT"ENTER THE INITIAL VALUE OF VOLTAGE, IF
    RC, OR CURRENT, IF RL. ";I
150 INPUT"ENTER THE TRANSIENT TIME (t) IN SEC-
    ONDS. USE E NOTATION IF DESIRED. ";T
160 PRINT"ENTER THE FINAL VALUE ASSUMING THE
    CIRCUIT IS PERMITTED TO FULLY CHARGE,"
170 INPUT"DISCHARGE, BUILDUP, OR DECAY. ";F
180 PRINT""
200 IF A$ = "RL" THEN Y = -1 * T/(L/R):GOTO 220
210 Y = -1 * T/(R * C)
220 X = F + (1 - F) * EXP(Y)
230 PRINT"THE INSTANTANEOUS VALUE IS ";X
240 INPUT"ANOTHER PROBLEM? (Y/N)";A$
250 IF A$ = "Y" THEN CLEAR: GOTO 70
260 CLS:END
```

```
10  CLS
20  PRINT"TRANSIENT TIME"
30  PRINT""
40  PRINT"THIS PROGRAM WILL CALCULATE THE
    TIME REQUIRED FOR A CAPACITOR TO ARRIVE
    AT A"
50  PRINT"SPECIFIED CHARGE VOLTAGE OR FOR AN
    INDUCTOR TO ARRIVE AT A SPECIFIED BUILDUP"
60  PRINT"OR DECAY CURRENT."
70  PRINT""
80  INPUT"IS THE TRANSIENT CIRCUIT RC OR RL?
    (ENTER RC OR RL)";A$
90  IF A$ = "RL" THEN GOTO 120
100 INPUT"ENTER THE CAPACITANCE IN FARADS.
    USE E NOTATION. ";C
110 GOTO 130
120 INPUT"ENTER THE INDUCTANCE IN HENRIES. USE
    E NOTATION IF DESIRED. ";L
130 INPUT"ENTER THE RESISTANCE IN OHMS. ";R
140 INPUT"ENTER THE INITIAL VALUE OF VOLTAGE, IF
    RC, OR CURRENT, IF RL. ";I
150 INPUT"ENTER THE INSTANTANEOUS VALUE THAT
    EXISTS AT THE TRANSIENT TIME. ";X
160 PRINT"ENTER THE FINAL VALUE ASSUMING THE
    CIRCUIT IS PERMITTED TO FULLY CHARGE,"
170 INPUT"DISCHARGE, BUILDUP, OR DECAY. ";F
180 PRINT""
190 Y = LOG ((X - F)/(I - F))
200 IF A$ = "RL" THEN T = -1 * Y * L/R:GOTO 220
210 T = -1 * Y * R * C
220 PRINT"THE TRANSIENT TIME REQUIRED IS ";T;"
    SECONDS."
230 PRINT""
240 INPUT"ANOTHER PROBLEM? (Y/N)";A$
250 IF A$ = "Y" THEN CLEAR: GOTO 70
260 CLS:END
```

Practical Pulsed-DC Circuits

The circuit illustrated in Figure 14-6 is not practical, since a mechanical switch would not withstand rapid switching and continuous operation. The circuit is merely analogous to the operation of electronic pulse circuits that switch at rates much higher than would be practical for a mechanical switch. These electronic circuits vary in design and are categorized as timers, multivibrators, choppers, switchers, and function generators. In the discussions to follow, we will use a function generator to provide a pulsed DC signal at the desired pulse rate, pulse amplitude, and 50% duty cycle.

Square-Wave Integration

Integrated Waveform Across the Capacitor

See color photo insert page D4.

When a pulsed-DC square wave is applied to an *RC* circuit, the square waveform is modified according to the transient response of the *RC* combination. If the pulsed DC is applied to the resistor and a voltage output is taken across the capacitor as shown in Figure 14-7, a transient waveform called an **integrated waveform** is produced. The circuit, with resistor in series and capacitor to ground, is called an **integrator**. The word *integrate* means "to combine and average out." An integrator combines and averages out DC pulses. In so doing, the capacitor charges slowly during the DC pulse and discharges slowly when the DC pulse is removed. The capacitor's opposition to rapid changes in voltage becomes quite evident when observed with an oscilloscope. As shown in the figure, the capacitor's voltage waveform includes the charge curve on the rising edge of the pulse and the discharge curve on the falling edge of the pulse. The capacitor's voltage is slow to change with the very quick rising and falling edges of the applied pulsed-DC square wave.

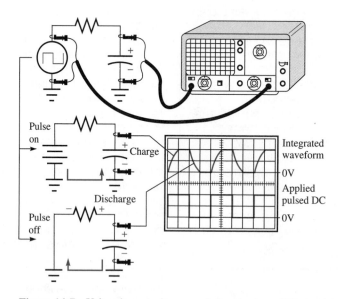

Figure 14-7   Using the capacitor as an integrator in an RC circuit.

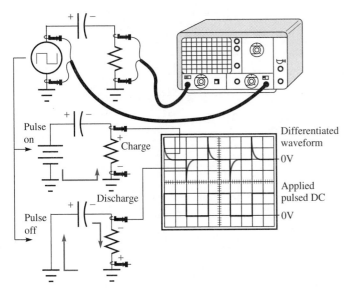

Figure 14-8   The voltage across the resistor is a differentiated waveform in an *RC* circuit.

Square-Wave Differentiation

Differentiated Waveform Across the Resistor

Figure 14-8 illustrates how a pulsed-DC square wave is differentiated across the resistor
of an *RC* transient circuit. The word "differentiate" means to separate, to change or alter,
and to make distinct. The **differentiated waveform** is characterized by two distinct pulses
of opposite polarity. Notice that the resistor voltage rises sharply in the positive direction
with the rising edge of the applied square wave and rises sharply in the negative direction
with the falling edge of the square wave. This is because the capacitor acts like a short at
the rising edge of the pulse, providing an instantaneous positive potential across the resistor. On the falling edge of the square-wave pulse, the positive side of the charged capacitor
is taken to ground. The capacitor's charge is neutralized through the resistor. Notice that
the voltage polarity across the resistor is suddenly reversed because the capacitor is acting
as a voltage source, temporarily reversing the direction of current through the resistor, as
it discharges.

See color photo insert
page D5.

Short, Medium, and Long Time Constants

5*RC* and Pulse Width

The integrated or differentiated transient waveforms may vary significantly in appearance
and characteristics depending on the length of the time constant as compared to the pulse
duration. The pulse duration is the time *allowed* for the capacitor to charge. Five times the
RC time constant is the time *required* by the capacitor to fully charge. If the time required
for a full charge (5*RC*) is *shorter* than the time allowed (pulse duration or pulse width),
the *RC* time constant is said to be SHORT (**SHORT TC** if 5*RC* < *PW*). If the time
required is much *longer* than the time allowed, the *RC* time constant is said to be LONG

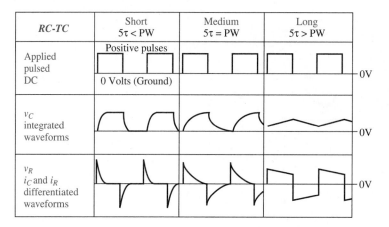

RC-TC	Short $5\tau < PW$	Medium $5\tau = PW$	Long $5\tau > PW$
Applied pulsed DC	Positive pulses⎍⎍ 0 Volts (Ground)		—0V
v_C integrated waveforms			—0V
v_R i_C and i_R differentiated waveforms			—0V

Figure 14-9   Short, medium, and long-time constants for RC transient circuits.

(**LONG TC** if $5RC > PW$). If the time required for a full charge is *equal* to the time allowed, the *RC* time constant is said to be MEDIUM (**MEDIUM TC** if $5RC = PW$).

Time Required vs Time Allowed

Transient waveforms for short, medium, and long time constants are illustrated in Figure 14-9. As you can see, integrated waveforms from short-time-constant transient circuits show a very rapid charge and discharge time compared to the time allowed by the applied pulsed-DC square wave. The short-time-constant differentiated waveform is composed of very narrow positive and negative pulses. Because of the time allowed, short-time-constant *RC* circuits are able to fully charge and discharge with each pulse cycle. The medium-time-constant circuits also have sufficient time allowed to completely charge and discharge, since the time allowed is equal to $5RC$, or $5\ \tau$. The long-time-constant circuit, however, does not have sufficient time allowed to fully charge and discharge. As a result, the integrated waveform does not reach a peak voltage equal to the peak pulse amplitude. The long-time-constant circuit is treated as a special case in the next section.

Significance of the Long Time Constant

A Steady-State Average DC Level

The waveform produced by the long-time-constant *RC* circuit is very different from that produced by the short- or medium-time-constant circuit. If the *RC* time constant is long compared to the time allowed by the square-wave pulse width, the capacitor will not fully charge during any pulse *on* time. Neither will the capacitor fully discharge the charge it received during the previous *on* time. With each applied-square wave pulse cycle, the capacitor will gain a little more voltage during the *on* time than it had gained in the previous *on* time and will discharge to a higher voltage during the *off* time than it had dropped to in the previous *off* time. This process continues until the long-time-constant integrated waveform reaches a steady-state condition varying slightly above and below the average pulse voltage (½ peak pulse amplitude for 50% duty cycle). As shown in Figure 14-10, it takes approximately five full time constants for the capacitor voltage to reach this average steady-state condition.

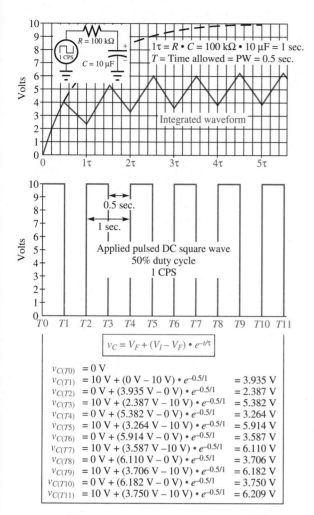

$$v_C = V_F + (V_I - V_F) \cdot e^{-t/\tau}$$

$v_{C(T0)}$ = 0 V
$v_{C(T1)}$ = 10 V + (0 V – 10 V) • $e^{-0.5/1}$ = 3.935 V
$v_{C(T2)}$ = 0 V + (3.935 V – 0 V) • $e^{-0.5/1}$ = 2.387 V
$v_{C(T3)}$ = 10 V + (2.387 V – 10 V) • $e^{-0.5/1}$ = 5.382 V
$v_{C(T4)}$ = 0 V + (5.382 V – 0 V) • $e^{-0.5/1}$ = 3.264 V
$v_{C(T5)}$ = 10 V + (3.264 V – 10 V) • $e^{-0.5/1}$ = 5.914 V
$v_{C(T6)}$ = 0 V + (5.914 V – 0 V) • $e^{-0.5/1}$ = 3.587 V
$v_{C(T7)}$ = 10 V + (3.587 V –10 V) • $e^{-0.5/1}$ = 6.110 V
$v_{C(T8)}$ = 0 V + (6.110 V – 0 V) • $e^{-0.5/1}$ = 3.706 V
$v_{C(T9)}$ = 10 V + (3.706 V – 10 V) • $e^{-0.5/1}$ = 6.182 V
$v_{C(T10)}$ = 0 V + (6.182 V – 0 V) • $e^{-0.5/1}$ = 3.750 V
$v_{C(T11)}$ = 10 V + (3.750 V – 10 V) • $e^{-0.5/1}$ = 6.209 V

Figure 14-10   Long time constant integrated transient response.

Full Integration

As long as the *RC* time constant is long, the average integrated waveform voltage (average capacitor voltage) will always reach a steady-state value equal to half of the peak pulse amplitude (assuming a 50% duty cycle). However, the difference between the peak and valley voltages of the integrated waveform will vary widely, depending on how long a time constant the *RC* circuit really has. If the *RC* time constant is extremely long compared to the pulse width, the integrated waveform will have peak and valley voltages very close to each other, varying only slightly above and below the average DC level. If the time constant is extremely long, the integrated waveform will appear flat and the pulsed-DC square wave applied is said to approach **full integration**. The example in Figure 14-10 is far from being fully integrated. The long time constant must be lengthened further by increasing the value of *C* or *R* or both in order to approach full integration. While it might be argued that full integration is never really achieved in a practical circuit, we can design the *RC* time constant to provide a level of integration that is more than adequate for a given application, where a peak and valley variation of ± 1% of the average voltage may be considered to be a fully integrated condition for all practical purposes. In summary, the overall effect of full integration achieved using an extremely long-time-constant circuit is that of converting a pulsed-DC square wave to a smooth DC level equal to the average of the peak pulse amplitude. The extremely long TC integrator is actually a pulsed-DC to pure DC converter.

The Long-TC Differentiated Waveform

While the capacitor is producing an integrated waveform, a differentiated waveform is being developed across the resistor. Figure 14-11 illustrates the development of the differentiated waveform using the same circuit as used in the previous discussion. One of the first things you should notice in the figure is that as the integrated waveform's average voltage increases to 5 VDC, the average voltage of the resistor's differentiated wave decreases to 0 V. During the first 5 τ period, the differentiated wave shifts downward in step

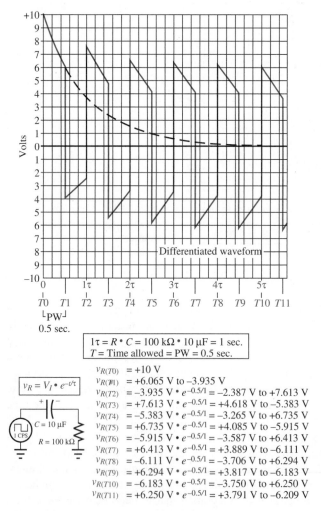

$1\tau = R \cdot C = 100 \text{ k}\Omega \cdot 10 \text{ }\mu\text{F} = 1 \text{ sec.}$
$T = \text{Time allowed} = \text{PW} = 0.5 \text{ sec.}$

$v_R = V_I \cdot e^{-t/\tau}$

$v_{R(T0)} = +10 \text{ V}$
$v_{R(T1)} = +6.065 \text{ V to } -3.935 \text{ V}$
$v_{R(T2)} = -3.935 \text{ V} \cdot e^{-0.5/1} = -2.387 \text{ V to } +7.613 \text{ V}$
$v_{R(T3)} = +7.613 \text{ V} \cdot e^{-0.5/1} = +4.618 \text{ V to } -5.383 \text{ V}$
$v_{R(T4)} = -5.383 \text{ V} \cdot e^{-0.5/1} = -3.265 \text{ V to } +6.735 \text{ V}$
$v_{R(T5)} = +6.735 \text{ V} \cdot e^{-0.5/1} = +4.085 \text{ V to } -5.915 \text{ V}$
$v_{R(T6)} = -5.915 \text{ V} \cdot e^{-0.5/1} = -3.587 \text{ V to } +6.413 \text{ V}$
$v_{R(T7)} = +6.413 \text{ V} \cdot e^{-0.5/1} = +3.889 \text{ V to } -6.111 \text{ V}$
$v_{R(T8)} = -6.111 \text{ V} \cdot e^{-0.5/1} = -3.706 \text{ V to } +6.294 \text{ V}$
$v_{R(T9)} = +6.294 \text{ V} \cdot e^{-0.5/1} = +3.817 \text{ V to } -6.183 \text{ V}$
$v_{R(T10)} = -6.183 \text{ V} \cdot e^{-0.5/1} = -3.750 \text{ V to } +6.250 \text{ V}$
$v_{R(T11)} = +6.250 \text{ V} \cdot e^{-0.5/1} = +3.791 \text{ V to } -6.209 \text{ V}$

Figure 14-11   Long time constant differentiated transient response.

with each cycle of the applied pulsed-DC square wave. (Take time to study the diagrams and calculations to see how this is done.) After 5 τ, the differentiated wave will have settled to a steady-state average voltage of 0 V with positive and negative peaks of approximately 6.2 V. You should recall that the negative peaks are created as the capacitor discharges through the resistor. As the capacitor begins to discharge, the resistor current is reversed and so is the resistor's voltage drop.

Full Differentiation

It is interesting to note here that if the *RC* time constant is extremely long, the differentiated waveform across the resistor will be a close reproduction of the pulsed-DC square wave and can be considered to be **fully differentiated**. The main difference between the resistor's differentiated wave and the applied pulsed DC is that the applied pulsed DC has an average voltage equal to half the peak pulse amplitude (assuming a 50% duty cycle) and the differentiated wave has an average voltage of 0 V. If the *RC* time constant of the circuit used in Figure 14-11 were significantly increased, the differentiated wave would look like the applied pulsed DC displaced to an average of 0 V and with positive and negative peaks of ± 5 V—a fully differentiated condition. In effect, the pulsed DC is transformed into an AC square wave. Thus, the extremely-long-time-constant differentiator is a pulsed-DC to AC square-wave converter.

Time for a review. Take a few minutes to answer the questions of Self-Check 14-4 before continuing.

SELF-CHECK 14-4

1. Define *pulse duration*.

2. What does 50% duty cycle mean?

3. Is an integrated waveform produced across the (a) capacitor or the (b) resistor in an *RC* transient circuit?

4. In effect, what does a very long TC integrator do?

5. Describe a fully differentiated waveform.

6. Is a fully differentiated waveform a pulsed-DC or an AC square wave?

7. In effect, what does an extremely long TC differentiator do?

14-5 *RL* Transient Response and Waveforms

Square-Wave Integration Across the Resistor

In *RL* transient circuits, the integrated waveform is developed across the resistor. The voltage waveform across the resistor is the result of the current waveform established by the inductor during buildup and decay. Recall that an inductor will act like an open for a brief instant the moment a DC is applied. This is due to the large CEMF that is instantly

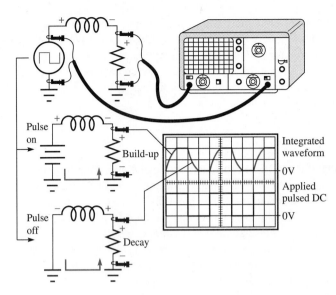

Figure 14-12   The voltage across the resistor is an integrated waveform in an RL circuit.

established. Over the 5 τ buildup period, the CEMF decreases and inductor current increases. As the inductor current increases, so does the resistor current. The increasing inductor and resistor current establishes an increasing voltage drop across the resistor. During decay, the process is reversed, with inductor and resistor current decreasing. Note that the resistor's voltage polarity does not reverse during decay, the inductor's voltage does. This maintains the current in the same direction as the buildup current, and therefore the resistor's voltage drop polarity does not change. The resulting voltage and current waveform is the integrated waveform shown in Figure 14-12.

Square-Wave Differentiation Across the Inductor

While the integrated waveform is developed across the resistor, the differentiated waveform is developed across the inductor. Differentiation takes place across the inductor because the inductor's voltage polarity reverses at the start of decay. An instantaneous positive voltage pulse is created at the start of buildup, and an instantaneous negative voltage pulse is created at the start of decay. The negative pulse starts at a maximum negative amplitude because the inductor decay current starts at maximum and self-induces a large voltage at the beginning of decay. Figure 14-13 illustrates the differentiated inductor waveform.

Short, Medium, and Long Time Constants

As in *RC* transient circuits, *RL* transient circuits produce short-, medium-, and long-time-constant transient waveforms. The definition for each is the same as that for *RC* circuits: short TC is when 5 τ < *PW*; medium TC is when 5 τ = *PW*; and long TC is when 5 τ > *PW* (where 5 τ = 5 · *L/R* and *PW* is the pulse width, or pulse *on* time). Five time constants is the required buildup and decay time, while the pulse width is the time allowed

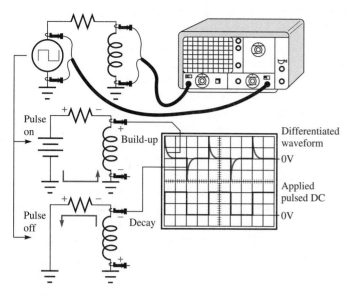

Figure 14-13   The voltage across the inductor is a differentiated waveform in an RL circuit.

for the buildup and the pulse *off* time is the time allowed for decay. These three different conditions are illustrated in Figure 14-14. Note that whether a TC is short, medium, or long is determined by the ratio of L/R and, of course, the pulsed-DC *on* and *off* times. A short TC can be made long by increasing the inductance or decreasing the resistance. Conversely, a long TC can be made short by decreasing the inductance or increasing the resistance. Changing the applied pulse width, duty cycle, and/or cycle time will also change the classification of a particular circuit as being short, medium, or long. If the applied pulsed-DC square wave is not 50% duty cycle (*on* time and *off* time not equal), it is possible for the circuit to have a long TC on buildup and a short TC on decay or a short TC on buildup and a long TC on decay (this also applies to *RC* transient circuits).

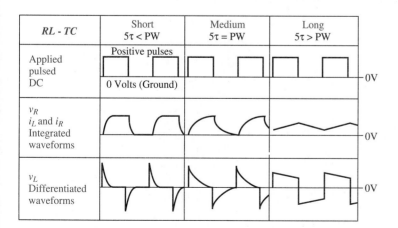

Figure 14-14   Short, medium, and long time constants for RL transient circuits.

Significance of the Long Time Constant

What makes the long-TC *RL* transient circuit more significant than the short- and medium-TC circuits? The inductor is not permitted to fully build up or fully decay, since 5 τ (5 · *L/R*) is longer than the *on* and *off* times of the applied pulsed DC. The result is an integrated waveform across the resistor that rises to an average DC level of some value other than 0 VDC. The differentiated waveform across the inductor is a wide-pulse differentiated waveform that looks like an AC square wave with sloped crowns. These are illustrated in Figure 14-15.

As shown in Figure 14-15, the integrated waveform rises to an average steady-state DC level equal to half the peak pulse amplitude after 5 τ (assuming the applied pulsed DC has a 50% duty cycle). During the same 5-τ period, the differentiated waveform settles to an average steady-state DC level of 0 V. As with long-TC *RC* circuits, if the *RL* time constant is made very long (by increasing *L* or decreasing *R*), the integrated waveform across the resistor will approach a smooth DC (full integration) and the differentiated waveform will approach a perfect AC square wave (full differentiation).

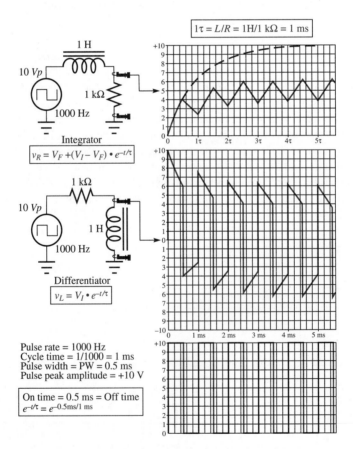

Figure 14-15   Long time constant RL transient waveforms.

Significance of Long Buildup and Short Decay

If in an *RL* transient circuit the inductor decay time is made shorter than the buildup time, the inductor will generate an initial decay voltage that is higher than the applied source voltage. This higher self-induced voltage can be, and often is, harmful to other circuit components. But how is it possible for an inductor to generate a higher voltage than what was originally applied? Figure 14-16 shows a practical example. The circuit includes a voltage source, a switch, a 1 H inductor, and a 1 kΩ resistor. The resistor is used to help limit the maximum current to 10 mA when the switch is closed (10 mA = 10 V/1 kΩ). As you can see, the inductor will store 50 μJ of energy with a current of 10 mA.

What happens to that stored energy when the switch is opened? Does it simply vanish or disappear? Absolutely not. The energy is concentrated into a higher-voltage pulse of very short duration. Let's see how. For our investigation, let's assume that the open-contact resistance of the switch is 1 GΩ (1 GΩ = resistance through the air between contacts). When the switch is opened, the circuit time constant drops from 1 ms (1 ms = 1 H/1 kΩ) to 1 ns (1 ns = 1 H/1 GΩ). In other words, one time constant on buildup equals 1 ms, while one time constant on decay equals 1 ns. Thus, because of the sudden and extreme increase in resistance due to the opened switch, the decay time drops to one millionth of the buildup time. The inductor must release the energy one million times faster than it collected the energy (1 ms/1 ns = 1,000,000). Recall that Faraday's Law for induced voltage states that the amount of induced voltage is in direct proportion to the rate of change in flux ($\Delta\Phi/\Delta t$). If the rate of change is increased one million times, the induced voltage will increase a million times. Also, since the inductor will attempt to maintain the circuit current at 10 mA the instant the switch is opened, the self-induced voltage will jump to 10,000,000 V (10 MV = 10 mA · 1 GΩ). During the first nanosecond of decay, the inductor voltage will theoretically decay from 10 MV to 3.7 MV (63% decay). In practice, an electric arc will occur between the switch contacts the moment the switch

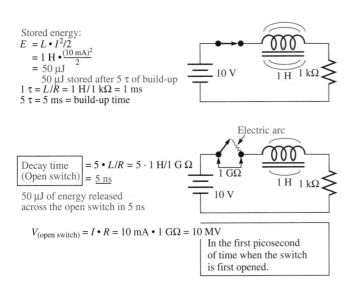

Stored energy:
$$E = L \cdot I^2/2$$
$$= 1\ \text{H} \cdot \frac{(10\ \text{mA})^2}{2}$$
$$= 50\ \mu\text{J}$$
50 μJ stored after 5 τ of build-up
$$1\ \tau = L/R = 1\ \text{H}/1\ \text{k}\Omega = 1\ \text{ms}$$
$$5\ \tau = 5\ \text{ms} = \text{build-up time}$$

10 V   1 H   1 kΩ

Electric arc

| Decay time | $= 5 \cdot L/R = 5 \cdot 1\ \text{H}/1\ \text{G}\,\Omega$ |
| (Open switch) | $= \underline{5\ \text{ns}}$ |

1 GΩ 1 H   1 kΩ

50 μJ of energy released
across the open switch in 5 ns

10 V

$$V_{(\text{open switch})} = I \cdot R = 10\ \text{mA} \cdot 1\ \text{G}\Omega = 10\ \text{MV}$$

In the first picosecond
of time when the switch
is first opened.

Figure 14-16   High self-induced voltage with short decay time.

is opened. This arc produces a relatively low-resistance decay path through the air, which significantly reduces the decay voltage. At the same time, the electric arc produces minor damage (contact pitting) to the switch contacts, damage that becomes significant over many switch operations.

Time for a self-check.

SELF-CHECK 14-5

1. In an *RL* transient circuit, across which component is the integrated waveform produced?

2. In an *RL* transient circuit, across which component is the differentiated waveform produced?

3. How many time constants does it take for a long TC *RL* transient circuit to reach a steady-state condition?

4. Describe what happens when a fully built-up inductor decays through a short time constant.

14-6 Some Practical Applications for Transient Circuits

Pulsed-DC to DC Converter

Figure 14-17 illustrates a **pulsed-DC to DC converter** that uses two extremely-long-time-constant transient circuits to convert a positive pulsed DC into a smooth negative DC. As you can see, the converter is made up of three basic sections: (1) a differentiator to convert the positive pulsed DC into an AC square wave, (2) a diode used to rectify the AC square wave by blocking the positive alternation and passing only the negative alternation, and (3) an integrator to convert the negative pulsed DC into a smooth negative DC. Current flows through the diode only during the negative alternation of the AC square wave (a

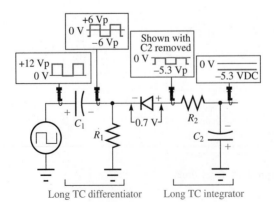

Long TC differentiator   Long TC integrator

Figure 14-17   A simple pulsed-DC to DC converter.

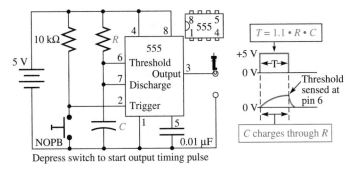

Figure 14-18 The 555 integrated circuit timer.

diode is like a check valve, allowing current to flow in only one direction). In so doing, a voltage drop of approximately 0.6 to 0.7 V is left across the diode. The remaining negative voltage is applied to the extremely-long-TC integrator. The result is a pure negative DC voltage. A converter such as this is normally used only when a negative-voltage, very-low-current, supply is needed.

Timer Circuit

Figure 14-18 illustrates the 555 timer integrated circuit being used to generate a timing pulse of a specific duration. The 555 IC is a very popular and useful circuit, but here we will only emphasize the fact that the 555 makes use of an RC time constant for its operation. As shown in the figure, R and C form an RC time constant that will determine the length of time pin 3 of the timer remains high. A momentary contact pushbutton switch is used to trigger the timer into action by momentarily shorting pin 2 to ground. When pin 2 is shorted to ground, pin 3 immediately goes high ($+5$ V) and the capacitor C begins to charge through the resistor R. When the voltage on the capacitor reaches a threshold voltage equal to about $\frac{2}{3}$ of the supply voltage, the voltage at pin three drops sharply to zero and the capacitor rapidly discharges through pin 7. Pin 3 remains low and the capacitor is held discharged (pin 7 to ground) until S1 is pressed again. Accurate output timing pulses can be obtained, ranging from milliseconds to many minutes depending on the values of R and C. For accuracy and stability, tantalum electrolytics (low-leakage) should be used for long time periods and film capacitors for short periods.

Pulse-Width DC Motor-Speed Control

Figure 14-19 illustrates a very efficient, and therefore popular, method of controlling the speed of a DC motor. This motor is powered by a pulsed-DC supply, the pulse duration of which is adjustable and will cause the motor to increase or decrease its speed according to the duty cycle. The armature of a DC motor is an inductor. The current in an inductor is an integrated waveform when a pulsed DC is applied. Thus the armature integrates, or smooths and averages out, the applied pulsating current. If the duty cycle of the applied pulsed-DC is high, the average current is high and the motor runs at a high speed. If the duty cycle is low, the average armature current is low and the motor runs at a low speed. This method of speed control is popular in DC motor applications ranging from model electric trains to full-size electric vehicles.

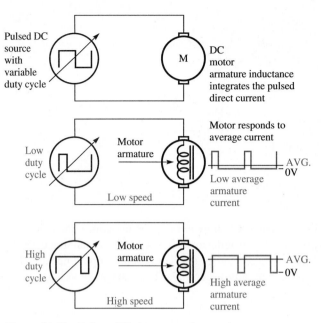

Figure 14-19   Pulse-width motor speed control.

SELF-CHECK 14-6

1. Briefly explain how a positive pulsed-DC square wave can be converted to a smooth negative DC voltage.

2. How long is the output pulse of a 555 timer circuit that has a 25 kΩ timing resistor and a 0.01 μF timing capacitor? (See Figure 14-18.)

3. Explain how a DC motor integrates and operates from a pulsed DC supply.

Summary

FORMULAS

(14.1) 1 TC $= 1\,\tau = R \cdot C$

   (for *RC* transient circuits)

(14.2) 1 TC $= 1\,\tau = L/R$

   (for *RL* transient circuits)

(14.3) $v = V_F + (V_I - V_F) \cdot e^{-t/\tau}$

   where v is the instantaneous voltage at any instant in time t for a transient *RC* or *RL* circuit starting from any initial voltage V_I heading toward a final voltage V_F.

(14.4) $i = I_F + (I_I - I_F) \cdot e^{-t/\tau}$

where i is the instantaneous current at any instant in time t for a transient RC or RL circuit starting from any initial current I_I heading toward a final current I_F.

CONCEPTS

- A capacitor will charge and discharge in five time constants ($5 \cdot R \cdot C$).
- A capacitor will charge or discharge 63% of any remaining voltage with each time constant.
- Charge and discharge current and resistor voltage will always start at maximum and decrease to zero over the $5\ \tau$ transient period.
- The time required for a capacitor to fully charge or discharge depends only on the value of the capacitor and the amount of circuit resistances ($5 \cdot R \cdot C$).
- An inductor will build up and decay in five time constants ($5 \cdot L/R$).
- In each time period the inductor's current will build up an additional 63% of any remaining current (63% of the difference in current from each time constant to the maximum value).
- Inductor buildup current starts at 0 A at t_0 and increases to a maximum at t_5.
- During buildup, the inductor voltage starts at maximum and decreases to a very low voltage at t_5.
- Inductor decay current starts at a maximum value at t_0 and decreases to 0 A at t_5 along with inductor voltage and resistor voltage.
- The buildup and decay times depend solely on the value of the inductor and the amount of circuit resistance ($5 \cdot L/R$).
- An integrated waveform is developed across the capacitor in an RC circuit and across the resistor in an RL circuit.
- A differentiated waveform is developed across the resistor in an RC circuit and across the inductor in an RL circuit.
- A short-time-constant waveform is produced if $5\ \tau$ is less than the applied pulse width.
- A medium-time-constant waveform is produced if $5\ \tau$ is equal to the applied pulse width.
- A long-time-constant waveform is produced if $5\ \tau$ is greater than the applied pulse width.
- Full integration or full differentiation occurs if the transient circuit's time constant is extremely long.
- Full differentiation produces an AC square wave from a pulsed DC.
- Full integration produces a smooth DC from a pulsed DC.

PROCEDURES

Using the Universal Time-Constant Chart to Find Values of Transient Voltage and Current

1. Convert the specified length of time into an equivalent number of time constants.

2. Read the percentage of current or voltage from the appropriate curve and convert the percentage to actual voltage or current.

Using the Universal Time-Constant Chart to Find the Time (*t*) Required for Specific Values of Transient Voltage and Current

1. Convert the specific voltage or current to a percentage of the maximum value.

2. Use the appropriate curve on the time-constant chart to translate the percentage to the corresponding number of time constants.

3. Convert the number of time constants to actual time by multiplying the number of time constants by the time for one time constant.

SPECIAL TERMS

- Time constant (TC = τ)
- Transient
- Charge, discharge, buildup, decay
- Universal time-constant chart
- Exponential curves
- Natural logarithm (e^x)
- Pulsed-DC square wave
- Peak pulse amplitude
- Pulse duration (*PD*), pulse width (*PW*)
- Pulse cycle, cycle time, pulse crown
- Pulse duty cycle
- Integrated waveform, integrator
- Differentiated waveform, differentiator
- Short TC, medium TC, long TC
- Full integration, full differentiation
- Pulsed-DC to DC converter
- Pulse-width motor control

Need to Know Solution

DC motor-speed control—see Figure 14-19.

Traffic light control—a resistor and a capacitor can be used to create a time delay to determine the length of time a light stays red, yellow, or green (see Figure 14-18).

Pulsed-DC to DC converter—Figure 14-17.

Negative pulses—see Figure 14-13.

Extremely high inductor decay voltage—see Figure 14-16.

Questions

14-1 The *RC* Time Constant

1. What is a time constant for an *RC* transient circuit?

2. How many time constants does it take for a capacitor to fully charge or fully discharge?

3. To what percentage of the applied DC voltage will the capacitor charge in the first time constant?

4. Describe, in terms of percentage, how a capacitor charges with each time constant.

5. What happens to resistor voltage and charge current as the capacitor charges in an *RC* transient circuit?

6. When the capacitor is discharging in an *RC* transient circuit, what is happening to resistor voltage and circuit current?

7. Does the capacitor act like (a) a short or (b) an open when a DC voltage is first applied?

14-2 The *L/R* Time Constant

8. What factors affect the buildup or decay time of an inductor in an *RL* transient circuit?

9. Will increasing resistance (a) increase or (b) decrease the buildup and decay times?

10. What determines the maximum, or final, buildup current in an *RL* transient circuit?

11. Describe what is happening to inductor voltage and resistor voltage during the buildup period.

12. Does the inductor act like (a) an open or (b) a short at the start of buildup?

13. Describe what is happening to inductor voltage, resistor voltage, and inductor and resistor current during decay.

14-3 The Universal Time-Constant Chart

14. Curve A of the universal time-constant chart is the rising curve and curve B is the falling curve.
    a. Which curve represents the charge current for a capacitor?
    b. Which curve represents resistor current during capacitor discharge?
    c. Which curve represents inductor voltage during buildup?
    d. Which curve represents resistor voltage during inductor decay?

15. What is the general equation for finding instantaneous values of voltage in any *RC* and *RL* transient circuit?

16. What does t/τ represent?

14-4 *RC* Transient Response and Waveforms

17. What is another name for pulse duration?

18. What is a 50% duty cycle pulsed DC?

19. The integrated voltage waveform is developed across which component in an *RC* transient circuit?

20. Are integrated waveforms produced by applying (a) a DC or (b) a pulsed-DC voltage to a transient circuit?

21. The differentiated voltage waveform is developed across which component in an *RC* transient circuit?

22. Define a short-time-constant waveform in terms of pulse width and time constants.

23. Define a long-time-constant waveform in terms of pulse width and time constants.

24. What is the difference between time *required* and time *allowed* in reference to charging and discharging capacitors?

25. How does the long-time-constant integrated waveform obviously differ from the medium- and short-time-constant waveforms?

26. In terms of number of time constants, how long does it take for a long-time-constant waveform to reach a steady-state condition?

27. Describe the long-time-constant differentiated voltage waveform in terms of average voltage.

14-5 *RL* Transient Response and Waveforms

28. In an *RL* transient circuit, which component is the integrated voltage waveform developed across?
29. Why is the differentiated voltage waveform developed across the inductor of an *RL* transient circuit?
30. Is the inductor current in an *RL* transient circuit (a) an integrated waveform or (b) a differentiated waveform?
31. Define the short-, medium-, and long-time-constant waveforms for *RL* transient circuits in terms of number of time constants and pulse width.
32. If an inductor builds up through one resistance and decays through a second larger resistance, will the initial decay voltage be (a) higher or (b) lower than the voltage source used during buildup? Explain.
33. Explain what occurs when a switch is opened that controls DC to a relay solenoid.

14-6 Some Practical Applications for Transient Circuits

34. Describe how a differentiator and an integrator can be used in converting a positive pulsed DC to a pure negative DC.
35. What is the purpose for the diode in a pulsed-DC to DC converter (Figure 14-17)?
36. What determines the *on* time of the output from a 555 timer circuit?
37. Does the armature of a DC motor (a) integrate or (b) differentiate the applied pulsed *current* from a variable-duty-cycle pulse-DC source?
38. If the duty cycle of a pulsed DC, applied to a DC motor, is increased, will the speed of the DC motor increase or decrease? Explain.

Problems

14-1 The *RC* Time Constant

1. How long will it take for a 10 μF capacitor to fully charge through a 27 kΩ resistor?
2. How long will it take for a 10 μF capacitor to fully discharge through a 1 kΩ resistor?
3. If 10 VDC is applied to a 10 kΩ resistor in series with a 100 μF capacitor, how long will it take for the capacitor to reach a charge of 6.3 V?
4. Using a 100 kΩ resistor, what value capacitor would you need if you wanted the capacitor to take 6 seconds to fully charge?
5. Using a 0.1 μF capacitor, what value resistor would you need if you wanted the capacitor to fully charge in 25 ms?

14-2 The *L/R* Time Constant

6. What is the percentage of maximum buildup current in an *RL* transient circuit after 2 τ?
7. How long will it take for a 250 mH inductor to fully decay through a 10 kΩ resistor?
8. If you want an inductor to fully buildup through a 4.7 kΩ resistor in 2.7 ms, what value must the inductor be?

9. If you want a 2 mH inductor to fully decay through a resistor in 1 μs, what value must the resistor be?

14-3 The Universal Time-Constant Chart

10. Using the universal time-constant chart, determine the charge voltage across a 47 μF capacitor after 254 ms of charge time. The capacitor charges from 0 V through a 2.7 kΩ resistor. The applied voltage is $+12$ V.
11. Using the universal time-constant chart, determine the instantaneous charge current to a 0.1 μF capacitor after 1 ms of charge time. The capacitor charges from 0 V through a 10 kΩ resistor. The applied voltage is $+10$ V.
12. Using the universal time-constant chart, determine the buildup current in a 5 mH inductor after 0.2 μs of buildup time. The inductor builds up from 0 A through a 33 kΩ resistor. The applied voltage is $+5$ V. (Assume inductor winding resistance to be negligible.)
13. Using the universal time-constant chart, determine the decay voltage of a 100 μH inductor after 1.5 ns of decay time. The inductor current just before decay begins is 500 μA and the inductor decays through a 100 kΩ resistor. (Assume inductor winding resistance to be negligible.)
14. Given an RC transient circuit with $C = 10$ μF, $R = 1.5$ MΩ, and a 10 volt source, how long will it take for the capacitor to charge to 8 V after the 10 volt source is applied? (The capacitor begins to charge from 0 V.)
15. Given an RC transient circuit with $C = 50$ μF, $R = 150$ kΩ, and a 24 volt source, how long will it take for the capacitor to discharge to 5 V through the 150 kΩ resistor starting from a full charge?
16. Given an RL transient circuit with $L = 1$ H, $R = 68$ kΩ, and a 10 volt source, how long will it take for the resistor voltage to reach 3 V during buildup after the 10 volt source is applied? (Assume inductor winding resistance to be negligible.)
17. Given an RL transient circuit with a 15 V source, $R = 1100$ Ω and $L = 50$ mH (winding resistance is negligible), use the general equation for instantaneous current to find the instantaneous inductor current after 100 μs of decay. (Inductor decays from full buildup.)
18. Given an RC transient circuit with an 18 V source, R $= 68$ kΩ and C $= 4.7$ μF, use the general equation for instantaneous voltage to determine the resistor voltage after one second of charge time. (Assume the capacitor is charging from 0 V.)
19. Rearrange the general equation for instantaneous voltage to determine the length of time (t) it will take for a 250 μF capacity to charge to 6.5 V through a 47 kΩ resistor from a 12 V source.
20. Rearrange the general equation for instantaneous current to determine the length of time (t) it will take for the current in a 1.8 H inductor to build up to 18 mA through a 1 kΩ resistor from a 24 V source.

14-5 RL Transient Response and Waveforms

21. If an inductor builds up through a 1 kΩ resistor from a 10 V source, what is the initial decay voltage if the inductor decays through a 100 kΩ resistor? Explain.

Answers to Self-Checks

Self-Check 14-1

1. 235 s
2. It decreases. Charge current decreases as the capacitor charges, thus reducing the resistor voltage drop.
3. 12.9 V

Self-Check 14-2

1. L/R
2. 5
3. 0.106 ms
4. Inductor voltage decreases.
5. No. The inductor voltage polarity changes to maintain the circuit current in the same direction as that of buildup.

Self-Check 14-3

1. 86%
2. 23%
3. For *RC* circuits—capacitor voltage on discharge, resistor voltage, and circuit current on charge and discharge. For *RL* circuits—inductor voltage on buildup and decay, resistor voltage and circuit current on decay.
4. Approximately 3 V
5. Maximum
6. Maximum

Self-Check 14-4

1. The length of time the DC pulse is on, also known as the *pulse width*
2. The pulse width is half of the total cycle time.
3. Capacitor
4. Converts pulsating DC top a smooth DC
5. A differentiated waveform that appears as a perfect AC square wave having an average voltage of 0 V
6. AC square wave
7. Converts a pulsed-DC to an AC square wave

Self-Check 14-5

1. The resistor
2. The inductor
3. 5 time constants
4. A very high voltage is self-induced across the inductor.

Self-Check 14-5

1. The pulsed DC is fully differentiated (converted to an AC square wave) and rectified (positive half-cycle is blocked with a diode). The resulting negative pulsed DC is then integrated producing a smooth negative DC.

2. 275 μs
3. The motor's armature is an inductor that integrates the current pulses of the pulsed DC supply. The DC motor responds to the average current.

SUGGESTED PROJECTS

1. Add some of the main concepts, formulas, and procedures from this chapter to your Electricity and Electronics Notebook.

2. Construct a pulsed-DC to DC converter like that shown in Figure 14-17. Use a bench function generator to supply a 1 kHz, +12 V peak pulse amplitude DC square wave to your converter ($PW = 0.5$ ms). $C_1 \cdot R_1$ and $C_2 \cdot R_2$ of the converter should be extremely-long-time-constant circuits. $5 \cdot R_1 \cdot C_1$ should be greater than $100 \cdot PW$. $100 \cdot PW = 50$ ms, therefore, $5 \cdot R_1 \cdot C_1$ should be greater than 50 ms. $5 \cdot R_2 \cdot C_2$ should also be greater than 50 ms. For example, R_1 and R_2 could be 47 kΩ and C_1 and C_2 could be 1 μF ($5 \cdot R_1 \cdot C_1 = 235$ ms). You will notice that the $-$DC voltage at the converter output is equal to the peak amplitude of the negative pulsed DC and not the average of the negative pulsed DC. The reason for this is that C_2 has no discharge path. Thus, C_2 charges to the full negative peak voltage in the first 5 time constants of time.

Part V

AC Theory and Circuit Analysis

On December 6, 1877, Edison recorded and played back ''Mary Had a Little Lamb'' on a simple hand-cranked machine composed of a tinfoil-covered cylinder mounted on a shaft and a needle and diaphragm structure riding on the foil. The machine worked the first time without modification. He applied for a patent on his cylindrical phonograph on December 24 of the same year and was granted his patent only seven weeks later on February 19, 1878. From 1887 to 1892 he patented over 80 improvements to the cylindrical phonograph and began manufacturing and selling them. Edison made significant improvements to disc-type phonographs between 1910 and late 1914. His record-cutting machine produced records of relatively high quality suitable for music reproduction. The large photograph to the left shows Edison's experimental recording studio in which he recorded music on cylinders in 1892. Edison developed and produced hundreds of varieties of phonographs, many of which can be seen today at the Edison Winter Home Museum in Fort Myers, Florida.

Edison—The Inventor of the Phonograph

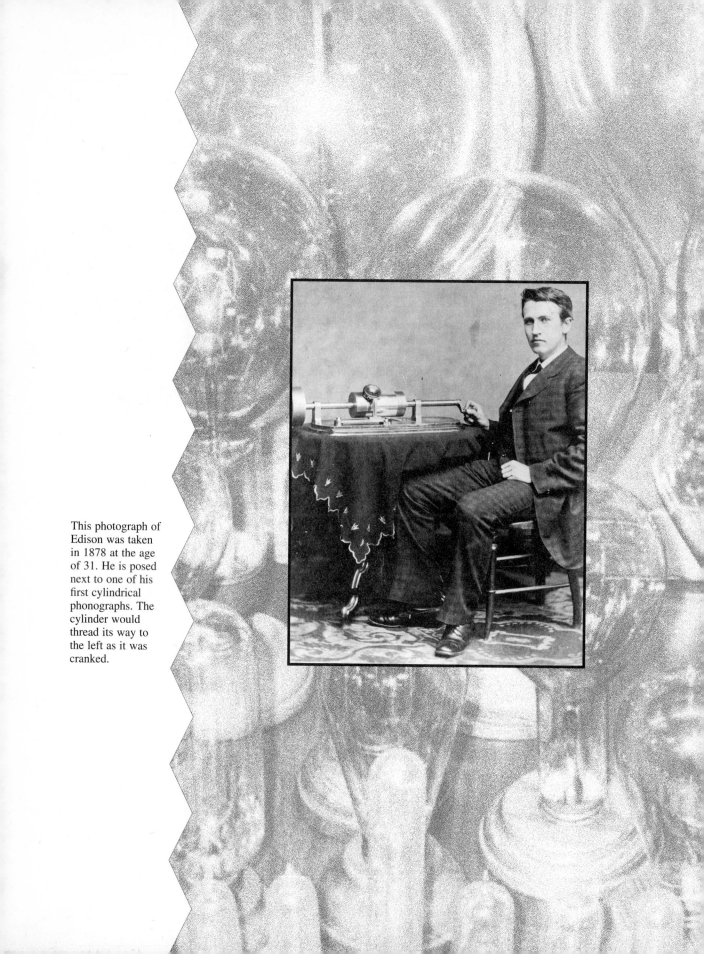

This photograph of
Edison was taken
in 1878 at the age
of 31. He is posed
next to one of his
first cylindrical
phonographs. The
cylinder would
thread its way to
the left as it was
cranked.

Chapter 15

AC

OBJECTIVES

After studying this chapter, you will be able to
- explain and calculate the following AC values: period, frequency, time, angle, instantaneous values of voltage and current, peak and peak-to-peak voltage and current, RMS voltage and current, average voltage and current, and power.
- analyze AC resistive circuits and solve for voltage drops, branch currents, and power dissipations.
- understand what phasors are and how to use them to represent the relative phase and amplitude of AC voltages and currents.
- understand and use voltage and power decibels.
- use a digital frequency counter to measure frequency.

- use the oscilloscope to measure AC voltage and frequency.
- use DMMs and clamp-on ammeters to measure AC voltage and current.
- define the square and sawtooth waves in terms of harmonic content.
- describe the difference between AC and pulsating DC.
- determine the duty cycle of a square wave.
- explain the difference between ideal and nonideal square waves.
- determine period and frequency for sine waves, square waves, sawtooth waves, and triangle waves.

INTRODUCTION

This chapter will serve as an introduction to the study of alternating current and voltage. As you study this chapter, you will become familiar with AC theory, AC circuit calculations, AC measurements—including phase relationships, decibels, and the use of test equipment—and you will investigate some of the more common complex waveforms. A thorough understanding of these areas will provide you with the necessary foundation upon which you will build as you continue your experiences in electricity and electronics.

NEED TO KNOW

You might be wondering what else you would possibly have to know about alternating current besides the fact that AC is current that frequently changes direction in a circuit. Maybe you thought you already learned just about all there is to know about AC from a previous chapter that discussed AC generation. Surprise! There's more! In fact, there are many facts about AC, considered fundamental and essential by today's technicians and engineers, to which you have not yet been introduced. Consider the following modest sampling of basic facts that you will be expected to know as an electronics technician or engineer:

- RMS values of voltage and current are about 7/10 of the peak values.
- Phasors are used to graphically demonstrate the relative amplitudes and phase of AC voltages, currents, and impedances.
- An AC voltage that leads a second AC voltage by 270° also lags the second voltage by 90°.
- A power level 3 dB higher than another power level is 2 times greater.
- A gain of −6 dB is actually a loss of 6 dB.
- An oscilloscope can be used to measure peak amplitude, period, and phase of AC voltages.
- A sawtooth voltage is often referred to as a *ramp voltage* or a *sweep voltage*.

How many of these facts did you already know? Not many huh? That's OK. At this point, you are not expected to know any of these facts. But it does demonstrate, once again, that you do have a need to know. As you study this chapter, you will be carefully guided through the terminology, formulas, use of test equipment, and theory that will enhance your understanding of AC and add to your experience in the world of electricity and electronics. Enjoy!

Heinrich Rudolph Hertz (1857 – 1894)

Heinrich Hertz was a German physicist born in Hamburg. He studied at the University of Berlin and, in the late years of his life, taught physics at the University of Bonn. His greatest contribution to the fields of physics and electricity was his discovery of electromagnetic waves. Hertz built and demonstrated the first spark-gap transmitter, which consisted of a high voltage, a high-frequency tuned circuit, and a spark gap. The high-frequency electric arc radiated electromagnetic energy to a distant tuned circuit. He was also the first to demonstrate that light is electromagnetic energy. (Photo courtesy of Culver Pictures, Inc.)

15-1 AC and the Sine Wave

Alternating Current

A Periodic Waveform

Alternating current (AC) is electrical current that is caused to periodically change direction in the conductors and components through which it flows. Each current reversal is known as an **alternation**. Two consecutive alternations form one complete cycle. A **cycle** can be defined as a recurring **period** of time during which certain events repeat themselves in the same order and intervals. Thus, an **AC cycle** is a periodic repetition of current reversals, or alternations. Because the AC waveform is periodically repeated in the manner previously described, it is naturally called a **periodic waveform**.

Figure 15-1 illustrates the concept of alternating current and the AC terminology we have discussed thus far. As you can see, an AC source, or generator, is supplying alternating current and voltage to a device or piece of equipment. During the positive alternation, bus A is made positive and the ground bus is made negative. As the AC source reverses polarity, circuit current reverses direction; the ground bus becomes positive and bus A becomes negative. Each alternation, or voltage and current reversal, is defined in terms of degrees, time, peak voltage, and waveform.

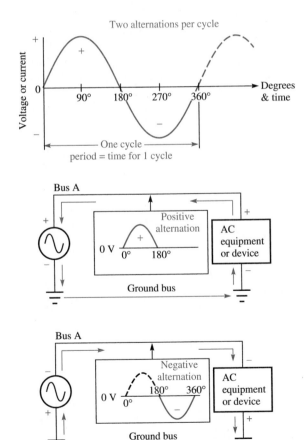

Figure 15-1   Alternating current AC is current that periodically changes direction in a circuit.

Waveform and Degrees

In many cases, each cycle of an AC waveform is **symmetrical** (equal in size, form, and/or time on opposite sides of a plane, line, or point). The positive alternation is a mirror image of the negative alternation—containing the same number of electrical degrees, lasting the same amount of time, and having the same peak voltage. Each alternation of a **symmetrical waveform** contains 180 electrical degrees with each cycle totaling 360°. Thus, a full cycle is 360°, a half cycle is 180°, a quarter cycle is 90°, an eighth cycle is 45°, etc. This degree-cycle relationship is easily understood when related to the operation of a simple electromechanical AC generator in which one complete 360° rotation of the armature produces one complete AC cycle. (See Chapter 9, Section 10 for a discussion of AC power generation.) However, this degree-cycle relationship is applied to all periodic, symmetrical, AC waveforms regardless of whether the AC signal is produced by an electromechanical or an electronic generator (bench signal generator or oscillator).

Time (T) and Frequency (f)

Just as an AC alternation is defined in terms of electrical degrees, it is also defined in terms of time (T). Each alternation requires a certain amount of time to increase in amplitude, reach a peak, and decrease in amplitude, returning to zero volts and zero current (referred to as **zero crossing**) to begin the next alternation. The overall time for both alternations, or one cycle, is determined by the frequency, or rate of repetition, of each cycle. The **frequency** (f) of an AC waveform is determined by the number of cycles repeated every second (the second being a standard SI unit of measure). As an example, the AC power frequency used in the United States is 60 cycles per second (60 cps). If 60 cycles occur every second, each individual cycle will require 1/60 of a second (16.7 ms). Thus, the time for each cycle of an AC waveform can be determined using the following formula if frequency is known:

$$T = 1/f \qquad (15.1)$$

Conversely, frequency can be found when the time for one cycle is known by rearranging Formula 15.1 as follows:

$$f = 1/T \qquad (15.2)$$

EXAMPLE   15-1

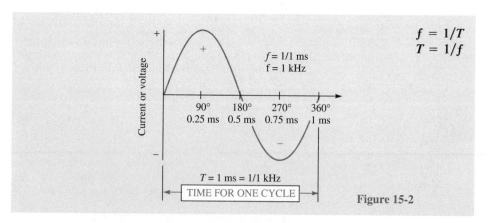

Figure 15-2

Example 15-1 illustrates the time–frequency relationship of a periodic waveform. Notice the time for one complete cycle of this symmetrical and periodic waveform is 1 ms. Thus, according to Formula 15.2, the frequency of this waveform, or signal, is 1,000 cps. In this example, however, the frequency unit is not shown as cps but, rather, kHz. As stated earlier, frequency is measured in cycles per second (cps, or often simply c). The cps unit is adequate and useful, since it is a clear description of what frequency is (a certain number of cycles per second of time). Nevertheless, a more modern and internationally accepted SI unit for frequency is the **hertz (Hz)** named in honor of Heinrich Hertz. The Hz, cps, and c are interchangeable: i.e., 1,000 cps = 1 kc = 1 kHz; 60 cps = 60 Hz; and 800 kc = 800 kHz.

The Sine Wave

A Sinusoidal Waveform

The most common and fundamental of all AC voltage or current waveforms is the sine wave. The **sine wave** is the waveform of AC power applied to homes and factories, AM and FM radio-frequency carriers, television carriers, satellite broadcast and communications signals, pure audio frequencies, and more. It is symmetrical and periodic. Figure 15-3 illustrates how the sine wave is so named. One cycle of a sine wave is, in fact, a graph of the trigonometric sine function. Current, or voltage, is said to vary sinusoidally throughout each cycle.

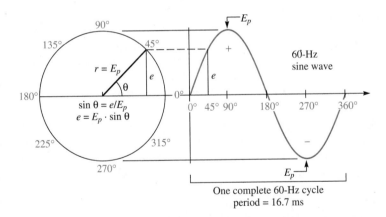

$e = E_p \cdot \sin \theta$			$E_p = 100$ V
Time	Angle	Sine θ	e
0 ms	0°	0	0
2.08 ms	45°	0.707	70.7 V
4.17 ms	90°	1.0	100 V
6.25 ms	135°	0.707	70.7 V
8.33 ms	180°	0	0
10.42 ms	225°	−0.707	−70.7 V
12.5 ms	270°	−1.0	−100 V
14.58 ms	315°	−0.707	−70.7 V
16.67 ms	360°	0	0

Figure 15-3   How the AC sine wave is related to degrees of rotation.

Angles and Instantaneous Values

The voltage or current amplitude that exists on or in an AC carrying conductor at any instant in time is referred to as the **instantaneous voltage** or **current**. This instantaneous voltage and current is expressed mathematically as follows:

$$e = E_\text{p} \cdot \sin \theta \tag{15.3}$$

where e represents the instantaneous voltage of a periodic waveform, E_p represents the maximum or peak value of voltage for the periodic waveform, and θ is the angle, and

$$i = I_\text{p} \cdot \sin \theta \tag{15.4}$$

where i represents the instantaneous current of a periodic waveform, I_p represents the maximum or peak current for the periodic waveform, and θ is the angle.

Figure 15-3 uses a table to illustrate the sinusoidal relationship between instantaneous voltages and various instants of time and angle. The peak voltage (E_p) of 100 V was chosen arbitrarily for simplification. As you can see, as long as peak voltage, or peak current, is known and the angle (θ) is known, the instantaneous voltage (e), or current (i), can be determined using Formulas 15.3 or 15.4. (Note: Lowercase characters such as e and i are used to designate time-related AC values or parameters and to distinguish them from DC parameters.)

Angular Velocity and Instantaneous Values

The instantaneous voltage, or current, can also be determined if the **angular velocity** of the sine wave and a specific period of time is given. But what is angular velocity? Just as cycles are repeated at some definite rate called *frequency*, every angle from 0 to 360° within a cycle is also repeated. Whereas frequency defines the number of cycles per second, angular velocity defines the number of degrees per second (dps, or d/s, or °/s). Thus, angular velocity is an expression of the total number of electrical degrees repeated every second of time. For example, a 60 Hz sine wave will repeat 60 cycles every second and each cycle contains 360 electrical degrees. Thus, the angular velocity is 21,600°/s (60 cps · 360°/1 cycle = 21,600°/s). As you can see, the angular velocity is very easily determined if frequency is known.

If the angular velocity is known, instantaneous values of current and voltage can be determined at any instant in time. Consider the following formulas:

$$e = E_\text{p} \cdot \sin(v \cdot t + \theta) \tag{15.5}$$

where $v = 360° \cdot f$ and is the angular velocity in degrees per second, t is elapsed time in seconds, and θ is any offset angle, and

$$i = I_\text{p} \cdot \sin(v \cdot t + \theta) \tag{15.6}$$

Study Examples 15-2 and 15-3 to see how Formulas 15.5 and 15.6 are used.

EXAMPLE   15-2

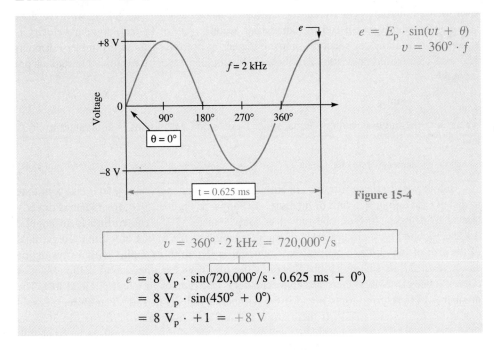

$$e = E_p \cdot \sin(vt + \theta)$$
$$v = 360° \cdot f$$

Figure 15-4

$$v = 360° \cdot 2 \text{ kHz} = 720,000°/s$$

$$e = 8 \text{ V}_p \cdot \sin(720,000°/s \cdot 0.625 \text{ ms} + 0°)$$
$$= 8 \text{ V}_p \cdot \sin(450° + 0°)$$
$$= 8 \text{ V}_p \cdot +1 = +8 \text{ V}$$

EXAMPLE   15-3

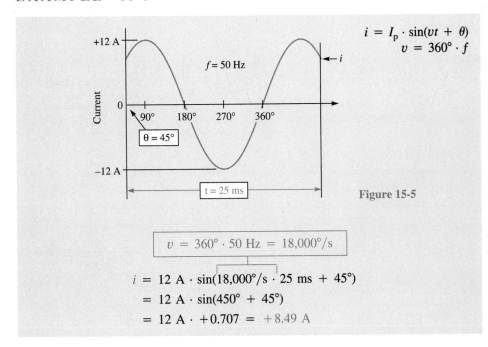

$$i = I_p \cdot \sin(vt + \theta)$$
$$v = 360° \cdot f$$

Figure 15-5

$$v = 360° \cdot 50 \text{ Hz} = 18,000°/s$$

$$i = 12 \text{ A} \cdot \sin(18,000°/s \cdot 25 \text{ ms} + 45°)$$
$$= 12 \text{ A} \cdot \sin(450° + 45°)$$
$$= 12 \text{ A} \cdot +0.707 = +8.49 \text{ A}$$

π and Radians

Even though angular velocity can be expressed in degrees per second, as we have done above, it is usually expressed in radians per second (rad/s). In fact, angular velocity is expressed in radians per second in most engineering formulas. As an example, Formulas 15.5 and 15.6 would be written as follows using angular velocity expressed in radians per second:

$$e = E_p \cdot \sin(\omega \cdot t + \theta) \tag{15.7}$$

where ω is the angular velocity equal to $2\pi f$ rad/s and 2π radians is equivalent to $360°$, and

$$i = I_p \cdot \sin(\omega \cdot t + \theta) \tag{15.8}$$

Before we use these formulas in solving for instantaneous values, let's take a minute to discuss the radian as a unit of measure. As shown in Figure 15-6a, the radian is directly related to degrees. In fact, one radian is equivalent to $57.3°$. The **radian** is an angular distance defined by a circumferal arc equal in length to the radius of a circle. Regardless of the size of the circle, a circumferal arc equal in length to its radius will always define an angle of $57.3°$. Figure 15-6b illustrates the origin and mathematical definition of π (Greek letter pi). Notice there are approximately 3.14 radians in a half circle ($180°$). The number 3.14 is referred to as π and is the number of radians in $180°$. Therefore, it is said that there are π rad in $180°$. If there are π rad in $180°$, there must be 2π rad in $360°$, as shown in Figure 15-6c. A complete cycle of an AC waveform can therefore be labeled in degrees or radians, as illustrated in Figure 15-6d.

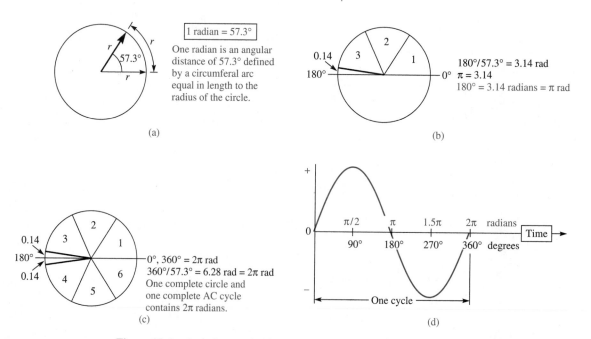

Figure 15-6   A circle contains 2π radians equal to $360°$.

Angular Velocity in Radians per Second

This brings us back to angular velocity measured in radians per second (rad/s). Since an AC waveform contains 2π rad per cycle (2π rad/c) and the cycles are being repeated at some rate, or frequency (c/s), the angular velocity (ω) must be 2π rad multiplied by the frequency ($\omega = 2\pi f$). As an example, the angular velocity for a 50 Hz waveform is 314 rad/s ($\omega = 2\pi f = 6.28$ rad/c $\cdot$ 50 c/s = 314 rad/s). Now that you have a basic understanding of radian measure, let's consider an example that makes use of Formula 15.7. Formula 15.8 is worked in a similar manner.

The object of Example 15-4 is to determine the instantaneous voltage of an 80 kHz sine wave 23 μs from a designated starting point. The starting point is an offset angle of 270°. The angular velocity in rad/s must be determined, and the offset angle must be converted from degrees to radians. It is very important to note here that Formulas 15.7 and 15.8 cannot be worked with your scientific calculator in the normal or DEG mode. You must change the calculator to the RAD mode, since you will be taking the sine of a quantity of radians, not degrees. Consult your calculator's manual to determine how to place it in the RAD mode. Once in the RAD mode, Formula 15.7 may be used to solve for the instantaneous voltage.

Take a few minutes to go back and review what you have studied. Pay particular attention to terms and definitions and the examples that illustrate the use of formulas. Also, take a moment to study Design Note 15-1 (page 571). Then, test your understanding by answering the questions of Self-Check 15-1.

SELF-CHECK   15-1

1. What is an alternation?

2. What is a symmetrical waveform?

3. What is the period of a 250 cps signal?

4. What is the accepted SI unit of measure for frequency?

5. Why is a sine wave called a sine wave?

6. If a sine wave has a peak current of 100 mA, what is the instantaneous current at the 160° point in the waveform?

7. What is the angular velocity, in degrees per second, of a 100 kHz signal?

8. Use degree angular velocity to find the instantaneous voltage of a 10 kHz sine wave 180 μs from an offset angle of 30°. The peak voltage is 300 mV.

9. What is a radian?

10. Use radian angular velocity to find the instantaneous current of a 400 Hz sine wave 1.5 ms from an offset angle of 90°. The peak current is 750 mA.

EXAMPLE 15-4

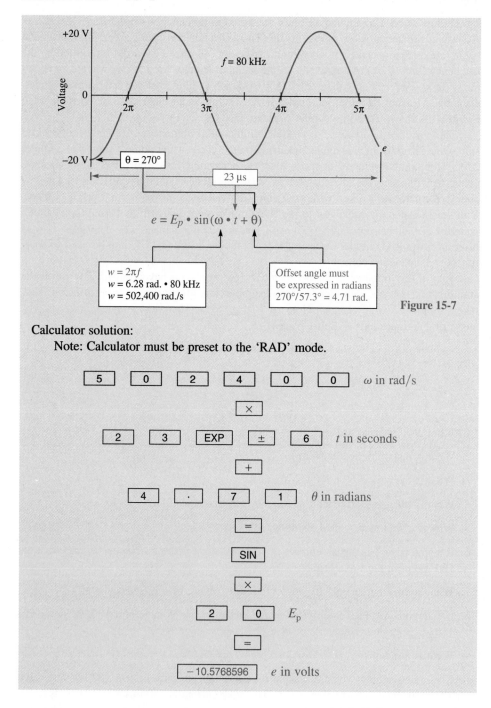

Figure 15-7

Calculator solution:
   Note: Calculator must be preset to the 'RAD' mode.

$$e = E_p \cdot \sin(\omega \cdot t + \theta)$$

| 5 | 0 | 2 | 4 | 0 | 0 | ω in rad/s |

×

| 2 | 3 | EXP | ± | 6 | t in seconds |

+

| 4 | . | 7 | 1 | θ in radians |

=

SIN

×

| 2 | 0 | E_p |

=

| −10.5768596 | e in volts |

D E S I G N   N O T E   1 5 - 1 :   Instantaneous Values

$$i = I_p \cdot \sin(\omega t + \theta)$$
$$e = E_p \cdot \sin(\omega t + \theta)$$

e, i = the instantaneous value of voltage or current
E_p, I_p = the peak value of voltage or current
$\omega = 2\pi f$ = the angular velocity in rad/s
t = length of time in seconds
θ = the offset angle

"BASIC" PROGRAM SOLUTION

THIS PROGRAM WILL SOLVE FOR THE INSTANTANEOUS VALUE OF VOLTAGE OR CURRENT IN A SINE WAVE.

```
10 CLS
20 PRINT"INSTANTANEOUS VALUES"
30 PRINT""
40 PRINT"THIS PROGRAM WILL CALCULATE THE INSTANTANEOUS VALUE OF VOLTAGE OR CURRENT OF A"
50 PRINT"SINE WAVE AT ANY INSTANT IN TIME STARTING FROM ANY INITIAL OFFSET ANGLE."
60 PRINT""
70 INPUT"ENTER THE PEAK VALUE OF VOLTAGE OR CURRENT. "; P
80 INPUT"ENTER THE FREQUENCY. ";F
90 INPUT"ENTER THE LENGTH OF TIME (t) IN SECONDS. ";T
100 INPUT"ENTER THE OFFSET ANGLE IN DEGREES. ";O
110 A = O/57.29578
120 W = 6.2831853# * F
130 I = P * SIN(W * T + A)
140 PRINT""
150 PRINT"THE INSTANTANEOUS VALUE IS ";I
160 PRINT""
170 INPUT"ANOTHER PROBLEM?   (Y/N)";A$
180 IF A$ = "Y" THEN CLEAR:GOTO 60
190 CLS:END
```

15-2 AC Voltage, Current, and Power

Values of Voltage and Current

Peak and Peak-to-Peak Values

As you know, the **peak** value of voltage and current is the maximum value, or maximum point, in each alternation corresponding to 90° and 270°. Two such peaks occur with every cycle, a positive peak and a negative peak. For this reason, a voltage sine wave having a peak voltage of 10 V may be described as having a peak voltage of ± 10 V. However, since most sine waves are symmetrical, it is sufficient to state the peak voltage as simply 10 V_p.

Another way to describe the maximum positive and negative extremes of the sine waveshape is in terms of **peak-to-peak** values. Figure 15-8 illustrates the peak and peak-

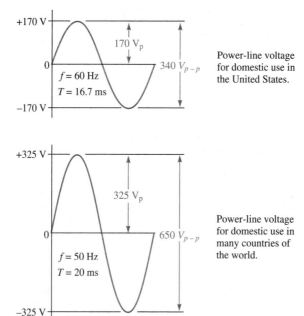

Power-line voltage for domestic use in the United States.

Power-line voltage for domestic use in many countries of the world.

Figure 15-8 Peak and peak-to-peak values of AC.

to-peak power-line voltages used in the United States and other countries. The peak-to-peak value (V_{p-p} or A_{p-p}) of voltage or current is simply twice the peak value ($V_{p-p} = 2 \cdot V_p$ and $A_{p-p} = 2 \cdot A_p$). While the peak value (V_p or A_p) may be used in circuit analysis to solve for peak current, peak voltage, or even peak power, the peak-to-peak values cannot. This is because the two peaks (positive and negative) do not occur, or exist, at the same time. Each peak has an independent effect on the circuit, although the effect of each peak is the same for symmetrical waveforms. The peak value of voltage and current is often used in determining AC circuit component ratings, since the peak value represents the highest value that occurs at any one time.

Average Value

Every waveform has what is known as an **average** value of voltage or current. However, the manner in which that average value is determined depends on the waveform that is being considered. For one complete cycle of a sine wave, the average voltage or current is zero (0), since the positive alternation mathematically cancels the negative alternation. For example, the average of -10 V_p and $+10$ V_p is zero $(-10 + 10)/2 = 0/2 = 0$). The average voltage available at the wall outlets in our homes is 0 V, since it is a symmetrical sine waveform. Thus, the average voltage for *one complete cycle* of a sine wave is 0 V.

However, *each individual alternation* of a sine wave has its own average voltage. The alternation rises to a peak and falls back to zero over some period of time. The average voltage, or current, during that period of time is determined from a large number of samples of instantaneous values taken from one alternation (0° to 180° or 180° to 360°). This is a simple arithmetic average, found by dividing the number of samples into the sum of all samples taken. Figure 15-9 shows how this is done using a positive alternation. Each sample is an instantaneous value found using the sine function and the angle corresponding

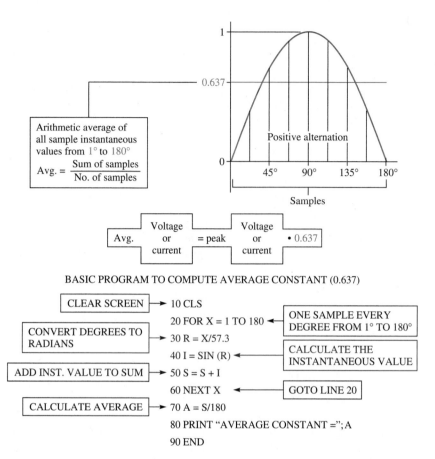

Figure 15-9 Computing the average value for one alternation (180°).

to the instantaneous value ($e = E_p \cdot \sin \theta$). Ideally, the number of samples should be infinite between 0° and 180° for perfect accuracy. From a practical standpoint, 180 samples taken every degree from 1° to 180° will provide more than adequate accuracy. The BASIC program provided in Figure 15-9 uses the speed of a computer to sample and average 180 instantaneous values. The result is an average relative amplitude of 0.6366 . . . , rounded off to a practical 0.637. Table 15-1 also demonstrates how the average value is determined. Notice that the sine values of 18 sample angles from 10° to 180° are listed and totaled. The total is then divided by the number of samples to determine the arithmetic average (11.43078/18 = 0.635). As you can see, the average in Table 15-1 is slightly low due to the few number of samples used.

As an example of determining an average value from a peak value, consider a 170 V_p power-line voltage. The average voltage of a 170 V_p half cycle is 108 V_{avg} (170 $V_p \cdot 0.637 = 108\ V_{avg}$). A general formula may be expressed as follows:

$$E_{avg} = E_p \cdot 0.637 \qquad \text{or} \qquad I_{avg} = I_p \cdot 0.637 \tag{15.9}$$

To convert average values back to peak values, Formula 15.9 is rearranged as follows:

$$E_p = \frac{E_{avg}}{0.637} = E_{avg} \cdot 1.57 \qquad \text{or} \qquad I_p = \frac{I_{avg}}{0.637} = I_{avg} \cdot 1.57 \tag{15.10}$$

TABLE 15-1

Sample	Angle	Sin θ	(Sin θ)^2
1	10°	0.1736	0.0301
2	20°	0.3420	0.1170
3	30°	0.5	0.2500
4	40°	0.6427	0.4131
5	50°	0.7660	0.5868
6	60°	0.8660	0.7499
7	70°	0.9397	0.8830
8	80°	0.9848	0.9698
9	90°	1.0	1.0
10	100°	0.9848	0.9698
11	110°	0.9397	0.8830
12	120°	0.8660	0.7499
13	130°	0.7660	0.5868
14	140°	0.6427	0.4131
15	150°	0.5	0.2500
16	160°	0.3420	0.1170
17	170°	0.1736	0.0301
18	180°	0	0
	TOTALS	11.4296	8.9994
	AVERAGES	0.6350	0.5
		RMS = $\sqrt{0.5}$ = 0.707	

We are only concerned with the average voltage or current when the AC sine wave has been rectified. *Rectified* means the AC has been converted to pulsating DC, as in a laboratory power supply. This is a topic you will study in detail in another course.

Effective Values (RMS)

The **effective** values of voltage and current are the most commonly used values for AC circuit calculations. An AC voltage or current will have the same heating effect as a certain value of DC voltage or current in a resistive circuit. That certain value of DC voltage or current is the effective value of the AC voltage or current. It is this effective value of current and voltage that actually performs work or supplies power to a load. Since an AC voltage or current is constantly changing in amplitude and only momentarily reaches a peak, the peak value contributes very little to powering a load. As a result, the overall effect of the sine wave is averaged out over the duration of each alternation in a special way. This special averaging is known as the **root mean square (RMS)**. In other words, the effective or RMS value is found by taking the square root of the average (mean) of a large number of squared instantaneous values as illustrated in Figure 15-10.

Table 15-1 also demonstrates the manner in which the RMS value is determined. Notice that the sine of each sample angle is *squared*, then all samples are totaled. The *mean* (average) of the squared values is determined by dividing the total by the number of samples (0.5). Finally, the square *root* of the mean is taken and found to be 0.707. Thus, the effective, or RMS, value of an AC signal is determined using the following formula:

$$E_{rms} = E_p \cdot 0.707 \quad \text{or} \quad I_{rms} = I_p \cdot 0.707$$

(15.11)

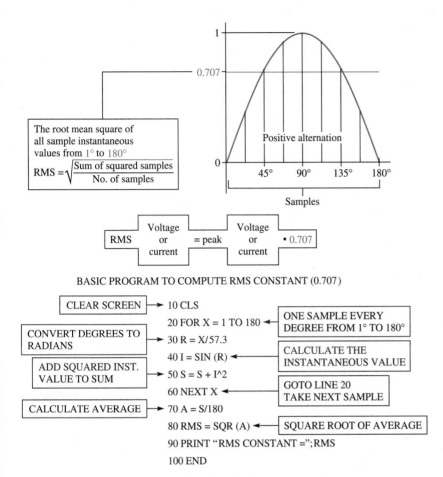

Figure 15-10   Computing the RMS value of AC.

To convert effective values to peak values, Formula 15.11 is rearranged as follows:

$$E_p = E_{rms}/0.707 = E_{rms} \cdot 1.414 \quad \text{or} \quad I_p = I_{rms}/0.707 = I_{rms} \cdot 1.414 \quad (15.12)$$

The power-line voltage for domestic use in the United States is 170 V_p or 120 V_{rms} (170 $V_p \cdot 0.707 = 120$ V_{rms}). In Europe, and many other parts of the world, the power-line voltage is 325 V_p or 230 V_{rms} (325 $V_p \cdot 0.707 = 230$ V_{rms}). There are many ways to designate effective (RMS) values of AC. For example, 120 V_{rms} = 120 V_{eff} = 120 V_{AC} = 120 VAC = 120 V, and 10 A_{rms} = 10 A_{eff} = 10 A_{AC} = 10 A. Capital letters are used here to indicate that the AC quantity is equivalent to the same DC quantity. As a general rule, simply V and A are used to denote effective (RMS) values. Peak, peak-to-peak, and average AC values are clearly designated with an appropriate subscript (V_p, V_{p-p}, V_{avg}, A_p, A_{p-p}, A_{avg}). Figure 15-11 summarizes the AC voltage values.

AC Power

Earlier it was mentioned that it is the effective, or RMS, values of AC voltage and current that actually supply power to a load. As shown in Figure 15-12, a 170 V_p sine wave voltage has the same heating effect as 120 VDC. What we are actually saying is that all

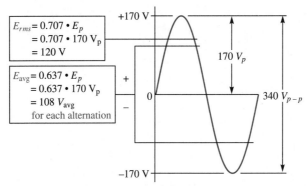

NOTE: The average voltage for the entire cycle is 0 V since the positive and negative alternations are equal.

Figure 15-11   A summary of AC voltage and current values.

power dissipation calculations must be made using effective values of voltage and current. While it is true we can calculate peak power using peak values, it is the effective, or RMS, power that is actually doing the work.

AC power is calculated using the same formulas you used to calculate DC power in earlier chapters: $P = IE$, $P = I^2R$, and $P = E^2/R$. To calculate peak power, peak values of current and voltage must be used. To calculate effective power, effective values of current and voltage must be used. Consider the following Example Set.

EXAMPLE SET   15-5

1. 120 V is applied to a 240 Ω light bulb. The resulting effective power dissipation is 60 W.

   $$P = E^2/R: (120 \text{ V})^2/240 \text{ } \Omega = 60 \text{ W}$$

2. 500 V_{p-p} is applied to a 20 Ω resistive heating element. The resulting power dissipation is 1,562 W.

   $$E_p = 500 \text{ V}_{p-p}/2 = 250 \text{ V}_p$$
   $$E_{rms} = 250 \text{ V}_p \cdot 0.707 = 176.75 \text{ V}$$
   $$P = E^2/R: (176.75 \text{ V})^2/20 \text{ } \Omega = 1,562 \text{ W}$$

3. A peak current of 20 A_p is supplied to a radio station antenna having a radiation resistance of 50 Ω. The resulting effective power applied to the antenna is 10 kW while the peak power is 20 kW.

   $$I_{rms} = 20 \text{ A}_p \cdot 0.707 = 14.14 \text{ A}$$
   $$P = I^2R = (14.14 \text{ A})^2 \cdot 50 \text{ } \Omega = 9,996.98 \text{ W}$$
   $$(20 \text{ A}_p)^2 \cdot 50 \text{ } \Omega = 20,000 \text{ W}_p$$

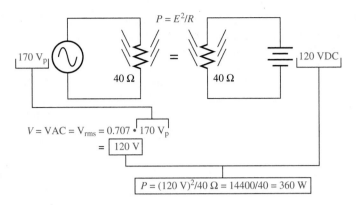

$P = E^2/R$

170 V$_p$

40 Ω

=

40 Ω

120 VDC

$V = \text{VAC} = V_{rms} = 0.707 \cdot 170\ V_p$

$= \boxed{120\ V}$

$P = (120\ V)^2/40\ \Omega = 14400/40 = 360\ W$

Figure 15-12 AC power calculations.

From Example Set 15-5, you should have seen the importance of converting peak and peak-to-peak values of current and voltage to RMS values before calculating effective, or actual dissipated, power. From the third example you were made aware that peak power is twice the effective power. However, it is the effective power that is actually doing work in an AC circuit.

Take a few minutes now to review this section. Then, test your understanding by solving the problems of Self-Check 15-2.

1. Given a 200 V$_{p-p}$ sine wave, find E_p, E_{rms}, and E_{avg} (average for one alternation).

2. Given a 55 V$_p$ sine wave, find E_{p-p}, E_{rms}, and E_{avg} (average for one alternation).

3. Given a 62 V sine wave, find E_p, E_{p-p}, and E_{avg} (average for one alternation).

4. Determine the effective power dissipated by a 30 Ω load with 160 V$_{p-p}$ applied to its terminals.

5. Determine the effective power dissipated by a 470 Ω load with 12 V applied to its terminals.

15-3 AC and Resistive Circuits

In this section you will once again analyze series and parallel resistor circuits, solving for voltage drops, currents, and power dissipations. The difference here is that AC sources will be used instead of DC. The purpose of this section is to demonstrate that AC resistive circuits are analyzed in the same manner, with the same formulas, as DC resistive circuits. The three versions of Ohm's Law and the three power formulas are used in AC resistive circuit analysis just as they are used in DC circuit analysis.

Before we begin what is in essence a brief review of fundamental circuit analysis, let's examine a couple of somewhat obvious considerations. First, if AC current is calculated using a peak value of voltage, the current will also be peak ($I_p = E_p/R$). Secondly, effective or RMS values are generally used for all AC circuit analysis, since power dissipation is calculated using RMS values and most AC equipment is rated according to RMS values of voltage and power. Therefore, peak or peak-to-peak values should immediately be converted to RMS in the analysis of an AC circuit. Now, take time to carefully study Examples 15-6, 15-7, and 15-8.

EXAMPLE   15-6

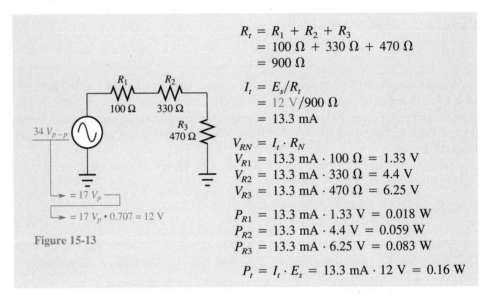

$$R_t = R_1 + R_2 + R_3$$
$$= 100\ \Omega + 330\ \Omega + 470\ \Omega$$
$$= 900\ \Omega$$

$$I_t = E_s/R_t$$
$$= 12\ \text{V}/900\ \Omega$$
$$= 13.3\ \text{mA}$$

$$V_{RN} = I_t \cdot R_N$$
$$V_{R1} = 13.3\ \text{mA} \cdot 100\ \Omega = 1.33\ \text{V}$$
$$V_{R2} = 13.3\ \text{mA} \cdot 330\ \Omega = 4.4\ \text{V}$$
$$V_{R3} = 13.3\ \text{mA} \cdot 470\ \Omega = 6.25\ \text{V}$$

$$P_{R1} = 13.3\ \text{mA} \cdot 1.33\ \text{V} = 0.018\ \text{W}$$
$$P_{R2} = 13.3\ \text{mA} \cdot 4.4\ \text{V} = 0.059\ \text{W}$$
$$P_{R3} = 13.3\ \text{mA} \cdot 6.25\ \text{V} = 0.083\ \text{W}$$

$$P_t = I_t \cdot E_s = 13.3\ \text{mA} \cdot 12\ \text{V} = 0.16\ \text{W}$$

Figure 15-13

EXAMPLE   15-7

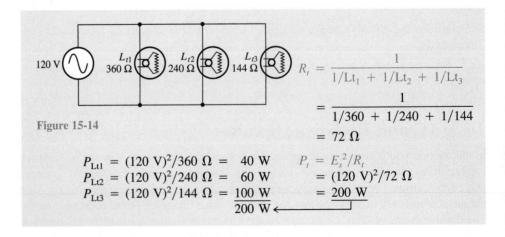

Figure 15-14

$$R_t = \cfrac{1}{1/Lt_1 + 1/Lt_2 + 1/Lt_3}$$
$$= \cfrac{1}{1/360 + 1/240 + 1/144}$$
$$= 72\ \Omega$$

$$P_{Lt1} = (120\ \text{V})^2/360\ \Omega = 40\ \text{W}$$
$$P_{Lt2} = (120\ \text{V})^2/240\ \Omega = 60\ \text{W}$$
$$P_{Lt3} = (120\ \text{V})^2/144\ \Omega = \underline{100\ \text{W}}$$
$$\underline{200\ \text{W}} \leftarrow$$

$$P_t = E_s^2/R_t$$
$$= (120\ \text{V})^2/72\ \Omega$$
$$= 200\ \text{W}$$

EXAMPLE   15-8

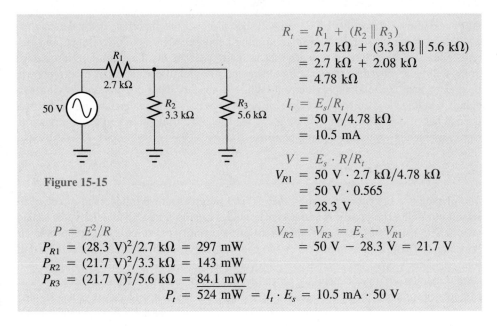

$$R_t = R_1 + (R_2 \parallel R_3)$$
$$= 2.7 \text{ k}\Omega + (3.3 \text{ k}\Omega \parallel 5.6 \text{ k}\Omega)$$
$$= 2.7 \text{ k}\Omega + 2.08 \text{ k}\Omega$$
$$= 4.78 \text{ k}\Omega$$

$$I_t = E_s/R_t$$
$$= 50 \text{ V}/4.78 \text{ k}\Omega$$
$$= 10.5 \text{ mA}$$

$$V = E_s \cdot R/R_t$$
$$V_{R1} = 50 \text{ V} \cdot 2.7 \text{ k}\Omega/4.78 \text{ k}\Omega$$
$$= 50 \text{ V} \cdot 0.565$$
$$= 28.3 \text{ V}$$

Figure 15-15

$$P = E^2/R$$
$$P_{R1} = (28.3 \text{ V})^2/2.7 \text{ k}\Omega = 297 \text{ mW}$$
$$P_{R2} = (21.7 \text{ V})^2/3.3 \text{ k}\Omega = 143 \text{ mW}$$
$$P_{R3} = (21.7 \text{ V})^2/5.6 \text{ k}\Omega = \underline{84.1 \text{ mW}}$$
$$P_t = 524 \text{ mW} = I_t \cdot E_s = 10.5 \text{ mA} \cdot 50 \text{ V}$$

$$V_{R2} = V_{R3} = E_s - V_{R1}$$
$$= 50 \text{ V} - 28.3 \text{ V} = 21.7 \text{ V}$$

Take a moment to test your skills by answering the questions of Self-Check 15-3.

SELF-CHECK   15-3

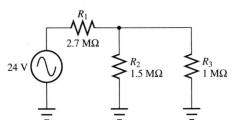

Figure 15-16

1. Solve for voltage drops across R_1, R_2, and R_3 in Figure 15-16.

2. Solve for power dissipated in R_1, R_2, and R_3 in Figure 15-16.

3. Solve for total power in Figure 15-16.

4. How much AC current is there flowing through R_3 of Figure 15-16?

15-4 Phasors and Phase Relationships

As you continue your study of AC, you will soon discover the importance of understanding and knowing the phase relationship and timing between various circuit voltages and currents. You will be surprised to learn that in many AC circuits the sum of series voltage drops will be greater than the source voltage and the sum of parallel branch currents will be greater than the total supply current. While it appears these circuits are defying Kirchhoff's voltage and current laws, in reality they are not. Your understanding of this will depend heavily upon your understanding of voltage and current phase, or timing, relationships graphically represented using phasors.

The Phasor

The **phasor** is a ray (graphic arrow) that is used to represent magnitude and direction in relation to time. The phasor represents magnitude, or amplitude, by its relative or scaled length. The length of the phasor may represent a quantity of current, voltage, power, or impedance. The direction indicated by the phasor is expressed in degrees and is related to time. In other words, the angular position of a particular phasor is directly related to a certain amount of time, depending on the frequency of the AC voltage or current. For example, if a phasor indicates an angle of 30° and the frequency of AC is 1,000 Hz, the time associated with the 30° angle will be 0.083 ms, since one 1 kHz cycle (360°) has a period of 1 ms (30°/360° · 1 ms = 0.083 ms). Often the term *vector* is used interchangeably with *phasor*. However, there is a technical difference between the two terms; a vector is like a phasor in that it indicates magnitude and direction, but a vector is related to physical space, not time.

Figure 15-17 shows three phasors, each indicating a certain magnitude at a particular angle. The upper polar chart indicates an angle of 0°. A phasor that is to indicate an angle of 0° is drawn in the same physical orientation as that indicated by the polar chart (0° is indicated in the right-hand horizontal position). As stated earlier, the length of the phasor is scaled to represent some electrical quantity. The middle and lower diagrams are further examples of phasors drawn to represent a specific angle and magnitude. Notice the lower phasor is at an angle of 315°, which may also be considered as −45°. Any phasor indicating an angle greater than 180° may be labeled as a negative angle less than 180°.

Phase Relationships and Phasors

Figure 15-18 demonstrates how waveform diagrams can be represented as **phasor diagrams**. Figure 15-18a illustrates the phase relationship between voltage and current in an AC resistive circuit. As you can see in the waveform diagram, the current waveform reaches zero crossings and peaks at the same time the voltage waveform reaches zero crossings and peaks. The phasor diagram indicates this in-phase condition by the use of overlapping phasors at 0°.

See color photo insert pages D7 through D9.

Figure 15-18b shows two quadrature (phase-related by 90°) voltage waveforms. In the waveform diagram, you can see that E_2 is already at 90° when E_1 is at zero crossing heading in the positive direction. Therefore, E_2 is leading E_1 by 90°. Remember here that 90° represents ¼ cycle and ¼ the time for one cycle. The corresponding phasor diagram represents the leading E_2 waveform as a phasor pointing upward at 90°. Both phasors, E_1 and E_2, are of the same length, since the voltages are of the same amplitude.

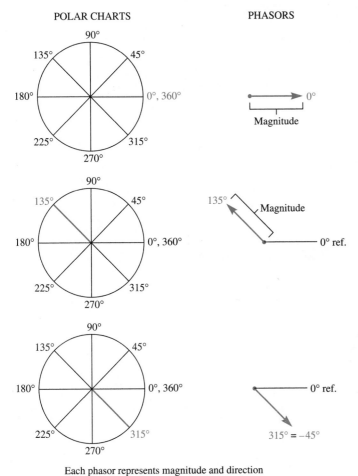

POLAR CHARTS   PHASORS

Each phasor represents magnitude and direction
(The direction in degrees is related to time)

Figure 15-17   Phasors represent magnitude and angular direction.

Two-phase (bi-phase) AC power-line voltage is common for domestic use in the United States. Figure 15-18c illustrates these two phases. As you can see, they are 180° out of phase as indicated in both the waveform diagram and the phasor diagram. While one phase is reaching a positive peak, the other is reaching its negative peak.

Another quadrature phase relationship is shown in Figure 15-18d. This time E_2 is leading E_1 by 270°. It can also be said that E_2 is lagging E_1 by 90°.

Figure 15-19 shows three AC waveforms that might supply a three-phase motor in an industrial application. The phasor diagram indicates very clearly that there are three phases equal in amplitude and evenly spaced 120 electrical degrees apart. Notice that I_1 is considered to be the reference phase at 0°, with I_2 at 120° and I_3 at 240°. It can also be said that I_3 is at $-120°$ or lagging I_1 by 120°.

In later chapters, you will further your understanding of phasor diagrams as you use them to analyze AC circuits. Now, take just a moment to test your knowledge by answering the questions of Self-Check 15-4.

WAVEFORM DIAGRAMS   PHASOR DIAGRAMS

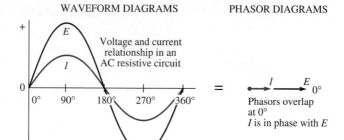

(a)

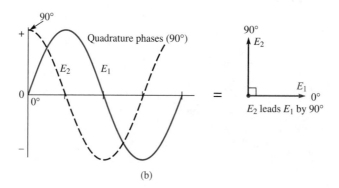

(b)

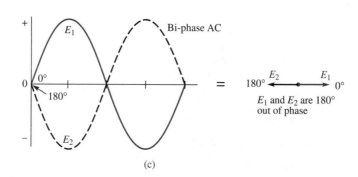

(c)

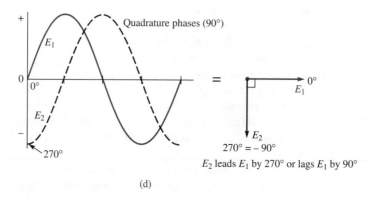

(d)

Figure 15-18   AC phase relationships.

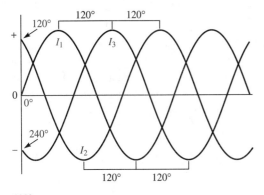

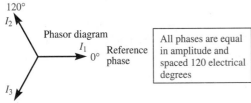

Figure 15-19   Three-phase AC.

1. What is the difference between a vector and a phasor?

2. Draw a phasor diagram that represents two AC voltages as follows: E_1 is 120 V, E_2 is 60 V, E_2 is leading E_1 by 90°.

3. Draw a phasor diagram that represents three AC currents as follows: I_1 is 5 A, I_2 is 3 A, I_3 is 2 A, I_2 is leading I_1 by 90° and I_3 is lagging I_1 by 90°.

15-5 Decibels (dB)

The Bel and Decibel

The electrical and electronics industry the world over has adopted a mathematical means by which voltage levels can be compared and power levels can be compared for analysis and evaluation. This means of comparison is through the use of decibels (dB). Decibels are used for voltage- and power-level comparisons in the fields of telephone communications, AM and FM broadcasting, television, satellite communications, professional and domestic audio equipment, two-way radio communications—and the list goes on and on. The subject of decibels is therefore a very important one, since the technician or engineer is inevitably required to converse in terms of decibels.

The Bel

The **Bel**, named in honor of Alexander Graham Bell, is a logarithmic relationship or function. It is related to the manner in which human senses function; particularly the manner in which sound levels are perceived. The response of the human ear in perceiving loudness, or sound pressure levels, is logarithmic. In order for a person to perceive a

Alexander Graham Bell (1847 – 1922)

Alexander Graham Bell was born in Edinburgh, Scotland. In 1871, he moved to Boston to open a school for teachers of the deaf. Later, in 1873, he became a professor at the University of Boston. Although much work had already been done in the area of nonvoice communications over a pair of wires, Bell was the first scientist to devise instruments capable of transmitting voice. He re- ceived his telephone patent in 1876 at the age of 29. Interestingly, Bell wanted to be remembered for his work with the deaf instead of as the inventor of the telephone. (Photo courtesy of The Bettmann Archive.)

doubling of loudness, the acoustical power, or sound pressure, must be increased 10 times. In the field of acoustics, 1 Bel is a tenfold increase in acoustical power and is perceived as a doubling of loudness. As an example, if the audio power applied to a loudspeaker is increased from 100 mW to 1 W, the power will have increased by 1 Bel and the sound is perceived as being twice as loud. If the power is further increased to 10 W, the power is 2 Bels above the 100 mW level and sounds four times as loud. Since the Bel is a logarithmic function, the formula used to determine the number of Bels from a given power ratio is as follows:

$$\# \text{ Bels } = \log_{10}(P_1/P_2) \tag{15.13}$$

The Decibel (dB)

The decibel is used far more than the Bel in nearly all areas of sound and electronics. The decibel is one tenth of a Bel (10 dB = 1 Bel and 1 dB = 0.1 Bel).

Power Decibels

The Power dB Formula

From the previous discussion, you learned that 1 Bel = 10 dB and, since 1 Bel represents a power ratio of 10/1, 10 dB also represents a power ratio of 10/1. Thus, a power ratio of 100/1 is 20 dB and a power ratio of 1,000/1 equals 30 dB, etc. But how many dB is a power ratio of 1,580/1? The following formula and your calculator will make dB calculations fairly simple.

$$\# \text{ dB } = 10 \cdot \log_{10}(P_1/P_2) \tag{15.14}$$

Using the Power dB Formula

Decibels may be used in comparing a high power level to a low power level, in comparing a low power level to a high power level, or to indicate an increase or decrease in power level. When the power ratio of P_1/P_2, in Formula 15.14, is greater than 1, the number of dB is expressed as a positive number. When the power ratio is less than 1, the number of dB is expressed as a negative number. If an amplifier raises the power level of an input signal 10 times, it is said that the amplifier has a gain of 10 dB. However, if an attenuator circuit decreases an input signal in power level by 10 times, it is said the circuit has a loss of 10 dB or a gain of − 10 dB. The illustrated examples in Example Set 15-9 will help you understand how power decibels are used in many very practical applications.

EXAMPLE SET   15-9

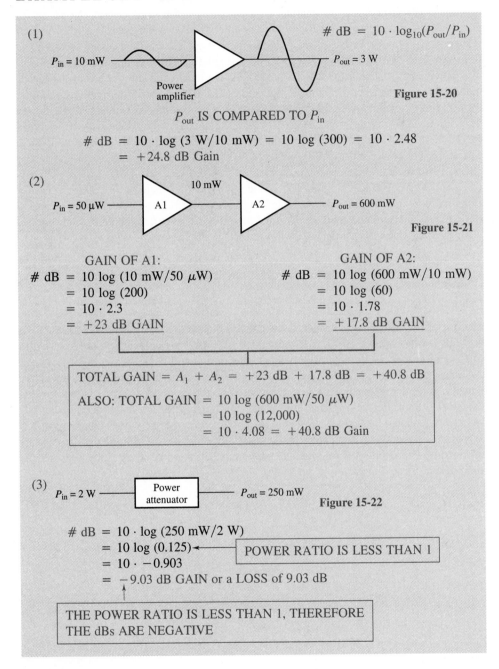

(1)

$$\# \text{ dB} = 10 \cdot \log_{10}(P_{\text{out}}/P_{\text{in}})$$

$P_{\text{in}} = 10 \text{ mW}$

$P_{\text{out}} = 3 \text{ W}$

Power amplifier

Figure 15-20

P_{out} IS COMPARED TO P_{in}

$$\# \text{ dB} = 10 \cdot \log (3 \text{ W}/10 \text{ mW}) = 10 \log (300) = 10 \cdot 2.48$$
$$= +24.8 \text{ dB Gain}$$

(2)

10 mW

$P_{\text{in}} = 50 \text{ μW}$ A1   A2 $P_{\text{out}} = 600 \text{ mW}$

Figure 15-21

GAIN OF A1:

$$\# \text{ dB} = 10 \log (10 \text{ mW}/50 \text{ μW})$$
$$= 10 \log (200)$$
$$= 10 \cdot 2.3$$
$$= +23 \text{ dB GAIN}$$

GAIN OF A2:

$$\# \text{ dB} = 10 \log (600 \text{ mW}/10 \text{ mW})$$
$$= 10 \log (60)$$
$$= 10 \cdot 1.78$$
$$= +17.8 \text{ dB GAIN}$$

TOTAL GAIN $= A_1 + A_2 = +23 \text{ dB} + 17.8 \text{ dB} = +40.8 \text{ dB}$

ALSO: TOTAL GAIN $= 10 \log (600 \text{ mW}/50 \text{ μW})$
$= 10 \log (12,000)$
$= 10 \cdot 4.08 = +40.8 \text{ dB Gain}$

(3) $P_{\text{in}} = 2 \text{ W}$ Power attenuator $P_{\text{out}} = 250 \text{ mW}$

Figure 15-22

$$\# \text{ dB} = 10 \cdot \log (250 \text{ mW}/2 \text{ W})$$
$$= 10 \log (0.125) \longleftarrow \boxed{\text{POWER RATIO IS LESS THAN 1}}$$
$$= 10 \cdot -0.903$$
$$= -9.03 \text{ dB GAIN or a LOSS of } 9.03 \text{ dB}$$

THE POWER RATIO IS LESS THAN 1, THEREFORE THE dBs ARE NEGATIVE

Voltage Decibels

The Voltage dB Formula

Decibels may also be used to compare two voltage levels. The formula for voltage dB is very similar to the power dB formula but has one very important difference: the formula constant is 20 instead of 10.

$$\# \text{ dB} = 20 \cdot \log_{10}(V_1/V_2) \tag{15.15}$$

Why is the formula constant 20 instead of 10? Simply stated, it is because the number of voltage dB must equal the number of power dB in the same circuit. As an aid to understanding this, consider the following example. The AC voltage across a 10 Ω resistor is increased from 1 V to 10 V. This increase in voltage is a 20 dB increase according to Formula 15.15 (# dB = $20 \cdot \log_{10}(10 \text{ V}/1 \text{ V})$ = 20 dB). The resulting increase in power dissipation should also be 20 dB. To prove this, we must calculate the power dissipated in the 10 Ω resistor when 1 V and 10 V are applied. The two power dissipations will then form the ratio in the power dB formula. With 1 V across the 10 Ω resistor, the power dissipation is 100 mW ($P = E^2/R = (1 \text{ V})^2/10 \text{ } \Omega = 100 \text{ mW}$). Then, with 10 V across the 10 Ω resistor, the power dissipation increases to 10 W [$(10 \text{ V})^2/10 \text{ } \Omega = 10 \text{ W}$]. Therefore, the power ratio is 100: 1 (10 W/100 mW = 100/1) and, according to the power dB formula, the increase in power is, in fact, 20 dB [# dB = $10 \cdot \log_{10}(P_1/P_2)$ = $10 \cdot \log_{10}(10 \text{ W}/100 \text{ mW})$ = 20 dB]. Notice that the 10:1 increase in voltage caused a 100:1 increase in power dissipation. This is because power is a function of voltage squared ($P = E^2/R$). Thus, to keep voltage dB equal to power dB, as pertaining to the same circuit, it is necessary to use a constant of 20 in the voltage dB formula.

Using the Voltage dB Formula

The use of the voltage dB formula is very much the same as the power dB formula. As you can see in Formula 15.15, V_1 is compared to V_2 and the comparison is converted to dB. The ratio of V_1 to V_2 may be the voltage gain of an amplifier where $V_1/V_2 = V_{out}/V_{in}$. Assuming the amplifier's gain is greater than 1, the number of dB is expressed as a positive number. If V_1/V_2 represents the performance of a voltage divider, V_1 (V_{out}) will be less than V_2 (V_{in}) and the ratio will be less than 1. In the case of a voltage divider, the AC signal is attenuated instead of being amplified. Therefore, the gain of a voltage divider is expressed as negative dB. A better understanding of the use of voltage dB can best be obtained by examining several practical applications. Take time to carefully consider the examples illustrated in Example Set 15-10.

EXAMPLE SET   15-10

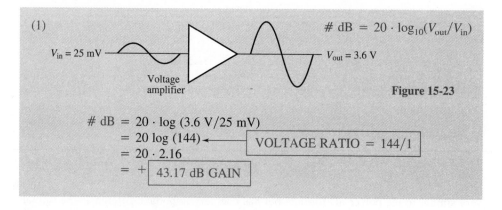

(1)

$V_{in} = 25 \text{ mV}$ Voltage amplifier $V_{out} = 3.6 \text{ V}$

dB = $20 \cdot \log_{10}(V_{out}/V_{in})$

Figure 15-23

dB = $20 \cdot \log (3.6 \text{ V}/25 \text{ mV})$
= 20 log (144) ← VOLTAGE RATIO = 144/1
= $20 \cdot 2.16$
= + 43.17 dB GAIN

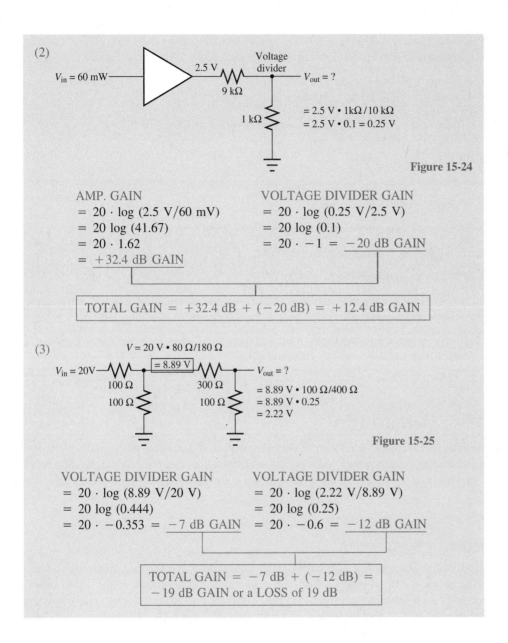

(2)

$V_{in} = 60$ mW

2.5 V

9 kΩ

Voltage divider

$V_{out} = ?$

1 kΩ

= 2.5 V • 1kΩ / 10 kΩ
= 2.5 V • 0.1 = 0.25 V

Figure 15-24

AMP. GAIN
= 20 · log (2.5 V/60 mV)
= 20 log (41.67)
= 20 · 1.62
= +32.4 dB GAIN

VOLTAGE DIVIDER GAIN
= 20 · log (0.25 V/2.5 V)
= 20 log (0.1)
= 20 · −1 = −20 dB GAIN

TOTAL GAIN = +32.4 dB + (−20 dB) = +12.4 dB GAIN

(3)

$V = 20$ V • 80 Ω/180 Ω

$V_{in} = 20$V

100 Ω

100 Ω

= 8.89 V

300 Ω

100 Ω

$V_{out} = ?$

= 8.89 V • 100 Ω/400 Ω
= 8.89 V • 0.25
= 2.22 V

Figure 15-25

VOLTAGE DIVIDER GAIN
= 20 · log (8.89 V/20 V)
= 20 log (0.444)
= 20 · −0.353 = −7 dB GAIN

VOLTAGE DIVIDER GAIN
= 20 · log (2.22 V/8.89 V)
= 20 log (0.25)
= 20 · −0.6 = −12 dB GAIN

TOTAL GAIN = −7 dB + (−12 dB) =
−19 dB GAIN or a LOSS of 19 dB

Comparing Voltage and Power Decibels

To insure that you have a clear understanding of the differences between voltage dB and power dB, consider the following relationships:

• A voltage ratio of 2:1 = +6 dB   (20 · log 2 = 20 · 0.3 = +6 dB)
• A power ratio of 2:1 = +3 dB   (10 · log 2 = 10 · 0.3 = +3 dB)

• A voltage ratio of 1:2 = −6 dB   (20 · log 0.5 = 20 · −0.3 = −6 dB)
• A power ratio of 1:2 = −3 dB   (10 · log 0.5 = 10 · −0.3 = −3 dB)

- A voltage ratio of 10:1 = +20 dB   (20 · log 10 = 20 · 1 = +20 dB)
- A power ratio of 10:1 = +10 dB   (10 · log 10 = 10 · 1 = +10 dB)

- A voltage ratio of 1:10 = −20 dB   (20 · log 0.1 = 20 · −1 = −20 dB)
- A power ratio of 1:10 = −10 dB   (10 · log 0.1 = 10 · −1 = −10 dB)

Take time now to test your understanding with Self-Check 15-5.

SELF-CHECK 15-5

1. 5 Bels is equal to how many dB?

2. Express a change in power level from 5 mW to 50 W in dB.

3. 4,500 W is how many dB above 20 W?

4. 5 W is how many dB below 10,000 W?

5. A 6 V signal is applied to a voltage divider that has an output of 100 mV. Express the signal attenuation in dB.

6. 30 mV is applied to the input of an amplifier and 5 V is obtained at the amplifier's output. Express the amplifier gain in dB.

15-6 Measuring AC

See color photo insert pages D6 through D9.

An introduction to AC would not be complete without a discussion of the practical aspects of measuring AC frequency, voltage, and current. In this section, you will be introduced to a variety of pieces of test equipment used to measure these important parameters.

Measuring Frequency

Digital Frequency Counters

Modern digital test instruments have brought convenience and accuracy to the technician's work bench and the engineer's R&D lab. Among the most common of the digital instruments is the digital frequency counter. Many of these instruments can accurately measure frequencies of AC signals from a few hertz to past 1 GHz ($1 · 10^9$ Hz). Figure 15-26 shows a sampling of digital frequency counters available from several leading manufacturers. Some DMMs even have a frequency counter function built in.

Using the Frequency Counter

Many digital frequency counters have the following features: a standard BNC connector for a test cable/probe; an attenuator selector switch marked X1, X10, X100; range switches; gate-time or resolution switches; and a frequency/period function switch. The test probe is usually a straight-through insulated probe with a ground clip that is placed in parallel across the AC source or load device to which the AC frequency is being applied. The input impedance to the frequency counter is very high, usually 1 MΩ, and therefore

(a)

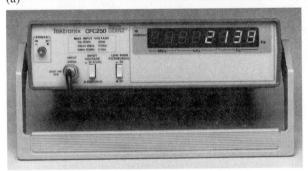

(b)

(c)

Figure 15-26   (a) The FC130A Frequency Counter—courtesy of Beckman Industrial, Inc. (b) The CFC250 Frequency Counter—courtesy of Tektronix, Inc. (c) The B&K 1805 Frequency Counter—courtesy of B&K Precision/Dynascan Corp.

will not load down most circuits. For very sensitive circuits, a 10:1 probe, which increases the probe/counter impedance to 10 MΩ, may be used. If the amplitude of the signal is very high, it may be necessary to select an X10 or X100 attenuator by using the attenuator selector switch on the counter. If the amplitude of the AC signal is too high or too low, the digital reading will be unstable and inaccurate. This becomes obvious in practice.

The Oscilloscope

The **oscilloscope** is a very versatile and important piece of test equipment. Although not as accurate as a digital frequency counter, the oscilloscope has the advantage of allowing you to view the waveform as you measure it. By being able to view the AC waveform, it

See color photo insert page D6.

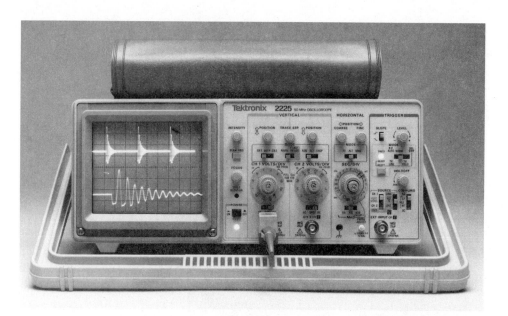

Figure 15-27   This is the very popular Tektronix 2225 dual-trace, 50 MHz oscilloscope—courtesy of Tektronix, Inc.

is possibly to observe signal distortion, measure peak and peak-to-peak voltage, measure period and frequency, and compare the phase of two or more waveforms (depending on the capabilities of the oscilloscope being used). Figure 15-27 clearly shows the controls and screen of a good-quality medium-priced dual-trace oscilloscope.

Measuring Period and Frequency with the Oscilloscope

In order to determine the frequency of an AC waveform, at least one complete cycle of the signal must be displayed on the screen. To do this, connect the oscilloscope test probe to the signal source and select a time-scale factor, using the time/division control, that will cause one to three cycles to be displayed as shown in Figure 15-28. The volts/division

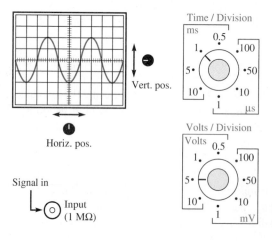

Figure 15-28   Oscilloscope measurements.

control should also be set for a peak-to-peak display covering about half, or more, of the screen vertically. Next, count the number of horizontal gradicules (1 cm divisions) contained in one complete cycle (subdivisions within a gradicule are 0.2 each). The displayed signal can be positioned as desired on the screen using the horizontal and vertical position controls. This may be useful in measuring the cycle. Multiply the number of gradicules by the time/division setting to find the period for one cycle. For example, one cycle may take 4.2 divisions and the time/division control may be set at 1 ms/div. The period is 4.2 div · 1 ms/div = 4.2 ms. The frequency is the reciprocal of the period ($f = 1/T$). In this case, the frequency is $1/4.2$ ms = 238 Hz.

The accuracy of your frequency measurement will not in any way come close to the performance of a digital frequency counter because of reading errors due to parallax, and poor adjustment and calibration of the oscilloscope. However, in many applications, extreme accuracy is not necessary and a ball park measurement ($\pm 5\%$) is adequate.

Measuring AC Voltage

The Oscilloscope

As stated above, the oscilloscope is also very useful in measuring peak and peak-to-peak voltages. Again, one to three cycles should be displayed on the screen. This time, the gradicules, or divisions, are counted vertically from the tip of the positive peak to the tip of the negative peak, as shown in Figure 15-28. The number of divisions is then multiplied by the setting of the volts/division control to find the peak-to-peak voltage. In this case, the waveform occupies 4 vertical divisions and the volt/division control is set at 5 V/div. Therefore, the signal voltage is 20 V_{p-p}. All other voltage values (E_p, E_{avg}, E_{rms}) can be found using formulas discussed earlier in this chapter. As before, the horizontal and vertical position controls may be used to place the displayed signal for convenient measurement. Accuracy is again affected by parallax and calibration.

See color photo insert page D6.

The Multimeter

Analog and digital multimeters are very useful in the measurement of AC voltages due to their extreme portability and ruggedness. As shown in Figure 15-29a, it is a good practice to use the one-hand rule when measuring lethal voltages, especially when working under adverse conditions. In case you are wondering why the meter in photo (b) has a reading being displayed even though the test leads are not connected to a voltage source, it is because this particular Fluke meter has sample and hold capability. While a measurement is taken, a button can be pressed that freezes the reading on the display.

When measuring AC, you need not be concerned with the proper polarity of the test leads as we are when measuring DC. The voltage is constantly changing polarity anyway. However, you do need to consider the approximate amount of voltage to be measured and preset the meter to the appropriate range. Start with the highest range if you are not sure, but keep in mind that the multimeter will have a maximum safe AC voltage limit of about 650 to 750 VAC in most cases.

The reading displayed on the meter is an RMS value. Some digital multimeters are only accurate for pure sine wave measurement while others will provide a true RMS reading for even complex waveforms. The accuracy of the reading also depends on the frequency of the AC being measured. Usually, digital multimeters are limited to measuring

(a)

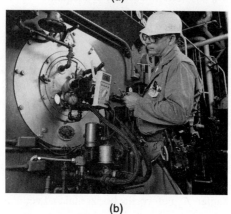

(b)

Figure 15-29   (a) Multimeter portability in measuring AC voltage. (b) Multimeter ruggedness for industrial applications. (Courtesy of John Fluke Mfg. Co., Inc.)

AC voltage and current whose frequency is below 10 kHz. However, most digital multimeters are far more accurate in the frequency range below 500 Hz. In many cases, analog multimeters are more accurate than digital multimeters at frequencies from 1 kHz to about 20 kHz. The previous statements are generalizations. Naturally, the specific performance of a particular multimeter will be clearly indicated on a specification sheet by the manufacturer. Make sure you are familiar with the limitations of your particular multimeter before you use it.

Measuring Alternating Current

The Multimeter

The multimeter may also be used to measure alternating current. As you know, to measure current the meter must be connected in series with the voltage source and the load. As in DC measurements, make sure power is removed before attempting to insert the meter in the circuit. For circuits of lethal voltage, the meter leads should be secured for hands-free operation. Usually multimeters are very limited in choices of AC ranges. Many DMMs have only one low AC range (200 or 300 mA) and a high range of 10 A maximum. Also, the accuracy of the current measurements is usually not as high as voltage or DC measurements. Even so, the accuracy of current measurements is very adequate from a practical standpoint (usually in the area of 1% to 3% up to 500 Hz).

The Clamp-on Ammeter

Perhaps the most common and useful instrument for measuring AC power-line currents is the **clamp-on ammeter**. These instruments are particularly useful to electricians since many of them, such as those shown in Figure 15-30, have both voltage and current measurement capability.

There are two main advantages to the clamp-on ammeter: (1) the AC circuit is not broken or interrupted when making current measurements, and (2) very high AC can be measured with reasonable accuracy (1 to 5% at <500 Hz). The clamp-on ammeter is very convenient to use, since measurements are made without interrupting the service of the piece of equipment whose AC you are measuring. The clamp is an inductive loop that is opened and placed around one current-carrying conductor. The constantly changing magnetic field surrounding the conductor will induce a current in the clamp-on loop. The induced AC is converted to DC and applied to the D'Arsonval meter movement through a selector switch and range resistors. Often, moving-iron-vane meter movements are used, since they respond to AC directly and there is no need for AC to DC conversion.

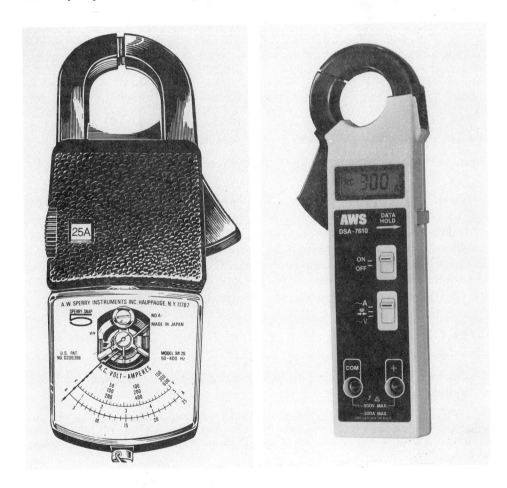

Figure 15-30   (a) Analog clamp-on ammeter with manual range select. (b) Digital clamp-on volt-ammeter with autoranging and data hold. (Courtesy of A.W. Sperry Instruments, Inc.)

Figure 15-31   An AC clamp-on probe for a DMM. (Courtesy of John Fluke Mfg. Co., Inc.)

Figure 15-31 illustrates the use of a clamp-on probe for use with a DMM. As you can see, the clamp is placed around the insulated conductor and the reading is obtained from the DMM. In this particular case, the current in each phase of a three-phase AC power installation is being measured to see if the phases are closely balanced.

In later chapters, we will continue to explore AC circuits and AC measurements using some of the test equipment discussed in this section. For now, take time to review this section. Then, test your knowledge by answering the questions of Self-Check 15-6.

SELF-CHECK 15-6

1. Is an oscilloscope as accurate as a digital frequency counter for measuring frequency? Explain.

2. Are you actually measuring (a) frequency or (b) period with the oscilloscope?

3. What value of AC voltage do you actually measure with the oscilloscope; E_{rms}, E_{avg}, or $E_{\text{p-p}}$?

4. What value of AC voltage do digital multimeters usually display; E_{rms}, E_{avg}, or $E_{\text{p-p}}$?

5. Describe the use of a clamp-on ammeter for measuring AC current.

15-7 Complex Waveforms

Square Waveforms

Harmonics and the Square Wave

The perfect square wave, shown in Figure 15-32, is a complex waveform consisting of an infinite number of odd harmonics. In order to gain a clear understanding of square waves, it is necessary to first understand harmonics. A **harmonic** is a frequency that is a member of a group of related frequencies called a **harmonic series**. The frequencies of the harmonic series are all related in that they all share the same equal spacing from each other as determined by the first (lowest) frequency in the series. Thus, the first harmonic in the series is called the **fundamental** frequency. As an example, consider the following harmonic series:

2 kHz, 4 kHz, 6 kHz, 8 kHz, 10 kHz, 12 kHz, and 14 kHz

The fundamental, or first harmonic, is 2 kHz. All other harmonics in the series are multiples of 2 kHz: the second harmonic is $2 \cdot 2$ kHz, the third harmonic is $3 \cdot 2$ kHz, the fourth harmonic is $4 \cdot 2$ kHz, etc.

As stated earlier, a perfect **square wave** consists of an infinite number of odd harmonics. More specifically, the instantaneous voltage at any point on a perfect, or ideal, square wave is in direct proportion to the sum of the instantaneous voltages, at that same point in time (t), of an infinite number of odd harmonics. Each of the odd harmonics is a perfect sine wave whose peak amplitude is inversely proportional to its harmonic number. For the square wave shown in Figure 15-32, the fundamental frequency is 1 kHz. The next harmonic contained in this square wave is the third harmonic ($3 \cdot 1$ kHz = 3 kHz). The peak amplitude of this third harmonic is $\frac{1}{3}$ of the peak amplitude of the fundamental. The next harmonic is the fifth, with a peak amplitude $\frac{1}{5}$ of the fundamental. The next is the seventh, with a peak amplitude $\frac{1}{7}$ that of the fundamental, etc.

Figure 15-33 demonstrates how a three-harmonic square wave might be created in the lab. Using an oscilloscope, each of the three signal generators is carefully adjusted for the appropriate frequency and amplitude as shown. The resulting square wave is available at the output of the summing amplifier and is shown in relation to the three harmonics that created it. The number of peaks on the crown of each alternation of the square wave is

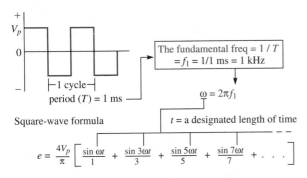

Figure 15-32 The square wave and harmonics.

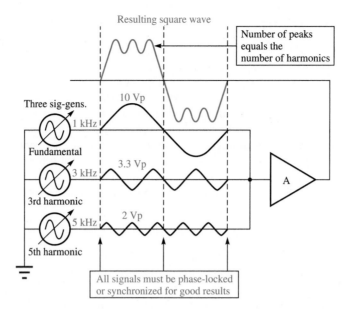

Figure 15-33 A three-harmonic square wave.

the same as the number of harmonics used to generate the wave (in this case three). Naturally, the more harmonics, the smoother the square wave will become. One practical problem in actually performing this experiment is that of keeping all three signals synchronized in proper phase. This may be impossible without a special phase-locked generator circuit that creates all three frequencies at once. It is far more practical to use computer simulation to graphically generate square waves of any desired number of harmonics.

AC vs Pulsating DC Square Waves

Are all square waves AC waveforms? No. Figure 15-34 illustrates three types of square waves: (a) the symmetrical AC square wave; (b) the asymmetrical AC square wave; and (c) the pulsating DC square wave. Notice that the **pulsating DC** square wave represents a DC voltage or current that is switching on and off but never changes polarity or direction. The AC square waves, (a) and (b), do periodically change polarity or direction.

Square-Wave Duty Cycle and Repetition Rate

The **pulse repetition rate (PRR)** of a square wave is the same as the fundamental frequency of the square wave. It is found by taking the reciprocal of the time (period) for one square wave cycle (PRR $= f = 1/T$). The PRR is measured in Hertz or pulses per second (pps). As shown in Example Set 15-11, the period for a cycle is the time from the rising edge of one pulse to the rising edge of the next pulse, or from falling edge to falling edge.

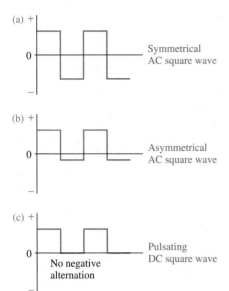

Symmetrical AC square wave

Asymmetrical AC square wave

Pulsating DC square wave

No negative alternation

Figure 15-34 AC or pulsating DC?

The % **duty cycle** of a square wave is a percentage based on the duration of the pulse (**pulse width**) as compared to the cycle time (T). The following formula is used to find the % duty cycle:

$$\text{\% Duty Cycle} = (PW/T) \cdot 100\% \tag{15.16}$$

where PW is the pulse width measured in seconds and T is the cycle period measured in seconds.

Example Set 15-11 illustrates a variety of PRRs, duty cycles, and pulse polarities. Pulses shown at (1) and (2) are called positive going pulses because the pulses are heading in a positive direction from the designated reference level. At (1) the reference level is zero while at (2) the reference level is a negative potential. In like manner, negative going pulses are shown at (3) and (4). In these cases, the pulses are heading in a negative direction from the established reference levels.

EXAMPLE SET 15-11

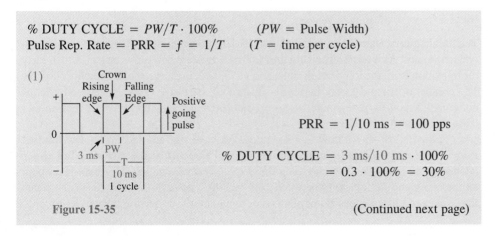

$\text{\% DUTY CYCLE} = PW/T \cdot 100\%$ (PW = Pulse Width)
Pulse Rep. Rate = PRR = f = $1/T$ (T = time per cycle)

(1)

Crown
Rising | Falling
edge   Edge
+
Positive going pulse
0
3 ms
PW
—T—
10 ms
1 cycle

$\text{PRR} = 1/10 \text{ ms} = 100 \text{ pps}$

$$\text{\% DUTY CYCLE} = 3 \text{ ms}/10 \text{ ms} \cdot 100\%$$
$$= 0.3 \cdot 100\% = 30\%$$

Figure 15-35   (Continued next page)

EXAMPLE SET   15-11   (Continued)

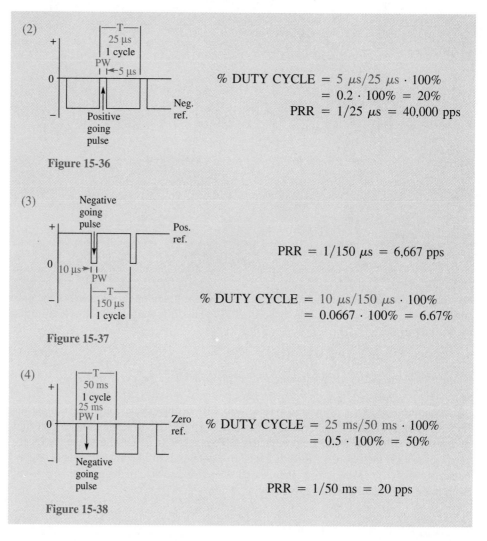

(2)

% DUTY CYCLE $= 5\ \mu s/25\ \mu s \cdot 100\%$
$= 0.2 \cdot 100\% = 20\%$

PRR $= 1/25\ \mu s = 40{,}000$ pps

Figure 15-36

(3)

PRR $= 1/150\ \mu s = 6{,}667$ pps

% DUTY CYCLE $= 10\ \mu s/150\ \mu s \cdot 100\%$
$= 0.0667 \cdot 100\% = 6.67\%$

Figure 15-37

(4)

% DUTY CYCLE $= 25\ ms/50\ ms \cdot 100\%$
$= 0.5 \cdot 100\% = 50\%$

PRR $= 1/50\ ms = 20$ pps

Figure 15-38

Ideal vs Nonideal Square Waves

The **ideal square wave** is shown in Figure 15-39. It is composed of an infinite number of odd harmonics. As a result, its **rising and falling edges** are perfectly vertical, which means the transition from the low to high state, and vice versa, is instantaneous (requires no time). The corners of the pulses are perfectly square and the **pulse crown** is perfectly flat and horizontal. Practical, real square waves are **nonideal** and fall short of perfection in all of these categories.

Figure 15-39b shows how real square waves have rise and fall times. The transition from one level to the other is not instantaneous. Notice that the **rise and fall time** is defined as the transition time between the $0.1 \cdot$ max level and the $0.9 \cdot$ max level points on the rising and falling edge (between 10% and 90% points). In many cases, the slant in the rising and falling edges is not obvious or apparent when viewing the waveform on the

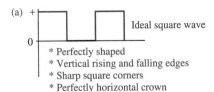

(a) Ideal square wave

* Perfectly shaped
* Vertical rising and falling edges
* Sharp square corners
* Perfectly horizontal crown

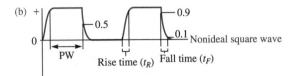

(b) 0.9 / 0.5 / 0.1 Nonideal square wave

PW   Rise time (t_R)   Fall time (t_F)

Rounded corners indicates attenuated
high-frequency harmonics

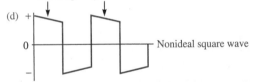

(c) Nonideal square wave

Sloped crown indicates attenuated
low-frequency harmonics

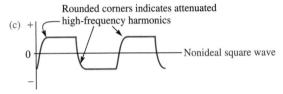

(d) Nonideal square wave

Figure 15-39   Ideal and nonideal square waves.

oscilloscope. However, if a wide-bandwidth oscilloscope is used and the time per division switch is set at a very low time setting, it is possible to view and measure the rise and fall times accurately. Notice also that the pulse width is specified as the time between the $0.5 \cdot$ max points on the rising and falling edge of the pulse (50% points). This specification is adequate for many applications. However, in many practical applications, especially those dealing with digital circuits, the pulse width will be specified by the operating thresholds of the electronic devices. These thresholds may be higher or lower than $0.5 \cdot$ max.

Figure 15-39c shows a square wave that is severely rounded on its corners. This is an indication that the higher-order harmonics are attenuated more than would be the case for an ideal square wave. Again, all practical square waves have some rounding of the corners, even though it is not always obvious on the screen of an oscilloscope. This is because real circuits are very much limited in bandwidth and cannot conduct very-high-order harmonics without greatly attenuating them. No practical circuit is capable of handling harmonics all the way to infinity. One practical application of the rounded corners characteristic is in testing and optimizing the high-frequency response of a broad-banded amplifier. A good-quality square wave is applied to the input of the amplifier and the amplifier is adjusted to minimize the rounding effect in the square wave at the output of the amplifier. The sharper the corners are, the better the high-frequency response of the amplifier.

Figure 15-39 shows a square wave whose low-order harmonics are attenuated more than that of an ideal square wave. The result is sloping crowns, often referred to as **tilt**. Again, this characteristic can be put to good use in testing and optimizing the low-

Slope is measured in V / s (or A / s)
(Sometimes a smaller unit of time such as ms or μs is used)

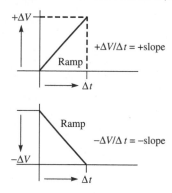

Figure 15-40   Ramp voltage slope is a function of change in voltage with time.

frequency response of an amplifier. The better the low-frequency response, the flatter the crown (less tilt).

Ramp Waveforms

Ramp Slope

A **ramp** waveform is one whose voltage, or current, increases and/or decreases linearly with time. Figure 15-40 illustrates two ramps: one with a positive slope and the other having a negative slope. A ramp that increases in amplitude with time has a **positive slope** and a ramp that decreases in amplitude with time has a **negative slope**. The actual **slope** of the ramp is expressed in volts per second and is found by dividing a change in amplitude by the corresponding change in time ($\Delta V/\Delta t$ or $\Delta A/\Delta t$). For example, suppose a ramp increases from 0 V to a positive peak of 10 V as time increases from 0 to 10 μs. The change in voltage is 10 V and the change in time is 10 μs. Thus, the slope is 1 MV/s (10 V/10 μs = 1 MV/s). This does not mean the voltage will reach 1 million volts in 1 second. It simply means the rate of change is 1 MV/s. In many practical cases, the slope is expressed in terms of μs instead of seconds. In this example, the slope could be expressed as 1 V/μs instead of 1 MV/s.

Sawtooth Waveform

The **sawtooth** waveform is a ramp waveform. As such, a sawtooth signal may be referred to as a ramp voltage or a ramp current. Figure 15-41 shows two forms of the sawtooth: one is a negative-slope sawtooth and the other is a positive-slope sawtooth. Both can be mathematically defined as the sum of an infinite number of all harmonics. Note the subtle difference in mathematical expressions for the negative-slope and the positive-slope sawtooths. For the positive-slope sawtooth, all even harmonics are preceded with a negative sign. This indicates a 180° phase reversal for all even harmonics.

The frequency of the sawtooth is determined in the same way frequency is determined for sine waves or square waves. Once the period for one cycle is determined, the frequency is found by taking the reciprocal of time ($f = 1/T$).

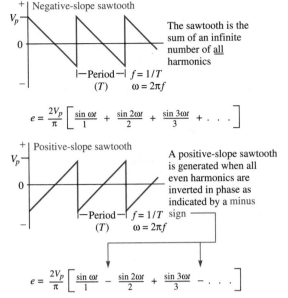

The sawtooth is the sum of an infinite number of <u>all</u> harmonics

$$e = \frac{2V_p}{\pi} \left[\frac{\sin \omega t}{1} + \frac{\sin 2\omega t}{2} + \frac{\sin 3\omega t}{3} + \cdots \right]$$

A positive-slope sawtooth is generated when all even harmonics are inverted in phase as indicated by a minus sign

$$e = \frac{2V_p}{\pi} \left[\frac{\sin \omega t}{1} - \frac{\sin 2\omega t}{2} + \frac{\sin 3\omega t}{3} - \cdots \right]$$

Figure 15-41   Sawtooth waveforms.

Triangle Waveform

The **triangle wave**, shown in Figure 15-42, is another member of the ramp family. As you can see, the triangle differs from the sawtooth in that the triangle has rising and falling ramps whose slopes are equal in magnitude yet opposite in sign. The triangle wave is one of three waveforms commonly available from function-generator integrated circuits found in many low-priced shop signal generators. (The other two waveforms are the sine wave and the square wave.) The triangle wave can be used as a sweep voltage, in like manner as the sawtooth, or to test the linearity of amplifiers and frequency-modulated transmitters.

The frequency of the triangle wave is determined in the same way frequency is determined for other waveforms. The period for one cycle is first determined, then the frequency is found by taking the reciprocal of time ($f = 1/T$).

Take a few minutes to go back and review this section on complex waveforms before answering the questions of Self-Check 15-7.

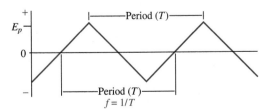

Figure 15-42   The triangle waveform.

SELF-CHECK 15-7

1. If the duration of a positive going pulse is 5 μs and the time for one cycle is 60 μs, what is the % duty cycle?

2. If the duration of a negative going pulse is 0.1 μs and the pulse repetition rate is 3 MHz, what is the % duty cycle?

3. What is the mathematical definition of a square wave?

4. What is the difference between an AC square wave and a pulsating DC square wave?

5. A square wave with rounded corners is an indication of what?

6. What is the slope of a ramp voltage that increases by 250 mV in 25 μs?

Summary

FORMULAS

(15.1) $T = 1/f$ (period)

(15.2) $f = 1/T$ (frequency)

(15.3) $e = E_p \cdot \sin \theta$ (instantaneous voltage)
where e is the instantaneous voltage, E_p is the peak voltage, and θ is the angle.

(15.4) $i = I_p \cdot \sin \theta$ (instantaneous current)
where i is the instantaneous current.

(15.5) $e = E_p \cdot \sin (v \cdot t + \theta)$ (instantaneous voltage)
where e is the instantaneous voltage, v is the angular velocity in $°/s$, t is the time in seconds, and θ is the angle.

(15.6) $i = I_p \cdot \sin (v \cdot t + \theta)$ (instantaneous current)

(15.7) $e = E_p \cdot \sin (\omega \cdot t + \theta)$ (instantaneous voltage)
where ω is the angular velocity in rad/s.

(15.8) $i = I_p \cdot \sin (\omega \cdot t + \theta)$ (instantaneous current)

(15.9) $E_{avg} = E_p \cdot 0.637$ or $I_{avg} = I_p \cdot 0.637$

(15.10) $E_p = E_{avg}/0.637 = E_{avg} \cdot 1.57$ or $I_p = I_{avg}/0.637 = I_{avg} \cdot 1.57$

(15.11) $E_{rms} = E_p \cdot 0.707$ or $I_{rms} = I_p \cdot 0.707$

(15.12) $E_p = E_{rms}/0.707 = E_{rms} \cdot 1.414$ or $I_p = I_{rms}/0.707 = I_{rms} \cdot 1.414$

(15.13) # Bels $= \log_{10}(P_1/P_2)$ (Bels from power ratio)
where P_1/P_2 is a power ratio.

(15.14) # dB $= 10 \cdot \log_{10}(P_1/P_2)$ (decibels from power ratio)

(15.15) # dB $= 20 \cdot \log_{10}(V_1/V_2)$ (decibels from voltage ratio)
where V_1/V_2 is a voltage ratio.

(15.16) % Duty Cycle $= (PW/T) \cdot 100\%$
　　　　where PW is the pulse width in seconds and T is the period for a cycle.

CONCEPTS

- An AC sine wave is a symmetrical, periodic waveform.
- Each AC cycle contains 360° or 2π radians (6.28 rad).
- Frequency is the inverse of time and time is the inverse of frequency.
- The Hertz (Hz) is the standard SI unit of measure for frequency equal to cycles per second.
- Angular velocity may be expressed in degrees per second or radians per second, where 1 radian is equal to 57.3 degrees.
- The average value for a complete symmetrical AC cycle is zero (0).
- The average value of a sine wave alternation is the arithmetic average of all instantaneous values from 0° to 180° or 180° to 360°.
- The effective value of a sine wave alternation is the square root of the arithmetic average (mean) of all squared instantaneous values from 0° to 180° or 180° to 360°.
- The effective value (RMS) is the DC equivalent in terms of power or heating effect.
- AC power is normally calculated using effective values unless otherwise specified.
- AC resistive circuit calculations are performed in the same manner as DC calculations. Effective (RMS) AC values are used.
- The phasor is a ray that is used to represent magnitude and direction in relation to time.
- One Bel equals 10 decibels.
- The digital frequency counter is the most accurate instrument used to measure frequency.
- The oscilloscope is used to measure peak values, peak-to-peak values, period, and phase relationships.
- The clamp-on ammeter offers a safe and convenient means to measure alternating current without interrupting the circuit.
- A perfect square wave consists of the sum of an infinite number of odd harmonics.
- Pulsating DC rapidly changes in amplitude but not in direction or polarity.
- Pulse repetition rate (PRR) is the same as frequency ($f = 1/T$).
- Duty cycle is the comparison of pulse width (duration) to total cycle time.
- Nonideal square waves have rise and fall times, have rounded corners, indicating attenuation of high-frequency harmonics, and may have sloped crowns (tilt), indicating attenuation of low-frequency harmonics.
- A ramp that increases with time has a positive slope; a ramp that decreases with time has a negative slope.
- A sawtooth is a ramp waveform.
- A triangle wave is a ramp waveform with equal rising and falling, positively and negatively sloped edges.

PROCEDURES

Using the Digital Frequency Counter

1. Apply power to the frequency counter and connect the test probe/cable to the counter and the circuit whose frequency is to be measured.

2. Select the appropriate attenuator setting (X1, X10, X100) until a stable reading is displayed.

3. Select the appropriate gate time, or range, for the frequency being measured.

Using the Oscilloscope to Measure Period and Calculate Frequency

1. Apply power to the oscilloscope and connect the test probe/cable to the oscilloscope and the circuit whose frequency is to be measured.

2. Select a scale factor using the time/division control that will cause at least one complete cycle to be displayed.

3. Select a scale factor using the volts/division control that will cause the waveform to occupy about half of the screen vertically.

4. Count the number of horizontal divisions contained in one cycle and multiply by the time/division setting to find the time for one cycle. Calculate the frequency using the measured period ($f = 1/T$).

Using the Oscilloscope to Measure Peak-to-Peak Voltage

1. Follow Steps 1 – 3 above.

2. Count the number of vertical divisions between the positive and negative peak of a cycle and multiply by the volts/division setting to find the peak-to-peak voltage.

Using the Clamp-on Ammeter

1. Beware of any lethal voltages in the vicinity where measurements are to be taken.

2. If the meter is not autoranging, preset it to the highest range.

3. Open the clamp and place it around a single conductor whose current is to be measured. (Insulated conductors only!)

4. Switch down in range until an up-scale reading is obtained.

SPECIAL TERMS

- AC, alternation, cycle
- Periodic waveform, symmetrical waveform
- Period, zero crossing, frequency, hertz
- Sine wave, instantaneous voltage and current
- Angular velocity (v or ω), π (pi), radians
- Peak to peak, average, effective, RMS
- Phasor, phasor diagram
- Decibel (dB), Bel
- Frequency counter, oscilloscope, clamp-on ammeter
- Harmonic, harmonic series, fundamental
- Square wave, pulsating DC, pulse repetition rate (PRR)
- Duty cycle, % duty cycle, pulse width (*PW*)
- Ideal square wave, nonideal square wave
- Rising and falling edges, rise and fall time

- Pulse crown, tilt
- Ramp, slope, positive and negative slope
- Sawtooth wave, triangle wave

Need to Know Solution

Since this particular "Need to Know" did not present a specific problem, there is no need for a specific solution. However, after having studied this chapter, you should be even more keenly aware of the importance of a good understanding of the subjects covered. As you complete your formal education and begin a career, you will see how much of this information really is fundamental, foundational, and common knowledge. Perhaps an additional review of this chapter wouldn't be a bad idea!

Questions

15-1 AC and the Sine Wave

1. What is the technical name for current that periodically changes direction in a circuit?
2. What is the technical name for a half cycle?
3. What is a symmetrical waveform?
4. How many electrical degrees are there in one cycle?
5. What is the relationship between period and frequency?
6. What is the standard SI unit for measuring frequency?
7. Why are many AC waveforms called *sine waves*?
8. How many degrees are there in one radian?
9. How many radians are there in one cycle?

15-2 AC Voltage, Current, and Power

10. Define the effective value of voltage or current.
11. Which value of voltage and current should you use to calculate effective power?

15-3 AC and Resistive Circuits

12. Which values of voltage and current should you normally use in AC circuit analysis?

15-4 Phasors and Phase Relationships

13. What is the difference between a vector and a phasor?
14. The direction of a phasor is expressed in degrees and related to what?
15. What are quadrature AC voltages?

15-5 Decibels (dB)

16. What is the Bel?
17. What is a decibel in relation to a Bel?
18. If audio power increases by a factor of 10, what is the perceived change in loudness?
19. One Bel is equal to what power ratio?
20. A doubling of power is expressed as how many dB?

15-6 Measuring AC

21. What is the most accurate instrument used to measure frequency?
22. Write a brief description of the use of the oscilloscope to measure frequency.
23. Why is the oscilloscope not as accurate as the digital frequency counter for measuring frequency?
24. Explain how the oscilloscope is used to measure AC voltage.
25. What value of AC voltage does the oscilloscope display; average, RMS, or peak-to-peak?
26. List two advantages of the clamp-on ammeter compared to a digital or analog in-line ammeter.
27. Explain how a clamp-on ammeter is used.

15-7 Complex Waveforms

28. What is a harmonic series?
29. What is another name for the first harmonic?
30. Define a square wave in terms of harmonics.
31. What is the difference between an AC square wave and a pulsating DC square wave?
32. Define "pulse repetition rate" (PRR).
33. How is the duty cycle of a square wave determined?
34. Define pulse width (*PW*).
35. Describe an ideal square wave.
36. Define rise and fall time as they pertain to the rising and falling edge of a nonideal square wave pulse.
37. How is the pulse width of a nonideal pulse usually specified?
38. Rounded corners on a square-wave pulse indicate what?
39. Square-wave pulses with sloped crowns indicate what?
40. A sawtooth is sometimes referred to as what?
41. How does a triangle wave differ from a sawtooth wave?
42. How is the frequency of a sawtooth wave or a triangle wave determined?

Problems

15-1 AC and the Sine Wave

1. What fraction of a cycle is represented by 45°?
2. How many electrical degrees are there in ¾ cycle?
3. What is the period for a frequency of 10 kcps?
4. If a cycle has a period of 20 μs, what is its frequency?
5. If a sine wave has a peak voltage of 20 V, what is the instantaneous voltage at 60 electrical degrees?
6. If a sine wave has a peak current of 300 mA, what is the instantaneous current at 140 electrical degrees?
7. What is the angular velocity, in degrees per second, of a 10 MHz frequency?
8. What is the angular velocity, in degrees per second, of a 0.5 MHz frequency?
9. If a 3 kHz sine wave has a peak amplitude of 8 V and an initial offset angle of 30°, what is the instantaneous voltage after an elapsed time of 1.5 ms?

10. If a 60 Hz sine wave has a peak amplitude of 170 V and an initial offset angle of 90°, what is the instantaneous voltage after an elapsed time of 20 ms?

11. If a 50 Hz sine wave has a peak amplitude of 30 A and an initial offset angle of 180°, what is the instantaneous current after an elapsed time of 5 ms?

12. What is the angular velocity, in radians per second, of a 5 kHz frequency?

13. What is the angular velocity, in radians per second, of a 600 kHz frequency?

14. If a sine wave has a peak amplitude of 30 A, an angular velocity of 1,257 rad/s, and an initial offset angle of 45°, what is the instantaneous current after an elapsed time of 1 ms?

15. If a sine wave has a peak amplitude of 50 V, an angular velocity of 6,000 rad/s, and an initial offset angle of 0°, what is the instantaneous voltage after an elapsed time of 50 μs?

15-2 AC Voltage, Current, and Power

16. A symmetrical sine wave that has a voltage of 6 V_p has a peak-to-peak voltage of how much?

17. A symmetrical sine wave that has a peak current of 8 A has a peak-to-peak current of how much?

18. What is the average value for a 10 V_p voltage alternation?

19. What is the average value for a 500 mA_p current alternation?

20. If $E_{avg} = 700$ mV, what is E_p for a single alternation?

21. If $I_{avg} = 3$ A, what is I_p for a single alternation?

22. Calculate the RMS value of 14.14 V_p.

23. Calculate the effective value of 50 A_p.

24. Calculate the peak value of 21 V_{rms}.

25. Calculate the peak value of 120 VAC.

26. If a resistor has a voltage drop of 30 V_p and a current of 100 mA_p, what is the effective power dissipation?

27. If a circuit has a total current of 4 A and an applied voltage of 20 V_{p-p}, what is the total effective power applied to the circuit?

15-3 AC and Resistive Circuits

28. The following three resistors are placed in series with each other and a 45 V_p voltage source: 1 kΩ, 2.7 kΩ, and 3.3 kΩ. Find the RMS voltage drops, the total RMS current, and the power dissipation of each resistor.

29. The following three resistors are placed in series with each other and a 15 V_{p-p} voltage source: 220 Ω, 430 Ω, and 560 Ω. Find the RMS voltage drops, the total RMS current, and the power dissipation of each resistor.

30. The following three resistors are placed in parallel with each other and a 340 V_{p-p} voltage source: 10 kΩ, 47 kΩ, and 51 kΩ. Find the RMS branch currents, the total RMS current, and the power dissipation of each resistor.

31. The following three resistors are placed in parallel with each other and a 36 V_p voltage source: 1 MΩ, 470 kΩ, and 820 kΩ. Find the RMS branch currents, the total RMS current, and the power dissipation of each resistor.

15-4 Phasors and Phase Relationships

32. Draw a waveform diagram showing two sine waves separated by 90°. One sine wave shall be twice the amplitude of the other.
33. Draw the phasor diagram that represents the sine waves of Problem 32.
34. Draw a waveform diagram of two sine waves 180° out of phase yet equal in amplitude.
35. A voltage that leads another by 200° can also be said to lag the other by how many degrees?
36. Draw a phasor diagram to represent the following: a reference voltage of 20 VAC, 40 VAC leading the reference by 90°, and 80 VAC lagging the reference voltage by 120°.
37. Draw a phasor diagram that represents the voltage and current in a resistive AC circuit.
38. If a phasor diagram shows two AC voltages separated by 90° and the frequency of the AC voltages is 50 Hz, what is the actual difference in time between the two voltages?

15-5 Decibels (dB)

39. 50 dB is equal to how many Bels?
40. A 2,100 W amplifier is how many dB above a 1 W amplifier?
41. If a voltage divider has 20 VAC applied and an output voltage of 0.5 VAC, what is the voltage divider gain in dB?
42. A loss of 40 dB can be expressed as a gain of how many dB?
43. If a 200 Ω resistor has a 14 dB increase in voltage across it, what will the dB power increase be in the resistor?
44. If a 4 VAC audio signal is increased by 6 dB, what will the increased voltage be?
45. If a 50 mW power level is increased by 20 dB, what will the increased power level be?
46. If a 10 VAC signal is decreased by 12 dB, what will the decreased voltage be?
47. If a 100,000 W power level is decreased to 100 W, what is the decrease in terms of dB?
48. What is the dB level of 1 μW compared to 5 W?

15-7 Complex Waveforms

49. Find the 11th harmonic of 5 kHz.
50. Find the 6th harmonic of 300 Hz.
51. If the pulse rate of a square wave is 400 pps and the *PW* is 0.3 ms, what is the duty cycle in percent?

Answers to Self-Checks

Self-Check 15-1

1. ½ cycle = 180° = 1 alternation
2. The positive and negative alternations are equal in amplitude and time.
3. 4 ms
4. The hertz (Hz)
5. The voltage, or current, changes in amplitude according to the sine of the angle.
6. 34.2 mA

7. $36 \cdot 10^6 \, °/s$
8. -200.7 mV
9. The radian is an angular distance defined by a circumferal arc equal in length to the radius of a circle.
10. -606.8 mA

Self-Check 15-2

1. $100 \, V_p$; 70.7 VAC; $63.7 \, V_{avg}$
2. $110 \, V_{p-p}$; 38.89 VAC; $35.04 \, V_{avg}$
3. $87.67 \, V_p$; $175.34 \, V_{p-p}$; $55.84 \, V_{avg}$
4. 106.63 W
5. 306 mW

Self-Check 15-3

1. $V_{R2} = V_{R3} = 4.36$ V; $V_{R1} = 19.64$ V
2. $P_{R2} = 12.67 \, \mu W$; $P_{R3} = 19 \, \mu W$; $P_{R1} = 143 \, \mu W$
3. $P_t = 174.67 \, \mu W$
4. $I_{R3} = 4.36 \, \mu A$

Self-Check 15-4

1. The vector is space related, and the phasor is time related.

2.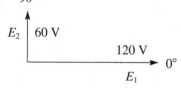

3.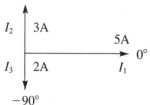

Self-Check 15-5

1. 50 dB
2. 40 dB
3. 23.5 dB
4. 33 dB below or at a level of -33 dB compared to 10 kW
5. 35.6 dB of attenuation or a gain of -35.6 dB
6. 44.4 dB gain

Self-Check 15-6

1. No. The chance for reading errors is greatly increased using the oscilloscope. There is also the chance of control-adjustment error and the oscilloscope being out of calibration.
2. Period
3. Peak to peak
4. RMS
5. Select the highest range. Open the clamp and place it around a single current-carrying conductor. Switch down in ranges until a satisfactory reading is obtained. Beware of any lethal voltages.

Self-Check 15-7

1. 8.33%
2. 30%
3. The sum of an infinite number of odd harmonics.
4. The current of an AC square wave periodically changes direction while the current of a pulsating DC square wave does not.
5. Attenuation of higher-order harmonics.
6. 10,000 V/s or 10 mV/μs

SUGGESTED PROJECTS

1. Add some of the main concepts, formulas, and procedures from this chapter to your personal Electricity and Electronics Notebook.

2. Begin now to familiarize yourself with as many pieces of AC test equipment as you have access to. A thorough understanding of the use of frequency counters, oscilloscopes, and other pieces of test equipment will be invaluable to you as you continue in your education and career.

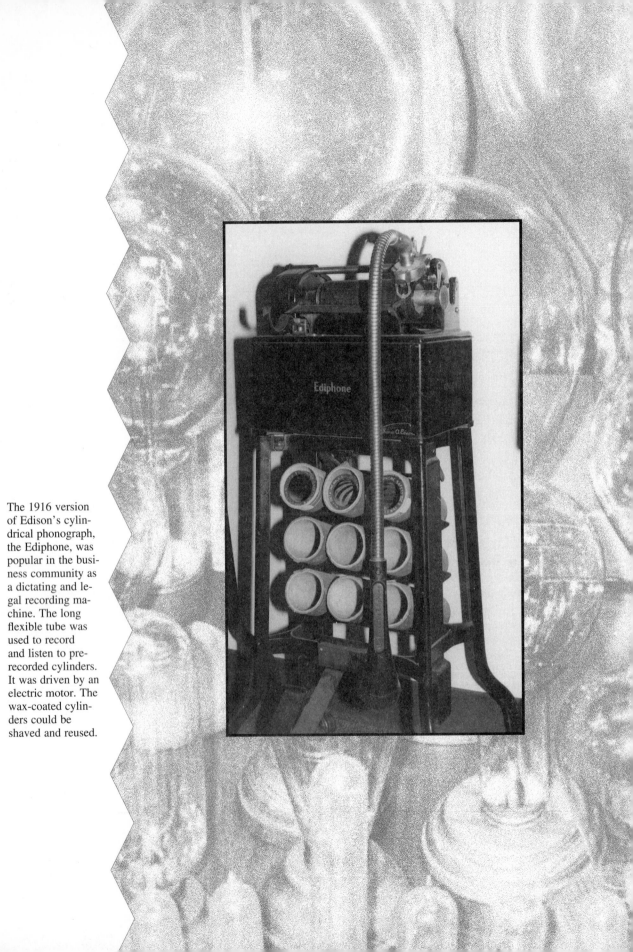

The 1916 version of Edison's cylindrical phonograph, the Ediphone, was popular in the business community as a dictating and legal recording machine. The long flexible tube was used to record and listen to pre-recorded cylinders. It was driven by an electric motor. The wax-coated cylinders could be shaved and reused.

Chapter 16

AC and Reactance

CHAPTER OUTLINE

OBJECTIVES

After studying this chapter, you will be able to

- explain the meaning and use of the phrase, "ELI the ICEman."
- calculate inductive reactance (X_L) using Ohm's Law or the inductive reactance formula when signal frequency and inductance are known.
- solve for applied-signal frequency when inductance and inductive reactance are known, or inductance when frequency and inductive reactance are known.

- calculate capacitive reactance (X_C) using Ohm's Law or the capacitive reactance formula when signal frequency and capacitance are known.
- solve for applied-signal frequency when capacitance and capacitive reactance are known, or capacitance when frequency and capacitive reactance are known.
- calculate all voltages and currents in series and parallel capacitive and inductive circuits.

INTRODUCTION

In this chapter, you will begin to see the inductor and capacitor as AC components. You will see that the inductor and capacitor react much differently to AC sine wave signals than they do to DC or pulsed DC. Capacitors and inductors are as common in electrical and electronic circuits as transistors, diodes, or their integrated-circuit counterparts. In the following pages, you will begin to understand the role these reactive components play in electrical and electronic circuitry. In particular, you will discover that the reactance of a capacitor or inductor is measured in ohms, just like resistance. You will also discover how capacitive and inductive reactances limit current and develop voltage drops in AC circuits. As you explore this chapter, be careful to identify the similarities and differences between resistance, inductive reactance, and capacitive reactance. The similarities are many and so are the differences. At the close of the chapter, you will be introduced to some important applications for capacitors and inductors in AC circuits.

NEED TO KNOW

Consider the following "black box." It puts out a 1 kHz AC signal along with a DC voltage on the same wire (using a common ground reference). Using as few components as possible, how might you separate the 1 kHz AC from the DC? In other words, how can the signal output be converted to two outputs: one a 1 kHz AC signal, and the other a DC voltage? Explain how and why.

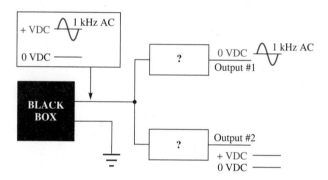

Not sure? That's OK. But wait. Don't look at the solution yet. Wait until you finish this chapter and try to solve this puzzle again. I think you will be able to do it without looking at the solution. Good luck!

16-1 AC and the Inductor

AC Inductor Voltage

As we begin to investigate the inductor as an AC component, it is important to recall the inductor's basic characteristics. First, recall that an inductor is made of a low-resistance conductor: copper, aluminum, and so on. Thus, the DC voltage drop across an inductor, after a buildup transient period is past, is very small, near zero volts. The inductor only has a significant voltage drop during the transient time and this voltage is induced. Remember, inductor current is changing (increasing or decreasing) during any transient time. The change in current with time ($\Delta i/\Delta t$) causes a corresponding change in magnetic flux with time ($\Delta\Phi/\Delta t$). This changing flux induces a voltage across the inductor and is expressed in Faraday's formula: $v_L = N \cdot \Delta\Phi/\Delta t$. Note that the rate of change in flux will determine the amount of induced voltage in an inductor of a given number of turns.

Recall, from an earlier chapter introducing inductance, that a useful formula for induced voltage can be derived from Faraday's formula. This formula states that the amount of induced voltage is the product of inductance and the rate of change of inductor current: $v_L = L \cdot \Delta i/\Delta t$. Figure 16-1 shows the derivation of this formula and illustrates the relationship between rate of change in current and rate of change in magnetic flux. At this point it is important for you to clearly understand that the amount of induced instantaneous voltage is directly related to the instantaneous rate of change in magnetic flux, which is

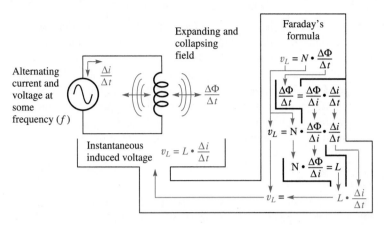

Figure 16-1 Induced inductor voltage depends upon the amount of inductance and rate of change of current in the inductor's windings.

directly related to the instantaneous rate of change of inductor current. In an AC inductor circuit, the current applied to the inductor is continually changing in amplitude, direction, and rate of change. Therefore, there is a continual change in magnetic flux and induced voltage. The alternating current produces an alternating flux that induces an alternating voltage.

ELI

Voltage Maximum

Now that you have reviewed and have a clear understanding of the relationship between changing current, flux, and induced voltage, you are ready to discover a very interesting relationship between the alternating inductor current and the alternating induced voltage: The phase relationship between the current and voltage is 90°. Figure 16-2 illustrates how

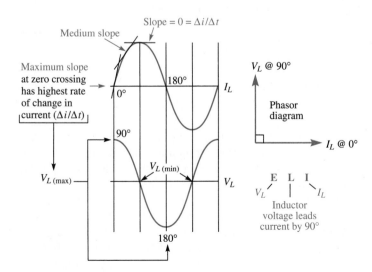

Figure 16-2 Inductor voltage leads inductor current by 90° (ELI).

the induced inductor voltage actually leads the inductor current by 90° (a quarter of a cycle). In other words, the induced inductor voltage is at a maximum value when the inductor current is changing at its highest rate. Notice, from Figure 16-2, that the current exhibits its highest rate of change as it passes through zero crossing (0° and 180°). It is at zero crossing that the slope of the sine curve is maximum, and therefore the rate of change in current is maximum ($\Delta i/\Delta t$ = maximum). Thus, when inductor current is minimum (0° or 180°), the inductor voltage is maximum (90° or 270°) and it is said that the inductor voltage *leads* its current by 90° or the inductor current *lags* its voltage by 90°.

Voltage Minimum

At 90° and 270° points in the sine curve for inductor current, the current reaches positive and negative peak values. At these peaks, the rate of change in current is zero ($\Delta i/\Delta t$ = 0), since the slope of a line drawn tangent to a peak is zero. Since the rate of change in current is zero, the induced inductor voltage is zero ($v_L = L \cdot \Delta i/\Delta t = 0$). Thus, when inductor current is maximum (90° or 270°), the inductor voltage is minimum (0° or 180°). Again, we can see the 90° voltage and current phase relationship (**ELI** = voltage leads current by 90° in an inductor). Notice the phasor diagram, in Figure 16-2, illustrating the 90° phase relationship. The current phasor is drawn horizontally as a reference at 0°. The reason for using current as the reference instead of voltage will become obvious later.

Take a moment now to review this short section and test your knowledge by answering the questions of Self-Check 16-1.

SELF-CHECK 16-1

1. Besides the inductance of a coil, what other factor determines the amount of voltage induced across a coil?

2. At what points in a sine curve for inductor current will induced inductor voltage be maximum? Why?

3. What is the phase relationship between inductor voltage and current?

16-2 AC and the Capacitor

AC Capacitor Current

How is it possible for a capacitor to pass an alternating current even though an insulator is placed between its plates? That is an important question and it is important that you understand the answer. A capacitor passes AC because an alternating voltage is applied to its plates. Here's what I mean. Recall that, in an *RC* transient circuit, capacitor charge current will flow until the capacitor is fully charged and discharge current will flow until the capacitor is fully discharged. When an alternating voltage is applied to the plates of a capacitor, the capacitor is forced to follow repeated cycles of charge and discharge. As the applied AC voltage increases, decreases, and changes polarity, the quantity of charge on the capacitor's plates will also increase, decrease, and change polarity. Therefore, the

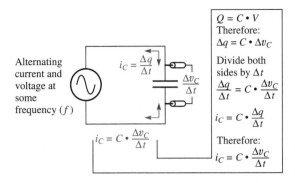

Figure 16-3   Alternating capacitor current.

alternating capacitor current at any instant in time depends on the rate of change in quantity of charge at that instant in time ($i_C = \Delta q / \Delta t$).

Don't lose sight of the fact that the rate of change in charge ($\Delta q / \Delta t$) is established by the rate of change in applied alternating voltage ($\Delta v / \Delta t$). It is the changing difference in potential that creates the changing quantity of charge. This is illustrated in Figure 16-3. Notice too, that since $Q = C \cdot V$ (from Chapter 13), a change in quantity of charge is equal to capacitance times a change in voltage ($\Delta q = C \cdot \Delta v_C$). Also, a particular rate of change in voltage ($\Delta v_C / \Delta t$) produces a corresponding rate of change in quantity of charge ($\Delta q / \Delta t = C \cdot \Delta v_C / \Delta t$). Since $i_C = \Delta q / \Delta t$, i_C must also equal $C \cdot \Delta v_C / \Delta t$ by substitution. Thus, the rate of change in voltage at any instant in time determines the instantaneous capacitor current at that instant in time.

ICE

Current Maximum

Just as we were able to explain the phase relationship between current and voltage of an inductor, we may now explain the current/voltage phase relationship of a capacitor. We can begin by determining the conditions under which the capacitor current is maximum. According to the formula derived above, the capacitor current at any instant in time is determined by the value of the capacitor and the rate of change in capacitor voltage ($i_C = C \cdot \Delta v_C / \Delta t$). Thus, when the rate of change in capacitor voltage is maximum, the instantaneous capacitor current is maximum. As shown in Figure 16-4, the rate of change in voltage is maximum at the points of zero crossing (0° and 180°) in the sine curve for alternating voltage. As the capacitor voltage sweeps through the points of zero crossing (0° and 180°), the capacitor current sweeps through maximum positive and negative peaks (90° and 270°). As you can see, this results in the capacitor current leading the capacitor voltage by 90°.

Current Minimum

Current minimums occur as capacitor voltage reaches points of maximum. It is at these points of voltage maximum that the rate of change in voltage is zero ($\Delta v_C / \Delta t = 0$). If the rate of change in capacitor voltage is zero, the current at that instant must also be zero $i_C = C \cdot 0 = 0$). Thus, the points of voltage maximum (90° and 270°) correspond to

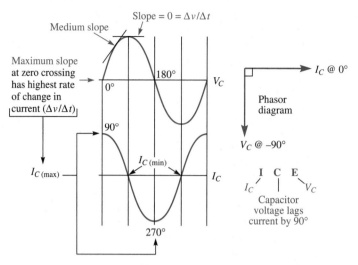

Figure 16-4   Capacitor current leads capacitor voltage by 90° (ICE).

points of current minimum (0° and 180°). Again, we see that the phase relationship between capacitor voltage and current is 90° (**ICE** = capacitor current leads capacitor voltage by 90°). This is also illustrated with the phasor diagram shown in Figure 16-4.

Remember This   HELPFUL HINT: Remember the phrase "ELI the ICEman." This will help you remember the voltage and current phase relationship for the inductor and capacitor.

Once again, time for a review and a self-check.

S E L F - C H E C K   1 6 - 2

1. Explain how the rate of change in capacitor voltage affects the capacitor current.

2. What is the rate of change in capacitor voltage when the instantaneous capacitor current is zero?

3. Does capacitor current lead or lag capacitor voltage?

16-3 Inductive Reactance (X_L)

Defining X_L

An inductor produces an opposition to alternating current that limits the amount of current flowing through the inductor. This opposition is the inductor's reaction, or reactance, to the alternating current. Reactance in general is symbolized with a capital X, and **inductive**

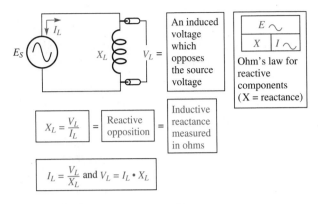

Figure 16-5 Inductive reactance X_L.

reactance (X_L) is symbolized with a subscript. As you know, a changing current in an inductor produces a changing magnetic flux, which in turn induces a counterelectromotive force (CEMF) across the inductor. This CEMF opposes the applied EMF and thus limits the amount of AC flowing in the inductor. This opposition is the inductor's reaction to the AC. Since this reaction (reactance) opposes the applied EMF and limits inductor current, it is measured in ohms.

Figure 16-5 illustrates inductive reactance showing how it can be calculated using Ohm's Law. Notice that, if the inductor voltage (V_L) and the inductor current (I_L) are known, the inductive reactance can be calculated in the same way resistance is calculated, using Ohm's Law. Also note that Ohm's Law may be used in any of its three forms to solve for any of the three variables when the other two are known. Consider the examples shown in Example Set 16-1.

EXAMPLE SET 16-1

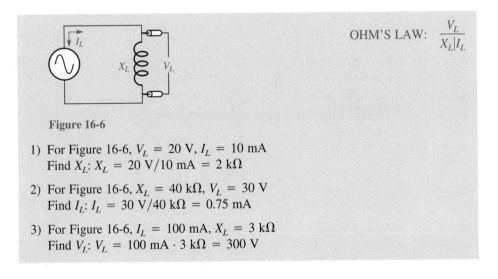

Figure 16-6

OHM'S LAW: $\dfrac{V_L}{X_L | I_L}$

1) For Figure 16-6, $V_L = 20$ V, $I_L = 10$ mA
   Find X_L: $X_L = 20$ V/10 mA $= 2$ kΩ

2) For Figure 16-6, $X_L = 40$ kΩ, $V_L = 30$ V
   Find I_L: $I_L = 30$ V/40 k$\Omega = 0.75$ mA

3) For Figure 16-6, $I_L = 100$ mA, $X_L = 3$ kΩ
   Find V_L: $V_L = 100$ mA $\cdot$ 3 k$\Omega = 300$ V

The X_L Formula

The inductive reactance of a particular inductor, in an AC circuit, can be calculated using a formula that takes into account the inductance L of the inductor and the frequency f of the applied AC as follows:

$$X_L = \omega \cdot L = 2\pi f L \tag{16.1}$$

where the inductive reactance is measured in ohms, and the angular velocity ($\omega = 2\pi f$) of the AC and the inductance L are known.

As you can see, the reactance of a particular inductor is dependent upon the inductor's inductance L and the angular velocity ($\omega = 2\pi f$) of the applied AC. Recall that the angular velocity of an AC voltage or current is the radian rate of change in amplitude of the sine waveform measured in radians per second (rad/s). Naturally, the higher the frequency (Hz or cps) the higher the angular velocity. If the frequency of the alternating current applied to the inductor is high, the rate of change in inductor current will be high ($\Delta i/\Delta t$). Recall that the rate of change in current determines the rate of change in flux ($\Delta\Phi/\Delta t$), which determines the amount of induced CEMF (Faraday's Voltage Law). Therefore, a higher AC frequency means a higher rate of change in current, which results in a greater induced CEMF or opposition to the alternating current. Thus, the amount of inductive reactance is directly related to the frequency of the applied AC. Naturally, the amount of inductance affects the amount of induced CEMF [$v_L = L \cdot (\Delta i/\Delta t)$] and is also directly related to the inductive reactance. In summary, an increase in frequency (f) and/or inductance (L) will result in an increase in X_L, and a decrease in frequency and/or inductance will result in a decrease in X_L. Study the examples in Example Set 16-2 to gain a clear understanding of the use of the X_L formula.

E X A M P L E   S E T   1 6 - 2

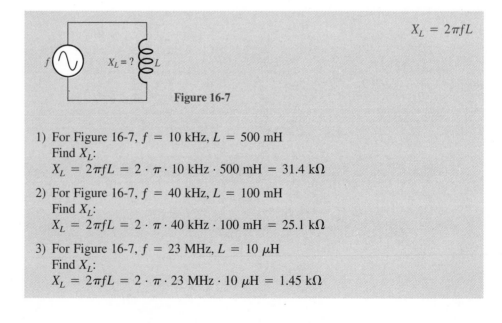

$$X_L = 2\pi f L$$

Figure 16-7

1) For Figure 16-7, $f = 10$ kHz, $L = 500$ mH
   Find X_L:
   $X_L = 2\pi f L = 2 \cdot \pi \cdot 10 \text{ kHz} \cdot 500 \text{ mH} = 31.4 \text{ k}\Omega$

2) For Figure 16-7, $f = 40$ kHz, $L = 100$ mH
   Find X_L:
   $X_L = 2\pi f L = 2 \cdot \pi \cdot 40 \text{ kHz} \cdot 100 \text{ mH} = 25.1 \text{ k}\Omega$

3) For Figure 16-7, $f = 23$ MHz, $L = 10$ μH
   Find X_L:
   $X_L = 2\pi f L = 2 \cdot \pi \cdot 23 \text{ MHz} \cdot 10 \text{ }\mu\text{H} = 1.45 \text{ k}\Omega$

Solving for L

Let's suppose you have an unmarked inductor and you want to determine its inductance. This can be done by first placing the inductor in a simple AC circuit to see how the inductor reacts in the circuit. The AC applied to the inductor can be measured with an AC ammeter (RMS value), and the voltage across the inductor can be measured with an AC voltmeter or oscilloscope (RMS value). The inductive reactance of the inductor can then be calculated using Ohm's Law ($X_L = V_L/I_L$). Once the inductive reactance of the unknown inductor is calculated and the frequency of the applied AC is known, the inductance L can be determined by rearranging the X_L formula as follows:

$$L = X_L/2\pi f \qquad\qquad (16.2)$$

Study Example Set 16-3 to see how this formula might actually be used.

EXAMPLE SET 16-3

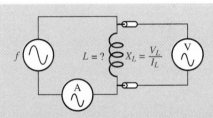

Figure 16-8

$$L = \frac{X_L}{2\pi f}$$

NOTE: Many AC voltmeters and ammeters are accurate only below 500 Hz. The meter's manual should be consulted to determine its actual frequency limitations.

1) For Figure 16-8, X_L is found to be 1,200 Ω.
   The frequency is 100 Hz.
   Find L:
   $$L = \frac{X_L}{2\pi f} = \frac{1,200 \ \Omega}{2 \cdot \pi \cdot 100 \ \text{Hz}} = 1.91 \ \text{H}$$

2) For Figure 16-8, X_L is found to be 200 Ω.
   The frequency is 60 Hz.
   Find L:
   $$L = \frac{X_L}{2\pi f} = \frac{200 \ \Omega}{2 \cdot \pi \cdot 60 \ \text{Hz}} = 0.531 \ \text{H}$$

Solving for f

Just as the X_L formula can be rearranged to solve for an unknown inductance, the formula may also be rearranged to solve for frequency (f). As long as L and X_L are known, the X_L formula may be rearranged as follows:

$$f = X_L/2\pi L \qquad\qquad (16.3)$$

Consider the examples of Example Set 16-4 before testing your understanding in Self-Check 16-3.

EXAMPLE SET   16-4

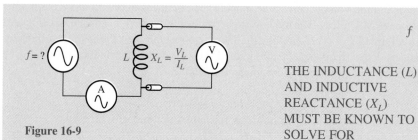

Figure 16-9

$$f = \frac{X_L}{2\pi L}$$

THE INDUCTANCE (L)
AND INDUCTIVE
REACTANCE (X_L)
MUST BE KNOWN TO
SOLVE FOR
FREQUENCY (f)

1) For Figure 16-9, $X_L = 350\ \Omega$ and $L = 1$ mH
   Find the frequency (f):
   $$f = \frac{X_L}{2\pi L} = \frac{350\ \Omega}{2 \cdot \pi \cdot 1\ \text{mH}} = 55.7\ \text{kHz}$$

2) For Figure 16-9, $X_L = 350\ \Omega$ and $L = 1$ H
   Find the frequency (f):
   $$f = \frac{X_L}{2\pi L} = \frac{350\ \Omega}{2 \cdot \pi \cdot 1\ \text{H}} = 55.7\ \text{Hz}$$

SELF-CHECK   16-3

1. If the voltage across an inductor is 8 V and the inductor's current is 100 mA, what is the inductor's reactance?

2. Determine the inductive reactance of a 100 μH inductor at a frequency of 200 kHz.

3. If an inductor produces 2,500 Ω of reactance in response to an applied frequency of 400 kHz, what is the inductance of the inductor?

4. If the frequency applied to an inductor is doubled, what will happen to its inductive reactance?

16-4 X_L in Series and Parallel

X_L in Series

Total Series X_L

The total inductive reactance produced by two or more inductors placed in a series arrangement is simply the sum of all individual inductive reactances. The total may be found by adding all individual inductive reactances or by applying the total inductance to the X_L formula.

$$X_{Lt} = X_{L1} + X_{L2} + X_{L3} + \cdots \qquad (16.4)$$

Examine Example Set 16-5 to see how total series inductive reactance is determined.

EXAMPLE SET   16-5

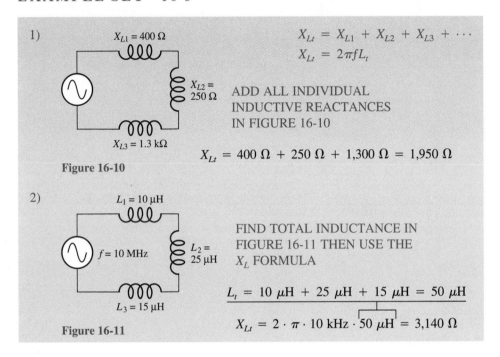

1) $X_{L1} = 400\ \Omega$

$X_{L2} = 250\ \Omega$

$X_{L3} = 1.3\ k\Omega$

Figure 16-10

$$X_{Lt} = X_{L1} + X_{L2} + X_{L3} + \cdots$$
$$X_{Lt} = 2\pi f L_t$$

ADD ALL INDIVIDUAL INDUCTIVE REACTANCES IN FIGURE 16-10

$$X_{Lt} = 400\ \Omega + 250\ \Omega + 1,300\ \Omega = 1,950\ \Omega$$

2) $L_1 = 10\ \mu H$

$f = 10\ MHz$

$L_2 = 25\ \mu H$

$L_3 = 15\ \mu H$

Figure 16-11

FIND TOTAL INDUCTANCE IN FIGURE 16-11 THEN USE THE X_L FORMULA

$$L_t = 10\ \mu H + 25\ \mu H + 15\ \mu H = 50\ \mu H$$

$$X_{Lt} = 2 \cdot \pi \cdot 10\ kHz \cdot 50\ \mu H = 3,140\ \Omega$$

Distributed Voltage

Inductors in series in an AC circuit will self-induce voltages that are in proportion to each individual inductor's inductance and inductive reactance. This is similar to voltage drops across resistors in a series circuit, inasmuch as the sum of all induced voltages will equal the source voltage. Remember, however, that the inductor voltages are leading the circuit current by 90° (ELI), unlike the resistor voltage drops, which are in phase with the circuit current. Each inductor voltage may be calculated just as though the inductor were a resistor. In other words, you may use Ohm's Law or the proportion formula as follows:

$$V_L = I_t \cdot X_L \qquad \text{(Ohm's Law)} \qquad (16.5)$$
$$V_L = E_S \cdot L/L_t = E_s \cdot X_L/X_{Lt} \qquad \text{(proportion formula)} \qquad (16.6)$$

where, for series inductors, any individual inductor's voltage will be in proportion to its inductance or inductive reactance as compared to total inductance or total inductive reactance.

Study Example Set 16-6 to see how these formulas can be used. Also, notice the phasor diagrams showing the phasor sum of all individual inductor voltages. All inductor voltages are in phase with each other and add up to the source voltage (Kirchhoff's Voltage Law for series circuits).

EXAMPLE SET   16-6

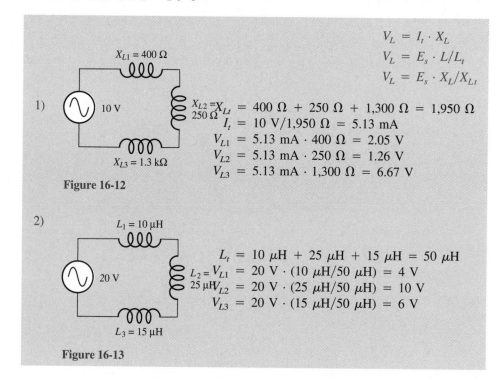

$$V_L = I_t \cdot X_L$$
$$V_L = E_s \cdot L/L_t$$
$$V_L = E_s \cdot X_L/X_{Lt}$$

1)   10 V $X_{L1} = 400\ \Omega$

$X_{L2} = 250\ \Omega$

$X_{L3} = 1.3\ \text{k}\Omega$

$X_{Lt} = 400\ \Omega + 250\ \Omega + 1{,}300\ \Omega = 1{,}950\ \Omega$
$I_t = 10\ \text{V}/1{,}950\ \Omega = 5.13\ \text{mA}$
$V_{L1} = 5.13\ \text{mA} \cdot 400\ \Omega = 2.05\ \text{V}$
$V_{L2} = 5.13\ \text{mA} \cdot 250\ \Omega = 1.26\ \text{V}$
$V_{L3} = 5.13\ \text{mA} \cdot 1{,}300\ \Omega = 6.67\ \text{V}$

Figure 16-12

2)   20 V $L_1 = 10\ \mu\text{H}$

$L_2 = 25\ \mu\text{H}$

$L_3 = 15\ \mu\text{H}$

$L_t = 10\ \mu\text{H} + 25\ \mu\text{H} + 15\ \mu\text{H} = 50\ \mu\text{H}$
$V_{L1} = 20\ \text{V} \cdot (10\ \mu\text{H}/50\ \mu\text{H}) = 4\ \text{V}$
$V_{L2} = 20\ \text{V} \cdot (25\ \mu\text{H}/50\ \mu\text{H}) = 10\ \text{V}$
$V_{L3} = 20\ \text{V} \cdot (15\ \mu\text{H}/50\ \mu\text{H}) = 6\ \text{V}$

Figure 16-13

X_L in Parallel

Total Parallel X_L

Inductive reactances in parallel are similar to resistances in parallel. Total inductive reactance is found using one of the parallel formulas for resistors: the same values formula, the product over the sum formula, or the reciprocal formula.

$$X_{Lt} = X_{L1}/N \qquad \text{(same values formula)} \tag{16.7}$$

where all parallel inductances and inductive reactances are the same value and the total inductive reactance is found by dividing the reactance of one inductor (X_{L1}) by the number N of parallel inductors.

$$X_{Lt} = \frac{X_{L1} \cdot X_{L2}}{X_{L1} + X_{L2}} \qquad \text{(product over the sum formula)} \tag{16.8}$$

where two parallel inductors have different values of inductance and inductive reactance.

$$X_{Lt} = \frac{1}{1/X_{L1} + 1/X_{L2} + 1/X_{L3} + \cdots} \qquad \text{(reciprocal formula)} \tag{16.9}$$

where more than two inductors of different value are placed in parallel.

Example Set 16-7 illustrates the use of these formulas in calculating total inductive reactance for parallel inductor circuits.

EXAMPLE SET　16-7

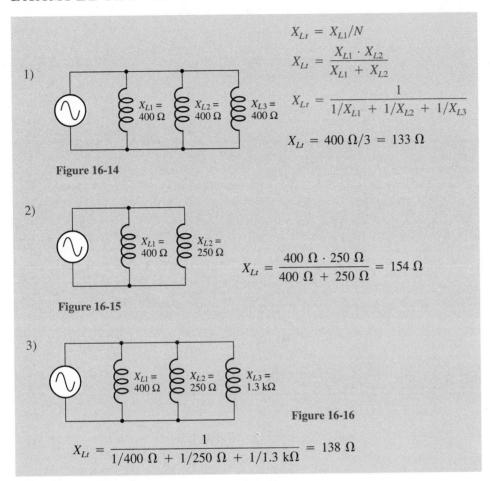

$$X_{Lt} = X_{L1}/N$$

$$X_{Lt} = \frac{X_{L1} \cdot X_{L2}}{X_{L1} + X_{L2}}$$

1)

$$X_{Lt} = \frac{1}{1/X_{L1} + 1/X_{L2} + 1/X_{L3}}$$

$$X_{Lt} = 400\ \Omega/3 = 133\ \Omega$$

Figure 16-14

2)

$$X_{Lt} = \frac{400\ \Omega \cdot 250\ \Omega}{400\ \Omega + 250\ \Omega} = 154\ \Omega$$

Figure 16-15

3)

Figure 16-16

$$X_{Lt} = \frac{1}{1/400\ \Omega + 1/250\ \Omega + 1/1.3\ k\Omega} = 138\ \Omega$$

Distributed Current

Once again, as in parallel resistor circuits, the total current supplied to a parallel section is distributed among the section's branches. The current is distributed to each parallel inductor in accordance with Ohm's Law and in accordance with the current divider formula as follows (see Chapter 7, Section 7-1 for a review of current division):

$$I_L = I_t \cdot X_{Lt}/X_L = I_t \cdot L_t/L \qquad \text{(current divider formula)} \tag{16.10}$$

where the current in any branch is in proportion to the ratio of total inductive reactance to the inductive reactance of the branch or in proportion to the ratio of total inductance to the inductance of the branch.

Example Set 16-8 illustrates the use of Ohm's Law and the current divider formula in determining individual branch currents of parallel inductive circuits.

EXAMPLE SET 16-8

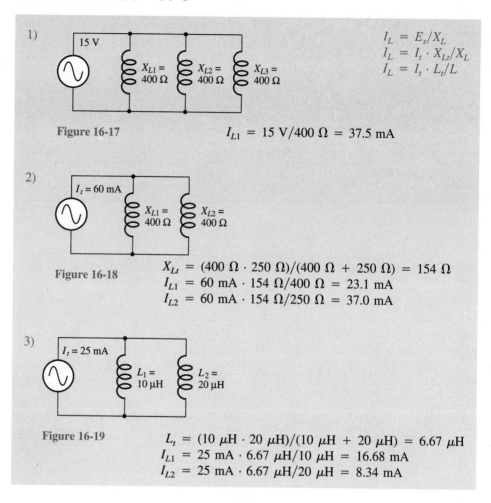

1)
15 V

$X_{L1} = 400\ \Omega$ $X_{L2} = 400\ \Omega$ $X_{L3} = 400\ \Omega$

$I_L = E_s/X_L$
$I_L = I_t \cdot X_{Lt}/X_L$
$I_L = I_t \cdot L_t/L$

Figure 16-17

$I_{L1} = 15\ V/400\ \Omega = 37.5\ mA$

2)
$I_t = 60\ mA$

$X_{L1} = 400\ \Omega$ $X_{L2} = 400\ \Omega$

Figure 16-18

$X_{Lt} = (400\ \Omega \cdot 250\ \Omega)/(400\ \Omega + 250\ \Omega) = 154\ \Omega$
$I_{L1} = 60\ mA \cdot 154\ \Omega/400\ \Omega = 23.1\ mA$
$I_{L2} = 60\ mA \cdot 154\ \Omega/250\ \Omega = 37.0\ mA$

3)
$I_t = 25\ mA$

$L_1 = 10\ \mu H$ $L_2 = 20\ \mu H$

Figure 16-19

$L_t = (10\ \mu H \cdot 20\ \mu H)/(10\ \mu H + 20\ \mu H) = 6.67\ \mu H$
$I_{L1} = 25\ mA \cdot 6.67\ \mu H/10\ \mu H = 16.68\ mA$
$I_{L2} = 25\ mA \cdot 6.67\ \mu H/20\ \mu H = 8.34\ mA$

SELF-CHECK 16-4

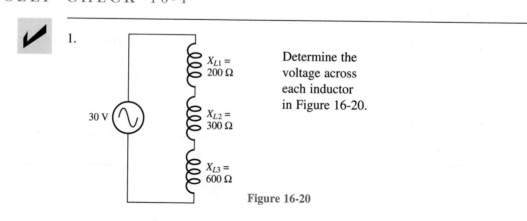

1.

$X_{L1} = 200\ \Omega$

30 V

$X_{L2} = 300\ \Omega$

$X_{L3} = 600\ \Omega$

Determine the voltage across each inductor in Figure 16-20.

Figure 16-20

2.

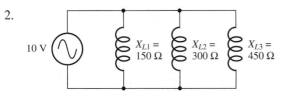

Figure 16-21

Determine the current through each inductor of Figure 16-21. Also, determine total current and total inductive reactance.

16-5 Capacitive Reactance (X_C)

Defining X_C

A capacitor produces an opposition to alternating current that limits the amount of current flowing through the capacitor. This opposition is the capacitor's reactance. Like inductors, capacitors react to the AC and produce a **capacitive reactance** measured in ohms. The capacitive reactance variable is symbolized with a subscript as follows: X_C.

Figure 16-22 illustrates capacitive reactance, showing how it can be calculated using Ohm's Law. Notice that, if the capacitor voltage V_C and the capacitor current I_C are known, the capacitive reactance can be calculated in the same way resistance, or inductive reactance, is calculated using Ohm's Law. Also note that Ohm's Law may be used in any of its three forms to solve for any of the three variables when the other two are known. Consider the examples shown in Example Set 16-9.

The X_C Formula

Capacitive reactance is calculated using a formula that takes into account the capacitance C of the capacitor and the frequency f of the applied AC as follows:

$$X_C = \frac{1}{\omega \cdot C} = \frac{1}{2\pi f C} \tag{16.11}$$

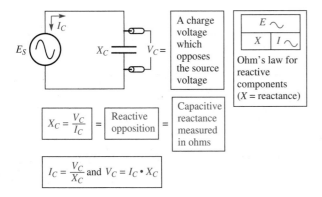

Figure 16-22   Capacitive reactance X_C.

EXAMPLE SET   16-9

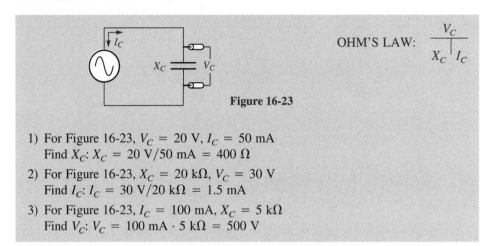

OHM'S LAW: $\dfrac{V_C}{X_C \mid I_C}$

Figure 16-23

1) For Figure 16-23, $V_C = 20$ V, $I_C = 50$ mA
   Find X_C: $X_C = 20$ V/50 mA $= 400\ \Omega$

2) For Figure 16-23, $X_C = 20$ kΩ, $V_C = 30$ V
   Find I_C: $I_C = 30$ V/20 kΩ $= 1.5$ mA

3) For Figure 16-23, $I_C = 100$ mA, $X_C = 5$ kΩ
   Find V_C: $V_C = 100$ mA $\cdot$ 5 kΩ $= 500$ V

where the capacitive reactance is measured in ohms and the angular velocity ($\omega = 2\pi f$) of the AC and the capacitance C are known.

As you can see, the reactance of a particular capacitor depends upon C and the angular velocity ($\omega = 2\pi f$) of the applied AC. Naturally, the higher the frequency (Hz or cps) the higher the angular velocity. If the frequency of the alternating current applied to the capacitor is high, the rate of change in capacitor charge ($\Delta q/\Delta t$) will also be high. Recall that the rate of change in charge determines the amount of capacitor current. Therefore, a higher AC frequency means a higher rate of change in charge, which results in a higher current. Thus, the amount of capacitive reactance is inversely related to the frequency of the applied AC (the higher the frequency, the lower the X_C). Naturally, the amount of capacitance is also inversely related to the amount of capacitive reactance. In summary, an increase in frequency and/or capacitance results in a decrease in X_C, and a decrease in frequency and/or capacitance results in an increase in X_C. Study the examples in Example Set 16-10 to gain a clear understanding of the use of the X_C formula.

EXAMPLE SET   16-10

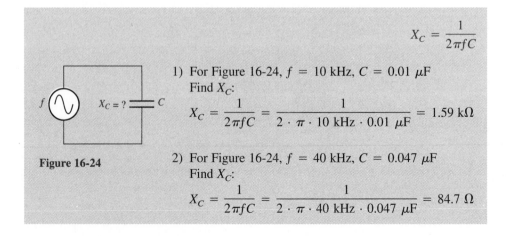

$$X_C = \frac{1}{2\pi f C}$$

1) For Figure 16-24, $f = 10$ kHz, $C = 0.01\ \mu$F
   Find X_C:
   $$X_C = \frac{1}{2\pi f C} = \frac{1}{2 \cdot \pi \cdot 10\ \text{kHz} \cdot 0.01\ \mu\text{F}} = 1.59\ \text{k}\Omega$$

Figure 16-24

2) For Figure 16-24, $f = 40$ kHz, $C = 0.047\ \mu$F
   Find X_C:
   $$X_C = \frac{1}{2\pi f C} = \frac{1}{2 \cdot \pi \cdot 40\ \text{kHz} \cdot 0.047\ \mu\text{F}} = 84.7\ \Omega$$

3) For Figure 16-24, $f = 23$ MHz, $C = 100$ pF
Find X_C:

$$X_C = \frac{1}{2\pi fC} = \frac{1}{2 \cdot \pi \cdot 23 \text{ MHz} \cdot 100 \text{ pF}} = 69.2 \text{ }\Omega$$

Solving for C

Let's suppose you have an unmarked capacitor and you want to determine its capacitance. As with unknown inductors, this can be done by first placing the capacitor in a simple AC circuit to see how the capacitor reacts in the circuit. The alternating current, applied to the capacitor, can be measured with an AC ammeter (RMS value), and the voltage across the capacitor can be measured with an AC voltmeter or an oscilloscope (RMS value). The capacitive reactance of the capacitor can then be calculated using Ohm's Law ($X_C = V_C/I_C$). Once the capacitive reactance of the unknown capacitor is calculated, and the frequency of the applied AC is known, C can be determined by rearranging the X_C formula as follows:

$$C = \frac{1}{2\pi fX_C} \qquad\qquad (16.12)$$

Study Example Set 16-11 to see how this formula is actually used.

EXAMPLE SET   16-11

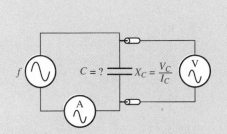

$$C = \frac{1}{2\pi fX_C}$$

Figure 16-25

1) For Figure 16-25, X_C is found to be 200 Ω.
The frequency is 100 Hz.
Find C:

$$C = \frac{1}{2\pi fX_C} = \frac{1}{2 \cdot \pi \cdot 100 \text{ Hz} \cdot 200 \text{ }\Omega} = 7.96 \text{ }\mu\text{F}$$

2) For Figure 16-25, X is found to be 80 Ω.
The frequency is 60 Hz.
Find C:

$$C = \frac{1}{2\pi fX_C} = \frac{1}{2 \cdot \pi \cdot 60 \text{ Hz} \cdot 80 \text{ }\Omega} = 33.2 \text{ }\mu\text{F}$$

Solving for f

Just as the X_C formula can be rearranged to solve for an unknown capacitance, it may also be rearranged to solve for frequency f, as long as C and X_C are known.

$$f = \frac{1}{2\pi CX_C} \tag{16.13}$$

Consider Example Set 16-12 before testing your understanding in Self-Check 16-5.

EXAMPLE SET   16-12

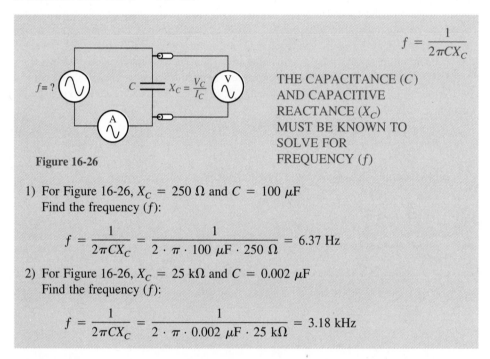

$$f = \frac{1}{2\pi CX_C}$$

THE CAPACITANCE (C)
AND CAPACITIVE
REACTANCE (X_C)
MUST BE KNOWN TO
SOLVE FOR
FREQUENCY (f)

Figure 16-26

1) For Figure 16-26, $X_C = 250\ \Omega$ and $C = 100\ \mu F$
   Find the frequency (f):

$$f = \frac{1}{2\pi CX_C} = \frac{1}{2 \cdot \pi \cdot 100\ \mu F \cdot 250\ \Omega} = 6.37\ \text{Hz}$$

2) For Figure 16-26, $X_C = 25\ k\Omega$ and $C = 0.002\ \mu F$
   Find the frequency (f):

$$f = \frac{1}{2\pi CX_C} = \frac{1}{2 \cdot \pi \cdot 0.002\ \mu F \cdot 25\ k\Omega} = 3.18\ \text{kHz}$$

SELF-CHECK   16-5

1. If the voltage across a capacitor is measured as 9 V and the capacitor's current is measured as 10 mA, what is the capacitor's reactance?

2. Determine the capacitive reactance of a 100 μF capacitor at a frequency of 20 kHz.

3. If a capacitor produces 4,500 Ω of reactance in response to an applied frequency of 200 kHz, what is the capacitance of the capacitor?

4. If the frequency applied to a capacitor is doubled, what will happen to its capacitive reactance?

16-6 X_C in Series and Parallel

X_C in Series

Total Series X_C

The total capacitive reactance produced by two or more capacitors placed in a series arrangement is simply the sum of all individual capacitive reactances. The total may be found by adding all individual capacitive reactances or by applying the total capacitance to the X_C formula.

$$X_{Ct} = X_{C1} + X_{C2} + X_{C3} + \cdots \qquad \text{(for series capacitors)} \qquad (16.14)$$

Examine Example Set 16-13 to see how total series capacitive reactance is determined.

EXAMPLE SET   16-13

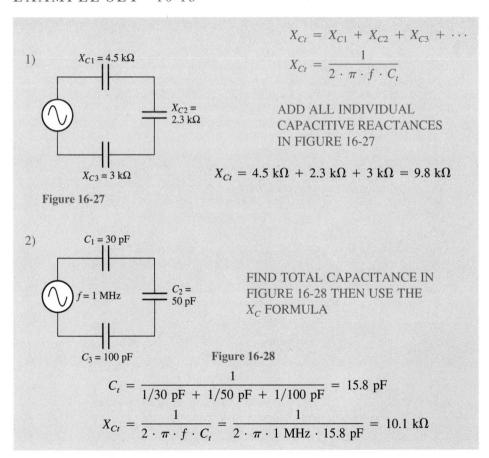

$$X_{Ct} = X_{C1} + X_{C2} + X_{C3} + \cdots$$

$$X_{Ct} = \frac{1}{2 \cdot \pi \cdot f \cdot C_t}$$

1) $X_{C1} = 4.5 \text{ k}\Omega$

$X_{C2} = 2.3 \text{ k}\Omega$

ADD ALL INDIVIDUAL CAPACITIVE REACTANCES IN FIGURE 16-27

$X_{C3} = 3 \text{ k}\Omega$

$$X_{Ct} = 4.5 \text{ k}\Omega + 2.3 \text{ k}\Omega + 3 \text{ k}\Omega = 9.8 \text{ k}\Omega$$

Figure 16-27

2) $C_1 = 30 \text{ pF}$

$f = 1 \text{ MHz}$

$C_2 = 50 \text{ pF}$

FIND TOTAL CAPACITANCE IN FIGURE 16-28 THEN USE THE X_C FORMULA

$C_3 = 100 \text{ pF}$ **Figure 16-28**

$$C_t = \frac{1}{1/30 \text{ pF} + 1/50 \text{ pF} + 1/100 \text{ pF}} = 15.8 \text{ pF}$$

$$X_{Ct} = \frac{1}{2 \cdot \pi \cdot f \cdot C_t} = \frac{1}{2 \cdot \pi \cdot 1 \text{ MHz} \cdot 15.8 \text{ pF}} = 10.1 \text{ k}\Omega$$

Distributed Voltage

Capacitors in series in an AC circuit will develop voltages that are in proportion to each individual capacitor's reactance. This is similar to voltage drops across resistors and inductor voltages in series circuits, inasmuch as the sum of all capacitor voltages will equal the source voltage. Remember, however, that the capacitor voltages are lagging the circuit current by 90° (ICE), unlike the resistor voltage drops, which are in phase with the circuit current. Each capacitor voltage may be calculated just as though the capacitor were a resistor. In other words, you may use Ohm's Law or the proportion formula as follows:

$$V_C = I_t \cdot X_C \quad \text{(Ohm's Law)} \tag{16.15}$$

$$V_C = E_S \cdot C_t/C = E_s \cdot X_C/X_{Ct} \quad \text{(proportion formula)} \tag{16.16}$$

where, for series capacitors, any individual capacitor's voltage will be in proportion to the ratio of C_t to the individual capacitance (C) or in proportion to the ratio of the capacitor's reactance to total capacitive reactance.

Study Example Set 16-14 to see how these formulas can be used. Also, notice the phasor diagrams showing the phasor sum of all individual capacitor voltages. All capacitor voltages are in phase with each other and add up to being equal to the source voltage (Kirchhoff's Voltage Law for series circuits).

EXAMPLE SET   16-14

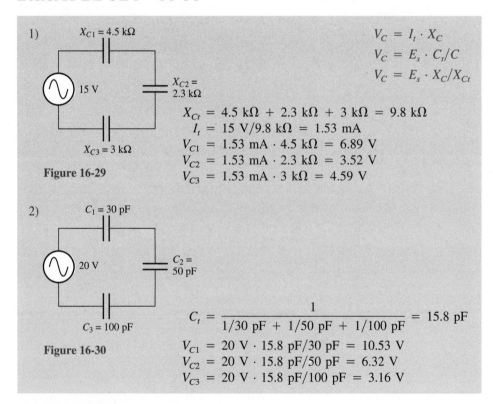

1) $X_{C1} = 4.5$ kΩ

$\bigcirc$ 15 V

$X_{C2} = 2.3$ kΩ

$X_{C3} = 3$ kΩ

Figure 16-29

$$V_C = I_t \cdot X_C$$
$$V_C = E_s \cdot C_t/C$$
$$V_C = E_s \cdot X_C/X_{Ct}$$

$$X_{Ct} = 4.5 \text{ k}\Omega + 2.3 \text{ k}\Omega + 3 \text{ k}\Omega = 9.8 \text{ k}\Omega$$
$$I_t = 15 \text{ V}/9.8 \text{ k}\Omega = 1.53 \text{ mA}$$
$$V_{C1} = 1.53 \text{ mA} \cdot 4.5 \text{ k}\Omega = 6.89 \text{ V}$$
$$V_{C2} = 1.53 \text{ mA} \cdot 2.3 \text{ k}\Omega = 3.52 \text{ V}$$
$$V_{C3} = 1.53 \text{ mA} \cdot 3 \text{ k}\Omega = 4.59 \text{ V}$$

2) $C_1 = 30$ pF

$\bigcirc$ 20 V

$C_2 = 50$ pF

$C_3 = 100$ pF

Figure 16-30

$$C_t = \frac{1}{1/30 \text{ pF} + 1/50 \text{ pF} + 1/100 \text{ pF}} = 15.8 \text{ pF}$$

$$V_{C1} = 20 \text{ V} \cdot 15.8 \text{ pF}/30 \text{ pF} = 10.53 \text{ V}$$
$$V_{C2} = 20 \text{ V} \cdot 15.8 \text{ pF}/50 \text{ pF} = 6.32 \text{ V}$$
$$V_{C3} = 20 \text{ V} \cdot 15.8 \text{ pF}/100 \text{ pF} = 3.16 \text{ V}$$

X_C in Parallel

Total Parallel X_C

Capacitive reactances in parallel are similar to resistances in parallel. Total capacitive reactance is found using one of the parallel formulas for resistors: the same values formula, the product over the sum formula, or the reciprocal formula.

$$X_{Ct} = X_{C1}/N \qquad \text{(same values)} \tag{16.17}$$

where all parallel capacitances and capacitive reactances are the same value and the total capacitive reactance is found by dividing the reactance of one capacitor (X_{C1}) by the number N of parallel capacitors.

$$X_{Ct} = \frac{X_{C1} \cdot X_{C2}}{X_{C1} + X_{C2}} \qquad \text{(product over the sum)} \tag{16.18}$$

where two parallel capacitors have different values of capacitance and capacitive reactance.

$$X_{Ct} = \frac{1}{1/X_{C1} + 1/X_{C2} + 1/X_{C3} + \cdots} \qquad \text{(reciprocal)} \tag{16.19}$$

where more than two capacitors of different value are placed in parallel.

Example Set 16-15 illustrates the use of these formulas in calculating total capacitive reactance for parallel capacitor circuits.

EXAMPLE SET   16-15

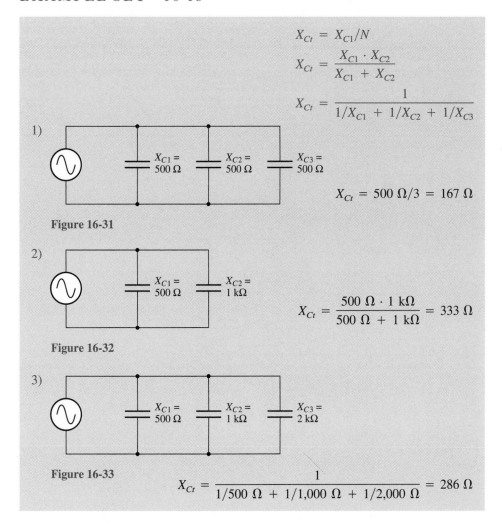

$$X_{Ct} = X_{C1}/N$$

$$X_{Ct} = \frac{X_{C1} \cdot X_{C2}}{X_{C1} + X_{C2}}$$

$$X_{Ct} = \frac{1}{1/X_{C1} + 1/X_{C2} + 1/X_{C3}}$$

1)

$X_{C1} = 500\ \Omega$ $X_{C2} = 500\ \Omega$ $X_{C3} = 500\ \Omega$

$$X_{Ct} = 500\ \Omega/3 = 167\ \Omega$$

Figure 16-31

2)

$X_{C1} = 500\ \Omega$ $X_{C2} = 1\ k\Omega$

$$X_{Ct} = \frac{500\ \Omega \cdot 1\ k\Omega}{500\ \Omega + 1\ k\Omega} = 333\ \Omega$$

Figure 16-32

3)

$X_{C1} = 500\ \Omega$ $X_{C2} = 1\ k\Omega$ $X_{C3} = 2\ k\Omega$

Figure 16-33

$$X_{Ct} = \frac{1}{1/500\ \Omega + 1/1{,}000\ \Omega + 1/2{,}000\ \Omega} = 286\ \Omega$$

Distributed Current

Once again, as in parallel resistor circuits, the total current supplied to a parallel section is distributed among the branches of the parallel section. The current is distributed to each parallel capacitor in accordance with Ohm's Law and in accordance with the current divider formula as follows:

$$I_C = I_t \cdot X_{Ct}/X_C = I_t \cdot C/C_t \qquad \text{(current divider formula)} \qquad (16.20)$$

where the current in any branch is in proportion to the ratio of the total capacitive reactance to the individual capacitor's reactance or in proportion to the ratio of the capacitor's capacitance to total capacitance.

Example Set 16-16 illustrates the use of Ohm's Law and the current divider formula in determining individual branch currents of parallel capacitive circuits.

EXAMPLE SET   16-16

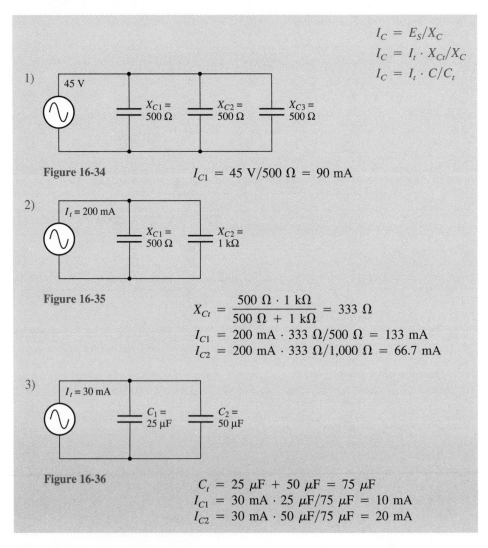

$$I_C = E_S/X_C$$
$$I_C = I_t \cdot X_{Ct}/X_C$$
$$I_C = I_t \cdot C/C_t$$

1)

45 V

$X_{C1} = 500\ \Omega$ $X_{C2} = 500\ \Omega$ $X_{C3} = 500\ \Omega$

Figure 16-34

$$I_{C1} = 45\ \text{V}/500\ \Omega = 90\ \text{mA}$$

2)

$I_t = 200$ mA

$X_{C1} = 500\ \Omega$ $X_{C2} = 1\ \text{k}\Omega$

Figure 16-35

$$X_{Ct} = \frac{500\ \Omega \cdot 1\ \text{k}\Omega}{500\ \Omega + 1\ \text{k}\Omega} = 333\ \Omega$$
$$I_{C1} = 200\ \text{mA} \cdot 333\ \Omega/500\ \Omega = 133\ \text{mA}$$
$$I_{C2} = 200\ \text{mA} \cdot 333\ \Omega/1{,}000\ \Omega = 66.7\ \text{mA}$$

3)

$I_t = 30$ mA

$C_1 = 25\ \mu\text{F}$ $C_2 = 50\ \mu\text{F}$

Figure 16-36

$$C_t = 25\ \mu\text{F} + 50\ \mu\text{F} = 75\ \mu\text{F}$$
$$I_{C1} = 30\ \text{mA} \cdot 25\ \mu\text{F}/75\ \mu\text{F} = 10\ \text{mA}$$
$$I_{C2} = 30\ \text{mA} \cdot 50\ \mu\text{F}/75\ \mu\text{F} = 20\ \text{mA}$$

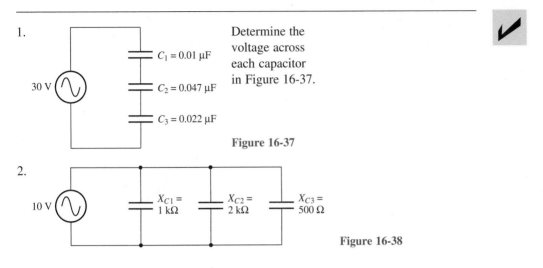

1. Determine the voltage across each capacitor in Figure 16-37.

$C_1 = 0.01\ \mu F$

$C_2 = 0.047\ \mu F$

$C_3 = 0.022\ \mu F$

30 V

Figure 16-37

2.

10 V

$X_{C1} = 1\ k\Omega$

$X_{C2} = 2\ k\Omega$

$X_{C3} = 500\ \Omega$

Figure 16-38

Determine the current through each capacitor of Figure 16-38. Also, determine total current and total capacitive reactance.

16-7 Applications of Reactance

Inductive Coupling

One of the primary applications for inductive reactance is found in circuits that employ **inductive coupling**. The purpose of inductive coupling is to pass low frequencies more easily than higher frequencies from one circuit or source to another circuit, or to a load. The inductor automatically increases its opposition (X_L) to higher frequencies in accordance to the X_L formula. This is particularly useful when it is desired to eliminate, or reduce, a high-frequency noise from a low-frequency audio signal. The inductor actually produces a high X_L in reaction to the high-frequency noise, while at the same time producing a low X_L for the low-frequency audio signal. This is illustrated in Figure 16-39. Such a circuit is called a *lowpass filter*.

Capacitive Coupling

Capacitive coupling is very much the opposite of inductive coupling. The capacitor blocks DC (0 Hz) and passes AC. Also, a capacitor passes higher frequencies more easily than lower frequencies. In other words, X_C decreases as frequency increases. Figure 16-40 illustrates the use of a capacitor to pass high frequencies while blocking DC and greatly attenuating very low frequencies (*attenuate* means "to reduce in amplitude or amount"). The size, or value, of the capacitor must be chosen to have a relatively low X_C at the lowest frequency the circuit is to handle (pass and amplify). The size of coupling capacitor shown in Figure 16-40 has a low X_C for supersonic and radio-frequency coupling applications. The X_C of the coupling capacitor forms a series voltage divider with the input resistance of the next stage. Notice, from the graph of Figure 16-40, at 10 kHz the

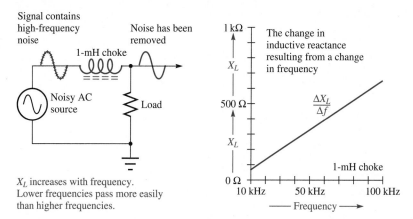

Signal contains
high-frequency
noise

Noise has been
removed

1-mH choke

Noisy AC
source

Load

X_L increases with frequency.
Lower frequencies pass more easily
than higher frequencies.

The change in
inductive reactance
resulting from a change
in frequency

$\frac{\Delta X_L}{\Delta f}$

X_L

X_L

1-mH choke

Frequency

Figure 16-39   Inductive coupling can separate high-frequency noise from a low-frequency signal or greatly attenuate AC while passing DC.

capacitor's X_C is about 1,600 Ω, while at 100 kHz its X_C is only about 160 Ω. As you can see, the 100 kHz frequency passes more easily than 10 kHz.

More will be discussed about capacitor and inductor applications in later chapters. For the moment, it is sufficient for you to understand the relationship between frequency and reactance for inductors and capacitors. Take time to review this section, then answer the questions of Self-Check 16-7.

SELF-CHECK 16-7

1. Will inductive coupling block DC (see Fig. 16-39)? Why?

2. Will inductive coupling pass high frequencies more easily than low frequencies? Why?

3. Which would a coupling capacitor pass more easily, 50 Hz or 5,000 Hz? Why?

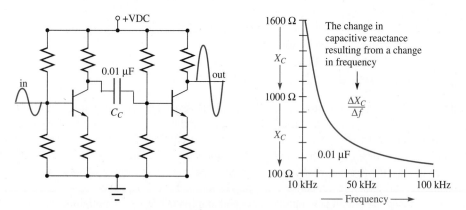

+VDC

0.01 μF

in

out

C_C

The change in
capacitive reactance
resulting from a change
in frequency

X_C

$\frac{\Delta X_C}{\Delta f}$

X_C

0.01 μF

Frequency

Coupling capacitor blocks DC while passing AC. X_C decreases as frequency increases.
Higher frequencies pass more easily than lower frequencies.

Figure 16-40   Capacitive coupling blocks DC and offers less opposition as the frequency of AC is increased.

Summary

FORMULAS

(16.1) $X_L = \omega \cdot L = 2\pi f L$

(16.2) $L = X_L / 2\pi f$

(16.3) $f = X_L / 2\pi L$

(16.4) $X_{Lt} = X_{L1} + X_{L2} + X_{L3} + \cdots$ (total series X_L)

(16.5) $V_L = I_t \cdot X_L$

(16.6) $V_L = E_S \cdot L / L_t = E_s \cdot X_L / X_{Lt}$

(16.7) $X_{Lt} = X_{L1} / N$ (total parallel X_L—same values)

(16.8) $X_{Lt} = \dfrac{X_{L1} \cdot X_{L2}}{X_{L1} + X_{L2}}$ (total parallel X_L—two different values)

(16.9) $X_{Lt} = \dfrac{1}{1/X_{L1} + 1/X_{L2} + 1/X_{L3} + \cdots}$ (total parallel X_L)

(16.10) $I_L = I_t \cdot X_{Lt} / X_L = I_t \cdot L_t / L$

(16.11) $X_C = \dfrac{1}{\omega \cdot C} = \dfrac{1}{2\pi f C}$

(16.12) $C = \dfrac{1}{2\pi f X_C}$

(16.13) $f = \dfrac{1}{2\pi C X_C}$

(16.14) $X_{Ct} = X_{C1} + X_{C2} + X_{C3} + \cdots$ (total series X_C)

(16.15) $V_C = I_t \cdot X_C$

(16.16) $V_C = E_S \cdot C_t / C = E_S \cdot X_C / X_{Ct}$

(16.17) $X_{Ct} = X_{C1} / N$ (total parallel X_C—same values)

(16.18) $X_{Ct} = \dfrac{X_{C1} \cdot X_{C2}}{X_{C1} + X_{C2}}$ (total parallel X_C—two different values)

(16.19) $X_{Ct} = \dfrac{1}{1/X_{C1} + 1/X_{C2} + 1/X_{C3} + \cdots}$ (total parallel X_C)

(16.20) $I_C = I_t \cdot X_{Ct} / X_C = I_t \cdot C / C_t$

CONCEPTS

- The self-induced voltage of an inductor leads the inductor current by 90 electrical degrees (ELI) in an AC circuit.
- Capacitor charge current leads capacitor voltage by 90 electrical degrees (ICE) in an AC circuit.
- An inductor will react to an applied AC and produce an opposition to the AC that is measured in ohms. This opposition is known as inductive reactance (X_L).
- A capacitor's opposition to AC is known as capacitive reactance (X_C), and is measured in ohms.
- Inductive reactance increases with frequency, while capacitive reactance decreases as frequency increases.

- Voltage and current distribution in series and parallel inductor circuits are determined using the same formulas and procedures as are used for resistor circuits: Ohm's Law, Kirchhoff's laws, and voltage and current divider formulas are applied in the same way. The same is true for AC capacitor circuits.
- Inductive signal coupling, in AC electrical or electronic circuits, is used to pass any DC and low-frequency AC while attenuating high-frequency AC.
- Capacitive coupling is used to block DC and low-frequency AC while passing high-frequency AC.

SPECIAL TERMS

- ELI, ICE
- Inductive reactance (X_L)
- Capacitive reactance (X_C)
- Inductive coupling
- Capacitive coupling

Need to Know Solution

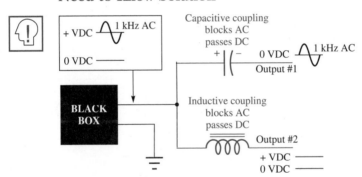

Questions

16-1 AC and the Inductor

1. What determines the amount of voltage induced across an inductor besides the number of turns?
2. What determines the rate of change in flux about the windings of an inductor in an AC circuit?
3. What is the meaning of ELI?
4. Explain why the inductor voltage is maximum when the inductor current is minimum (passing through zero crossing).
5. Draw the phasor diagram that represents the phase relationship between inductor voltage and current. The current phasor should be drawn as the reference at 0°.

16-2 AC and the Capacitor

6. What determines the capacitor current at any instant in time in an AC circuit?
7. What is the meaning of ICE?

8. Why is the capacitor current minimum when the capacitor voltage reaches points of maximum in an AC circuit?
9. Is it correct to say that capacitor voltage lags capacitor current by 90° in an AC circuit?
10. Draw the phasor diagram that represents the phase relationship between capacitor voltage and current. The current phasor should be drawn as the reference at 0°.

16-3 Inductive Reactance (X_L)

11. The opposition an inductor develops to an AC signal is called what?
12. The inductor's opposition to AC is measured in what units?
13. Explain what the AC frequency has to do with the reactance of an inductor.

16-5 Capacitive Reactance (X_C)

14. The opposition a capacitor develops to an AC signal is called what?
15. The capacitor's opposition to AC is measured in what units?
16. Explain what the AC frequency has to do with the reactance of a capacitor.

16-7 Applications of Reactance

17. Which would an inductor pass more easily, a high-frequency signal or a low-frequency signal? Why?
18. Which would provide more opposition to an AC signal, a 1 mH coil or a 10 mH coil? Why?
19. Does inductive reactance vary directly or inversely with frequency? Why?
20. Which would a capacitor pass more easily, a high-frequency signal or a low-frequency signal? Why?
21. Which would provide more opposition to an AC signal, a 1 μF capacitor or a 1 pF capacitor? Why?
22. Does capacitive reactance vary directly or inversely with frequency? Why?
23. Which component would you choose to block DC and pass AC? (R, C, or L)
24. Which component would you choose to pass DC and greatly oppose high frequencies? (R, C, or L)
25. Which component opposes DC and all frequencies equally? (R, C, or L)

Problems

16-3 Inductive Reactance (X_L)

1. If an inductor has 100 mA of AC flowing through its windings and 40 V across its windings, what is its reactance?
2. What is the inductive reactance of an inductor whose current is 450 μA and whose voltage is 300 mV?
3. Determine the reactance of a 100 μH inductor at (a) 100 Hz; (b) 100 kHz; (c) 100 MHz.
4. Determine the reactance of a 1 mH inductor at a frequency of 10 kHz.
5. Determine the inductance (L) of an inductor that exhibits 2,300 Ω of reactance at a frequency of 50 kHz.
6. Determine the inductance (L) of an inductor that exhibits 45 Ω of reactance at a frequency of 2 MHz.

7. Determine the frequency (f) at which a 20 μH inductor will exhibit 30 Ω of reactance.
8. Given: $X_L = 40$ kΩ; $L = 25$ mH; find $f = $?

16-4 X_L in Series and Parallel

9. Given the following *series* inductors and AC source voltage, find the total inductive reactance, total current, and all inductor voltages:

$$E_S = 10 \text{ V}; X_{L1} = 400 \ \Omega; X_{L2} = 320 \ \Omega; X_{L3} = 100 \ \Omega$$
$$X_{Lt} = ?, I_t = ?, V_{L1} = ?, V_{L2} = ?, V_{L3} = ?$$

10. Given the following *series* inductors and AC source voltage, find the total inductive reactance, total current, and all inductor voltages:

$$E_S = 150 \text{ mV}; X_{L1} = 30 \text{ k}\Omega; X_{L2} = 40 \text{ k}\Omega; X_{L3} = 10 \text{ k}\Omega$$
$$X_{Lt} = ?, I_t = ?, V_{L1} = ?, V_{L2} = ?, V_{L3} = ?$$

11. Given the following *series* inductors and AC source voltage, find the total inductance, and all inductor voltages:

$$E_S = 20 \text{ V}; L_1 = 200 \ \mu\text{H}; L_2 = 400 \ \mu\text{H}; L_3 = 300 \ \mu\text{H}$$
$$V_{L1} = ?, V_{L2} = ?, V_{L3} = ?, L_t = ?$$

12. Given the following *parallel* inductors and AC source voltage, find the total inductive reactance, total current, and all inductor currents:

$$E_S = 15 \text{ V}; X_{L1} = 2.2 \text{ k}\Omega; X_{L2} = 4 \text{ k}\Omega; X_{L3} = 1 \text{ k}\Omega$$
$$X_{Lt} = ?, I_t = ?, I_{L1} = ?, I_{L2} = ?, I_{L3} = ?$$

13. Given the following *parallel* inductors and AC source voltage, find the total inductive reactance, total current, and all inductor currents:

$$E_S = 30 \text{ V}; X_{L1} = 2.2 \text{ k}\Omega; X_{L2} = 2.2 \text{ k}\Omega; X_{L3} = 1.1 \text{ k}\Omega$$
$$X_{Lt} = ?, I_t = ?, I_{L1} = ?, I_{L2} = ?, I_{L3} = ?$$

14. Given the following *parallel* inductors and AC source voltage, find the total inductive reactance, total inductance, and all inductor currents:

$$E_S = 5 \text{ V}; L_1 = 1 \text{ mH}; L_2 = 0.5 \text{ mH}; L_3 = 0.75 \text{ mH}; L_4 = 1.5 \text{ mH};$$
$$I_t = 30 \text{ mA}; X_{L1} = ?, L_t = ?, I_{L1} = ?, I_{L2} = ?, I_{L3} = ?, I_{L4} = ?$$

16-5 Capacitive Reactance (X_C)

15. If a capacitor has 80 mA of AC flowing to and from its plates and 40 V across its plates, what is its reactance?
16. What is the capacitive reactance of a capacitor whose current is 50 μA and voltage is 100 mV?
17. Determine the reactance of a 100 μF capacitor at (a) 1 Hz; (b) 100 Hz; (c) 10 kHz.
18. Determine the reactance of a 10 pF capacitor at a frequency of 10 kHz.
19. Determine the capacitance of a capacitor that exhibits 500 Ω of reactance at a frequency of 100 kHz.
20. Determine the capacitance of a capacitor that exhibits 25 Ω of reactance at a frequency of 5 MHz.

21. Determine the frequency at which a 20 pF capacitor will exhibit 10 Ω of reactance.
22. Given: $X_C = 40$ kΩ; $C = 25$ pF; find f.

16-6 X_C in Series and Parallel

23. Given the following *series* capacitors and AC source voltage, find the total capacitive reactance, total current, and all capacitor voltages:

$E_S = 150$ V, $X_{C1} = 70$ kΩ, $X_{C2} = 30$ kΩ, $X_{C3} = 10$ kΩ

$X_{Ct} = ?$, $I_t = ?$, $V_{C1} = ?$, $V_{C2} = ?$, $V_{C3} = ?$

24. Given the following *series* capacitors and AC source voltage, find the total capacitance, and all capacitor voltages:

$E_S = 40$ V; $C_1 = 200$ pF; $C_2 = 400$ pF; $C_3 = 300$ pF

$V_{C1} = ?$, $V_{C2} = ?$, $V_{C3} = ?$, $C_t = ?$

25. Given the following *parallel* capacitors and AC source voltage, find the total capacitive reactance, total current, and all capacitor currents:

$E_S = 25$ V, $X_{C1} = 2$ kΩ, $X_{C2} = 4$ kΩ, $X_{C3} = 1$ kΩ

$X_{Ct} = ?$, $I_t = ?$, $I_{C1} = ?$, $I_{C2} = ?$, $I_{C3} = ?$

26. Given the following *parallel* capacitors and AC source voltage, find the total capacitive reactance, total current, and all capacitor currents:

$E_S = 30$ V; $X_{C1} = 2.2$ kΩ; $X_{C2} = 2.2$ kΩ; $X_{C3} = 1.1$ kΩ

$X_{Ct} = ?$, $I_t = ?$, $I_{C1} = ?$, $I_{C2} = ?$, $I_{C3} = ?$

Answers to Self-Checks

Self-Check 16-1

1. The rate of change in flux and current
2. The inductor voltage will reach points of maximum as the inductor current passes through points of zero crossing, since it is at these zero crossing points that the rate of change in inductor current is greatest.
3. The inductor voltage leads the inductor current by 90 electrical degrees in an AC circuit.

Self-Check 16-2

1. The rate of change in capacitor voltage is directly related to the rate of change in capacitor current. If the rate of change in capacitor voltage is increased, the rate of change in current will also be increased.
2. The capacitor voltage is at a maximum peak. Since the rate of change in voltage is zero at any peak, the current will be zero.
3. Capacitor current leads capacitor voltage by 90 electrical degrees in an AC circuit.

Self-Check 16-3

1. 80 Ω
2. 126 Ω

3. 0.995 mH
4. It will also double.

Self-Check 16-4

1. $V_{L1} = 5.45$ V; $V_{L2} = 8.18$ V; $V_{L3} = 16.4$ V
2. $I_{L1} = 66.7$ mA; $I_{L2} = 33.3$ mA; $I_{L3} = 22.2$ mA; $I_t = 122$ mA; $X_{Lt} = 81.8$ Ω

Self-Check 16-5

1. 900 Ω
2. 0.080 Ω
3. 177 pF
4. Its reactance will be cut in half

Self-Check 16-6

1. $V_{C1} = 18$ V; $V_{C2} = 3.83$ V; $V_{C3} = 8.18$ V
2. $I_{C1} = 10$ mA; $I_{C2} = 5$ mA; $I_{C3} = 20$ mA, $I_t = 35$ mA, $X_{Ct} = 286$ Ω

Self-Check 16-7

1. No; the inductor has no reactance at 0 Hz (DC)
2. No; the higher the frequency, the greater the inductive reactance
3. 5,000 Hz; the higher the frequency, the lower the capacitive reactance

SUGGESTED PROJECTS

1. Add some of the main concepts, formulas, and procedures from this chapter to your Electricity and Electronics Notebook.

2. Obtain a 100 mH choke (laminated iron-core type with a current rating greater than 250 mA) and a 6 VAC source (50 or 60 Hz) such as a small step-down transformer (115/6 VAC or 230/6 VAC at greater than 250 mA). Apply the 6 VAC to the choke and measure the AC current and actual voltage across the choke with an AC volt/ammeter. Use Ohm's Law to calculate the X_L of the choke from the measured values of current and voltage. Then verify the inductance of the choke using Formula 16.2. Your findings may not be exact due to component tolerance and meter accuracy.

3. Perform the same experiment using a 1 μF, 25 V nonpolarized electrolytic capacitor in place of the choke.

Here Edison sits in his library in later years musing over his earlier works. His first cylindrical phonograph is to his left on the desk and an early business phonograph is on his right.

Chapter 17

Transformers

CHAPTER OUTLINE

OBJECTIVES

After studying this chapter, you will be able to

- explain how mutual inductance can be used to increase or decrease the total inductance between two coils.
- explain the importance of mutual inductance in transformer action.
- calculate primary or secondary transformer voltage and current as related to the transformer's turns ratio.
- explain the theory of reflected impedance between the primary and secondary, or secondaries, of transformers.
- calculate the impedance ratio for a transformer with a given turns ratio.
- calculate reflected impedance given a transformer's turns ratio and secondary load impedance.
- explain various transformer ratings, such as voltage, current, power, impedance, frequency, and efficiency.
- explain various transformer losses, such as winding losses, and core losses.
- discuss a variety of transformer types and applications.
- troubleshoot a transformer using an ohmmeter and/or a voltmeter.

INTRODUCTION

You have probably seen many different transformers used in many common applications—such as power-line transformers, model train transformers, battery eliminators for portable radios and tape players, and household doorbell transformers. More than likely you have also wondered how they are made and how they work. Why are they called *transformers*? What do they transform? Why do some transformers get very hot, while others are barely warm? What makes a transformer ''burn up''? How are transformers rated? These are only a few of the many questions that will be answered in this chapter. You will soon discover how important transformers are. The purpose for this chapter is to give you a good practical knowledge of transformers. There is still much more that can be learned about transformer design, types, and applications that go beyond the scope and purpose of this text.

NEED TO KNOW

This "Need to Know" is very simple and straightforward. It centers around a very simple question:

How do you know a particular transformer will satisfy the requirements for a particular application?

A good technician will know the answer to this question and have a good working knowledge of transformers.

Consider this simple application: A transformer is needed to connect a 12.6 V light bulb, that draws 2 A of current, to a 115 VAC source. Describe the needed transformer. Read on and satisfy your need to know in this important area. Keep the above question and application in mind as you read. The "Need to Know Solution" at the end of the chapter will confirm your findings.

17-1 Mutual Inductance (L_M)

Mutual Inductance Explained

Mutual Inductance (L_M)

In previous chapters, total inductance and total inductive reactance were calculated under the assumption that neighboring inductors did not interact with one another magnetically. In practice, many inductors do interact with one another when mounted close to one another in a circuit. The expanding and collapsing magnetic field generated by one inductor may cut into or link the windings of a neighboring inductor, thus inducing a current in, and voltage across, the neighboring inductor. The linking flux is mutual to (in common with) both inductors and induces a voltage across the windings of the host inductor as well as the neighboring inductor. The property of one inductor inducing a voltage across a neighboring inductor is known as **mutual inductance** (L_M).

Like self-inductance, mutual inductance is measured in henries (H). Recall that a current changing at the rate of 1 A/s will induce 1 V across a 1 H coil. A mutual inductance of 1 H is defined in very much the same way:

> Two inductors have a mutual inductance of 1 H when a rate of change in current of 1 A/s in one inductor induces a voltage of 1 V across the second inductor.

Coefficient of Coupling (k)

The actual amount of mutual inductance between two coils is largely determined by the percentage of flux from a host inductor that links the windings of a neighboring inductor. If 100% of the flux from the host inductor links the windings of the neighboring inductor, every individual turn of the neighboring inductor will receive the same amount of induced voltage as each turn of the host inductor. In other words, the changing current in the host inductor induces the same amount of voltage across each turn of the neighboring inductor as it does across each of its own turns. However, if the flux linkage is less than 100%, the induced voltage in the neighboring coil is also less than 100%.

Mathematically, it is more convenient to express the percentage of flux linkage in decimal form so it may be used in a formula as a coefficient. Thus, 100% is expressed as 1, 80% as 0.80, 75% as 0.75, and so on. Because this coefficient refers to the coupling of flux between inductors, it is called the **coefficient of coupling (k)**. It is found from the following ratio:

$$k = \frac{\text{amount of flux between } L_1 \text{ and } L_2}{\text{total flux produced by } L_1 \text{ or } L_2} = \frac{\Phi_M}{\Phi_t} \qquad (17.1)$$

where the amount of flux is expressed in Webers (Wb) or Maxwells (Mx) and the coefficient of coupling k has no units.

Thus, if the total produced flux (Φ_t) and the amount of mutual flux (Φ_M) are known, the coefficient of coupling can easily be determined. Consider the following simple examples:

1. $\Phi_t = 25 \ \mu\text{Wb}$ and $\Phi_M = 15 \ \mu\text{Wb}$, so $k = 15 \ \mu\text{Wb}/25 \ \mu\text{Wb} = 0.6$
2. $\Phi_t = 140 \ \mu\text{Wb}$ and $\Phi_M = 85 \ \mu\text{Wb}$, so $k = 85 \ \mu\text{Wb}/140 \ \mu\text{Wb} = 0.607$
3. $\Phi_t = 10,000 \ \text{Mx}$ and $\Phi_M = 800 \ \text{Mx}$, so $k = 800 \ \text{Mx}/10,000 \ \text{Mx} = 0.08$

Mutual Inductance—Desired and Undesired

Figure 17-1 illustrates various ways of arranging coils to give different values of k and, therefore, varying amounts of mutual inductance. In many applications, mutual inductance is desired and coils are deliberately placed to take advantage of the mutual coupling. In such cases, an AC signal in one coil is transferred, or magnetically coupled, to a second coil by mutual induction (a voltage is induced in the second coil as a voltage is self-induced in the first coil). Special components called **transformers** are designed to take advantage of the mutual inductance between two or more coils wound on the same core. The remaining sections of this chapter are dedicated to the study of transformers and their applications.

In cases where mutual coupling between coils is not desired, neighboring inductors are shielded with a high permeability material, mounted at right angles to each other, separated from one another as far as is practical, or wound on toroidal cores for near total flux containment. Many modern circuits today use toroidal inductors almost exclusively. Shielded tin can inductors and intermediate-frequency transformers (IF xfmrs.) are commonly found in radio receivers of all types.

Calculating Mutual Inductance

The mutual inductance of two coils can be calculated as long as the self-inductance of each coil is known and the coefficient of coupling k between the coils is known. Iron-core transformers, used in audio- and power-frequency applications, have a coefficient of coupling equal to, or very close to, unity ($k = 1$). Ferrite-slug radio-frequency transformers generally have coefficients of coupling in the range of 0.4 to about 0.7 and air-core radio-frequency transformers range from about 0.1 to approximately 0.4. Naturally, the actual k

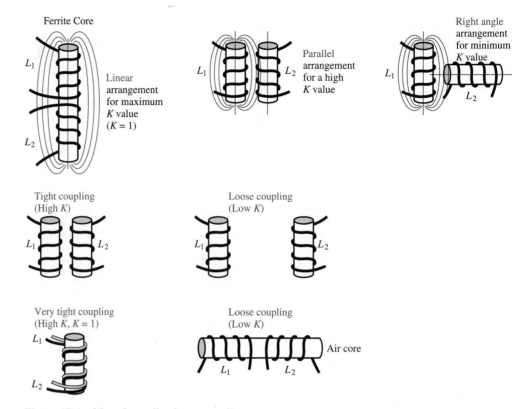

Figure 17-1 Mutual coupling between coils.

value depends on the particular transformer design and is usually specified by the manufacturer.

Assuming that the self-inductance of each coil (L_1 and L_2) and the k value are known, the mutual inductance L_M may be calculated by first finding the geometric mean of the two self-inductances and multiplying the result by the k value, as shown in the following formula:

$$L_M = k \cdot \sqrt{L_1 \cdot L_2} \tag{17.2}$$

Example Set 17-1 offers several examples of the use of Formula 17.2. As you can see, the calculations are very straightforward.

EXAMPLE SET 17-1

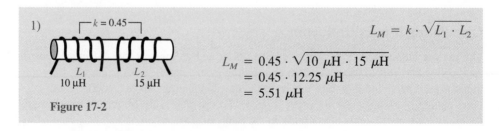

1)

$$L_M = k \cdot \sqrt{L_1 \cdot L_2}$$

$$L_M = 0.45 \cdot \sqrt{10\ \mu H \cdot 15\ \mu H}$$
$$= 0.45 \cdot 12.25\ \mu H$$
$$= 5.51\ \mu H$$

Figure 17-2

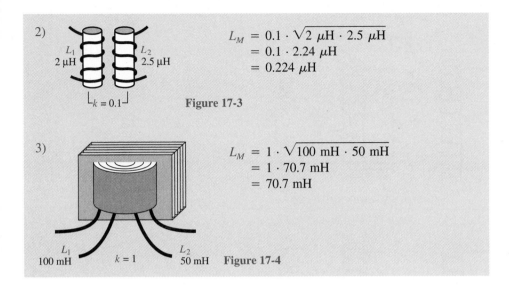

$$L_M = 0.1 \cdot \sqrt{2 \ \mu H \cdot 2.5 \ \mu H}$$
$$= 0.1 \cdot 2.24 \ \mu H$$
$$= 0.224 \ \mu H$$

Figure 17-3

$$L_M = 1 \cdot \sqrt{100 \ mH \cdot 50 \ mH}$$
$$= 1 \cdot 70.7 \ mH$$
$$= 70.7 \ mH$$

Figure 17-4

The Effects of Mutual Inductive Coupling

Considering the Effects

Inductors do not need to be wound on the same core in order for flux to be shared mutually. Coils that are mounted side by side or end to end, as shown in Figure 17-1, also have mutual inductance in proportion to their coefficient of coupling. The mutual inductance of the coils definitely affects their total inductance. In cases where the coefficient of coupling is greater than 0.05, the mutual inductance of the coils should not be ignored. The mutual inductance has an effect not only on total inductance, but also on total inductive reactance, since the total inductive reactance is found using L_t in the X_L formula.

Inductors With Aiding Fields (Boosting)

As shown in Figure 17-5a, inductors may be connected in series so as to cause an overall increase in the total inductance due to the contribution of the mutual inductance. The mutual inductance can increase, or **boost**, the total inductance significantly, depending on the k value of the two coils. However, this boost in total inductance will only occur if the magnetic fields of the two coils are aiding each other (north end of one coil facing the south end of the other when placed end to end, or the north and south ends aligned when placed in parallel).

In order to determine whether the inductors are aiding or opposing, the manner, or sense, in which the coils are wound and arranged must be known. Often, in schematics, **sense dots** are used to indicate the magnetic polarity or phase of the coils, since actual pictorials of each inductor are impractical. Figure 17-5a uses the sense dots. Pictorials and schematic symbols are used to help you understand what it means for coils to be wound and arranged in the same or opposite sense with fields aiding or opposing.

When inductors are wound and arranged in the same sense, the total inductance is boosted according to the following formula:

$$L_t = L_1 + L_2 + 2L_M \qquad \text{(fields aiding)} \tag{17.3}$$

(a) BOOST ARRANGEMENT

Magnetic fields aiding
$L_t = L_1 + L_2 + 2L_M$

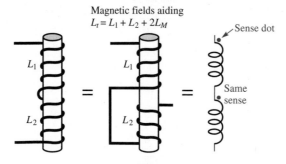

(b) BUCK ARRANGEMENT

Magnetic fields opposing
$L_t = L_1 + L_2 - 2L_M$

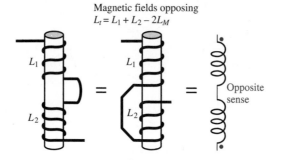

Figure 17-5   The effect of mutual inductance on total inductance.

Naturally, the individual coil inductances and the mutual inductance must be known or determined before the total inductance is determined. Again, the coefficient of coupling is the significant factor in determining the mutual inductance and, therefore, the total inductance. Example Set 17-2 illustrates the boosting effects of mutual inductance.

EXAMPLE SET 17-2

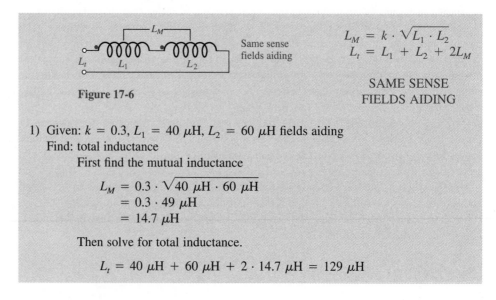

Figure 17-6

Same sense
fields aiding

$$L_M = k \cdot \sqrt{L_1 \cdot L_2}$$
$$L_t = L_1 + L_2 + 2L_M$$

**SAME SENSE
FIELDS AIDING**

1) Given: $k = 0.3$, $L_1 = 40 \ \mu H$, $L_2 = 60 \ \mu H$ fields aiding
    Find: total inductance
      First find the mutual inductance

$$L_M = 0.3 \cdot \sqrt{40 \ \mu H \cdot 60 \ \mu H}$$
$$= 0.3 \cdot 49 \ \mu H$$
$$= 14.7 \ \mu H$$

Then solve for total inductance.

$$L_t = 40 \ \mu H + 60 \ \mu H + 2 \cdot 14.7 \ \mu H = 129 \ \mu H$$

2) Given: $k = 0.8$, $L_1 = 10$ mH, $L_2 = 40$ mH, fields aiding
   Find: total inductance
      First find the mutual inductance

$$L_M = 0.8 \cdot \sqrt{10 \text{ mH} \cdot 40 \text{ mH}}$$
$$= 0.8 \cdot 20 \text{ mH}$$
$$= 16 \text{ mH}$$

   Then solve for total inductance.

$$L_t = 10 \text{ mH} + 40 \text{ mH} + 2 \cdot 16 \text{ mH} = 82 \text{ mH}$$

Inductors With Opposing Fields (Bucking)

When inductors are placed and connected in such a way that the magnetic fields of the two coils oppose each other (opposite sense), the total inductance is actually reduced. Once again, if the coefficient of coupling is significant, the resulting mutual inductance will significantly reduce the total inductance. In such a case, one coil is **bucking** the other. This condition is also illustrated in Figure 17-5b. To determine the total inductance for coils whose fields are bucking, the following simple equation is used:

$$L_t = L_1 + L_2 - 2L_M \quad \text{(fields opposing)} \tag{17.4}$$

   Example Set 17-3 illustrates the use of this formula in a couple of examples. Take a moment to study these examples before continuing.

EXAMPLE SET   17-3

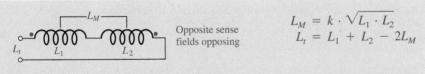

$$L_M = k \cdot \sqrt{L_1 \cdot L_2}$$
$$L_t = L_1 + L_2 - 2L_M$$

Opposite sense
fields opposing

Figure 17-7

NOTE: VALUES GIVEN ARE THE SAME AS EXAMPLE SET 17-2.
1) Given: $k = 0.3$, $L_1 = 40$ μH, $L_2 = 60$ μH, fields opposing
   Find: total inductance
      First find the mutual inductance

$$L_M = 0.3 \cdot \sqrt{40 \ \mu\text{H} \cdot 60 \ \mu\text{H}}$$
$$= 0.3 \cdot 49 \ \mu\text{H}$$
$$= 14.7 \ \mu\text{H}$$

   Then solve for total inductance.

$$L_t = 40 \ \mu\text{H} + 60 \ \mu\text{H} - 2 \cdot 14.7 \ \mu\text{H} = 70.6 \ \mu\text{H}$$

(Continued next page)

E X A M P L E S E T 17-3 (Continued)

2) Given: $k = 0.8$, $L_1 = 10$ mH, $L_2 = 40$ mH, fields opposing
   Find: total inductance
   First find the mutual inductance

$$L_M = 0.8 \cdot \sqrt{10 \text{ mH} \cdot 40 \text{ mH}}$$
$$= 0.8 \cdot 20 \text{ mH}$$
$$= 16 \text{ mH}$$

   Then solve for total inductance.
$$L_t = 10 \text{ mH} + 40 \text{ mH} - 2 \cdot 16 \text{ mH} = 18 \text{ mH}$$

Determining L_M and k Through Measurement

Figure 17-8 illustrates a technique by which the mutual inductance and coefficient of coupling between two coils may be determined. First, the actual self-inductance of each coil and the total inductance of the two series coils is measured using a standard laboratory impedance bridge. The dial on the impedance bridge is adjusted to zero the bridge's meter and the inductance is read from the dial itself (usually the dial reading times a range setting). The two inductors should be connected so that they are series aiding or boosting one another when making the measurement for total inductance. Also, the two inductors must remain in the positions in which they will normally function. Otherwise, the coefficient of coupling between the coils will be disturbed. With this measured information, Formula 17.3 can be rearranged to solve for mutual inductance as follows:

$$L_M = (L_t - L_1 - L_2)/2 \tag{17.5}$$

Once the mutual inductance is determined, the coefficient of coupling may be calculated. Formula 17.2 may be rearranged to solve for k as follows:

$$k = L_M/\sqrt{L_1 \cdot L_2} \tag{17.6}$$

IMPEDANCE BRIDGE USED TO MEASURE INDUCTANCE

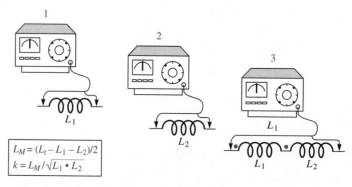

Figure 17-8 Determining mutual inductance and coefficient of coupling experimentally.

The following examples will illustrate the use of Formulas 17.5 and 17.6 in determining mutual inductance and coefficient of coupling:

EXAMPLE   17-4

The inductance of two series-aiding inductors is measured and found to be 50 μH and 80 μH. The total inductance is found to be 170 μH. Therefore, according to Formula 17.5, the mutual inductance is found to be 20 μH:

$$L_M = (170\ \mu H - 50\ \mu H - 80\ \mu H)/2 = 20\ \mu H$$

Using Formula 17.6, the k value is found to be 0.316:

$$k = 20\ \mu H/\sqrt{50\ \mu H \cdot 80\ \mu H} = 0.316$$

EXAMPLE   17-5

The inductance of two series-aiding inductors is measured and found to be 1 mH and 0.5 mH. The total inductance is found to be 1.783 mH. Therefore, according to Formula 17.5, the mutual inductance is found to be 0.142 mH:

$$L_M = (1.783\ mH - 1\ mH - 0.5\ mH)/2 = 0.142\ mH$$

Using Formula 17.6, the k value is found to be 0.2:

$$k = 0.142\ mH/\sqrt{1\ mH \cdot 0.5\ mH} = 0.2$$

SELF-CHECK   17-1

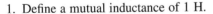

1. Define a mutual inductance of 1 H.

2. If the total flux generated by one coil is 25 μWb and 10 μWb of flux links a neighboring coil, what is the coefficient of coupling?

3. List four techniques that can be used to prevent mutual inductance between two inductors.

4. If the k values between two coils is 0.6 and the inductances of the two coils are 5 μH and 7.5 μH, calculate their mutual inductance.

5. Determine the total inductance of two series-aiding inductors with the following parameters: $L_1 = 200\ \mu H$; $L_2 = 100\ \mu H$; and $k = 0.9$.

6. The inductance of two series-aiding inductors is measured and found to be 18 mH and 4.5 mH. The total inductance is found to be 28 mH. Calculate the mutual inductance and the coefficient of coupling.

17-2 Voltage and Current Transformation

The Transformer

Transformers are special components that are used to transfer AC power from one circuit to another through the principle of mutual inductance. A basic transformer will have two coils, or windings. One coil is called the **primary winding** and receives AC power from some source. The second coil is called the **secondary winding** and receives an induced voltage and current due to mutual coupling of the primary winding's flux. AC power is transferred from the primary to the secondary by mutual induction. The secondary winding is usually connected to some load where the transferred AC power is converted to work, or dissipated as heat.

In the process of transferring AC power from primary to secondary, the AC voltage may be increased (stepped up) or decreased (stepped down). Therefore, a transformer can be used to reduce or increase AC voltage. In many cases, a transformer has more than one secondary winding in order to obtain a variety of voltages higher or lower than the primary source voltage.

Voltage Transformation

Figure 17-9 illustrates step-up and step-down voltage transformation. The amount of instantaneous voltage induced across the secondary winding is in accordance with Faraday's Voltage Law and formula: $v_L = N \cdot \Delta\Phi/\Delta t$. The amount of induced voltage depends on the number of turns (N) and the rate of change in flux ($\Delta\Phi/\Delta t$). Thus, the induced voltage across the secondary winding is determined by the number of turns in the secondary (N_S) and the rate of change in mutual flux ($\Delta\Phi_m/\Delta t$). (In most laminated iron-core transformers and toroidal transformers, the k value is 1, which means all flux generated by the primary is mutual to, or links, the secondary.) Likewise, the self-induced voltage of the primary is determined by the number of turns in the primary winding (N_P) and the same rate of change in mutual flux ($\Delta\Phi_m/\Delta t$). Faraday's formula is used to express the instantaneous primary and secondary voltages as follows:

$$v_S = N_S \cdot \Delta\Phi_m/\Delta t \qquad \text{(instantaneous secondary voltage)} \tag{17.7}$$

$$v_P = N_P \cdot \Delta\Phi_m/\Delta t \qquad \text{(instantaneous primary voltage)} \tag{17.8}$$

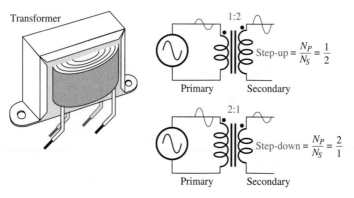

Figure 17-9   Voltage step-up and step-down transformers.

Formulas 17.7 and 17.8 can be combined to reveal a very useful formula that can be used to determine secondary, or primary, voltage based on the ratio of the number of turns in the primary and secondary windings. To do this, each of the formulas is rearranged to isolate the $\Delta\Phi_m/\Delta t$ terms on one side of the equations. Then, the two equations can be set equal to each other as follows:

$$v_S/N_S = \Delta\Phi_m/\Delta t \quad\text{and}\quad v_P/N_P = \Delta\Phi_m/\Delta t$$

therefore

$$v_S/N_S = v_P/N_P$$

since the rate of change in mutual flux is the same for the primary and the secondary (assuming $k = 1$).

Evaluating this relationship further, we see that

$$v_S/v_P = N_S/N_P \quad\text{and}\quad v_P/v_S = N_P/N_S$$

The previous relationships are not only true for instantanous voltages. They are also true for peak-to-peak, peak, and RMS values of AC voltage and can be expressed as follows:

$$\frac{V_S}{V_P} = \frac{N_S}{N_P} \quad\text{and}\quad \frac{V_P}{V_S} = \frac{N_P}{N_S}$$

These relationships, or ratios, are termed **transformation ratios** because they mathematically express the transformation of voltage from the primary to the secondary and vice versa. The following formulas make use of the turns transformation ratios to determine unknown primary or secondary voltages:

$$V_S = \frac{V_P N_S}{N_P} \tag{17.9}$$

$$V_P = \frac{V_S N_P}{N_S} \tag{17.10}$$

The transformation ratio is often shown in a schematic above the transformer diagram as two numbers separated by a colon. The first number represents the primary winding and the second number represents the secondary winding (i.e., 1:2 indicates a step-up ratio where the primary winding might have 500 turns and the secondary has 1,000 turns). Now, take a few minutes to consider Example Set 17-6.

EXAMPLE SET 17-6

1)

115 V   200 T   1:3   600 T

Primary   Secondary

$$V_S = \frac{V_P N_S}{N_P}$$

$$V_P = \frac{V_S N_P}{N_S}$$

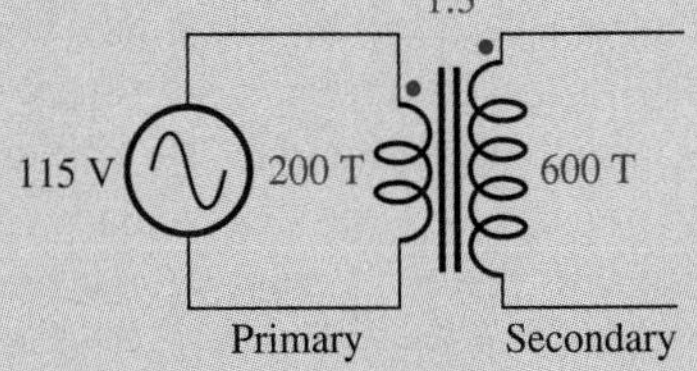

$$V_S = \frac{115\ \text{V} \cdot 600\ \text{T}}{200\ \text{T}} = 115\ \text{V} \cdot 3$$

$$= 345\ \text{V}$$

Figure 17-10

(Continued next page)

EXAMPLE SET 17-6 (Continued)

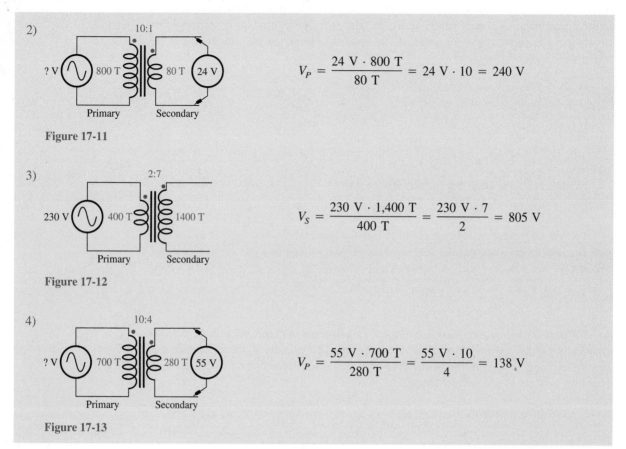

2)

Figure 17-11

$$V_P = \frac{24 \text{ V} \cdot 800 \text{ T}}{80 \text{ T}} = 24 \text{ V} \cdot 10 = 240 \text{ V}$$

3)

Figure 17-12

$$V_S = \frac{230 \text{ V} \cdot 1{,}400 \text{ T}}{400 \text{ T}} = \frac{230 \text{ V} \cdot 7}{2} = 805 \text{ V}$$

4)

Figure 17-13

$$V_P = \frac{55 \text{ V} \cdot 700 \text{ T}}{280 \text{ T}} = \frac{55 \text{ V} \cdot 10}{4} = 138 \text{ V}$$

Current Transformation

If the voltage is stepped up from primary to secondary in a transformer, what happens to the current? If the voltage is stepped up, the current is stepped down. Why?

Power In = Power Out

The simplest approach to explaining the inverse current–voltage relationship is found in understanding the primary and secondary power relationship. For most practical purposes, the power in the primary winding is equal to the power in the secondary winding for all transformers with $k = 1$. In other words, the power demanded by the secondary winding circuit is essentially the same as the power supplied to the primary winding from some power source. This, of course, assumes the transformer is 100% efficient. We will discuss this more later. Since $P_P = P_S$, $E_P \cdot I_P = E_S \cdot I_S$. As you can see, if the secondary voltage E_S is greater than the primary voltage E_P, the secondary current I_S must be less than the primary current I_P to maintain the balance in primary and secondary power. The inverse must also be true: If voltage is stepped down, current is stepped up. The primary to secondary current ratio is always the exact inverse of the primary to secondary voltage ratio and turns ratio. If the primary to secondary turns ratio is given as 1:3, the secondary voltage is tripled, while the secondary current is cut to one third of the primary current. Thus, the primary and secondary power balance is maintained.

In other words: $N_P/N_S = I_S/I_P$, where the secondary to primary current ratio is equal to the primary to secondary turns ratio. From this, the following useful formulas are obtained:

$$I_S = \frac{I_P N_P}{N_S} \tag{17.11}$$

$$I_P = \frac{I_S N_S}{N_P} \tag{17.12}$$

Study Example Set 17-7 and Example 17-8 to gain a clear understanding of the relationship between the voltage and current of step-up and step-down transformers. Keep in mind that the terms "step-up" and "step-down" refer to the voltage, not the current.

EXAMPLE SET   17-7

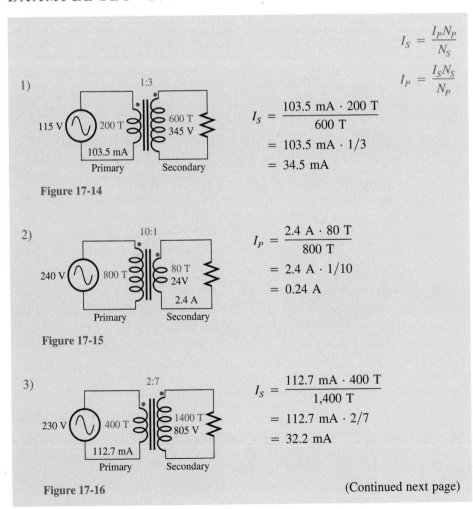

$$I_S = \frac{I_P N_P}{N_S}$$

$$I_P = \frac{I_S N_S}{N_P}$$

1)

115 V   200 T   1:3   600 T / 345 V   103.5 mA   Primary   Secondary

$$I_S = \frac{103.5 \text{ mA} \cdot 200 \text{ T}}{600 \text{ T}}$$
$$= 103.5 \text{ mA} \cdot 1/3$$
$$= 34.5 \text{ mA}$$

Figure 17-14

2)

240 V   800 T   10:1   80 T / 24V   2.4 A   Primary   Secondary

$$I_P = \frac{2.4 \text{ A} \cdot 80 \text{ T}}{800 \text{ T}}$$
$$= 2.4 \text{ A} \cdot 1/10$$
$$= 0.24 \text{ A}$$

Figure 17-15

3)

230 V   400 T   2:7   1400 T / 805 V   112.7 mA   Primary   Secondary

$$I_S = \frac{112.7 \text{ mA} \cdot 400 \text{ T}}{1,400 \text{ T}}$$
$$= 112.7 \text{ mA} \cdot 2/7$$
$$= 32.2 \text{ mA}$$

Figure 17-16

(Continued next page)

EXAMPLE SET 17-7 (Continued)

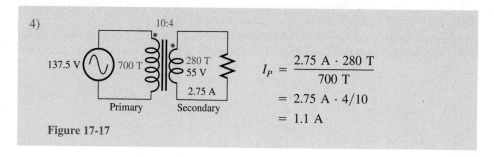

4)

$$I_P = \frac{2.75 \text{ A} \cdot 280 \text{ T}}{700 \text{ T}}$$

$$= 2.75 \text{ A} \cdot 4/10$$

$$= 1.1 \text{ A}$$

137.5 V · 700 T · 10:4 · 280 T / 55 V / 2.75 A · Primary · Secondary

Figure 17-17

EXAMPLE 17-8

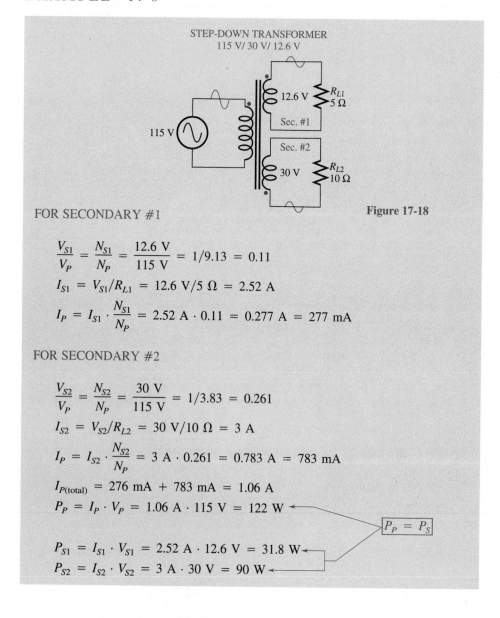

STEP-DOWN TRANSFORMER
115 V/ 30 V/ 12.6 V

115 V · 12.6 V · R_{L1} 5 Ω · Sec. #1 · Sec. #2 · 30 V · R_{L2} 10 Ω

Figure 17-18

FOR SECONDARY #1

$$\frac{V_{S1}}{V_P} = \frac{N_{S1}}{N_P} = \frac{12.6 \text{ V}}{115 \text{ V}} = 1/9.13 = 0.11$$

$$I_{S1} = V_{S1}/R_{L1} = 12.6 \text{ V}/5 \text{ Ω} = 2.52 \text{ A}$$

$$I_P = I_{S1} \cdot \frac{N_{S1}}{N_P} = 2.52 \text{ A} \cdot 0.11 = 0.277 \text{ A} = 277 \text{ mA}$$

FOR SECONDARY #2

$$\frac{V_{S2}}{V_P} = \frac{N_{S2}}{N_P} = \frac{30 \text{ V}}{115 \text{ V}} = 1/3.83 = 0.261$$

$$I_{S2} = V_{S2}/R_{L2} = 30 \text{ V}/10 \text{ Ω} = 3 \text{ A}$$

$$I_P = I_{S2} \cdot \frac{N_{S2}}{N_P} = 3 \text{ A} \cdot 0.261 = 0.783 \text{ A} = 783 \text{ mA}$$

$$I_{P\text{(total)}} = 276 \text{ mA} + 783 \text{ mA} = 1.06 \text{ A}$$

$$P_P = I_P \cdot V_P = 1.06 \text{ A} \cdot 115 \text{ V} = 122 \text{ W} \longleftarrow$$

$$\boxed{P_P = P_S}$$

$$P_{S1} = I_{S1} \cdot V_{S1} = 2.52 \text{ A} \cdot 12.6 \text{ V} = 31.8 \text{ W} \longleftarrow$$

$$P_{S2} = I_{S2} \cdot V_{S2} = 3 \text{ A} \cdot 30 \text{ V} = 90 \text{ W} \longleftarrow$$

Take time now to test your understanding of voltage and current transformation by answering the questions of Self-Check 17-2.

1. A certain transformer is marked with a turns ratio of 2:5. If the primary voltage is 115 V, what is the secondary voltage? Is this a step-up or step-down transformer?

2. If the power supplied to the primary winding of the transformer of Question 1 is 250 W, how much current is in the secondary winding?

3. If a certain transformer has a turns ratio of 3:2, the secondary voltage is 60 V, and the secondary power is 150 W, what is the primary voltage and current?

4. Determine the secondary currents and powers and the primary current and power for the transformer of Figure 17-19.

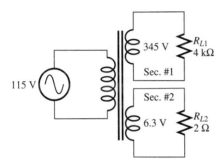

Figure 17-19

17-3 Impedance Transformation

Reflected Impedance

Impedance (Z)

Transformers not only transfer power by transforming voltage and current, but they also transform impedance. **Impedance (Z)** is the opposition offered by a circuit to an AC voltage. The impedance of an AC circuit may include inductive reactance, capacitive reactance, and resistance. The use of the term *impedance* generally implies the presence of some reactance and resistance. However, the term may be used loosely as a substitute for total resistance in a purely resistive AC circuit. The impedance for any AC circuit is measured in ohms and may be calculated using Ohm's Law as follows:

$$Z = V/I \qquad\qquad (17.13)$$

where Z is the impedance of any AC circuit whose voltage V and current I are known.

Determining Reflected Impedance

When a transformer transforms impedance, the secondary load impedance is **reflected** back to the primary circuit, and may be stepped up or stepped down in proportion to the square of the primary to secondary turns ratio. We may demonstrate this mathematically as follows:

- First, the impedance of the secondary load circuit is expressed using Ohm's Law: $Z_S = V_S/I_S$
- The reflected impedance is the primary winding impedance and is also expressed using Ohm's Law: $Z_P = V_P/I_P$
- The ratio of the primary impedance to the secondary impedance is, naturally, the impedance transformation ratio: Z_P/Z_S
- Since $Z_P = V_P/I_P$ and $Z_S = V_S/I_S$, $Z_P/Z_S = (V_P/I_P)/(V_S/I_S)$
- Also, $Z_P/Z_S = (V_P/I_P) \cdot (I_S/V_S) = (V_P/V_S) \cdot (I_S/I_P)$
- Recall that $V_P/V_S = N_P/N_S$ and $I_S/I_P = N_P/N_S$
- Therefore, by substitution $Z_P/Z_S = (N_P/N_S) \cdot (N_P/N_S) = N_P^2/N_S^2 = (N_P/N_S)^2$

Thus, the primary to secondary impedance ratio is equal to the primary to secondary turns ratio squared. (This is assuming the transformer is 100% efficient.)

$$Z_P/Z_S = (N_P/N_S)^2 \tag{17.14}$$

From this relationship, we derive the reflected impedance formula as follows:

$$Z_P = Z_S \cdot (N_P/N_S)^2 \tag{17.15}$$

If the impedance ratio is known, the turns ratio can be determined by rearranging Formula 17.15 as follows:

$$N_P/N_S = \sqrt{Z_P/Z_S} \tag{17.16}$$

Example Set 17-9 illustrates the use of Formulas 17.14, 17.15, and 17.16. Take time to study the examples carefully before continuing. Also, take a moment to study Design Note 17-1 as a review and summary of transformer analysis.

EXAMPLE SET 17-9

1)

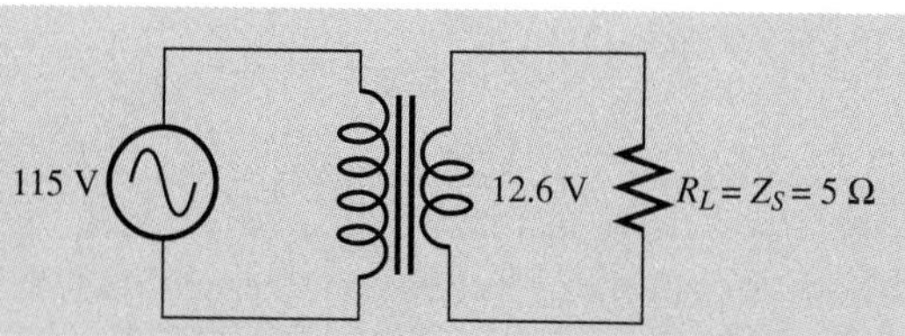

Figure 17-20

$$Z_P/Z_S = (N_P/N_S)^2$$
$$Z_P = Z_S \cdot (N_P/N_S)^2$$
$$N_P/N_S = \sqrt{Z_P/Z_S}$$

$$\frac{V_P}{V_S} = \frac{N_P}{N_S} = \frac{115\ \text{V}}{12.6\ \text{V}} = \frac{9.13}{1} \quad \therefore \frac{N_S}{N_P} = 1/9.13 = 0.11$$

$$I_P = I_S \cdot \frac{N_S}{N_P} = 2.52\ \text{A} \cdot 0.11 = 0.277\ \text{A} = 277\ \text{mA}$$

$$Z_P = \frac{V_P}{I_P} = \frac{115\ \text{V}}{277\ \text{mA}} = 415\ \Omega \longleftarrow$$

$$Z_P = Z_S \cdot (N_P/N_S)^2 = 5\ \Omega \cdot (9.13/1)^2 = 417\ \Omega \longleftarrow$$

SLIGHT ERROR DUE TO ROUNDING

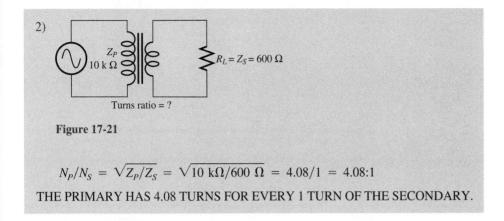

Figure 17-21

$$N_P/N_S = \sqrt{Z_P/Z_S} = \sqrt{10 \text{ k}\Omega/600 \text{ }\Omega} = 4.08/1 = 4.08{:}1$$

THE PRIMARY HAS 4.08 TURNS FOR EVERY 1 TURN OF THE SECONDARY.

How the Impedance Is Transformed

The Inductive Reactance of the Primary

How is the impedance that is connected to the secondary winding transformed, or reflected, back to the primary winding? How does the transformer know if a large load (heavy current demand), a light load (low current demand), or no load at all is connected to the secondary winding? To answer these questions, let's first consider a transformer with no load. In other words, the secondary winding is not connected to any load device. Since no current is demanded in the secondary circuit, does that mean there is no current delivered to the primary winding from the AC source? No. Some current will flow in the primary winding, limited by the inductive reactance of the primary winding itself. Since there is no secondary load, the primary current is very small and the inductive reactance of the primary is high. When a load device is connected to the secondary winding, the current in the primary winding increases to meet the demand for power (power delivered to the secondary circuit). The increase in primary winding current must be the result of a decrease in the primary winding's impedance.

CEMF Is the Key

But how can an increase in secondary winding current cause a decrease in the primary winding impedance? The answer to this question also explains reflected impedance. When the secondary load is increased (load impedance is decreased and load current increases), the magnetic flux produced in the core by current in the secondary winding increases as well. This secondary flux opposes the primary winding flux, which in turn reduces the CEMF normally developed across the primary winding. The CEMF of the primary should nearly equal the applied primary voltage. When the secondary flux opposes the primary flux and reduces the CEMF of the primary, the primary current must increase to restore the CEMF to its original level (nearly equal to the applied source voltage). Thus, the overall effect of an increased secondary current is the necessary increase in primary current, which amounts to a reduction of the primary's impedance.

In Case of a Short

If the secondary winding of the transformer becomes shorted, the extremely high demand for current causes a large quantity of secondary flux in the core. This generates a great

D E S I G N   N O T E   1 7 - 1:   Transformer Analysis

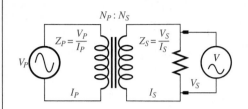

PROGRAM VARIABLES

VP = PRIMARY VOLTAGE
VS = SECONDARY VOLTAGE
IP = PRIMARY CURRENT
IS = SECONDARY CURRENT
TR = TURNS RATIO
ZP = PRIMARY IMPEDANCE
ZS = SECONDARY IMPED
PP = PRIMARY POWER
PS = SECONDARY POWER

"B A S I C"   P R O G R A M   S O L U T I O N

```
 10 CLS
 20 PRINT"TRANSFORMER ANALYSIS"
 30 PRINT"":PRINT"":CLEAR
 40 PRINT"(1) SECONDARY VOLTAGE, CURRENT, AND POWER"
 50 PRINT"(2) PRIMARY VOLTAGE, CURRENT, AND POWER"
 60 PRINT"(3) REFLECTED IMPEDANCE"
 70 PRINT"(4) PRIMARY TO SECONDARY TURNS RATIO"
 80 PRINT"(5) QUIT"
 85 PRINT""
 86 INPUT C:PRINT""
 87 IF C < 1 OR C > 5 THEN GOTO 30
 90 ON C GOTO 100,200,300,400,800
100 INPUT"ENTER THE PRIMARY VOLTAGE. ";VP
110 INPUT"ENTER THE PRIMARY TO SECONDARY TURNS RATIO AS A WHOLE NUMBER. ";TR
120 INPUT"ENTER THE SECONDARY LOAD RESISTANCE. ";RL
130 VS = VP/TR:IS = VS/RL:PS = IS * VS
140 PRINT"THE SECONDARY VOLTAGE IS ";ES;" VOLTS."
150 PRINT"THE SECONDARY CURRENT IS ";IS;" AMPS."
160 PRINT"THE SECONDARY POWER IS ";PS;" WATTS."
170 GOTO 30
200 INPUT"ENTER THE SECONDARY VOLTAGE. ";VS
210 INPUT"ENTER THE PRIMARY TO SECONDARY TURNS RATIO AS A WHOLE NUMBER. ";TR
220 INPUT"ENTER THE SECONDARY LOAD RESISTANCE. ";RL
230 VP = VS * TR:IP = VS/RL/TR:PP = IP * VP
240 PRINT"THE PRIMARY VOLTAGE IS ";VP;" VOLTS."
250 PRINT"THE PRIMARY CURRENT IS ";IP;" AMPS."
260 PRINT"THE PRIMARY POWER IS ";PP;" WATTS."
270 GOTO 30
300 INPUT"ENTER THE PRIMARY TO SECONDARY TURNS RATIO AS A WHOLE NUMBER. ";TR
310 INPUT"ENTER THE SECONDARY LOAD RESISTANCE. ";RL
320 ZP = RL * TR^2
330 PRINT"THE REFLECTED IMPEDANCE IS ";ZP;" OHMS."
340 GOTO 30
400 PRINT"DO YOU WANT TO USE:"
410 PRINT""
420 PRINT"(1) VOLTAGES"
```

```
430  PRINT"(2) CURRENTS"
440  PRINT"(3) IMPEDANCES"
450  PRINT""
460  INPUT"TO FIND THE TURNS RATIO";C:PRINT""
470  IF C<1 OR C>3 THEN GOTO 410
480  ON C GOTO 500,600,700
500  INPUT"ENTER THE PRIMARY VOLTAGE. ";VP
510  INPUT"ENTER THE SECONDARY VOLTAGE. ";VS
520  TR = VP/VS
530  PRINT"":PRINT"THE PRIMARY TO SECONDARY TURNS RATIO IS ";TR
540  GOTO 30
600  INPUT"ENTER THE PRIMARY CURRENT. ";IP
610  INPUT"ENTER THE SECONDARY CURRENT. ";IS
620  TR = IS/IP
630  PRINT"":PRINT"THE PRIMARY TO SECONDARY TURNS RATIO IS ";TR
640  GOTO 30
700  INPUT"ENTER THE PRIMARY IMPEDANCE. ";ZP
710  INPUT"ENTER THE SECONDARY IMPEDANCE. ";ZS
720  TR = SQR(ZP/ZS)
730  PRINT"":PRINT"THE PRIMARY TO SECONDARY TURNS RATIO IS ";TR
740  GOTO 30
800  CLS:END
```

opposition to primary winding flux. As a result, the CEMF of the primary is made to drop very low. Thus, the primary winding current increases dramatically to restore the CEMF. In a short time, this results in the primary windings burning open. It is for this reason that a fuse placed in the primary winding protects both the primary and secondary circuits from excessive damage due to shorts.

Time for a Self-Check. The problems of Self-Check 17-3 will reinforce the concept of reflected impedance and the reflected impedance calculations.

SELF-CHECK 17-3

1. Define *impedance*.

2. In what way is the transformer impedance ratio related to the transformer turns ratio?

3. If a transformer's turns ratio is 2:5, what is the impedance ratio?

4. If a transformer has a turns ratio of 3:7 and the secondary is connected to a 400 Ω resistor, what is the reflected primary impedance?

5. Explain how a short across the secondary of a transformer can burn the primary winding open.

17-4 Transformer Ratings and Losses

Transformer Ratings

Voltage Rating

Power transformers are rated according to primary and secondary voltage. The transformer specification plate, or silk-screen, will show the voltages for which the transformer was designed. As an example, a transformer may be labeled VOLTS: 230 CT/12.6. This indicates that this transformer has a center-tapped winding rated at 230 V. Using the center-tap lead, the transformer will also operate on 115 V. In this case, the high-voltage winding is obviously intended as the primary side of the transformer. The secondary winding is rated at 12.6 V. However, the 12.6 V winding may be used as a primary winding and the center-tapped high-voltage winding may be used as the secondary. In other words, any power transformer may be used as a step-up or step-down transformer as long as the windings voltage ratings are not exceeded along with other transformer ratings.

EXAMPLE   17-10

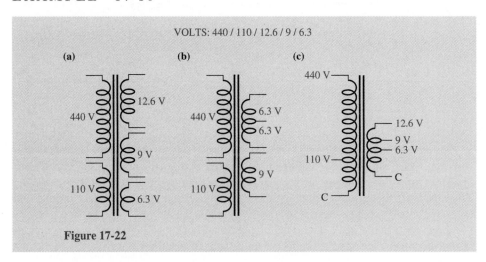

Figure 17-22

As another example, a transformer may be labeled as follows: VOLTS: 440/110/12.6/9/6.3. In this case, the transformer either has five separate windings or has two or more main windings that are tapped. A description or schematic is needed to know for sure (see Example 17-10). As shown in Figure 17-22c, it is possible that the high voltage windings (440/110) are actually one winding that has a tap in the winding one quarter of the way in from a common end ($110/440 = 1/4$). If this is the case, the transformer could actually be powered from a 110 V source, a 440 V source, or even a 330 V source ($440 - 110 = 330$, using the larger side of the tapped winding). The secondary windings may also be a single winding with two taps. The entire winding would supply 12.6 V. The 9 V tap would be seven tenths of the way from a common end and the 6.3 V tap would be a center tap.

Current Rating

Often the specification plate on the transformer indicates the maximum current that can safely be supplied to or demanded from a particular winding. For example, a particular transformer may be rated as follows: VOLTS: 115/12.6@1.5 A. This indicates that the transformer will supply an RMS voltage of 12.6 V to an RMS load of 1.5 A with 115 V applied to the primary. The 1.5 A rating is a maximum rating and should not be exceeded. If the rated current is exceeded, the I^2R losses of the windings will increase and excessive heating will occur. In this case, the primary current should not exceed about 164 mA, since the current transformation ratio is $115/12.6 = 9.13:1$ (164 mA $\cdot$ 9.13 = 1.5 A). To double-check our calculations, recall that the power in the primary will approximately equal the power in the secondary. Thus, 115 V $\cdot$ 164 mA $=$ 12.6 V $\cdot$ 1.5 A. A $\frac{3}{16}$ A or a $\frac{1}{4}$ A slow-blow fuse would be recommended for the primary of this transformer.

Power Rating

The power rating of any transformer is an important one. It is quite common for a transformer's power rating to be included on the transformer's specification plate. As an example, a particular power transformer may be labeled as follows: POWER: 250 VA. In this case, the transformer is rated as 250 volt-amperes (VA). Note that power is voltage times current (VA). The transformer is not rated in watts because it does not convert the power to heat or mechanical energy (real or true power). It merely transfers the power from a source to a load. The transformer power is *apparent power* (P_A) and is the power that is only apparently used by the transformer but is actually passed on to a load (with the exception of normal transformer losses). To distinguish apparent power from real power, the unit VA is used instead of watts. In this case, a rating of 250 VA means the product of the primary, or secondary, voltage and current must not exceed 250 VA ($V_P \cdot I_P \leq 250$ VA and $V_S \cdot I_S \leq 250$ VA). If this transformer's voltage rating is shown on the specification plate as: VOLTS: 115/6.3, the current in the 115 volt winding cannot exceed 2.17 A ($I = P/E = 250$ VA$/115$ V $= 2.17$ A) and the current in the 6.3 volt winding cannot exceed 39.7 A (250 VA$/6.3$ V $= 39.7$ A). In this case, the 6.3 volt winding is few turns of very heavy-gauge wire (AWG #8 or #6) and the 115 volt winding is 18.25 times more turns of a lighter-gauge wire (AWG #24 or #22).

Consider the power rating of a multiple secondary transformer with the following voltage and current ratings: VOLTS: 115/12.6@2 A/24@3 A. The maximum power in each of the low-voltage windings must be calculated and added to find the total power rating and 115 V winding current. The maximum power of the 12.6 V winding is 25.2 VA and the maximum power of the 24 V winding is 72 VA (2 A $\cdot$ 12.6 V $= 25.2$ VA and 3 A $\cdot$ 24 V $= 72$ VA). The total power mutually coupled to these two windings from the 115 V winding is 97.2 VA (25.2 VA $+$ 72 VA $= 97.2$ VA). Thus, the maximum current in the 115 V winding will be 850 mA ($I = P/E = 97.2$ VA$/115$ V $= 0.85$ A). A 1A fuse should be used in the primary.

Impedance Rating

Audio transformers are often rated according to primary and secondary impedance along with an output power rating. As an example, the specification plate on a particular audio transformer may read: Z: 10k/600 50 mW(max). This particular audio transformer is designed to match a 10 kΩ impedance to a 600 Ω impedance for maximum power transfer. The output power, or power delivered to a load, must not exceed 50 mW, and the load must be either 600 Ω or 10 kΩ. If the load is 600 Ω, the low-Z side of the transformer is connected to the load and the transformer reflects an impedance back to the primary that is 10 kΩ (the transformer's turns ratio is $\sqrt{10k/600} = 4.08:1$). If the load is 10 kΩ, the high-Z side of the transformer is connected to the load and the transformer reflects an impedance back to the primary that is 600 Ω.

Frequency Rating

The frequency rating of a transformer is just as important as any of the ratings already discussed. A transformer is specially designed to operate at a particular frequency or frequency range. Power transformers are commonly rated at 400 Hz, 60 Hz, or 50 Hz. The 400 Hz power transformers are commonly used on aircraft and are much smaller and lighter than a 50 Hz or 60 Hz transformer of the same power rating. The lower the frequency, the larger and heavier the transformer must become to maintain the inductive reactance of the windings at a level that will not allow excessive current to flow. A 400 Hz transformer would soon burn open if operated at 50 Hz or 60 Hz due to the low inductive reactance and resulting high current. Even a 60 Hz transformer becomes very warm when operated at 50 Hz and must, therefore, be derated in power.

Transformer Losses

Winding Losses

Winding loss is the conductor resistance loss (copper loss) in each winding. Any resistance converts current to heat. The amount of power dissipated as heat in the windings is found using the following power formula: $P = I^2R$. Naturally, this loss is often referred to as an I^2R loss. As the winding current increases, the I^2R loss naturally increases. This loss is particularly dangerous when the transformer is overloaded. The I^2R heat can quickly become severe enough to burn the enamel coating from the windings, causing them to short and blow a fuse or simply burn open. This winding loss can be fairly significant, depending on the resistance of the wire and amount of current drawn.

Eddy-Current Losses

The iron core of many transformers is another place where significant power losses can occur. It is possible for the iron core to act as a shorted secondary winding. Since the core would look like only one or two turns, the induced current would be extremely high. For example, suppose a primary winding has 350 turns and the iron core looks like a single very heavy turn. The current induced in the core would be 350 times greater than the primary current. To prevent such a thing from actually occurring, the iron core must be broken up, or segmented, to interrupt an organized induced current. See Figure 17-23. The unwanted induced currents are known as **eddy currents**. In many audio and power transformers, eddy currents are prevented, or greatly reduced, by laminating the iron core. A **laminated iron core** is made of thin layers of iron called **laminations**. Each lamination is enamel coated or oxidized to act as an insulating layer. When the laminations are placed together, the enamel or oxidation prevents organized current between plates. The eddy currents that do exist are very small. The eddy-current loss in ferrite cores, used in radio-frequency and some audio-frequency applications, is greatly reduced, since the core is made of millions of small individually insulated iron particles.

Hysteresis Losses

The **hysteresis loss** in an iron core takes into account the energy required to align and realign magnetic domains in the core with every AC cycle. This loss has a heating effect

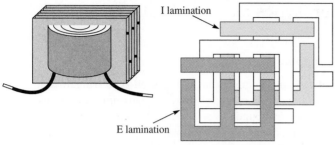

I lamination

E lamination

Insulated iron laminations form the core

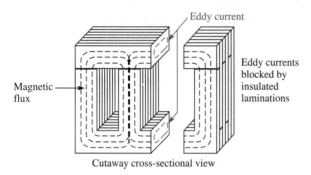

Eddy current

Magnetic flux

Eddy currents blocked by insulated laminations

Cutaway cross-sectional view

Figure 17-23   Reducing eddy-current losses by using a laminated iron core.

on the core, since there is some atomic stress and molecular friction involved with the alternating movement of the magnetic domains. This loss is controlled, or minimized, by using a core material with a low retentivity.

Altogether, the transformer losses discussed here can be very significant. Some small-power transformers have power efficiencies as low as 60% where the power delivered to the primary winding is much greater than the power used by the secondary load device. For example: 10 VA delivered to the primary and the secondary load uses only 6 W. 4 W is converted to heat in the transformer.

Time for a quick self-check.

SELF-CHECK   17-4

1. What is the meaning of the following information found stamped on a transformer? VOLTS: 120/24 CT

2. From the following transformer specification plate information, what is the transformer's maximum power rating? VOLTS: 115/12.6@2 A

3. What is the maximum high-voltage winding current for the transformer of Question 2?

4. From the following transformer specification plate information, what is the transformer's maximum power rating? VOLTS: 115/24@1.5 A/30@1 A/6.3@4 A

5. Why will a 400 Hz transformer burn up if it is operated at 60 Hz?

6. List two transformer losses.

17-5 Transformer Types and Applications

Power Transformers

Transformers for Power Supplies

See color photo insert
page D7.

Figure 17-24 shows a variety of power transformers. The **shell-type transformer** is encased in a soft iron housing to trap all flux leakage from the transformer windings. Many circuits used to amplify very low-level signals are sensitive to the presence of magnetic fields. **Open-core-type transformers** are for general-purpose power-supply applications where flux leakage is not a problem.

The AC to DC converter shown in Figure 17-25 is a bipolar supply using a center-tapped secondary winding. With the center tap to ground, the secondary voltage is cut in half and rectified (converted to pulsating DC) by the diode circuit. Two electrolytic filter capacitors integrate the dual polarity pulsating DC, converting it to a smooth ± 16.3 VDC. The 24 V CT secondary is rated in RMS voltage. The capacitors charge to the peak voltage of the pulsating DC (minus the approximate 0.7 V diode drop).

Transformers for Adjustable and Variable AC

AC voltages are often adjusted or varied using special single-winding transformers called **autotransformers**. Figure 17-26 illustrates a variety of autotransformers, both tapped and variable. Tapped autotransformers permit line-voltage adjustment over a wide voltage range. The taps on the autotransformer are labeled in actual voltage or, in some cases, percentage adjustment (i.e., $+5\%$, $+10\%$, -5%, etc.). The line voltage should be con-

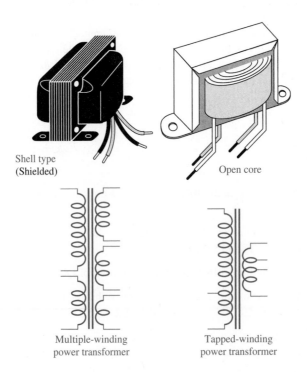

Shell type
(Shielded)

Open core

Multiple-winding
power transformer

Tapped-winding
power transformer

Figure 17-24 Power-supply transformers.

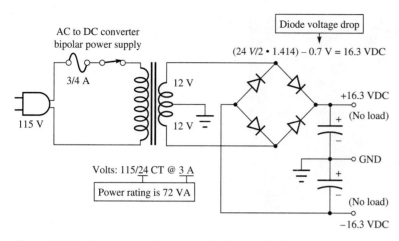

Figure 17-25   A power transformer application—a bipolar power supply.

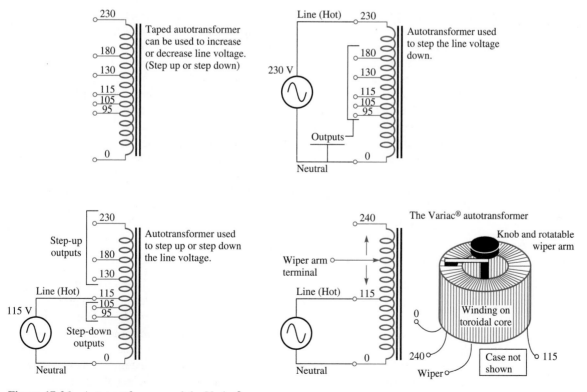

Figure 17-26   Autotransformers and the Variac®.

Figure 17-27 A Variac®.

nected to the correspondingly labeled transformer tap, then a higher or lower voltage can be selected from the remaining taps. Thus, the autotransformer may be used as a step-up or step-down transformer.

The **Variac**® is a **VARI**able **AC** autotransformer. It consists of a single transformer winding that is tapped to match a 110 – 120 V line voltage. The entire winding will accept or produce approximately 220 to 240 V. The Variac® is a single-layer winding wound on a very large toroidal core. A rotatable wiper arm slides across exposed windings (insulation removed) on one end of the transformer. The wiper arm is then connected to a terminal post for external use. This output terminal provides a range of AC from 0 – 240 V as the control knob is manually varied. The Variac® winding is rated at a maximum current level in accordance with the core size and windings wire gauge. Variacs® are used to permit manual line-voltage adjustment or to vary the brightness of a light or speed of some motors. A typical Variac® is shown in Figure 17-27.

Autotransformer action follows the same basic principles as multiple-winding-transformer action: principles of mutual inductance. Since the entire winding is on the one iron core, the coefficient of coupling k is 1 for the entire winding. This means that any voltage induced across any single turn of the winding is induced across each and all remaining turns. If an autotransformer has 300 turns and 1 V appears across one of its turns, the voltage across the entire 300 turn winding will be 300 V. Also, the amount of voltage step-up or step-down is in proportion to the ratio of the number of turns used for the output voltage to the number of turns used for the input voltage. For example, if 100 V is applied to 200 turns of a 500 turn autotransformer, the output voltage across the entire winding is 250 V (250 $V_{out} = V_{in} \cdot N_{out}/N_{in} =$ 100 V $\cdot$ 500 T/200 T).

Transformers for Isolation

Power transformers are also used to isolate a high-voltage source from the equipment chassis ground. Figure 17-28 illustrates this application. The idea here is to protect people and other pieces of equipment from the unpleasant effects of misplaced high voltage. It is definitely undesirable for the metal chassis of a piece of equipment to become ''hot.'' As shown in the figure, it is conceivable that the chassis could become hot if a two-pronged plug is plugged in the outlet the wrong way and an isolation transformer is not used. Using the **isolation transformer**, there is no chance that the chassis might accidentally become hot due to improper plug placement.

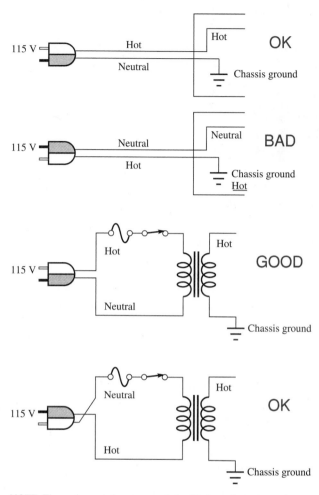

NOTE: The equipment chassis ground should always be connected to
      the utility earth ground for maximum protection from a high-voltage
      short circuit. This insures that the chassis can never become hot.

Figure 17-28   The use of the isolation transformer.

Audio Transformers

Audio Transformers for Impedance Matching

Recall that, according to the maximum-power-transfer theorem, maximum power is delivered to a load when the load impedance matches the source impedance. With this in mind, transformers are an almost ideal component used in matching impedances. The impedance of the secondary (load impedance) is reflected back to the primary while being stepped up or stepped down in proportion to the square of the turns ratio (see Section 17-3). Since most iron-core transformers have very high efficiencies, nearly all of the power applied to the primary is transferred to the secondary circuit.

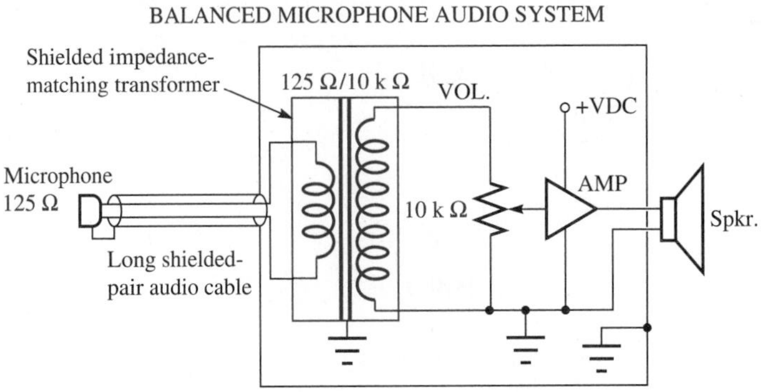

BALANCED MICROPHONE AUDIO SYSTEM

Shielded impedance-
matching transformer

125 Ω/10 k Ω   VOL.

+VDC

Microphone
125 Ω

AMP

10 k Ω

Spkr.

Long shielded-
pair audio cable

Figure 17-29   Using an audio transformer for microphone impedance matching.

Figure 17-29 illustrates two very common applications for audio **impedance-matching transformers**. In many instances, low-impedance microphones (125 Ω to 600 Ω) are connected to public address systems or audio mixers through long shielded-pair audio cables. The input impedance of the public address system should match the microphone impedance for optimum performance (maximum power transfer and flat frequency response over the entire audio range). In some cases, impedance matching is accomplished through the use of a specially designed and properly shielded audio transformer. The audio transformer serves more than a single purpose here. It not only provides for maximum power transfer and proper cable loading (impedance matching), it also provides a step-up in microphone signal voltage and a balanced input. As you can see from the impedance ratio (125 Ω/10 kΩ), the matching transformer is a voltage step-up transformer. Thus, the audio voltage across the 10 kΩ volume control is greater than the microphone voltage.

Figure 17-30 illustrates the use of impedance-matching transformers in some audio amplifiers. Many low-powered audio amplifiers (and some very-high-powered amplifiers) use an impedance-matching transformer to transfer maximum power to a speaker. Two transistors take turns driving each half of a center-tapped primary winding. Notice that the entire primary winding has an impedance of 200 Ω, while each half of the primary only has 50 Ω. This may not seem correct at first thought. However, impedance is a function

LOW-POWER AUDIO AMP

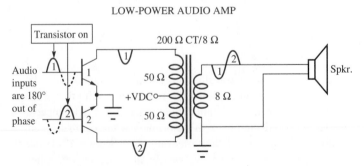

Figure 17-30   Using an audio transformer for power transistor impedance matching.

of the square of a turns ratio. In this case, the turns ratio is 2:1 since the entire primary winding is twice each half. Therefore, the impedance ratio for the center-tapped primary is $(2/1)^2$ or 4:1. Thus, the entire primary winding impedance is 200 Ω (50 Ω · 4 = 200 Ω).

Radio-Frequency Transformers

Transformers for Intermediate-Frequency Amplifiers

The circuitry of radio and television receivers includes a very important section called the **intermediate-frequency section (IF strip)**. In this section, special transformers are used to select and couple an IF signal from one amplifier stage to the next (usually 2 to 4 stages). These **IF transformers** are illustrated in Figure 17-31. Each transformer has a threaded ferrite core that is used to adjust the transformer circuit for maximum sensitivity (gain) at the intermediate frequency (455 kHz for AM and 10.7 MHz for FM). The transformer is shielded in a small metal can to prevent interaction with neighboring stages. The IF transformer plays an important role in communications circuitry.

Transformers for RF Power Applications

Many solid-state radio-frequency transmitters use RF transformers in the final amplifier circuit. Figure 17-32 illustrates a typical medium-power (25 to 200 W) transmitter power amplifier (PA). The output transformer is a ferrite-core toroidal transformer with a center-

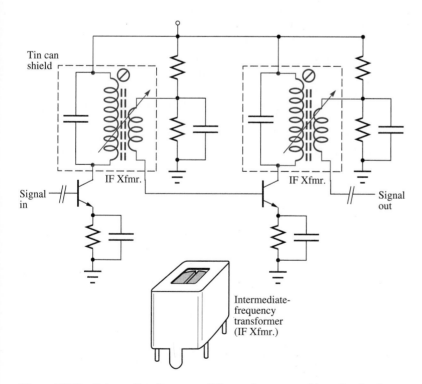

Figure 17-31   Intermediate frequency (IF) transformers used in radio circuits.

BROAD-BANDED RADIO-FREQUENCY AMPLIFIER

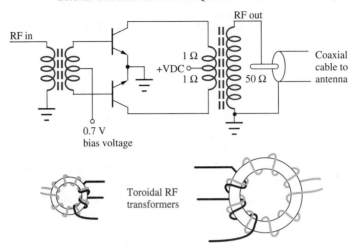

Figure 17-32   Radio-frequency power transformers.

tapped primary. The center tap on the primary is used to provide +VDC to each of the two transistors. The transistors take turns driving the output transformer, each conducting only for half a cycle. The half-cycle AC current in each half of the center-tapped primary winding is very high, as each transistor conducts. This high current develops a large primary power ($P_P = I_P = V_P$) which is transferred to the secondary circuit (transmission line and antenna). The step-up design of the output transformer matches the very low transistor impedance to the 50 Ω load impedance (communications antenna). Thus, maximum power is transferred from the low impedance transistor circuit to the antenna. Note that the transistors are also driven with a smaller center-tapped toroidal transformer. The 0.7 VDC bias voltage is supplied to each transistor through the center-tapped secondary of the input transformer. The entire circuit is very broad-banded (covers a large frequency range) and requires no tuning.

Time for a short review. Test your knowledge by answering the questions of Self-Check 17-5 before continuing.

SELF-CHECK 17-5

1. What is the difference between an open-core and a shell-type power transformer?

2. Briefly describe an autotransformer.

3. Describe the impedance-matching transformer needed to match a 250 Ω microphone to an amplifier that has a 5 kΩ input impedance.

4. What is an IF transformer? (describe its purpose or use)

17-6 Troubleshooting Transformers

Using the Ohmmeter

The ohmmeter can be used to troubleshoot transformers much as it can be used to troubleshoot an inductor. After all, a transformer is one or more inductors on the same core. These troubleshooting procedures are carefully examined in Chapter 12 (Section 12-6). We will summarize and review the ohmmeter troubleshooting procedure here.

TROUBLESHOOTING PROCEDURE

1. Remove power and isolate the winding from the circuit. To do this, disconnect at least one lead per winding from the circuit. If the winding to be tested has more than two leads (taps), all leads should be disconnected from the circuit.

2. Measure the winding resistance with the ohmmeter. An open indication, or extremely high reading, obviously means the winding has burned open. A very low resistance reading means the winding is either good or partially shorted. The actual winding resistance will depend on the number of turns and wire gauge of the winding. A high-voltage winding will have a larger resistance than a low-voltage, high-current winding. A low-voltage, high-current winding may appear as a short, while a high-voltage, low-current winding may have 10 Ω, 50 Ω, 70 Ω, or 100 Ω or so of wire resistance.

It is important for you to realize that the winding *resistance* and the primary or secondary *impedance* are two different things. You cannot measure the primary or secondary impedance with an ohmmeter. Winding impedance is an AC opposition. Resistance, however, is an opposition that exists for both DC and AC.

Using the AC and DC Voltmeter

Figure 17-33 illustrates how a DC and AC voltmeter can be used to troubleshoot a power transformer. In this case, measurements are made with power on. It is very important that you clearly understand what you want to do and how you are going to accomplish it before you start probing around. You must be aware of key test points and what voltages to expect at each point (AC or DC voltage and normal values). Also, make sure you are aware of any high-voltage points and take your measurements carefully. It is good practice to clip one test lead to a common ground and use only one hand to make all measurements. Keep the other hand away from the equipment to avoid accidentally completing a high-voltage circuit with your body. Finally, maneuver the test probe with a steady hand. Don't hurry and let the probe slip, shorting neighboring terminals or foils. A careless technician often further damages a circuit in attempting to repair it. Take your time and be deliberate in your actions.

Figure 17-33 illustrates how the DC voltmeter is used to verify the presence of a DC voltage at the output of the power supply and at the output of the diode-bridge circuit. If the primary or secondary winding of the transformer is open, there will be no DC voltage

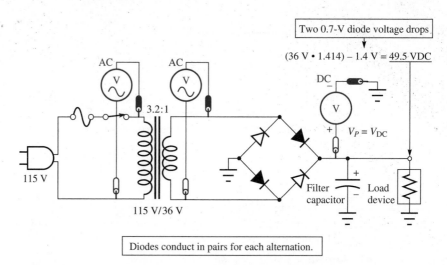

Figure 17-33 Using a voltmeter to troubleshoot a transformer circuit. Take care when measuring lethal voltages.

present at the supply output or the diode bridge. If a DC voltage is present but it is lower than it should be, it is possible that some of the turns of the secondary winding are shorted. It is also possible that there is a problem with the diode-bridge rectifier or filter capacitor. However, we will not involve ourselves with diode or filter-capacitor troubleshooting at this time. If a DC voltage is present but is higher than it should be, it is possible that some of the primary turns are shorted together. If this is the case, it is very likely that the transformer will get very warm, or even hot.

Once the transformer is suspected of having trouble, the AC voltmeter can be used for further diagnosis. The AC voltmeter should be placed across the transformer's primary leads to verify that a line voltage is present (be careful). You don't want to replace a transformer if a switch or fuse is bad. If the line voltage is in fact present at the primary, the secondary can then be measured. If no AC voltage is present at the secondary, the secondary or the primary is open. If the voltage is lower than normal, some of the secondary turns may be shorted (this increases the primary to secondary turns ratio). The following is a brief summary:

1. Use the AC voltmeter to confirm the presence of line voltage at primary leads (caution).

2. Use the AC voltmeter to measure the secondary (caution).

   - no or very low voltage = open secondary or open primary
   - lower than normal voltage = possible shorted secondary turns
   - higher than normal voltage = possible shorted primary turns (or a higher than normal line voltage)

3. Use the ohmmeter with power off to further identify the problem.

   - infinite or very high reading across winding = winding open
   - lower than normal winding resistance = some turns shorted (this may be difficult to determine, since the normal winding resistance may be very low)

Now, take a few minutes to test your troubleshooting skill by answering the questions of Self-Check 17-6.

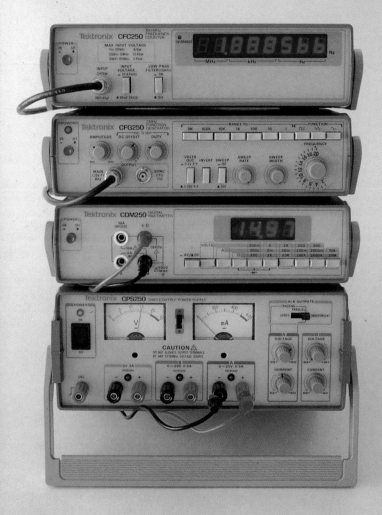

Tektronix offers the CXX250 series of test instruments—a set of match
ckable test instruments for the electronics laboratory. From top to bottom:
C250 Frequency Counter, the CFG250 Sweep Function Generator,
M250 Digital Multimeter, and the CPS250 Triple Power Supply. (equip

Troubleshooting an Inductor

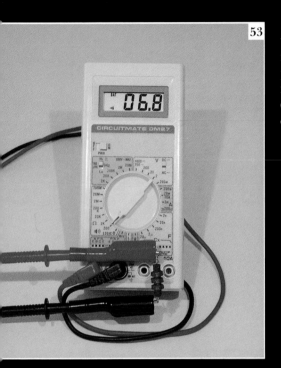

53

53) An ohmmeter (DMM) can be used to check an inductor to see if it is open. The inductor in this photograph is a 1 mH choke having a winding resistance of 6.8 Ω. (Section 12-6) (equipment courtesy of Beckman Industrial)

Troubleshooting a Film Capacitor

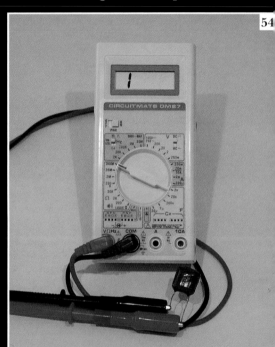

54

Measuring Capacitance

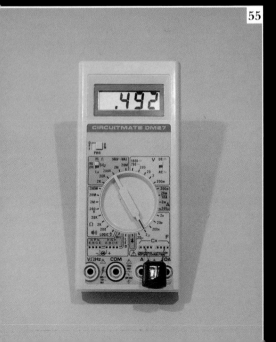

55

54) The ohmmeter can be used to test a capacitor to see if it is shorted. Here, a film capacitor checks as not being shorted. The meter indicates a resistance out of range on the 2000 MΩ range. This is typical for good film capacitors. (Section 13-8) (equipment courtesy of Beckman Industrial)

55) Here, the DM27 is used to measure the actual capacitance of this film capacitor. The reading is 0.492 μF. This 10% tolerance capacitor is marked as 0.47 μF. (Section 13-8) (equipment courtesy of Beckman Industrial)

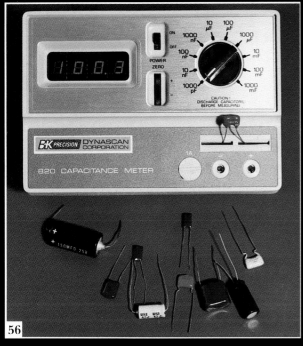

56

56) Here, a 100 pF capacitor is being tested on the B&K capacitance meter. Meters such as this are very useful to determine tolerance and value of unknown capacitors. Test leads are used for physically large capacitors, or capacitors that do not have wire leads. (Section 13-8) (test instruments courtesy of B&K PRECISION/Dynascan Corporation)

57) An LCR bridge such as this Leader Instruments bridge can be used to determine unknown values of capacitance. The capacitor is connected to the test terminals and the C function and range are chosen. The sensitivity control is set for a right scale meter reading of between 5 and 10. The D.Q dial and the R.C.L. dial are alternately adjusted to obtain maximum null on the meter. The actual value of capacitance is taken from the R.C.L. dial times the range multiplier. The dissipation factor can be read from the large D.Q dial. The capacitor being measured is marked as 0.0047 μF = 470 pF. The R.C.L. dial indicates 468 pF. (Sections 13-5 & 13-8) (test instruments courtesy of Leader Instruments Corp., Hauppauge, NY)

Checking Capacitance with an LCR Bridge

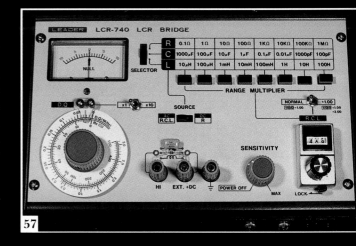

57

Checking Inductance with the LCR Bridge

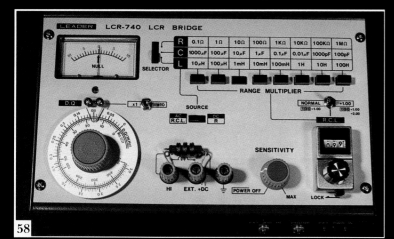

58

58) Inductance and quality factor are determined in similar manner as capacitance is determined using the LCR bridge. Here, a 1 mH inductor is tested and found to be nearly exactly 1 mH ($0.99 \cdot 1$ mH $= 0.99$ mH). The inductor Q can be obtained from the D.Q dial. (Sections 12-3 & 12-6) (test instruments courtesy of Leader Instruments Corp., Hauppauge, NY)

Short Time Constant Integrator

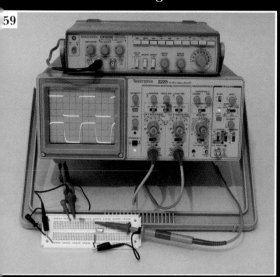

59) Channel 1 (upper trace) shows the applied 10 kHz square wave (5 horizontal divisions · 20 μs per division). Channel 2 shows the short time constant integrated waveform developed across the capacitor. R = 1 kΩ, C = 2500 pF, 5TC = 5 · 1 kΩ · 2500 pF = 12.5 μs time allowed = 50 μs = time for half cycle (Section 14-4) (test instruments courtesy of Tektronix, Inc.)

Medium Time Constant Integrator

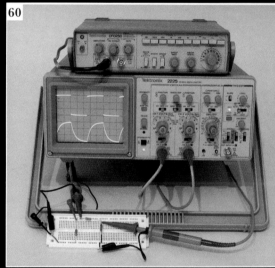

60) Channel 1 (upper trace) shows the applied 10 kHz square wave. Channel 2 shows the medium time constant integrated waveform developed across the capacitor. R = 1 kΩ, C = 0.01 μF, 5TC = 5 · 1 kΩ · 0.01 μF = 50 μs time allowed = 50 μs = time for half cycle (Section 14-4) (test instruments courtesy of Tektronix, Inc.)

Long Time Constant Integrator

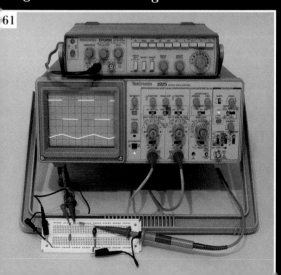

61) Channel 1 (upper trace) shows the applied 10 kHz square wave. Channel 2 shows the long time constant integrated waveform developed across the capacitor. R = 1 kΩ, C = 0.1 μF, 5TC = 5 · 1 kΩ · 0.1 μF = 500 μs time allowed = 50 μs = time for half cycle (Section 14-4) (test instruments courtesy of Tektronix, Inc.)

Short Time Constant Differentiator

62) Channel 1 (upper trace) shows the applied 10 kHz square wave (5 horizontal divisions · 20 µs per division). Channel 2 shows the short time constant differentiated waveform developed across the resistor.

R = 1 kΩ, C = 2500 pF, 5TC = 5 · 1 kΩ · 2500 pF = 12.5 µs

time allowed = 50 µs = time for half cycle (Section 14-4)

(test instruments courtesy of Tektronix, Inc.)

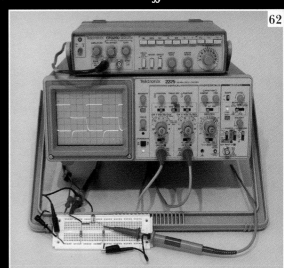

Medium Time Constant Differentiator

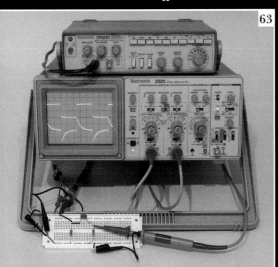

63) Channel 1 (upper trace) shows the applied 10 kHz square wave. Channel 2 shows the medium time constant differentiated waveform developed across the resistor.

R = 1 kΩ, C = 0.01 µF, 5TC = 5 · 1 kΩ · 0.01 µF = 50 µs

time allowed = 50 µs = time for half cycle (Section 14-4)

(test instruments courtesy of Tektronix, Inc.)

Long Time Constant Differentiator

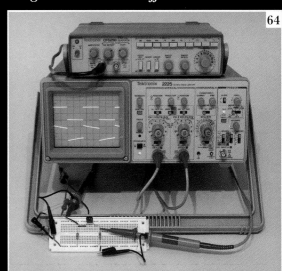

64) Channel 1 (upper trace) shows the applied 10 kHz square wave. Channel 2 shows the long time constant differentiated waveform developed across the resistor.

R = 1 kΩ, C = 0.1 µF, 5TC = 5 · 1 kΩ · 0.1 µF = 500 µs

time allowed = 50 µs = time for half cycle (Section 14-4)

(test instruments courtesy of Tektronix, Inc.)

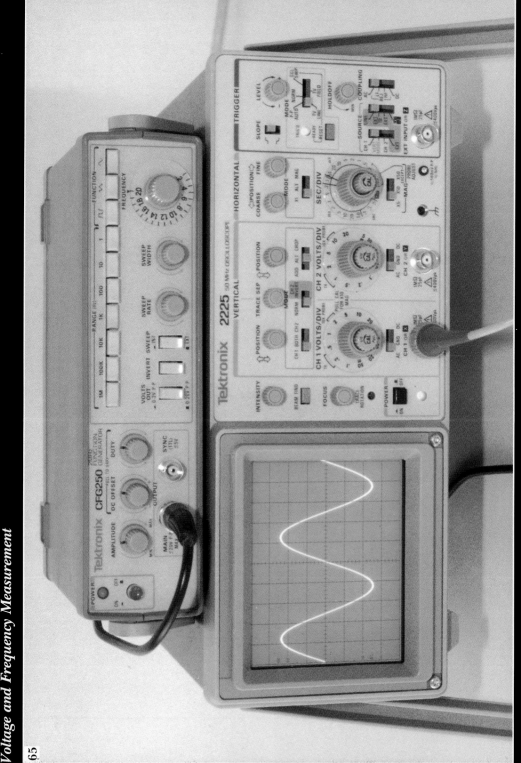

65) The peak-to-peak voltage of this sine wave is found by multiplying the number of vertical divisions from positive to negative peaks by the volts/div control setting— V_{P-P} = 5 div . 2 V/div = 10 V_{P-P}. (Note: a 10X probe is being used so the 2 is used on the volts/div dial.) The frequency of the sine wave is the reciprocal of cycle time f = 1T. The time for one cycle here is 5 horizontal divisions times 0.1 ms/div =

In Phase Transformer Windings

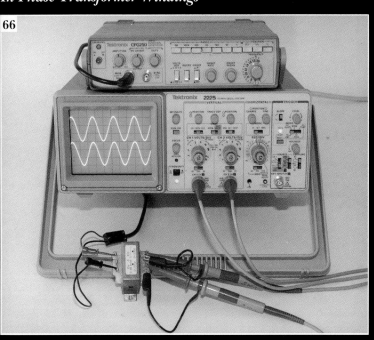

66) This photo demonstrates the in phase voltages of two identical transformer windings (winding terminals marked as 0-100V). Note the black sense dot above each 100V marking. A test probe is placed at each 100V terminal with its corresponding ground clip on the 0 terminal. (Section 17-5) (test instruments courtesy of Tektronix, Inc.)

180° Out of Phase Transformer Windings

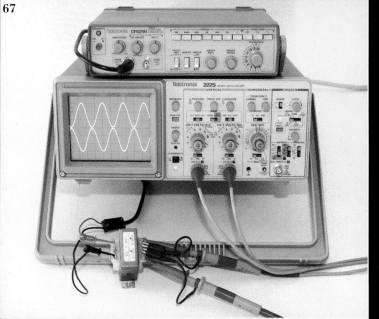

67) In this photo, the upper test probe is switched with its ground clip to obtain a 180° phase shift with the other winding. When dual phase AC is needed, the outer terminals or the two inner terminals can be connected to a common ground. (Section 17-5) (test instruments courtesy of Tektronix, Inc.)

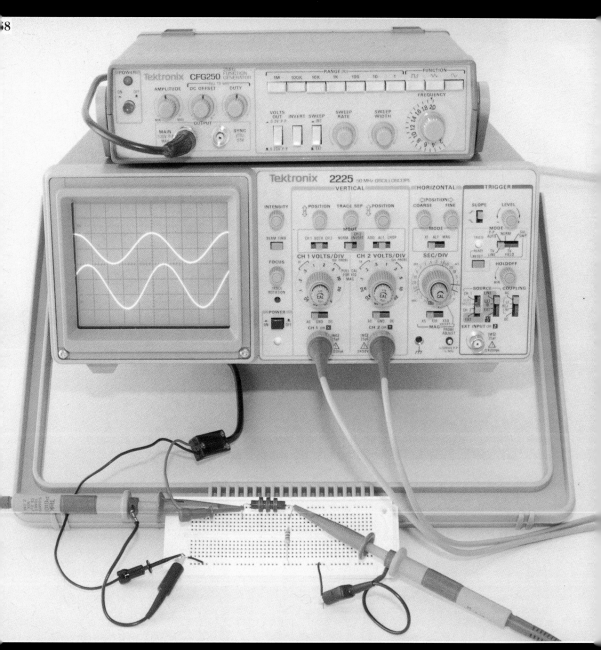

68) This photograph demonstrates that resistor voltage lags the source voltage for an RL circuit. The source voltage is the upper trace and the resistor voltage is the lower trace. Notice that the source voltage reaches its positive peak first (left side of screen). The actual phase angle for V_R is determined as follows: 1) Determine the horizontal displacement between positive peaks of the two signals. In this case it is about 0.7 horizontal divisions. 2) Determine the number of horizontal divisions for one cycle. Here, each cycle is about 5 horizontal divisions. 3) Phase Angle $= 360° \cdot 0.7/5 = 50.4°$

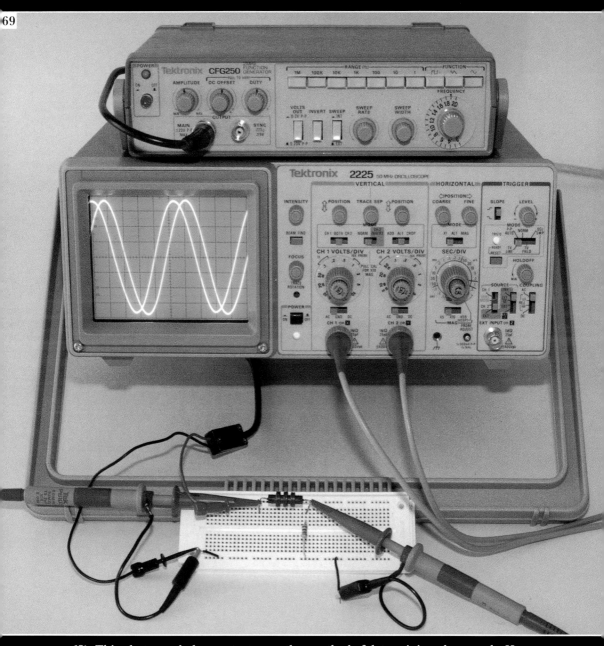

69) This photograph demonstrates another method of determining phase angle. Here, the horizontal displacement of the two signals is measured on the horizontal center line at zero crossing. The displacement is about 0.8 divisions and the number of divisions for one cycle is about 5.4. Phase angle = 360° · 0.8/5.4 = 53.3°. (Section 15-6) (equipment courtesy of Tektronix, Inc.)

"Q" Rise in Voltage

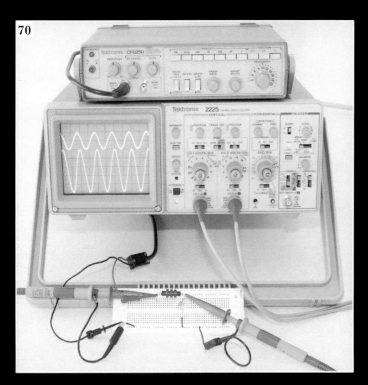

70) A "Q rise in voltage" can be seen across each reactive component of a series resonant circuit. The upper trace is the input voltage and the lower trace is the voltage across the capacitor. The input voltage is 2 V_{P-P} and the output voltage is 100 V_{P-P}. This indicates that the circuit Q is 100/2 = 50. (Section 20-3) (equipment courtesy of Tektronix, Inc.)

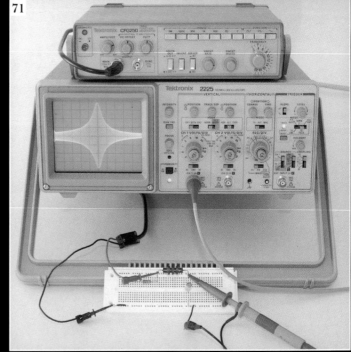

71) In this photograph, the function generator is sweeping through a wide range of frequencies while the capacitor voltage is being observed with the oscilloscope. At the resonant frequency, the capacitor voltage is very high, again demonstrating the "Q rise in voltage" at resonance. (Section 20-3) (equipment courtesy of Tektronix, Inc.)

72) This photo demonstrates the frequency response of a highpass filter. The output is taken across the inductor and load resistor. The function generator is swept internally over a wide range starting from 20 kHz. The resulting display resembles a typical highpass bode plot. (Section 21-2) (test instruments courtesy of Tektronix, Inc.)

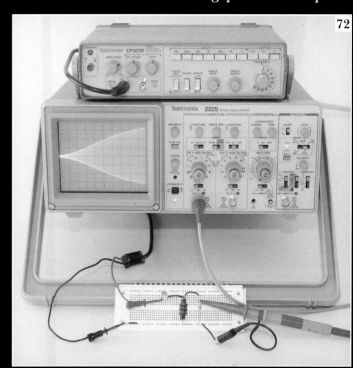

Lowpass Filter Response

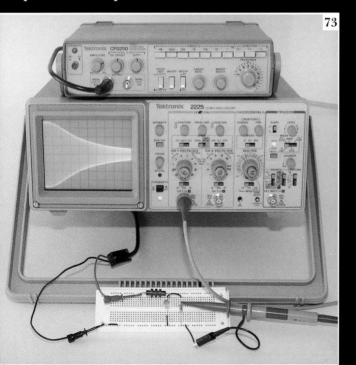

73) This photo demonstrates the frequency response of a lowpass filter. The output is taken across the capacitor and load resistor. The resulting display resembles a typical lowpass bode plot. This test setup can be used to tailor the filter to a desired frequency response. Values of capacitance and load resistance can be changed to modify the response. (Section 21-3) (test instruments courtesy of Tektronix, Inc.)

74) This photo demonstrates the frequency response of a bandpass filter. A series resonant circuit is placed in the inverted "L" configuration with the output taken across the load resistor. The resulting display resembles a typical bandpass bode plot. Relative frequency characteristics, such as center frequency and bandwidth, can be modified and observed by changing values of L, C, and R. (Section 21-4) (test instruments courtesy of Tektronix, Inc.)

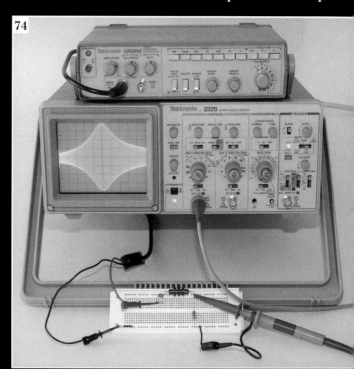

Bandstop Filter Response

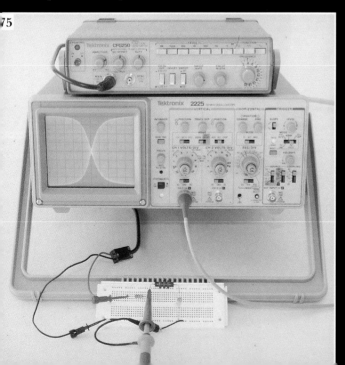

75) This photo demonstrates the frequency response of a bandstop filter (bandreject filter, notch filter, trap). The output is taken across the series resonant circuit (L&C) whose impedance is very low at resonance. The resulting oscilloscope display is a typical bandreject bode plot. Center frequency, bandstop bandwidth, and notch depth can be modified and observed by changing values of L, C, and R (Section 21-5) (test instruments courtesy of Tektronix, Inc.)

1. What reading should you expect when measuring an open winding with an ohmmeter?

2. If the AC voltage across a secondary winding is measured and found to be lower than normal, what might the trouble be? (Assume a normal line voltage across the primary winding.)

3. What are two possible troubles that would cause there to be no voltage across the secondary winding? (Assume a normal line voltage across the primary winding.)

4. Describe how you would troubleshoot further to pinpoint the actual trouble for Question 3.

Summary

FORMULAS

$$(17.1) \quad k = \frac{\text{amount of flux between } L_1 \text{ and } L_2}{\text{total flux produced by } L_1 \text{ or } L_2} = \frac{\Phi_M}{\Phi_t}$$

$(17.2) \quad L_M = k \cdot \sqrt{L_1 \cdot L_2}$

$(17.3) \quad L_t = L_1 + L_2 + 2L_M \qquad \text{(fields aiding)}$

$(17.4) \quad L_t = L_1 + L_2 - 2L_M \qquad \text{(fields opposing)}$

$(17.5) \quad L_M = (L_t - L_1 - L_2)/2$

$(17.6) \quad k = L_M/\sqrt{L_1 \cdot L_2}$

$(17.7) \quad v_S = N_S \cdot \Delta\Phi_m/\Delta t \qquad \text{(instantaneous secondary voltage)}$

$(17.8) \quad v_P = N_P \cdot \Delta\Phi_m/\Delta t \qquad \text{(instantaneous primary voltage)}$

$$(17.9) \quad V_S = \frac{V_P \cdot N_S}{N_P}$$

$$(17.10) \quad V_P = \frac{V_S \cdot N_P}{N_S}$$

$$(17.11) \quad I_S = I_P \cdot \frac{N_P}{N_S}$$

$$(17.12) \quad I_P = I_S \cdot \frac{N_S}{N_P}$$

$(17.13) \quad Z = V/I$

$(17.14) \quad Z_P/Z_S = (N_P/N_S)^2$

$(17.15) \quad Z_P = Z_S \cdot (N_P/N_S)^2$

$(17.16) \quad N_P/N_S = \sqrt{Z_P/Z_S}$

CONCEPTS

- Two inductors have a mutual inductance of 1 H when a rate of change in current of 1 A/s in one inductor induces a voltage of 1 V across the second inductor.
- The coefficient of coupling (k) is the ratio of the amount of mutual flux to the total flux.
- Transformers are described as step-up or step-down in accordance with the primary to secondary voltage ratio (step-up $= V_P/V_S < 1$, where $V_S > V_P$, and step-down $= V_P/V_S > 1$, where $V_P > V_s$)
- Power in the primary is equal to power in the secondary ($k = 1$), assuming 100% power efficiency.
- Primary and secondary impedances cannot be measured with an ohmmeter. They are AC impedances.
- Transformer ratings of voltage, current, power, and frequency cannot be violated without excessive transformer heating and possible burnout.
- The winding I^2R loss is due to winding resistance.
- Laminated iron cores and ferrite cores greatly reduce eddy-current loss.
- Hysteresis loss is due to atomic stress and molecular friction caused by continually reversing reluctant magnetic domains in the core.

PROCEDURES

Ohmmeter Transformer Troubleshooting Procedure

1. Remove power and isolate the winding from the circuit.
2. Measure the winding resistance with the ohmmeter. An open indication, or extremely high reading, obviously means the winding has burned open. A very low resistance reading means the winding is either good or partially shorted.

Voltmeter/Ohmmeter Transformer Troubleshooting Procedure

1. Use AC voltmeter to confirm presence of line voltage at primary leads (caution).
2. Use AC voltmeter to measure the secondary under normal operating conditions with power on (caution).

   - no or very low voltage = open secondary or open primary
   - lower than normal voltage = possible shorted secondary
   - higher than normal voltage = possible shorted primary

3. Use the ohmmeter with power off to further identify the problem.

   - infinite or very high reading = winding open
   - lower than normal winding resistance = some turns shorted (this may be difficult to determine, since the normal winding resistance may be very low)

SPECIAL TERMS

- Mutual inductance (L_M)
- Coefficient of coupling (k)
- Transformer, boost, bucking, sense dot
- Primary winding, secondary winding

- Transformation ratios
- Impedance (Z), reflected impedance
- Eddy current, laminations
- Laminated iron core
- Hysteresis loss
- Shell-type transformer, open-core-type transformer
- Autotransformer, Variac®
- Isolation transformer
- Impedance-matching transformer
- IF transformer

Need to Know Solution

A transformer's ratings must be known in order to determine its suitability for a particular application. The transformer secondary must be rated at the needed voltage and have a maximum current rating that is greater than the amount of current demanded by the load. In any case, the transformer's power rating must not be exceeded. The transformer must also be operated at the rated frequency or poor performance and/or overheating will occur. For the application presented at the beginning of this chapter, we will need a 115 V to 12.6 V step-down transformer whose secondary is rated for at least 2 A continuous current. A rating of 3 or 4 A would be better to provide a safe operating margin.

Questions

17-1 Mutual Inductance (L_M)

1. Define mutual inductance.
2. How does mutual inductance differ from self-inductance?
3. What does it mean for two inductors to have a mutual inductance of 1 H?
4. What does it mean for two inductors to have a coefficient of coupling of 0.5?
5. What do sense dots indicate?

17-2 Voltages and Current Transformation

6. What determines whether a transformer winding is called a primary winding or a secondary winding?
7. What is a step-up transformer?

17-3 Impedance Transformation

8. What does *impedance* mean and what is its unit of measure?
9. What is reflected impedance?
10. How is a transformer's turns ratio related to its impedance ratio?
11. Express the impedance transformation ratio in terms of voltage, turns, and current.
12. Briefly explain how a change in current in the secondary winding of a transformer can cause a change in current in the primary winding.
13. Briefly explain how a fuse in the primary circuit can protect the entire transformer from severe damage due to a short in the secondary circuit.

17-4 Transformer Ratings and Losses

14. Explain what the following transformer label indicates: VOLTS: 115/24 CT.
15. A transformer is labeled as follows: VOLTS: 230/12.6. Can the 12.6 V winding be used as the primary? Explain.
16. An audio transformer is rated as follows: Z: 600/2,500, 10 mW. Explain what this rating indicates.
17. What does it mean to say a power transformer has an efficiency of 90%?
18. List three losses common to transformers.
19. Briefly explain the eddy-current loss.
20. How is the hysteresis loss minimized in transformers?

17-5 Transformer Types and Applications

21. What is the purpose of the iron-shell housing on shell-type transformers?
22. What is an autotransformer?
23. Describe the use and operation of a Variac®.
24. What is an isolation transformer? (its purpose or use)
25. Explain the purposes for using impedance-matching transformers to match low-Z microphones to high-Z amplifiers.
26. What is an IF transformer?

17-6 Troubleshooting Transformers

27. When using an ohmmeter to test a low-voltage, high-current transformer winding, what reading should you expect for a good winding?
28. When using an ohmmeter to test a high-voltage, low-current transformer winding, what reading should you expect for a good winding?
29. If an ohmmeter is used to measure the windings of a 10k/1k impedance-matching transformer, will the ohmmeter indicate 10 kΩ for one winding and 1 kΩ for the other? Explain.
30. If the primary voltage is normal and the secondary voltage is lower than it should be, what might be the trouble?

Problems

17-1 Mutual Inductance (L_M)

1. If the total flux generated by one inductor is 120 μWb and 70 μWb links a neighboring inductor, what is the coefficient of coupling?
2. If 200 μWb of flux links two inductors and the coefficient of coupling is 0.4, what is the total generated flux?
3. If a 10 mH and 20 mH inductor have a k value of 0.6, what is their mutual inductance?
4. If a 45 μH and a 30 μH inductor have a k value of 0.4, what is their mutual inductance?

5. Determine the total inductance of the following two *series-aiding* inductors: $L_1 = 0.59$ mH; $L_2 = 0.3$ mH; $k = 0.25$

6. Determine the total inductance of the following two *series-opposing* inductors: $L_1 = 0.59$ mH; $L_2 = 0.3$ mH; $k = 0.25$

7. The inductance of each of two series-aiding inductors is measured and found to be 150 μH and 325 μH. The total inductance is found to be 580 μH. Calculate the mutual inductance and the coefficient of coupling.

17-2 Voltage and Current Transformation

8. Determine the instantaneous voltage across a 150-turn secondary winding when the mutual flux is changing at the rate of 66.7 mWb/s.

9. If a transformer has 450 turns in the primary winding and 90 turns in the secondary winding, and 115 V is applied to the primary, what is the secondary voltage?

10. If a transformer has 300 turns in the primary winding and 60 turns in the secondary winding, and the secondary voltage is 30 V, what is the primary voltage?

11. If a transformer has 120 V applied to the primary winding and the resulting secondary voltage is 24 V, what is the *current* transformation ratio (primary to secondary)?

12. If a transformer has 250 turns in the primary winding and 650 turns in the secondary winding, what is the *current* transformation ratio (primary to secondary)?

13. If a certain transformer has 200 turns in the primary winding and 1,400 turns in the secondary winding, and the secondary current is 100 mA, what is the primary current?

14. If a certain transformer has 200 turns in the primary winding and 40 turns in the secondary winding, and the secondary current is 10 A, what is the primary current?

15. If a certain transformer has 325 turns in the primary winding and 140 turns in the secondary winding, and the primary current is 100 mA, what is the secondary current?

16. Determine the primary and secondary power for the transformer of Problem 15. The secondary load resistance is 213 Ω.

17-3 Impedance Transformation

17. If the turns ratio for a given transformer is 3:2 (3/2), what is the transformer's impedance ratio?

18. If the voltage ratio for a given transformer is 120 V/20 V (6:1 or 6/1), what is the transformer's impedance ratio?

19. If a transformer's primary to secondary current ratio is 5 A/600 mA (8.33:1 or 8.33/1), what is the transformer's impedance ratio?

20. If a certain transformer has a primary to secondary turns ratio of 4:1 and the secondary load impedance is 1 kΩ, what is the impedance reflected to the primary?

21. If a certain transformer has a primary to secondary turns ratio of 3.5:1 and the secondary load impedance is 600 Ω, what is the impedance reflected to the primary?

22. If a certain transformer has a primary to secondary turns ratio of 5:2 and the impedance reflected to the primary is 1 kΩ, what is the load impedance of the secondary?

23. If a certain transformer has a primary to secondary turns ratio of 4:9 and the impedance reflected to the primary is 600 Ω, what is the load impedance of the secondary?

17-4 Transformer Ratings and Losses

24. A transformer is labeled as follows: VOLTS: 115/24 @ 3A. What is the maximum current that can be permitted to flow in the primary winding?
25. A transformer is labeled as follows: VOLTS: 110/450 @ 100mA/6.3 @ 4 A. If the maximum current is demanded from each secondary winding, what is the maximum power delivered to the primary winding from the AC source?
26. What size fuse (current rating) would you place in the primary circuit of the transformer of Problem 25?
27. Explain why power transformers are rated in apparent-power units of volt-amperes (VA) instead of watts.
28. If a transformer is rated at 350 VA and the primary voltage is 115 V, what is the maximum current that can be permitted to flow in the primary winding?

Answers to Self-Checks

Self-Check 17-1

1. Two inductors have a mutual inductance of 1 H when a rate of change in current of 1 A/s in one inductor induces a voltage of 1 V across the second inductor.
2. $k = 0.4$
3. Use a shield; mount coils at right angles to each other; separate coils as far as possible; use toroidal cores for total flux containment.
4. 3.67 μH
5. 554.6 μH
6. 2.75 mH and $k = 0.306$

Self-Check 17-2

1. 288 V; step-up
2. 0.87 A
3. 90 V and 1.67 A
4. $I_{S1} = 86.3$ mA; $P_{S1} = 29.8$ W; $I_{S2} = 3.15$ A; $P_{S2} = 19.8$ W; $I_P = 0.431$ A; $P_P = 49.6$ W

Self-Check 17-3

1. The opposition a component or circuit offers to AC (usually includes resistance and reactance)
2. The impedance ratio is the square of the turns ratio.
3. 4:25 or $1:6.25 = 0.16$
4. 73.5 Ω
5. The short in the secondary circuit demands a large amount of current. The large amount of current develops a large quantity of flux, which opposes the primary flux. This reduces the CEMF of the primary. The primary current then increases to restore the CEMF to its original level (nearly equal to the applied source voltage). The excessive current burns the primary winding open.

Self-Check 17-4

1. One transformer winding is rated at 120 V and the second winding is rated at 24 V. The 24 V winding is center tapped.
2. 25.2 VA
3. 219 mA
4. 91.2 VA
5. The 400 Hz transformer is very small and will have a much lower reactance at 60 Hz. As a result, winding current will be much higher than normal.
6. Winding resistance and core eddy currents.

Self-Check 17-5

1. The shell-type transformer is fully shielded to prevent flux leakage.
2. An autotransformer is a single winding transformer used to step voltage up or down.
3. The matching transformer must have a turns ratio of 1:4.47. This might mean there are 100 turns on the low-Z side and 447 turns on the high-Z side.
4. The IF transformer is an intermediate-frequency transformer used in radio and television receivers. They are tuned to a particular intermediate frequency.

Self-Check 17-6

1. Very high or infinite
2. Some of the turns of the secondary winding may be shorted.
3. Open primary or open secondary
4. Turn power off; isolate the primary and secondary winding; measure each with the ohmmeter.

SUGGESTED PROJECTS

1. Add some of the main concepts, formulas, and procedures from this chapter to your Electricity and Electronics Notebook.

2. Visit your local library and do a little research into how power transformers are designed and constructed. Try to find out how the size of the core is determined for a specific power rating and frequency rating.

3. Obtain a small power transformer and conduct some simple experiments to confirm some of the things you have learned in this chapter. Measure the DC resistance of the windings with an ohmmeter to see the difference in resistance between high-voltage and low-voltage windings. Apply AC power to the transformer and connect different resistive loads to the secondary. Measure the primary and secondary currents with each different load resistor connected to the secondary. Verify that the primary to secondary current ratio is the inverse of the primary to secondary voltage ratio. Be careful not to exceed the power rating of the transformer by loading the secondary too heavily. Also, be aware of proper resistor power rating. Be very careful when working around high voltage. Make sure the high-voltage connections are well insulated and secured before power is applied.

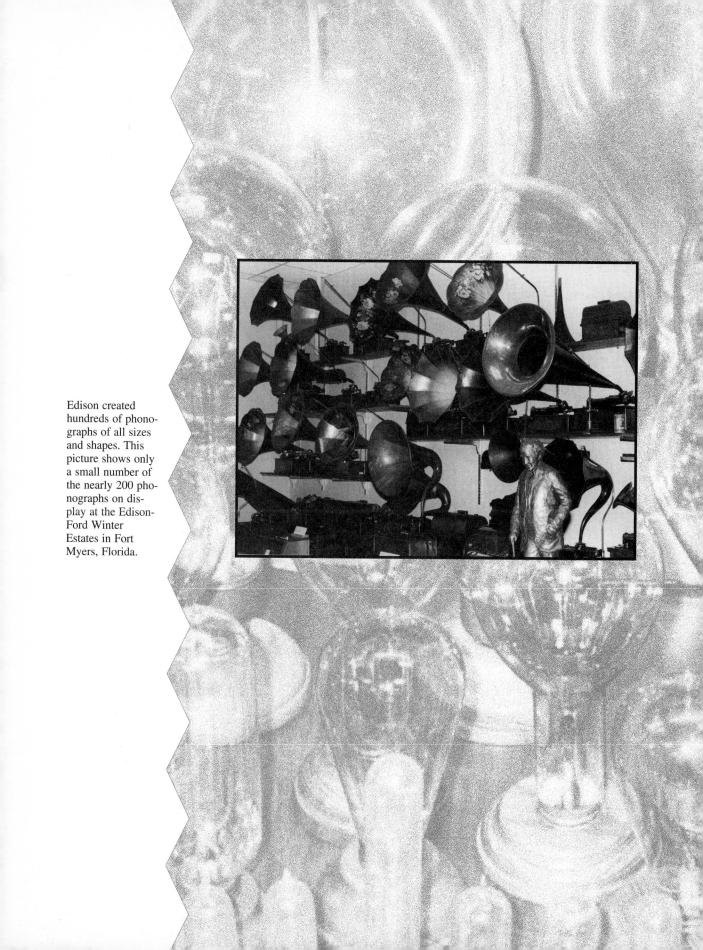

Edison created hundreds of phonographs of all sizes and shapes. This picture shows only a small number of the nearly 200 phonographs on display at the Edison-Ford Winter Estates in Fort Myers, Florida.

Chapter 18

Series *RL*, *RC*, and *RCL* Circuits

OBJECTIVES

After studying this chapter you will be able to

- perform algebraic operations with complex numbers in rectangular and polar form.
- convert complex numbers in rectangular form to complex numbers in polar form and vice versa.
- use basic trigonometric functions and the Pythagorean theorem for right triangles in the analysis of series AC circuits.
- calculate impedance, current, voltages, and power for series *RL*, *RC*, and *RCL* circuits.
- represent series AC circuits with voltage, impedance, and power phasors in phasor diagrams.

- use an *RL* or *RC* circuit as a lead or lag circuit.
- express phase relationships in terms of time.
- understand and calculate AC circuit efficiency.
- use a dual-trace oscilloscope for sine-wave-voltage phase comparison.
- use a vector impedance meter to measure an unknown AC circuit impedance.

INTRODUCTION

What is an AC circuit? How does an AC circuit differ from a DC circuit? Can an AC circuit be analyzed in the same way as a DC circuit? Do Ohm's Law and Kirchoff's Voltage Law apply to series AC circuits? In this chapter, you will discover the answers to these questions and more. You will learn how to use old, familiar circuit-analysis tools in new ways as they are applied in AC circuit analysis. Also, you will be introduced to some new tools and techniques that must be used in AC circuit analysis. The approach in this chapter is to build on and expand upon the knowledge you already have of DC circuit analysis. You will soon discover that AC circuit analysis is really no more difficult than DC circuit analysis.

However, there are new things to learn and special procedures to follow. You will learn about *complex* numbers and the special mathematical operations of *complex algebra* that are used to manipulate them. *Imaginary* numbers will become very real to you as you learn of their importance in simple AC circuit analysis. Happily, you will discover that the so-called "complex" algebra is not complex. Finally, as you learn to analyze series *RL*, *RC*, and *RCL* circuits, you will become keenly aware of the very different and unique personalities of the three main actors in AC circuits: the resistor, the capacitor, and the inductor.

NEED TO KNOW

Let's suppose you construct a simple AC circuit consisting of a resistor, a capacitor, and an inductor all placed in series. You connect your bench signal generator to the circuit as an AC voltage source, setting it to 10 V_{p-p} as viewed on your oscilloscope. Then you use the oscilloscope to measure the voltage across each component. You measure the following voltages: $V_R = 8.66$ V_{p-p}; $V_C = 13$ V_{p-p}; and $V_L = 18$ V_{p-p}. Immediately, you notice that the capacitor voltage and the inductor voltage are both greater than the applied source voltage (10 V_{p-p}). It is also obvious that the sum of the three series voltages is much greater than the applied source voltage (8.66 + 13 + 18 = 39.66). This is something you had never seen in series DC circuits where the sum of all series voltages equals the applied voltage. There must be a reason for these unusual voltages. Or, are they really unusual for an AC circuit? That is a very important question, the answer to which you need to know.

18-1 Rectangular and Polar Notation

The Rectangular System

Special Notation

As we continue in the study of AC circuits and circuit analysis, it is necessary for you to have a fundamental knowledge of **complex algebra**. The numbers that you will work with in AC analysis involve **complex numbers** (terms that include resistance and reactance, "real" and "imaginary" numbers). An AC circuit impedance may be expressed as a complex number that includes resistance and inductive reactance, or capacitive reactance (R and X_L or R and X_C). Such an impedance is referred to as a **complex impedance**. For example, the impedance of a particular circuit might be 100 Ω of resistance plus 240 Ω

of inductive reactance. These two values *cannot* simply be added together for a total of 340 Ω, because the reactance is not in phase, or in time, with the resistance. There is a fixed and definite time delay (phase shift) between the two values. Thus, the total impedance of the two must be expressed in a way that takes into account this time, or phase, difference.

There are two forms of notation that you will learn to use to express complex impedances: rectangular notation and polar notation. In **rectangular notation**, the aforementioned impedance is expressed as $100 + j240$ or $100\ \Omega + j240\ \Omega$ or $100 + j240\ \Omega$. In **polar notation**, the same impedance is expressed as $260\ \angle 67.4°\ \Omega$ ($260\ \Omega$ at an angle of 67.4°). Both forms of the impedance are complex in that they represent resistive and reactive (real and imaginary) values that are added phasorally (using phasors).

The Rectangular Coordinate System

Let's begin with the **rectangular coordinate system** (Cartesian coordinate system). You may already be somewhat familiar with this system from previous math courses. The basic version of the rectangular system is often referred to as the X, Y coordinate system since it consists of two perpendicular axes: the horizontal X axis and the vertical Y axis. Figure 18-1 illustrates this system. As you can see, it is a simple two-dimensional, or planar, coordinate system. The point of intersection for the two axes is known as the **origin** (the point at which all numbers originate). Starting at the origin and moving to the right on the X axis, numbers increase in a positive direction. Moving to the left on the X axis from the origin, numbers increase in the negative direction. Starting again at the origin and moving upward on the Y axis, numbers increase in the positive direction; moving downward from the origin, numbers increase in the negative direction. Any point within the limits of the plane can be located and plotted as long as X and Y values are given. The X axis is known as the **abscissa** and the Y axis is the **ordinate**. Any point of interest is specified by an abscissa value first, then an ordinate value; the two values are separated by a comma and contained in parentheses: (X, Y).

The rectangular coordinate system is used in many areas of math and science. Its basic purpose is to graphically demonstrate the result of interacting magnitudes and directions. For example, a mathematical formula can be plotted to demonstrate the effects on the Y value that results with changes in the X value. As the X value changes in magnitude and direction (sign $\pm$), the Y value also changes as defined by the formula.

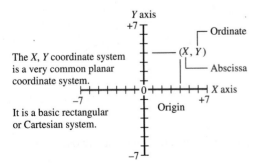

The X, Y coordinate system is a very common planar coordinate system.

It is a basic rectangular or Cartesian system.

Figure 18-1   The X,Y coordinate system.

EXAMPLE   18-1

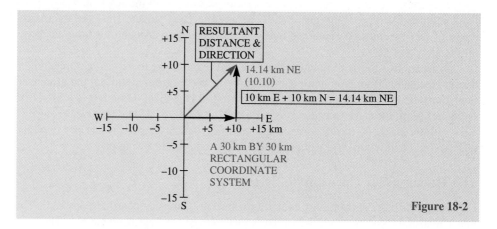

Figure 18-2

Let's consider a simple application for the *X, Y* coordinate system. Example 18-1 illustrates the use of the *X, Y* coordinate system in determining the resulting direction and distance you have traveled from a particular point. In this case, you start at a particular location, called your *point of origin*, and proceed east a distance of 10 km. Then, you head north a distance of 10 km. The result of your trek, 10 km east and 10 km north, is a straight-line distance of 14.14 km in a northeasterly direction (10 km E + 10 km N = 14.14 km NE). The same result would be achieved if you had gone north first, then east. You would arrive at the same point, the same direction and distance from the origin.

The Pythagorean Theorem

Figure 18-3 illustrates how the Pythagorean theorem is used to determine the resultant distance from the point of origin. Notice that the resultant is actually the hypotenuse of a right triangle formed by the two interacting vectors (rays or arrows that indicate magnitude and direction). In Example 18-1, the two vectors are 10 km E and 10 km N.

Pythagorean Theorem

IN A RIGHT TRIANGLE, THE SUM OF THE SQUARES OF THE TWO SIDES ADJACENT TO THE RIGHT ANGLE IS EQUAL TO THE SQUARE OF THE SIDE OPPOSITE THE RIGHT ANGLE ($a^2 + b^2 = c^2$).

Therefore:

$$c = \sqrt{a^2 + b^2}$$

(18.1)

where *c* is the resultant magnitude opposite the right angle and *a* and *b* are two magnitudes adjacent to the right angle.

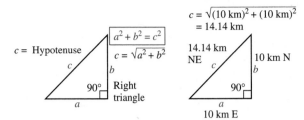

Figure 18-3   The Pythagorean theorem.

Figure 18-3 shows the Pythagorean theorem in equation form. As you can see, the hypotenuse (side c or the resultant) is found by taking the square root of the sum of the squares of the two sides adjacent to the right angle ($c = \sqrt{a^2 + b^2}$). In Example 18-1, the distance from the origin to the destination (resultant) is found by applying the two vector values to the Pythagorean formula (resultant distance $= \sqrt{(10 \text{ km})^2 + (10 \text{ km})^2} = 14.14$ km).

The Pythagorean Theorem in AC Analysis

Now, let's apply the Pythagorean theorem to our opening discussion regarding AC impedance. It was stated that circuit resistance must be added phasorally (using phasor addition). In AC circuit analysis, resistance and reactance are represented as phasors, each having a magnitude and direction related to time (phase). In this case, the AC impedance consists of 100 Ω of resistance and 240 Ω of inductive reactance. Example 18-2 illustrates how these phasors are placed on the rectangular coordinate system. Resistance values are plotted on the positive side of the X axis (in a direction of 0°), and inductive reactance values are plotted on the positive half of the Y axis (in a direction of $+90°$). The right triangle is formed, as shown, by adding the two phasors (the tail of the X_L phasor is placed at the head of the R phasor). The resultant of this phasor addition is the AC impedance Z. As you can see, the resulting impedance is the square root of the sum of $R^2 + X_L^2$. Thus, the resultant impedance Z is (100 Ω R + 240 Ω X_L) = $\sqrt{(100 \text{ Ω})^2 + (240 \text{ Ω})^2} = 260$ Ω.

EXAMPLE 18-2

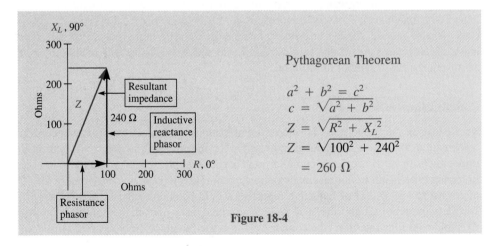

Pythagorean Theorem

$$a^2 + b^2 = c^2$$
$$c = \sqrt{a^2 + b^2}$$
$$Z = \sqrt{R^2 + X_L^2}$$
$$Z = \sqrt{100^2 + 240^2}$$
$$= 260 \text{ Ω}$$

Figure 18-4

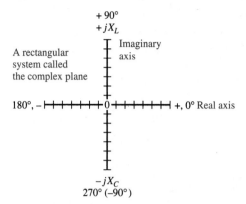

Figure 18-5   The complex plane.

Imaginary and Real Numbers

The Complex Plane

Earlier, it was indicated that a complex number is a mathematical term that consists of a real and imaginary number. But, what are real and imaginary numbers? First of all, do not try to conjure up a definition for real and imaginary numbers on your own. You probably won't even come close to the actual mathematical definition of these two terms. Figure 18-5 illustrates real and imaginary numbers as they relate to each other in a planar system called the **complex plane**. **Real numbers** are all numbers (both positive and negative) that lie on the *X* axis (**real axis**). **Imaginary numbers** are numbers (both positive and negative) that lie on the *Y* axis (**imaginary axis**). In electronics, real numbers are resistance values and imaginary numbers are reactive values (inductive and capacitive reactance, or reactive voltages or currents). Does this mean reactive values are imaginary and do not exist? No! The term *imaginary* is another exotic term from the world of pure mathematical theory. It simply indicates that another separate set of numbers does exist, which interact with "real" numbers in a two-dimensional coordinate system (planar system) called the complex plane.

When the complex plane is used in AC circuit analysis, the imaginary axis is referred to as the *j* axis. In pure or unapplied math, the imaginary axis is labeled with the lowercase letter *i* (for imaginary). To avoid the obvious confusion with electrical current the lowercase *j* is used instead. Since our primary interest does not lie in the pursuit of unapplied math, we will make the change to *j* instead of *i* immediately. Again, this is only to avoid confusing the *i* as a symbol for current.

Complex Impedance

Recall that an AC impedance is often a combination of resistance and reactance. This combination forms a mathematical expression for impedance that is called a **complex impedance**. In other words, a complex impedance is expressed as a complex number. Referring once again to our original example, a complex impedance consisting of 100 Ω of resistance and 240 Ω of inductive reactance would be expressed as 100 + j240. The 100 Ω is the real part and the +j240 Ω is the imaginary part of the complex number. Example 18-3 illustrates how the complex plane is used to graphically represent this complex impedance. Notice that the complex impedance can be expressed in rectangular notation or resolved using the Pythagorean theorem (260 Ω). In rectangular notation, the $+j$ immediately indicates that the reactive part of the complex impedance is inductive. The $+j$ also indicates that the inductive reactance is phase-shifted by $+90$ electrical degrees (as compared to resistance).

EXAMPLE   18-3

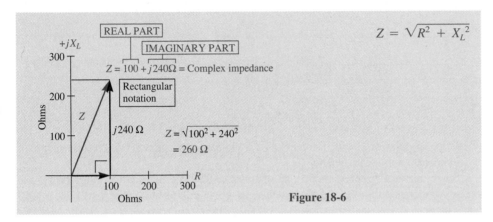

Figure 18-6

j Is an Operator

Any quantity represented by a phasor on the complex plane has magnitude and direction related to time (phase). The phase angle for any phasor is known as the **operator** for the phasor's magnitude. In other words, the phasor *operates* at a certain angle with respect to a reference. The 0° reference is assigned to the right half of the real axis. The operator for quantities of resistance is 0°, since resistance phasors lie on the right half of the real axis. Thus, all positive numbers on the real axis have an operator of 0°. All positive numbers on the imaginary axis have an operator of $+90°$ with respect to the positive half of the real axis. Inductive-reactance phasors are plotted in accordance with the positive half of the imaginary axis. The $+j$ indicates this positive 90° angle in rectangular notation ($100 + j240$). Since the $+90°$ is the operator for the X_L phasor, and $+j$ represents the $+90°$, the j is referred to as **operator j**. If operator j is positive, the angle is $+90°$. If operator j is negative ($-j$), the angle is $-90°$. (Later, you will see that capacitive reactance is represented by a $-j$, meaning $\angle -90°$.)

j and the Complex Plane

Figure 18-7 demonstrates how operator j is related to the complex plane, or rectangular coordinate system. First, notice that all positive real numbers have an operator of 0° and all negative real numbers have an operator of 180°. The 0° operator is implied by the positive sign of the real number. The positive real number can be represented as a $+1$ times the number ($+100 \ \Omega = +1 \cdot 100 \ \Omega$). The 180° operator is implied by the negative sign of the real number (i.e.: $-12 \ \text{V} = -1 \cdot 12 \ \text{V} = 12 \ \text{V}$ with a 180° phase reversal

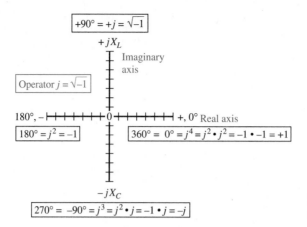

Figure 18-7 Operator j and the complex plane.

or change in polarity). Thus, a factor of $+1$ indicates an operator of $0°$ and a factor of -1 indicates an operator of $180°$.

Since operator j represents $90°$, an operator of $180°$ represents two operator j phase shifts. Mathematically, $180°$ is represented as j^2. It has been shown that $180°$ is also mathematically represented as a factor of -1. Therefore, $j^2 = -1 = 180°$. Now we are at a point to discover that j not only represents $90°$, it is also mathematically equivalent to the square root of negative one ($j = \sqrt{-1}$). Since $j^2 = -1$, j must equal $\sqrt{-1}$. The square root of negative one is imaginary indeed, since there is no real root for a negative number. Thus, all numbers on the imaginary axis are imaginary, since there is no real root for these numbers (e.g., $\sqrt{-49} = \sqrt{-1} \cdot \sqrt{49} = \sqrt{-1} \cdot 7 = j7 =$ an imaginary number).

With each additional counterclockwise rotation of $90°$, operator j's exponent is incremented by 1. At an operator of $270°$ ($270° = \angle -90°$), j becomes negative because $270° = -90° = j^3 = j^2 \cdot j = -1 \cdot j = -j = -1\sqrt{-1}$. At an operator of $360°$, operator j is replaced by a $+1$ because $360° = 0° = j^4 = j^2 \cdot j^2 = -1 \cdot -1 = +1 = j^0$. If counterclockwise rotation is continued in $90°$ increments the cycle of j operations will be repeated again and again ($+j$ @ $\angle +90°$ to -1 @ $\angle 180°$ to $-j$ @ $\angle -90°$ back to $+1$ @ $\angle 0° \ldots$).

The Polar System

Close Cousins

A close cousin to the rectangular system is the **polar system**. Figure 18-8 illustrates this system, which is sometimes referred to as the **polar plane** or **polar chart**. Like the rectangular system, this system also permits the plotting of magnitude and direction (vectors and phasors). The biggest difference between the rectangular system and the polar system is appearance and the manner in which a resultant vector or phasor is expressed. As is shown in the figure, polar charts are used for many applications and are found in a variety of forms. In any case, the polar system is used to graphically express a magnitude and direction.

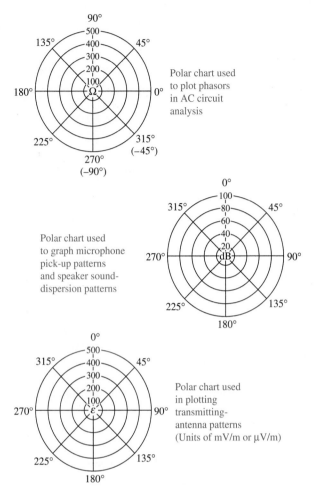

Polar chart used to plot phasors in AC circuit analysis

Polar chart used to graph microphone pick-up patterns and speaker sound-dispersion patterns

Polar chart used in plotting transmitting-antenna patterns (Units of mV/m or µV/m)

Figure 18-8 The polar system.

Example 18-4 shows how an impedance phasor is plotted on a polar chart. To conform with the rectangular coordinate system, this polar chart has its reference at the 3 o'clock position (0° on the right half of the real axis). The angle of the impedance ($\angle\theta$, **angle theta**) is the phase angle between the total impedance and the resistance. In this case, the impedance is shown as 260 Ω at an angle of 67.4°. The impedance expressed as 260 Ω $\angle 67.4°$, or $260\angle 67.4°$ Ω, is polar notation conforming to the polar-chart system.

Polar or Rectangular

Any AC impedance may be expressed in polar form if it is mathematically convenient or necessary. Some pieces of test equipment are designed to show magnitude and angle when used to measure the AC impedance of a circuit. The pieces of test equipment are often called *vector impedance meters*, or *vector impedance bridges*. The reading they produce,

EXAMPLE 18-4

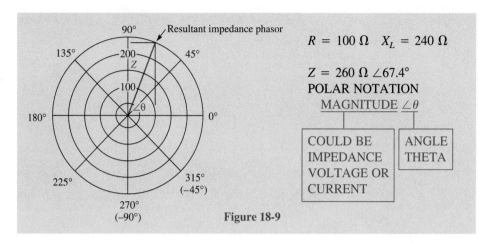

Figure 18-9

using analog meters or digital displays, is an impedance expressed in polar notation (a magnitude and an angle). When using such equipment, we often must convert the polar form of the impedance to rectangular form to determine the resistive and reactive parts. Likewise, it is often necessary to convert the rectangular form of an impedance to its polar form for ease of analysis and calculation. But how are these conversions performed?

Rectangular to Polar Conversion

How can a complex number expressed in rectangular notation be converted to polar notation? There are basically two methods that can be used to perform such a conversion. One method might be called the *manual method*, while the other might be termed the *automatic method*. The automatic method is handled very quickly by many scientific calculators, but the manual method requires a little more work on your part. If your scientific calculator does not have R→P and P→R capability, you will have to use the manual method. Both methods are illustrated in Example 18-5.

EXAMPLE 18-5

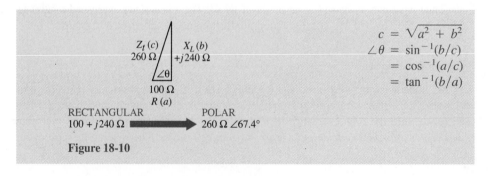

Figure 18-10

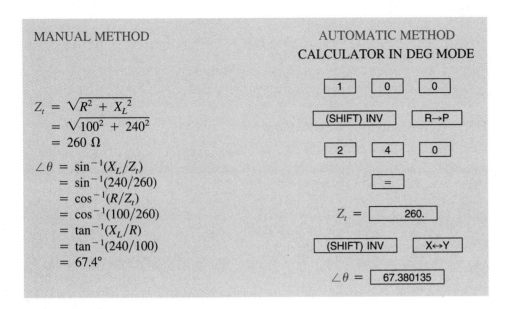

MANUAL METHOD

$$Z_t = \sqrt{R^2 + X_L^2}$$
$$= \sqrt{100^2 + 240^2}$$
$$= 260\ \Omega$$

$$\angle\theta = \sin^{-1}(X_L/Z_t)$$
$$= \sin^{-1}(240/260)$$
$$= \cos^{-1}(R/Z_t)$$
$$= \cos^{-1}(100/260)$$
$$= \tan^{-1}(X_L/R)$$
$$= \tan^{-1}(240/100)$$
$$= 67.4°$$

AUTOMATIC METHOD
CALCULATOR IN DEG MODE

| 1 | 0 | 0 |

(SHIFT) INV   R→P

| 2 | 4 | 0 |

=

$Z_t =$ 260.

(SHIFT) INV   X↔Y

$\angle\theta =$ 67.380135

Manual R→P Conversion

- Use the Pythagorean theorem ($c = \sqrt{a^2 + b^2}$, where a is the real magnitude and b is the imaginary magnitude) to find the resultant magnitude.
- Use a trigonometric function to find the angle ($\angle\theta$) (see Appendix D, "Basic Trigonometry"). *Note:* $\sin^{-1}(a/b) = \text{invsin}(a/b) = \arcsin(a/b)$

$$\angle\theta = \sin^{-1}(b/c) \tag{18.2}$$

where $\sin^{-1}$ is the inverse sine, b is the side opposite $\angle\theta$, and c is the hypotenuse or resultant side.

$$\angle\theta = \cos^{-1}(a/c) \tag{18.3}$$

where $\cos^{-1}$ is the inverse cosine, a is the side adjacent to $\angle\theta$, and c is the hypotenuse or resultant side.

$$\angle\theta = \tan^{-1}(b/a) \tag{18.4}$$

where $\tan^{-1}$ is the inverse tangent, b is the side opposite $\angle\theta$, and a is the side adjacent to $\angle\theta$.

Automatic R→P Conversion

- Use the R→P function of your scientific calculator. The following general procedure can be applied to many scientific calculators that have the R→P and P→R functions (consult the operations manual for your calculator):

    Note: Your calculator should be in the DEG mode.

    —Enter the real magnitude (side a)
    —press the [R→P] key (you may have to press the INV, 2nd, or SHIFT key first)
    —enter the imaginary magnitude (side b)
    —press the [=] key

—the resultant magnitude is displayed (side *c*)
—press the [X↔Y] key (you may have to press the INV, 2nd, or SHIFT key first)
—the angle is displayed (∠θ)
—press the [X↔Y] key to see the magnitude again
—press the [X↔Y] key to see the angle again.

Polar to Rectangular Conversion

As in rectangular to polar conversion, two basic methods can be employed when converting from polar to rectangular notation: the manual method and the automatic method. These methods are illustrated in Example 18-6.

Manual P→R Conversion

• Use the cosine function to find the magnitude of the side adjacent to the angle.

$$a = \cos \angle\theta \cdot c \tag{18.5}$$

where *a* is the real magnitude and *c* is the given polar magnitude (resultant, hypotenuse).
• Use the sine function to find the magnitude of the side opposite to the angle. This magnitude will be expressed as a ±*j* value in rectangular notation.

$$b = \sin \angle\theta \cdot c \tag{18.6}$$

where *b* is the imaginary magnitude and *c* is the given polar magnitude.

EXAMPLE   18-6

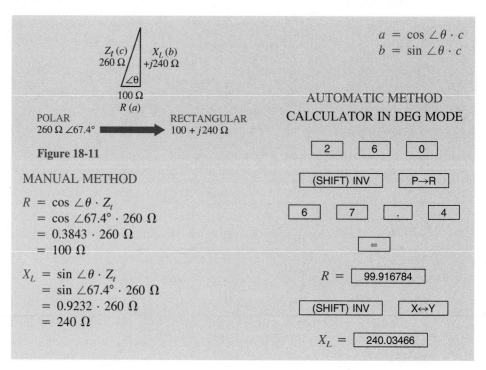

$$a = \cos \angle\theta \cdot c$$
$$b = \sin \angle\theta \cdot c$$

Z_t (*c*)
260 Ω

X_L (*b*)
+*j*240 Ω

∠θ

100 Ω
R (*a*)

POLAR
260 Ω ∠67.4° ⟶ RECTANGULAR
100 + *j*240 Ω

Figure 18-11

MANUAL METHOD

$R = \cos \angle\theta \cdot Z_t$
$\quad = \cos \angle 67.4° \cdot 260 \ \Omega$
$\quad = 0.3843 \cdot 260 \ \Omega$
$\quad = 100 \ \Omega$

$X_L = \sin \angle\theta \cdot Z_t$
$\quad = \sin \angle 67.4° \cdot 260 \ \Omega$
$\quad = 0.9232 \cdot 260 \ \Omega$
$\quad = 240 \ \Omega$

AUTOMATIC METHOD
CALCULATOR IN DEG MODE

| 2 | | 6 | | 0 |

| (SHIFT) INV | | P→R |

| 6 | | 7 | | . | | 4 |

| = |

$R = $ | 99.916784 |

| (SHIFT) INV | | X↔Y |

$X_L = $ | 240.03466 |

Automatic P→R Conversion

• Use the P→R function of your scientific calculator. (Consult the operations manual for your calculator):

   Note: calculator should be in the DEG mode.

   —Enter the resultant magnitude (side c)
   —press the [P→R] key (you may have to press the INV, 2nd, or SHIFT key first)
   —enter the angle ($\angle\theta$)
   —press the [=] key
   —the real magnitude is displayed (side a)
   —press the [X↔Y] key (you may have to press the INV, 2nd, or SHIFT key first)
   —the imaginary magnitude is displayed (side b)
   —press the [X↔Y] key to see the real magnitude again
   —press the [X↔Y] key to see the imaginary magnitude again.

   Take a moment now to review this section and answer the questions of Self-Check 18-1. Review any areas that seem unclear to you. It is very important that you have a good understanding of the material covered here.

SELF-CHECK 18-1

1. What is a complex impedance? (describe and give an example)

2. What is an operator?

3. If a magnitude's operator is $-j$, what is its angle?

4. A j^6 represents what factor and angle?

5. Convert $200 - j450$ to polar notation.

6. Convert $50 + j75$ to polar notation.

7. Convert $300 \angle 45°$ to rectangular notation.

8. Convert $25 \angle -60°$ to rectangular notation.

18-2 Mathematical Operations with Complex Numbers

Since AC impedances are often complex numbers, it is very important that you understand how to perform basic mathematical operations with them. In this section, you will learn by rule and example how to manipulate complex numbers expressed in rectangular and polar notation.

Operations with Rectangular Notation

Addition and Subtraction

> **Rule**
>
> When adding or subtracting complex numbers in rectangular form, add or subtract all real numbers and add or subtract all imaginary numbers until all real terms are combined and all imaginary terms are combined to yield one complex number.

EXAMPLE SET
18-7

1. $(100 - j200) + (25 + j30) + (60 + j75)$
   $= (100 + 25 + 60) + (-j200 + j30 + j75) = 185 - j95$
2. $(50 + j30) + (-40 + j60) + (70 - j20) + (10 - j10)$
   $= (50 - 40 + 70 + 10) + (j30 + j60 - j20 - j10) = 90 + j60$
3. $(5 - j4) - (8 + j20) - (12 - j6) - (10 + j4)$
   $= (5 - 8 - 12 - 10) + (-j4 - j20 + j6 - j4) = -25 - j22$
4. $(-6 + j12) - (3 - j6) - (-21 + j10)$
   $= (-6 - 3 + 21) + (j12 + j6 - j10) = 12 + j8$

Multiplying *j* Terms

> **Rule**
>
> Multiply the magnitudes and the *j* terms and resolve to the simplest form keeping in mind that: $j = \sqrt{-1}; j^2 = -1; j^3 = -j = -\sqrt{-1}; j^4 = j^2 \cdot j^2 = -1 \cdot -1 = 1 = j^0$.

EXAMPLE SET
18-8

1. $j40 \cdot j20 = j^2 800 = -1 \cdot 800 = -800$
2. $-j30 \cdot j10 \cdot -j5 = j^2 \cdot j1,500 = -j1,500$
3. $-j^2 \cdot j^3 40 = j^3 40 = -j40$
4. $j^3 20 \cdot j10 \cdot -j^2 10 = -j20 \cdot j10 \cdot 10 = 2,000$
5. $j^2 6 \cdot j^4 3 \cdot j^6 4 = -6 \cdot 3 \cdot -4 = 72$

Multiplying Complex Numbers in Rectangular Form

> **Rule**
>
> Multiply the complex numbers as you would binomials and reduce to the simplest rectangular form.

EXAMPLE SET
18-9

1. $(60 + j30)(20 - j10) = 1,200 - j600 + j600 - j^2 300 = 1,200 - j^2 300$
   $= 1,200 + 300 = 1,500$

2. $(200 + j400)(100 - j500) = 20,000 - j100,000 + j40,000 - j^2200,000$
   $= 20,000 - j60,000 + 200,000 = 220,000 - j60,000$
3. $(3 - j4)(20 - j50) = 60 - j150 - j80 + j^2200 = 60 - j230 - 200$
   $= -140 - j230$
4. $(4 + j6)^3 = (4 + j6)(4 + j6)(4 + j6) = (16 + j24 + j24 + j^236)(4 + j6)$
   $= (16 + j48 - 36)(4 + j6) = (-20 + j48)(4 + j6)$
   $= -80 - j120 + j192 + j^2288 = -368 + j72$
5. $(70 - j30)(30 - j40) = 2,100 - j2,800 - j900 + j^21,200$
   $= 2,100 - j3,700 - 1,200 = 900 - j3,700$

Dividing by j Terms

> **Rule**
>
> When any number is to be divided by a single j term, change the sign of the j and multiply. The j's magnitude is divided into the number in a normal fashion.

1. $1/j = 1/j \cdot j/j = j/-1 = -j$
2. $10/j5 = -j2$
3. $100/-j20 = j5$
4. $(20 - j40)/j4 = -j \cdot (5 - j10) = -j5 + j^210 = -j5 - 10 = -10 - j5$
5. $(30 + j60 - j^280)/-j5 = (110 + j60)/-j5 = j \cdot (22 + j12) = j22 - 12$
   $= -12 + j22$
6. $(36 - j12)/-j4 = j \cdot (9 - j3) = j9 - j^23 = 3 + j9$

EXAMPLE SET
18-10

Dividing Complex Numbers in Rectangular Form

> **Rule #1**
>
> To divide by a complex number in rectangular form, multiply both the numerator and the denominator by the conjugate of the complex denominator. This converts the denominator to a real number.

This process of conversion is known as **rationalization** and is a legal process. The value of the term is not changed, because you are actually multiplying the term by 1. Both the numerator and the denominator are multiplied by the same conjugate term. Once rationalization is accomplished, division can proceed in a normal manner to render the simplest form of the expression.

The **conjugate** of a complex number is another complex number containing the same magnitudes but separated by the opposite sign. Some examples are: the conjugate of $40 - j30$ is $40 + j30$, the conjugate of $-100 - j20$ is $-100 + j20$, the conjugate of $3 + j3$ is $3 - j3$, etc.

Rule #2

The product of a binomial complex number and its conjugate is always equal to the sum of the squares of the real and imaginary magnitudes where $(a + jb)(a - jb) = a^2 + b^2$.

E X A M P L E   S E T
1 8 - 1 1

1. $\dfrac{1}{1-j} = \dfrac{1}{1-j} \cdot \dfrac{1+j}{1+j} = \dfrac{1+j}{2} = 0.5 + j0.5$

2. $\dfrac{10 - j20}{2 + j4} = \dfrac{10 - j20}{2 + j4} \cdot \dfrac{2 - j4}{2 - j4} = \dfrac{20 - j40 - j40 - 80}{20} = \dfrac{-60 - j80}{20}$

   $= -3 - j4$

3. $\dfrac{30 + j60}{5 - j5} = \dfrac{30 + j60}{5 - j5} \cdot \dfrac{5 + j5}{5 + j5} = \dfrac{-150 + j450}{50} = -3 + j9$

4. $\dfrac{2 - j8}{-2 - j4} = \dfrac{2 - j8}{-2 - j4} \cdot \dfrac{-2 + j4}{-2 + j4} = \dfrac{28 + j24}{20} = 1.4 + j1.2$

Operations with Polar Notation

Addition and Subtraction

Rule

Convert the polar notation to rectangular notation and follow the rule for addition and subtraction of complex numbers in rectangular form. Complex numbers in polar form cannot directly be added or subtracted. The final answer can be converted back to polar form if necessary or desired.

E X A M P L E   S E T
1 8 - 1 2

1. $5 \angle 53.1° + 8.6 \angle -54.5° = (3 + j4) + (5 - j7) = (3 + 5) + (j4 - j7)$
   $= 8 - j3 = 8.54 \angle -20.6°$
2. $8.1 \angle 119.7° - 10 \angle -53.1° = (-4 + j7) - (6 - j8) = -10 + j15 = 18 \angle 123.7°$

Multiplication of Complex Numbers in Polar Form

Rule

Multiply the magnitudes and add the angles.

$$a \angle \theta \cdot b \angle \phi = a \cdot b \angle \theta + \phi$$

1. $200 \angle 30° \cdot 20 \angle 45° = 200 \cdot 20 \angle 30° + 45° = 4{,}000 \angle 75°$
2. $80 \angle 90° \cdot 30 \angle -20° = 80 \cdot 30 \angle 90° - 20° = 2{,}400 \angle 70°$
3. $5 \angle 60° \cdot 3 \angle 45° = 5 \cdot 3 \angle 60° + 45° = 15 \angle 105°$

Dividing Complex Numbers in Polar Form

> **Rule**
>
> Divide the magnitudes and subtract the angles.
>
> $a \angle \theta / b \angle \phi = a/b \angle \theta - \phi$

1. $20 \angle 60°/2 \angle 45° = 20/2 \angle 60° - 45° = 10 \angle 15°$
2. $750 \angle 45°/100 \angle 90° = 750/100 \angle 45° - 90° = 7.5 \angle -45°$
3. $450 \angle 65°/200 \angle 50° = 450/200 \angle 65° - 50° = 2.25 \angle 15°$

Which Notation When? (A Summary)

If complex numbers are to be added or subtracted, they must be expressed in rectangular form. If necessary, convert the complex numbers to rectangular notation and follow the rule for addition and subtraction of complex numbers in rectangular form.

If complex numbers are to be multiplied or divided, it is often more convenient to perform these operations using the polar form of the complex numbers. If the complex numbers are not already in polar form, convert them to polar notation and follow the rule for multiplication or division of complex numbers in polar form.

Self-Check 18-2 reinforces what you have learned in this section. Compare your work to the examples that have been presented.

SELF-CHECK   18-2

1. $(35 + j25) + (60 - j22)$

2. $(12 - j4) - (20 + j5)$

3. $j6 \cdot j4$

4. $-j^3 50 \cdot j30$

5. $(3 + j8)(4 - j4)$

6. $(22 - j44)(10 - j5)$

7. $460/-j10$

8. $(33 - j11)/j11$

9. $(30 - j5)/(40 + j8)$

10. $(4 + j9)/(3 - j2)$

(Continued next page)

11. $50 \angle 20° + 30 \angle 45°$

12. $100 \angle 90° - 65 \angle -30°$

13. $400 \angle 25° \cdot 300 \angle 40°$

14. $6 \angle 110°/2 \angle -50°$

18-3 Series *RL* Circuit Analysis

In this section, you will learn how to analyze a simple series AC circuit composed of a resistor and an inductor as shown in Figure 18-12. The purpose of the analysis is to show the time (phase) relationship between all voltages and impedances in the circuit. You will be able to calculate the circuit's total impedance (Z_t) and total current (I_t), and all voltage drops (V_R and V_L). All of this, of course, is in preparation for more advanced circuit analysis later. As we proceed, you will discover some similarities and some differences between DC and AC circuit analysis. Make sure you recognize and understand the differences in particular.

At this time, it might be helpful for you to go back and briefly review Chapter 16, "AC and Reactive Components." Here, we will make use of the inductive reactance formula and the ELI concept. It is important for you to recall that voltage leads current by 90° for an inductor in an AC circuit (ELI). This is the key to understanding the voltage and impedance phase relationships.

RL Circuit Voltage Phasors

Voltages Out of Phase

See color photo insert pages D8 and D9.

When a sinusoidal AC voltage is applied to an *RL* circuit, as shown in Figure 18-13, the AC voltage drops produced across each component do not reach peak values all at the same time. In fact, the source voltage E_S is also out of phase with the component voltage drops. Since all voltages are out of phase with each other, the component voltage drops

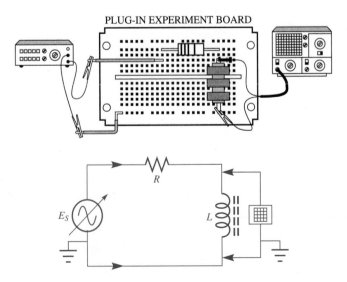

Figure 18-12

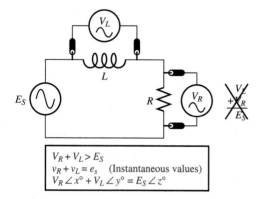

$V_R + V_L > E_S$
$v_R + v_L = e_s$ (Instantaneous values)
$V_R \angle x° + V_L \angle y° = E_S \angle z°$

Figure 18-13   Series *RL* circuit voltages.

cannot be added arithmetically to equal the applied source voltage. If this is done, the sum of the resistor and inductor voltage drops will be greater than the source voltage ($V_R + V_L > E_S$). Does this mean that Kirchhoff's Voltage Law does not apply to AC circuits? No. Kirchhoff's Voltage Law does apply. But, it applies to the sum of the instantaneous voltages or the **phasor sum** of all voltages.

Kirchhoff's Voltage Law for AC circuits can be stated in two ways:

1. The sum of all instantaneous voltages in a closed loop of an AC circuit will equal zero volts. For a series *RL* circuit, this means $v_R + v_L - e_s = 0$ or $e_s = v_R + v_L$.

2. The phasor sum of all voltages in a closed loop of an AC circuit will equal zero volts.

Since the circuit voltages are out of phase, they must be added using phasor addition. In other words, the phase angle of each voltage must be included in the calculations. According to Kirchoff's Voltage Law, for AC circuits the sum of the phasors that represent each voltage drop will equal the applied source voltage. For a series *RL* circuit, this means $V_R \angle x° + V_L \angle y° = E_S \angle z°$.

RL Voltage Phasors

But how are voltage phasors represented and added? To answer this, we will first develop the voltage-phasor diagram for a series *RL* circuit. Let's begin with what you already know. First, you know that current in a simple series circuit is the same at any point in the circuit. Therefore, the current (I_t) is used as a reference in a polar system that contains all of the voltage phasors for the circuit. Thus, we can place current at 0° and compare the phase of all voltage drops to it. Next, you already know that there is no phase shift between resistor current and resistor voltage. So the phasor that represents resistor voltage will also be plotted at 0° along with the current. Also, you know that the inductor voltage leads the current by 90° (ELI). This means the inductor-voltage phasor is plotted in a vertical position above and at a right angle to the resistor-voltage phasor. Thus we see that inductor voltage leads resistor voltage by 90°. This is demonstrated in Figure 18-14.

Phase Angle Represents Time

Remember, a phase angle represents time as related to the AC cycle time. A phase angle of 90°, for example, represents ¼ of the cycle time for the particular AC frequency that is being applied to the circuit. If the frequency is 1 kHz, the time for a full cycle is 1 ms

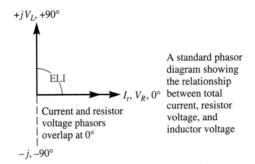

A standard phasor diagram showing the relationship between total current, resistor voltage, and inductor voltage

Figure 18-14 The series *RL* voltage-phasor diagram.

$(T = 1/f)$. Thus, 90° of a cycle represents $T/4$ $(T \cdot 90°/360° = T \cdot \frac{1}{4} = T/4)$. At a frequency of 1 kHz, 90° represents 0.25 ms of time. If the frequency of the applied source voltage is increased, cycle time and the time represented by 90° will decrease and vice versa. What does all of this mean? It means that a phase shift represents a definite time delay in accordance with the frequency of the source. AC time delays are very important in many areas of electricity and electronics. However, we will not take the liberty of pursuing such applications at this time.

Adding the Resistor and Inductor Voltage Phasors

Now that you know how the voltages are represented as phasors, let's see how they are added. Remember, we said the *R* and *L* voltages must be added using phasor addition. When the voltages are added in this way, their sum is equal to the applied source voltage. Figure 18-15 shows how this is done. When phasors are added, the tail of one phasor is placed at the head of the other. In this case, the inductor-voltage phasor $(V_L \angle 90° = +jV_L)$ is added to the resistor-voltage phasor $(V_R \angle 0° = V_R)$. As you can see, the result is a right triangle formed by a resultant phasor extending from the origin to the head of the inductor-voltage phasor. This resultant is the applied source voltage, the phasor sum of the two circuit voltages $(E_s \angle \theta = V_R \angle 0° + V_L \angle 90° = V_R + jV_L)$. The Pythagorean theorem is used to find the magnitude of the source voltage and $\tan^{-1}(V_L/V_R)$ can be used to find the angle (use Formulas 18.7 and 18.8). Also, if you have a scientific calculator with the R→P and P→R functions, the real and imaginary rectangular magnitudes of resistor voltage and inductor voltage can be entered into the calculator to find the resultant

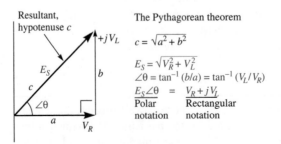

The Pythagorean theorem

$$c = \sqrt{a^2 + b^2}$$

$$E_S = \sqrt{V_R^2 + V_L^2}$$
$$\angle\theta = \tan^{-1}(b/a) = \tan^{-1}(V_L/V_R)$$

$$\frac{E_S \angle \theta}{\text{Polar notation}} \quad = \quad \frac{V_R + jV_L}{\text{Rectangular notation}}$$

Figure 18-15 *RL* circuit voltage phasor addition.

polar value of the source voltage using the R→P function. If the source voltage and angle are already known ($E_s \angle \theta$), the resistor voltage and inductor voltage can be found by performing a P→R conversion (see Section 18-1).

$$E_S = \sqrt{V_R^2 + V_L^2} \tag{18.7}$$

also: $V_R = \sqrt{E_S^2 - V_L^2}$ and $V_L = \sqrt{E_S^2 - V_R^2}$

$$\angle \theta = \tan^{-1}(V_L/V_R) \tag{18.8}$$

Voltage Lag and Lead

Figures 18-16 and 18-17 illustrate two series *RL* circuit configurations. Figure 18-16 is a *lag* network and Figure 18-17 is a *lead* network. When the *RL* circuit is arranged so an output is taken across the resistor, as in Figure 18-16, the circuit is referred to as a *lag network*. It is so named because the resistor **voltage lags** the source voltage in phase and time (because current lags voltage—ELI). This is illustrated in the dual-trace oscilloscope display and the phasor diagrams. The electron beam in the cathode ray tube (CRT) of the oscilloscope impinges on the back side of the CRT face and draws the sine wave voltage waveforms from left to right (time increases from left to right). As you can see, the resistor-voltage waveform reaches a positive peak after the source voltage does. The phase angle theta ($\angle \theta$) represents this lag in time. Often, the voltage-phasor diagram is rotated to place the source-voltage phasor at 0°. This establishes the source voltage as a reference to which the resistor and inductor voltages may be compared. Rotating the phasor diagram places the resistor-voltage phasor at a negative angle, indicating a voltage lag. In a circuit

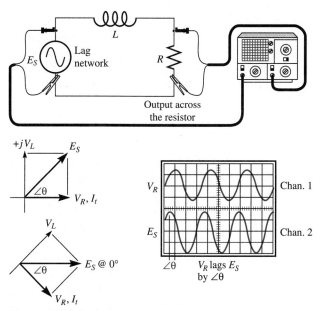

Phasor diagram is rotated to place E_S @ 0°

Figure 18-16   The resistor voltage lags the source voltage in a series *RL* circuit.

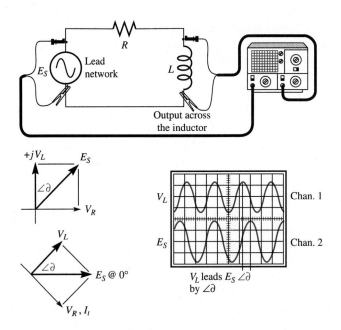

Figure 18-17   The inductor voltage leads the source voltage in a series *RL* circuit.

such as this, the source voltage is delayed in time as an output is taken across the resistor. As you continue your studies in electricity and electronics, you will discover many applications where such a delay is needed.

When the *RL* circuit is arranged so the output is taken across the inductor, as in Figure 18-17, the circuit becomes a lead network. The **voltage** across the inductor **leads** the source voltage in phase and time. Again, this is represented in the dual-trace oscilloscope display and the phasor diagrams. Notice that the inductor voltage reaches a positive peak before the source voltage does. This is indicated by the positive phase angle between the inductor-voltage phasor and the source-voltage phasor ($\angle \delta$). The positive angle of the inductor voltage indicates voltage lead. In other words, the self-induced voltage of the inductor actually leads the source voltage.

Determining Resistor and Inductor Voltages

You now have some idea of how the series *RL* voltages are represented and added as phasors. But how are the individual resistor and inductor voltages determined? In practice, the actual voltages can be measured. For AC frequencies less than 500 Hz, a DMM may be used. In any case, an oscilloscope may be used to measure peak and peak-to-peak voltages, which can then be converted to RMS values ($V_{\text{peak}} \cdot 0.707$). In theoretical analysis, the voltages must be determined through calculation. If the total current can be determined, the individual circuit voltages can be determined using Ohm's Law ($E = I \cdot R$ or $E = I \cdot Z$ or $E = I \cdot X_{L \text{ or } C}$, so $V_R = I_t \cdot R$ and $V_L = I_t \cdot X_L$). The total current must be determined from the total circuit impedance, again using Ohm's Law ($I_t = E_s / Z_t$). But, how is total circuit impedance determined for a series *RL* circuit? The solution to this is readily illustrated through impedance phasor analysis.

$$I_t = E_S/Z_t \tag{18.9}$$
$$V_R = I_t \cdot R \tag{18.10}$$
$$V_L = I_t \cdot X_L \tag{18.11}$$

RL Circuit Impedance Phasors

As shown in Figure 18-18, an impedance-phasor diagram is readily obtained by dividing each of the voltage phasors by the total current (Ohm's Law: $R = V_R/I_t$; $X_L \angle 90° = V_L \angle 90°/I_t \angle 0°$; $Z_t \angle \theta = E_s \angle \theta/I_t \angle 0°$). The result is an inductive reactance at 90° ($X_L \angle 90° = +jX_L$) and a total impedance at $\angle \theta$ ($Z_t \angle \theta$) with respect to the resistance at 0°. The complete impedance-phasor diagram looks much like the voltage-phasor diagram, except for the labeling (impedances instead of voltages).

From our previous discussion and Figure 18-18, you can see that the total impedance is the result of the phasor sum of the resistance and the inductive reactance ($Z_t \angle \theta = R \angle 0° + X_L \angle +90° = R + jX_L$). Since a right triangle is formed from the phasor sum of resistance and inductive reactance, the total impedance phasor can be determined using the Pythagorean theorem and the inverse tangent (arctan) of X_L/R. If the total impedance and angle are already known ($Z_t \angle \theta$), the resistance and inductive reactance may be determined using P→R conversion.

$$Z_t = \sqrt{R^2 + X_L^2} \tag{18.12}$$

also: $R = \sqrt{Z_t^2 - X_L^2}$ and $X_L = \sqrt{Z_t^2 - R^2}$

$$\angle \theta = \tan^{-1}(X_L/R) = \tan^{-1}(V_L/V_R) \tag{18.13}$$

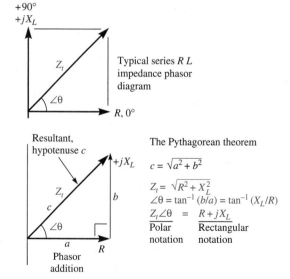

+90°
+jX_L

Z_t

$\angle \theta$

R, 0°

Typical series *R L* impedance phasor diagram

Resultant, hypotenuse c

+jX_L

Z_t

b

c

$\angle \theta$

a R

Phasor addition

The Pythagorean theorem

$c = \sqrt{a^2 + b^2}$

$Z_t = \sqrt{R^2 + X_L^2}$
$\angle \theta = \tan^{-1}(b/a) = \tan^{-1}(X_L/R)$

$\underline{Z_t \angle \theta}$ = $\underline{R + jX_L}$
Polar   Rectangular
notation   notation

Figure 18-18   A complex *RL* circuit impedance-phasor diagram.

EXAMPLE   18-15

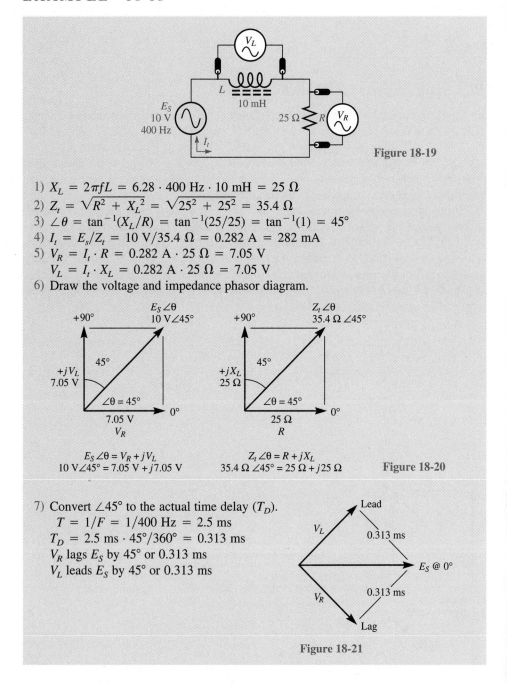

Figure 18-19

1) $X_L = 2\pi fL = 6.28 \cdot 400 \text{ Hz} \cdot 10 \text{ mH} = 25 \ \Omega$
2) $Z_t = \sqrt{R^2 + X_L^2} = \sqrt{25^2 + 25^2} = 35.4 \ \Omega$
3) $\angle\theta = \tan^{-1}(X_L/R) = \tan^{-1}(25/25) = \tan^{-1}(1) = 45°$
4) $I_t = E_s/Z_t = 10 \text{ V}/35.4 \ \Omega = 0.282 \text{ A} = 282 \text{ mA}$
5) $V_R = I_t \cdot R = 0.282 \text{ A} \cdot 25 \ \Omega = 7.05 \text{ V}$
   $V_L = I_t \cdot X_L = 0.282 \text{ A} \cdot 25 \ \Omega = 7.05 \text{ V}$
6) Draw the voltage and impedance phasor diagram.

$E_S \angle\theta = V_R + jV_L$
$10 \text{ V}\angle 45° = 7.05 \text{ V} + j7.05 \text{ V}$

$Z_t \angle\theta = R + jX_L$
$35.4 \ \Omega \angle 45° = 25 \ \Omega + j25 \ \Omega$

Figure 18-20

7) Convert $\angle 45°$ to the actual time delay (T_D).
   $T = 1/F = 1/400 \text{ Hz} = 2.5 \text{ ms}$
   $T_D = 2.5 \text{ ms} \cdot 45°/360° = 0.313 \text{ ms}$
   V_R lags E_S by 45° or 0.313 ms
   V_L leads E_S by 45° or 0.313 ms

Figure 18-21

RL Circuit Calculations

Now, let's take what has been discussed and apply it to actual series *RL* circuits. Consider Example 18-15. The theoretical analysis is performed in a systematic step-by-step method.

The following step-by-step summary corresponds to those steps used in Example 18-15:

1. Determine the inductive reactance at the applied frequency.
2. Use the Pythagorean theorem to find the total impedance.
3. Use the inverse tangent of X_L/R to find $\angle\theta$. (If Z_t and $\angle\theta$ are already known, use P→R conversion to find R and X_L.)
4. Use Ohm's Law to find the total current.
5. Use Ohm's Law to find the resistor and inductor voltages. Resistor and inductor voltages may also be found using P→R conversion when E_S and $\angle\theta$ are known.
6. Draw and label the impedance- and voltage-phasor diagrams.
7. If you wish, convert phase angles to actual time delay, taking into account the time for one cycle at the applied frequency.

   The next two examples reinforce this analysis procedure. As you evaluate them, the following basic concepts should become evident:

1. Component voltages in AC circuits can be found the same way they are found in DC circuits—using Ohm's Law.
2. Kirchhoff's Voltage Law for AC circuits is different than for DC circuits. Resistor and inductor voltages in AC circuits must be added phasorally using the Pythagorean theorem.
3. All impedances and voltages are out of phase.
4. The voltage-phasor diagram can be rotated to place the source voltage phasor at 0°. The source voltage, instead of the resistor voltage and circuit current, then becomes the reference voltage.

EXAMPLE 18-16

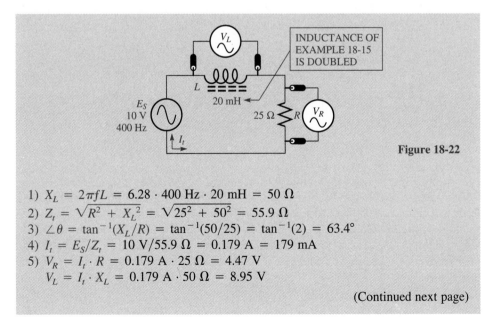

Figure 18-22

1) $X_L = 2\pi fL = 6.28 \cdot 400 \text{ Hz} \cdot 20 \text{ mH} = 50 \ \Omega$
2) $Z_t = \sqrt{R^2 + X_L^2} = \sqrt{25^2 + 50^2} = 55.9 \ \Omega$
3) $\angle\theta = \tan^{-1}(X_L/R) = \tan^{-1}(50/25) = \tan^{-1}(2) = 63.4°$
4) $I_t = E_S/Z_t = 10 \text{ V}/55.9 \ \Omega = 0.179 \text{ A} = 179 \text{ mA}$
5) $V_R = I_t \cdot R = 0.179 \text{ A} \cdot 25 \ \Omega = 4.47 \text{ V}$
   $V_L = I_t \cdot X_L = 0.179 \text{ A} \cdot 50 \ \Omega = 8.95 \text{ V}$

(Continued next page)

EXAMPLE 18-16 (Continued)

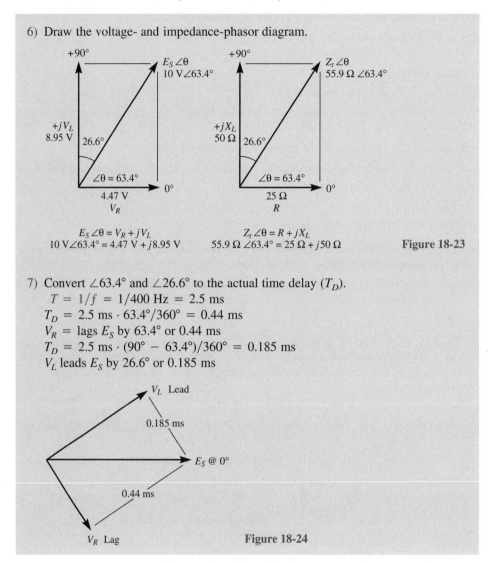

6) Draw the voltage- and impedance-phasor diagram.

$$E_S \angle\theta = V_R + jV_L$$
$$10\ V\angle63.4° = 4.47\ V + j8.95\ V$$

$$Z_t \angle\theta = R + jX_L$$
$$55.9\ \Omega\ \angle63.4° = 25\ \Omega + j50\ \Omega$$

Figure 18-23

7) Convert $\angle63.4°$ and $\angle26.6°$ to the actual time delay (T_D).

$T = 1/f = 1/400\ Hz = 2.5\ ms$
$T_D = 2.5\ ms \cdot 63.4°/360° = 0.44\ ms$
$V_R = $ lags E_S by $63.4°$ or $0.44\ ms$
$T_D = 2.5\ ms \cdot (90° - 63.4°)/360° = 0.185\ ms$
V_L leads E_S by $26.6°$ or $0.185\ ms$

Figure 18-24

Example 18-17 demonstrates how the total impedance, resistance, inductive reactance, resistor voltage, and inductor voltage can be found when only the source voltage, angle theta, and current are known. Notice the total impedance is found using Ohm's Law. The polar quantities of voltage and current are divided to find the total impedance and phase angle. The resistance and inductive reactance are then found using manual or automatic P→R conversion. The resistor and inductor voltages may be found using Ohm's Law or P→R conversion.

EXAMPLE 18-17

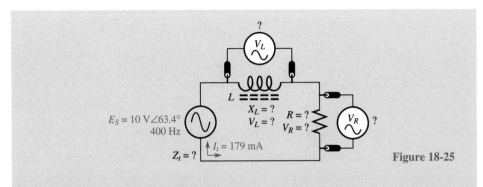

Figure 18-25

1) $Z_t = E_S/I_t = 10 \angle 63.4°/179 \angle 0°$ mA $= 55.9 \ \Omega \ \angle 63.4°$
2) Use $P{\rightarrow}R$ conversion to find R and X_L.

> FORMULA (18.5)
> $R = \cos \angle 63.4° \cdot Z_t = 0.448 \cdot 55.9 \ \Omega = 25 \ \Omega$
> FORMULA (18.6)
> $X_L = \sin \angle 63.4° \cdot Z_t = 0.894 \cdot 55.9 \ \Omega = 50 \ \Omega$

3) $V_R = I_t \cdot R = 0.179$ A $\cdot 25 \ \Omega = 4.48$ V
$V_L = I_t \cdot X_L = 0.179$ A $\cdot 50 \ \Omega = 8.95$ V
also, using $P{\rightarrow}R$ conversion to find V_R and V_L:

> FORMULA (18.5)
> $V_R = \cos \angle 63.4° \cdot E_s = 0.448 \cdot 10$ V $= 4.48$ V

> FORMULA (18.6)
> $V_L = \sin \angle 63.4° \cdot E_s = 0.894 \cdot 10$ V $= 8.94$ V

4) Draw voltage- and impedance-phasor diagrams.

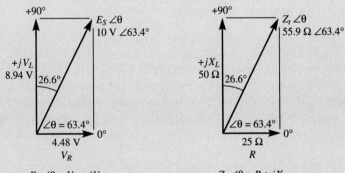

Figure 18-26

Now, try your skills by analyzing the circuits of Self-Check 18-3.

SELF-CHECK 18-3

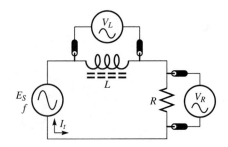

Figure 18-27

Fill in the blanks and draw the voltage-phasor diagram for each problem.

#	E_S	f	L	X_L	R	Z_t	I_t	V_R	V_L	$\angle\theta$
1.	5 V	60 Hz	0.531 H		100 Ω					
2.	20 V	1 MHz	31.8 μH		400 Ω					
3.	15 mV	200 kHz		500 Ω	300 Ω					
4.	4 V		50 mH	25 kΩ	10 kΩ					
5.*	18 V	100 Hz			1 kΩ					30°

Hint: Use the source voltage and the angle to find resistor and inductor voltage. Then, find total current using resistance and resistor voltage.

18-4 Series *RC* Circuit Analysis

In this section, we discuss series *RC* circuit analysis. See Figure 18-28. You will discover very quickly the distinct similarities and differences between *RC* and *RL* circuits. Much of what we cover here will be so similar, it will seem like a review of *RL* circuit analysis. But don't be fooled. There are some important differences. The main differences you will see are negative angle thetas ($\angle-\theta$) and V_C and X_C phasors heading downward at the $-90°$ position ($-jX_C$ and $-jV_C$). Here, we will make use of the capacitive reactance formula and the ICE concept. It is important for you to recall that current leads voltage by 90° for a capacitor in an AC circuit (ICE). As with *RL* circuits, this is the key to understanding the voltage and impedance phase relationships in *RC* circuits.

RC Circuit Voltage Phasor

Voltages Out of Phase

When a sinusoidal AC voltage is applied to an *RC* circuit, as shown in Figure 18-29, the AC voltage drops produced across each component do not reach peak values all at the same time. In fact, the source voltage (E_S) is also out of phase with the component voltage

PLUG-IN EXPERIMENT BOARD

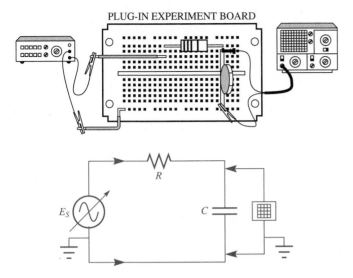

Figure 18-28

drops. Doesn't that sound familiar? It should. As in the *RL* circuits, all voltages are out of phase with each other. Again, the component voltage drops cannot be added arithmetically to equal the applied source voltage, since the sum of the resistor and capacitor voltages will be greater than the source voltage ($V_R + V_C > E_S$). They must be added phasorally. As before, Kirchhoff's Voltage Law does apply. But it applies only to the instantaneous voltages and the phasor sum of all voltages ($e_S = v_R + v_C$ and $E_S \angle z° = V_R \angle x° + V_C \angle y°$).

RC Voltage Phasors

How are the *RC* voltage phasors represented and added? As before, the total circuit current (I_t) is used as a reference and its phasor is plotted at 0° ($I_t \angle 0°$). Since there is no phase shift between resistor current and resistor voltage, the phasor that represents resistor voltage will also be plotted at 0° ($V_R \angle 0°$). Recall that the capacitor voltage lags the current by

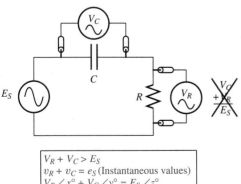

$$V_R + V_C > E_S$$
$$v_R + v_C = e_S \text{ (Instantaneous values)}$$
$$V_R \angle x° + V_C \angle y° = E_S \angle z°$$

Figure 18-29   Series *RC* circuit voltages.

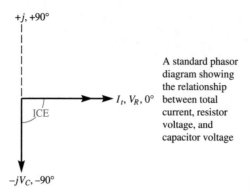

A standard phasor diagram showing the relationship between total current, resistor voltage, and capacitor voltage

Figure 18-30   The series *RC* circuit voltage-phasor diagram.

90° (ICE). This means the capacitor voltage phasor is plotted in a vertical position below and at a right angle to the resistor voltage phasor. Thus, as shown in Figure 18-30, capacitor voltage lags resistor voltage by 90° ($V_C \angle -90° = -jV_C$).

Adding the Resistor and Capacitor Voltage Phasors

Now that you know how the voltages are represented as phasors, let's see how they are added. Remember, we said the *R* and *C* voltages must be added using phasor addition. Also, when the voltages are added phasorally, their sum will equal the applied source voltage. Figure 18-31 shows how this is done. In this case, the capacitor-voltage phasor ($V_C \angle -90° = -jV_C$) is added to the resistor-voltage phasor ($V_R \angle 0° = V_R$). As you can see, the result is a right triangle formed by a resultant phasor extending from the origin to the head of the capacitor-voltage phasor. This resultant is the applied source voltage, the sum of the two circuit voltages ($E_s \angle \theta = V_R \angle 0° + V_C \angle -90° = V_R - jV_C$). The Pythagorean theorem is used once again to find the magnitude of the source voltage and $\tan^{-1}(-V_C/V_R)$ can be used to find the angle (Formulas 18.14 and 18.15). Also, the R→P function of your scientific calculator may be used to find $E_s \angle \theta$ by entering resistor voltage, INV R→P, then capacitor voltage. If the source voltage and angle are already known, the resistor and capacitor voltage may be found using P→R conversion.

$$E_s = \sqrt{V_R^2 + V_C^2} \tag{18.14}$$

$$\text{also: } V_R = \sqrt{E_s^2 - V_C^2} \quad \text{and} \quad V_C = \sqrt{E_s^2 - V_R^2}$$

$$\angle\theta = \tan^{-1}(-V_C/V_R) \tag{18.15}$$

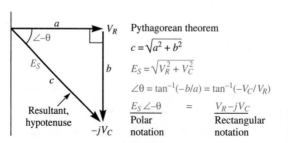

Pythagorean theorem

$c = \sqrt{a^2 + b^2}$

$E_S = \sqrt{V_R^2 + V_C^2}$

$\angle\theta = \tan^{-1}(-b/a) = \tan^{-1}(-V_C/V_R)$

$E_S \angle -\theta$	$V_R - jV_C$
Polar notation	Rectangular notation

Figure 18-31   *RC* circuit voltage phasor addition.

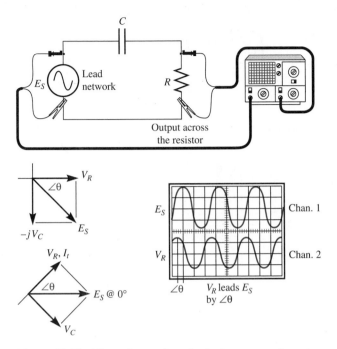

Figure 18-32   The resistor voltage leads the source voltage in a series *RC* circuit.

Voltage Lead and Lag

Figures 18-32 and 18-33 illustrate two series *RC* circuit configurations. Figure 18-32 shows a *lead* network and Figure 18-33 shows a *lag* network. When the *RC* circuit is arranged so an output is taken across the resistor, the circuit is called a *lead network*—because the resistor voltage leads the source voltage in phase and time (because current leads voltage—ICE). The oscilloscope display shows the phase shift between resistor voltage and source voltage that is represented in the phasor diagrams. As with *RL* phasor diagrams, the entire diagram may be rotated to place the source voltage phasor at 0° as a reference. Angle theta ($\angle\theta$) is then shown as a positive angle to emphasize the fact that resistor voltage leads the source voltage.

When the *RC* circuit is arranged so the output is taken across the capacitor, as shown in Figure 18-33, the circuit becomes a lag network. The voltage across the capacitor lags the source voltage in phase and time. As before, the two voltages are shown on a dual trace oscilloscope display and are represented by the phasor diagrams. The rotated diagram once again places the source voltage at 0° as a reference. The capacitor voltage is at a negative angle indicating it lags the source voltage ($\angle\delta$).

Determining Resistor and Capacitor Voltages

You now know how series *RC* voltages are represented and added as phasors. But, how are the individual resistor and capacitor voltages determined? The same way individual voltages are determined in *RL* circuits—using Ohm's Law ($E = I \cdot R$ or $E = I \cdot Z$ or

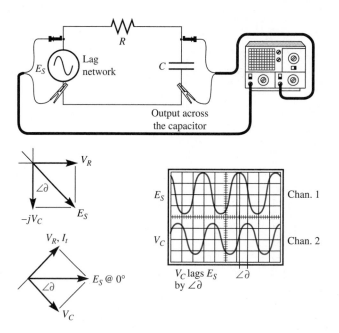

Figure 18-33 The capacitor voltage lags the source voltage in a series *RC* circuit.

$E = I \cdot X_{L \text{ or } C}$, so $V_R = I_t \cdot R$ and $V_C = I_t \cdot X_C$). The total current must be determined from the total circuit impedance, again using Ohm's Law ($I_t = E_S/Z_t$). The total impedance for series *RC* circuits is found the same way as for *RL* circuits, using the Pythagorean theorem or R→P conversion.

$$V_C = I_t \cdot X_C \tag{18.16}$$

RC Circuit Impedance Phasors

As shown in Figure 18-34, the impedance-phasor diagram is obtained by dividing each of the voltage phasors by the total current (Ohm's Law: $R = V_R/I_t$); ($X_C \angle -99° = V_C \angle -90°/I_t \angle 0°$), ($Z_t \angle \theta = E_S \angle \theta/I_t \angle 0°$). The result is a capacitive reactance at $-90°$ ($X_C \angle -90° = -jX_C$) and a total impedance at $\angle -\theta$ ($Z_t \angle -\theta$) with respect to the resistance at $0°$. The complete impedance-phasor diagram looks much the same as the voltage-phasor diagram except for the labels (impedances instead of voltages).

As you can see, the total impedance is the result of the phasor sum of the resistance and the capacitive reactance ($Z_t \angle -\theta = R + X_C \angle -90° = R - jX_C$). Since a right triangle is formed from the phasor sum of resistance and inductive reactance, the total impedance phasor can be determined using the Pythagorean theorem and the inverse tangent of $-X_C/R$. If the total impedance and angle are already known, the resistance and capacitive reactance may be found using P→R conversion.

$$Z_t = \sqrt{R^2 + X_C^2} \tag{18.17}$$

also: $R = \sqrt{Z_t^2 - X_C^2}$ and $X_C = \sqrt{Z_t^2 - R^2}$

$$\angle \theta = \tan^{-1}(-X_C/R) = \tan^{-1}(-V_C/V_R) \tag{18.18}$$

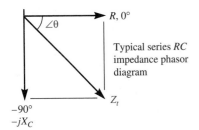

R, 0°

Typical series *RC*
impedance phasor
diagram

$-90°$
$-jX_C$

Z_t

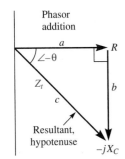

Phasor
addition

a

$\angle -\theta$

Z_t

c

Resultant,
hypotenuse

R

b

$-jX_C$

The Pythagorean theorem

$c = \sqrt{a^2 + b^2}$

$Z_t = \sqrt{R^2 + X_C^2}$

$\angle\theta = \tan^{-1}(-b/a) = \tan^{-1}(-X_C/R)$

$$\underset{\substack{\text{Polar} \\ \text{notation}}}{Z_t \angle -\theta} = \underset{\substack{\text{Rectangular} \\ \text{notation}}}{R - jX_C}$$

Figure 18-34 A complex series *RC*
circuit impedance-phasor diagram.

EXAMPLE   18-18

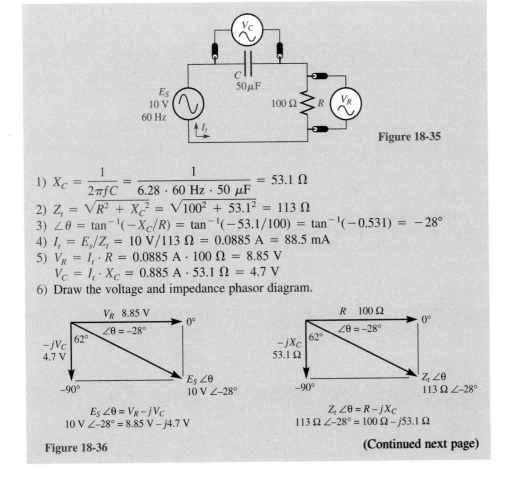

Figure 18-35

1) $X_C = \dfrac{1}{2\pi fC} = \dfrac{1}{6.28 \cdot 60 \text{ Hz} \cdot 50 \ \mu\text{F}} = 53.1 \ \Omega$

2) $Z_t = \sqrt{R^2 + X_C^2} = \sqrt{100^2 + 53.1^2} = 113 \ \Omega$

3) $\angle\theta = \tan^{-1}(-X_C/R) = \tan^{-1}(-53.1/100) = \tan^{-1}(-0.531) = -28°$

4) $I_t = E_s/Z_t = 10 \text{ V}/113 \ \Omega = 0.0885 \text{ A} = 88.5 \text{ mA}$

5) $V_R = I_t \cdot R = 0.0885 \text{ A} \cdot 100 \ \Omega = 8.85 \text{ V}$
   $V_C = I_t \cdot X_C = 0.885 \text{ A} \cdot 53.1 \ \Omega = 4.7 \text{ V}$

6) Draw the voltage and impedance phasor diagram.

V_R 8.85 V   0°

$\angle\theta = -28°$

62°

$-jV_C$
4.7 V

$-90°$

$E_S \angle\theta$
10 V $\angle -28°$

R 100 Ω   0°

$\angle\theta = -28°$

62°

$-jX_C$
53.1 Ω

$-90°$

$Z_t \angle\theta$
113 Ω $\angle -28°$

$E_S \angle\theta = V_R - jV_C$
10 V $\angle -28° = 8.85$ V $- j4.7$ V

$Z_t \angle\theta = R - jX_C$
113 Ω $\angle -28° = 100$ Ω $- j53.1$ Ω

Figure 18-36

(Continued next page)

EXAMPLE   18-18   (Continued)

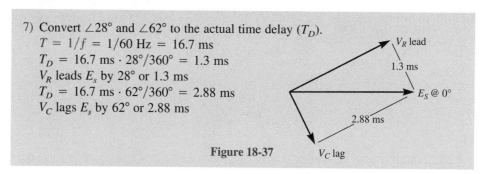

7) Convert $\angle 28°$ and $\angle 62°$ to the actual time delay (T_D).
$T = 1/f = 1/60 \text{ Hz} = 16.7 \text{ ms}$
$T_D = 16.7 \text{ ms} \cdot 28°/360° = 1.3 \text{ ms}$
V_R leads E_s by 28° or 1.3 ms
$T_D = 16.7 \text{ ms} \cdot 62°/360° = 2.88 \text{ ms}$
V_C lags E_s by 62° or 2.88 ms

Figure 18-37

RC Circuit Calculations

Now, let's take what has been discussed and apply it to actual series *RC* circuits. Consider Example 18-18. The theoretical analysis is performed in a systematic step-by-step method just as we did with the series *RL* circuits. The following step-by-step summary corresponds to those steps used in Example 18-18:

1. Determine the capacitive reactance at the applied frequency.

2. Use the Pythagorean theorem to find the total impedance.

3. Use the inverse tangent of $-X_C/R$ to find $\angle \theta$. (If Z_t and $\angle \theta$ are already known, use P→R conversion to find R and X_C.)

4. Use Ohm's Law to find the total current.

5. Use Ohm's Law to find the resistor and capacitor voltages. Resistor and capacitor voltages may also be found using P→R conversion when E_S and $\angle \theta$ are known.

6. Draw and label the impedance- and voltage-phasor diagrams.

7. If desired, convert phase angles to actual time delay, taking into account the time for one cycle at the applied frequency.

The next two examples help reinforce this analysis procedure. Example 18-20 demonstrates once again how R, X_C, V_R, and V_C can be found using P→R conversion. The manual method of P→R conversion is shown. If your scientific calculator has the P→R function, the values may be obtained a little more quickly using the automatic method.

EXAMPLE   18-19

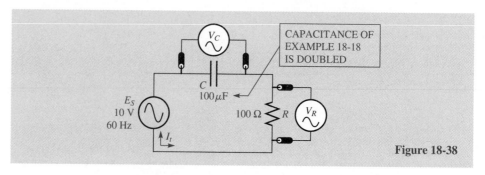

Figure 18-38

1) $X_C = \dfrac{1}{2\pi f C} = \dfrac{1}{6.28 \cdot 60 \text{ Hz} \cdot 100 \ \mu\text{F}} = 26.54 \ \Omega$

2) $Z_t = \sqrt{R^2 + X_C{}^2} = \sqrt{100^2 + 26.54^2} = 103.5 \ \Omega$

3) $\angle\theta = \tan^{-1}(-X_C/R) = \tan^{-1}(-26.54/100) = \tan^{-1}(-0.265) = -14.9°$

4) $I_t = E_S/Z_t = 10 \text{ V}/103.5 \ \Omega = 0.0966 \text{ A} = 96.6 \text{ mA}$

5) $V_R = I_t \cdot R = 0.0966 \text{ A} \cdot 100 \ \Omega = 9.66 \text{ V}$

   $V_C = I_t \cdot X_C = 0.0966 \text{ A} \cdot 26.54 \ \Omega = 2.56 \text{ V}$

6) Draw the voltage- and impedance-phasor diagram.

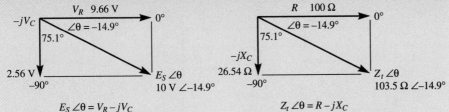

$$E_S \angle\theta = V_R - jV_C$$
$$10 \text{ V} \angle -14.9° = 9.66 \text{ V} - j2.56 \text{ V}$$

$$Z_t \angle\theta = R - jX_C$$
$$103.5 \ \Omega \angle -14.9° = 100 \ \Omega - j26.54 \ \Omega$$

Figure 18-39

7) Convert $\angle 14.9°$ and $\angle 75.1°$ to actual time delay (T_D).

   $T = 1/f = 1/60 \text{ Hz} = 16.7 \text{ ms}$

   $T_D = 16.7 \text{ ms} \cdot 14.9°/360° = 0.691 \text{ ms}$

   V_R leads E_S by 14.9° or 0.691 ms (691 μs)

   $T_D = 16.7 \text{ ms} \cdot 75.1°/360° = 3.48 \text{ ms}$

   V_C lags E_S by 75.1° or 3.48 ms

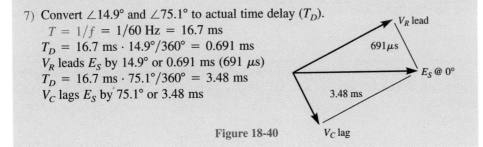

Figure 18-40

EXAMPLE   18-20

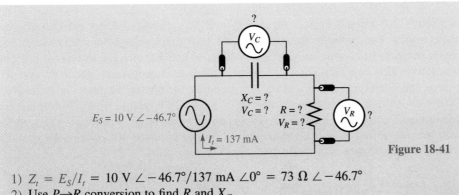

Figure 18-41

1) $Z_t = E_S/I_t = 10 \text{ V} \angle -46.7°/137 \text{ mA} \angle 0° = 73 \ \Omega \angle -46.7°$

2) Use $P{\rightarrow}R$ conversion to find R and X_C.

   $R = \cos \angle -46.7° \cdot Z_t = 0.686 \cdot 73 \ \Omega = 50 \ \Omega$

   $X_C = \sin \angle -46.7° \cdot Z_t = -0.728 \cdot 73 \ \Omega = -53.1 \ \Omega \ (-j53.1 \ \Omega)$

   (Continued next page)

EXAMPLE 18-20 (Continued)

3) $V_R = I_t \cdot R = 0.137 \text{ A} \cdot 50 \text{ }\Omega = 6.85 \text{ V}$
$V_C = I_t \cdot X_C = 0.137 \text{ A} \cdot 53.1 \text{ }\Omega = 7.27 \text{ V}$
also: using $P \rightarrow R$ conversion

$$V_R = \cos \angle -46.7° \cdot E_S = 0.686 \cdot 10 \text{ V} = 6.86 \text{ V}$$
$$V_C = \sin \angle -46.7° \cdot E_S = -0.728 \cdot 10 \text{ V} = -7.28 \text{ V} \ (-j7.28 \text{ V})$$

4) Draw the voltage- and impedance-phasor diagrams.

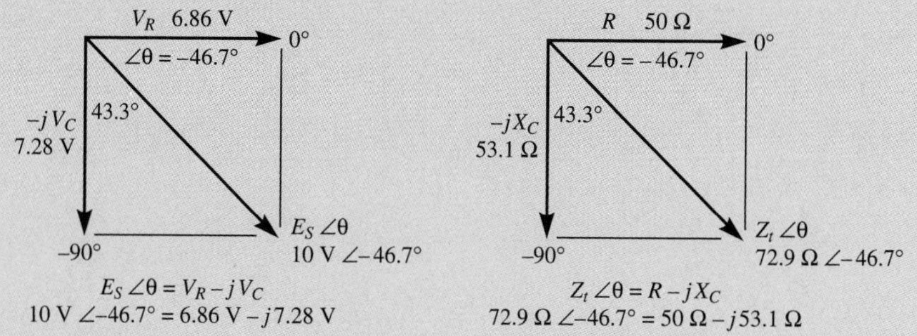

$$E_S \angle \theta = V_R - jV_C$$
$$10 \text{ V} \angle -46.7° = 6.86 \text{ V} - j7.28 \text{ V}$$

$$Z_t \angle \theta = R - jX_C$$
$$72.9 \text{ }\Omega \angle -46.7° = 50 \text{ }\Omega - j53.1 \text{ }\Omega$$

Figure 18-42

Once again, the problems of Self-Check 18-4 will reinforce what you have learned thus far. It would be best that you not continue until you understand how to do these problems.

18-5 Series *RCL* Circuit Analysis

RCL Circuit Voltage Phasors

RCL Voltage Phasors

In a series *RCL* circuit, such as Figure 18-44, three different voltages exist at different phase angles. You probably have a good idea how these voltages are represented in a phasor diagram from your knowledge of series *RL* and *RC* circuits. Instead of learning new information, we now simply combine what you already know about *RL* and *RC* circuits and circuit analysis. Figure 18-45 illustrates the phasor diagrams for series *RCL* circuits under three different conditions: $V_L > V_C$, $V_C > V_L$ and $V_L = V_C$. Notice that the phasor representing inductor voltage is still plotted in the $+90°$ position ($+jV_L$) and the voltage representing capacitor voltage is still plotted in the $-90°$ position ($-jV_C$). Of course, resistor voltage is in phase with circuit current at $0°$.

Adding *RCL* Voltage Phasors

Once again, the resistor, capacitor, and inductor voltages cannot be added arithmetically. They must be added using phasor addition. A simple arithmetic sum of the three voltages

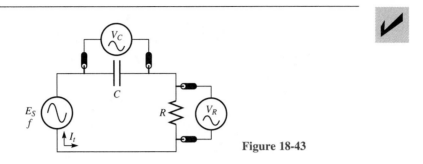

Figure 18-43

Fill in the blanks and draw the voltage-phasor diagram for each problem.

#	E_S	f	C	X_C	R	Z_t	I_t	V_R	V_C	$\angle\theta$
1.	30 V	50 Hz	20 μF		220 Ω					
2.	115 V	60 Hz	10 μF		180 Ω					
3.	25 μV		100 pF	3 kΩ	0.5 kΩ					
4.	50 mV	650 kHz		300 Ω	200 Ω					
5.	100 mV		0.1 μF	700 Ω			119 μA			$-56.1°$

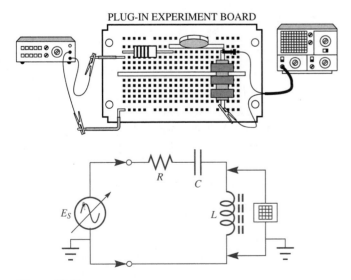

PLUG-IN EXPERIMENT BOARD

Figure 18-44

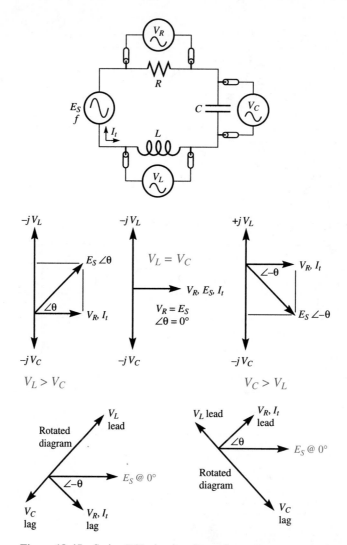

Figure 18-45   Series *RCL* circuit voltage phasors.

would be much greater than the source voltage. As before, Kirchhoff's Voltage Law applies only to the instantaneous voltages. The voltage phasors can be added following the rules for real and imaginary number addition. In this case, the resultant phasor, representing the source voltage, is found as follows:

$$E_S \angle \theta = V_R \angle 0° + V_L \angle +90° + V_C \angle -90° = V_R + jV_L - jV_C \qquad (18.19)$$

The magnitude of the sum of the three voltages (V_R, V_C, V_L) can be found using the Pythagorean theorem. As you can see, in Formula 18.20, the difference between the inductor and capacitor voltages ($V_L - V_C$) is used. This is because the inductor and capacitor voltages are 180° out of phase ($+90°$ and $-90°$). The result of the phasor sum of V_L and V_C is the total cancellation of one or both of the reactive voltages. If V_C is greater than V_L, a portion of V_C will remain, showing that the circuit is acting capacitively. If V_C is less than V_L, a portion of V_L will remain, showing that the circuit is acting inductively. And if V_C is equal to V_L, the reactive voltages will have cancelled each other, causing the circuit to appear as only a resistance (acting resistively—no phase shift between E_S and V_R, since $E_S = V_R$). The phase angle of the source voltage can be found using the inverse tangent as shown in Formula 18.21. As before, the R→P function of your scientific calculator can be used to quickly determine the source voltage magnitude and phase angle. However, when you enter the imaginary value, you must enter the difference between V_L and V_C ($V_L - V_C$).

$$E_S = \sqrt{V_R^2 + (V_L - V_C)^2} \qquad\qquad (18.20)$$
$$\angle\theta = \tan^{-1}[(V_L - V_C)/V_R] \qquad\qquad (18.21)$$

Cancelled Voltage?

What does it mean when we say the capacitor voltage cancels the inductor voltage and vice versa? Does it mean that the cancelled voltage no longer exists? Does it mean that, if the inductor voltage cancels the capacitor voltage, the capacitor will have no voltage across it? No. Each of the components (R, L, and C) will have its own voltage. If the voltage of each individual component is measured, it will have a voltage value as predicted by Ohm's Law (total current times the impedance of the component). Cancellation occurs when a voltage is taken across both the capacitor and the inductor at the same time. The voltmeter, or oscilloscope, will measure the result of the two reactive component voltages being 180° out of phase. The overall effect of the two reactive components working in opposition is to electrically eliminate one or both of the components. This should become clearer as we continue.

The Total *RCL* Picture

The total picture for *RCL* circuits can be seen in three different perspectives. In other words, any series *RCL* circuit will operate in one of three electrical states: (1) $V_L > V_C$, (2) $V_C > V_L$, or (3) $V_L = V_C$. As shown in Figure 18-46, this will cause $\angle\theta$ to be positive, negative, or at 0°. To the signal generator (voltage source), the *RCL* circuit appears electrically as a simple *RL* circuit, an *RC* circuit, or a resistor circuit having no reactance at all.

What determines whether an *RCL* circuit will act inductively ($\angle+\theta$), capacitively ($\angle-\theta$), or resistively ($\angle\theta = 0°$)? As has already been stated, it is determined by the amount of voltage on the capacitor and the inductor. But what determines if the capacitor voltage is greater than the inductor voltage and vice versa? This is determined by the inductive and capacitive reactance. If the inductive reactance is greater than the capacitive reactance, the inductor will have a greater voltage. Conversely, if the capacitive reactance is greater, the capacitor voltage is greater. If the reactances are equal, their voltages are equal. From the inductive and capacitive reactance formulas, we see that reactance is determined by component value and frequency ($X_L = 2\pi fL$ and $X_C = 1/2\pi fC$). Thus, the component value and applied frequency determine the reactance of the component, which in turn affects the component's voltage. (We will soon see how a change in the frequency of the applied source voltage can radically change how the circuit is acting.) In the analysis process, the actual component voltages are determined using Ohm's Law [$V_R = I_t \cdot R$, $V_L = I_t \cdot X_L$ (Formula 18.11), and $V_C = I_t \cdot X_C$ (Formula 18.16)].

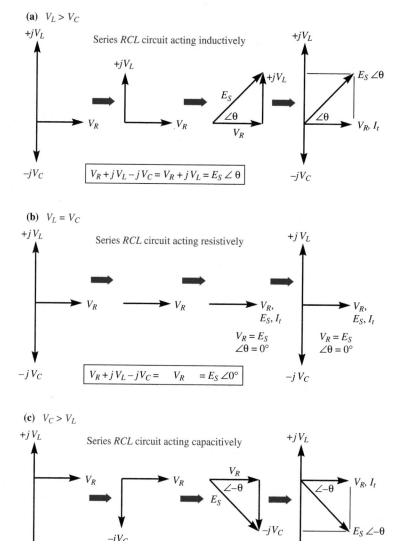

Figure 18-46   Phasor addition for the three electrical states of the series *RCL* circuit.

RCL Circuit Impedance Phasors

The series *RCL* impedance-phasor diagram is obtained in the same manner as for *RL* and *RC* circuits. Each voltage phasor is divided by the total current. The resulting impedance-phasor diagram appears to be the same as the voltage-phasor diagram except for labeling. Again, the impedance-phasor diagram shows that the circuit is acting inductively, capacitively, or resistively as determined by frequency, inductance, and capacitance (see Figure 18-47). From the impedance-phasor diagrams, it becomes evident that the total circuit

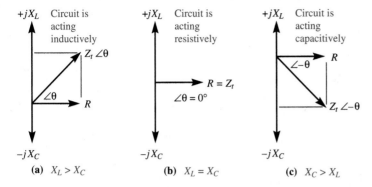

Figure 18-47   The series *RCL* circuit impedance-phasor diagrams.

impedance (Z_t) is the result of the phasor sum of resistance and reactances. The inductive and capacitive reactances are 180° out of phase and have a mutual cancelling effect. The larger of the two reactances determines how the circuit is acting. The Pythagorean theorem is used once again to determine the resultant total impedance and inverse tangent can be used to determine $\angle \theta$.

$$Z_t \angle \theta = R \angle 0° + X_L \angle +90° + X_C \angle -90° = R + jX_L - jX_C \qquad (18.22)$$

$$Z_t = \sqrt{R^2 + (X_L - X_C)^2} \qquad (18.23)$$

$$\angle \theta = \tan^{-1}[(X_L - X_C)/R] = \tan^{-1}[(V_L - V_C)/V_R] \qquad (18.24)$$

RCL Circuit Calculations

Enough discussion. It's time to apply what you have learned here and in the previous sections. We reinforce what has been covered by analyzing the same circuit at three different frequencies. This creates three totally different analysis problems even though the applied source voltage and component values are not changed. At the lower applied frequency, the capacitive reactance is greater than the inductive reactance, causing the circuit to act capacitively (as though the circuit were a simple *RC* circuit ($\angle -\theta$, Example 18-21). Conversely, at the higher frequency the inductive reactance is greater than the capacitive reactance, causing the circuit to act inductively (as though the circuit were a simple *RL* circuit ($\angle +\theta$, Example 18-23). At the center frequency, the inductive and capacitive reactances are equal, causing the circuit to appear purely resistive (Example 18-22). As you follow the examples, use your scientific calculator to verify the answers and to make sure you are using your calculator properly. Use the R→P function if it is available on your calculator. To find $Z_t \angle \theta$, enter the resistance, press INV and R→P (or 2nd and R→P), enter the difference between the reactive values ($X_L - X_C$), press = to read the magnitude (Z_t), then, press X↔Y to read the angle ($\angle \theta$).

Things to Recognize

1. The circuit impedance is minimum at a frequency where $X_L = X_C$ and $V_L = V_C$ because equal and opposite reactances cancel, leaving only resistance as the impedance in the circuit.

$$Z_t = \sqrt{R^2 + (X_L - X_C)^2} = \sqrt{R^2 + (0)^2} = \sqrt{R^2} = R$$

2. The total current is maximum when $X_L = X_C$ and $V_L = V_C$ because circuit impedance is minimum and equal to circuit resistance ($Z_t = R$). Thus, by Ohm's Law, current is maximum ($\uparrow I_t = E_S/Z_t\downarrow$).

3. The reactive voltages are maximum when $X_L = X_C$ and $V_L = V_C$ because the inductor and capacitor voltages are determined by the product of their reactance and current. When the current is maximum, the voltages will be maximum.

4. The resistor voltage is equal to the source voltage when $X_L = X_C$ and $V_L = V_C$ because the inductor and capacitor voltages effectively cancel each other and, in a real sense, disappear. This means the resistor appears electrically to be the only component in the circuit. Also, when $X_L = X_C$ and $V_L = V_C$ the product of the total current and the resistance will equal the applied source voltage.

EXAMPLE   18-21

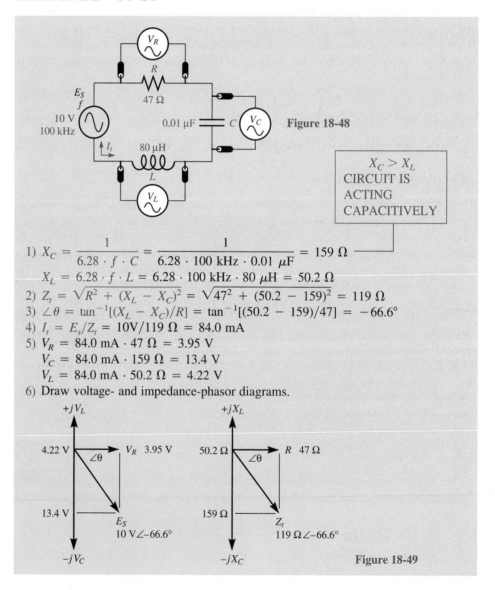

Figure 18-48

$X_C > X_L$
CIRCUIT IS
ACTING
CAPACITIVELY

1) $X_C = \dfrac{1}{6.28 \cdot f \cdot C} = \dfrac{1}{6.28 \cdot 100 \text{ kHz} \cdot 0.01\ \mu\text{F}} = 159\ \Omega$

   $X_L = 6.28 \cdot f \cdot L = 6.28 \cdot 100 \text{ kHz} \cdot 80\ \mu\text{H} = 50.2\ \Omega$

2) $Z_t = \sqrt{R^2 + (X_L - X_C)^2} = \sqrt{47^2 + (50.2 - 159)^2} = 119\ \Omega$

3) $\angle\theta = \tan^{-1}[(X_L - X_C)/R] = \tan^{-1}[(50.2 - 159)/47] = -66.6°$

4) $I_t = E_s/Z_t = 10\text{V}/119\ \Omega = 84.0 \text{ mA}$

5) $V_R = 84.0 \text{ mA} \cdot 47\ \Omega = 3.95 \text{ V}$

   $V_C = 84.0 \text{ mA} \cdot 159\ \Omega = 13.4 \text{ V}$

   $V_L = 84.0 \text{ mA} \cdot 50.2\ \Omega = 4.22 \text{ V}$

6) Draw voltage- and impedance-phasor diagrams.

Figure 18-49

7) All voltages may be compared in phase to the applied source voltage.

$T = 1/f = 1/100 \text{ kHz} = 10 \ \mu s$

V_L leads E_S by 156.6° or 4.35 μs

$4.35 \ \mu s = 10 \ \mu s \cdot 156.6°/360°$

V_R leads E_S by 66.6° or 1.85 μs

$1.85 \ \mu s = 10 \ \mu s \cdot 66.6°/360°$

V_C lags E_S by 23.4° or 0.65 μs

$0.65 \ \mu s = 10 \ \mu s \cdot 23.4°/360°$

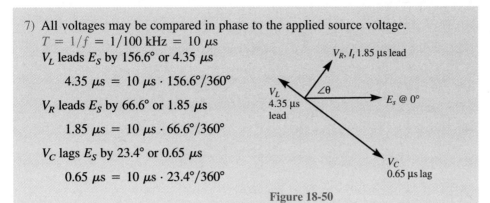

Figure 18-50

EXAMPLE 18-22

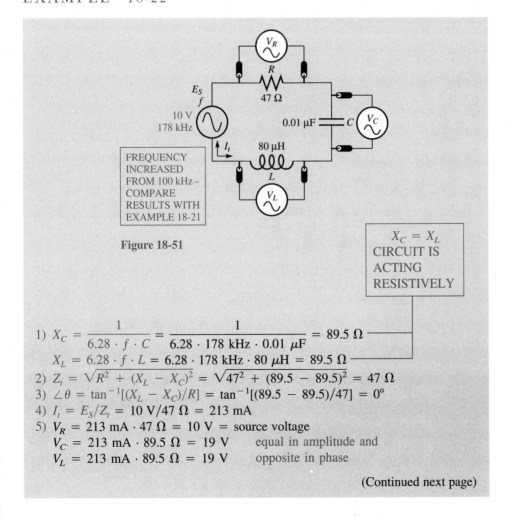

Figure 18-51

FREQUENCY INCREASED FROM 100 kHz– COMPARE RESULTS WITH EXAMPLE 18-21

$X_C = X_L$
CIRCUIT IS ACTING RESISTIVELY

1) $X_C = \dfrac{1}{6.28 \cdot f \cdot C} = \dfrac{1}{6.28 \cdot 178 \text{ kHz} \cdot 0.01 \ \mu F} = 89.5 \ \Omega$

$X_L = 6.28 \cdot f \cdot L = 6.28 \cdot 178 \text{ kHz} \cdot 80 \ \mu H = 89.5 \ \Omega$

2) $Z_t = \sqrt{R^2 + (X_L - X_C)^2} = \sqrt{47^2 + (89.5 - 89.5)^2} = 47 \ \Omega$

3) $\angle\theta = \tan^{-1}[(X_L - X_C)/R] = \tan^{-1}[(89.5 - 89.5)/47] = 0°$

4) $I_t = E_S/Z_t = 10 \text{ V}/47 \ \Omega = 213 \text{ mA}$

5) $V_R = 213 \text{ mA} \cdot 47 \ \Omega = 10 \text{ V} = $ source voltage

$V_C = 213 \text{ mA} \cdot 89.5 \ \Omega = 19 \text{ V}$ equal in amplitude and

$V_L = 213 \text{ mA} \cdot 89.5 \ \Omega = 19 \text{ V}$ opposite in phase

(Continued next page)

EXAMPLE 18-22 (Continued)

6) Draw voltage- and impedance-phasor diagrams.

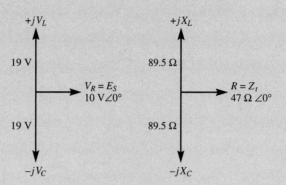

Figure 18-52

7) All voltages may be compared in phase to the applied source voltage.
$T = 1/f = 1/178$ kHz $= 5.62$ μs
V_L leads E_S by 90° or 1.4 μs (5.62 μs · 90°/360°).
V_R is in phase with E_S (both at 0°).
V_C lags E_S by 90° or 1.4 μs (5.62 μs · 90°/360°).

EXAMPLE 18-23

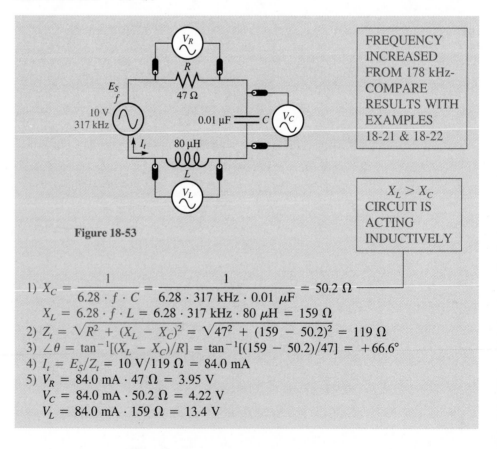

Figure 18-53

FREQUENCY INCREASED FROM 178 kHz-COMPARE RESULTS WITH EXAMPLES 18-21 & 18-22

$X_L > X_C$ CIRCUIT IS ACTING INDUCTIVELY

1) $X_C = \dfrac{1}{6.28 \cdot f \cdot C} = \dfrac{1}{6.28 \cdot 317 \text{ kHz} \cdot 0.01 \text{ } \mu\text{F}} = 50.2 \text{ } \Omega$
   $X_L = 6.28 \cdot f \cdot L = 6.28 \cdot 317 \text{ kHz} \cdot 80 \text{ } \mu\text{H} = 159 \text{ } \Omega$
2) $Z_t = \sqrt{R^2 + (X_L - X_C)^2} = \sqrt{47^2 + (159 - 50.2)^2} = 119 \text{ } \Omega$
3) $\angle\theta = \tan^{-1}[(X_L - X_C)/R] = \tan^{-1}[(159 - 50.2)/47] = +66.6°$
4) $I_t = E_S/Z_t = 10 \text{ V}/119 \text{ } \Omega = 84.0 \text{ mA}$
5) $V_R = 84.0 \text{ mA} \cdot 47 \text{ } \Omega = 3.95 \text{ V}$
   $V_C = 84.0 \text{ mA} \cdot 50.2 \text{ } \Omega = 4.22 \text{ V}$
   $V_L = 84.0 \text{ mA} \cdot 159 \text{ } \Omega = 13.4 \text{ V}$

6) Draw voltage- and impedance-phasor diagrams.

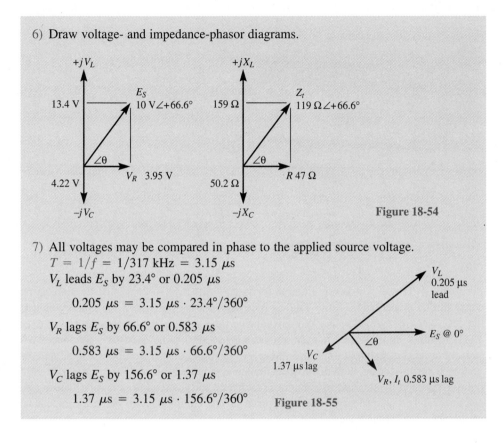

Figure 18-54

7) All voltages may be compared in phase to the applied source voltage.

$T = 1/f = 1/317 \text{ kHz} = 3.15 \ \mu s$

V_L leads E_S by 23.4° or 0.205 μs

$\quad 0.205 \ \mu s = 3.15 \ \mu s \cdot 23.4°/360°$

V_R lags E_S by 66.6° or 0.583 μs

$\quad 0.583 \ \mu s = 3.15 \ \mu s \cdot 66.6°/360°$

V_C lags E_S by 156.6° or 1.37 μs

$\quad 1.37 \ \mu s = 3.15 \ \mu s \cdot 156.6°/360°$

Figure 18-55

The analysis procedure used in the previous examples is very much the same as that used for *RL* and *RC* circuits. In fact, this procedure can be used to analyze series *RL*, *RC*, or *RCL* circuits. If a component is missing, a zero is simply entered in its place or the formula is adjusted to eliminate that component's variable (X_L or X_C and V_L or V_C). Here is a summary:

General Series AC Circuit Analysis Procedure

1. Determine the capacitive and inductive reactances at the applied frequency.
2. Use the Pythagorean theorem to find the total impedance (Formula 18.23).
3. Use the inverse tangent of $(X_L - X_C)/R$ to find $\angle\theta$ (Formula 18.24).
4. Use Ohm's law to find the total current ($I_t = E_S/Z_t$).
5. Use Ohm's Law to find the resistor, inductor, and capacitor voltages.
6. Draw and label the impedance- and voltage-phasor diagrams.
7. If desired, convert phase angles to actual time delay, taking into account the time for one cycle at the applied frequency and the phase angles.

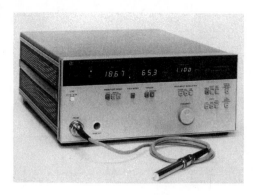

Figure 18-56 The vector impedance meter. (Courtesy of the Hewlett-Packard Company.)

Measuring an Unknown Impedance

Often, Measurement Is Necessary

For many AC circuits, it is not possible (or is very difficult) to determine the total impedance through calculation. Some kind of measurement is often necessary. Such would be the case when stray inductance and capacitance have a significant effect on circuit operation. In other applications, it may not be easy to determine inductive reactances and capacitive reactances in a circuit or system. An example of this would be a large antenna system for a broadcast station. As the antenna is adjusted for operation on a particular radio frequency, the amount of antenna capacitance and inductance will vary, causing the antenna's impedance to vary. The right piece of test equipment can make an impedance measurement quick and accurate.

The Right Equipment

One of the nicest pieces of test equipment used to measure an unknown impedance is the **vector impedance meter**. Figure 18-56 shows a digital display vector impedance meter manufactured by Hewlett-Packard. Notice the vector impedance meter has three digital displays: the first display is for magnitude (Z_t), the second is for phase angle ($\angle \theta$), and the third is for frequency. The vector impedance meter has a broad-spectrum radio-

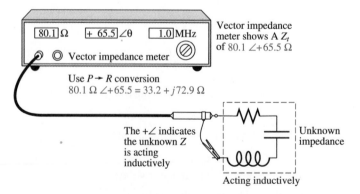

Figure 18-57 Using the vector impedance meter.

frequency generator built in. The frequency of the generator may be selected manually, or the generator can be swept (scanned) automatically between any two preset frequencies. The frequency is applied to the circuit under test through the test probe. Thus, the impedance of any circuit can be measured at any frequency within the range of the signal generator. (Recall that the impedance of an AC circuit depends on the frequency applied to the circuit—Examples 18-21, 18-22, and 18-23.) Figure 18-57 illustrates how a vector impedance meter is used to measure an unknown impedance.

Now, here's your chance to test your analysis skills. Solve the problems of Self-Check 18-5. Check your answers with those provided at the end of the chapter.

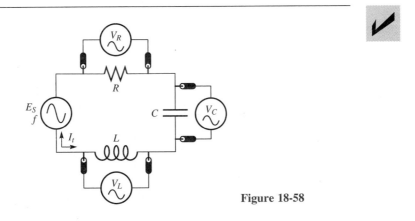

Figure 18-58

Fill in the blanks and draw the voltage-phasor diagram for each problem.

#	E_S	f	R	C	L	X_C	X_L	Z_t	I_t	V_R	V_C	V_L	$\angle\theta$
1.	10 V	1 kHz	1 kΩ	0.47 μF	0.1 H								
2.	25 mV	455 kHz	20 Ω	0.008 μF	10 μH								
3.	50 mV	10 MHz							1.02 mA	40.8 mV	67.6 mV	96.1 mV	
4.		120 MHz	5 Ω	50 pF	0.01 μH			19.6 Ω	4.07 μA				
5.				10 μF	1 H	133 Ω			95.3 mA	9.53 V			

18-6 Power in AC Circuits

Real Power (P_R)

Real power (P_R) (sometimes called **true power** or **average power**) in an AC circuit is the electrical power that is actually converted into heat, radiation (radio waves, light, etc.), or work (mechanical motion). In an *RCL* AC circuit, the resistor is, ideally, the only component that converts electrical power to an unrecoverable form of energy (heat). It should be noted, however, that some power is converted to heat in inductors and capacitors. The winding resistance and core losses of inductors and the dielectric losses of capacitors account for very small real power conversions (I^2R losses). In some applications these small losses are not important and need not be considered. In other applications the quality of the components is very important. More will be said about this in a later chapter on resonance. Calculating real power is very straightforward. Just remember that real power is only dissipated in resistance. Therefore, real power calculations deal with the resistor or resistors, as follows:

$$P_R = I_R \cdot V_R = I_R{}^2 \cdot R = V_R{}^2/R \quad \text{(W)} \tag{18.25}$$

Reactive Power (P_r)

Reactive power (P_r) is the power that a capacitor or inductor *seems* to be using. However, in a sense, capacitors and inductors do not use power. That is, they do not convert electrical power to some other form of unrecoverable energy (except for the small losses previously mentioned). Reactive components store electrical energy in the form of a charge or force field (in an electric or magnetic energy field). This energy is recovered and returned to the circuit during the next half cycle. Even though the reactive component does not use the energy, the charge and discharge current or buildup and decay current *is* current. Any current multiplied by a voltage qualifies as being some kind of power according to the power formula ($P = IE$). To distinguish this reactive power from real power, the watt power unit is not used. Instead, **voltamperes reactive (VAR)** is used. The normal power formulas are used to calculate reactive power with VAR used for the units. Only the current through the reactive components and the voltage across the reactive components are used to calculate reactive power.

$$P_r = P_C = I_C \cdot V_C = I_C{}^2 \cdot X_C = V_C{}^2/X_C \quad \text{(VAR)}$$
$$= P_L = I_L \cdot V_L = I_L{}^2 \cdot X_L = V_L{}^2/X_L \quad \text{(VAR)} \tag{18.26}$$

Apparent Power (P_A)

Apparent power (P_A) is the total power that the entire AC circuit is apparently using. It is the total power supplied to the circuit from the source. Since an AC circuit may consist of reactances (capacitance and inductance), some of the total apparent power may be reactive power. Apparent power is a combination of real and reactive power. Some of the total applied power is being converted to heat and work, while some is being stored and returned to the circuit. Therefore, apparent power must be labeled in **volt-ampere (VA)** units to distinguish it from real power (recall transformer power ratings, Section 17-4). Apparent power is the phasor sum of all real and reactive powers in a circuit (more on this later). It is calculated using the normal power formulas. However, total circuit parameters must be used as follows:

$$P_A = P_{\text{total}} = I_t \cdot E_S = I_t^2 \cdot Z_t = E_S^2/Z_t \quad \text{(VA)} \tag{18.27}$$

The Power Factor (*PF*)

The **power factor** (*PF*) is a decimal number, equal to or less than 1, that indicates the relationship between real power and apparent power. It is also known as an efficiency factor, since it indicates what portion of the total applied power is actually converted to heat or work. The power conversion efficiency in percent is found by multiplying the power factor by 100% (% eff = $PF \cdot 100\%$). The power factor itself is the decimal expression of the ratio of real power to apparent power as follows:

$$PF = P_R/P_A \quad (\leq 1, \text{ no units}) \tag{18.28}$$

Power Phasors

Creating the Power Phasor Diagrams

Recall that to create the impedance phasors and phasor diagram, we simply divided all voltage phasors representing a series *RL*, *RC*, or *RCL* circuit by the total current. In like manner, we may multiply each voltage phasor by total current to create power phasors and the power-phasor diagram: $V_L \angle 90° \cdot I_t \angle 0° = P_L \angle 90°$; $V_C \angle -90° \cdot I_t \angle 0° = P_C \angle -90°$; $V_R \angle 0° \cdot I_t \angle 0° = P_R \angle 0°$ and $E_S \angle \theta \cdot I_t \angle 0° = P_A \angle \theta$. We can do this because current times voltage equals power ($P = IE$). The result is the trio of power-phasor diagrams shown in Figure 18-59. As you can see, the three typical diagrams are for $X_L > X_C$, $X_L = X_C$, and $X_C > X_L$.

Apparent Power Is a Phasor Sum

The phasor diagrams show very clearly that the apparent power is the phasor sum of the real and reactive power in the circuit. Notice how the reactive powers cancel each other

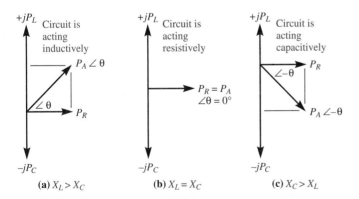

Figure 18-59   Power phasor diagrams for series *RCL* circuits.

just as reactive voltages and reactances do. This is because the capacitive and inductive powers are 180° out of phase ($+jP_L$ and $-jP_C$).

$$P_A = P_R \angle 0° + P_L \angle +90° + P_C \angle -90° = P_R + jP_L - jP_C \quad \text{(VA)} \quad (18.29)$$

Power Factor and Cos $\angle\theta$

Something else becomes evident as we study the power-phasor diagrams further. The power factor is actually equal to the cosine of $\angle\theta$. This is because the cosine function is the ratio of the side adjacent to $\angle\theta$ to the hypotenuse (cos $\angle\theta$ = adjacent/hypotenuse). In this case, the adjacent side is the real power and the hypotenuse is the apparent power. In the impedance-phasor diagram, the adjacent side is resistance and the hypotenuse is the total impedance. In the voltage-phasor diagram, the adjacent side is resistor voltage and the hypotenuse is the source voltage. Thus, any one of the following formulas may be used to find the power factor of a series *RC*, *RL*, or *RCL* circuit:

$$PF = \cos \angle\theta = P_R/P_A = V_R/E_S = R/Z_t \quad (\leq 1, \text{ no units}) \quad (18.30)$$

When Apparent Power Equals Real Power

Notice one more very important fact. When $X_L = X_C$, causing $V_L = V_C$ and $P_L = P_C$, $\angle\theta$ will be 0°. If $\angle\theta$ is 0°, the apparent power and the real power become one and the same (the same phasor at 0°). In other words, apparent power equals real power when $X_L = X_C$ and $\angle\theta = 0°$. Since the power factor is equal to the cosine of $\angle\theta$, the power factor is 1 when $\angle\theta$ is 0° (cos $\angle 0° = 1$). A power factor of 1 indicates a power efficiency of 100%, where all of the applied power is converted to heat or work. In some applications, it is important for the AC circuit to have as high a power factor as possible for the most efficient conversion of electricity. A high power factor means there is little or no reactance in the electrical circuit (or at least the reactance has been cancelled out by an equal and opposite reactance). For example, a large industrial factory looks like a large, yet simple, AC circuit to the power company. In most cases, the factory looks like an *RL* circuit (the inductance contributed by motors, etc.). If the factory has too much inductance, the power factor will be low and the efficiency of the factory will be low. In effect, the factory is demanding more electrical energy than it is converting to heat or work. In such cases, large banks of capacitors are added to the power grid to correct the power factor and increase the efficiency of the factory. The capacitors cancel the inductance. In many cases, a surcharge penalty may be imposed on the factory by the power company if the power factor is less than 0.85 (<85% efficiency). Thus, the power factor, and power-factor correction, can be very important.

Series *RCL* Power Calculations

As we examine actual AC circuit power calculations, we will revisit three previous examples: Example 18-21, Example 18-22, and Example 18-23, shown here as Examples 18-24, 18-25, and 18-26. In each example, the power formulas are applied as previously discussed. Once again, the Pythagorean theorem can be used to calculate the resultant phasor and the inverse tangent can be used to find the phase angle ($\angle\theta$). In this case, the resultant phasor is the apparent power. Recall that apparent power is the phasor sum of the real and reactive powers (Formula 18.29). Applying reactive power and real power to the Pythagorean threorem and the tangent function yields the following formulas:

$$P_A = \sqrt{P_R^2 + (P_L - P_C)^2} \tag{18.31}$$

$$\angle\theta = \tan^{-1}\left[(P_L - P_C)/P_R\right] \tag{18.32}$$

Note: you should recognize that these formulas can be used for any *RL*, *RC*, or *RCL* circuit. Simply delete the variable of the missing reactive component from the formula.

EXAMPLE   18-24

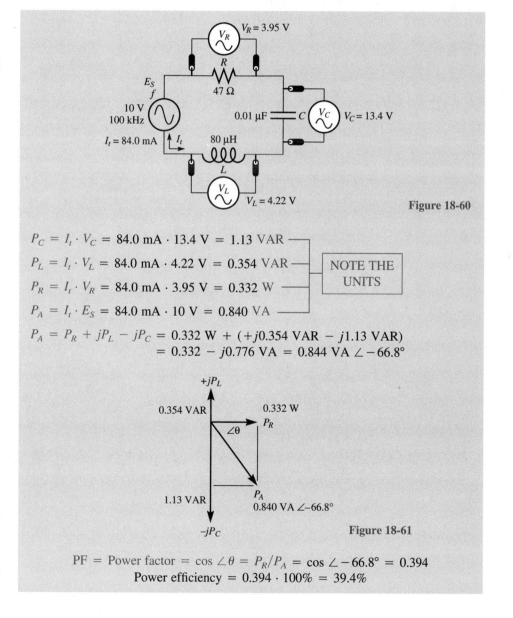

Figure 18-60

$P_C = I_t \cdot V_C = 84.0 \text{ mA} \cdot 13.4 \text{ V} = 1.13 \text{ VAR}$ ⎤

$P_L = I_t \cdot V_L = 84.0 \text{ mA} \cdot 4.22 \text{ V} = 0.354 \text{ VAR}$ ⎤ NOTE THE

$P_R = I_t \cdot V_R = 84.0 \text{ mA} \cdot 3.95 \text{ V} = 0.332 \text{ W}$ ⎦ UNITS

$P_A = I_t \cdot E_S = 84.0 \text{ mA} \cdot 10 \text{ V} = 0.840 \text{ VA}$ ⎦

$P_A = P_R + jP_L - jP_C = 0.332 \text{ W} + (+j0.354 \text{ VAR} - j1.13 \text{ VAR})$
$\qquad = 0.332 - j0.776 \text{ VA} = 0.844 \text{ VA} \angle -66.8°$

Figure 18-61

$\text{PF} = \text{Power factor} = \cos\angle\theta = P_R/P_A = \cos\angle -66.8° = 0.394$

$\text{Power efficiency} = 0.394 \cdot 100\% = 39.4\%$

EXAMPLE 18-25

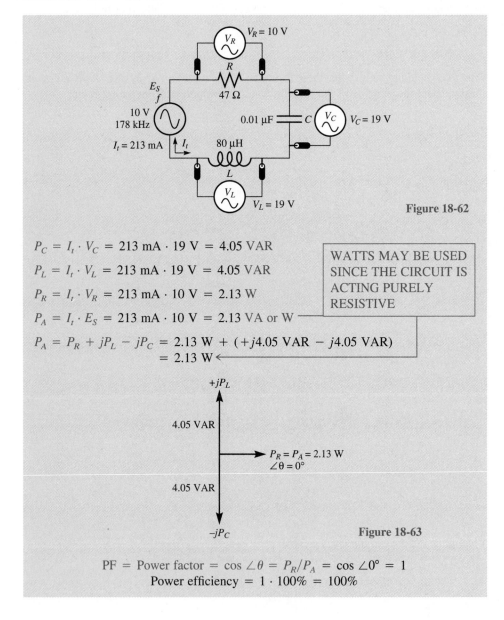

Figure 18-62

$P_C = I_t \cdot V_C = 213 \text{ mA} \cdot 19 \text{ V} = 4.05 \text{ VAR}$

$P_L = I_t \cdot V_L = 213 \text{ mA} \cdot 19 \text{ V} = 4.05 \text{ VAR}$

$P_R = I_t \cdot V_R = 213 \text{ mA} \cdot 10 \text{ V} = 2.13 \text{ W}$

$P_A = I_t \cdot E_S = 213 \text{ mA} \cdot 10 \text{ V} = 2.13 \text{ VA or W}$

$P_A = P_R + jP_L - jP_C = 2.13 \text{ W} + (+j4.05 \text{ VAR} - j4.05 \text{ VAR})$
$ = 2.13 \text{ W}$

WATTS MAY BE USED SINCE THE CIRCUIT IS ACTING PURELY RESISTIVE

$+jP_L$

4.05 VAR

$P_R = P_A = 2.13 \text{ W}$
$\angle\theta = 0°$

4.05 VAR

$-jP_C$

Figure 18-63

$$PF = \text{Power factor} = \cos \angle\theta = P_R/P_A = \cos \angle 0° = 1$$
$$\text{Power efficiency} = 1 \cdot 100\% = 100\%$$

EXAMPLE 18-26

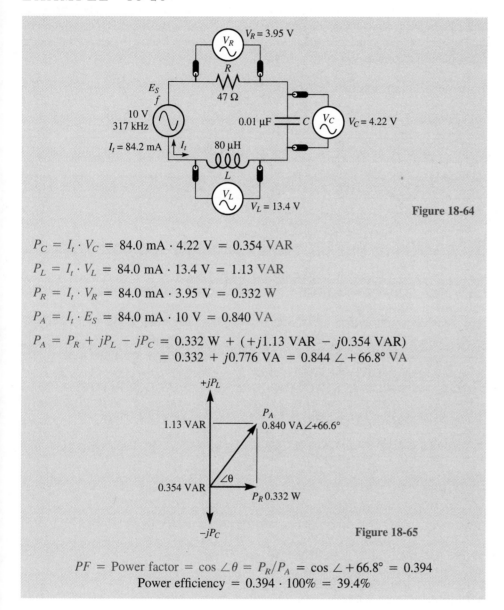

Figure 18-64

$P_C = I_t \cdot V_C = 84.0 \text{ mA} \cdot 4.22 \text{ V} = 0.354 \text{ VAR}$

$P_L = I_t \cdot V_L = 84.0 \text{ mA} \cdot 13.4 \text{ V} = 1.13 \text{ VAR}$

$P_R = I_t \cdot V_R = 84.0 \text{ mA} \cdot 3.95 \text{ V} = 0.332 \text{ W}$

$P_A = I_t \cdot E_S = 84.0 \text{ mA} \cdot 10 \text{ V} = 0.840 \text{ VA}$

$P_A = P_R + jP_L - jP_C = 0.332 \text{ W} + (+j1.13 \text{ VAR} - j0.354 \text{ VAR})$
$= 0.332 + j0.776 \text{ VA} = 0.844 \angle +66.8° \text{ VA}$

Figure 18-65

$PF = \text{Power factor} = \cos \angle \theta = P_R/P_A = \cos \angle +66.8° = 0.394$
$\text{Power efficiency} = 0.394 \cdot 100\% = 39.4\%$

As you considered these examples, the following should have been apparent:

1. The inductive and capacitive reactive powers have a cancelling effect on one another $(+jP_L$ and $-jP_C)$.

2. Make sure you can see the relationship between power factor, $\angle \theta$, apparent power and real power, and circuit efficiency.

3. The circuit is most efficient when all reactance is cancelled $(X_L = X_C)$.

Now, take a few minutes to study Design Note 18-1. Then test your understanding by solving the problems of Self-Check 18-6. *Note:* A series RCL circuit analysis program called Phasepro is available on diskette for the IBM® and compatibles, along with all of the Design Note programs throughout this text. Ask your instructor to contact the author or the publisher for a master copy.

SELF-CHECK 18-6

Go back to Self-Check 18-5 and calculate P_L, P_C, P_R, P_A, and PF for each problem. Check your answers at the end of the chapter.

DESIGN NOTE 18-1:   Series *RCL* Analysis

PROGRAM VARIABLES	
ES = SOURCE VOLTAGE	TH = ANGLE THETA
F = FREQUENCY	VR = RESISTOR VOLTAGE
R = RESISTANCE	VC = CAPACITOR VOLTAGE
C = CAPACITANCE	VL = INDUCTOR VOLTAGE
L = INDUCTANCE	PR = REAL POWER
XC = CAPACITIVE REACTANCE	PC = CAPACITOR POWER
XL = INDUCTIVE REACTANCE	PL = INDUCTOR POWER
ZT = TOTAL IMPEDANCE	PA = APPARENT POWER
I = TOTAL CURRENT	T = CYCLE TIME

"BASIC" PROGRAM SOLUTION

```
10 CLS
20 PRINT"SERIES RCL ANALYSIS"
30 PRINT"":PRINT""
40 PRINT"THIS PROGRAM WILL ANALYZE ANY SERIES RC, RL, OR RCL CIRCUIT."
50 PRINT"ENTER A ZERO (0) FOR L OR C IF IT DOES NOT EXIST IN YOUR CIRCUIT."
60 PRINT""
70 INPUT"ENTER THE INDUCTANCE IN HENRIES. ";L
80 INPUT"ENTER THE CAPACITANCE IN FARADS. ";C
90 INPUT"ENTER THE TOTAL SERIES RESISTANCE. ";R
100 INPUT"ENTER THE SOURCE VOLTAGE. ";ES
110 INPUT"ENTER THE FREQUENCY OF THE APPLIED AC. ";F
120 IF C = 0 THEN GOTO 140
130 XC = 1/(6.2831853# * F * C)
140 IF L=0 THEN GOTO 160
150 XL = 6.2831853# * F * L
160 ZT = SQR(R^2 + (XL − XC)^2)
170 I = ES/ZT
180 TH = ATN((XL − XC)/R) * 57.29578
190 VR = I * R:VC = I * XC:VL = I * XL
200 PR = I * VR:PC = I * VC:PL = I * VL
```

```
210 PA = I * ES:PF = PR/PA
220 T = 1/F
300 IF C = 0 THEN GOTO 320
310 PRINT"XC = ";XC;" OHMS."
320 IF L = 0 THEN GOTO 340
330 PRINT"XL = ";XL;" OHMS."
340 PRINT"ZT = ";ZT;" OHMS."
350 PRINT"I = ";I;" AMPS."
360 PRINT"ANGLE THETA = ";TH;" DEGREES."
370 PRINT"VR = ";VR;" VOLTS. PR = ";PR;" WATTS."
380 IF C = 0 THEN GOTO 400
390 PRINT"VC = ";VC;" VOLTS. PC = ";PC;" VOLT-AMPERES REACTIVE"
400 IF L = 0 THEN GOTO 420
410 PRINT"VL = ";VL;" VOLTS. PL = ";PL;" VOLT-AMPERES REACTIVE."
420 PRINT"PA = ";PA;" VOLT AMPERES. PF = ";PF
520 IF XL > XC THEN GOTO 590
530 PRINT"VR LEADS ES BY ";-TH;" DEGREES OR ";T * -TH/360;" SECONDS."
540 IF L = 0 THEN GOTO 560
550 PRINT"VL LEADS ES BY ";90 - TH;" DEGREES OR ";T * (90 - TH)/360;" SECONDS."
560 IF C = 0 THEN GOTO 580
570 PRINT"VC LAGS ES BY ";90 + TH;" DEGREES OR ";T * (TH + 90)/360;" SECONDS."
580 GOTO 640
590 PRINT"VR LAGS ES BY ";TH;" DEGREES OR ";T * TH/360;" SECONDS."
600 IF C = 0 THEN GOTO 620
610 PRINT"VC LAGS ES BY ";90 + TH;" DEGREES OR ";T * (90 + TH)/360;" SECONDS."
620 IF L = 0 THEN GOTO 640
630 PRINT"VL LEADS ES BY ";90 - TH;" DEGREES OR ";T * (90 - TH);" SECONDS."
640 PRINT""
650 INPUT"ANOTHER PROBLEM? (Y/N)";A$
660 IF A$ = "Y" THEN CLEAR:GOTO 30
670 CLS:END
```

Summary

FORMULAS

Trigonometry Formulas

(18.1) $c = \sqrt{a^2 + b^2}$ (Pythagorean theorem)

where c is the resultant magnitude opposite the right angle and a and b are two magnitudes adjacent to the right angle.

(18.2) $\angle\theta = \sin^{-1}(b/c)$

where $\sin^{-1}$ is the inverse sine, b is the side opposite $\angle\theta$, and c is the hypotenuse or resultant side.

(18.3) $\angle\theta = \cos^{-1}(a/c)$

where $\cos^{-1}$ is the inverse cosine, a is the side adjacent to $\angle\theta$, and c is the hypotenuse or resultant side.

(18.4) $\angle\theta = \tan^{-1}(b/a)$

where $\tan^{-1}$ is the inverse tangent, b is the side opposite $\angle\theta$, and a is the side adjacent to $\angle\theta$.

(18.5) $a = \cos\angle\theta \cdot c$

where a is the real magnitude and c is the given polar magnitude (resultant, hypotenuse).

(18.6) $b = \sin\angle\theta \cdot c$

where b is the imaginary magnitude and c is the given polar magnitude.

Formulas for Series *RL* Circuit Analysis

(18.7) $E_S = \sqrt{V_R^2 + V_L^2}$

   also: $V_R = \sqrt{E_s^2 - V_L^2}$ and $V_L = \sqrt{E_s^2 - V_R^2}$

(18.8) $\angle\theta = \tan^{-1}(V_L/V_R)$

(18.9) $I_t = E_S/Z_t$

(18.10) $V_R = I_t \cdot R$

(18.11) $V_L = I_t \cdot X_L$

(18.12) $Z_t = \sqrt{R^2 + X_L^2}$

   also: $R = \sqrt{Z_t^2 - X_L^2}$ and $X_L = \sqrt{Z_t^2 - R^2}$

(18.13) $\angle\theta = \tan^{-1}(X_L/R) = \tan^{-1}(V_L/V_R)$

Formulas for Series *RC* Circuit Analysis

(18.14) $E_S = \sqrt{V_R^2 + V_C^2}$

   also: $V_R = \sqrt{E_s^2 - V_C^2}$ and $V_C = \sqrt{E_s^2 - V_R^2}$

(18.15) $\angle\theta = \tan^{-1}(-V_C/V_R)$

(18.16) $V_C = I_t \cdot X_C$

(18.17) $Z_t = \sqrt{R^2 + X_C^2}$

   also: $R = \sqrt{Z_t^2 - X_C^2}$ and $X_C = \sqrt{Z_t^2 - R^2}$

(18.18) $\angle\theta = \tan^{-1}(-X_C/R) = \tan^{-1}(-V_C/V_R)$

Formulas for Series *RCL* Circuit Analysis

(18.19) $E_S \angle\theta = V_R \angle 0° + V_L \angle +90° + V_C \angle -90° = V_R + jV_L - jV_C$

(18.20) $E_S = \sqrt{V_R^2 + (V_L - V_C)^2}$

(18.21) $\angle\theta = \tan^{-1}[(V_L - V_C)/V_R]$

(18.22) $Z_t \angle\theta = R \angle 0° + X_L \angle +90° + X_C \angle -90° = R + jX_L - jX_C$

(18.23) $Z_t = \sqrt{R^2 + (X_L - X_C)^2}$

(18.24) $\angle\theta = \tan^{-1}[(X_L - X_C)/R] = \tan^{-1}[(V_L - V_C)/V_R]$

Power Formulas for Series *RCL* Circuit Analysis

(18.25) $P_R = I_R \cdot V_R = I_R^2 \cdot R = V_R^2/R$ (W)

(18.26) $P_r = P_C = I_C \cdot V_C = I_C^2 \cdot X_C = V_C^2/X_C$ (VAR
$= P_L = I_L \cdot V_L = I_L^2 \cdot X_L = V_L^2/X_L$ (VAR)

(18.27) $P_A = P_{\text{total}} = I_t \cdot E_S = I_t^2 \cdot Z_t = E_S^2/Z_t$ (VA)

(18.28) $PF = P_R/P_A$ (≤ 1, no units)

(18.29) $P_A = P_R \angle 0° + P_L \angle +90° + P_C \angle -90° = P_R + jP_L - jP_C$ (VA)

(18.30) $PF = \cos \angle \theta = P_R/P_A = V_R/E_S = R/Z_t$ (≤ 1, no units)

(18.31) $P_A = \sqrt{P_R^2 + (P_L - P_C)^2}$ (VA)

(18.32) $\angle \theta = \tan^{-1}[(P_L - P_C)/P_R]$

CONCEPTS

• A complex impedance is an opposition to AC that includes resistance and reactance.
• Operator j is a $+90°$ angle and is mathematically equivalent to $\sqrt{-1}$.
• Rectangular notation is composed of a real and an imaginary part (using operator j) and polar notation is composed of a magnitude and an angle.
• If complex numbers are to be added or subtracted, they must be in rectangular form.
• If complex numbers are to be multiplied or divided, it is more convenient to perform these operations using the polar form of the complex numbers.
• Phase angles, or phase shifts, actually represent a lead or lag in time.
• Inductive reactance and capacitive reactance are opposing reactances that have an overall cancelling effect on one another in an AC circuit. The entire circuit will take on the characteristics of the larger of the two reactances (X_L or X_C).
• Resistor power is real, true, or average power and is expressed in units of watts (W).
• Reactive power is the power stored in a capacitor or inductor and is expressed in units of volt-amperes reactive (VAR).
• Apparent power is the total applied power and is the power that an AC circuit is apparently using. It is expressed in units of volt-amperes (VA).
• Power factor is the ratio of real power to apparent power expressed in decimal form and indicates the efficiency of an AC circuit.

PROCEDURES

General Series AC Circuit-Analysis Procedure

1. Determine the capacitive and inductive reactances at the applied frequency.
2. Use the Pythagorean theorem to find the total impedance (Formula 18.23).
3. Use the inverse tangent of $(X_L - X_C)/R$ to find $\angle \theta$ (Formula 18.24).
4. Use Ohm's Law to find the total current ($I_t = E_S/Z_t$).
5. Use Ohm's Law to find the resistor, inductor, and capacitor voltages.
6. Draw and label the impedance- and voltage-phasor diagrams.
7. If desired, convert phase angles to actual time delay, taking into account the time for one cycle at the applied frequency and the phase angles.

SPECIAL TERMS

- Complex numbers, complex impedance, complex algebra
- Rectangular notation, polar notation
- Rectangular coordinate system
- Origin, abscissa, ordinate
- The Pythagorean theorem
- The complex plane
- Real numbers, imaginary numbers
- Real axis, imaginary axis
- Operator, operator *j*
- Polar system, polar plane, polar chart
- Angle theta, $\angle\theta$
- Conjugate, rationalization
- Phasor sum
- Voltage lead, voltage lag
- Vector impedance meter
- Real power (P_R), reactive power (P_L and P_C)
- Apparent power (P_A), power factor (*PF*)
- Volt-ampere (VA), volt-amperes reactive (VAR)

Need to Know Solution

As you have seen during your study of this chapter, it is possible and normal for the arithmetic sum of all voltages in a series *RCL* circuit to be much greater than the source voltage. The reason, of course, is because these voltages do not all reach a peak at the same time. They are all out of phase. If all voltages in a closed AC loop are added using phasor addition, their sum is zero—Kirchhoff's Law for AC circuits.

Perhaps you are still wondering how the inductor and the capacitor can each have a larger voltage than the source voltage. If only 10 V_{p-p} is applied, how can the capacitor have 13 V_{p-p} and the inductor have 18 V_{p-p}? Recall that inductive and capacitive reactances have a cancelling effect on one another. That is, the total impedance of the circuit is actually reduced when a capacitor is added to a series *RL* circuit or an inductor is added to a series *RC* circuit. This lower impedance allows a relatively high current to flow. This high current produces relatively large voltages across the inductor and capacitor according to Ohm's Law ($V_L = I_t \cdot X_L$ and $V_C = I_t \cdot X_C$). The only way V_L or V_C can exceed E_s is if both components (*L* and *C*) are in the same circuit, reducing Z_t and increasing I_t as previously described.

Questions

18-1 Rectangular and Polar Notation

1. What is a complex impedance?
2. What is the complex plane?
3. Name two values, or parameters, in electronics that are considered to be imaginary.

4. What is an operator?
5. Why is $\sqrt{-30}$ imaginary?
6. List two applications for polar systems (polar charts).

18-2 Mathematical Operations with Complex Numbers

7. Complex numbers must be expressed in which form (notation) if they are to be added or subtracted?
8. Complex numbers should be expressed in which form (notation) if they are to be multiplied or divided?

18-3 Series *RL* Circuit Analysis

9. How does Kirchoff's Voltage Law apply to AC circuits?
10. In a series *RL* circuit, which component voltage is in phase with the circuit current?
11. If the output voltage is taken across the resistor of a series *RL* circuit, is the circuit considered to be a voltage *lead* or *lag* network? Why?
12. How is the impedance-phasor diagram similar to the voltage-phasor diagram for a particular series *RL* circuit?

18-4 Series *RC* Circuit Analysis

13. How does the *RC* voltage-phasor diagram differ from the *RL* voltage-phasor diagram?
14. Is $\angle\theta$ positive or negative for a series *RC* circuit? Why?
15. In a series *RC* circuit, which component voltage is in phase with the circuit current?
16. If the output voltage is taken across the resistor of a series *RC* circuit, is the circuit considered to be a voltage *lead* or *lag* network? Why?
17. How is the impedance phasor diagram similar to the voltage-phasor diagram for a particular series *RC* circuit?

18-5 Series *RCL* Circuit Analysis

18. What does it mean to say the inductor voltage cancels capacitor voltage in a series *RCL* circuit?
19. Under what conditions will $V_R = E_s$ in a series *RCL* circuit?
20. Under what conditions will V_L and V_C be maximum in a series *RCL* circuit?
21. Briefly describe the purpose and use of a vector impedance meter.

18-6 Power in AC Circuits

22. What is the unit of measure for reactive power? Why is this unit different than the unit for real power?
23. What is apparent power and what is its unit of measure?
24. What does the power factor for an *RL*, *RC*, or *RCL* circuit indicate and how is it calculated?
25. How is the power factor related to $\angle\theta$ in an AC circuit?
26. If a circuit is acting inductively, how can the circuit's power factor and efficiency be improved?

Problems

18-1 Rectangular and Polar Notation

1. Use the Pythagorean theorem to find the hypotenuse for right triangles having the following given sides a and b:

	side a	side b	side c
a.	3 ft	4 ft	?
b.	10 in.	5 in.	?
c.	6 cm	8 cm	?

2. How many degrees does operator j represent?
3. What is j^2 equal to? j^3? j^4? j^5?
4. Express 270° in terms of operator j.
5. Convert the following complex impedances to polar notation:
   a. 600 Ω + j300
   b. 150 Ω + j250
   c. 120 Ω + j75
   d. 50 Ω − j100 (*Hint:* the 100 must be entered as − 100, $\angle - \theta$.)
6. Convert the following complex impedances to rectangular notation:
   a. 3,000 Ω $\angle$60°
   b. 15 Ω $\angle$30°
   c. 50 Ω $\angle$45°
   d. 250 Ω $\angle -25°$ (*Hint:* the − 25° must be entered as − 25.)

18-2 Mathematical Operations with Complex Numbers

7. Solve the following:
   a. $(140 - j50) - (100 + j60) - (300 + j55) =$
   b. $(-4 - j6) + (-2 + j4) - (+3 + j4) =$
8. Solve the following:
   a. $j20 \cdot j40 =$
   b. $-j10 \cdot j30 =$
   c. $j8 \cdot j^24 \cdot -j^25 =$
9. Solve the following:
   a. $(60 - j30)(30 + j60) =$
   b. $(5 + j4)(5 - j4) =$
   c. $(10 - j40)^2 =$
10. Solve the following:
    a. $1/j =$
    b. $(10 + j30 - j40)/j2 =$
11. Solve the following:
    a. $(12 + j6)/(6 - j3) =$
    b. $1/(30 - j10) =$
12. Solve the following:
    a. 30 $\angle$40° + 20 $\angle$30°
    b. 60 $\angle$25° + 25 $\angle -30° =$
13. Solve the following:
    a. 80 $\angle -40° \cdot 4 \angle$40° $=$
    b. 200 $\angle -120° \cdot 30 \angle -30° =$

14. Solve the following:
    a. $8 \angle 10° / 4 \angle 70° =$
    b. $100 \angle -20° / 50 \angle -30° =$

18-3 Series *RL* Circuit Analysis

15. If the frequency of the applied source voltage is 20 kHz, how much time does 90° represent? 45°? 60°?
16. Express the source voltage in polar form from the following:
    a. $10 + j30$ V $=$
    b. $30 + j50$ mV $=$
17. Solve for resistor and inductor voltage from the following source voltages expressed in polar form (use P→R conversion).
    a. 200 V $\angle 45° =$
    b. 70 V $\angle 60° =$
18. Find the total impedance for series *RL* circuits that have the following resistances and inductive reactances:
    a. $R = 10$ kΩ and $X_L = 20$ kΩ
    b. $30 + j20$ Ω $=$
19. Find the phase angle ($\angle \theta$) for each of the total impedances of Problem 18.

Analyze the following *RL* circuits and draw a voltage-phasor diagram for each: Figure 18-66 is a circuit model for Problems 20 through 24.

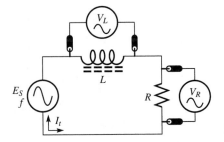

Figure 18-66

Fill in the blanks.

#	E_S	f	L	X_L	R	Z_t	I_t	V_R	V_L	$\angle \theta$
20.	8 V	400 Hz	100 mH		220 Ω					
21.	10 mV	1 kHz	1 H		1 kΩ					
22.	15 V	50 kHz		3.14 kΩ	2.7 kΩ					
23.	12 V		25 mH	10 kΩ	4.7 kΩ					
24.	115 V	60 Hz			1 kΩ					60°

18-4 Series *RC* Circuit Analysis

25. Find the capacitive reactance for a 0.001 μF capacitor at the following frequencies: 1 kHz, 50 kHz, 8 MHz, and 40 MHz.

26. Express the source voltage in polar form from the following:
    a. $5 - j30$ V =
    b. $30 - j60$ μV =
27. Solve for resistor and capacitor voltage from the following source voltages expressed in polar form (use P→R conversion).
    a. 20 V $\angle -40°$ =
    b. 30 mV $\angle -40°$ =
28. Find the total impedance for series *RC* circuits that have the following resistances and capacitive reactances:
    a. $R = 5$ kΩ and $X_C = 15$ kΩ
    b. $8 - j5$ Ω =
29. Find the phase angle ($\angle \theta$) for each of the total impedances of Problem 28.

Analyze the following *RC* circuits and draw a voltage-phasor diagram for each: Figure 18-67 is a circuit model for Problems 30 through 34.

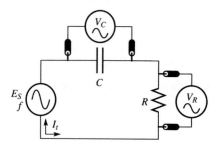

Figure 18-67

Fill in the blanks.

#	E_S	f	C	X_C	R	Z_t	I_t	V_R	V_C	$\angle \theta$
30.	70 μV	15 MHz	28 pF		100 Ω					
31.	14 mV	90 kHz	0.01 μF		150 Ω					
32.	10 V		0.47 μF	50 Ω	100 Ω					
33.	30 mV	1.5 MHz		300 Ω	100 Ω					
34.	80 V		0.001 μF	1.48 kΩ			50 mA			$-68.2°$

18-5 Series *RCL* Circuit Analysis

35. Solve for E_S using the following given information:
    a. $V_R = 12$ V; $V_L = 14$ V; $V_C = 16$ V; $E_S = ?$
    b. $E_S = 10$ V $+ j9$ V $- j18$ V = ?
    c. $E_S = 1.5$ V $+ j1$ V $- j2$ V = ?
36. Find $\angle \theta$ for each of the source voltages of Problem 35 and state whether the circuit is acting inductively or capacitively.

37. Of the problems in #35, which is the closest to being purely resistive?
38. Given the following information, determine the phase angle and time delay for each voltage with respect to the source voltage: $f = 10$ kHz; $V_R = 5$ V; $V_L = 5$ V; $V_C = 10$ V.
39. Given the following information, determine the phase angle and time delay for each voltage with respect to the source voltage: $f = 80$ kHz; $V_R = 20$ V; $V_L = 14$ V; $V_C = 13$ V.
40. From the following information, find the total impedance and phase angle for each case ($Z_t \angle \theta$).
    a. $R = 1$ kΩ; $X_L = 3$ kΩ; $X_C = 2$ kΩ; $Z_t \angle \theta = ?$
    b. $Z_t \angle \theta = 5\ \Omega + j4\ \Omega - j9\ \Omega = ?$
    c. $Z_t \angle \theta = 47\ \Omega + j100\ \Omega - j70\ \Omega = ?$
41. At a frequency of 100 kHz, what size inductor would it take to cancel 2.4 kΩ of capacitive reactance in a series *RCL* circuit? (*Hint*: rearrange the X_L formula to solve for *L*.)
42. At a frequency of 4.5 MHz, what size capacitor would it take to cancel 45 Ω of inductive reactance in a series *RCL* circuit?

Analyze the following *RCL* circuits and draw a voltage-phasor diagram for each: Figure 18-68 is a circuit model for Problems 43 through 47.

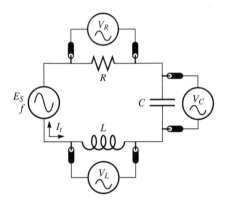

Figure 18-68

Fill in the blanks and draw the voltage phasor diagram for each problem.

#	E_S	f	R	C	L	X_C	X_L	Z_t	I_t	V_R	V_C	V_L	$\angle \theta$
43.	20 V	3 kHz	330 Ω	0.3 μF	50 mH								
44.	8 μV	45 kHz	10 kΩ	0.001 μF	10 mH								
45.		5 MHz							1 mA	30 mV	40 mV	50 mV	
46.		600 kHz	5.6 kΩ			12 kΩ	4 kΩ		25 mA				
47.				5 μF	0.5 H	200 Ω			100 mA	10 V			

18-6 Power in AC Circuits

48. Calculate P_R, P_L, P_C, P_A, and *PF* for Problem 43.
49. Calculate P_R, P_L, P_C, P_A, and *PF* for Problem 44.
50. Calculate P_R, P_L, P_C, P_A, and *PF* for Problem 45.
51. Calculate P_R, P_L, P_C, P_A, and *PF* for Problem 46.
52. Calculate P_R, P_L, P_C, P_A, and *PF* for Problem 47.

Answers to Self-Checks

Self-Check 18-1

1. A complex impedance is an AC impedance composed of resistance and reactance (i.e., $50 + j40\ \Omega$).
2. The operator is the angle of a magnitude in a polar system.
3. $-j = -90°$
4. $+j^6 = j^2 = 180° = -1$
5. $200 - j450 = 492.4\ \angle -66°$
6. $50 + j75 = 90.1\ \angle 56.3°$
7. $300\ \angle 45° = 212 + j212$
8. $25\ \angle -60° = 12.5 - j21.7$

Self-Check 18-2

1. $95 + j3$
2. $-8 - j9$
3. -24
4. $-1,500$
5. $12 - j12 + j32 + 32 = 44 + j20$
6. $220 - j110 - j440 - 220 = -j550$
7. $j46$
8. $-j \cdot (3 - j) = -j3 \cdot -1 = j3$
9. $\dfrac{30 - j5}{40 + j8} \cdot \dfrac{40 - j8}{40 - j8} = (1,200 - j240 - j200 - 40)/(1,600 + 64)$
   $= (1,160 - j440)/1,664 = 0.697 - j0.264$
10. $\dfrac{4 + j9}{3 - j2} \cdot \dfrac{3 + j2}{3 + j2} = (12 + j8 + j27 - 18)/(9 + 4)$
    $= (-6 + j35)/13 = -0.462 + j2.69$
11. $(47 + j17.1) + (21.2 + j21.2) = 68.2 + j38.3 = 78.2\ \angle 29.3°$
12. $(0 + j100) - (56.3 - j32.5) = -56.3 + j132.5 = 144\ \angle 113°$
13. $120,000\ \angle 65°$
14. $3\ \angle 160°$

Self-Check 18-3

Fill in the blanks and draw the voltage-phasor diagram for each problem.

#	E_S	f	L	X_L	R	Z_t	I_t	V_R	V_L	$\angle \theta$
1.	5 V	60 Hz	0.531 H	**200 Ω**	100 Ω	**224 Ω**	22.3 mA	2.23 V	4.47 V	63.5°
2.	20 V	1 MHz	31.8 μH	**200 Ω**	400 Ω	**447 Ω**	44.7 mA	17.9 V	8.94 V	26.5°
3.	15 mV	200 kHz	**0.398 mH**	500 Ω	300 Ω	583 Ω	25.7 μA	7.72 mV	12.9 mV	59°
4.	4 V	**79.6 kHz**	50 mH	25 kΩ	10 kΩ	26.9 kΩ	149 μA	1.49 V	3.73 V	68.2°
5.	18 V	100 Hz	**0.918 H**	**577 Ω**	1 kΩ	1.15 kΩ	15.6 mA	15.6 V	9 V	30°

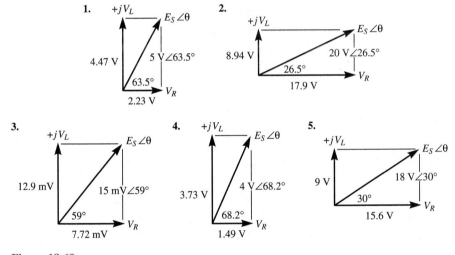

Figure 18-69

Self-Check 18-4

Fill in the blanks and draw the voltage-phasor diagram for each problem.

#	E_S	f	C	X_C	R	Z_t	I_t	V_R	V_C	$\angle\theta$
1.	30 V	50 Hz	20 μF	**159 Ω**	220 Ω	**271 Ω**	110 mA	**24.3 V**	**17.6 V**	**$-35.9°$**
2.	115 V	60 Hz	10 μF	**265 Ω**	180 Ω	**321 Ω**	359 mA	**64.6 V**	**95.2 V**	**$-55.8°$**
3.	25 μV	**531 kHz**	100 pF	3 kΩ	0.5 kΩ	**3.04 kΩ**	8.22 nA	**4.11 μV**	**24.7 μV**	**$-80.5°$**
4.	50 mV	650 kHz	**816 pF**	300 Ω	200 Ω	**361 Ω**	139 μA	**27.7 mV**	**41.6 mV**	**$-56.3°$**
5.	100 mV	**2.27 kHz**	0.1 μF	700 Ω	**465 Ω**	**840 Ω**	119 μA	**55.3 mV**	**83.3 mV**	**$-56.1°$**

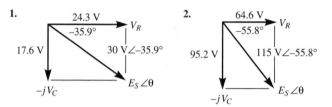

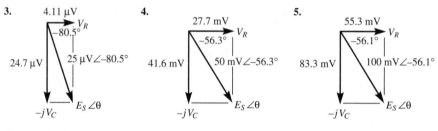

Figure 18-70

Self-Check 18-5

Fill in the blanks and draw the voltage-phasor diagram for each problem.

#	E_S	f	R	C	L	X_C	X_L	Z_t	I_t	V_R	V_C	V_L	$\angle\theta$
1.	10 V	1 kHz	1 kΩ	0.47 μF	0.1 H	339 Ω	628 Ω	1,041 Ω	9.61 mA	9.61 V	3.26 V	6.04 V	16.1°
2.	25 mV	455 kHz	20 Ω	0.008 μF	10 μH	43.7 Ω	28.6 Ω	25.1 Ω	996 μA	19.9 mV	43.5 mV	28.5 mV	−37.1°
3.	50 mV	10 MHz	40 Ω	240 pF	1.5 μH	66.3 Ω	94.2 Ω	48.8 Ω	1.02 mA	40.8 mV	67.6 mV	96.1 mV	34.9°
4.	79.8 μV	120 MHz	5 Ω	50 pF	0.01 μH	26.5 Ω	7.54 Ω	19.6 Ω	4.07 μA	20.4 μV	108 μV	30.7 μV	−75°
5.	59.7 V	120 Hz	100 Ω	10 μF	1 H	133 Ω	752 Ω	627 Ω	95.3 mA	9.53 V	12.7 V	71.7 V	80.8°

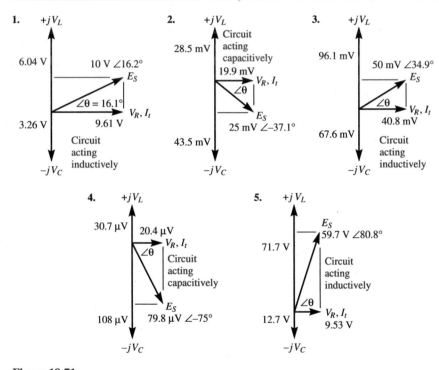

Figure 18-71

Self-Check 18-6

1. P_C = 31.3 mVAR; P_L = 58 mVAR; P_A = 96 mVA; P_R = 92.4 mW; *PF* = 0.96
2. P_C = 43.3 μVAR; P_L = 28.4 μVAR; P_A = 24.9 μVA; P_R = 19.8 μW; *PF* = 0.8
3. P_C = 69 μVAR; P_L = 98 μVAR; P_A = 51 μVA; P_R = 41.6 μW; *PF* = 0.82
4. P_C = 439.6 pVAR; P_L = 125 pVAR; P_A = 324.8 pVA; P_R = 83 pW; *PF* = 0.255
5. P_C = 1.21 VAR; P_L = 6.83 VAR; P_A = 5.69 VA; P_R = 0.91 W; *PF* = 0.16

SUGGESTED PROJECTS

1. Add some of the main concepts, formulas, and procedures from this chapter to your Electricity and Electronics Notebook.

2. Construct a series *RCL* circuit using a 1 kΩ resistor, a 0.01 μF capacitor, and a 10 mH coil. Apply a 10 V$_{p-p}$, 10 kHz signal to the circuit and use an oscilloscope to measure the voltage across each component. As you measure the voltage across each component, rearrange the order of the components so that the component that is being measured is connected to the common ground of your circuit. The scope probe ground must stay connected to your circuit's common ground with the signal generator's ground or there is the possibility the circuit will be shorted by the probe's ground. Calculate all voltages and compare them to the measured values. You can use the 10 V$_{p-p}$ to calculate the peak-to-peak total current and use this to calculate the peak-to-peak resistor, capacitor, and inductor voltages. These can be compared directly with the peak-to-peak oscilloscope measurements.

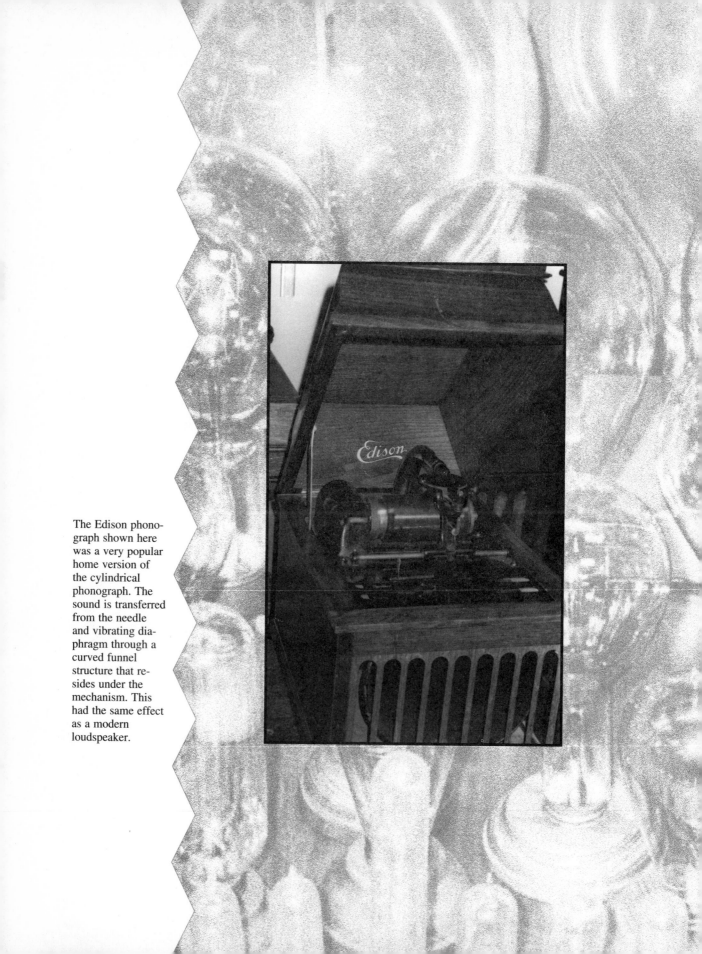

The Edison phonograph shown here was a very popular home version of the cylindrical phonograph. The sound is transferred from the needle and vibrating diaphragm through a curved funnel structure that resides under the mechanism. This had the same effect as a modern loudspeaker.

Chapter 19

Parallel *RL*, *RC*, and *RCL* Circuits

CHAPTER OUTLINE

OBJECTIVES

After studying this chapter, you will be able to
- calculate branch currents and total current for parallel *RL*, *RC*, and *RCL* circuits.
- calculate the phase angle for each branch current and total current of a parallel AC circuit.
- calculate real power, reactive power, apparent power, and the power factor for parallel AC circuits.
- calculate the power efficiency of a parallel AC circuit.
- correct the power factor for parallel AC circuits.

INTRODUCTION

In this chapter, you will learn how to analyze simple parallel AC circuits. You will discover many similarities between parallel circuit analysis and series circuit analysis. Many of the formulas (such as the Pythagorean theorem, Ohm's Law, the power formulas, and the arctan function) that are used in series analysis are also used in parallel analysis. The biggest difference between parallel analysis and series analysis is this: In series circuits, voltage and impedance phasors are analyzed, while in parallel circuits, current phasors are analyzed. In a series circuit, there is one current and many voltage drops at various phase angles ($0°$, $\pm 90°$, etc.). In a parallel circuit, there is one voltage and many branch currents at various phase angles. Enjoy your experience here.

NEED TO KNOW

Consider the AC circuit shown here. The voltage source shown is to be connected to this circuit. The problem here is to determine the impedance that the circuit will offer and the amount of current the circuit will demand from the voltage source. Can you do this? OK, here's what you do. Take a crack at it anyway. Take a sheet of paper and try to establish a procedure to follow in determining the impedance and current. Go ahead and perform all calculations according to what you think might be correct. Then fold the paper and place it in this book for safekeeping until you finish this chapter. When you finish this chapter, recover the folded paper from your book and check your work. Revise your procedure and redo your calculations if you discover it is necessary (don't be disappointed if it's necessary, it probably will be). You'll find this little exercise to be symbolic and rewarding. At this point in time, this problem represents your need to know, as you may feel a little unsure of how to proceed. Later, this same problem will represent the significant progress you have made in expanding your experiences in electricity and electronics.

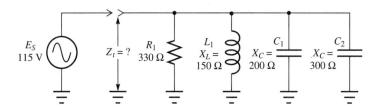

19-1 Parallel *RL* Circuit Analysis

RL Circuit Current Phasors

Currents Out of Phase

In this section, we discuss the analysis of parallel *RL* AC circuits as shown in Figure 19-1. The entire analysis process is based on the fact that total current, the resistor current, and the inductor current are all out of phase. Each current reaches a peak independent of the other two. The inductor current is not in phase with the resistor current and neither of the two is in phase with the resulting total current. This may sound familiar to you, since this out-of-phase condition is also true of voltage drops in series AC circuits. In simple parallel AC circuits, it is the currents that are out of phase, not the voltages, since the voltages are all the same across each branch of a parallel circuit (same amplitude and phase).

Since all currents are out of phase, they must be added using phasor addition, taking into consideration the phase angle of each branch current. Just as series voltage drops cannot be added arithmetically, parallel currents cannot be added arithmetically. In a parallel *RL* circuit, the simple arithmetic sum of the resistor branch current and the inductor branch current is much greater than the total applied current. Does this mean Kirchhoff's Current Law does not apply to AC circuits? No. Kirchhoff's Current Law does apply to parallel AC circuits. However, Kirchhoff's Current Law only applies to *instantaneous* current values or *phasor* current values. As shown in Figure 19-2, the simple arithmetic sum of the instantaneous resistor current and inductor current equals the instantaneous total current. Also, the phasor sum of all branch currents equals the total current.

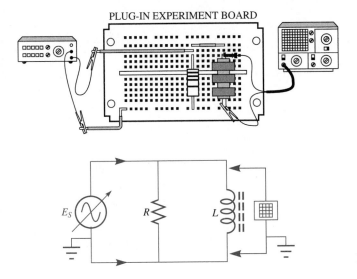

Figure 19-1

Kirchhoff's Current Law for AC circuits can be stated in two ways:

1. The algebraic sum of all instantaneous currents entering and leaving any point in an AC circuit must equal zero. For a parallel *RL* circuit this means; $i_R + i_L - i_t = 0$ or $i_R + i_L = i_t$

2. The phasor sum of all currents entering and leaving any point in an AC circuit must equal zero. For a parallel *RL* circuit this means: $I_R\angle x° + I_L\angle y° - I_t\angle z° = 0$, or $I_R\angle x° + I_L\angle y° = I_t\angle z°$

RL Current Phasors

Figure 19-3 illustrates a standard phasor diagram showing the phase relationship between the branch currents and voltages of a parallel *RL* circuit. Notice that the applied source voltage, the resistor voltage, and the inductor voltage are all the same, since this is a simple parallel circuit. The source voltage, or component voltages, can be used as a reference with which to compare all current phase angles. Since the resistor is a nonreactive component, its current is in phase with its voltage. Therefore, the resistor current phasor is

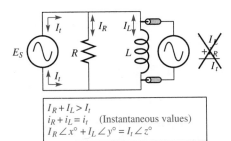

Figure 19-2   Adding parallel *RL* circuit currents.

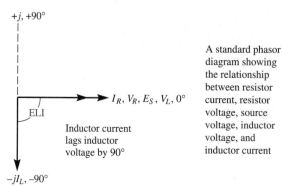

A standard phasor diagram showing the relationship between resistor current, resistor voltage, source voltage, inductor voltage, and inductor current

Figure 19-3 Parallel *RL* circuit current phasor diagram.

placed at 0° along with voltage. The inductor current is found to be at −90° due to ELI. Recall that inductor current lags inductor voltage by 90° (see Section 16-1). Thus, inductor current has an operator of −*j* or −90°.

Adding Resistor- and Inductor-Current Phasors

Earlier, it was stated that parallel currents in AC circuits must be added using phasor addition. From your experience with series AC circuit analysis, you already know how this is done (except you were adding voltage phasors for series circuits). Figure 19-4 illustrates the method used to add current phasors. The Pythagorean theorem is used once again. Notice that the Pythagorean theorem is used to find the magnitude (amplitude) of the resultant total current phasor and arctan ($\tan^{-1}$) is used to find the total current's phase angle ($\angle \theta$). For parallel *RL* circuits, the phase angle is negative ($\angle -\theta$) when the resistor current is used as a reference at 0° (recall that all voltages are at 0° as well). The rectangular form of the complex total current is $I_R - jI_L$. Instead of manually applying the Pythagorean theorem and arctan, automatic R→P conversion may be used to find the polar form of the complex total current: $I_R - jI_L = I_t \angle -\theta$ (see R→P conversion in Section 18-1). Enter the resistor current first then INV R→P followed by the inductor current (enter the inductor current as a negative value). Depress the = button and I_t is displayed. Depress X↔Y and the negative angle is displayed (see your calculator's operating manual to verify the R→P conversion procedure).

$$I_t = \sqrt{I_R^2 + I_L^2} \tag{19.1}$$

also: $I_R = \sqrt{I_t^2 - I_L^2}$ and $I_L = \sqrt{I_t^2 - I_R^2}$

$$\angle \theta = \tan^{-1}(-I_L/I_R) \tag{19.2}$$

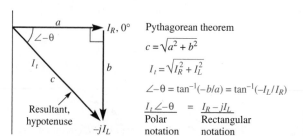

Pythagorean theorem

$c = \sqrt{a^2 + b^2}$

$I_t = \sqrt{I_R^2 + I_L^2}$

$\angle -\theta = \tan^{-1}(-b/a) = \tan^{-1}(-I_L/I_R)$

$I_t \angle -\theta$	=	$I_R - jI_L$
Polar notation		Rectangular notation

Figure 19-4 *RL* circuit current phasor addition.

Determining the Branch Currents (I_R and I_L)

Calculating the branch currents for parallel AC circuits can be as simple as using Ohm's Law. The current flowing through a branch is determined by the voltage across the branch and the impedance of the branch. Thus, the general form of Ohm's Law for branch currents is as follows:

$$I_B = E_B/Z_B \qquad\qquad (19.3)$$

(where the subscript B stands for *Branch* and, in a simple parallel circuit, $E_B = E_S$).

When a certain branch current is calculated, the phase angle of the current may also be calculated by applying the rules for division of complex numbers in polar form. For example, the current flowing through a parallel branch composed only of resistance may be calculated using Ohm's Law as follows: $I_R\angle 0° = E_R\angle 0°/R\angle 0°$. Notice the phase angle for resistor current is $0°$ since the phase angle for resistance and the resistor voltage is $0°$ (in a simple parallel circuit $E_S = E_R = E_L$). In a parallel AC circuit, the applied source voltage is often used as the reference parameter ($E_S\angle 0°$). All branch currents are then compared in phase to the source voltage. With this in mind, we discover that current flowing in a purely inductive branch has a $-90°$ phase angle as follows: $I_L\angle -90° = E_L\angle 0°/X_L\angle +90°$. Once again, $E_S = E_R = E_L$ for a simple parallel circuit. Thus E_L is at the reference angle of $0°$. It may be useful at this point to go back to Section 18-2 and review the rules of complex algebra. Here, you should recall that, when dividing these angles, the denominator is subtracted from the numerator (i.e., $\angle 0°/\angle +90° = \angle 0° - \angle +90° = \angle -90°$).

Current Lead and Lag

Figure 19-5 illustrates the total picture for a simple parallel *RL* AC circuit. Notice that the total current for the circuit can be represented in polar or rectangular form ($I_t\angle -\theta =$

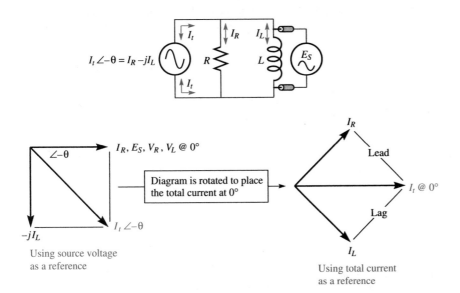

Figure 19-5   The total picture for parallel *RL* circuits.

polar form, and $I_R - jI_L$ = rectangular form). Later, you will see how each form of the complex current may be used for various calculations in the analysis process. Also note that two phasor diagrams represent currents in parallel *RL* circuits. The phasor diagram on the left conventionally places I_R and E_S at 0° to be used as reference parameters. In this case, the total current and the inductor current are shown to lag the resistor current and source voltage. In some cases, however, it may be useful to represent the circuit currents as shown in the diagram on the right. In this case, the entire phasor diagram has been rotated to place the total current phasor at 0°. The total current is then used as the reference parameter. As you can see, the resistor current is shown to lead by $\angle\theta$ while the inductor current lags by $\angle(90° - \theta°)$.

Determining Total Impedance (Z_t) for Parallel *RL*

Using Ohm's Law to Find Z_t

The total impedance for any circuit, AC or DC, may be calculated using Ohm's Law as long as the source voltage and total current are known. The following general form of Ohm's Law is used

$$Z_t = E_S/I_t \qquad\qquad (19.4)$$

If we wish to calculate the phase angle for total impedance, the phase angle for total current must be known and the rules of complex algebra must be followed as shown in Figure 19-6. In the case of a parallel *RL* circuit, the phase angle for total current is always negative ($I_t\angle -\theta$). Thus, the phase angle for total impedance is positive ($Z_t\angle +\theta = E_S\angle 0°/I_t\angle -\theta$), which indicates the circuit is acting inductively.

Using Parallel Formulas to Find Z_t

Figure 19-6 also illustrates how the familiar product over the sum formula for parallel components may be used to find the total impedance. Notice that the numerator and denominator of the equation should both be in polar form before division can be performed.

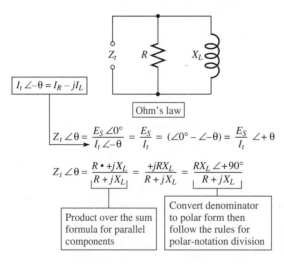

Figure 19-6   Total impedance for parallel *RL* circuits.

Thus, $+jRX_L$ must be expressed as $RX_L\angle +90°$ and $R + jX_L$ must be converted to its polar form using R→P conversion ($R + jX_L = Z\angle +\theta$). The $\angle +\theta$ in the denominator may then be subtracted from the $\angle +90°$ in the numerator to find the phase angle for the total impedance. This is demonstrated in examples to follow.

Any of the following standard formulas may be used to find total impedance in parallel AC circuits. However, the rules of complex algebra must be followed. Recall that complex numbers are multiplied or divided in polar form, and added or subtracted in rectangular form.

$$Z_t = Z_B/N \qquad \text{(same values formula)} \tag{19.5}$$

where all branches have the same impedance Z_B and N is the number of branches.

$$Z_t = \frac{Z_{B1} \cdot Z_{B2}}{Z_{B1} + Z_{B2}} \qquad \text{(product over the sum formula)} \tag{19.6}$$

where there are two parallel branches having different impedances.

$$Z_t = \frac{1}{1/Z_{B1} + 1/Z_{B2} + 1/Z_{B3} + \cdots} \qquad \text{(reciprocal formula)} \tag{19.7}$$

where there are more than two branches having different impedances.

Parallel *RL* Circuit Calculations

Now that the formulas have been identified and parallel *RL* currents have been defined in terms of phase angle, it is time to consider some practical examples. As in series AC circuit analysis, parallel AC circuit analysis involves a step-by-step process. Examples 19-1 and 19-2 illustrate two different step-by-step approaches to the analysis of parallel *RL* circuits. In each example, a different set of given information is provided for analysis.

EXAMPLE   19-1

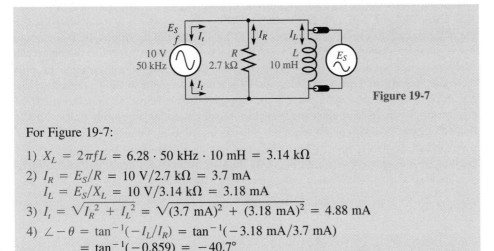

Figure 19-7

For Figure 19-7:

1) $X_L = 2\pi fL = 6.28 \cdot 50 \text{ kHz} \cdot 10 \text{ mH} = 3.14 \text{ k}\Omega$

2) $I_R = E_S/R = 10 \text{ V}/2.7 \text{ k}\Omega = 3.7 \text{ mA}$
   $I_L = E_S/X_L = 10 \text{ V}/3.14 \text{ k}\Omega = 3.18 \text{ mA}$

3) $I_t = \sqrt{I_R^2 + I_L^2} = \sqrt{(3.7 \text{ mA})^2 + (3.18 \text{ mA})^2} = 4.88 \text{ mA}$

4) $\angle -\theta = \tan^{-1}(-I_L/I_R) = \tan^{-1}(-3.18 \text{ mA}/3.7 \text{ mA})$
   $= \tan^{-1}(-0.859) = -40.7°$

(Continued next page)

EXAMPLE 19-1 (Continued)

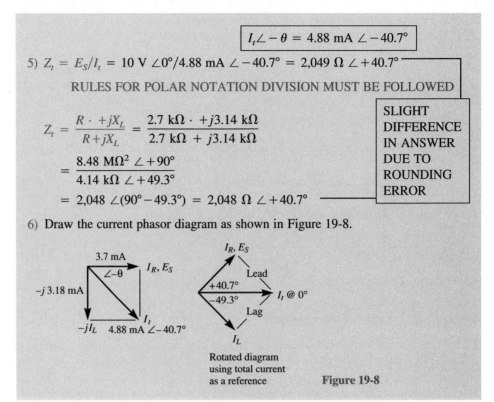

$$I_t \angle -\theta = 4.88 \text{ mA} \angle -40.7°$$

5) $Z_t = E_S/I_t = 10 \text{ V} \angle 0°/4.88 \text{ mA} \angle -40.7° = 2{,}049 \text{ } \Omega \angle +40.7°$

RULES FOR POLAR NOTATION DIVISION MUST BE FOLLOWED

SLIGHT DIFFERENCE IN ANSWER DUE TO ROUNDING ERROR

$$Z_t = \frac{R \cdot +jX_L}{R+jX_L} = \frac{2.7 \text{ k}\Omega \cdot +j3.14 \text{ k}\Omega}{2.7 \text{ k}\Omega + j3.14 \text{ k}\Omega}$$

$$= \frac{8.48 \text{ M}\Omega^2 \angle +90°}{4.14 \text{ k}\Omega \angle +49.3°}$$

$$= 2{,}048 \angle (90° - 49.3°) = 2{,}048 \text{ } \Omega \angle +40.7°$$

6) Draw the current phasor diagram as shown in Figure 19-8.

Rotated diagram
using total current
as a reference

Figure 19-8

In Example 19-1, the resistance R, the inductance L, the source voltage E_S, and the frequency f of the applied source voltage are given. Starting with this set of given information, all currents (I_R, I_L, I_t), total current's phase angle ($\angle -\theta$), and total impedance with phase angle ($Z_t \angle \theta$) must be found. The inductive reactance is first calculated so that the inductor current can be determined. Ohm's Law is used to determine the inductor and resistor branch currents. The Pythagorean theorem is then used to determine the magnitude of the total current and arctan ($\tan^{-1}$) is used to find the total current's phase angle ($\angle -\theta$). Note that the inductor current must be entered as a negative value to yield a negative phase angle. The total current and its phase angle may also be determined using automatic R→P conversion. Remember to enter the inductor branch current as a negative value. Finally, the total impedance with phase angle may be calculated using Ohm's Law or the product over the sum formula. Note that the rules of complex algebra must be followed (review Section 18-2 if necessary). Carefully study the two methods for determining $Z_t \angle \theta$ until you are able to obtain the same results using your calculator.

EXAMPLE   19-2

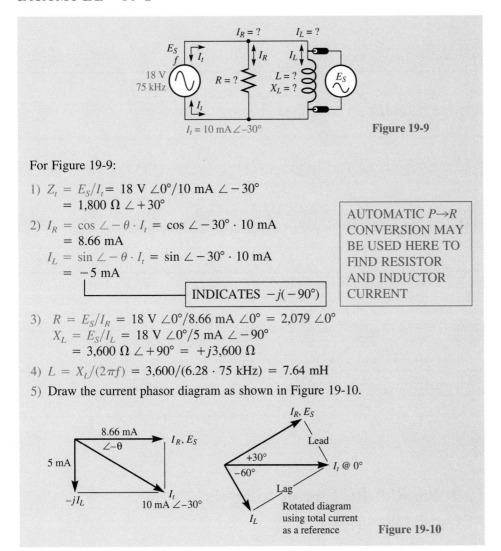

Figure 19-9

For Figure 19-9:

1) $Z_t = E_S/I_t = $ 18 V $\angle 0°$/10 mA $\angle -30°$
   $= 1,800\ \Omega \angle +30°$

2) $I_R = \cos \angle -\theta \cdot I_t = \cos \angle -30° \cdot$ 10 mA
   $= 8.66$ mA
   $I_L = \sin \angle -\theta \cdot I_t = \sin \angle -30° \cdot$ 10 mA
   $= -5$ mA

   └─────── | INDICATES $-j(-90°)$ |

> AUTOMATIC $P{\to}R$
> CONVERSION MAY
> BE USED HERE TO
> FIND RESISTOR
> AND INDUCTOR
> CURRENT

3) $R = E_S/I_R =$ 18 V $\angle 0°$/8.66 mA $\angle 0° = 2,079 \angle 0°$
   $X_L = E_S/I_L =$ 18 V $\angle 0°$/5 mA $\angle -90°$
   $= 3,600\ \Omega \angle +90° = +j3,600\ \Omega$

4) $L = X_L/(2\pi f) = 3,600/(6.28 \cdot 75$ kHz$) = 7.64$ mH

5) Draw the current phasor diagram as shown in Figure 19-10.

Figure 19-10

In Example 19-2, the only given information is: the source voltage (E_S), the frequency (f), and the total current with phase angle ($I_t\angle -\theta$). Since the source voltage and total current are already known, the total impedance can be calculated immediately using Ohm's Law. The individual branch currents I_R and I_L are found using either manual or automatic P→R conversion. If automatic P→R conversion is used, for many calculators, the current magnitude is entered first then [SHIFT] [P→R] is pressed. The phase angle is then entered, in this case as a negative value, and the = key is pressed to display the magnitude of the resistor current. When X↔Y is pressed, the magnitude of the inductor current is displayed. With all currents determined, the unknown resistance and inductive reactance is found using Ohm's Law. Finally, the unknown inductance is determined using the rearranged X_L formula.

After you have carefully studied Examples 19-1 and 19-2, test your skills by solving the problems of Self-Check 19-1. Check your answers at the end of the chapter.

SELF-CHECK 19-1

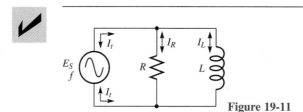

Figure 19-11

Figure 19-11 is a model circuit for Problems 1 through 5.
Fill in the blanks and draw the current phasor diagram for each problem.

#	E_S	f	L	X_L	R	I_L	I_R	I_t	Z_t	$\angle\theta$
1.	50 V	60 Hz	0.5 H		100 Ω					
2.	20 V	1 MHz	32 μH		400 Ω					
3.	15 mV	200 kHz		500 Ω	300 Ω					
4.	10 V		50 mH	25 kΩ	10 kΩ					
5.		100 Hz			1.04 kΩ			20 mA		−30°

19-2 Parallel *RC* Circuit Analysis

RC Circuit Current Phasors

Currents Out of Phase

As you may have already guessed, currents in parallel *RC* circuits, such as Figure 19-12, are also out of phase. Again, the entire analysis process is based on the fact that the resistor current, the capacitor current, and the total current all reach peak amplitudes at different times (out of phase).

As shown in Figure 19-13, the resistor and capacitor branch currents must be added using phasor addition, taking into consideration the phase angle of each current. The instantaneous currents may be added arithmetically to obtain the instantaneous total current. Kirchhoff's Current Law for AC circuits once again applies to *instantaneous* or *phasor* current values only: $i_R + i_C - i_t = 0$ or $i_R + i_C = i_t$, and $I_R\angle x° + I_C\angle y° - I_t\angle z° = 0$ or $I_R\angle x° + I_C\angle y° = I_t\angle z°$.

PLUG-IN EXPERIMENT BOARD

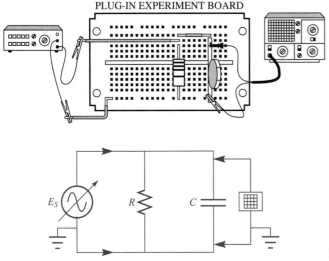

Figure 19-12

RC Current Phasors

Figure 19-14 illustrates a standard phasor diagram, showing the relationship between the branch currents and voltages of a parallel *RC* circuit. Once again the source voltage, which is also the resistor and capacitor voltage, is placed at 0° as a reference along with the resistor current. The capacitor current is shown to lead the resistor current and source voltage by 90 electrical degrees due to ICE. For a capacitor, current always leads voltage by 90° (see Section 16-2). Thus the operator for capacitor current is $+j$ or $+90°$.

Adding Resistor and Capacitor Current Phasors

The Pythagorean theorem and the arctan function are used once again to determine the magnitude and angle of the resultant total current. Figure 19-15 illustrates the use of these formulas for the phasor addition of resistor and capacitor current. This phasor addition is represented in rectangular form as $I_R + jI_C$, which is equal to the total current expressed in polar form as $I_t \angle \theta$. Note that the phase angle for total current in a parallel *RC* circuit is positive. Also, the capacitor current value is entered into the formulas, and your cal-

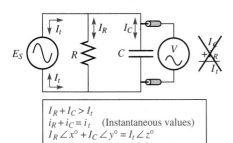

$I_R + I_C > I_t$
$i_R + i_C = i_t$ (Instantaneous values)
$I_R \angle x° + I_C \angle y° = I_t \angle z°$

Figure 19-13   Adding parallel *RC* circuit currents.

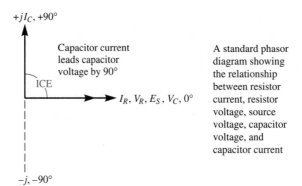

A standard phasor diagram showing the relationship between resistor current, resistor voltage, source voltage, capacitor voltage, and capacitor current

Figure 19-14 Parallel *RC* circuit current phasor diagram.

culator, as a positive value (recall that inductor current is entered as a negative value in the arctan formula). As before, the total current and phase angle may be obtained automatically using the R→P function of your calculator (see R→P conversion in Section 18-1 and your calculator's operating manual).

$$I_t = \sqrt{I_R^2 + I_C^2} \tag{19.8}$$

also: $I_R = \sqrt{I_t^2 - I_C^2}$ and $I_C = \sqrt{I_t^2 - I_R^2}$

$$\angle\theta = \tan^{-1}(I_C/I_R) \tag{19.9}$$

Determining the Branch Currents I_R and I_C

Determining the resistor and capacitor branch currents is again as simple as using Ohm's Law. Using Ohm's Law, we can demonstrate mathematically that the resistor current is in phase with the source voltage and that the capacitor current is leading the source voltage by 90°:

$$I_R \angle 0° = E_S \angle 0°/R \angle 0° \qquad (\angle 0°/\angle 0° = \angle 0° - \angle 0° = \angle 0°)$$

and

$$I_C \angle +90° = E_S \angle 0°/X_C \angle -90° \quad (\angle 0°/\angle -90° = \angle 0° - \angle -90° = \angle +90°)$$

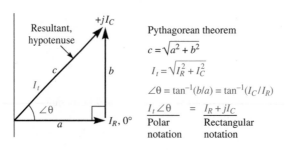

Pythagorean theorem

$c = \sqrt{a^2 + b^2}$

$I_t = \sqrt{I_R^2 + I_C^2}$

$\angle\theta = \tan^{-1}(b/a) = \tan^{-1}(I_C/I_R)$

$I_t\angle\theta \quad = \quad I_R + jI_C$

Polar notation — Rectangular notation

Figure 19-15 *RC* circuit current phasor addition.

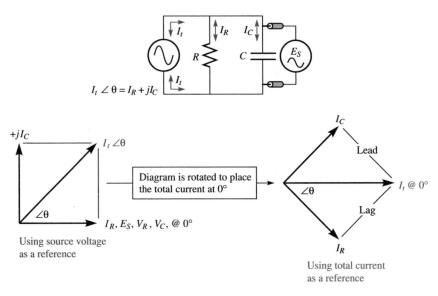

Figure 19-16   The total picture for parallel *RC* circuits.

Current Lead and Lag

Figure 19-16 summarizes much of what has been discussed regarding parallel *RC* currents and phase. Notice that the phasor diagram on the left represents the current-phase relationships in a conventional sense, using the source voltage and resistor current as a reference. The capacitor current and total current both lead the resistor current in time, or phase. As in parallel *RL* analysis, the current phasor diagram may be rotated to place the total current at 0° as a reference. This may be done to emphasize the fact that capacitor current leads the total current while resistor current lags.

Determining Total Impedance (Z_t) for Parallel *RC*

Using Ohm's Law to Find Z_t

Ohm's Law can always be used to find the total impedance of a circuit if the source voltage and total current are known. The magnitude of total current can be divided into the source voltage to find the magnitude of the total impedance ($Z_t = E_S/I_t$, Formula 19.4). If we wish to find the phase angle for the total impedance, all phase angles may be included in the formula and the rules of complex algebra must be followed ($Z_t \angle -\theta = E_S \angle 0°/I_t \angle \theta$). For a parallel *RC* circuit, the phase angle of the total impedance is negative. See Figure 19-17.

Using Parallel Formulas to Find Z_t

Once again, the total impedance of any parallel circuit may be found using one or more of the three standard parallel formulas (19.5, 19.6, or 19.7). For AC circuits, each branch impedance must be expressed in rectangular or polar form, whichever is appropriate.

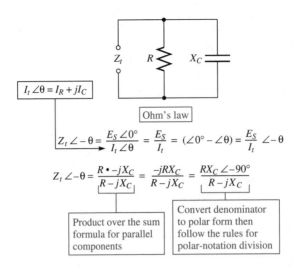

$$I_t \angle \theta = I_R + jI_C$$

Ohm's law

$$Z_t \angle -\theta = \frac{E_S \angle 0°}{I_t \angle \theta} = \frac{E_S}{I_t} = (\angle 0° - \angle \theta) = \frac{E_S}{I_t} \angle -\theta$$

$$Z_t \angle -\theta = \frac{R \bullet -jX_C}{R - jX_C} = \frac{-jRX_C}{R - jX_C} = \frac{RX_C \angle -90°}{R - jX_C}$$

| Product over the sum formula for parallel components | Convert denominator to polar form then follow the rules for polar-notation division |

Figure 19-17 Total impedance for parallel *RC* circuits.

Again, recall that complex numbers are multiplied or divided in polar form, and added or subtracted in rectangular form. Figure 19-17 illustrates the use of the product over the sum formula in finding the total impedance for parallel *RC* circuits. Note that both the numerator and the denominator must be expressed in polar form before a final solution can be obtained.

Parallel *RC* Circuit Calculations

The analysis process for parallel *RC* circuits is very similar to that used for parallel *RL* circuits: Ohm's Law is used to find the branch currents I_R and I_C, the Pythagorean theorem may be used to find the magnitude of the total current, the arctan function may be used to find the phase angle for the total current, and Ohm's Law or a parallel formula may be used to find the total impedance and phase angle. Also, automatic R→P conversion may be used to find the total current and its phase angle ($I_R + jI_C = I_t \angle \theta$). Study Example 19-3 to see how these formulas are applied. You will, of course, immediately recognize the many similarities between this example and Example 19-1 (*RL* analysis). The only real differences between *RC* and *RL* analysis are: the $+jI_C$ instead of $-jI_L$; a positive total-current phase angle for *RC* circuits ($I_t \angle + \theta$) instead of the negative total-current phase angle for *RL* circuits ($I_t \angle - \theta$), and a negative total-impedance phase angle for *RC* circuits ($Z_t \angle - \theta$) instead of the positive total-impedance phase angle for *RL* circuits ($Z_t \angle + \theta$).

EXAMPLE 19-3

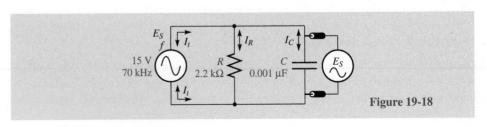

Figure 19-18

For Figure 19-18:

1) $X_C = \dfrac{1}{2\pi f C} = \dfrac{1}{6.28 \cdot 70 \text{ kHz} \cdot 0.001 \ \mu\text{F}} = 2{,}275 \ \Omega$

2) $I_R = E_S/R = 15 \text{ V}/2.2 \text{ k}\Omega = 6.82 \text{ mA}$
   $I_C = E_S/X_C = 15 \text{ V}/2{,}275 \ \Omega = 6.59 \text{ mA}$

3) $I_t = \sqrt{I_R^2 + I_C^2} = \sqrt{(6.82 \text{ mA})^2 + (6.59 \text{ mA})^2} = 9.48 \text{ mA}$

4) $\angle\theta = \tan^{-1}(I_C/I_R) = \tan^{-1}(6.59 \text{ mA}/6.82 \text{ mA}) = 44°$

$$\boxed{I_t \ \angle\theta = 9.48 \text{ mA} \ \angle 44°}$$

5) $Z_t = E_S/I_t = 15 \text{ V} \ \angle 0°/9.48 \text{ mA} \ \angle 44° = 1{,}582 \ \Omega \ \angle -44°$

RULES FOR POLAR NOTATION DIVISION MUST BE FOLLOWED

$Z_t = \dfrac{R \cdot -jX_C}{R - jX_C} = \dfrac{2.2 \text{ k}\Omega \cdot -j2{,}275 \ \Omega}{2.2 \text{ k}\Omega - j2{,}275 \ \Omega}$

$\quad = \dfrac{5{,}005 \text{ k}\Omega \ \angle -90°}{3{,}165 \ \Omega \ \angle -46°}$

$\quad = 1{,}581 \ \Omega \ \angle(-90° - (-46°)) = 1{,}581 \ \Omega \ \angle -44°$

> SLIGHT DIFFERENCE IN ANSWER DUE TO ROUNDING ERROR

6) Draw the current phasor diagram as shown in Figure 19-19.

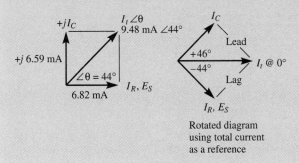

Rotated diagram
using total current
as a reference

Figure 19-19

Example 19-4 is very similar to Example 19-2 in that only the source voltage, the frequency, and the total current with phase angle are given. Unknown circuit parameters are found using the total current and phase angle as the key information. First, the total current and phase angle are applied to Ohm's Law to find the total impedance and phase angle. Then the individual branch currents I_R and I_C are found using manual or automatic P→R conversion (the manual method is shown in step 2, using the cosine and sine functions). Once the branch currents are determined, R and X_C are found using Ohm's Law. The unknown capacitance is found by rearranging the capacitive reactance formula.

EXAMPLE   19-4

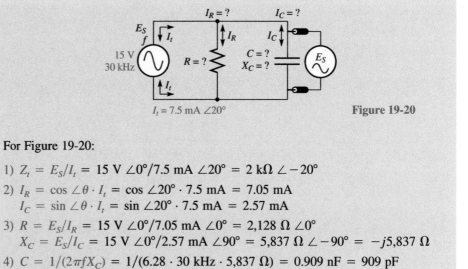

$I_t = 7.5$ mA $\angle 20°$

Figure 19-20

For Figure 19-20:

1) $Z_t = E_S/I_t = 15$ V $\angle 0°/7.5$ mA $\angle 20° = 2$ k$\Omega \angle -20°$

2) $I_R = \cos \angle \theta \cdot I_t = \cos \angle 20° \cdot 7.5$ mA $= 7.05$ mA
   $I_C = \sin \angle \theta \cdot I_t = \sin \angle 20° \cdot 7.5$ mA $= 2.57$ mA

3) $R = E_S/I_R = 15$ V $\angle 0°/7.05$ mA $\angle 0° = 2,128 \Omega \angle 0°$
   $X_C = E_S/I_C = 15$ V $\angle 0°/2.57$ mA $\angle 90° = 5,837 \Omega \angle -90° = -j5,837 \Omega$

4) $C = 1/(2\pi f X_C) = 1/(6.28 \cdot 30$ kHz $\cdot 5,837 \Omega) = 0.909$ nF $= 909$ pF

5) Draw the current phasor diagram as shown in Figure 19-21.

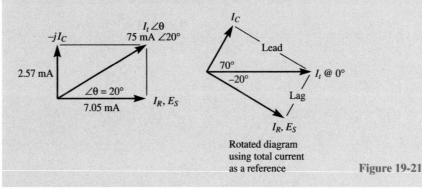

Rotated diagram
using total current
as a reference

Figure 19-21

Take a moment now to reinforce your understanding by analyzing the circuits of Self-Check 19-2.

19-3 Parallel *RCL* Circuit Analysis

RCL Circuit Current Phasors

RCL Currents Out of Phase

In this section, all that you have learned regarding parallel *RL* and *RC* circuits will be combined in parallel *RCL* circuit analysis. Figure 19-23 demonstrates the parallel *RCL* circuit. There is really very little new information for you to learn. As before, the parallel branch currents must be added using phasor addition, since all currents are out of phase. Each of the branch currents and the total current reach a voltage peak independent of all other currents. Kirchhoff's Current Law for AC circuits again applies to instantaneous

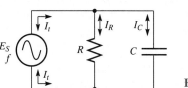

Figure 19-22

Figure 19-22 is a model circuit for Problems 1 through 5.
Fill in the blanks and draw the current phasor diagram for each problem.

#	E_S	f	C	X_C	R	I_C	I_R	I_t	Z_t	$\angle\theta$
1.	3 V	50 Hz	5 μF		470 Ω					
2.	11 V	60 Hz	5 μF		180 Ω					
3.	25 V		100 pF	3 kΩ	4.7 kΩ					
4.	50 mV	650 kHz		300 Ω	200 Ω					
5.			0.1 μF	900 Ω		0.111 A				40°

current values and phasor values. For parallel *RCL* circuits, Kirchhoff's Current Law is
mathematically represented as follows:

$i_R + i_C + i_L - i_t = 0$ or $i_R + i_C + i_L = i_t$ (instantaneous values)

and $I_R \angle 0° + I_C \angle +90° + I_L \angle -90° - I_t \angle \theta = 0$ (phasor values)

or $I_R \angle 0° + I_C \angle +90° + I_L \angle -90° = I_t \angle \theta$.

PLUG-IN EXPERIMENT BOARD

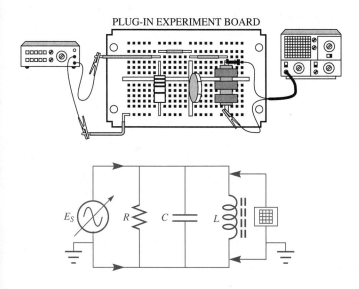

Figure 19-23

RCL Current Phasors

Figure 19-24 illustrates the three possible conditions of parallel *RCL* circuits. Depending on the frequency of the applied source voltage and the values of inductance and capacitance, the parallel *RCL* circuit will act in one of three ways:

1. Capacitively. The *RCL* circuit acts capacitively if X_C is less than X_L. This permits I_C to be greater than I_L as determined by Ohm's Law ($I_C = E_S/X_C$ and $I_L = E_S/X_L$). The greater current determines how the circuit is acting. In this case, $\angle\theta$ for I_t is always positive ($I_t\angle + \theta$).

2. Resistively. The *RCL* circuit acts resistively if $X_C = X_L$ and $I_C = I_L$. The capacitor and inductor current cancel each other, since they are equal and opposite in phase. This results in the entire circuit looking like a pure resistance ($Z_t = R$) to the source voltage. In this case, the resistor branch current and the total current are one and the same and $\angle\theta$ is 0°.

3. Inductively. The *RCL* circuits acts inductively if X_L is less than X_C, making I_L greater than I_C. In this case, $\angle\theta$ for I_t is always negative ($I_t\angle - \theta$).

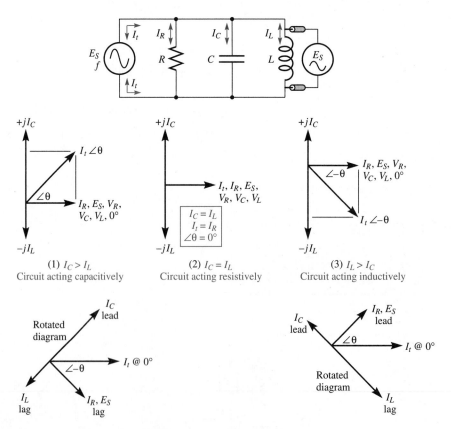

Figure 19-24   Parallel *RCL* circuit current phasor diagrams.

Figure 19-24 also shows a rotated phasor diagram for the capacitive and inductive conditions. In each case, the total current is placed at $0°$ as a reference. All other currents are said to lead or lag the total current.

Adding *RCL* Current Phasors

The *RCL* current phasors must be added using phasor addition. This is mathematically represented as follows:

$$I_t \angle \theta = I_R \angle 0° + I_C \angle +90° + I_L \angle -90° \quad \text{(values in polar form)}$$
$$= I_R + jI_C - jI_L \quad \text{(values in rectangular form)}$$

The actual phasor addition can be performed using manual or automatic R→P conversion. As before, the manual method employs the Pythagorean theorem and the arctan function. For parallel *RCL* analysis, the formulas show the cancelling effect of capacitor and inductor branch currents as I_L is subtracted from I_C. The individual branch currents used in these formulas are found using Ohm's Law as before: $I_R = E_S/R$; $I_C = E_S/X_C$; $I_L = E_S/X_L$.

$$I_t = \sqrt{I_R^2 + (I_C - I_L)^2} \qquad (19.10)$$
$$\angle \theta = \tan^{-1}[(I_C - I_L)/I_R] \qquad (19.11)$$

Determining Total Impedance for Parallel *RCL*

As in parallel *RC* and *RL* circuits, the total impedance for the parallel *RCL* circuit may be determined using Ohm's Law ($Z_t = E_S \angle 0°/I_t \angle \theta$) or any one or more of the standard parallel formulas. Of the three Z_t formulas for parallel circuits, the reciprocal formula (19.7) is the most practical to use, since a parallel *RCL* circuit has three branches containing different impedances (R, $-jX_C$ and $+jX_L$). The reciprocal formula would be applied in the following form, which requires complex algebra:

$$Z_t = \frac{1}{(1/R) + (1/-jX_C) + (1/+jX_L)}$$

Note: Each term in the denominator is a reciprocal of impedance measured in ohms. Thus, the units used for each term must be siemens, millisiemens, or microsiemens. For example, $1/270 \ \Omega = 3.7$ mS.

Parallel *RCL* Circuit Calculations

Just as we did with series *RCL* circuits, we will analyze a parallel *RCL* circuit operating at three different frequencies. Thus, Examples 19-5, 19-6, and 19-7 show the same circuit acting **inductively** at 25,165 Hz ($I_L > I_C$), **resistively** at 50,330 Hz ($I_L = I_C$), and **capacitively** at 100,660 Hz ($I_C > I_L$). It is important for you to understand that the way an AC circuit behaves greatly depends on the frequency of the applied source voltage. Also, it is important for you to realize there is only one frequency at which the *RCL* circuit will appear purely resistive. It is the frequency at which $I_C = I_L$ and is determined by the amount of capacitance and inductance in the circuit. We will investigate this further in later studies on resonance.

EXAMPLE 19-5

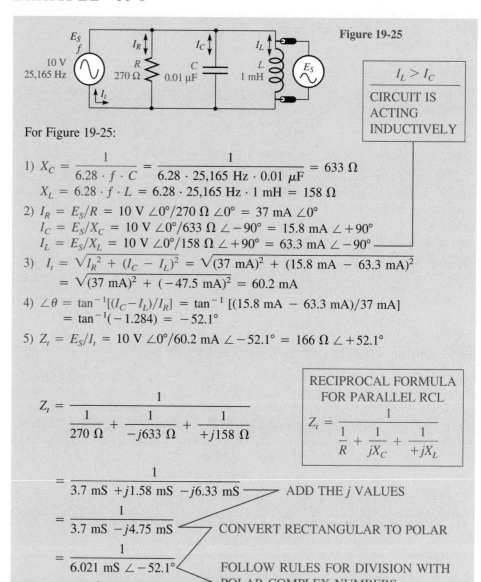

Figure 19-25

For Figure 19-25:

1) $X_C = \dfrac{1}{6.28 \cdot f \cdot C} = \dfrac{1}{6.28 \cdot 25{,}165 \text{ Hz} \cdot 0.01 \ \mu\text{F}} = 633 \ \Omega$

$X_L = 6.28 \cdot f \cdot L = 6.28 \cdot 25{,}165 \text{ Hz} \cdot 1 \text{ mH} = 158 \ \Omega$

2) $I_R = E_S/R = 10 \text{ V} \angle 0°/270 \ \Omega \angle 0° = 37 \text{ mA} \angle 0°$

$I_C = E_S/X_C = 10 \text{ V} \angle 0°/633 \ \Omega \angle -90° = 15.8 \text{ mA} \angle +90°$

$I_L = E_S/X_L = 10 \text{ V} \angle 0°/158 \ \Omega \angle +90° = 63.3 \text{ mA} \angle -90°$

3) $I_t = \sqrt{I_R{}^2 + (I_C - I_L)^2} = \sqrt{(37 \text{ mA})^2 + (15.8 \text{ mA} - 63.3 \text{ mA})^2}$

$= \sqrt{(37 \text{ mA})^2 + (-47.5 \text{ mA})^2} = 60.2 \text{ mA}$

4) $\angle\theta = \tan^{-1}[(I_C - I_L)/I_R] = \tan^{-1}[(15.8 \text{ mA} - 63.3 \text{ mA})/37 \text{ mA}]$

$= \tan^{-1}(-1.284) = -52.1°$

5) $Z_t = E_S/I_t = 10 \text{ V} \angle 0°/60.2 \text{ mA} \angle -52.1° = 166 \ \Omega \angle +52.1°$

$Z_t = \dfrac{1}{\dfrac{1}{270 \ \Omega} + \dfrac{1}{-j633 \ \Omega} + \dfrac{1}{+j158 \ \Omega}}$

> **RECIPROCAL FORMULA FOR PARALLEL RCL**
> $$Z_t = \dfrac{1}{\dfrac{1}{R} + \dfrac{1}{jX_C} + \dfrac{1}{+jX_L}}$$

$= \dfrac{1}{3.7 \text{ mS} + j1.58 \text{ mS} - j6.33 \text{ mS}}$ ⟶ ADD THE j VALUES

$= \dfrac{1}{3.7 \text{ mS} - j4.75 \text{ mS}}$ ⟶ CONVERT RECTANGULAR TO POLAR

$= \dfrac{1}{6.021 \text{ mS} \angle -52.1°}$ ⟶ FOLLOW RULES FOR DIVISION WITH POLAR COMPLEX NUMBERS

$= 166 \ \Omega \angle +52.1°$

6) Draw the current phasor diagram as shown in Figure 19-26.

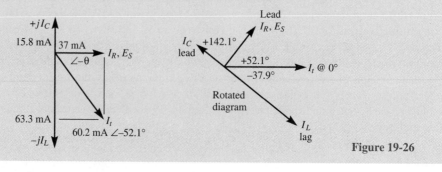

Figure 19-26

In Example 19-5, the *RCL* circuit is acting inductively. This is discovered immediately in the analysis process as the capacitive and inductive reactances are calculated in step 1. Naturally, the lower of the two reactances determines how the circuit is acting. The actual branch currents are determined in step 2 using Ohm's Law. The phase angles are included in the calculations to show the resulting phase angle for each current. This verifies the fact that capacitor current is at $+90°$ and inductor current is at $-90°$. In step 3, the total current is found manually using the Pythagorean theorem. The phase angle for total current is found in step 4 using the arctan function. Notice that the inductor current must be subtracted from the capacitor current to yield a negative phase angle when inductor current is greater than capacitor current. The findings of steps 2 to 4 are graphically related to the phasor diagrams of step 6.

The total current and phase angle may also be found using automatic R→P conversion. The resistor current is entered first, then press [SHIFT] [R→P] followed by entering the difference between capacitor and inductor current ($I_C - I_L$). Depress = to display the total current magnitude, then depress X↔Y to display the phase angle.

Step 5 shows how the total impedance may be calculated using Ohm's Law or the reciprocal formula. Ohm's Law, of course, is the easier of the two methods. However, Ohm's Law can only be used if a voltage is given or assumed (a voltage may be assumed to allow the use of Ohm's Law in finding total impedance). If the reciprocal formula is used, the basic rules of complex algebra must be followed. Note that addition is accomplished in rectangular form and division is accomplished in polar form.

Example 19-6 shows the analysis of the same circuit operating at a higher frequency (50,330 Hz). This is the single frequency at which this circuit will act purely resistively. Any increase in frequency will cause the circuit to act capacitively, and any decrease will cause the circuit to act inductively. At 50,330 Hz, the inductive reactance and capacitive reactance are both 316 Ω. The capacitor and inductor currents are shown to be equal in step 2 (each 31.6 mA but opposite in phase). Notice that the total current, found in step 3, is the same as the resistor current. This indicates that the total impedance in this circuit is simply the resistor. Step 5 verifies this fact ($Z_t = R\angle0°$).

EXAMPLE   19-6

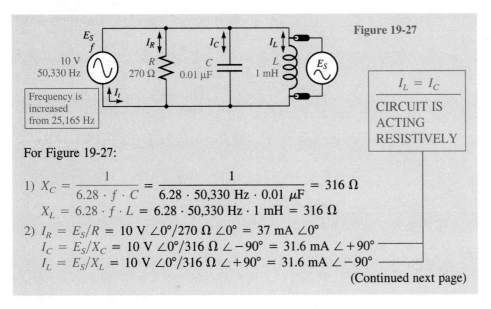

Figure 19-27

$$I_L = I_C$$

CIRCUIT IS
ACTING
RESISTIVELY

For Figure 19-27:

1) $X_C = \dfrac{1}{6.28 \cdot f \cdot C} = \dfrac{1}{6.28 \cdot 50,330 \text{ Hz} \cdot 0.01 \ \mu\text{F}} = 316 \ \Omega$

   $X_L = 6.28 \cdot f \cdot L = 6.28 \cdot 50,330 \text{ Hz} \cdot 1 \text{ mH} = 316 \ \Omega$

2) $I_R = E_S/R = 10 \text{ V} \angle0°/270 \ \Omega \angle0° = 37 \text{ mA} \angle0°$

   $I_C = E_S/X_C = 10 \text{ V} \angle0°/316 \ \Omega \angle-90° = 31.6 \text{ mA} \angle+90°$

   $I_L = E_S/X_L = 10 \text{ V} \angle0°/316 \ \Omega \angle+90° = 31.6 \text{ mA} \angle-90°$

(Continued next page)

EXAMPLE 19-6 (Continued)

3) $I_t = \sqrt{I_R^2 + (I_C - I_L)^2} = \sqrt{(37\text{ mA})^2 + (31.6\text{ mA} - 31.6\text{ mA})^2}$
   $= \sqrt{(37\text{ mA})^2 + (0)^2} = \sqrt{(37\text{ mA})^2} = 37\text{ mA}$

4) $\angle\theta = \tan^{-1}[(I_C - I_L)/I_R] = \tan^{-1}[(0)/37\text{ mA}] = \tan^{-1}0 = 0°$

5) $Z_t = E_S/I_t = 10\text{ V }\angle0°/37\text{ mA }\angle0° = 270\ \Omega\ \angle0°$

$$Z_t = \cfrac{1}{\cfrac{1}{270\ \Omega} + \cfrac{1}{-j316\ \Omega} + \cfrac{1}{+j316\ \Omega}}$$

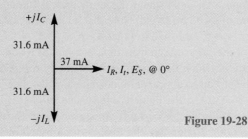

RECIPROCAL FORMULA
FOR PARALLEL RCL

$$Z_t = \cfrac{1}{\cfrac{1}{R} + \cfrac{1}{jX_C} + \cfrac{1}{+jX_L}}$$

$$= \frac{1}{3.7\text{ mS} + j3.16\text{ mS} - j3.16\text{ mS}}$$

$$= \frac{1}{3.7\text{ mS}} = 270\ \Omega\ \angle0°$$

6) Draw the current phasor diagram as shown in Figure 19-28.

+jI_C

31.6 mA

37 mA → I_R, I_t, E_S, @ 0°

31.6 mA

−jI_L

Figure 19-28

EXAMPLE 19-7

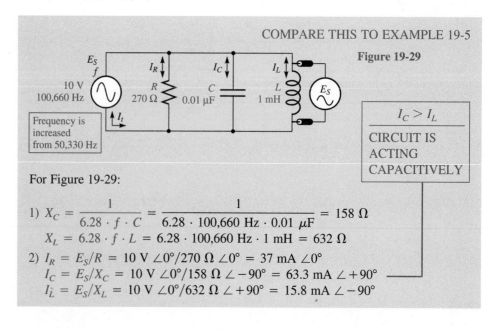

COMPARE THIS TO EXAMPLE 19-5

Figure 19-29

E_S
f

10 V
100,660 Hz

I_R I_C I_L

R
270 Ω

C
0.01 μF

L
1 mH

E_S

Frequency is
increased
from 50,330 Hz

I_t

$I_C > I_L$

CIRCUIT IS
ACTING
CAPACITIVELY

For Figure 19-29:

1) $X_C = \dfrac{1}{6.28 \cdot f \cdot C} = \dfrac{1}{6.28 \cdot 100,660\text{ Hz} \cdot 0.01\ \mu F} = 158\ \Omega$

$X_L = 6.28 \cdot f \cdot L = 6.28 \cdot 100,660\text{ Hz} \cdot 1\text{ mH} = 632\ \Omega$

2) $I_R = E_S/R = 10\text{ V }\angle0°/270\ \Omega\ \angle0° = 37\text{ mA }\angle0°$

$I_C = E_S/X_C = 10\text{ V }\angle0°/158\ \Omega\ \angle-90° = 63.3\text{ mA }\angle+90°$

$I_L = E_S/X_L = 10\text{ V }\angle0°/632\ \Omega\ \angle+90° = 15.8\text{ mA }\angle-90°$

3) $I_t = \sqrt{I_R^2 + (I_C - I_L)^2} = \sqrt{(37 \text{ mA})^2 + (63.3 \text{ mA} - 15.8 \text{ mA})^2}$
$= \sqrt{(37 \text{ mA})^2 + (47.5 \text{ mA})^2} = 60.2 \text{ mA}$

4) $\angle\theta = \tan^{-1}[(I_C - I_L)/I_R] = \tan^{-1}[(63.3 \text{ mA} - 15.8 \text{ mA})/37 \text{ mA}]$
$= \tan^{-1}(1.284) = 52.1°$

5) $Z_t = E_S/I_t = 10 \text{ V} \angle0°/60.2 \text{ mA} \angle52.1° = 166 \text{ } \Omega \angle-52.1°$

$$Z_t = \cfrac{1}{\cfrac{1}{270 \text{ } \Omega} + \cfrac{1}{-j158 \text{ } \Omega} + \cfrac{1}{+j632 \text{ } \Omega}}$$

RECIPROCAL FORMULA
FOR PARALLEL RCL
$Z_t = \cfrac{1}{\cfrac{1}{R} + \cfrac{1}{-jX_C} + \cfrac{1}{+jX_L}}$

$$= \cfrac{1}{3.7 \text{ mS} + j6.33 \text{ mS} - j1.58 \text{ mS}} \quad \text{ADD THE } j \text{ VALUES}$$

$$= \cfrac{1}{3.7 \text{ mS} + j4.75 \text{ mS}} \quad \text{CONVERT RECTANGULAR TO POLAR}$$

$$= \cfrac{1}{6.021 \text{ mS} \angle52.1°} \quad \begin{array}{l}\text{FOLLOW RULES FOR DIVISION WITH} \\ \text{POLAR COMPLEX NUMBERS}\end{array}$$

$$= 166 \text{ } \Omega \angle-52.1°$$

6) Draw the current phasor diagram as shown in Figure 19-30.

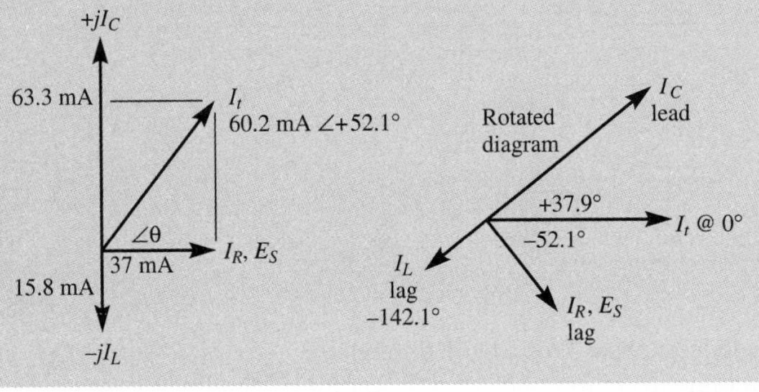

Figure 19-30

Finally, Example 19-7 shows the same circuit acting capacitively at 100,660 Hz. Compare Example 19-7 to Example 19-5. Notice that the frequency of Example 19-5 is half the frequency of Example 19-6 and the frequency of Example 19-7 is twice that of Example 19-6. As a result, the calculations of Example 19-7 are the opposite, or reverse, of Example 19-5.

Things to Recognize

1. The parallel *RCL* circuit's total impedance is maximum at a frequency where $I_C = I_L$. The reason: total current is minimum at this one frequency since I_C and I_L cancel. If total current is minimum, total impedance must be maximum and equal to R.

2. The parallel *RCL* circuit is acting capacitively if $I_C > I_L$ and inductively if $I_L > I_C$.

General Parallel AC Circuit Analysis Procedure

The following procedure is a general step-by-step procedure that may be used to analyze parallel *RL*, *RC*, and *RCL* circuits. This procedure is utilized in Design Note 19-1.

1. Calculate the inductive, and/or capacitive, reactance at the applied frequency.
2. Calculate all branch currents using Ohm's Law.
3. Calculate the total current and phase angle using the Pythagorean theorem and the arctan function, or use automatic R→P conversion.
4. Calculate the total impedance using Ohm's Law or one of the standard parallel formulas.
5. Draw and label the current phasor diagram.

Take a few minutes now to exercise your skills on Self-Check 19-3.

SELF-CHECK 19-3

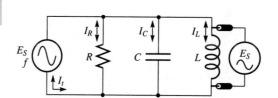

Figure 19-31

Figure 19-31 is a model circuit for Problems 1 through 5.
Fill in the blanks and draw the current phasor diagram for each problem.

#	E_S	f	R	C	L	X_C	X_L	Z_t	I_R	I_C	I_L	I_t	$\angle\theta$
1.	1 V	1 kHz	1 kΩ	0.47 μF	0.1 H								
2.	50 mV	455 kHz	20 Ω	0.002 μF	20 μH								
3.	5 V	10 MHz							2 mA	4 mA	6 mA		
4.		120 MHz	5 Ω	50 pF	0.01 μH				15 mA				
5.	10 V			10 μF	1 H	150 Ω			100 mA				

D E S I G N   N O T E   1 9 - 1 :   Parallel *RCL* Analysis

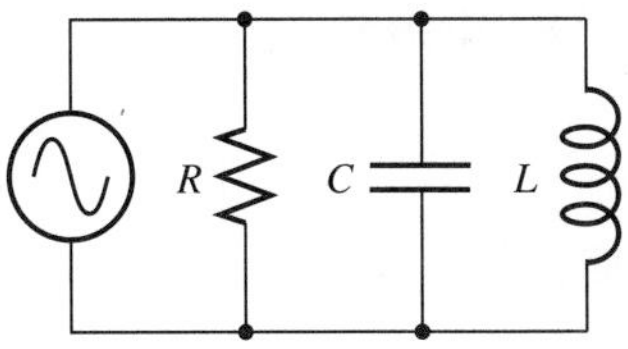

PROGRAM VARIABLES

ES = SOURCE VOLTAGE	TH = ANGLE THETA
F = FREQUENCY	IR = RESISTOR CURRENT
R = RESISTANCE	IC = CAPACITOR CURRENT
C = CAPACITANCE	IL = INDUCTOR CURRENT
L = INDUCTANCE	PR = REAL POWER
XC = CAPACITIVE REACTANCE	PC = CAPACITOR POWER
XL = INDUCTIVE REACTANCE	PL = INDUCTOR POWER
ZT = TOTAL IMPEDANCE	PA = APPARENT POWER
IT = TOTAL CURRENT	T = CYCLE TIME

"BASIC" PROGRAM SOLUTION

```
 10 CLS
 20 PRINT"PARALLEL RCL ANALYSIS"
 30 PRINT"":PRINT""
 40 PRINT"THIS PROGRAM WILL ANALYZE ANY PARALLEL RC, RL, OR RCL CIRCUIT."
 50 PRINT"ENTER A ZERO (0) FOR L OR C IF IT DOES NOT EXIST IN YOUR CIRCUIT."
 60 PRINT""
 70 INPUT"ENTER THE INDUCTANCE IN HENRIES. ";L
 80 INPUT"ENTER THE CAPACITANCE IN FARADS. ";C
 90 INPUT"ENTER THE TOTAL PARALLEL RESISTANCE. ";R
100 INPUT"ENTER THE SOURCE VOLTAGE. ";ES
110 INPUT"ENTER THE FREQUENCY OF THE APPLIED AC. ";F
120 IF C = 0 THEN GOTO 140
130 XC = 1/(6.2831853# * F * C)
135 IC = ES/XC
140 IF L = 0 THEN GOTO 160
150 XL = 6.2831853# * F * L
155 IL = ES/XL
160 IR = ES/R
170 IT = SQR(IR^2 + (IC-IL)^2)
175 ZT = ES/IT
180 TH = ATN((IC − IL)/IR) * 57.29578
200 PR = IR * ES:PC = IC * ES:PL = IL * ES
210 PA = IT * ES:PF = PR/PA
220 T = 1/F
300 IF C = 0 THEN GOTO 320
310 PRINT"XC = ";XC;" OHMS. IC = ";IC;" AMPS."
320 IF L = 0 THEN GOTO 340
330 PRINT"XL = ";XL;" OHMS. IL = ";IL;" AMPS."
340 PRINT"ZT = ";ZT;" OHMS."
350 PRINT"IT = ";IT;" AMPS."
360 PRINT"ANGLE THETA = ";TH;" DEGREES."
370 PRINT"IR = ";IR;" AMPS. PR = ";PR;" WATTS."
380 IF C = 0 THEN GOTO 400
390 PRINT"PC = ";PC;" VOLT-AMPERES REACTIVE."
400 IF L = 0 THEN GOTO 420
```

```
410  PRINT"PL = ";PL;" VOLT-AMPERES REACTIVE."
420  PRINT"PA = ";PA;" VOLT AMPERES. PF = ";PF
520  IF IC > IL THEN GOTO 590
530  PRINT"IR LEADS IT BY ";-TH;" DEGREES OR ";T * -TH/360;" SECONDS."
540  IF C = 0 THEN GOTO 560
550  PRINT"IC LEADS IT BY";90 - TH;" DEGREES OR "T * (90 - TH)/360;" SECONDS."
560  IF L = 0 THEN GOTO 580
570  PRINT"IL LAGS IT BY ";90 + TH;" DEGREES OR ";T * (TH + 90)/360;" SECONDS."
580  GOTO 640
590  PRINT"IR LAGS IT BY ";TH;" DEGREES OR ";T * TH/360;" SECONDS."
600  IF L = 0 THEN GOTO 620
610  PRINT"IL LAGS IT BY ";90 + TH;" DEGREES OR ";T * (90 + TH)/360;" SECONDS."
620  IF C = 0 THEN GOTO 640
630  PRINT"IC LEADS IT BY ";90 - TH;" DEGREES OR ";T * (90 - TH)/360;" SECONDS."
640  PRINT""
650  INPUT"ANOTHER PROBLEM? (Y/N)";A$
660  IF A$ = "Y" THEN CLEAR:GOTO 30
670  CLS:END
```

19-4 Power in Parallel AC Circuits

The power formulas for parallel AC circuits are the same as those used for series circuits. Real power (P_R) is the power dissipated by the resistor. Reactive power ($P_r = P_C$ and/or P_L) is the power stored in a reactive component. Apparent power (P_A) is the total power that is apparently being used by the circuit and is usually a combination of real and reactive power. The basic formulas for these powers were introduced in Chapter 18 and are summarized as follows:

$$(18.25) \quad P_R = I_R \cdot V_R = I_R^2 \cdot R = V_R^2/R \ \text{(W, watt)}$$

$$(18.26) \quad P_r = P_C = I_C \cdot V_C = I_C^2 \cdot X_C = V_C^2/X_C$$

$$= P_L = I_L \cdot V_L = I_L^2 \cdot X_L = V_L^2/X_L \ \text{(VAR, volt-amperes reactive)}$$

$$(18.27) \quad P_A = P_{\text{total}} = I_t \cdot E_S = I_t^2 \cdot Z_t = E_S^2/Z_t \ \text{(VA, volt-ampere)}$$

Power Phasors for Parallel AC Circuits

Creating Power Phasors for Parallel RCL

The power-phasor diagrams for parallel AC circuits are created from the current-phasor diagrams. Each of the current phasors is simply multiplied by the common source voltage: $I_R \angle 0° \cdot E_S \angle 0° = P_R \angle 0°$; $I_C \angle +90° \cdot E_S \angle 0° = P_C \angle +90° = +jP_C$; $I_L \angle -90° \cdot E_S \angle 0° = P_L \angle -90° = -jP_L$; $I_t \angle \pm \theta° \cdot E_S \angle 0° = P_A \angle \pm \theta°$. The resulting power-phasor diagrams for the three possible conditions of a parallel *RCL* circuit are shown in Figure 19-32. In each of the three conditions, we can see once again that the apparent power is the phasor sum of the real and reactive powers in the circuit.

$$P_A \angle \theta = P_R \angle 0° + P_C \angle +90° + P_L \angle -90° = P_R + jP_C - jP_L \quad (19.12)$$

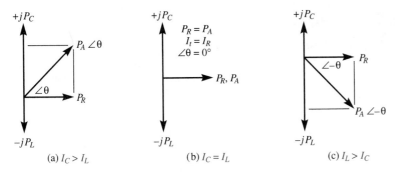

Figure 19-32 Parallel *RCL* circuit power phasor diagrams.

Power Factor for Parallel AC Circuits

Calculating Power Factor (*PF*)

The power factor for parallel AC is the same as for series AC circuits. It is the decimal expression of the ratio of real power to apparent power ($PF = P_R/P_A$). The power factor is a power efficiency factor. When multiplied by 100%, the power factor becomes the percent efficiency for the AC circuit (percent efficiency $= PF \cdot 100\%$). Any of the following ratios may be used to calculate power factor:

$$PF = P_R/P_A = I_R/I_t = \cos \angle \theta \qquad (\leq 1, \text{ no units}) \qquad (19.13)$$

Maximum Power Efficiency

Under what conditions is a parallel AC circuit operating at maximum efficiency? Maximum efficiency occurs when $I_C = I_L$, which makes $P_R = P_A$ and $\angle \theta = 0°$. Consider Figure 19-32b. When $I_C = I_L$, then $P_C = P_L$. Under these circumstances, the total current equals the resistor current (I_C and I_L cancel). As a result, $P_R = P_A = I_R \cdot E_S = I_t \cdot E_S$.

Power Factor Correction

As in series AC circuits, the power factor can be corrected and efficiency improved if one reactance is used to balance out another. In most practical applications, such as electric motors, it is impossible to physically remove unwanted reactance. In such a case, the reactance can be neutralized with an equal amount of the opposite reactance. Figure 19-32a indicates the circuit is acting capacitively ($I_C > I_L$). Therefore, the inductive reactance should be reduced (by reducing *L*) or the capacitive reactance should be increased (by reducing *C*). This will balance the reactances and reactive currents to improve efficiency. Figure 19-32c indicates the circuit is acting inductively ($I_L > I_C$). In this case, the inductive reactance should be increased (by increasing *L*) or the capacitive reactance should be decreased (by increasing *C*) to correct the power factor and improve efficiency. In some cases, it is possible to change the frequency of the applied source voltage to improve circuit efficiency. For the phasor diagram on the left, the frequency should be decreased [$\downarrow X_L = 2\pi \downarrow fL, \uparrow X_C = 1/(2\pi \downarrow fC)$], and for the diagram on the right, the frequency should be increased [$\uparrow X_L = 2\pi \uparrow fL, \downarrow X_C = 1/(2\pi \uparrow fC)$].

Parallel *RCL* Power Calculations

Examples 19-8, 19-9, and 19-10 illustrate how the various powers are calculated for parallel *RCL* circuits. You will immediately recognize these examples as a further analysis of Examples 19-5, 19-6, and 19-7. The examples are designed to demonstrate the effects of frequency on the apparent power and reactive power distribution. As you study these examples, you should realize the following:

1. Real power (P_R) is independent of frequency. The resistor dissipates 370 mW in each example.

2. Apparent power (P_A) is minimum only at one frequency (50,330 Hz).

3. $P_R = P_A$ at one frequency (50,330 Hz). P_C and P_L cancel, leaving the circuit purely resistive.

4. At any frequency below the center frequency, $P_L > P_C$. Therefore, an imbalance exists between the two.

5. At any frequency above the center frequency, $P_C > P_L$. Therefore, an imbalance exists between the two.

6. The circuit is 100% efficient ($PF = 1$) at only one frequency (the center frequency, in this case 50,330 Hz).

7. The way the circuit is acting depends heavily on the applied frequency and the size of the inductor (L) and capacitor (C).

$$\uparrow X_L = 2\pi \cdot \uparrow f \cdot L \qquad \text{or} \qquad \uparrow X_L = 2\pi \cdot f \cdot \uparrow L$$
$$\downarrow X_L = 2\pi \cdot \downarrow f \cdot L \qquad \text{or} \qquad \downarrow X_L = 2\pi \cdot f \cdot \downarrow L$$
$$\uparrow X_C = 1/(2\pi \cdot \downarrow f \cdot C) \qquad \text{or} \qquad \uparrow X_C = 1/(2\pi \cdot f \cdot \downarrow C)$$
$$\downarrow X_C = 1/(2\pi \cdot \uparrow f \cdot C) \qquad \text{or} \qquad \downarrow X_C = 1/(2\pi \cdot f \cdot \uparrow C)$$

EXAMPLE   19-8

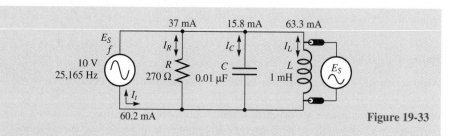

Figure 19-33

For Figure 19-33:

$P_R = I_R \cdot E_S = 37\ \text{mA} \cdot 10\ \text{V} = 370\ \text{mW}$

$P_C = I_C \cdot E_S = 15.8\ \text{mA} \cdot 10\ \text{V} = 158\ \text{mVAR}$

$P_L = I_L \cdot E_S = 63.3\ \text{mA} \cdot 10\ \text{V} = 633\ \text{mVAR}$

$P_A = I_t \cdot E_S = 60.2\ \text{mA} \cdot 10\ \text{V} = 602\ \text{mVA}$

$P_A = P_R + jP_C - jP_L = 370\ \text{mW} + j158\ \text{mVAR} - j633\ \text{mVAR}$
$\quad = 370\ \text{mW} - j475\ \text{mVAR} = 602\ \text{mVA}\ \angle -52.1°$

Figure 19-34 is the phasor diagram for Figure 19-33.

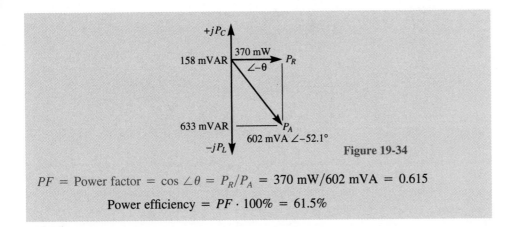

Figure 19-34

$$PF = \text{Power factor} = \cos \angle \theta = P_R/P_A = 370 \text{ mW}/602 \text{ mVA} = 0.615$$

$$\text{Power efficiency} = PF \cdot 100\% = 61.5\%$$

EXAMPLE   19-9

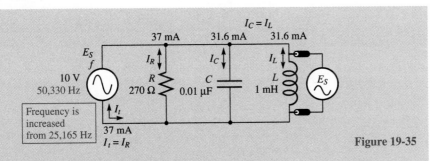

Figure 19-35

For Figure 19-35:

$$P_R = I_R \cdot E_S = 37 \text{ mA} \cdot 10 \text{ V} = 370 \text{ mW}$$

$$P_C = I_C \cdot E_S = 31.6 \text{ mA} \cdot 10 \text{ V} = 316 \text{ mVAR}$$

$$P_L = I_L \cdot E_S = 31.6 \text{ mA} \cdot 10 \text{ V} = 316 \text{ mVAR}$$

$$P_A = I_t \cdot E_S = 37 \text{ mA} \cdot 10 \text{ V} = 370 \text{ mVA or mW}$$

$$P_A = P_R + jP_C - jP_L = 370 \text{ mW} + j316 \text{ mVAR} - j316 \text{ mVAR}$$
$$= 370 \text{ mVA or mW}$$

Figure 19-36 is the phasor diagram for Figure 19-35.

$$PF = \text{Power factor} = \cos \angle \theta = P_R/P_A = 370 \text{ mW}/370 \text{ mVA} = 1$$

$$\text{Power efficiency} = PF \cdot 100\% = 100\%$$

EXAMPLE   19-10

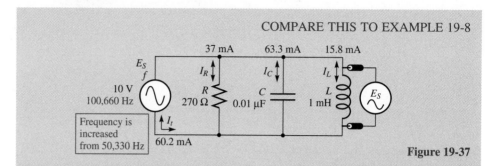

COMPARE THIS TO EXAMPLE 19-8

Figure 19-37

For Figure 19-37:

$P_R = I_R \cdot E_S = 37 \text{ mA} \cdot 10 \text{ V} = 370 \text{ mW}$

$P_C = I_C \cdot E_S = 63.3 \text{ mA} \cdot 10 \text{ V} = 633 \text{ mVAR}$

$P_L = I_L \cdot E_S = 15.8 \text{ mA} \cdot 10 \text{ V} = 158 \text{ mVAR}$

$P_A = I_t \cdot E_S = 60.2 \text{ mA} \cdot 10 \text{ V} = 602 \text{ mVA}$

$P_A = P_R + jP_C - jP_L = 370 \text{ mW} + j633 \text{ mVAR} - j158 \text{ mVAR}$
$\quad = 370 \text{ mW} + j475 \text{ mVAR} = 602 \text{ mVA} \angle +52.1°$

Figure 19-38 is the phasor diagram for Figure 19-37.

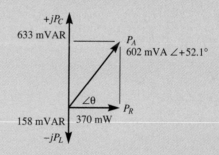

Figure 19-38

$PF = \text{Power factor} = \cos \angle\theta = P_R/P_A = 370 \text{ mW}/602 \text{ mVA} = 0.615$

Power efficiency $= PF \cdot 100\% = 61.5\%$

SELF-CHECK   19-4

Calculate P_R, P_C, P_L, P_A, and PF for each of the parallel analysis problems of Self-Check 19-3.

Summary

FORMULAS

Parallel *RL*

(19.1) $I_t = \sqrt{I_R^2 + I_L^2}$

   also: $I_R = \sqrt{I_t^2 - I_L^2}$ and $I_L = \sqrt{I_t^2 - I_R^2}$

(19.2) $\angle \theta = \tan^{-1}(-I_L/I_R)$

General Parallel Analysis

(19.3) $I_B = E_B/Z_B$

   where the subscript *B* stands for *B*ranch and, in a simple parallel circuit, $E_B = E_S$.

(19.4) $Z_t = E_S/I_t$

(19.5) $Z_t = Z_B/N$ (same values formula)

   where all branches have the same impedance (Z_B) and *N* is the number of branches

(19.6) $Z_t = \dfrac{Z_{B1} \cdot Z_{B2}}{Z_{B1} + Z_{B2}}$ (product over the sum formula)

   where there are two parallel branches having different impedances

(19.7) $Z_t = \dfrac{1}{1/Z_{B1} + 1/Z_{B2} + 1/Z_{B3} + \cdots}$ (reciprocal formula)

   where there are more than two branches having different impedances

Parallel *RC*

(19.8) $I_t = \sqrt{I_R^2 + I_C^2}$

   also: $I_R = \sqrt{I_t^2 - I_C^2}$ and $I_C = \sqrt{I_t^2 - I_R^2}$

(19.9) $\angle \theta = \tan^{-1}(I_C/I_R)$

Parallel *RCL*

(19.10) $I_t = \sqrt{I_R^2 + (I_C - I_L)^2}$

(19.11) $\angle \theta = \tan^{-1}[(I_C - I_L)/I_R]$

(19.12) $P_A \angle \theta = P_R \angle 0° + P_C \angle +90° + P_L \angle -90° = P_R + jP_C - jP_L$

(19.13) $PF = P_R/P_A = I_R/I_t = \cos \angle \theta$ (≤ 1, no units)

CONCEPTS

• A parallel *RCL* circuit is acting inductively if the inductor branch current is greater than the capacitor branch current.
• A parallel *RCL* circuit is acting capacitively if the capacitor branch current is greater than the inductor branch current.

- A parallel *RCL* circuit is acting resistively if the inductor and capacitor branch currents are equal.
- There is only one frequency at which $I_C = I_L$ for given values of inductance and capacitance.
- When $I_C = I_L$ for a simple parallel *RCL* circuit, the following relationships are also true: $P_C = P_L$; $Z_t = R$; $P_A = P_R$; $PF = 1$; efficiency = 100%; $\angle\theta = 0°$; $I_t = I_R$; Z_t is maximum, I_t is minimum.
- Circuit efficiency is improved by correcting the power factor.

PROCEDURES

General Parallel AC Circuit Analysis Procedure

1. Calculate the inductive, and/or capacitive, reactance at the applied frequency.

2. Calculate all branch currents using Ohm's Law.

3. Calculate the total current and phase angle using the Pythagorean theorem and the arctan formula, or use automatic R→P conversion.

4. Calculate the total impedance using Ohm's Law or one of the standard parallel formulas.

5. Draw and label the current phasor diagram.

Need to Know Solution

First, determine individual branch currents.

$$I_{R1} = 115 \text{ V}/330 \ \Omega = 348 \text{ mA}$$

$$I_{L1} = 115 \text{ V}/+j150 \ \Omega = -j767 \text{ mA}$$

$$I_{C1} = 115 \text{ V}/-j200 \ \Omega = +j575 \text{ mA}$$

$$I_{C2} = 115 \text{ V}/-j300 \ \Omega = +j383 \text{ mA}$$

Next, represent I_t in rectangular form.

$$I_t = 348 \text{ mA} - j767 \text{ mA} + j575 \text{ mA} + j383 \text{ mA}$$
$$= 348 \text{ mA} - j767 \text{ mA} + j958 \text{ mA}$$
$$= 348 \text{ mA} + j191 \text{ mA}$$

Now, convert to polar form.

$$348 \text{ mA} + j191 \text{ mA} = 397 \text{ mA} \angle +28.8°$$

Use the polar form of I_t to find Z_t.

$$Z_t = E_S/I_t = \frac{115 \text{ V}}{397 \text{ mA} \angle +28.8°} = 290 \ \Omega \ \angle -28.8°$$

Note: the circuit is acting capacitively and is equivalent to a 254 Ω resistor in series with a capacitor with an X_C of 140 Ω.

$$Z_t = 290 \ \Omega \ \angle -28.8° = 254 \ \Omega \ -j140 \ \Omega$$

Questions

19-1 Parallel *RL* Circuit Analysis

1. Explain how Kirchhoff's Current Law is applied to parallel AC circuits.
2. How is inductor current expressed in rectangular notation? Polar notation?
3. Explain what it means to add branch currents using phasor addition.
4. If the resistor current and source voltage are used as the reference at 0°, does the total current (a) lead or (b) lag in a parallel *RL* circuit?
5. Standard parallel formulas may be used to calculate the total impedance of a parallel AC circuit, but what mathematical rules must you follow?

19-2 Parallel *RC* Circuit Analysis

6. What is the phase angle for capacitor current if the source voltage is used as the reference at 0° for a parallel *RC* circuit? Explain.
7. How is capacitor current expressed in rectangular notation? Polar notation?
8. If the resistor current and source voltage are used as the reference at 0°, does the total current (a) lead or (b) lag in a parallel *RC* circuit?

19-3 Parallel *RCL* Circuit Analysis

9. How do you determine if a parallel *RCL* circuit is acting inductively or capacitively?
10. What is the phase relationship between inductor branch current and capacitor branch current in a simple parallel *RCL* circuit?
11. Does the total current (a) lead or (b) lag the resistor current in a parallel *RCL* circuit that is acting capacitively?
12. What does the frequency of the applied source voltage have to do with the way a circuit is acting? Explain.
13. How many frequencies are there at which a parallel *RCL* circuit will act purely resistively? Explain.

19-4 Power in Parallel AC Circuits

14. Write a mathematical expression showing the phasor sum of all powers of a simple parallel *RCL* circuit. Express the powers in rectangular and polar form.
15. Under what conditions is a parallel *RCL* circuit operating at maximum efficiency?
16. If a parallel *RCL* circuit is acting capacitively, explain how the circuit's power factor can be corrected for increased efficiency.
17. Under what conditions does real power equal apparent power in a parallel *RCL* circuit?

Problems

19-1 Parallel *RL* Circuit Analysis

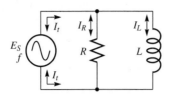

Figure 19-39

Figure 19-39 is a model circuit for Problems 1 through 5.
Fill in the blanks and draw the current phaser diagram for each problem.

#	E_S	f	L	X_L	R	I_L	I_R	I_t	Z_t	$\angle \theta$
1.	40 mV	150 kHz	0.2 mH		1 kΩ					
2.	150 μV	600 kHz	40 μH		50 Ω					
3.	115 V		2 H	754 Ω	1.8 kΩ					
4.	230 V		5 H			146 mA	411 mA	436 mA		
5.		1 MHz			100 Ω			1.13 mA		−28°

19-2 Parallel *RC* Circuit Analysis

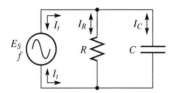

Figure 19-40

Figure 19-40 is a model circuit for Problems 6 through 10.
Fill in the blanks and draw the current phasor diagram for each problem.

#	E_S	f	C	X_C	R	I_C	I_R	I_t	Z_t	$\angle \theta$
6.	20 V	1 kHz	0.01 μF		4.7 kΩ					
7.	35 mV	80 kHz	0.01 μF		330 Ω					
8.	4 V		0.05 μF	26.5 kΩ	56 kΩ					
9.	140 V		1 μF			0.352 A	1.27 A	1.32 A		
10.			0.5 μF	5,305 Ω		1.89 mA				46.6°

19-3 Parallel *RCL* Circuit Analysis

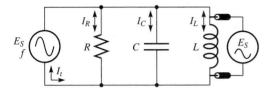

Figure 19-41

Figure 19-41 is a model circuit for Problems 11 through 15.
Fill in the blanks and draw the current phasor diagram for each problem.

#	E_S	f	R	C	L	X_C	X_L	Z_t	I_R	I_C	I_L	I_t	$\angle\theta$
11.	5 V	60 Hz	300 Ω	10 μF	1 H								
12.	12 V	50 Hz	25 Ω	100 μF	100 mH								
13.	200 μV	10 MHz							4 μA	0.251 μA	1.59 μA		
14.		170 kHz	100 Ω	0.02 μF	0.06 mH				600 μA				
15.	80 V			20 μF	3 H	159 Ω			667 mA				

19-4 Power in Parallel AC Circuits

16. Calculate all powers and power factor for Problem 14 (P_R, P_C, P_L, P_A, PF).
17. Calculate all powers and power factor for Problem 15 (P_R, P_C, P_L, P_A, PF).

Answers to Self-Checks

Self-Check 19-1

#	E_S	f	L	X_L	R	I_L	I_R	I_t	Z_t	$\angle\theta$
1.	50 V	60 Hz	0.5 H	**188 Ω**	100 Ω	**266 mA**	0.5 A	**566 mA**	88.3 Ω	**−28°**
2.	20 V	1 MHz	32 μH	**201 Ω**	400 Ω	**99.5 mA**	50 mA	**111 mA**	180 Ω	**−63.3°**
3.	15 mV	200 kHz	**398 μH**	500 Ω	300 Ω	**30 μA**	50 μA	**58.3 μA**	257 Ω	**−31°**
4.	10 V	**79.6 kHz**	50 mH	25 kΩ	10 kΩ	**0.4 mA**	1 mA	**1.08 mA**	9.26 kΩ	**−21.8°**
5.	**18 V**	100 Hz	**2.87 H**	**1,800 Ω**	1.04 kΩ	**10 mA**	17.3 mA	20 mA	900 Ω	**−30°**

1.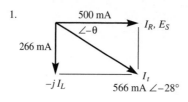

2.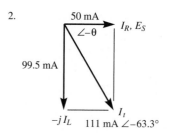

3.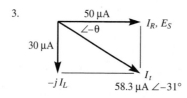

4.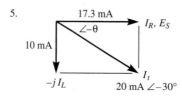

5.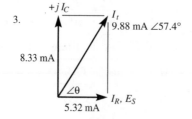

Figure 19-42

Self-Check 19-2

#	E_S	f	C	X_C	R	I_C	I_R	I_t	Z_t	$\angle\theta$
1.	3 V	50 Hz	5 μF	**637 Ω**	470 Ω	**4.71 mA**	**6.38 mA**	**7.93 mA**	**378 Ω**	**36.4°**
2.	11 V	60 Hz	5 μF	**531 Ω**	180 Ω	**20.7 mA**	**61.1 mA**	**64.5 mA**	**171 Ω**	**18.7°**
3.	25 V	**531 kHz**	100 pF	3 kΩ	4.7 kΩ	**8.33 mA**	**5.32 mA**	**9.88 mA**	**2.53 kΩ**	**57.4°**
4.	50 mV	650 kHz	**817 pF**	300 Ω	200 Ω	**167 μA**	**250 μA**	**301 μA**	**166 Ω**	**33.7°**
5.	**100 V**	**1.77 kHz**	0.1 μF	900 Ω	**758 Ω**	0.111 A	**132 mA**	**173 mA**	**578 Ω**	40°

1.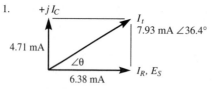

2.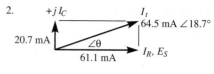

3.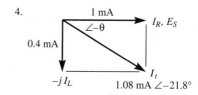

4.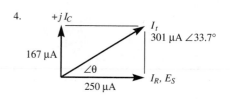

5.

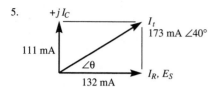

Figure 19-43

Self-Check 19-3

#	E_S	f	R	C	L	X_C	X_L	Z_t	I_R	I_C	I_L	I_t	$\angle\theta$
1.	1 V	1 kHz	1 kΩ	0.47 μF	0.1 H	339 Ω	628 Ω	592 Ω	1 mA	2.95 mA	1.59 mA	1.69 mA	53.7°
2.	50 mV	455 kHz	20 Ω	0.002 μF	20 μH	1.75 Ω	57.1 Ω	19.5 Ω	2.5 mA	286 μA	876 μA	2.57 mA	−13.3°
3.	5 V	10 MHz	2.5 kΩ	12.7 pF	13.3 μH	1.25 kΩ	833 Ω	1.77 kΩ	2 mA	4 mA	6 mA	2.83 mA	−45°
4.	**75 mV**	120 MHz	5 Ω	50 pF	0.01 μH	26.5 Ω	7.54 Ω	4.52 Ω	15 mA	2.83 mA	9.95 mA	16.6 mA	−25.4°
5.	10 V	**106 Hz**	**100 Ω**	10 μF	1 H	150 Ω	666 Ω	88.8 Ω	100 mA	66.7 mA	15 mA	112.6 mA	27.3°

1.

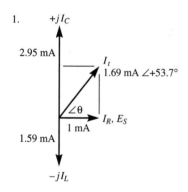

2.

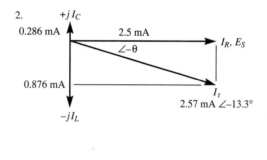

3.

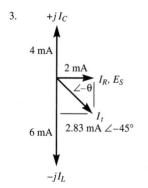

4.

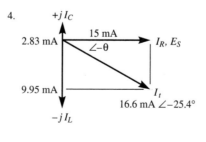

5.

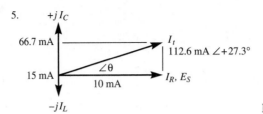

Figure 19-44

Self-Check 19-4

	P_R	P_C	P_L	P_A	PF
1.	1 mW	2.95 mVAR	1.59 mVAR	1.69 mVA	0.592
2.	125 μW	14.3 μVAR	43.8 μVAR	128.5 μVA	0.973
3.	10 mW	20 mVAR	30 mVAR	14.2 mVA	0.707
4.	1.125 mW	212 μVAR	746 μVAR	1.245 mVA	0.903
5.	1 W	0.667 VAR	0.15 VAR	1.126 VA	0.888

SUGGESTED PROJECTS

1. Add some of the main concepts, formulas, and procedures from this chapter to your Electricity and Electronics Notebook.

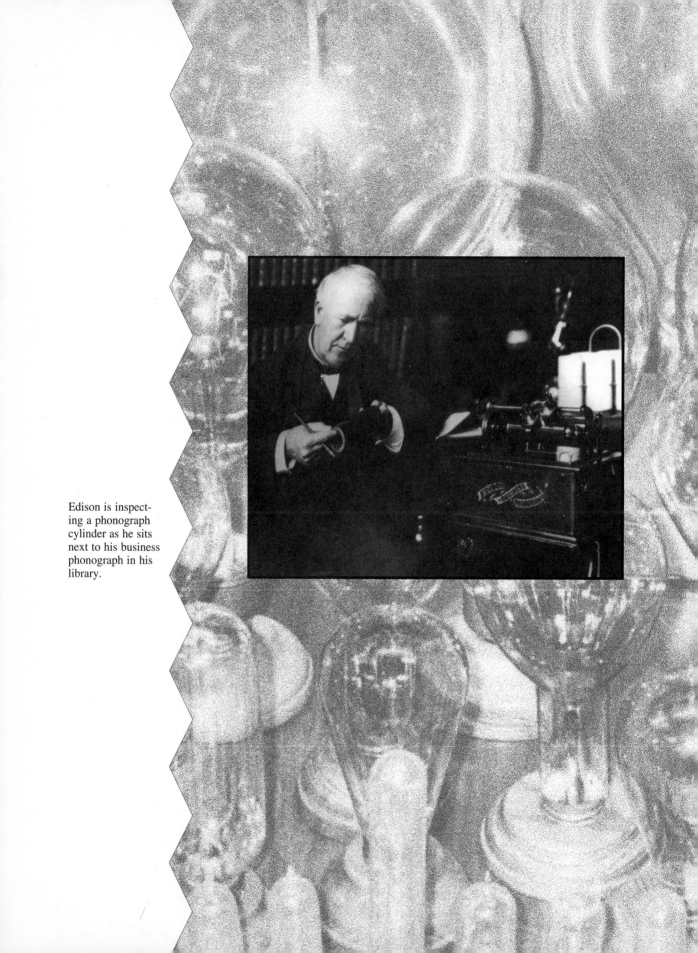

Edison is inspecting a phonograph cylinder as he sits next to his business phonograph in his library.

Chapter 20

Resonance

CHAPTER OUTLINE

OBJECTIVES

After studying this chapter you will be able to

- name several applications for series and parallel resonant circuits.
- list all of the significant parameters and characteristics of series and parallel resonant circuits.
- understand and read the characteristic graphs for series and parallel resonant circuits.
- calculate the resonant frequency for series and parallel resonant circuits.
- calculate circuit Q and bandwidth when the resonant frequency and total circuit resistance are known.
- calculate bandpass when the resonant frequency and bandwidth are known.
- calculate the proper size capacitor to resonate with a given inductor at a specified resonant frequency.
- calculate the proper amount of total resistance needed to provide a specified bandwidth for a given series resonant circuit.
- describe a high-Q inductor and capacitor in terms of construction and characteristics.
- accurately test series and parallel resonant circuits using a variable-frequency generator and an oscilloscope.
- explain similarities and differences between series and parallel resonance.

INTRODUCTION

One of the most common *RCL* circuits found in electronic equipment is the resonant circuit. In this chapter, you will learn what resonant circuits are, how and where they are used, how they are analyzed, how they can be designed, and how a series resonant circuit is tested using basic laboratory test equipment. In these pages, you will discover how very simple series and parallel resonant circuits can be used to select and pass, or reject and eliminate a band, or group, of frequencies.

NEED TO KNOW

Let's suppose you have just finished your formal training as an electronics technician or engineer and have just landed an exciting job with an international shortwave broadcasting company. You've been given the grand tour of the high-power shortwave facility, and introduced to the big transmitters for which you will eventually be responsible. As days go by, you are briefed on government rules and regulations, transmitter operation, transmitter maintenance, and emergency procedures.

Several weeks go by and all seems well. You are feeling a little more comfortable with your duties and you have your first paycheck in your pocket. Then, the boss puts you on the night shift with the skeleton crew (you and the night janitor). He says he has confidence in your abilities! So do you! Well, don't you? After all, what could possibly go wrong? All you have to do is periodically read voltage and current meters and make occasional frequency changes (which is no problem, since the transmitters retune themselves automatically using servo motors). So, what could go wrong?

Then, on your very first night shift, at exactly 1:59 a.m., you prepare to make a routine frequency change from 11.750 MHz to 15.315 MHz on transmitter #1. You power the transmitter down, push the auto-tune button, and wait a few seconds for the servos to automatically retune the transmitter to 15.315 MHz. (During this tuning cycle, several tuned circuits are servo-motor tuned to the new frequency.) You fully expect to be back on the air on 15.315 MHz in less than 10 seconds. But not this time! The final PA (power-amplifier) tuning trouble light comes on, and the transmitter will not come up. Your thoughts begin to race and you feel the pink warmth of panic on your face.

Somehow, you control yourself and try to remember what to do in this type of situation. You remember a quick briefing about reaching behind a front panel and feeling a belt-driven wheel that can be manually adjusted. This hidden wheel is connected to the final PA tuning capacitor. You were told to watch the PA-plate-tuning meter as you manually adjust the wheel and bring power up on the transmitter.

You're sweating, aren't you? You can't remember if the PA plate meter is supposed to be peaked or dipped! Should you tune for maximum or minimum current in the plate circuit? At that moment, all you can remember is that someone mentioned that the PA tube was temperamental and cost about $10,000 to replace. That's about six long months of going without your salary as you live out of a luxurious neighborhood trash dumpster.

You desperately need advice. Just then the janitor strolls by waving his feather duster. "Something wrong?" he asks. Whatever gave him that idea? You smile and wipe sweat from your chin. The station identification has already been relayed to you and the next program has already begun. You're losing air time and that costs a lot of money in lost revenue from the broadcaster.

Again, you calm yourself and try to think. The PA-plate-tuning network is a parallel resonant circuit. The plate current feeds the parallel resonant circuit. Should you tune the plate for minimum or maximum current? You swallow hard!

I am going to leave you in that predicament for now and come back to it with a solution at the end of this chapter. In the meantime, let's begin the study of resonance to discover what you will need to know to quickly get out of a situation such as this. You will find this to be a very practical study. Enjoy.

20-1 Series Resonance and Applications

The Series Resonant Circuit

The series resonant circuit is a series RCL circuit that is intended to operate at, or near, the frequency at which X_C and X_L are equal and cancel. In other words, the series resonant circuit is a series RCL circuit that is acting resistively. Thus, the series resonant circuit is operating at, or near, the frequency of maximum efficiency where $PF = 1$ and angle theta is $0°$ ($\angle \theta = 0°$, see Section 18-5).

It is at this frequency that the series RCL circuit is said to resonate. The word **resonate** means to respond in such a way as to reproduce or echo a stimulus. In a series resonant circuit, the capacitor and inductor will respond to, and echo, alternating current most efficiently at the frequency where a perfect, or near-perfect, balance exists between X_C and X_L, and V_C and V_L.

Applications for the Series Resonant Circuit

Series Resonant Tuned Circuits

Series resonant circuits are commonly used in radio and television receivers and in transmitters. They exhibit some very basic and desirable characteristics that make them useful as tuned or tunable circuits. A **tuned circuit** is one that will resonate at a particular frequency or narrow range of frequencies. A **tunable circuit** is one that can be varied or changed to resonate at many different frequencies. A series resonant circuit can be tuned to a particular frequency using fixed components or it can be made tunable by using a variable capacitor and/or variable inductor. In either case, the series resonant circuit can be used to either select a desired band (small group) of frequencies or reject an undesired band of frequencies.

Series Resonant Preselection

Figures 20-1a and 20-1b are examples of preselection at the front end of a radio receiver. An antenna receives many radio stations at the same time. The series resonant circuit selects only a few of the many stations arriving at the antenna, letting them enter the receiver for further selection and amplification. The series resonant characteristics that you will learn about in this chapter make this preselection possible. The circuit of Figure 20-1b not only preselects radio stations, it also very significantly increases the strength of the station or stations selected. In this study of series resonance, you will see that the inductor

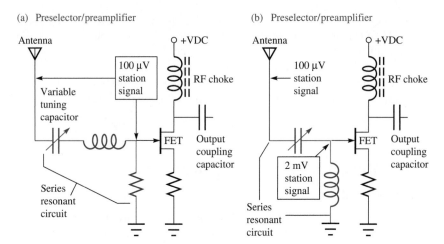

Figure 20-1   Series resonant circuits can be used for preselection in radio receivers.

and capacitor voltages are normally much higher than the source voltage at resonance. (In this case, the source voltage is the weak voltage induced on the antenna from the radio station.)

Series Resonant Trap

Figure 20-2 illustrates the use of the series resonant circuit as a frequency trap at the output of a transmitter. The series resonant trap will act like a short at a specific resonant frequency. In this application, the trap is tuned to the second harmonic of the transmitter frequency. If the station is authorized to transmit on 620 kHz, the transmitter will produce

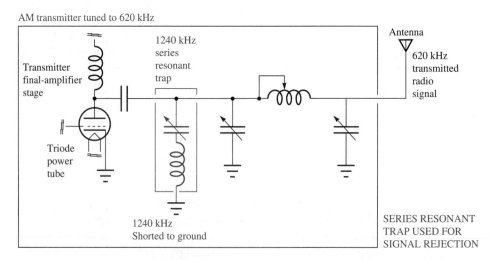

Figure 20-2   A series resonant trap is used to prevent illegal radiation of the second harmonic from broadcast transmitters.

a high-power signal at 620 kHz and a low-power signal at 1240 kHz, the second harmonic. Since the station is not authorized to broadcast on 1240 kHz, the low-power 1240 kHz must be removed before it reaches the antenna. The trap will act like a short only at 1240 kHz. Thus, the series resonant circuit removes the second harmonic before it can be radiated into the air.

As you can see, the series resonant circuit performs some important tasks in the world of electronics. As you continue in this chapter, you will learn about all the series resonance characteristics that make the series resonant circuit so very valuable. You will also be able to predict the resonant frequency and other important parameters by using a handful of simple formulas. Take a few minutes now to test your understanding by answering the questions of Self-Check 20-1.

1. How will a series resonant circuit that is operating at resonance act (resistively, inductively, or capacitively?)

2. What does *resonate* mean?

3. The resonant frequency for any combination of L and C depends on what?

4. What is a tunable series resonant circuit?

5. List two common applications for series resonant circuits.

20-2 Series Resonance Characteristics

Phase Relationships

One of the first characteristics of series resonant circuits to understand is the phase relationship of all voltages in the circuit—for the most part, a review of material covered in Section 18-5. From your previous studies, you know the inductor voltage leads the source voltage by 90° ($+jV_L$, ELI) and the capacitor voltage lags the source voltage by 90° ($-jV_C$, ICE). Also, at resonance, when the circuit is acting purely resistively $\angle\theta$ is equal to 0° and the source voltage and resistor voltage are the same. The phasor diagram of Figure 20-3 summarizes the phase relationships.

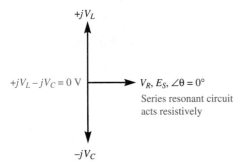

$+jV_L$

$+jV_L - jV_C = 0$ V ⟶ V_R, E_S, $\angle\theta = 0°$
Series resonant circuit
acts resistively

$-jV_C$

Figure 20-3   Voltage phase relationships at resonance.

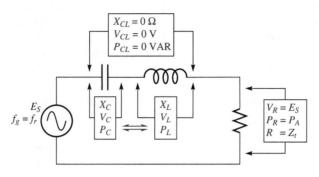

Figure 20-4 A series resonant circuit with all parameters shown.

Circuit Parameter Summary

This is a summary of the circuit parameters as they exist when the series *RCL* circuit is at resonance (when the resonant frequency is applied to the series *RCL* circuit). See Figure 20-4.

$X_L = X_C$ When the series circuit is operating at resonance, the inductive reactance and capacitive reactance are equal in magnitude (Ω) and opposite in phase ($+jX_L = X_L\angle +90°$, $-jX_C = X_C\angle -90°$). This means the two reactances have a cancelling effect on each other. In other words, the impedance measured across the inductor and capacitor together is 0 Ω ($+jX_L - jX_C = 0\ \Omega$). At resonance, *L* and *C* combined appear as a short ($X_{CL} = 0\ \Omega$ = short).

$V_L = V_C$ Since the reactances are equal at resonance, their voltages will also be equal. Also, just as the reactances are opposite in phase, the reactive voltages are opposite in phase ($+jV_L = V_L\angle +90°$, $-jV_C = V_C\angle -90°$). This, of course, means the two reactive voltages have a cancelling effect on each other. Thus, the voltage measured across the inductor and capacitor together is 0 V ($V_{CL} = +jV_L - jV_C = 0$ V). This makes sense since 0 V is always measured across a short.

$E_S = V_R$ Since the two reactances cancel each other and appear as a short at the resonant frequency, the entire source voltage appears across the resistance in the circuit.

$Z_t = R$ = minimum   Once again, since the reactances cancel at resonance, the resistance is the only remaining impedance in the circuit. Therefore, the total impedance is minimum and equal to the resistance in the circuit.

$I_t = E_S/Z_t = E_S/R$ = maximum   The total current of a series *RCL* circuit is maximum at resonance since the total impedance, equal to the resistance, is minimum.

$P_A = P_R = I_t \cdot E_S = I_t \cdot V_R$ When the series *RCL* circuit is operating at resonance, the apparent power (total applied power) equals the real power (resistor power), since the reactive powers are equal in magnitude and opposite in phase ($+jP_L = P_L\angle +90°$, $-jP_C = P_C\angle -90°$). The reactive powers cancel ($+jP_L - jP_C = 0$ VAR).

$\angle\theta = 0°$, $PF = 1$, %eff = 100%   At resonance, the phase angle between P_A and P_R, E_S and V_R, and Z_t and R is 0° since X_L and X_C cancel. Thus, P_A and P_R are the same magnitude and phase. The ratio of P_R to P_A, therefore, is simply 1 ($PF = 1$) and the circuit efficiency is 100% (all of the applied power is converted to heat in the resistance).

Series Resonance Graphs

The series resonance characteristics are summarized once again in Figure 20-5 in the form of overlapping graphs. The graphs show how the circuit parameters change with frequency above and below resonance. The horizontal scale is frequency and the vertical scale is

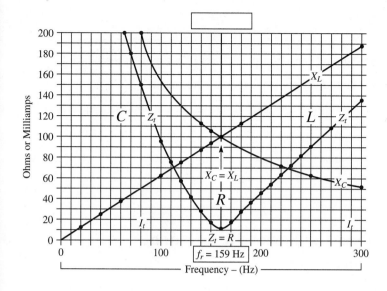

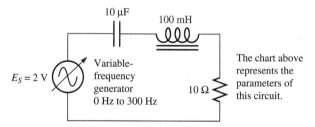

These graphs were created from these formulas

Figure 20-5   The series resonance graphs.

magnitude—representing amperes, volts, and ohms. Notice the total current is maximum and the total impedance is minimum only at resonance. Also, at resonance the X_C and X_L graphs intersect, indicating they are equal in magnitude. Below resonance, the circuit acts capacitively ($X_C > X_L$); at resonance, the circuit acts resistively ($X_C = X_L$); and above resonance, the circuit acts inductively ($X_L > X_C$).

Realize also that the inductor and capacitor voltages reach a maximum peak with current. At first it might seem that the inductor voltage would continue to increase with inductive reactance above resonance. However, the total circuit current drops sharply above resonance, cancelling out the effects of the increasing inductive reactance. According to Ohm's Law, $V_L = I_t \cdot X_L$. Since I_t decreases at a faster rate than X_L increases, the voltage across the inductor decreases. The same reasoning is true for the capacitor voltage as the frequency is decreased below resonance. Instead of the capacitor voltage increasing with capacitive reactance as the frequency is decreased, the capacitor voltage actually decreases with the sharply decreasing total current.

Time for a review and self-check.

SELF-CHECK 20-2

1. Explain why the inductor and capacitor together act like a short at resonance.

2. Explain why the total circuit current is maximum at resonance.

3. Explain why the inductor voltage actually decreases as the applied frequency is increased above resonance.

20-3 Calculations for Series Resonance

The Resonant Frequency (f_r)

As you know, the resonant frequency for any values of inductance and capacitance is the frequency at which X_L and X_C are equal. With this fact in mind, it is a simple matter to derive a formula to calculate the resonant frequency. We begin with the X_L and X_C formulas. Since X_L and X_C are equal, the two formulas can be set equal to each other. The frequency in each formula is the common frequency at which $X_L = X_C$, the resonant frequency (f_r).

$$X_L = 2\pi f_r L = 1/(2\pi f_r C) = X_C$$

$$2\pi f_r L = \frac{1}{2\pi f_r C}$$

Now divide both sides by $2\pi L$ to leave f_r on the left of the equation:

$$f_r = \frac{1}{2\pi f_r C \cdot 2\pi L} = \frac{1}{4 \cdot \pi^2 \cdot f_r \cdot L \cdot C}$$

Next, remove f_r from the denominator by multiplying both sides of the equation by f_r.

$$f_r^2 = \frac{1}{4 \cdot \pi^2 \cdot L \cdot C}$$

Finally, take the square root of both sides of the equation.

$$f_r = \frac{\sqrt{1}}{\sqrt{4 \cdot \pi^2 \cdot L \cdot C}} = \frac{1}{2\pi\sqrt{LC}} \qquad (20.1)$$

$$f_r = f_c = f_o = \frac{1}{2\pi\sqrt{LC}}$$

where the resonant frequency f_r is the same as the center frequency f_c, which is the same as the frequency of operation f_o.

Consider the examples of Example Set 20-1.

EXAMPLE SET   20-1

1. Given $L = 1$ H and $C = 1$ μF, find f_r.

$$f_r = \frac{1}{2\pi\sqrt{LC}} = \frac{1}{2\pi\sqrt{1 \text{ H} \cdot 1 \text{ } \mu\text{F}}} = 159 \text{ Hz}$$

2. Given $L = 1$ mH and $C = 0.01$ μF, find f_r.

$$f_r = \frac{1}{2\pi\sqrt{LC}} = \frac{1}{2\pi\sqrt{1 \text{ mH} \cdot 0.01 \text{ } \mu\text{F}}} = 50{,}329 \text{ Hz}$$

3. Given $L = 1$ μH and $C = 30$ pF, find f_r.

$$f_r = \frac{1}{2\pi\sqrt{LC}} = \frac{1}{2\pi\sqrt{1 \text{ } \mu\text{H} \cdot 30 \text{ pF}}} = 29.06 \text{ MHz}$$

Notice, from Example Set 20-1, how the resonant frequency is inversely related to the size of the inductor and capacitor. Large values of inductance and/or capacitance result in a low resonant frequency and small values of inductance and/or capacitance result in a high resonant frequency.

Designing for Resonance—Finding C or L

A series LC or RCL circuit can be designed to resonate at any desired frequency if one of the reactive components is known. For example, a 10 μH inductor can be made to resonate at 4 MHz with a capacitor if the correct value of capacitance is chosen. What is the correct value of capacitance? Or, perhaps you have a 0.001 μF capacitor and you want it to resonate at 200 kHz with an inductor. What size inductor do you need? The solution to each of these cases is found by rearranging the resonant-frequency formula as follows:

$$f_r = \frac{1}{2\pi\sqrt{LC}}$$

$$f_r\sqrt{LC} = \frac{1}{2\pi}$$

$$\sqrt{LC} = \frac{1}{2\pi f_r}$$

$$LC = \frac{1}{4 \cdot \pi^2 \cdot f_r^2}$$

$$C = \frac{1}{4 \cdot \pi^2 \cdot f_r^2 \cdot L} \tag{20.2}$$

$$L = \frac{1}{4 \cdot \pi^2 \cdot f_r^2 \cdot C} \tag{20.3}$$

The examples of Example Set 20-2 demonstrate the use of these formulas.

EXAMPLE SET   20-2

1. What value capacitor will you need to resonate with a 20 mH inductor at a frequency of 10 kHz?

$$C = \frac{1}{4 \cdot \pi^2 \cdot f_r^2 \cdot L} = \frac{1}{4 \cdot \pi^2 \cdot (10 \text{ kHz})^2 \cdot 20 \text{ mH}} = 0.013 \ \mu\text{F}$$

2. What value inductor will you need to resonate with a 400 pF capacitor at a frequency of 3.5 MHz?

$$L = \frac{1}{4 \cdot \pi^2 \cdot f_r^2 \cdot C} = \frac{1}{4 \cdot \pi^2 \cdot (3.5 \text{ MHz})^2 \cdot 400 \text{ pF}} = 5.17 \ \mu\text{H}$$

The Quality Factor (Q)

Quality Factor?

Earlier it was stated that series resonant circuits are used to select or reject a desired or undesired band of frequencies. The effectiveness with which the resonant circuit can accomplish these tasks depends greatly on the **quality factor (Q)** of the resonant circuit. The quality factor is a very important "figure of merit" for tuned circuits. (A figure of merit is an important characteristic that is used to compare or judge quality or worth.) If the band of frequencies to be selected or rejected is very narrow, the quality factor of the tuned circuit must be very high. If the band of frequencies is fairly wide, the Q must be low.

What Determines Circuit Q?

The obvious question here then is: What determines the Q of a series resonant circuit? The quality factor for a series resonant circuit (Q_{ckt}) is simply the ratio of inductive, or capacitive, reactance to the total amount of resistance in the circuit.

$$Q_{ckt} = X_L/R_t = X_C/R_t \qquad \text{(no units)} \tag{20.4}$$

The total resistance in the circuit is not just the sum of all real resistors; other resistances affect the Q of the series resonant circuit. The inductor has resistance in its windings (R_W) and must be considered for high-Q circuits. The generator used as a frequency and voltage source also has internal resistance (impedance) that is in series with the circuit. Ideally, for a high-Q series resonant circuit, the generator resistance R_g should be 0 Ω (a short). As shown in Figure 20-6, the total resistance in a series resonant circuit is actually the sum of the inductor resistance, the generator resistance, and any real resistors. As you solve various series resonance problems in this chapter, keep in mind that all of these resistances affect circuit Q. However, inductor and generator resistances can be ignored if they are small compared to any series resistors in the circuit (ignored if $R_g + R_W < 0.1 \ R_S$).

If, for example, the inductive reactance of an inductor in a series resonant circuit is 1.5 kΩ and a series resistor is 100 Ω, the Q of the circuit is 15 (1.5 kΩ/100 Ω = 15, R_g and R_W being considered insignificant here). If the 100 Ω resistor is replaced with a 10 Ω

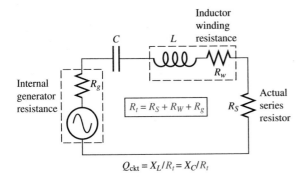

Figure 20-6   Total resistance affecting circuit Q.

$$Q_{ckt} = X_L/R_t = X_C/R_t$$

resistor, the Q is 150 (1.5 kΩ/10 Ω = 150, R_g and R_W may now be significant). On the extreme side, if the series resistance is totally removed (replaced with a wire), the Q will, theoretically, be infinite (1.5 kΩ/0 Ω = ∞, assuming R_g and R_W to be 0 Ω). In practice, however, conductors of a conventional, practical circuit have resistance.

The Significance of Conductor Resistance

What about the resistance in the conductors of a practical circuit? Is this resistance significant? Does it really affect the Q of a circuit? The answer to this is: It depends. If a very-high-Q resonant circuit is needed, the conductor resistance will become a significant factor. For the most part, the conductor resistance of concern (R_W) will be found in the windings of the inductor. Therefore, it is important, in high-Q applications, to select a high-quality inductor.

The High-Q Inductor

What is a high-Q inductor? A **high-Q inductor** is a coil that has very little winding resistance R_W. The Q of the inductor itself (Q_L) is found using the same basic formula.

$$Q_L = X_L/R_W \tag{20.5}$$

where X_L is determined using the resonant frequency.

As you can see, inductor Q is the ratio of the inductor's reactance to the inductor's resistance. This ratio may range from less than 10 to over 1,000 for very-high-Q inductors. Notice that, for any given inductor, the Q is low for low frequencies and high for high frequencies, because X_L increases with frequency. This is desirable, since higher Q values are normally needed at higher frequencies of operation (more on this later).

Figure 20-7 schematically illustrates a practical inductor with resistance in its windings (R_W). A high-Q inductor has a relatively high X_L compared to the winding resistance. The design objective for high-Q inductors is to make the inductor look as ideal as possible. This means the inductor should have an impedance of $Z_t\angle +90°$ ($+jX_L$). The inductor can only have an angle of $+90°$ if there is no winding resistance. The phasor diagrams of Figure 20-7 illustrate this. In all of our previous circuit-analysis examples (Chapters 18 and 19) the inductors were assumed to be perfect or ideal. For most applications this

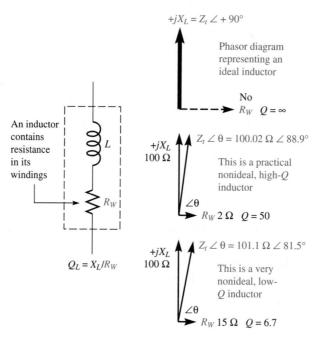

Figure 20-7   Inductor *Q*.

assumption is acceptable, since most inductors will have a phase angle of about $+89°$ or better. The difference between $+89°$ and $90°$ is insignificant in most cases.

It is interesting to note here that the *Q* of an inductor is actually the reciprocal of the cosine of the inductor's phase angle ($\angle\theta$). Also, recall that the power factor of an AC circuit is equal to the cosine of $\angle\theta$. The inductor itself is, in fact, a series *RL* circuit having a certain power factor and phase angle. Recall also that power factor is the ratio of real power to apparent power. Thus, for any inductor, the Q_L can be determined as follows:

$$Q_L = 1/\cos\angle\theta = 1/PF = 1/(P_R/P_A) \tag{20.6}$$
$$= 1/(R_W/Z_t) = 1/(R_W/X_L) = X_L/R_W$$

(*Note:* Z_t and X_L are considered equal, since X_L is extremely large compared to R_W.)

If the phase angle of an inductor operating at a certain frequency is $+89°$, its *Q* is the reciprocal of the $\cos\angle +89°$. The $\cos\angle +89°$ is 0.01745. Therefore, the inductor's *Q* at that frequency is 57.3 ($1/0.01745 = 57.3$). Conversely, if the inductor's *Q* is known to be 57.3 at a given operating frequency, its phase angle can be determined by finding the inverse cosine of the reciprocal of 57.3 ($\angle\theta = \cos^{-1}(1/Q_L) = \cos^{-1}(1/57.3) = +89°$). The *Q* of 57.3 indicates that the inductor's reactance is 57.3 times greater than its winding resistance ($Q = X_L/R_W$ and $X_L = Q \cdot R_W$).

Skin Effect

There is more that must be said about the resistance of the inductor and its effect on inductor *Q*. The resistance of the inductor is more than simply the DC resistance that can be measured with an ohmmeter. In practice, it is the amount of resistance that the windings exhibit at a particular frequency. The resistance of the windings will not be the same at 30 MHz as it is at 30 kHz. Why? This is due to a conductive property called **skin effect**. As frequency is increased, current tends to flow closer and closer to the surface (skin) of the windings. The overall effect is that of reducing the cross-sectional area of the conductor. The effective resistance of the windings is increased as the charges crowd at the skin of the windings. As a result, the increase in X_L is somewhat counteracted with an increase in R_W as frequency is increased. This prevents the inductor from exhibiting as high a Q as otherwise would be possible if it were not for skin effect.

Reducing Skin Effect to Obtain a High Q

The skin effect is reduced, or minimized, in high-Q inductors using several techniques. First, the diameter of the inductor is made large to reduce the number of turns needed for a particular value of inductance. This is particularly true for inductors used in transmitters or even VHF and UHF receivers. Second, the windings of the inductor are made of large-diameter wire, or, in the case of transmitters, copper tubing ($\frac{1}{4}''$ to $3''$, roughly 0.5 cm to 7.5 cm). Third, the windings are plated with silver to reduce surface resistance. Thus, skin effect is reduced by designing inductors with few turns of large-diameter silver-plated conductors.

High-Q Capacitors

What about capacitors? Is there such a thing as a high-Q capacitor? Yes. **High-Q capacitors** are capacitors that have highly conductive plates, usually silver plating or silver-plated copper. Also, the dielectric is of the highest quality to minimize leakage current between plates (mica and vacuum capacitors). Because of the very low conductor resistance of high-Q capacitors, their Q is very much higher than that of high-Q inductors. Capacitor Q will range into the thousands depending on design.

Often, high-Q capacitors are rated in terms of power factor. Recall from the discussion of inductor Q that Q is the reciprocal of power factor. (It is also true that power factor is the reciprocal of Q.) An ideal capacitor has a phase angle of $-90°$ and a cosine and power factor equal to 0.00000 ($PF = \cos\angle\theta = \cos\angle -90° = 0.00000$). High-quality capacitors will have power factors ranging below 0.0005. This means the Q values range above 2,000 ($Q = 1/PF = 1/0.0005 = 2,000$).

Insuring a High-Q Resonant Circuit—A High L/C Ratio

A high-Q series resonant circuit is one in which high-Q capacitors and inductors are used and the total series resistance is very low. Also, inductance can be increased and capacitance can be decreased to maintain the same desired resonant frequency ($\leftrightarrow f_r = 0.159/\sqrt{\uparrow L \downarrow C}$, note: $\leftrightarrow$ means "remains the same"). Increasing L causes X_L to increase, which causes the circuit Q to increase ($\uparrow Q_{ckt} = \uparrow X_L/R_t$). However, a very-high-$Q$ inductor must be used to achieve this.

Q Rise in Voltage—More Than the Source Voltage!

At or near the resonant frequency, a series resonant circuit exhibits a **Q rise in voltage** across the inductor and capacitor. This means the inductor and capacitor voltages are Q

See color photo insert page D10.

times greater than the resistor voltage. At resonance, the circuit current is limited by the series resistance. This relatively high circuit current produces high voltages across the reactive components ($V_C = I_t \cdot X_C$ and $V_L = I_t \cdot X_L$). The ratio of this high reactive component voltage to the resistor voltage is equal to the circuit Q.

$$Q_{ckt} = V_C/E_S = V_L/E_S \quad \text{and} \quad V_C = V_L = Q_{ckt} \cdot E_S \tag{20.7}$$

As an example, suppose in a series resonant circuit the X_L and X_C are 3 kΩ at a certain resonant frequency. Let's assume the total series circuit resistance to be 300 Ω. The Q for this circuit is 10 (3 kΩ/300 Ω = 10, Formula 20.4). If 5 V is applied to the circuit, the voltage across the inductor and capacitor will be 50 V each ($V_L = V_C = Q \cdot E_s = 10 \cdot 5$ V = 50 V, a total of 0 V since $V_{LC} = +j50$ V $- j50$ V = 0 V). Thus, the 50 V inductor and capacitor voltages represent a Q rise in voltage.

Bandwidth (*BW*) and Bandpass (*BP*)

Bandwidth (*BW*)

Another important calculation that is made at resonance for a series resonant circuit is the **bandwidth (*BW*)**. Bandwidth tells you the width of the group of frequencies being selected or rejected by the resonant circuit. It is measured in Hertz (Hz) just as frequency is. For example, many common AM radio receivers have a bandwidth of about 10 kHz in the most selective section of the radio. This means a band of frequencies 10 kHz wide is permitted to pass through the receiver at a certain resonant frequency. For most AM radios, this resonant frequency is a standard 455 kHz (more on this in the next chapter). The 10 kHz bandwidth is centered on 455 kHz. This means half of the bandwidth extends above the resonant frequency and half extends below as illustrated with the frequency response graph of Figure 20-8.

But, how is the bandwidth for a particular circuit determined? Bandwidth is inversely related to circuit Q. If the circuit Q is high, the bandwidth is small or narrow and vice

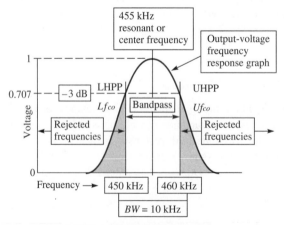

$Q_{ckt} = f_r / BW = 455$ kHz / 10 kHz = 45.5
$BW = f_r / Q_{ckt} = 455$ kHz / 45.5 = 10 kHz
$BP \cong [f_r - (BW / 2)] \rightarrow [f_r + (BW / 2)] = 450 - 460$ kHz

Figure 20-8   Bandwidth (*BW*) and bandpass (*BP*).

versa. Thus, the bandwidth for any resonant circuit can be calculated as follows:

$$BW = f_r/Q_{ckt} \qquad \text{(also:} \quad f_r = BW \cdot Q_{ckt} \quad \text{and} \quad Q_{ckt} = f_r/BW) \qquad (20.8)$$

where Q_{ckt} for a series resonant circuit is X_L/R_t or X_C/R_t.

If a resonant circuit has a Q of 25 and a center frequency, or resonant frequency, of 455 kHz, the circuit's bandwidth is 18.2 kHz ($BW = f_r/Q_{ckt} = 455$ kHz$/25 = 18.2$ kHz). Rearranging Formula 20.8, we can determine the circuit Q by dividing the resonant frequency by the measured bandwidth ($Q_{ckt} = f_r/BW$). For example, if an actual bandwidth of 10 kHz exists at a center frequency of 455 kHz, as in Figure 20-8, the circuit Q must be 45.5 (455 kHz$/10$ kHz $= 45.5$).

From this discussion you can see that the actual Q of a given circuit can be determined through measurement and simple calculation. The bandwidth is carefully measured and the circuit Q is calculated by dividing bandwidth into the resonant frequency. But how is the bandwidth of a resonant circuit measured? The actual use of test equipment will be discussed later. Here, we will continue to prepare for that discussion by defining the boundaries that determine the bandpass of the resonant circuit.

Bandpass (*BP*)

The **bandpass (*BP*)** information tells you how wide the band is and exactly where in the frequency spectrum it lies. The bandpass extends from a frequency below resonance, through the center resonant frequency, to a frequency above resonance. For high-Q circuits, this means about half of the bandwidth lies below resonance and about half lies above. In Figure 20-8, the bandwidth centered at 455 kHz starts at 450 kHz and extends up to 460 kHz. Thus the bandpass is 450 to 460 kHz. Mathematically, the bandpass is approximated as follows:

$$BP \cong [f_r - (BW/2)] \rightarrow [f_r + (BW/2)] \qquad \text{(for high-}Q\text{ circuits)} \qquad (20.9)$$

The two bandpass frequencies define the outer limits of the band. As a result, the bandpass frequencies are referred to as **cutoff frequencies** (f_{co}), since all frequencies beyond cutoff are considered to be excluded from the band. These cutoff frequencies are further defined as **−3 dB points**, **0.707 voltage points**, and **lower** and **upper half-power points** (**LHPP** and **UHPP**). As shown in Figure 20-8, the voltage amplitudes of the lower and upper cutoff frequencies are 0.707 times the highest voltage amplitude at the resonant frequency. If the peak-to-peak voltage at the resonant frequency is 10 V_{p-p}, the peak-to-peak voltages at the two cutoff frequencies will be 7.07 V_{p-p} ($V_{fco} = 0.707 \cdot V_{fr}$).

Why are the cutoff frequency points called half-power points? When the voltage across a load resistance is decreased to $7/10$, the power dissipated in the load resistance drops in half. This is easily proven. Let's suppose 10 V is applied to a 100 Ω resistor. The resulting dissipated power in the resistor is 1 W ($P = E^2/R = (10$ V$)^2/100$ $\Omega = 1$ W). If the applied voltage is reduced to 7.07 volts, the dissipated power drops to 0.5 W ($P = E^2/R = (7.07$ V$)^2/100$ $\Omega = 0.5$ W). Thus, a 0.707 voltage point is also a half-power point. The half-power points are also called -3 dB points, since -3 dB indicates a drop in half of power [-3 dB $= 10 \log (\frac{1}{2})$]. As a review, consider Example Set 20-3.

EXAMPLE SET 20-3

1. Analysis problem—Given: $R_t = 10\ \Omega$, $L = 0.5\ \text{mH}$, $C = 0.05\ \mu\text{F}$
   Find: f_r, Q_{ckt}, BW, and BP

$$f_r = \frac{1}{2\pi\sqrt{LC}} = \frac{1}{2 \cdot \pi \cdot \sqrt{0.5\ \text{mH} \cdot 0.05\ \mu\text{F}}} = 31.8\ \text{kHz}$$

$Q_{ckt} = X_L/R = 2\pi f L/R = 2 \cdot \pi \cdot 31.8\ \text{kHz} \cdot 0.5\ \text{mH}/10\ \Omega$
$\qquad = 10$

$BW = f_r/Q_{ckt} = 31.8\ \text{kHz}/10 = 3.18\ \text{kHz}$

$BP = [f_r - (BW/2)] \rightarrow [f_r + (BW/2)]$
$\qquad = [31.8\ \text{kHz} - 1.59\ \text{kHz}] \rightarrow [31.8\ \text{kHz} + 1.59\ \text{kHz}]$
$\qquad = 30.2\ \text{kHz} \rightarrow 33.4\ \text{kHz}$

30.2 kHz is the lower cutoff frequency and 33.4 kHz is the upper cutoff frequency.

2. Design problem—Given: $L = 0.25\ \text{mH}$, $R_W = 1\ \Omega$, $f_r = 50\ \text{kHz}$, $BW = 2.5\ \text{kHz}$
   Find: C, Q_{ckt}, R_t

$$C = \frac{1}{4 \cdot \pi^2 f_r^2 \cdot L} = \frac{1}{4 \cdot \pi^2 \cdot (50\ \text{kHz})^2 \cdot 0.25\ \text{mH}} = 0.041\ \mu\text{F}$$

$X_L = 2\pi f_r L = 2\pi \cdot 50\ \text{kHz} \cdot 0.25\ \text{mH} = 78.5\ \Omega$

$Q_{ckt} = f_r/BW = 50{,}000/2{,}500 = 20$

$R_t = X_L/Q_{ckt} = 78.5\ \Omega/20 = 3.93\ \Omega$

$R_S = R_t - R_W = 3.93\ \Omega - 1\ \Omega = 2.93\ \Omega$

Before continuing, take time to study Design Note 20-1. Then, test your skills and understanding by answering the questions and solving the problems of Self-Check 20-3.

SELF-CHECK 20-3

1. Find the resonant frequency for a 300 pF capacitor and a 3.5 μH inductor.

2. If the capacitance of a series resonant circuit is increased, will the resonant frequency increase or decrease? Why?

3. What value capacitor would you need to resonate with a 250 μH inductor at 100 kHz?

4. What is the circuit Q for a series resonant circuit that has 50 Ω of resistance and an X_C of 2,400 Ω?

5. List three techniques used to make high-Q inductors.

6. If the power factor of a capacitor is 0.0002, what is its Q?

7. If the voltage across a capacitor is 350 mV at resonance and the source voltage is 20 mV, what is the circuit's Q?

8. Given $R = 3\ \Omega$; $L = 0.025\ \text{mH}$; $C = 0.001\ \mu\text{F}$, find f_r, Q_{ckt}, BW, and BP.

DESIGN NOTE 20-1:   Series Resonance

$$C \qquad L \qquad R$$

NOTE: R = The sum of all resistances in the circuit including winding resistance, output impedance of the generator, and any actual series resistors.

$$R = R_W + R_g + R_S$$

ANALYSIS	DESIGN
$f_r = 1/(2\pi\sqrt{LC})$ $Q_{ckt} = X_L/R$ $BW = f_r/Q_{ckt}$	$L = \dfrac{1}{4 \cdot \pi^2 \cdot f_r^2 \cdot C}$ $C = \dfrac{1}{4 \cdot \pi^2 \cdot f_r^2 \cdot L}$ $R = X_L/Q_{ckt} = X_L/(f_r/BW) = X_L \cdot BW/f_r$

ANALYSIS & DESIGN PROGRAM FOR SERIES RESONANCE

```
 10 CLS
 20 PRINT"SERIES RESONANCE"
 30 PRINT"":PRINT""
 40 PRINT"(1) ANALYZE A SERIES RESONANT CIRCUIT"
 50 PRINT"(2) DESIGN A SERIES RESONANT CIRCUIT"
 55 PRINT"(3) QUIT"
 60 PRINT""
 70 INPUT"SELECT ANALYSIS, DESIGN, OR QUIT — 1, 2 OR 3";S
 80 IF S<1 OR S>3 THEN GOTO 30
 90 ON S GOTO 200,400,600
200 INPUT"ENTER THE TOTAL SERIES CAPACITANCE IN FARADS. ";C
210 INPUT"ENTER THE TOTAL SERIES INDUCTANCE IN HENRIES. ";L
220 INPUT"ENTER THE TOTAL SERIES RESISTANCE IN OHMS. ";R
230 F = 1/(6.2831853# * SQR(L * C))
240 XL = 6.2831853# * F * L
250 Q = XL/R
260 BW = F/Q
270 LF = F − (BW/2):UF = F + (BW/2)
280 PRINT""
290 PRINT"THE RESONANT FREQUENCY IS ";F;" HERTZ."
300 PRINT"XC AND XL ARE ";XL;" OHMS."
310 PRINT"THE CIRCUIT Q IS ";Q;"."
320 PRINT"THE BANDWIDTH IS ";BW;" HERTZ."
330 PRINT"THE BANDPASS IS APPROXIMATELY ";LF;" HERTZ TO ";UF;" HERTZ."
390 GOTO 30
400 INPUT"ENTER THE DESIRED RESONANT FREQUENCY. ";F
410 INPUT"ENTER THE DESIRED BANDWIDTH. ";BW
420 INPUT"ENTER THE INDUCTOR VALUE IN HENRIES. ";L
430 PRINT""
440 XL = 6.2831853# * F * L
450 Q = F/BW
460 R = XL/Q
```

(Continued next page)

```
470 C = .02533/(F^2 * L)
480 PRINT"THE CAPACITOR MUST BE ";C;" FARADS."
490 PRINT"THE TOTAL RESISTANCE INCLUDING INDUCTOR WINDING RESISTANCE MUST BE "
500 PRINT R;" OHMS."
510 PRINT"THE CIRCUIT Q IS ";Q
520 GOTO 30
600 CLS:END
```

20-4 Parallel Resonance and Applications

The Parallel Resonant Circuit

The **parallel resonant circuit** is a parallel *RCL* circuit that is intended to operate at or near the frequency at which I_C and I_L are equal and cancel. At this frequency of resonance, the circuit acts resistively due to the reactance cancellation. This means the circuit is operating at maximum efficiency where $PF = I$ and the phase angle for total current is $0°$ ($\angle\theta = 0°$, see Sections 19-3 and 19-4).

Applications for the Parallel Resonant Circuit

Like the series resonant circuit, the parallel resonant circuit is used to select or reject a band of frequencies. However, you will soon discover that the parallel resonant circuit performs these functions while having characteristics that are quite the opposite of series resonant circuits. The characteristics will become evident as we continue.

It is safe to say that parallel resonant circuits are in wider use than series resonant circuits for applications of frequency selection or rejection. As indicated in Figure 20-9, parallel resonant circuits are perhaps the key circuits in radios, televisions, and transmitters of all kinds. Without them, these important communications systems could not function. The parallel resonant circuit is sensitive to, and resonant at, a narrow band of frequencies.

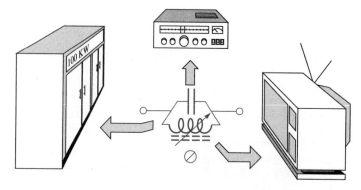

Figure 20-9   Parallel resonant circuits are used in all types of communications and broadcasting equipment.

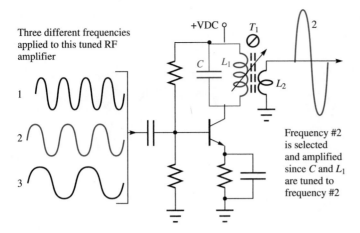

Figure 20-10   A tuned radio-frequency amplifier.

Thus, these simple circuits insure that only one station is received or that a well-defined band of frequencies is transmitted.

Figure 20-10 illustrates a typical tuned radio-frequency amplifier, found in many radio and television receivers. Notice that the amplifier has several different frequencies applied to its input and only one frequency appearing at its output. The parallel resonant circuit composed of C and L_1 is used to select frequency #2 and reject the others. C and L_1 are resonant at frequency #2. This means a large amount of current will be exchanged back and forth between C and L_1 at the resonant frequency (frequency #2). Transformer T_1 is formed by winding L_2 on the same core as L_1. The large resonant current exchanged between C and L_1 will induce a voltage across L_2, which is used to couple the signal to the next stage (not shown).

Figure 20-11 shows a typical transistor radio containing many of these small tuned radio-frequency transformers (**I**ntermediate **F**requency or **IF** transformers). As you can see, these parallel resonant circuits are contained in small metal enclosures ("tin cans") used to shield them from radiating or picking up electromagnetic energy from other transformers. It is interesting to note that the adjustable ferrite core is actually a ferrite cup that surrounds the coils. The cup is threaded on the outside, mating with threads in the wall of the tin can. The position of the cup with respect to the windings determines the inductance and, therefore, the resonant frequency.

As stated earlier, parallel resonance is also found in transmitters. Figure 20-12 illustrates the final power amplifier stage of a high-power broadcast transmitter. The parallel resonant circuit is formed from L_1, C_1, and C_B. The resonant frequency is determined by the size of L_1 and C_1. The radio-frequency bypass capacitor (C_B) is a relatively large-value capacitor that acts almost like a short at the radio frequency. Its purpose is to place the top end of L_1 at an AC ground while blocking the DC high voltage from being shorted to ground. Thus, C_B and the common ground are used to complete the parallel resonant circuit ($L_1 \| C_1$). The parallel resonant circuit is resonant at the desired radio frequency and a very large voltage is developed across L_1. This high-level radio signal is then coupled to the antenna.

As you can see, the parallel resonant circuit is very important. The applications presented here are only a few of the more common ones in which you will find the parallel

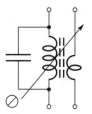

Figure 20-11 Tuned *IF* transformers are used in radios.

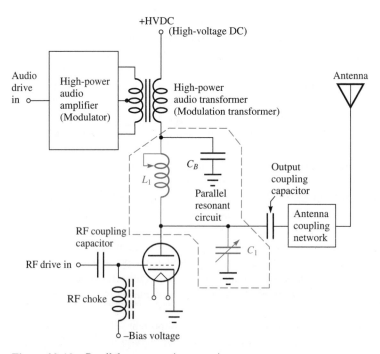

Figure 20-12 Parallel resonance in transmitters.

resonant circuit. Perhaps these few applications were enough to stimulate your curiosity as you continue in this chapter to see how these useful circuits work. Before continuing, take a moment to answer the questions of Self-Check 20-4.

SELF-CHECK 20-4

1. How is a parallel resonant circuit acting at resonance? Why?

2. In general, for what two purposes is the parallel resonant circuit used?

3. What is an IF transformer?

4. List at least three pieces of electronic equipment that use parallel resonant circuits.

20-5 Parallel Resonance Characteristics

Operational Theory

As we begin to explore parallel resonance, let's start with an investigation of operational theory. How does a parallel *LC* circuit resonate? How is energy stored in the parallel resonant circuit? How is the energy used up?

Ringing

The capacitor charges and then begins to discharge through the parallel inductor. The field of the inductor builds up, then collapses. The collapsing magnetic field induces current that recharges the parallel capacitor. This exchange of energy between the capacitor and the inductor does not end after only one cycle, but continues on its own for many cycles. This repeated exchange of energy in the parallel resonant circuit is known as **ringing** or **oscillation**. Ringing produces a sinusoidal voltage and current waveform that is equal to the resonant frequency of the circuit. The circuit will continue to ring until all of the exchanging energy is converted to heat. What converts the electrical energy to heat? Resistance. Resistance in the conductors, particularly the windings of the inductor (R_W), is responsible for this energy conversion. A high-Q inductor rings longer than a low-Q inductor since it has less resistance in its windings. With each cycle of ringing, a little more energy is converted to heat, an I^2R loss. Finally, the ringing is said to damp out. A **damped waveform** is shown in Figure 20-13.

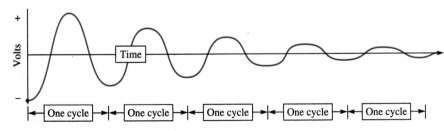

Figure 20-13 A damped waveform.

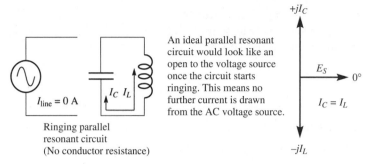

Figure 20-14   Ideal parallel resonant conditions.

Phase Relationships

Ideal Phase Relationships and Currents

Now that you have a basic understanding of the operation of the parallel resonant circuit, we can begin to examine more-practical circuits. Figure 20-14 illustrates a parallel resonant circuit that is being sustained by an AC signal source. The circuit operates at maximum efficiency when the generator frequency matches the resonant frequency of the LC circuit. At resonance, the inductor and capacitor currents are equal in amplitude and opposite in phase ($I_C \angle +90°$ and $I_L \angle -90°$, E_S is the reference at $0°$). Ideally, the parallel resonant circuit should look like an open, since the inductor and capacitor actually take turns sharing the same current. If the inductor and capacitor had no losses, no resistance (no R_W), the parallel resonant circuit would ring indefinitely, requiring no more current from the generator. The line current from the generator would then be 0 amps, as though the parallel circuit were no longer there (open).

Practical Phase Relationships and Currents

In practice, however, the parallel resonant circuit does require some line current to sustain the ringing. As you know, with each cycle some of the ringing current is converted to heat in the resistance of the inductor windings (damped by R_W). The line current resupplies the energy that is converted to heat. Thus the ringing is sustained at a constant level. If the resistance in the inductor windings is great, the I^2R energy conversion will be rapid, requiring a large amount of line current to make up for the loss. Thus, the higher the winding resistance, the higher the line current must be. At resonance, the parallel resonant circuit is acting resistively, since X_L and X_C cancel. With the reactances cancelled, the generator views the parallel resonant circuit as a large resistor. If the losses in the parallel resonant circuit are small, the circuit appears as a very large resistance that requires very little line current from the generator. Since the circuit is acting resistively, this line current is in phase with the applied source voltage. Therefore, as shown in Figure 20-15, a practical parallel resonant LC circuit has three currents at different phase angles: $I_{\text{line}} \angle 0°$; $I_C \angle +90°$; and $I_L \angle -90°$.

Circuit Parameter Summary

$I_C = I_L$ This is the main requirement and condition for parallel resonance.

Z_t = maximum   The total impedance of the parallel LC circuit is very high and maximum when it is operating at resonance. This is because the reactive currents cancel.

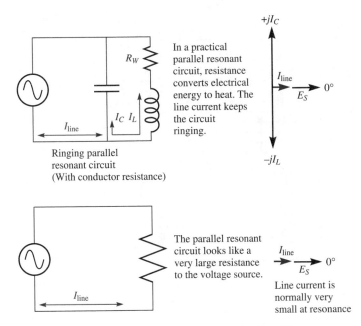

Figure 20-15   Practical and real parallel resonant conditions.

The parallel circuit looks like a very-high-value resistor demanding very little line current from the generator. If the reactive components had no resistance, the total impedance would be infinite (ideally an open).

$I_t = I_{\text{line}} = \text{minimum}$ Since the total impedance is maximum, the line current must be minimum ($I_{\text{line}} = E_S/Z_t$). The total current, or line current, is minimum at resonance since a balance exists between I_C and I_L and only a small amount of current is needed to sustain ringing. Ideally, there should be no line current at all, since ideally the impedance is infinite.

$P_A = P_R = I_t \cdot E_S$ The apparent power is equal to the real power since the reactive powers are equal and cancel ($P_C\angle +90° + P_L\angle -90° = 0$ VAR).

$\angle\theta = 0°$, $PF = 1$, $\%\text{eff} = 100\% = \text{maximum}$ At resonance, the parallel resonant circuit is operating at its maximum efficiency, which is usually very close to 100%. As indicated in Figure 20-15, the total current phase angle is 0°. The power factor is equal to $\cos\angle\theta$. Thus the power factor is 1.

Parallel Resonance Graphs

The chart of Figure 20-16 graphically summarizes the important characteristics of parallel resonant circuits. The graphs illustrate very clearly how the various parameters are affected by a change in frequency. Notice that the total impedance is maximum and the total current is minimum at resonance. The graphs indicate that as the generator frequency is increased or decreased above and below resonance, the total impedance decreases and the total current increases. Also, the capacitor and inductor currents are equal at resonance. Also note that at resonance the resonant circuit acts resistively ($I_L = I_C$), below resonance the circuit acts inductively ($I_L > I_C$), and above resonance the circuit acts capacitively ($I_C > I_L$).

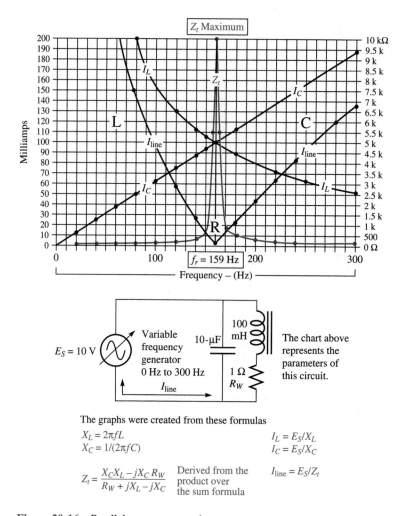

Figure 20-16   Parallel resonance graphs.

The graphs were created from these formulas

$$X_L = 2\pi f L$$
$$X_C = 1/(2\pi f C)$$

$$I_L = E_S/X_L$$
$$I_C = E_S/X_C$$

$$Z_t = \frac{X_C X_L - jX_C R_W}{R_W + jX_L - jX_C}$$

Derived from the product over the sum formula

$$I_{line} = E_S/Z_t$$

Take a few minutes now to review this section. Pay particular attention to the differences between ideal and practical parallel resonant circuits. Realize that the differences between ideal and practical are caused mainly by the inductor winding resistance (R_W). Test your understanding by answering the questions of Self-Check 20-5.

SELF-CHECK   20-5

1. What does *ringing* mean?

2. What is a damped waveform?

3. What causes damping?

4. Is the line current (a) minimum or (b) maximum at resonance? Explain.

20-6 Basic Circuit Calculations for Parallel Resonance

Tank-Circuit Calculations

Tank Impedance (Z_{tank})

In the previous discussions, it was shown that the line current is small but necessary to maintain the ringing, or oscillation, of the parallel resonant circuit. As the line current is being supplied, it is also being converted to heat in the windings of the inductor (R_W). Thus, the parallel resonant circuit can be compared to a fuel tank that contains energy in the form of a fuel. As the fuel is used up, it must be replaced. In a parallel resonant circuit, the dissipated electrical energy is replaced by the line current. Therefore, the parallel resonant circuit (L and C) is most often referred to as a **tank circuit**.

The tank circuit appears as a large resistance (impedance) at resonance. As shown in Figure 20-17, at resonance the generator sees a large resistor instead of an inductor or capacitor. Ideally, the resistance of this resistor would be infinite. In practice, the tank impedance is some large real value. To determine this **tank impedance**, two parameters must be known: (1) the resistance of the tank (usually just the winding resistance R_W), and (2) the inductive or capacitive reactance at the resonant frequency. Once these parameters are known, the tank impedance can be calculated by

$$Z_{\text{tank}} \cong X_L^2/R_W \cong X_C^2/R_W \qquad (20.10)$$

This is derived using either the product over the sum formula or the reciprocal formula for parallel components. Complex algebra must be used for the derivation. Here, the product over the sum formula is used to derive Formula 20.10:

$$Z_{\text{tank}} = \frac{-jX_C \cdot (R_W + jX_L)}{-jX_C + (R_W + jX_L)} = \frac{-jX_C \cdot (R_W + jX_L)}{-jX_C + R_W + jX_L} = \frac{-jX_C R_W + X_C X_L}{R_W}$$

$$\cong \frac{X_C X_L}{R_W} = \frac{X_L^2}{R_W} = \frac{X_C^2}{R_W} \quad (\cong Z_{\text{tank}} \text{ only at resonance where } X_C = X_L)$$

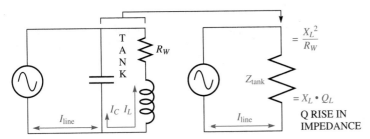

At resonance the generator sees the tank as a large resistive impedance.

$$I_{\text{tank}} = I_C = I_L = \frac{V_{\text{tank}}}{X_C} = \frac{V_{\text{tank}}}{X_L}$$

$$I_{\text{line}} = \frac{V_{\text{tank}}}{Z_{\text{tank}}}$$

Figure 20-17   Tank circuit calculations.

Note that the emphasized $-jX_CR_W$ term can be ignored in most cases, since R_W is usually very small compared to the reactances; $-jX_CR_W$ is considered insignificant compared to X_CX_L if the inductor Q is high ($Q_L \geq 10$).

If you look carefully, you will notice that Formula 20.10 contains the formula for the Q of an inductor. Recall that the Q of an inductor is the ratio of the inductor's reactance to its winding resistance ($Q_L = X_L/R_W$). Thus, Formula 20.10 can be expressed as:

$$Z_{\text{tank}} \cong X_L^2/R_W = X_L \cdot X_L/R_W = X_L \cdot Q_L = X_C \cdot Q_L \qquad (20.11)$$

where X_L or X_C is determined at the resonant frequency.

Formula 20.11 indicates that a Q *rise in impedance* exists in the tank circuit. In other words, the tank impedance is the product of the inductor Q and X_L or X_C.

Tank Current (I_{tank})

The **tank current** is calculated using Ohm's Law. It is simply the inductor branch current or the capacitor branch current. The inductor and capacitor share the same current back and forth as they ring. Remember, the two currents are 180° out of phase. Thus, the tank current is determined as follows:

$$I_{\text{tank}} = V_{\text{tank}}/X_C = V_{\text{tank}}/X_L \qquad (20.12)$$

[*Note:* The inductor branch impedance is actually $R_W + jX_L$. The inductor resistance is ignored for high-Q inductors (≥ 10).]

Line Current (I_{line})

The **line current** is determined by the source voltage (or tank voltage) and the tank impedance. Recall that the generator is supplying current to the tank which looks like a large resistance. Thus, the line current is found using Ohm's Law:

$$I_{\text{line}} = V_{\text{tank}}/Z_{\text{tank}} \qquad (20.13)$$

Calculations with Parallel Resistance

Total Current (I_t)

As shown in Figure 20-18, the total current for a parallel RCL circuit operating at resonance is simply the sum of the resistor branch current and the line current. The parallel resistor

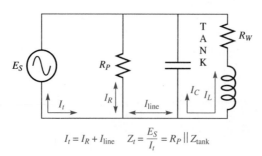

$$I_t = I_R + I_{\text{line}} \qquad Z_t = \frac{E_S}{I_t} = R_P \| Z_{\text{tank}}$$

Figure 20-18 The tank circuit with a parallel resistor.

(R_P) may be an actual resistor or may represent the resistance of some circuit that is connected to and loading the tank (i.e., an amplifier circuit or an antenna system). Since the overall circuit is acting resistively, and the tank looks like a large resistor, the two currents may be added arithmetically as follows:

$$I_t = I_R + I_{line} \tag{20.14}$$

Total Impedance (Z_t)

The total impedance for a parallel *RCL* circuit operating at resonance can be found using Ohm's Law (when E_S and I_t are known), or by using any of the parallel component formulas:

$$Z_t = E_S/I_t \tag{20.15}$$

$$Z_t = \frac{R_P \cdot Z_{tank}}{R_P + Z_{tank}} \qquad \text{(product over the sum)} \tag{20.16}$$

where R_P is a resistor in parallel with the tank.

Study Example 20-4 to see how these formulas are used to analyze a parallel *RCL* circuit that is operating at resonance.

EXAMPLE   20-4

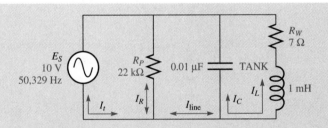

Figure 20-19

For Figure 20-19:

1) $f_r = \dfrac{1}{2\pi\sqrt{LC}} = \dfrac{1}{2 \cdot \pi \cdot \sqrt{1 \text{ mH} \cdot 0.01 \text{ }\mu F}} = 50.3 \text{ kHz}$

2) $X_C = X_L = 2 \cdot \pi \cdot f \cdot L = 2 \cdot \pi \cdot 50.3 \text{ kHz} \cdot 1 \text{ mH} = 316 \text{ }\Omega$

3) $I_C = I_L = I_{tank} = E_S/X_C = 10 \text{ V}/316 \text{ }\Omega = 31.6 \text{ mA}$

4) $Z_{tank} = X_L^2/R_W = 316^2/7 = 14.3 \text{ k}\Omega$

5) $I_{line} = E_S/Z_{tank} = 10 \text{ V}/14.3 \text{ k}\Omega = 0.7 \text{ mA}$

6) $I_R = E_S/R_P = 10 \text{ V}/22 \text{ k}\Omega = 0.454 \text{ mA}$

7) $I_t = I_R + I_{line} = 0.454 \text{ mA} + 0.7 \text{ mA} = 1.154 \text{ mA}$

8) $Z_t = E_S/I_t = 10 \text{ V}/1.154 \text{ mA} = 8.67 \text{ k}\Omega$

$$Z_t = \frac{R_P \cdot Z_{tank}}{R_P + Z_{tank}} = \frac{22 \text{ k}\Omega \cdot 14.3 \text{ k}\Omega}{22 \text{ k}\Omega + 14.3 \text{ k}\Omega} = 8.67 \text{ k}\Omega$$

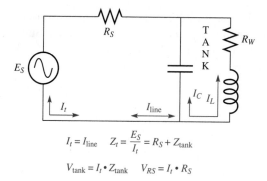

$$I_t = I_{\text{line}} \quad Z_t = \frac{E_S}{I_t} = R_S + Z_{\text{tank}}$$

$$V_{\text{tank}} = I_t \cdot Z_{\text{tank}} \quad V_{RS} = I_t \cdot R_S$$

Figure 20-20 The tank circuit with a series resistor.

Calculations With Series Resistance

Total Impedance

As shown in Figure 20-20, the total impedance of a parallel resonant circuit with a series resistor is simply the sum of the series resistor and the tank impedance. Once again, the tank looks like a large resistance at resonance. Thus, the tank impedance is added to the series resistance just as two resistors would be added.

$$Z_t = R_S + Z_{\text{tank}} \tag{20.17}$$

Total Current

Once the total impedance is known (Formula 20.17), the total current can be found using Ohm's Law. Note here, however, that the total current is also the line current supplying the tank.

$$I_t = I_{\text{line}} = E_S/Z_t \tag{20.18}$$

EXAMPLE   20-5

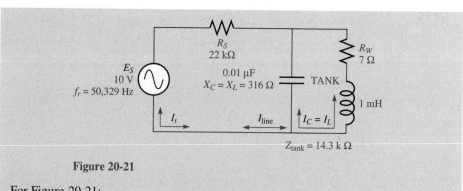

Figure 20-21

For Figure 20-21:

1) $Z_t = R_S + Z_{\text{tank}} = 22\ \text{k}\Omega + 14.3\ \text{k}\Omega = 36.3\ \text{k}\Omega$

2) $I_t = E_S/Z_t = 10\ \text{V}/36.3\ \text{k}\Omega = 275\ \mu\text{A}$

3) $V_{RS} = I_t = R_S = 275 \ \mu A \cdot 22 \ k\Omega = 6.05 \ V$

$V_{RS} = E_S \cdot \dfrac{R_S}{Z_t} = 10 \ V \cdot 22 \ k\Omega / 36.3 \ k\Omega = 6.06 \ V$ SLIGHT ROUNDING ERROR

4) $V_{tank} = I_t \cdot Z_{tank} = 275 \ \mu A \cdot 14.3 \ k\Omega = 3.93 \ V$

$V_{tank} = E_S \cdot \dfrac{Z_{tank}}{Z_t} = 10 \ V \cdot 14.3 \ k\Omega / 36.3 \ k\Omega = 3.94 \ V$ SLIGHT ROUNDING ERROR

Voltage Drops

Also shown in Figure 24-10, the voltage across the series resistor and the tank voltage are found using Ohm's Law or the voltage divider rule (proportion formula).

$$V_{RS} = I_t \cdot R_S = E_S \cdot R_S / Z_t \qquad (20.19)$$

$$V_{tank} = I_t \cdot Z_{tank} = E_S \cdot Z_{tank} / Z_t \qquad (20.20)$$

Study Example 20-5 to see how these formulas are applied.

It is very important for you to understand and be able to do the calculations shown in Examples 20-4 and 20-5. Review this section and solve the unknown parameters specified in the problems of Self-Check 20-6.

S E L F - C H E C K   2 0 - 6

1. Find Z_{tank}, I_{tank}, I_{line}, and I_t for Figure 20-22.

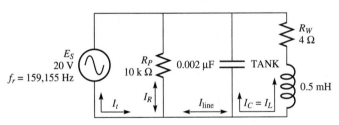

Figure 20-22

2. Find Z_t, I_t, V_{tank}, I_{tank}, and V_{RS} for Figure 20-23.

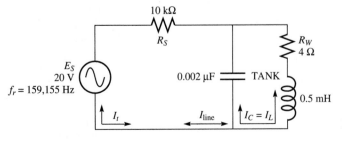

Figure 20-23

20-7 Calculations for Parallel Resonance

Quality Factors

Tank Quality Factor (Q_{tank})

The Q of an LC tank circuit is almost the same as the Q of the inductor itself (see Section 20-3 for a complete discussion of quality factor). For all practical purposes, Q_{tank} can be considered as equal to Q_L. In other words, the tank Q is equal to the ratio of inductive reactance to winding resistance (other conductor resistances are usually insignificant).

$$Q_{tank} = Q_L = X_L/R_W = X_C/R_W \tag{20.21}$$

Recall that the tank impedance is a Q rise in impedance, meaning tank Q times X_L, or tank Q times X_C, equals Z_{tank} ($Z_{tank} = Q_{tank} \cdot X_L = Q_{tank} \cdot X_C$). Rearranging this expression, we discover that:

$$Q_{tank} = Z_{tank}/X_L = Z_{tank}/X_C \tag{20.22}$$

Since Q_{tank} is actually the ratio of tank impedance to inductive, or capacitive, reactance, Q_{tank} must also be the ratio of tank current to line current. This is because the same voltage appears across the tank and reactive components. Note this simple proof:

$$\frac{I_{tank}}{I_{line}} = \frac{V_{tank}/X_L}{V_{tank}/Z_{tank}} = \frac{V_{tank}}{X_L} \cdot \frac{Z_{tank}}{V_{tank}} = \frac{Z_{tank}}{X_L} = Q_{tank}$$

Thus,

$$Q_{tank} = I_{tank}/I_{line} \tag{20.23}$$

From Formula 20.23, you can see that there is also a Q rise in *current*. The line current times the tank Q equals the tank current ($I_{tank} = Q_{tank} \cdot I_{line}$). Therefore, in a parallel resonant circuit, a Q rise in *current* is realized in the inductor and capacitor of the tank. (Recall that in a series resonant circuit a Q rise in *voltage* is obtained across the inductor or capacitor.)

Study Example Set 20-6 to see how the tank Q is found using Formulas 20.21, 20.22, and 20.23.

Overall Circuit Quality Factor (Q_{ckt})

The quality factor for an entire circuit that contains a tank circuit is very different from the quality factor for the tank itself. The circuit Q takes into consideration the loading effects of any circuits connected to the tank. A tank circuit cannot perform a useful task by itself. It must be connected to other circuitry or components. In the previous section, you saw examples of tank circuits with parallel and series resistors. These resistors actually act as a load on the tank circuit. This means the effective Q of the tank is reduced and the operating bandwidth of the circuit is increased. The tank, by itself, has a much higher Q and narrower bandwidth. The amount of reduction in the effective tank Q will depend on the amount of tank loading. The amount of tank loading depends on the size of resistor that appears in parallel with the tank impedance ($R_P \| Z_{tank}$). If the parallel resistance is large, the loading effect is small. If the parallel resistance is small, the loading effect will be great.

EXAMPLE   20-6

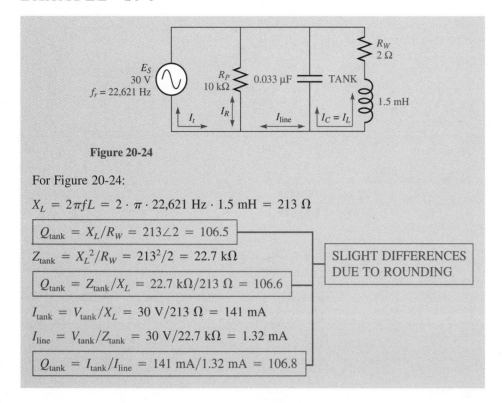

Figure 20-24

For Figure 20-24:

$$X_L = 2\pi fL = 2 \cdot \pi \cdot 22{,}621 \text{ Hz} \cdot 1.5 \text{ mH} = 213 \text{ }\Omega$$

$Q_{\text{tank}} = X_L/R_W = 213\angle 2 = 106.5$

$Z_{\text{tank}} = X_L^2/R_W = 213^2/2 = 22.7 \text{ k}\Omega$

$Q_{\text{tank}} = Z_{\text{tank}}/X_L = 22.7 \text{ k}\Omega/213 \text{ }\Omega = 106.6$

$I_{\text{tank}} = V_{\text{tank}}/X_L = 30 \text{ V}/213 \text{ }\Omega = 141 \text{ mA}$

$I_{\text{line}} = V_{\text{tank}}/Z_{\text{tank}} = 30 \text{ V}/22.7 \text{ k}\Omega = 1.32 \text{ mA}$

$Q_{\text{tank}} = I_{\text{tank}}/I_{\text{line}} = 141 \text{ mA}/1.32 \text{ mA} = 106.8$

SLIGHT DIFFERENCES
DUE TO ROUNDING

All circuitry, connected to the tank, can be reduced to a single equivalent parallel resistance (R_P). Thus, any circuit containing a tank can be reduced to a simple parallel *RCL* circuit for the purpose of analysis (to determine Q_{ckt}, *BW*, and *BP*). Example Set 20-7 illustrates several practical examples of circuit simplification. Notice the resistance of the generator (R_g) is also considered part of the load on the tank. In effect, everything that is connected to the tank circuit is converted to a single parallel resistor (R_P). Once the equivalent parallel resistance is determined, the circuit Q can be approximated as follows:

$$Q_{\text{ckt}} \cong R_P/X_L \cong R_P/X_C \qquad \text{(for } Z_{\text{tank}} \geq 10 \cdot R_P) \tag{20.24}$$

In most cases, Formula 20.24 will be very accurate since Z_{tank} is normally very high (much greater than 10 times R_P). However, for those cases where the tank impedance is less than $10 \cdot R_P$, another formula is needed. This formula must take into consideration the impedance of the tank, which is in parallel with the equivalent parallel resistance, as follows:

$$Q_{\text{ckt}} = (R_P\|Z_{\text{tank}})/X_L \cong (R_P\|Z_{\text{tank}})/X_C \qquad \text{(for } Z_{\text{tank}} < 10 \cdot R_P) \tag{20.25}$$

where $Z_{\text{tank}} = X_L^2/R_W$ and $R_P\|Z_{\text{tank}} = (R_P \cdot Z_{\text{tank}})/(R_P + Z_{\text{tank}})$

Example Set 20-8 illustrates how Formulas 20.24 and 20.25 are used to determine circuit Q. Notice that the equivalent parallel resistance R_P and the tank impedance must be found before it is decided which formula to use in determining circuit Q. Of course, accuracy is always assured if Formula 20.25 is always used to determine circuit Q.

EXAMPLE SET   20-7

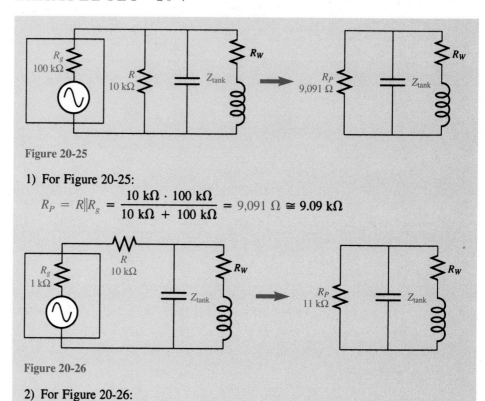

Figure 20-25

1) **For Figure 20-25:**

$$R_P = R \| R_g = \frac{10 \text{ k}\Omega \cdot 100 \text{ k}\Omega}{10 \text{ k}\Omega + 100 \text{ k}\Omega} = 9,091 \ \Omega \cong 9.09 \text{ k}\Omega$$

Figure 20-26

2) **For Figure 20-26:**
$$R_P = R + R_g = 10 \text{ k}\Omega + 1 \text{ k}\Omega = 11 \text{ k}\Omega$$

EXAMPLE SET   20-8

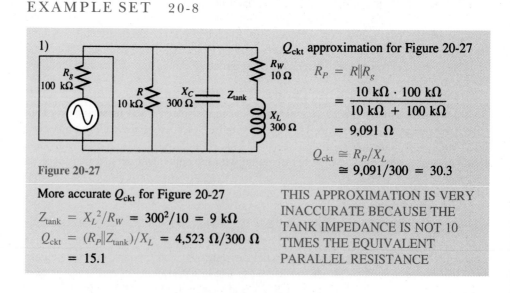

1)

Figure 20-27

Q_{ckt} approximation for Figure 20-27

$$R_P = R \| R_g$$

$$= \frac{10 \text{ k}\Omega \cdot 100 \text{ k}\Omega}{10 \text{ k}\Omega + 100 \text{ k}\Omega}$$

$$= 9,091 \ \Omega$$

$$Q_{\text{ckt}} \cong R_P / X_L$$
$$\cong 9,091/300 = 30.3$$

More accurate Q_{ckt} for Figure 20-27

$$Z_{\text{tank}} = X_L{}^2 / R_W = 300^2/10 = 9 \text{ k}\Omega$$
$$Q_{\text{ckt}} = (R_P \| Z_{\text{tank}})/X_L = 4,523 \ \Omega/300 \ \Omega$$
$$= 15.1$$

THIS APPROXIMATION IS VERY INACCURATE BECAUSE THE TANK IMPEDANCE IS NOT 10 TIMES THE EQUIVALENT PARALLEL RESISTANCE

2)

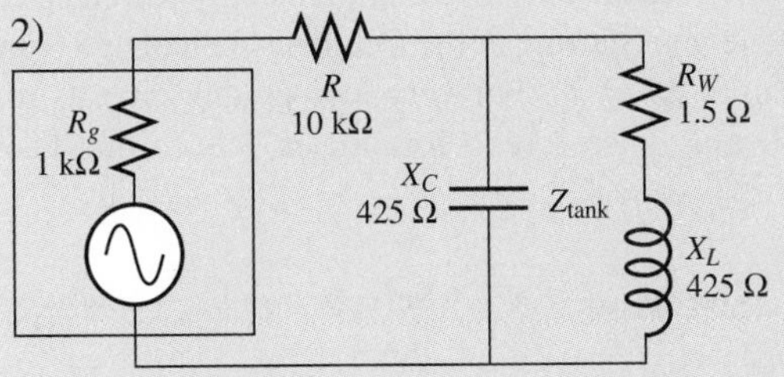

R_g 1 kΩ

R 10 kΩ

X_C 425 Ω

Z_{tank}

R_W 1.5 Ω

X_L 425 Ω

Figure 20-28

Q_{ckt} approximation for Figure 20-28

$$R_P = R + R_g$$
$$= 10\ k\Omega + 1\ k\Omega$$
$$= 11\ k\Omega$$

$$Q_{ckt} \cong R_P/X_L$$
$$\cong 11\ k\Omega/425\ \Omega$$
$$\cong 25.9\ (\text{CLOSE}$$
$$\text{APPROXIMATION)}$$

More accurate Q_{ckt} for Figure 20-28

$$Z_{tank} = X_L^{\,2}/R_W = 425^2/1.5 = 120.4\ k\Omega$$
$$Q_{ckt} = (R_P\|Z_{tank})/X_L = 10,079\ \Omega/425\ \Omega = 23.7$$

NOTE THAT THE TANK
IMPEDANCE IS $> 10 \cdot R_P$.

The Resonant Frequency and Q

High-Q Resonance

The resonant frequency for a high-Q (≥ 10) tank circuit is determined in the same way as for series resonant circuits, using the following resonance formula:

$$f_r = f_c = f_o = 1/(2\pi\sqrt{LC}) \qquad (\text{for } Q_{tank} \geq 10) \tag{20.1}$$

where f_r is the resonant frequency equal to f_c, the center frequency equal to f_o, the frequency of operation.

Low-Q Resonance

Formula 20.1 is not accurate enough for low-Q tank circuits. Tank circuits whose Q is less than 10 have a lower resonant frequency. Why? How can Q_{tank} affect the resonant frequency? Recall that two of the characteristics of parallel resonant circuits are: $X_C = X_L$ and $I_C = I_L$. Actually, these two characteristics only occur at the same frequency in high-Q tank circuits. You see, Formula 20.1 was derived from the X_C and X_L formulas, which were set equal to each other and solved for f. The resulting formula is based on the very fact that $X_C = X_L$. In any tank circuit the inductive and capacitive reactances are equal at a frequency found using Formula 20.1. However, the capacitor and inductor branch currents may not be equal at that frequency. In a low-Q tank circuit, the winding resistance of the inductor is significant. It is greater than one-tenth of X_L for a Q_{tank} less than 10 ($R_W > 0.1X_L$ for $Q_{tank} < 10$). This inductor resistance increases the inductor branch impedance and reduces the inductor branch current: $I_L = V_{tank}/(R_W + jX_L)$. Thus I_C is greater than I_L at the frequency at which $X_C = X_L$. The *true resonant frequency* for a parallel LC circuit is the frequency at which $I_C = I_L$. For I_C to equal I_L, the inductive reactance must be decreased and the capacitive reactance increased. This is accomplished by lowering the applied frequency to the true resonant frequency. Starting at the frequency calculated using Formula 20.1, the frequency is lowered until $I_C = I_L$. For very low-Q tank circuits, the true resonant frequency is significantly lower. Therefore, a formula is needed that takes into consideration Q_{tank} as follows:

$$f_r = \frac{1}{2\pi\sqrt{LC}} \cdot \sqrt{Q^2/(Q^2 + 1)} \qquad \text{(for } Q_{tank} < 10\text{, where } Q = Q_{tank}) \qquad (20.26)$$

Follow Example 20-9 to see how Formulas 20.1 and 20.26 are used. Note that the term $\sqrt{Q^2/(Q^2 + 1)}$ can be considered as a correction factor for the basic resonant-frequency formula. This correction factor is used when it is discovered the tank Q is less than 10.

EXAMPLE   20-9

Given a tank circuit with component values $L = 15\ \mu H$, $R_W = 204\ \Omega$, and $C = 40$ pF, find the true resonant frequency. To begin, Q_{tank} is not known. So, we do not know if the correction factor is needed in the resonant frequency formula. That doesn't matter. We will simply start with the basic formula (20.1). Once the resonant frequency is determined using Formula 20.1, the inductive reactance can be determined and a decision can be made as to whether the correction factor is needed.

$$f_r = \frac{1}{2\pi\sqrt{LC}} = \frac{1}{2 \cdot \pi \cdot \sqrt{15\ \mu H \cdot 40\ pF}} = 6.50\ \text{MHz}$$

$$X_L = 2\pi f L = 2\pi \cdot 6.50\ \text{MHz} \cdot 15\ \mu H = 613\ \Omega$$

$$Q_{tank} = X_L/R_W = 613\ \Omega/204\ \Omega = 3$$

The tank Q is very low. The correction factor is definitely needed.

$$f_r = 6.50\ \text{MHz} \cdot \sqrt{Q^2/(Q^2 + 1)} = 6.50\ \text{MHz} \cdot \sqrt{3^2/(3^2 + 1)}$$
$$= 6.50\ \text{MHz} \cdot \sqrt{0.9} = 6.50\ \text{MHz} \cdot 0.949 = 6.17\ \text{MHz}$$

The difference between 6.50 MHz and 6.17 MHz *is* significant (330 kHz)!

Bandwidth and Bandpass

Calculating *BW* and *BP*

Bandwidth and bandpass for a parallel resonant circuit are calculated in the same manner as for series resonant circuits. The bandwidth is the amount of frequencies that is selected or rejected by the parallel resonant circuit. The circuit Q and resonant frequency must be known to determine the bandwidth for a circuit: $BW = f_r/Q_{ckt}$ (Formula 20.8). Once the bandwidth is known, the bandpass may be calculated. Recall that the resonant frequency is in the center of the bandwidth (assuming a high-Q circuit, $Q \geq 10$). Thus $BP \cong [f_r - (BW/2)] \rightarrow [f_r + (BW/2)]$ (Formula 20.9). Consider Figure 20.29 and Example 20-10 and see Section 20-3.

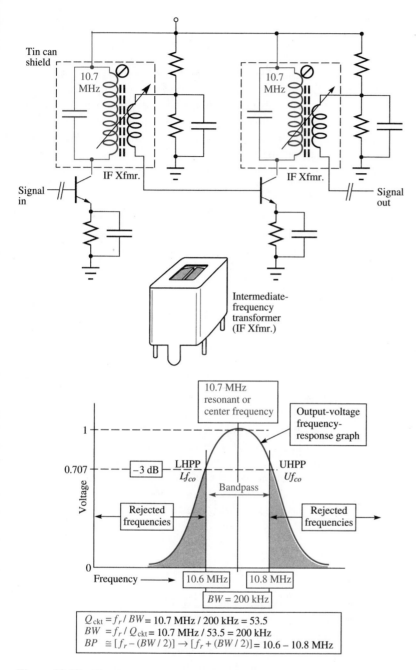

$$Q_{ckt} = f_r / BW = 10.7 \text{ MHz} / 200 \text{ kHz} = 53.5$$
$$BW = f_r / Q_{ckt} = 10.7 \text{ MHz} / 53.5 = 200 \text{ kHz}$$
$$BP \cong [f_r - (BW/2)] \rightarrow [f_r + (BW/2)] = 10.6 - 10.8 \text{ MHz}$$

Figure 20-29 Circuit Q, bandwidth, and bandpass.

EXAMPLE   20-10

The intermediate frequency section (IF section) of an FM receiver has a center frequency of 10.7 MHz and an overall circuit Q of 53.5. This section is the heart of the FM receiver where most of the selectivity and amplification takes place. The IF section usually consists of several tuned radio-frequency amplifiers. Of all the stations picked up by the receiver's antenna, only one at a time is allowed through the IF section to the audio amplifier and speaker. The bandwidth of the IF section is wide enough for only one station. Determine the bandwidth and bandpass for this important section.

$$BW = f_r/Q_{ckt} = 10.7 \text{ MHz}/53.5$$

$$= 200 \text{ kHz} \quad \text{(this is the channel width for a single FM station)}$$

$$BP = [f_r - (BW/2)] \rightarrow [f_r + (BW/2)]$$

$$= [10.7 \text{ MHz} - 100 \text{ kHz}] \rightarrow [10.7 \text{ MHz} + 100 \text{ kHz}]$$

$$= 10.6 \text{ MHz} \rightarrow 10.8 \text{ MHz}$$

Changing *BW* and *BP*

In order to change the *BW* and *BP* of a circuit, the circuit Q must be changed. To change the circuit Q, one or more of the resistive variables used in Formula 20.25 must be changed $(Q_{ckt} = (R_P \| Z_{tank})/X_L)$. In other words, the equivalent parallel resistance (R_P) or the tank impedance (Z_{tank}) must be changed. As you saw earlier, the tank impedance is largely dependent on the resistance of the inductor windings (R_W) or inductor Q $(Q_L = Q_{tank})$. Thus, if we wish to increase the tank Q and impedance, the inductor resistance must be decreased and vice versa $(Q_L = Q_{tank} = X_L/R_W)$. Increasing the tank impedance and/or the equivalent parallel resistance will increase the circuit Q and vice versa $(Q_{ckt} = (R_P \| Z_{tank})/X_L)$. Thus, if it is desired to increase the bandwidth and bandpass, the circuit Q must be decreased by decreasing R_P and/or Z_{tank} (see Figure 20-30). If we wish to decrease the bandwidth and bandpass, the circuit Q must be increased by increasing R_P and/or Z_{tank}.

Tunable-Range Calculations

Determining the Tuning Range

Often, a parallel resonant circuit can be tuned over a specific range of frequencies. For example, the ferrite-rod antenna circuit of a typical AM receiver can be tuned over a range of about 540 kHz to 1,610 kHz, covering the AM broadcast band. As shown in Figure 20-31, the inductance of the ferrite-rod antenna is 200 μH and a $48-435$ pF variable capacitor is used for tuning. The actual frequency-tuning range of this tunable circuit is determined using the resonant-frequency formula (20.1). The minimum capacitance of the variable capacitor will determine the highest frequency and the maximum capacitance the lowest frequency $(C_{min} = 48 \text{ pF}; C_{max} = 435 \text{ pF})$:

$$f_{max} = \frac{1}{2\pi\sqrt{LC_{min}}} = \frac{1}{2 \cdot \pi \cdot \sqrt{200 \ \mu\text{H} \cdot 48 \text{ pF}}}$$

$$= 1,624 \text{ kHz}$$

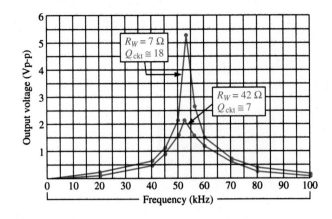

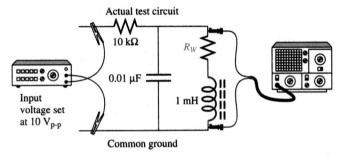

R_W represents the resistance in the inductor branch, in this experiment, R_W is a total of 7 Ω or 42 Ω.

Figure 20-30   How R_W affects Q and circuit performance.

$$f_{min} = \frac{1}{2\pi\sqrt{LC_{max}}}$$

$$= \frac{1}{2 \cdot \pi \cdot \sqrt{200\ \mu H \cdot 435\ pF}}$$

$$= 540\ kHz$$

Tunable range $= 540 \rightarrow 1{,}624$ kHz

Determining the Capacitance Range

Let's suppose you want to design a parallel resonant circuit that can be tuned over a range of $3 \rightarrow 3.5$ MHz. You will need a fixed inductor and a variable capacitor or a fixed capacitor and a variable inductor (both could be variable). Let's assume that you want to use a fixed-value 50 μH coil. How can you determine the range of capacitance needed to cover a frequency range of $3 \rightarrow 3.5$ MHz? The answer is Formula 20.2. The minimum capacitance (C_{min}) is calculated using the highest frequency and the maximum capacitance (C_{max}) is calculated using the lowest frequency.

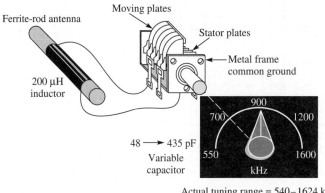

Ferrite-rod antenna

Moving plates

Stator plates

Metal frame common ground

200 μH inductor

48 ⟶ 435 pF Variable capacitor

Actual tuning range = 540–1624 kHz

$$f_{min} = \frac{1}{2\pi \sqrt{LC_{max}}} \quad \text{AND} \; f_{max} = \frac{1}{2\pi \sqrt{LC_{min}}}$$

Figure 20-31   Determining the tuning range of a parallel resonant circuit.

$$C_{min} = \frac{1}{4 \cdot \pi^2 \cdot f_{max}^2 \cdot L} = \frac{1}{4 \cdot \pi^2 \cdot (3.5 \text{ MHz})^2 \cdot 50 \text{ } \mu H}$$
$$= 41.4 \text{ pF}$$

$$C_{max} = \frac{1}{4 \cdot \pi^2 \cdot f_{min}^2 \cdot L} = \frac{1}{4 \cdot \pi^2 \cdot (3 \text{ MHz})^2 \cdot 50 \text{ } \mu H}$$
$$= 56.3 \text{ pF}$$

The range of needed capacitance is 41.4 → 56.3 pF.

As you can see, to obtain a tunable frequency range of 3 → 3.5 MHz, the capacitance must be variable, from 41.4 → 56.3 pF. The overall change in capacitance from minimum to maximum is about 15 pF. As shown in Figure 20-32, the required capacitance range

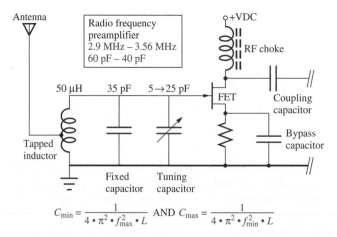

Antenna

Radio frequency preamplifier
2.9 MHz – 3.56 MHz
60 pF – 40 pF

+VDC

RF choke

50 μH   35 pF   5→25 pF

FET

Coupling capacitor

Bypass capacitor

Tapped inductor

Fixed capacitor   Tuning capacitor

$$C_{min} = \frac{1}{4 \cdot \pi^2 \cdot f_{max}^2 \cdot L} \quad \text{AND} \; C_{max} = \frac{1}{4 \cdot \pi^2 \cdot f_{min}^2 \cdot L}$$

Figure 20-32   Determining the needed range of capacitance for a tunable parallel resonant circuit.

can be obtained by placing a fixed-value 35 pF capacitor in parallel with a small 5 →
25 pF variable capacitor. This arrangement accommodates the required frequency range
with a little extra on each end.

Take a few minutes to study Design Note 20-2. Then test your skills by solving the
problems of Self-Check 20-7.

SELF-CHECK 20-7

1. If the line current feeding a tank circuit is 2 μA, the tank voltage is 1 V, and the
   reactance of the inductor is 3 kΩ, what is the tank Q?

2. If the Q of a tank is 240 and the line current feeding the tank is 150 μA, how much is
   the tank current?

3. List three factors that affect the overall circuit Q of a parallel resonant circuit.

4. If the tank impedance is only 4 times greater than a parallel resistance, which formula
   should you use to find Q_{ckt} and why?

5. Name the two parameters that are equal when a parallel resonant circuit is operating at
   its true resonant frequency.

6. Given a tank circuit with the values $L = 30$ μH, $R_W = 120$ Ω and $C = 50$ pF, find
   the true resonant frequency.

7. Given a tank circuit with a parallel resistor and $L = 500$ μH, $R_W = 5$ Ω, $C = 0.001$ μF
   and $R_P = 100$ kΩ, find f_r, Q_{ckt}, BW and BP.

8. Determine the range of capacitance needed to resonate with a 25 μH inductor over a
   frequency range of 7 → 7.5 MHz.

DESIGN NOTE 20-2: Parallel Resonance

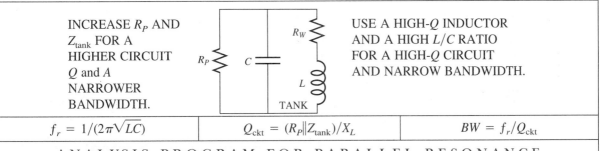

INCREASE R_P AND Z_{tank} FOR A HIGHER CIRCUIT Q and A NARROWER BANDWIDTH.

USE A HIGH-Q INDUCTOR AND A HIGH L/C RATIO FOR A HIGH-Q CIRCUIT AND NARROW BANDWIDTH.

$f_r = 1/(2\pi\sqrt{LC})$	$Q_{ckt} = (R_P\|Z_{tank})/X_L$	$BW = f_r/Q_{ckt}$

ANALYSIS PROGRAM FOR PARALLEL RESONANCE

```
10 CLS
20 PRINT"PARALLEL RESONANCE"
30 PRINT"":PRINT""
```

(Continued next page)

D E S I G N   N O T E   2 0 - 2:   (Continued)

```
 40  PRINT"(1) ANALYZE A PARALLEL RESONANT CIRCUIT"
 50  PRINT"(2) DESIGN A PARALLEL RESONANT CIRCUIT"
 55  PRINT"(3) QUIT"
 60  PRINT""
 70  INPUT"SELECT ANALYSIS, DESIGN, OR QUIT - 1, 2 OR 3";S
 80  IF S<1 OR S>3 THEN GOTO 30
 90  ON S GOTO 200,400,600
200  INPUT"ENTER THE TOTAL PARALLEL CAPACITANCE IN FARADS. ";C
210  INPUT"ENTER THE TOTAL PARALLEL INDUCTANCE IN HENRIES. ";L
220  INPUT"ENTER THE TOTAL PARALLEL RESISTANCE IN OHMS. ";R
225  INPUT"ENTER THE INDUCTOR WINDING RESISTANCE. ";RW
230  F = 1/(6.2831853# * SQR(L * C))
240  XL = 6.2831853# * F * L
242  IF RW = 0 THEN Q = R/XL:GOTO 260
245  QTANK = XL/RW
249  F = F * SQR(QTANK^2/(QTANK^2 + 1))
250  XL = 6.2831853# * F * L
252  ZTANK = XL^2/RW
255  RP = (R * ZTANK)/(R + ZTANK): Q = RP/XL
260  BW = F/Q
270  LF = F - (BW/2):UF = F + (BW/2)
280  PRINT""
290  PRINT"THE RESONANT FREQUENCY IS ";F;" HERTZ."
310  PRINT"THE CIRCUIT Q IS";Q;"."
320  PRINT"THE BANDWIDTH IS ";BW;" HERTZ."
330  PRINT"THE BANDPASS IS APPROXIMATELY ";LF;" HERTZ TO ";UF;" HERTZ."
390  GOTO 30
400  INPUT"ENTER THE DESIRED RESONANT FREQUENCY. ";F
410  INPUT"ENTER THE DESIRED BANDWIDTH. ";BW
420  INPUT"ENTER THE INDUCTOR VALUE IN HENRIES. ";L
425  PRINT"*** THIS DESIGN ASSUMES THE USE OF A HIGH-Q INDUCTOR ***"
430  PRINT""
440  XL = 6.2831853# * F * L
450  Q = F/BW
460  R = XL * Q
470  C = .02533/(F^2 * L)
480  PRINT"THE CAPACITOR MUST BE";C;" FARAD."
490  PRINT"THE TOTAL PARALLEL RESISTANCE MUST BE";R;" OHMS."
510  PRINT"THE CIRCUIT Q IS ";Q
520  GOTO 30
600  CLS:END
```

20-8 Testing and Troubleshooting Resonant Circuits

Measurements With the Oscilloscope

Prediction vs Measurement

Predicting the bandpass, bandwidth, and resonant frequency through calculation and application of theory is one thing, but determining these parameters by actual measurement is quite another. Not that the calculations and theory are wrong. The simple fact is we do not always bring all of the data into light for consideration and calculation when we apply formulas and theory. Consider these examples of overlooked data: stray capacitance, stray inductance, component tolerance, temperature effects, unknown generator resistance, and component Q. If all of these items are known, the theoretical analysis of a series resonant circuit will closely match the actual measurements. But the measurements must be performed carefully, with accurately calibrated equipment. And of course you must know what you are looking for. That's what you will learn here—the hows and whys of test and measurement.

A Measurement Problem

Before we begin our discussion on actual measurement, one very important consideration must be made known: the loading effects of a standard direct-scope probe. The oscilloscope probe, interconnecting coaxial cable, and scope input all have capacitance. The total capacitance of probe, cable, and scope input may be as high as 150 pF. When the probe is connected across a component to measure its voltage, this capacitance is placed in parallel with that component. This added capacitance will change the resonance of the circuit and render your measurements invalid. This probe/scope loading effect is more severe at higher frequencies. At frequencies in the kHz, the probe/scope capacitance is not normally a problem, since this capacitance is relatively small compared to circuit capacitances at these frequencies. At frequencies in the MHz, the probe/scope capacitance becomes very significant, since circuit capacitances are small. Connecting a normal direct probe to test a circuit operating in the MHz will detune the circuit.

A solution to this probe/scope loading problem is the **10:1 low-capacitance scope probe**. The 10:1 probe contains a large internal resistance (9 MΩ) connected directly to the probe tip and in series with the center conductor of the probe cable. This large resistance isolates the cable and scope input capacitance from the circuit under test. As a result, the scope impedance is raised from 1 MΩ to 10 MΩ and the total capacitance is decreased to about 10 pF. The volts-per-division control or controls on most oscilloscopes are calibrated to accommodate the 10:1 probe (some type of dual scaling, labeled 1X and 10X). Since the 10:1 probe has a large series resistance, the maximum voltage-measurement limit of the scope is raised ten times (i.e., instead of a maximum of 20 V/cm, the limit is raised to 200 V/cm). Figure 20-33 shows a typical 10:1 scope test probe.

A Bandpass Output

Figure 20-34 illustrates the use of the oscilloscope in measuring the bandpass and center frequency for resonant circuits. Notice the two ways in which a series circuit may be See color photo insert page D12.

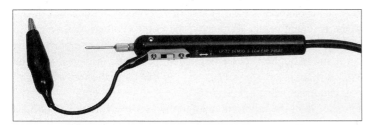

Figure 20-33   A 10:1 low capacitance test probe. (Courtesy of Leader Instruments Corporation, Hauppauge, NY)

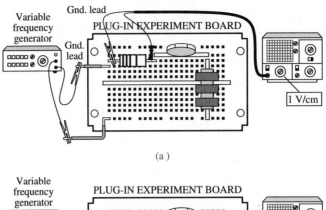

(a)

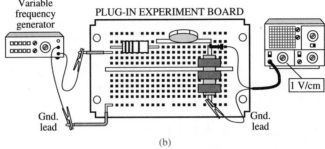

(b)

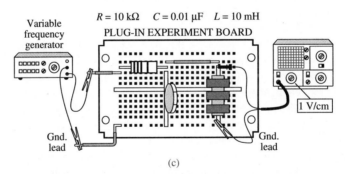

(c)

Figure 20-34   (a) $V_{out} = E_S = V_R$ (b) $V_{out} = V_L = Q_{ckt} \cdot E_S$ = a Q-rise in voltage (c) $V_{out} = V_{tank} = E_S - V_{RS}$.

arranged to produce a characteristic bell-shaped bandpass curve. The output may be taken across the series resistor (20-34a) or across the inductor or capacitor (20-34b). When the output is taken from across the resistor, the output voltage will be maximum and equal to the applied source voltage at resonance. As the generator frequency is increased or decreased above or below the resonant center frequency, the output voltage will decrease. When the output is taken from across one of the reactive components, the output voltage will be very much higher than the applied source voltage at resonance. This is due to the Q rise in voltage discussed earlier ($V_L = V_C = Q_{ckt} \cdot E_S$). In this case, the series resonant circuit not only provides bandpass selection, it also provides voltage gain—since the output voltage is much greater than the input voltage (applied source voltage). Again, as the generator frequency is increased or decreased above or below the resonant center frequency, the output voltage will decrease. Figure 20-34c shows how the generator and oscilloscope are connected to a parallel resonant circuit to observe a bandpass output.

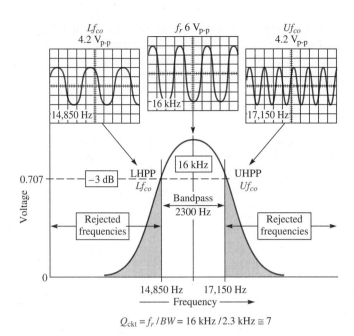

Figure 20-35   Using the oscilloscope to determine f_r, BP, BW, and Q_{ckt}.

Using the Oscilloscope

The inset oscilloscope screens of Figure 20-35 illustrate what you can expect to see as the oscilloscope is used to determine the bandpass and center resonant frequency of the circuit under test. The following procedure should be followed for these measurements:

1. Begin your measurements by determining the resonant frequency. The scope probe should be connected to the output of your circuit and the variable-frequency generator should be connected to its input to supply a voltage over a wide range of frequencies. Slowly increase the generator frequency through the various ranges until a maximum peak-to-peak voltage is seen on the oscilloscope. Carefully rock the frequency-adjust knob back and forth to find the exact center frequency (the frequency of maximum peak-to-peak voltage). Measure the time for one cycle and use this to determine the resonant frequency ($f = 1/T$).

2. Adjust the amplitude control on the frequency generator so the resonant frequency displayed on the scope is at some convenient reference level (a peak-to-peak voltage covering 6 to 8 vertical divisions is fine).

3. Determine the lower bandpass frequency by slowly lowering the generator frequency as you observe the circuit output on the oscilloscope. When the peak-to-peak voltage of the displayed frequency is 0.707 times the reference level (set in step 2), the lower bandpass frequency has been found. This lower bandpass frequency is the lower cutoff frequency (Lf_{co}) at the 0.707 voltage point or -3 dB point. It is also the frequency of the lower half-power point (LHPP). Measure the time for one cycle and find its frequency ($f = 1/T$).

4. Determine the upper bandpass frequency by slowly increasing the generator frequency above the resonant frequency until the displayed peak-to-peak voltage is 0.707 times

the reference level at resonance. Again, measure the time for one cycle and determine its frequency (upper cutoff frequency is at the UHPP, 0.707 point, and -3 dB point).

The bandpass for the resonant circuit lies between the lower cutoff frequency, found in step 3, and the upper cutoff frequency, found in step 4.

Determining BW and Q_{ckt}

With the information gathered using the oscilloscope and frequency generator (signal generator), the bandwidth and actual circuit quality factor can easily be determined. The bandwidth is simply the amount of frequencies that exists between the upper and lower bandpass frequencies (between the Uf_{co} and the Lf_{co}). Thus, the bandwidth is merely the difference between the two.

$$BW = Uf_{co} - Lf_{co} \tag{20.27}$$

Once the actual bandwidth for the circuit is determined, the actual circuit Q can be determined using Formula 20.8 ($Q_{ckt} = f_r/BW$).

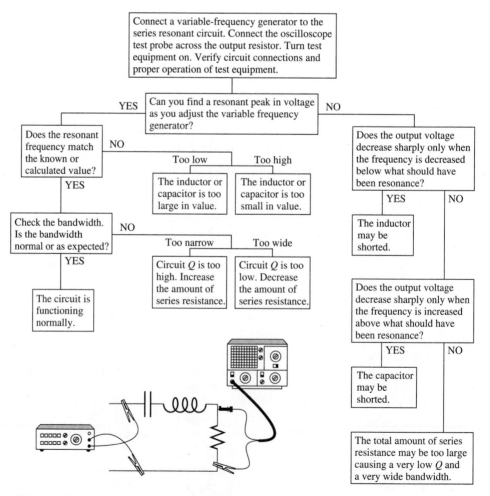

Figure 20-36 The series resonant circuit troubleshooting tree.

Troubleshooting Trees

Take time to study Figures 20-36 and 20-37 as an aid to troubleshooting and a review of series and parallel resonant circuits.

SELF-CHECK 20-8

1. Give at least two reasons why predicted values of bandwidth and circuit Q may not exactly match the actual measured values.

2. What is the advantage of using a 10:1 probe as compared to a regular direct probe?

3. Explain how the oscilloscope and frequency generator are used to determine the upper cutoff frequency for a series resonant circuit.

4. If the bandwidth of a series resonant circuit is too narrow, what might be wrong?

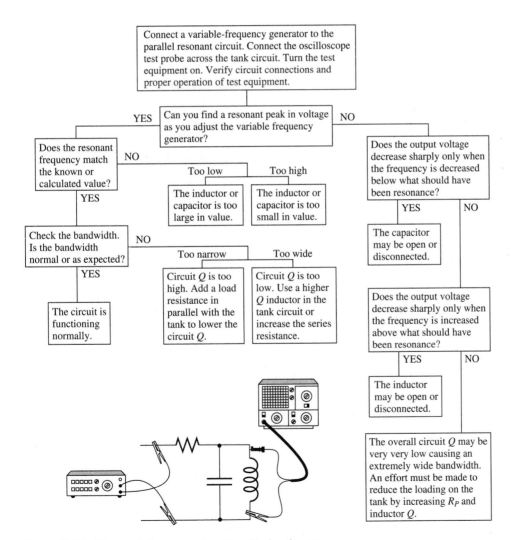

Figure 20-37   The parallel resonant circuit troubleshooting tree.

20-9 Comparing Series and Parallel Resonance

As we conclude this chapter, let's take a side-by-side look at the parameters and characteristics of series and parallel resonant circuits. Figure 20-38 will serve as a practical review and summary.

SELF-CHECK 20-9

1. List three major series and parallel resonance characteristics that are the same.

2. List three major series and parallel resonance characteristics that are opposite.

SERIES RESONANCE

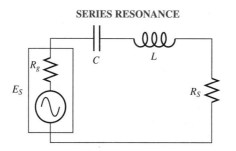

PARALLEL RESONANCE

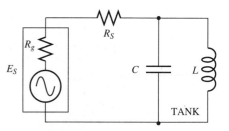

Resonance characteristics

IDEALLY: *L* AND *C* ACT LIKE A SHORT

$X_C = X_L$
$Z_t = R_S$ = minimum
$I_t = I_C = I_L$ = maximum
$V_C = V_L$ = maximum
V_{CL} = minimum (ideally 0 V)
$E_S = V_{RS}$ = maximum
$P_R = P_A = I_t \cdot E_S$
Power factor = 1
Phase angle = $\angle\theta = 0°$
The generator resistance
and the series resistance
must be as low as
possible for the highest
possible circuit Q.

Resonance characteristics

IDEALLY: THE TANK ACTS LIKE AN OPEN

$X_C = X_L$
$Z_t = R_S + Z_{tank}$ = maximum
$I_t = I_{line}$ = minimum
$I_C = I_L = I_{tank}$ = maximun
V_{tank} = maximum (ideally = E_S)
$E_S = V_{RS} + V_{tank}$
$P_R = P_A = I_t \cdot E_S$
Power factor = 1
Phase angle = $\angle\theta = 0°$
The generator resistance
and the series resistance
must be as high as
possible for the highest
possible circuit Q.

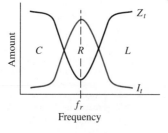

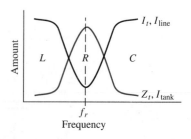

Figure 20-38   Comparing series and parallel resonance characteristics.

Summary

FORMULAS

(20.1) $f_r = f_c = f_o = \dfrac{1}{2\pi\sqrt{LC}}$

where the resonant frequency f_r is the same as the center frequency f_c, which is the same as the frequency of operation f_o.

(20.2) $C = \dfrac{1}{4 \cdot \pi^2 \cdot f_r^2 \cdot L}$

(20.3) $L = \dfrac{1}{4 \cdot \pi^2 \cdot f_r^2 \cdot C}$

(20.4) $Q_{ckt} = X_L/R_t = X_C/R_t$ (no units)

(20.5) $Q_L = X_L/R_W$

where X_L is determined using the resonant frequency.

(20.6) $Q_L = 1/\cos\angle\theta = 1/PF = 1/(P_R/P_A)$
$= 1/(R_W/Z_t) = 1/(R_W/X_L) = X_L/R_W$

(*Note:* Z_t and X_L are considered to be equal, since X_L is extremely large compared jto R_W.)

(20.7) $Q_{ckt} = V_C/E_S = V_L/E_S$ and $V_C = V_L = Q_{ckt} \cdot E_S$

(20.8) $BW = f_r/Q_{ckt}$

also: $f_r = BW \cdot Q_{ckt}$ and $Q_{ckt} = f_r/BW$
where Q_{ckt} for a series resonant circuit is X_L/R_t or X_C/R_t.

(20.9) $BP \cong [f_r - (BW/2)] \to [f_r + (BW/2)]$ (only for $Q > 10$)

(20.10) $Z_{tank} \cong X_L^2/R_W \cong X_C^2/R_W$

(20.11) $Z_{tank} \cong X_L^2/R_W = X_L \cdot Q_L = X_C \cdot Q_L$

where X_L or X_C is determined at the resonant frequency.

(20.12) $I_{tank} \cong V_{tank}/X_C \cong V_{tank}/X_L$

[The inductor branch impedance is actually $R_L + jX_L$. The inductor resistance is ignored for high-Q inductors (≥ 10).]

(20.13) $I_{line} = V_{tank}/Z_{tank}$

(20.14) $I_t = I_R + I_{line}$

(20.15) $Z_t = E_S/I_t$

(20.16) $Z_t = \dfrac{R_P \cdot Z_{tank}}{R_P + Z_{tank}}$ (product over the sum)

where R_P is a resistor in parallel with the tank.

(20.17) $Z_t = R_S + Z_{tank}$

where R_S is a resistor in series with the tank.

(20.18) $I_t = I_{line} = E_S/Z_t$

(20.19) $V_{RS} = I_t \cdot R_S = E_S \cdot R_S/Z_t$

(20.20) $V_{tank} = I_t \cdot Z_{tank} = E_S \cdot Z_{tank}/Z_t$

(20.21) $Q_{tank} = Q_L = X_L/R_W \cong X_C/R_w$

(20.22) $Q_{\text{tank}} = Z_{\text{tank}}/X_L \cong Z_{\text{tank}}/X_C$

(20.23) $Q_{\text{tank}} = I_{\text{tank}}/I_{\text{line}}$

(20.24) $Q_{\text{ckt}} \cong R_P/X_L \cong R_P/X_C$ (for $Z_{\text{tank}} \geq 10 \cdot R_P$)

(20.25) $Q_{\text{ckt}} = (R_P\|Z_{\text{tank}})/X_L \cong (R_P\|Z_{\text{tank}})/X_C$ (for $Z_{\text{tank}} < 10 \cdot R_P$)

where $Z_{\text{tank}} = X_L{}^2/R_W$ and $R_P\|Z_{\text{tank}} = (R_P \cdot Z_{\text{tank}})/(R_P + Z_{\text{tank}})$

(20.26) $f_r = \dfrac{1}{2\pi\sqrt{LC}} \cdot \sqrt{Q^2/(Q^2 + 1)}$ (for $Q_{\text{tank}} < 10$) where $Q = Q_{\text{tank}}$.

(20.27) $BW = Uf_{co} - Lf_{co}$

CONCEPTS

- For series resonance: $X_L = X_C$; $V_L = V_C$; $E_S = V_R$; $Z_t = R =$ minimum; $I_t = E_S/R =$ maximum; $P_A = P_R$; $\angle\theta = 0°$; $PF = 1$; % efficiency = 100%.
- For series resonance: $V_{LC} = 0$ V—the inductor and capacitor combined act as a short.
- The larger the values of capacitance and inductance, the lower the resonant frequency.
- The smaller the values of capacitance and inductance, the higher the resonant frequency.
- Bandwidth and Q are inversely proportional. If the circuit Q is high, bandwidth will be narrow and vice versa.
- A high-Q inductor has low-resistance windings—few turns, a coil of large diameter, plated with silver to reduce the skin effect.
- A Q rise in voltage occurs across the inductor and capacitor of a series RCL circuit at resonance.
- The bandpass frequencies tell you how wide and where the passed or rejected band of frequencies lie.
- The bandpass frequencies are also called the lower and upper cutoff frequencies (Lf_{co} and Uf_{co}) since they define the boundaries of selection or rejection.
- The cutoff frequencies are at the 0.707 voltage points, also known as the -3 dB points and the lower and upper half-power points (LHPP and UHPP).
- Actual circuit measurements may not match the predicted values, due to stray inductance and capacitance, component tolerance, temperature effects, unknown generator impedance, component Q, equipment accuracy, and type of scope test probe.
- The line current supplied to a ringing tank prevents the tank current from damping out. The resistance of the tank circuit (mainly the inductor) damps the ringing.
- At resonance, capacitor and inductor tank currents are equal and 180° out of phase.
- The impedance of the tank is maximum and looks resistive at resonance.
- Line current, supplying the tank, is minimum at resonance.
- The impedance of the tank and line current depend greatly on inductor resistance.
- The quality factor of the tank depends mainly on the Q of the inductor (X_L/R_W).
- The quality factor of the entire parallel resonant circuit determines the operating bandwidth of the circuit.
- The circuit Q is determined by the equivalent parallel resistance and the tank impedance. The higher the parallel resistance and tank impedance, the higher the circuit Q.
- Resistance in series with the tank should be considered as a parallel resistance for the purpose of circuit Q calculations only.
- The generator resistance cannot be ignored if the analysis of a parallel resonant circuit is to be accurate. Ideally, the generator resistance should be infinite for the highest possible circuit Q.

- Changing L or C will change the resonant frequency of a tank.
- Changing R_P or R_W will change the circuit Q and bandwidth.

SPECIAL TERMS

- Resonate, tuned circuit, tunable circuit
- Quality factor (Q), high-Q inductor, high-Q capacitor
- Skin effect
- Q rise in voltage
- Bandwidth (BW), bandpass (BP), cutoff frequency (f_{co})
- -3 dB point
- Lower half-power point (LHPP)
- Upper half-power point (UHPP)
- 10:1 low-capacitance scope probe
- Parallel resonance
- Ringing, damped waveform
- Tank circuit
- Tank impedance (Z_{tank}), tank current (I_{tank})
- Line current (I_{line})
- Tank quality factor (Q_{tank})

Need to Know Solution

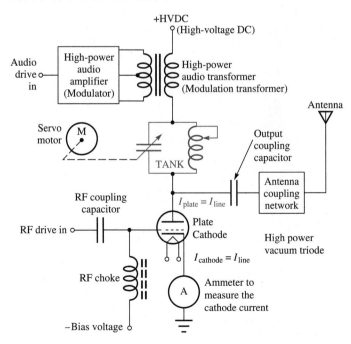

The above diagram is a typical final power amplifier for an AM broadcast transmitter. When you tune by hand, you manually adjust the plate-tuning capacitor to bring the plate tank circuit into resonance at the transmitter frequency. The ammeter that you must observe is in the cathode circuit as shown. The tank line current must go through the meter, through the high power triode, to the tank. The pointer on this ammeter must be dipped as you tune for resonance. Recall that the line current is minimum at resonance.

So, wipe the sweat from your chin for the last time and dip that meter. Bring the power up slowly and rock the belt-driven pulley back and forth to insure the tank is properly tuned. The indicated current should be at a minimum. OK, now bring the transmitter up to full power and apply modulation (apply the audio). You did it! You're a big shot now!

Questions

20-1 Series Resonance and Applications

1. What does the word *resonate* mean?
2. What determines the resonant frequency for any series resonant circuit?
3. What is the difference between a *tuned* circuit and a *tunable* circuit?
4. Describe an application for series resonance in which a band of frequencies is selected.
5. What is a trap?

20-2 Series Resonance Characteristics

6. What is the phase relationship between inductor and capacitor voltage in a series resonant circuit?
7. What is the total reactance (X_{LC}) of a series *RCL* circuit at resonance? Explain.
8. At resonance, would you expect to measure (a) a large voltage or (b) a very small voltage across the inductor and capacitor combined (V_{LC})? Explain.
9. Is the current in a series resonant circuit (a) minimum or (b) maximum (at resonance)? Explain.
10. What is the % efficiency of an *RCL* circuit at resonance? Explain.
11. Draw a graph that would typically represent the total impedance of a series *RCL* circuit below, at, and above resonance.
12. Explain why the inductor voltage decreases as the generator frequency is increased above resonance, even though the inductive reactance increases.

20-3 Calculations for Series Resonance

13. What is a "figure of merit"?
14. What is the relationship between bandwidth and Q?
15. What two factors determine the Q of a circuit?
16. What determines the Q of an inductor?
17. What is skin effect and what does it have to do with the Q of an inductor?
18. List three techniques used in making high-Q inductors.
19. What does the ratio of inductance to capacitance have to do with circuit Q?
20. Explain how total circuit resistance affects bandwidth.
21. Why is the upper cutoff frequency defined as the upper half-power point?

20-4 Parallel Resonance and Applications

22. Give two general purposes for the parallel resonant circuit.
23. List three different pieces of electronic equipment that use the parallel resonant circuit.
24. Briefly explain how an amplifier can amplify one frequency and reject all others.

25. What is an IF transformer? What is it used for?
26. How does the parallel resonant circuit act at resonance?

20-5 Parallel Resonance Characteristics

27. What is the meaning of the term *ringing* in regard to parallel resonant circuits?
28. What causes the ringing parallel resonant circuit to become damped?
29. Ideally, how should the parallel resonant circuit look (or act) at resonance?
30. How much line current would there be at resonance if the inductor and capacitor had no resistance or losses?
31. List six characteristics, or parameter relationships, for the parallel resonant circuit.

20-6 Basic Circuit Calculations for Parallel Resonance

32. Why is the parallel resonant circuit referred to as a *tank*, or *tank circuit*?
33. What is meant by a Q rise in impedance?

20-7 Calculations for Parallel Resonance

34. When should the parallel resonant-frequency correction factor definitely be used?

20-8 Testing and Troubleshooting Resonant Circuits

35. List four reasons why actual measured values of BP, BW, and Q_{ckt} may not match the predicted, or calculated, values.
36. Explain what will occur if a normal direct-scope probe is used to measure the voltage across a 200 pF capacitor.
37. What is the purpose for a 10:1 low-capacitance probe?
38. What is inside a 10:1 test probe?
39. Explain how the resonant frequency of a series RCL circuit can be determined using a variable-frequency generator and an oscilloscope.
40. Explain how the lower cutoff frequency of a series RCL circuit is determined using a variable-frequency generator and an oscilloscope.
41. If the actual resonant frequency of a series RCL circuit is measured and found to be too high, what could be the trouble?
42. If the actual bandwidth of a series RCL circuit is too wide, what is the trouble?
43. How many dB down is the upper cutoff frequency, compared to the resonant frequency?
44. What might be wrong with a parallel resonant circuit if its operating bandwidth is too narrow?
45. If the tank voltage does not decrease below the supposed resonant frequency but only decreases above the supposed resonant frequency, what might be wrong?

20-9 Comparing Series and Parallel Resonance

46. Ideally, the series LC resonant circuit acts like a(n) (short, open), while the tank circuit acts like a(n) (short, open).
47. Explain why the total impedance of a series resonant circuit is very low while the total impedance of a parallel resonant circuit is very high.
48. Should the generator resistance for a series resonant circuit be (a) high, or (b) low? Explain.

49. Should the generator resistance for a parallel resonant circuit be (a) high, or (b) low? Explain.
50. List three characteristics of series and parallel resonant circuits that are the same.

Problems

20-3 Calculations for Series Resonance

1. Given $L = 25$ μH and $C = 300$ pF, find f_r.
2. Given $L = 0.2$ μH and $C = 10$ pF, find f_r.
3. Given $L = 5$ mH and $C = 0.002$ μF, find f_r.
4. Given $L = 0.6$ H and $C = 1$ μF, find f_r.
5. Given $L = 130$ μH and $f_r = 6.4$ MHz, find C.
6. Given $L = 8$ μH and $f_r = 28$ MHz, find C.
7. Given $L = 3$ H and $f_r = 50$ Hz, find C.
8. Given $C = 0.022$ μF and $f_r = 6$ kHz, find L.
9. Given $C = 0.47$ μF and $f_r = 0.75$ kHz, find L.
10. Given $L = 1$ mH, $C = 0.003$ μF, and $R = 30$ Ω, find Q_{ckt}.
11. Given $L = 170$ μH, $C = 0.001$ μF, and $R = 200$ Ω, find Q_{ckt}.
12. Given $L = 150$ mH, $C = 2.2$ μF, and $R = 20$ Ω, find Q_{ckt}.
13. If a capacitor has a power factor of 0.0005, what is its Q? Explain.
14. If the voltage across the capacitor of a series resonant circuit is 1.58 V and the source voltage is 40 mV, what is the circuit Q?
15. If 100 mV is applied to a series resonant circuit that has a Q of 35, how much voltage will there be across the indicator?
16. Given $L = 40$ μH, $C = 200$ pF, and $R = 10$ Ω, find Q_{ckt} and BW.
17. Given $L = 0.24$ μH, $C = 4$ pF, and $R = 20$ Ω, find Q_{ckt} and BW.
18. Given $L = 20$ mH, $C = 0.022$ μF, and $R = 100$ Ω, find Q_{ckt}, BW, and BP.
19. Given $L = 0.9$ mH, $C = 0.003$ μF, and $R = 50$ Ω, find Q_{ckt}, BW, and BP.
20. If the output voltage at resonance is 20 $V_{p\text{-}p}$, what is the output voltage when the lower cutoff frequency is applied to the circuit? (assume the input voltage remained constant)
21. Given $L = 0.5$ mH, $R_W = 0.2$ Ω, $f_r = 3$ kHz and $BW = 200$ Hz, find C, Q_{ckt}, and R_t (*Note: R_t includes R_W and any other series resistance*).
22. Given $L = 60$ μH, $R_W = 0.1$ Ω, $f_r = 4$ MHz and $BW = 10$ kHz. Find C, Q_{ckt}, and R_t (*Note: R_t includes R_W and any other series resistance*).

20-6 Basic Circuit Calculations for Parallel Resonance

23. Determine the impedance of a tank that has a capacitive reactance of 600 Ω and an inductor resistance of 2.5 Ω.
24. Determine the impedance of a tank that has a capacitive reactance of 1.5 kΩ and an inductor resistance of 1.3 Ω.
25. Calculate the tank current when the tank voltage is 20 V, the inductive reactance is 750 Ω, and the inductor resistance is 5 Ω.
26. Calculate the line current for the tank described in Problem 25.
27. Calculate the tank current when the tank voltage is 150 mV, the inductive reactance is 80 Ω, and the inductor resistance is 0.5 Ω.
28. Calculate the line current for the tank described in Problem 27.

29. Given a tank circuit, with a parallel resistance, operating at resonance, and $X_C = 250\ \Omega$, $R_W = 2\ \Omega$, $R_P = 100\ k\Omega$, and $E_S = 10V$, find Z_{tank}, I_{line}, I_t, and Z_t (see Example 20-4).

30. Given a tank circuit, with a parallel resistance, operating at resonance, and $X_C = 440\ \Omega$, $R_W = 3\ \Omega$, $R_P = 270\ k\Omega$, and $E_S = 25\ mV$, find Z_{tank}, I_{line}, I_t, and Z_t (see Example 20-4).

31. Given a tank circuit, with a series resistance, operating at resonance, and $X_C = 1\ k\Omega$, $R_W = 13\ \Omega$, $R_S = 47\ k\Omega$, and $E_S = 100\ mV$, find Z_{tank}, Z_t, I_t, V_{tank}, and V_{RS} (see Example 20-5).

32. Given a tank circuit, with a series resistance, operating at resonance, and $X_C = 40\ \Omega$, $R_W = 0.08\ \Omega$, $R_S = 33\ k\Omega$, and $E_S = 1\ V$, find Z_{tank}, Z_t, I_t, V_{tank}, and V_{RS} (see Example 20-5).

20-7 Calculations for Parallel Resonance

33. Given a tank circuit with a parallel resistance operating at resonance and $X_C = 620\ \Omega$, $R_W = 5\ \Omega$, $R_P = 100\ k\Omega$, and $R_g = 250\ k\Omega$, find Q_{tank} and Q_{ckt} (see Example Set 20-7).

34. Given a tank circuit with a parallel resistance operating at resonance and $X_C = 30\ \Omega$, $R_W = 0.06\ \Omega$, $R_P = 150\ k\Omega$, and $R_g = 150\ k\Omega$, find: Q_{tank} and Q_{ckt} (see Example Set 20-7).

35. Given a tank circuit with a series resistance operating at resonance and $X_C = 65\ \Omega$, $R_W = 1\ \Omega$, $R_S = 10\ k\Omega$, and $R_g = 600\ \Omega$, find Q_{tank} and Q_{ckt} (see Example Set 20-7).

36. Given a tank circuit with a series resistance operating at resonance and $X_L = 95\ \Omega$, $R_w = 1.8\ \Omega$, $R_S = 22\ k\Omega$, and $R_g = 75\ \Omega$, find Q_{tank} and Q_{ckt} (see Example Set 20-7).

37. Given a tank circuit with the values $C = 48\ pF$, $L = 12\ \mu H$, and $R_W = 0.1\ \Omega$, find f_r.

38. Given a tank circuit with the values $C = 200\ pF$, $L = 40\ \mu H$, and $R_W = 0.7\ \Omega$, find f_r.

39. Given a tank circuit with the values $C = 0.001\ \mu F$, $L = 250\ \mu H$ and $R_W = 70\ \Omega$, find f_r.

40. Given a parallel resonant circuit with a series resistance operating at resonance and $L = 0.5\ mH$, $C = 0.01\ \mu F$, $R_W = 5\ \Omega$, $R_S = 10\ k\Omega$, and $R_g = 10\ \Omega$, find f_r, Q_{ckt}, BW, and BP (see Example Set 20-7 and Example 20-10).

41. Given a parallel resonant circuit with a series resistance operating at resonance and $L = 0.1\ mH$, $C = 0.022\ \mu F$, $R_W = 1.3\ \Omega$, $R_S = 27\ k\Omega$ and $R_g = 3\ k\Omega$, find f_r, Q_{ckt}, BW, and BP (see Example Set 20-7 and Example 20-10).

42. Given a parallel resonant circuit with a parallel resistance operating at resonance and $L = 1\ mH$, $C = 0.047\ \mu F$, $R_W = 30\ \Omega$, $R_P = 1\ k\Omega$ and $R_g = 1\ k\Omega$, find f_r, Q_{ckt}, BW, and BP (see Example Set 20-7 and Example 20-10).

43. List two ways in which the circuit Q for the circuit of Problem 42 can be increased.

44. What is the tuning range of a 50→200 pF capacitor that is placed in parallel with a 40 μH inductor?

45. What is the tuning range of a 12→20 pF capacitor that is placed in parallel with a 0.5 μH inductor?

46. What is the range of capacitance needed to resonate with a 10 μH inductor over a frequency range of 15→17 MHz?

47. What is the range of capacitance needed to resonate with a 350 μH inductor over a frequency range of 300→380 kHz?

Answers to Self-Checks

Self-Check 20-1

1. Resistively
2. To respond in such a way as to reproduce or echo a stimulus
3. The value of inductance and capacitance
4. One that includes a variable capacitor and/or variable inductor so the circuit can be tuned to a wide range of resonant frequencies
5. To select a band of frequencies, as in the front end of a radio receiver, and to eliminate a band of frequencies, such as the trap at the output of a broadcast transmitter

Self-Check 20-2

1. X_L and X_C cancel each other, since they are equal in magnitude and opposite in phase.
2. Because the total circuit impedance is minimum at resonance
3. The total current decreases at a much higher rate than the inductive reactance increases.

Self-Check 20-3

1. 4.91 MHz
2. Resonant frequency decreases, since it is inversely related to L and C.
3. 0.01 μF
4. 48
5. Large-conductor diameter, large-coil diameter with fewer turns, with silver plated windings
6. 5,000
7. 17.5
8. $f_r = 1.007$ MHz; $Q = 52.7$; $BW = 19,100$ Hz; $BP \cong 997$ kHz to 1,017 kHz

Self-Check 20-4

1. Resistively, due to reactance cancellation
2. To select or reject a narrow band of frequencies
3. Intermediate-frequency transformer used in radio receivers for selectivity
4. Transmitters, AM and FM radio receivers, televisions

Self-Check 20-5

1. Current is being exchanged back and forth between the inductor and capacitor in a parallel resonant circuit, producing a sinusoidal voltage and current waveform.
2. A waveform that decreases in amplitude with each cycle
3. The resistance in the parallel resonant circuit
4. Minimum, since $I_C = I_L$ and only a small amount of current is needed to sustain ringing

Self-Check 20-6

1. $Z_{\text{tank}} = 62.5$ kΩ; $I_{\text{tank}} = 40$ mA; $I_{\text{line}} = 320$ μA; $I_t = 2.32$ mA
2. $Z_t = 72.5$ kΩ; $I_t = 276$ μA; $V_{\text{tank}} = 17.25$ V; $I_{\text{tank}} = 34.5$ mA; $V_{RS} = 2.76$ V

Self-Check 20-7

1. 167
2. 36 mA
3. Inductor resistance, reactance, equivalent parallel resistance
4. $Q_{ckt} = (R_P \| Z_{tank})/X_L$; the tank impedance is not 10 times the parallel resistance.
5. I_C and I_L
6. 4.061 MHz
7. $f_r = 225.08$ kHz; $Q_{ckt} = 70.7$; $BW = 3{,}183$ Hz; $BP = 223{,}488 \rightarrow 226{,}670$ Hz
8. $18 \rightarrow 20.65$ pF

Self-Check 20-8

1. Stray capacitance and component tolerance
2. The 10:1 probe has far less capacitance than a direct probe—the probe capacitance changes the circuit operating conditions.
3. The frequency of the generator is increased above resonance until a frequency is reached whose peak-to-peak voltage is 0.707 times the maximum voltage measured at resonance.
4. The circuit Q is too high—it can be reduced by increasing the total series resistance.

Self-Check 20-9

1. $X_C = X_L$; $P_A = P_R$; $\angle \theta = 0°$
2. At resonance: for parallel, Z_t is maximum, I_t is minimum, ideally the tank looks like an open; for series, Z_t is minimum, I_t is maximum, ideally it looks like a short.

SUGGESTED PROJECTS

1. Add some of the main concepts, formulas, and procedures from this chapter to your Electricity and Electronics Notebook.

2. Construct and test the circuit of Figure 20-39. Make a graph of the output voltage. Determine the bandwidth and actual circuit Q. Compare your measurements to calculated values. An analog AC voltmeter can be used to set the input voltage and measure the output voltage at different frequencies. Take enough output measurements at different frequencies until you are able to draw a reasonably accurate frequency-response curve.

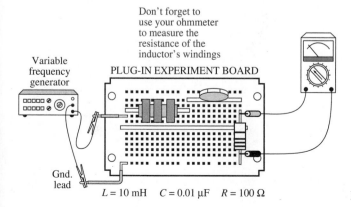

$L = 10$ mH $C = 0.01$ μF $R = 100\ \Omega$ **Figure 20-39**

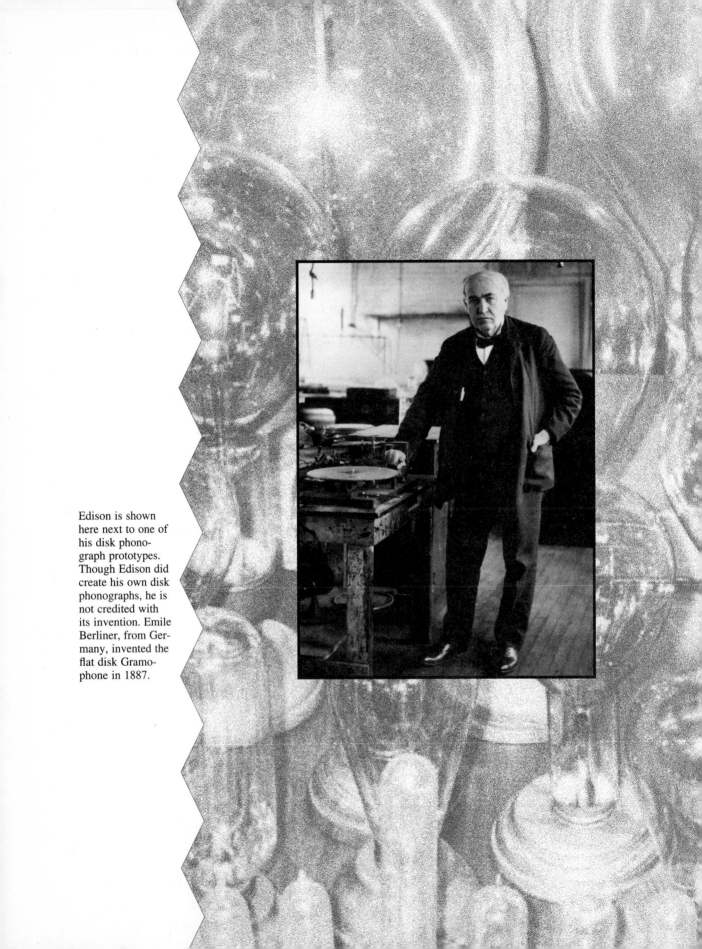

Edison is shown here next to one of his disk phonograph prototypes. Though Edison did create his own disk phonographs, he is not credited with its invention. Emile Berliner, from Germany, invented the flat disk Gramophone in 1887.

Chapter 21

Filters and Their Applications

CHAPTER OUTLINE

OBJECTIVES

After studying this chapter, you will be able to

- identify filters by type and configuration.
- discuss and analyze filter types in terms of frequency response, phase response, insertion loss, and rolloff slope.
- design many useful filters.

- list many practical applications for each of the basic filter types.
- predict and plot frequency response for many common filter types using the insertion loss formula.

INTRODUCTION

In this chapter, you will investigate electrical filters while applying much of the circuit theory covered in previous chapters. You should find this chapter interesting and useful since filters are examined in a very practical way with an emphasis on analysis and design. Here you will learn how filters are able to separate AC and DC, high frequencies and low frequencies, and a small band of frequencies from the frequency spectrum. You will discover that the filter's output voltage and phase angle are a function of frequency, and you will be able to analyze filters with this in mind. You will also be able to design practical filters for many real-world applications.

Though this chapter does go beyond the usual superficial introduction of filters, it is important for you to understand that it is impossible to present a thorough coverage in one chapter. The subject of electrical filters is very broad and can be very complex. Therefore, the purpose for this chapter is to give you an introductory knowledge of filters that can be built upon as needed. Much of the complex mathematical analysis has been avoided here so as not to cloud the overall practical intent of the material. To reinforce the practical side, this chapter emphasizes common filter terminology, filter characteristics, and analysis and design formulas.

NEED TO KNOW

Consider these practical situations:

1. Your new car radio is plagued with alternator whine when the engine is running. The whine is high frequency noise that is coming in on the $+\mathrm{VDC}$ supply line to the radio. Describe a filter that will effectively purify the DC and eliminate the whine.

2. Your CB radio transmits a clean-sounding signal when it is in your car but transmits a loud 120 Hz hum when it is in your house operating from an AC–DC converter. What might be wrong and how can it be corrected?

3. Your computer produces beautiful synthesized music but you notice an annoying low-frequency hum as you listen to the computer-generated music through your hi-fi sound system (the hum is generated inside the computer). Describe a filter that can be inserted between the computer and your hi-fi system to eliminate the hum.

4. You decide you want to add highrange audio-frequency tweeters to your stereo system. The midrange and lowrange audio frequencies must be prevented from getting to the tweeters lest they be damaged. Describe a filter that can be used to accomplish this.

These situations are practical applications for filters and help establish your need to know. The solution to each situation is based upon a good understanding of filter types and characteristics. When you finish studying this chapter, you should be able to offer solutions for each of these situations. As always, solutions are available at the end of the chapter.

21-1 Introduction to Filters

Frequency-Selective Filters

In this chapter, we take a new look at *RL*, *RC*, and *RCL* circuits functioning as frequency-selective filters. Here, you will realize the practical application of the AC circuit-analysis concepts that you have accumulated over the past ten chapters. A full understanding of frequency-selective filters will largely depend on your ability to understand and use decibels, apply phasor-analysis principles, apply X_C and X_L formulas, and relate to series and parallel resonance characteristics.

The frequency-selective filters covered in this chapter are called **passive filters**. They are made of resistors, inductors, and capacitors. There are no active devices directly involved in, and affecting the characteristics of, these filter circuits (i.e., transistors, FETs, integrated circuits, etc.) However, passive filters are used in circuits along with active devices. As the various types of passive filters are presented, you will be introduced to many practical applications in which these filters work together with active devices. Filters that include active devices that affect the overall operating characteristics of the filter are appropriately named **active filters**. However, the coverage of active filters is beyond the purpose of this text.

Basic Filter Types

I am sure the concept of filtering is not foreign to you. You already understand the purpose for any kind of filtering is to separate. In the realm of electricity and electronics, frequencies are separated. The frequency-selective filter separates low frequencies from high frequencies, highs from lows, AC from DC, and so on. Electrical filters are **categorized** in four basic **types**: **highpass**, **lowpass**, **bandpass**, and **bandstop**. The **highpass** filters are able to pass all frequencies above a specific cutoff frequency and reject, or greatly attenuate, all frequencies below cutoff. **Lowpass** filters are able to pass all frequencies below a specific cutoff frequency (all the way down to DC) and reject all frequencies above cutoff. **Bandpass** filters select out a group, or band, of frequencies to be passed on to another circuit while rejecting all frequencies above and below the desired band. Finally, **bandstop** filters, sometimes called **bandreject** filters, have the ability to reject a specific band of frequencies while passing all frequencies above and below the rejected band. Each of these filter types is examined in detail in the sections to follow.

Basic Filter Configurations

Filters are categorized not only by type but also by configuration. **Filter configuration** is the basic overall arrangement of the passive components. There are three basic filter configurations: inverted L, T, and π. For the inverted L configuration, the components are arranged in the shape of an inverted L. Likewise, components are arranged in the shape of a T or π as shown in Figure 21-1. The use of the different configurations depends largely upon the particular application. One configuration will pass DC, while another will block DC, or it may be necessary for the filter to provide a capacitive or inductive load at the input and output. Again, the application defines the configuration to be used. We will not take the time here to consider the appropriate use of the various configurations. The choice of configuration will in most cases be obvious to you as you ponder a particular application.

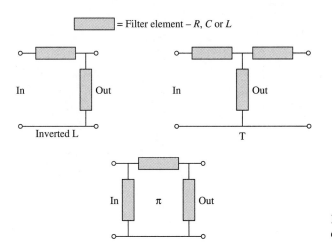

Figure 21-1 Basic filter configurations.

Filter Frequency Response

The most significant characteristic of any filter is its frequency response. The filter's **frequency response** is simply the filter's output voltage as a function of frequency. As the frequency applied to the filter changes, the voltage at the output of the filter changes. One of the key factors that characterizes a filter's frequency response is its cutoff frequency (or cutoff frequencies as in the case of bandpass and bandstop filters). The **cutoff frequency** marks the dividing line between frequencies that are passed and those that are rejected. It is defined as the frequency whose voltage (or, in some cases, current) is 0.707 times the maximum filter output voltage. It is also referred to as the **half-power frequency (HPF)** or the **half-power point (HPP)**, since 70.7% voltage across a load device produces only half the maximum power in the device (see Section 20-3).

The frequency response of a filter can be graphed to demonstrate the filter's operational characteristics such as: cutoff frequency and skirt slope. Figure 21-2 illustrates graphs representing the four filter types. In each case, the cutoff frequencies define the **passband** and **stopband** regions. The sloping curves of the graph in the stopband regions are called **skirts**; they roll off at a specific rate, or **slope**. The frequencies in the stopband regions are attenuated at some specific rate that is characteristic for each particular filter type and configuration. The rate of attenuation (slope) is expressed in units of decibels per octave (dB/octave) or decibels per decade (dB/decade). An octave is either double or half a reference frequency, and a decade is either 10 times or 1/10 a reference frequency (i.e., an octave is from 100 Hz to 200 Hz and a decade is from 100 Hz to 1,000 Hz). The rate of attenuation (slope) on the skirt for any filter is known as the filter's **rolloff slope** or **skirt slope**. We will make use of these terms when describing the performance of each filter type in the sections to follow.

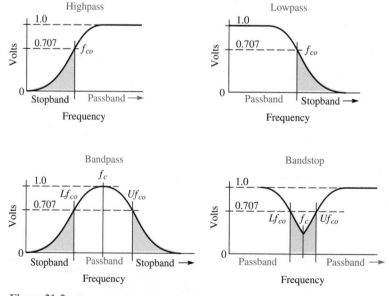

Figure 21-2 Frequency response graphs.

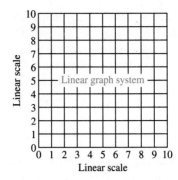

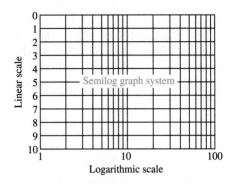

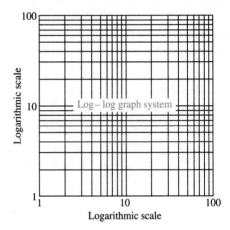

Figure 21-3 Types of graph systems.

The frequency response of a filter may be graphed using a variety of grids, or graph systems. As shown in Figure 21-3, the most common **graph systems** are **linear, semilog**, and **log–log**. Any of these may be used to plot the frequency response of a filter. However, the semilog system is most often used, since it provides a very wide frequency range along the horizontal axis. Each major division along this logarithmic axis is a decade in frequency ($10\times$ or an order of magnitude, i.e., 100 Hz, 1 kHz, 10 kHz, etc.). The vertical axis of a semilog system is linear, having evenly spaced divisions. This axis is most often scaled in decibels. Thus it is very easy to determine a filter's rolloff from a frequency-response graph plotted on semilog paper (semilog system) since rolloff slope can be expressed in dB/decade. When a semilog system like this is used, the plotted frequency response is known as a **Bode plot**.

Filter Phase Response

Another important characteristic of a filter is its **phase response**. Not only does the filter affect its output-voltage amplitude at each frequency, it also affects the voltage phase angle at each frequency. The voltage phase angle at the output of the filter, with respect to the filter's input voltage, is a function of frequency. As the frequency of the filter input voltage changes, the phase angle of the filter output voltage changes. Thus, voltage amplitude and phase angle change with frequency. Remember also that phase angle means time. Thus, a phase shift is a time delay. Therefore, it is important for you to realize that filters not only select and reject frequencies, they also introduce time delays through phase shift. Both of these important characteristics are demonstrated in the following sections which cover the four basic filter types.

Filter Insertion Loss

When a filter is inserted between a voltage source and a load, there is a certain amount of loss in output voltage and power referred to as **insertion loss**. The amount of loss, or attenuation, depends on filter design and the frequency of the AC voltage applied to the filter. It is important to know how much signals are attenuated by the filter. Therefore, insertion loss is an important figure of merit for filters. The insertion loss is measured in decibels and can be calculated using input and output, voltage or power levels. For our purposes here, we will use the output/input voltage levels and the voltage dB formula (see Section 15-5).

$$dB_{atten.} = 20 \cdot \log(V_{out}/V_{in}) \qquad \text{(insertion loss formula)} \qquad (21.1)$$

where V_{out}/V_{in} is a voltage divider expression for the particular filter.

For any filter, a voltage divider expression must be formulated to take the place of V_{out}/V_{in} in Formula 21.1. The voltage divider expression has a different value for every frequency, and a different amount of attenuation is calculated for every frequency. The Bode plot is actually a plot of the insertion loss of the filter over a wide range of frequencies. To make a theoretical Bode plot, you must derive a voltage divider expression for the filter. Then insert the expression in Formula 21.1 so the insertion loss can be calculated and plotted for any desired frequency.

S E L F - C H E C K   2 1 - 1

1. List the components used for passive filters.

2. List the four main filter types.

3. What is a T configuration?

4. What is a Bode plot?

5. What is insertion loss?

21-2 Highpass Filters

In this section, we examine common configurations of highpass filters. You will see how inductors or capacitors or both may be used to pass frequencies above a specified cutoff frequency and attenuate all frequencies below cutoff.

RL Highpass Filters

RL Highpass Frequency Response

Figure 21-4 illustrates the use of resistance and inductance in three highpass filter configurations (inverted L, T, and π). Notice, from the frequency response graph, that all frequencies above the cutoff frequency fall in the **passband**. This means the filter output voltage is greater than 0.707 times the input voltage for all frequencies above the cutoff frequency (for passband frequencies: $V_{out} > 0.707 \cdot E_S$). Those frequencies that are below the cutoff frequency fall in the **stopband** and are attenuated to a greater or lesser degree, depending on the frequency's distance below the cutoff frequency due to the frequency-dependent X_L ($\downarrow X_L = 2\pi \cdot \downarrow f \cdot L$). As X_L decreases with frequency, the output voltage across the inductor decreases. The word *attenuate* means "to reduce in amplitude" (reduce voltage). The frequencies of the stopband are attenuated more than 3 dB at the output of the highpass filter.

The actual cutoff frequency for the *RL* inverted L filter is determined by the values of resistance and inductance. The cutoff frequency formula is simply derived from the X_L formula. This is easily done since $X_L = R$ at the cutoff frequency (at f_{co}, $V_L = V_R = 0.707 \cdot E_S = \sin \angle 45° \cdot E_S$, $\angle 45° = \tan^{-1}(X_L/R) = \tan^{-1}(1/1)$ where $X_L = R$).

$$X_L = R = 2\pi f L$$

thus

$$f_{co} = R/(2\pi L) \qquad \text{(for } RL \text{ inverted L filter analysis)} \qquad (21.2)$$

Formula 21.2 is rearranged to solve for an unknown R or L value as follows:

$$R = 2\pi L f_{co} \qquad \text{(for design of } RL \text{ inverted L filters)} \qquad (21.3)$$
$$L = R/(2\pi f_{co}) \qquad \text{(for design of } RL \text{ inverted L filters)} \qquad (21.4)$$

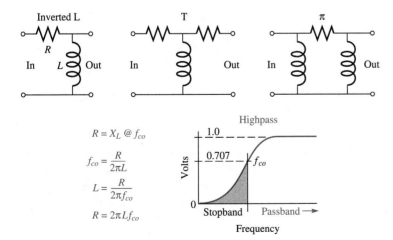

Figure 21-4 *RL* highpass filters.

EXAMPLE 21-1

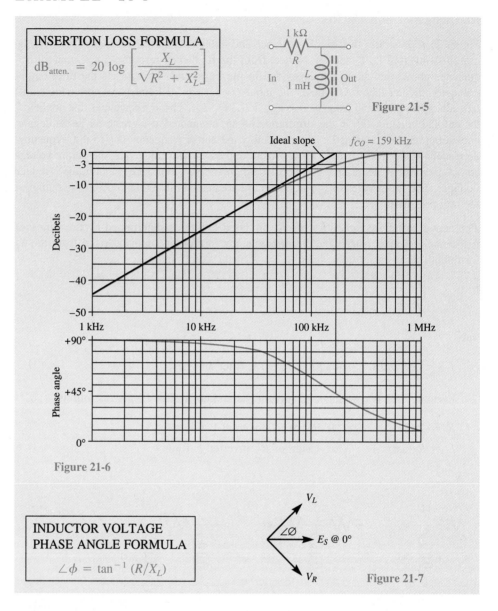

INSERTION LOSS FORMULA

$$dB_{atten.} = 20 \log \left[\frac{X_L}{\sqrt{R^2 + X_L^2}} \right]$$

Figure 21-5

Figure 21-6

INDUCTOR VOLTAGE
PHASE ANGLE FORMULA

$$\angle \phi = \tan^{-1} (R/X_L)$$

Figure 21-7

RL Highpass Phase vs Frequency

Example 21-1 demonstrates the complete analysis of a simple *RL* highpass inverted L filter. A Bode plot and phase plot are shown concurrently to help you understand the significance of phase shift with filter frequency response. Notice the output-voltage phase angle at the cutoff frequency is 45°. As frequency is increased above cutoff, the inductor-voltage phase angle decreases (for $f > f_{co}$, $V_L \angle \phi < 45°$). As frequency is decreased below cutoff, the inductor-voltage phase angle increases (for $f < f_{co}$, $V_L \angle \phi > 45°$ and $\leq 90°$). Also note that the plotted phase angle is the angle between the inductor voltage and the source voltage ($\angle \phi$), since the output is taken across the inductor. For this circuit, $\angle \phi = \tan^{-1} (R/X_L)$, where X_L is determined for each frequency.

RL Highpass Inverted L Bode Plot Analysis

The slope of the rolloff for Example 21-1 is found to be 18.57 dB/decade from 100 kHz down to 10 kHz. At 100 kHz, the output voltage is 5.48 dB below the input voltage, while at 10 kHz, the output is 24.05 dB below the input voltage ($V_{100\,kHz}$ at -5.48 dB and $V_{10\,kHz}$ at -24.05 dB). The difference between 24.1 dB and 5.48 dB is 18.57 dB, covering a decade of frequency (10–100 kHz). This slope of 18.57 dB/decade is also equal to a little over 5 dB/octave. The ideal slope for a filter such as this is 6 dB/octave, equal to 20 dB/decade. The entire Bode plot is a graph of the filter's insertion loss over a wide range of frequencies above and below cutoff. The voltage divider expression for V_{out}/V_{in} is inserted in Formula 21.1. The formula is then used to determine the amount of dB attenuation (insertion loss) at any frequency.

RC Highpass Filters

RC Highpass Frequency Response

Figure 21-8 illustrates the three basic configurations for *RC* highpass filters. The obvious difference between these filters and the *RL* filters is the use of capacitors instead of inductors. Also, the capacitors are placed in series with the signal path instead of in parallel. In these filters, the frequency response is controlled by X_C instead of X_L. Starting at the cutoff frequency, X_C increases as the frequency of the applied source voltage decreases: $\uparrow X_C = 1/(2\pi \cdot \downarrow f \cdot C)$. As X_C increases, the voltage across the capacitor increases leaving less voltage across the output resistor. Thus, the output voltage decreases as frequency decreases. The *RC* highpass filters are every bit as effective as *RL* highpass filters with one important advantage: lower cost (inductors are generally more costly than capacitors).

Determining the Cutoff Frequency for *RC* Inverted L

The actual cutoff frequency for the *RC* inverted L filter is determined by the values of resistance and capacitance. The cutoff-frequency formula is derived from the X_C formula. Resistance is substituted for X_C, since $X_C = R_L$ at the cutoff frequency [at f_{co}, $V_{RL} =$

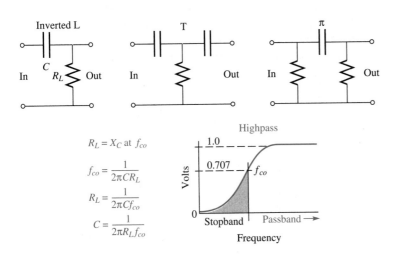

Figure 21-8 *RC* highpass filters.

$V_C = 0.707 \cdot E_S = \sin \angle 45° \cdot E_S$, also $\angle 45° = \tan^{-1}(X_C/R_L) = \tan^{-1}(1/1)$ where $X_C = R_L$].

$$X_C = R_L = 1/(2\pi f C)$$

thus

$$f_{co} = 1/(2\pi C R_L) \qquad \text{(for } RC \text{ inverted L filter analysis)} \tag{21.5}$$

Formula 21.5 is rearranged to solve for an unknown R_L or C value as follows:

$$R_L = 1/(2\pi C f_{co}) \qquad \text{(for design of } RC \text{ inverted L filters)} \tag{21.6}$$

$$C = 1/(2\pi R_L f_{co}) \qquad \text{(for design of } RC \text{ inverted L filters)} \tag{21.7}$$

RC Highpass Inverted L Phase vs Frequency

Example 21-2 demonstrates the complete analysis of a simple *RC* highpass inverted L filter. Like Example 21-1, a Bode plot and phase plot are shown concurrently to help you understand the significance of phase shift with filter frequency response. As with the *RL* highpass filter, the output-voltage phase angle at the cutoff frequency is +45°. As frequency is increased above cutoff, the resistor-voltage phase angle decreases (for $f > f_{co}$, $V_R \angle \theta < 45°$). As frequency is decreased below cutoff, the resistor-voltage phase angle increases (for $f < f_{co}$, $V_R \angle \theta > 45°$ and $\leq 90°$). Also note that the plotted phase angle is the angle between the resistor voltage and the source voltage ($\angle \theta$), since the output is taken across the resistor. For this circuit, $\angle \theta = \tan^{-1}(X_C/R_L)$, where X_C is determined for each frequency.

RC Highpass Inverted L Bode Plot Analysis

By now you have noticed that the Bode plot for the *RC* highpass filter of Example 21-2 is the same as the Bode plot for the *RL* highpass filter of Example 21-1. This demonstrates that the same filtering characteristics can be obtained with either an *RL* or *RC* filter. As you can see, the cutoff frequency and rolloff slope are the same for both filters ($f_{co} = 159$ kHz and slope equals 18.57 dB/decade from 10 kHz to 100 kHz). Again, the ideal slope for this filter is 20 dB/decade. For this filter, the voltage divider expression is $R_L/\sqrt{R_L^2 + X_C^2}$, since the filter output is the resistor voltage. Using this expression, the insertion loss (attenuation) can be calculated for any frequency. As in the *RL* filter, the insertion loss is less than 3 dB for all frequencies above cutoff.

E X A M P L E   2 1 - 2

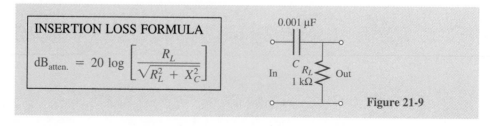

INSERTION LOSS FORMULA

$$dB_{atten.} = 20 \log \left[\frac{R_L}{\sqrt{R_L^2 + X_C^2}} \right]$$

0.001 μF

In C R_L Out
      1 kΩ

Figure 21-9

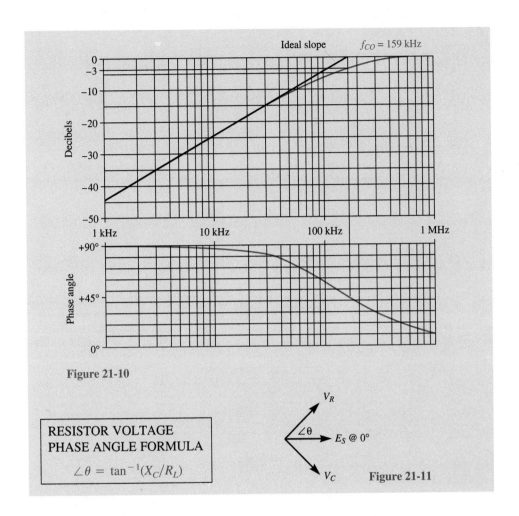

Figure 21-10

RESISTOR VOLTAGE
PHASE ANGLE FORMULA

$$\angle \theta = \tan^{-1}(X_C/R_L)$$

Figure 21-11

RCL Highpass Filters

RCL Highpass Frequency Response

Figure 21-12 shows the three basic configurations for the *RCL* highpass filters. As you can see, inductors *and* capacitors are used in these filters to improve the slope of stopband rolloff. These filters are much more decisive in separating high frequencies from low frequencies above and below the cutoff frequency. The reason for the improved response is the push-pull cooperative action of the inductor and capacitor. When the applied frequency is increased, X_L increases and X_C decreases, and vice versa. As frequency is decreased below cutoff, the series X_C increases and the parallel X_L decreases. This causes the filter output voltage across R_L to decrease rapidly, since most of the voltage is dropped across the series capacitor(s). Thus, the *RCL* filter is actually an *RC* and an *RL* filter working together for improved performance.

See color photo insert page D11, photograph 72.

Determining the Cutoff Frequency for *RCL* Inverted L

The design of these filters varies widely and goes well beyond the intent of this text. However, their general performance may be explored by examining the filters of Figure

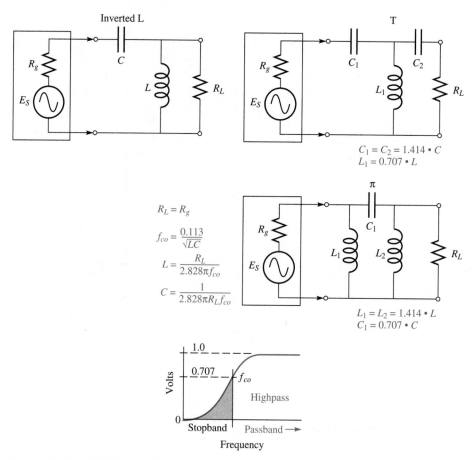

Figure 21-12   *RCL* highpass filters.

21-12. These are special filters that are designed to be inserted between a generator and a matched load for maximum power transfer. Recall that for maximum power transfer from a voltage source (generator) to a load, the load impedance must match the generator impedance. Thus, for these filters, R_L must equal R_g.

The derivation of the formula used to calculate the cutoff frequency is not as simple as for the *RL* and *RC* inverted L filters. In this case, the derivation is complicated by the fact that the output voltage is across $X_L \parallel R_L$. Since the derivation is somewhat involved, and beyond the purpose of this text, only the formula itself is presented here.

$$f_{co} = 1/(2.828\pi\sqrt{LC}) \cong 0.113/\sqrt{LC} \tag{21.8}$$

(for highpass *RCL* inverted L filters where $R_g = R_L$).

The design of an *RCL* highpass inverted L filter begins with the matched-load circuit in which the filter is to be inserted, where $R_g = R_L$. The values of L and C are determined based on the values of R_g and R_L and the desired cutoff frequency using the following equations:

$$L = R_L/(2.828\pi f_{co}) \cong 0.113\, R_L/f_{co} \tag{21.9}$$
$$C = 1/(2.828\pi R_L f_{co}) \cong 0.113/(R_L f_{co}) \tag{21.10}$$

(for highpass *RCL* inverted L filter design where $R_g = R_L$).

The filter shown in Example 21-3 was designed using these formulas. Use your calculator to verify that the values of L and C are correct ($R_g = R_L = 600\ \Omega$, $f_{co} = 300$ Hz).

EXAMPLE 21-3

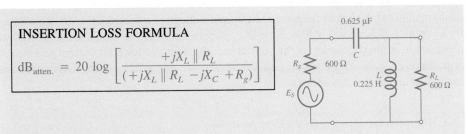

INSERTION LOSS FORMULA

$$dB_{atten.} = 20 \log \left[\frac{+jX_L \parallel R_L}{(+jX_L \parallel R_L - jX_C + R_g)} \right]$$

Figure 21-13

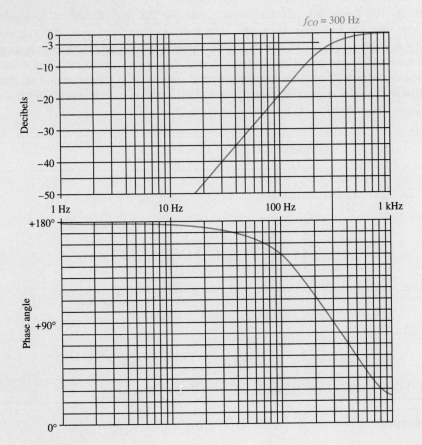

Figure 21-14

$V_{out} \angle x = I_t \angle y \cdot Z_{out} \angle z$
OUTPUT VOLTAGE PHASE ANGLE $= \angle x = \angle y + \angle z$

RCL Highpass Inverted L Phase vs Frequency

Example 21-3 illustrates the analysis of an *RCL* highpass inverted L filter. As with all filters, the phase angle of the output voltage changes with frequency. For this filter, the phase angle approaches 0° at some very high frequency, and approaches 180° as 0 Hz is approached. The phase plot shown in Example 21-3 is typical for this type of filter and configuration. At the cutoff frequency, the output-voltage phase angle is $+90°$. Recall that for a complex AC circuit $V \angle x = I \angle y \cdot Z \angle z$ where $\angle x = \angle y + \angle z$. The output voltage phase angle at any frequency is the sum of the phase angles for total current and output impedance ($V_{out} \angle x = I_t \angle y \cdot Z_{out} \angle z$ where $\angle x = \angle y + \angle z$; recall magnitudes are multiplied and angles are added).

RCL Highpass Inverted L Bode Plot Analysis

Examine the Bode plot of Example 21-3. Notice the slope of rolloff is twice the rate of that for *RL* and *RC* highpass filters. In this case the slope below cutoff is in the area of 37 to 40 dB/decade (40 dB/decade = 12 dB/octave). This means the output voltage at, say, 20 Hz, is 100 times less than the output voltage at 200 Hz [40 dB $=$ 20 log(100)]. For the filter of Example 21-3, the voltage divider expression is fairly complex: $(+jX_L \| R_L)/[(+jX_L \| R) - jX_C + R_g]$. This expression must be resolved following the rules for complex algebra. The values of X_L and X_C must first be determined at the frequency of interest. Then these reactance values are placed in the insertion-loss formula along with the load-resistance value. Obviously, the use of a computer is very desirable here. See Design Note 21-1.

DESIGN NOTE 21-1:   Highpass Filters

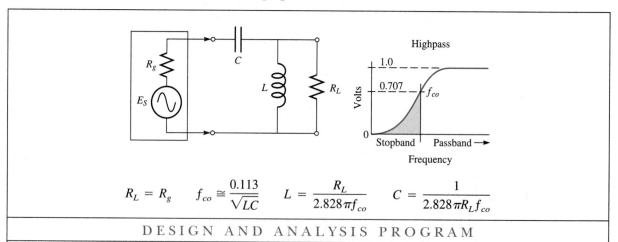

$$R_L = R_g \qquad f_{co} \cong \frac{0.113}{\sqrt{LC}} \qquad L = \frac{R_L}{2.828\pi f_{co}} \qquad C = \frac{1}{2.828\pi R_L f_{co}}$$

DESIGN AND ANALYSIS PROGRAM

```
10 CLS
20 PRINT"HIGHPASS FILTERS"
30 PRINT"":PRINT""
40 PRINT"THIS IS A DESIGN AND ANALYSIS PROGRAM FOR INVERTED L, RCL HIGHPASS FILTERS"
50 PRINT"** NOTE: GENERATOR AND LOAD RESISTANCES ARE EQUAL FOR THESE FILTERS **"
60 PRINT""
70 PRINT"(1) ANALYZE AN RCL HIGHPASS FILTER"
```

```
 80  PRINT"(2) DESIGN AN RCL HIGHPASS FILTER"
 90  PRINT"(3) QUIT"
100  PRINT""
110  INPUT"SELECT 1, 2, OR 3 - ";N
120  IF N<1 OR N>3 THEN GOTO 60
130  ON N GOTO 140,420,500
140  INPUT"ENTER THE INDUCTOR VALUE IN HENRIES. ";L
150  INPUT"ENTER THE CAPACITOR VALUE IN FARADS. ";C
160  FCO = .1125565 / SQR(L*C)
170  INPUT"ENTER THE LOAD RESISTANCE. ";RL
180  PRINT"THE SOURCE RESISTANCE IS ASSUMED TO BE ";RL;" OHMS."
200  PRINT"THE CUTOFF FREQUENCY IS ";FCO;" HZ."
210  PRINT""
220  INPUT"ENTER A TEST FREQUENCY. ";F
230  XL = 6.2832 * F * L
240  XC = 1/(6.2832 * F * C)
250  ZOUT = 1/SQR((1/RL)^2 + (1/XL)^2)
260  A1 = ATN(RL/XL) * 57.29578
270  R = COS(A1/57.29578) * ZOUT
280  X = SIN(A1/57.29578) * ZOUT
290  ZT = SQR((RL+R)^2 + (X-XC)^2)
300  A2 = ATN((X-XC)/(RL+R)) * 57.29578
310  A = A1 - A2
320  PRINT"THE OUTPUT VOLTAGE PHASE ANGLE = ";A;" DEGREES."
330  AF = 2 * ZOUT/ZT
340  PRINT"THE ATTENUATION FACTOR = ";AF
350  PRINT""
360  IF PP=1 THEN GOTO 390
370  FF = AF: PP=1
380  F = F/10:PRINT"AT ";F;" HZ.":GOTO 230
390  DB = 20 * ((LOG(FF/AF))/LOG(10))
400  PRINT"THE ROLLOFF SLOPE = ";DB;" dB/DECADE."
410  PP=0:PRINT"":GOTO 60
420  REM ** DESIGN SECTION **
430  INPUT"ENTER THE LOAD RESISTANCE. ";RL
440  INPUT"ENTER THE DESIRED CUTOFF FREQUENCY. ";FCO
450  C = .1125565/(FCO * RL)
460  L = .1125565 * RL/FCO
470  PRINT"THE REQUIRED INDUCTANCE IS ";L;" HENRIES."
480  PRINT"THE REQUIRED CAPACITANCE IS ";C;" FARADS."
490  GOTO 60
500  CLS:END
```

Highpass Filter Applications

Highpass filters are used for two general purposes: to separate AC signals from DC, and to separate high frequencies from low frequencies. Figure 21-15 shows an extremely common application for the *RC* inverted L highpass filter. In this application, the coupling capacitor blocks a DC voltage of one amplifier stage and prevents it from passing to the input of a second stage. While blocking the DC, the capacitor passes all frequencies above

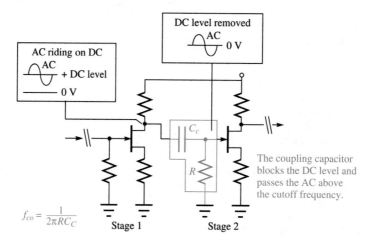

$$f_{co} = \frac{1}{2\pi RC_C}$$

Figure 21-15   The coupling capacitor as a highpass filter.

the cutoff frequency as determined by the value of coupling capacitor (C_C) and the input resistance (R) of the FET amplifier stage. Thus, the AC is separated from the DC, and all frequencies below the cutoff frequency are attenuated.

Figure 21-16 shows how the highpass filter is used to separate frequencies. In this case, an audio-frequency range of 20 Hz to 20 kHz is applied to a 5 kHz highpass *RCL* inverted L filter. All frequencies above 5 kHz are passed to the output of the filter. All frequencies below 5 kHz are attenuated at the rate of about 40 dB/decade. The output voltage at 500 Hz is approximately 40 dB below the output voltage at 5 kHz. In other words, if the output voltage at 5 kHz is 1 $V_{p\text{-}p}$, the output voltage at 500 Hz is about 0.01 $V_{p\text{-}p}$ [1 $V_{p\text{-}p}$/100, since 40 dB = 100, also −40 dB = 20 log (0.01/1)].

We have covered a fair amount of material in this section. Take time to go back and review the *RL*, *RC*, and *RCL* filters. As you do so, make a conscious effort to compare component placement and characteristics for the different filters. Then, test your understanding by answering the questions of Self-Check 21-2.

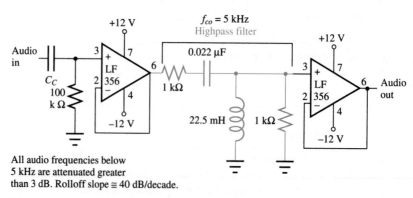

All audio frequencies below 5 kHz are attenuated greater than 3 dB. Rolloff slope ≅ 40 dB/decade.

Figure 21-16   A highpass filter for audio frequency separation.

1. All frequencies in the passband of a filter have voltage amplitudes greater than what fraction of the input voltage?

2. The frequencies in the stopband of a filter are attenuated more than how many dB?

3. Determine the cutoff frequency for an *RL* highpass inverted L filter if $R = 2.7 \text{ k}\Omega$ and $L = 10 \text{ mH}$.

4. Design a 10 kHz *RC* highpass inverted L filter using a 4.7 kΩ load resistor. ($C = ?$)

5. What is the ideal slope of rolloff for the filter of #4?

6. Is the output-voltage phase angle, for frequencies above cutoff, (a) greater or (b) less than 45° for an *RC* highpass inverted L filter?

7. Describe or draw a highpass filter that would provide a rolloff slope of 40 dB/decade.

21-3 Lowpass Filters

In this section, we examine common configurations of lowpass filters. You will see how inductors or capacitors or both may be used to pass frequencies below a specified cutoff frequency and attenuate all frequencies above cutoff.

RL Lowpass Filters

RL Lowpass Frequency Response

Figure 21-17 illustrates the three common *RL* lowpass filter configurations. As you can see, lowpass filters can be made by placing an inductor or inductors in series with the

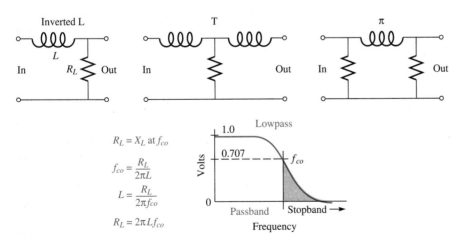

Figure 21-17   *RL* lowpass filters.

signal path. You are already aware that X_L varies directly with frequency. Thus, the higher the frequency, the greater the opposition of the inductor. Notice that the frequency-response graph is the opposite or flipped version of the highpass graph. The characteristics of the graph are the same, except now low frequencies are passed and high frequencies are attenuated.

Determining the Cutoff Frequency for *RL* Inverted L

As for highpass filters, the actual cutoff frequency for the lowpass *RL* inverted L filter is determined by the values of resistance and inductance. The cutoff-frequency design formulas used for highpass inverted L are also used for lowpass inverted L. Recall that $X_L = R_L$ at the cutoff frequency (at f_{co}, $V_L = V_R = 0.707 \cdot E_S = \sin \angle 45° \cdot E_S$, also $\angle 45° = \tan^{-1}(X_L/R_L) = \tan^{-1}(1/1)$ where $X_L = R_L$).

$$X_L = R_L = 2\pi f L$$

thus

(21.2) $f_{co} = R_L/(2\pi L)$

(21.3) $R_L = 2\pi L f_{co}$ (for design of *RL* inverted L filters)

(21.4) $L = R_L/2\pi f_{co}$ (for design of *RL* inverted L filters)

These formulas were used to design the filter of Example 21-4. Use your calculator to verify the values of *L* and *C*.

RL Lowpass Phase vs Frequency

Example 21-4 demonstrates the complete analysis of an *RL* lowpass inverted L filter. The Bode plot and phase plot are shown as before to graphically illustrate the output-voltage amplitude and phase-angle dependence on frequency. As you can see, the output-voltage phase angle at the cutoff frequency is $-45°$. Also note that the plotted phase angle is the angle between the resistor voltage and the source voltage ($\angle \theta$) since the output is taken across the resistor. For this circuit, $\angle \theta = \tan^{-1}(X_L/R_L)$, where X_L is determined for each frequency.

RL Lowpass Inverted L Bode Plot Analysis

The rolloff slope for Example 21-4 is found to be 17.89 dB/decade from 200 kHz up to 2 MHz. At 200 kHz, the output voltage is 4.12 dB below the input voltage, while at 2 MHz, the output is 22.01 dB below the input voltage ($V_{200\,kHz}$ at -4.12 dB and $V_{2\,MHz}$ at -22.01 dB). These attenuation figures were calculated using the insertion-loss formula shown in the example. The ideal rolloff slope for a filter such as this is 6 dB/octave, equal to 20 dB/decade. For the filter of Example 21-4, the voltage divider expression is: $V_{out}/V_{in} = R_L/\sqrt{R_L^2 + X_L^2}$. Thus, for the *RL* lowpass inverted L filter, the insertion loss at any frequency is: $dB_{atten.} = 20 \cdot \log(V_{out}/V_{in}) = 20 \cdot \log[R_L/\sqrt{R_L^2 + X_L^2}]$.

EXAMPLE   21-4

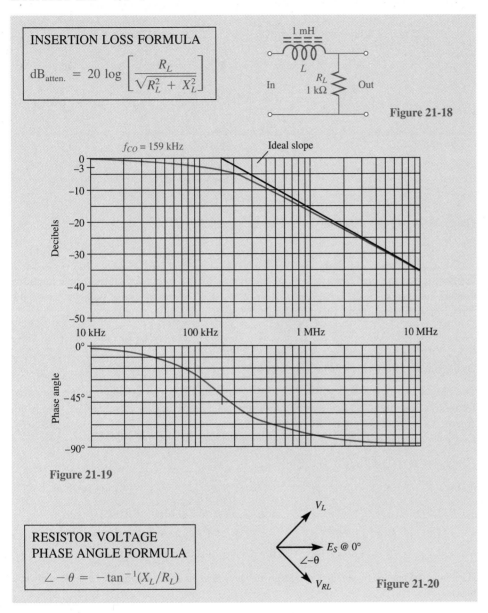

INSERTION LOSS FORMULA

$$dB_{atten.} = 20 \log \left[\frac{R_L}{\sqrt{R_L^2 + X_L^2}} \right]$$

Figure 21-18

$f_{CO} = 159$ kHz   Ideal slope

Figure 21-19

RESISTOR VOLTAGE
PHASE ANGLE FORMULA

$$\angle -\theta = -\tan^{-1}(X_L/R_L)$$

Figure 21-20

RC Lowpass Filters

RC Lowpass Frequency Response

Figure 21-21 illustrates the three basic configurations for *RC* lowpass filters. As you can see, capacitors are used instead of inductors, and they are placed in parallel with the output instead of in series. In these filters, the frequency response is controlled by X_C instead of

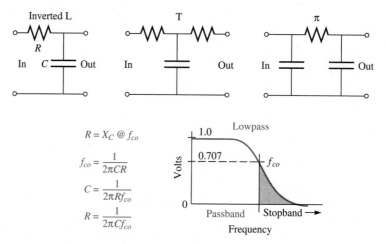

Figure 21-21   *RC* lowpass filters.

X_L. Starting at the cutoff frequency, X_C decreases as the frequency of the applied source voltage increases: $\downarrow X_C = 1/(2\pi \cdot \uparrow f \cdot C)$. As X_C decreases, the voltage across the capacitor decreases. Thus, the output voltage decreases as frequency increases.

Determining the Cutoff Frequency for *RC* Inverted L

The actual cutoff frequency for lowpass *RC* inverted L filters is found using the same formula as for highpass *RC* inverted L filters. The same design formulas are also used. Again, $X_C = R$ at the cutoff frequency (at f_{co}, $V_R = V_C = 0.707 \cdot E_S = \sin \angle 45° \cdot E_S$, also $\angle 45° = \tan^{-1}(R/X_C) = \tan^{-1}(1/1)$ where $X_C = R$).

$$X_C = R = 1/(2\pi f C)$$

thus

(21.5) $f_{co} = 1/(2\pi CR)$ (for *RC* inverted L filter analysis)

(21.6) $R = 1/(2\pi C f_{co})$ (for design of *RC* inverted L filters)

(21.7) $C = 1/(2\pi R f_{co})$ (for design of *RC* inverted L filters)

Use these formulas to verify the design of the filter given in Example 21-5 ($f_{co} = 159$ kHz).

RC Lowpass Inverted L Phase vs Frequency

Example 21-5 demonstrates the complete analysis of an *RC* lowpass inverted L filter. Again, the output voltage phase angle at the cutoff frequency is $-45°$, like the *RL* lowpass filter of Example 21-4. Also note, the plotted phase angle is the angle between the capacitor voltage and the source voltage ($\angle\phi$), since the output is taken across the capacitor. For this circuit $\angle\phi = \tan^{-1}(R/X_C)$, where X_C is determined for each frequency.

RC Lowpass Inverted L Bode Plot Analysis

Note that the Bode plot for the *RC* lowpass filter of Example 21-5 is the same as the Bode plot for the *RL* lowpass filter of Example 21-4. As with highpass filters, this demonstrates that the same filtering characteristics can be obtained with either filter. The cutoff frequency and rolloff slope are the same for both (f_{co} = 159 kHz and slope equals 17.89 dB/decade from 200 kHz to 2 MHz). Again, the ideal slope for this filter is 20 dB/decade. The voltage divider expression is $X_C/\sqrt{R^2 + X_C^2}$ since the filter output is the capacitor voltage. Using this expression, the insertion loss can be calculated for any frequency. As in the *RL* filter, the insertion loss is less than 3 dB for all frequencies below cutoff.

E X A M P L E   21-5

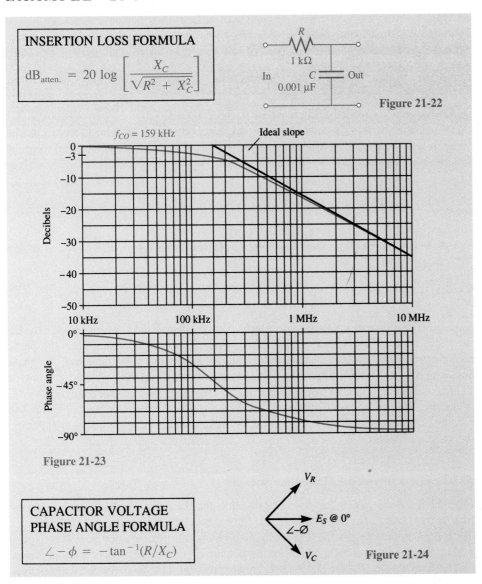

INSERTION LOSS FORMULA

$$dB_{atten.} = 20 \log \left[\frac{X_C}{\sqrt{R^2 + X_C^2}} \right]$$

Figure 21-22

fco = 159 kHz   Ideal slope

Figure 21-23

CAPACITOR VOLTAGE PHASE ANGLE FORMULA

$$\angle - \phi = -\tan^{-1}(R/X_C)$$

Figure 21-24

RCL Lowpass Filters

RCL Lowpass Frequency Response

See color photo insert
page D11, photograph 73.

Figure 21-25 shows the three basic configurations for *RCL* highpass filters. In these filters, the inductors and capacitors work together to improve the slope of stopband rolloff. As in highpass *RCL* filters, the improved response is due to the push-pull cooperative action of the inductor and capacitor. As frequency is increased above cutoff, the series X_L increases and the parallel X_C decreases. This causes the filter output voltage across R_L to decrease rapidly, since most of the voltage is dropped across the inductor(s).

Determining the Cutoff Frequency for *RCL* Inverted L

As with the *RCL* highpass filters previously discussed, these lowpass filters are placed between a matched source and load where $R_g = R_L$. Again, since the derivation is somewhat involved, and beyond the purpose of this text, only the formula itself is presented here.

$$f_{co} = 1/(1.414\pi\sqrt{LC}) \cong 0.225/\sqrt{LC} \qquad (21.11)$$

(for lowpass *RCL* inverted L filters where $R_g = R_L$)

The design of an *RCL* lowpass inverted L filter begins with the load resistance. The value of resistance must be known. Then, the values of L and C are determined from the following equations:

$$L = R_L/(1.414\pi f_{co}) \cong 0.225R_L/f_{co} \qquad (21.12)$$

$$C = 1/(1.414\pi R_L f_{co}) \cong 0.225/(R_L f_{co}) \qquad (21.13)$$

(for lowpass *RCL* inverted L filter design where $R_g = R_L$)

As before, take time to use these formulas to verify the values of L and C as shown in Example 21-6.

RCL Lowpass Inverted L Phase vs Frequency

Example 21-6 illustrates the analysis of an *RCL* lowpass inverted L filter. As you can see, the phase angle of the output voltage changes with frequency. For this filter, the phase angle approaches $0°$ as 0 Hz is approached, and, at some very high frequency, the phase angle approaches $-180°$. The phase plot shown in Example 21-6 is typical for this type of filter and configuration. At the cutoff frequency, the output-voltage phase angle is $-90°$. As before, for a complex AC circuit, $V \angle x = I \angle y \cdot Z \angle z$ where $\angle x = \angle y + \angle z$. The output-voltage phase angle at any frequency is the sum of the phase angles for total current and output impedance ($V_{out} \angle x = I_t \angle y \cdot Z_{out} \angle z$ where $\angle x = \angle y + \angle z$.

RCL Lowpass Inverted L Bode Plot Analysis

Examine the Bode plot of Example 21-6. Notice that the slope of rolloff for this lowpass filter is the same as for the *RCL* highpass filter (about 40 dB/decade, or 12 dB/octave). Thus, the output voltage at 10 kHz is approximately 40 dB below the output voltage at 1 kHz. For the filter of Example 21-6, the voltage divider expression is fairly complex: $(-jX_C\|R_L)/[(-jX_C\|R_L) + jX_L + R_g]$. This expression must be expanded and resolved following the rules of complex algebra. The values of X_L and X_C must first be determined

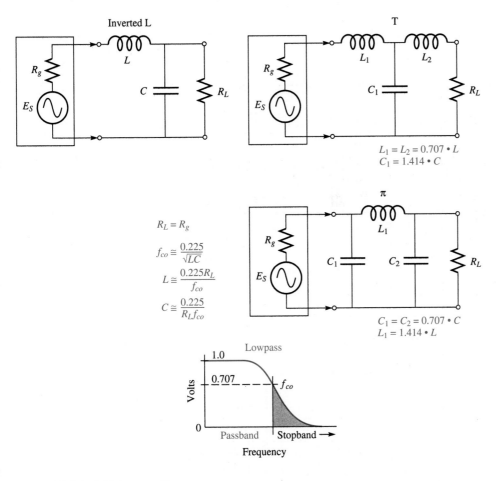

Figure 21-25 *RCL* lowpass filters.

at the frequency of interest. Then these reactance values are placed in the insertion-loss formula along with the load resistance ($R_g = R_L$) to determine the amount of attenuation at the frequency of interest. Again, a computer is very useful here. See Design Note 21-2.

EXAMPLE 21-6

INSERTION LOSS FORMULA

$$dB_{atten.} = 20 \log \left[\frac{-jX_C \parallel R_L}{(-jX_C \parallel R_L + jX_L + R_g)} \right]$$

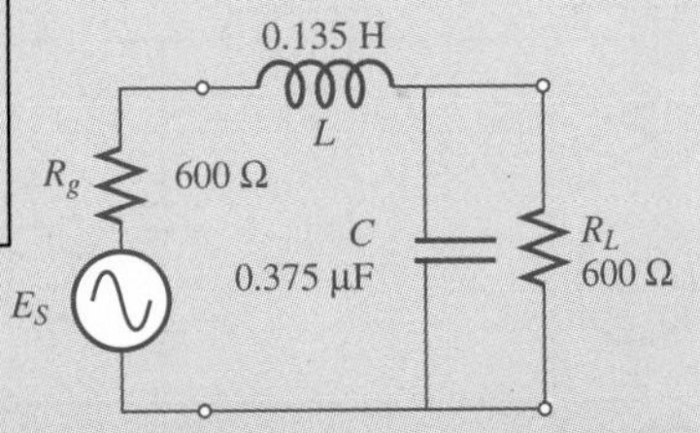

Figure 21-26

(Continued next page)

EXAMPLE   21-6   (Continued)

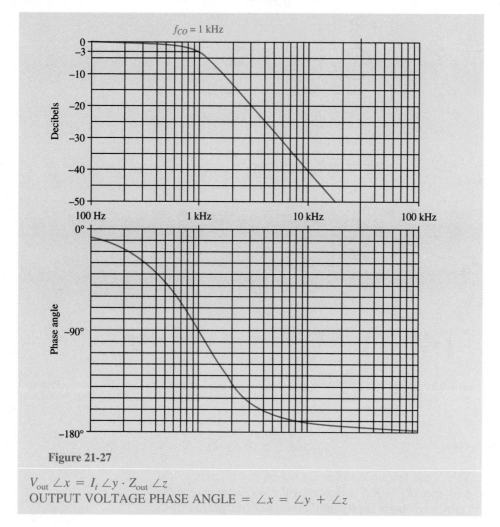

Figure 21-27

$$V_{\text{out}} \angle x = I_t \angle y \cdot Z_{\text{out}} \angle z$$

OUTPUT VOLTAGE PHASE ANGLE $= \angle x = \angle y + \angle z$

DESIGN NOTE 21-2:   Lowpass Filters

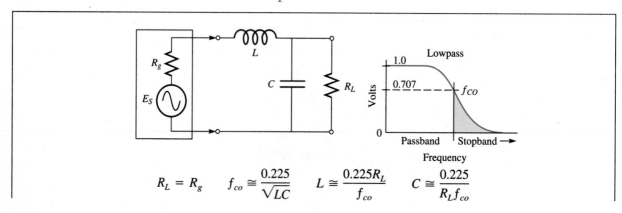

$$R_L = R_g \qquad f_{co} \cong \frac{0.225}{\sqrt{LC}} \qquad L \cong \frac{0.225 R_L}{f_{co}} \qquad C \cong \frac{0.225}{R_L f_{co}}$$

DESIGN AND ANALYSIS PROGRAM

```
10 CLS
20 PRINT"LOWPASS FILTERS"
30 PRINT"":PRINT""
40 PRINT"THIS IS A DESIGN AND ANALYSIS PROGRAM FOR INVERTED L, RCL LOWPASS FILTERS"
50 PRINT"** NOTE: GENERATOR AND LOAD RESISTANCES ARE EQUAL FOR THESE FILTERS **"
60 PRINT""
70 PRINT"(1) ANALYZE AN RCL LOWPASS FILTER"
80 PRINT"(2) DESIGN AN RCL LOWPASS FILTER"
90 PRINT"(3) QUIT"
100 PRINT""
110 INPUT"SELECT 1, 2, OR 3 – ";N
120 IF N<1 OR N>3 THEN GOTO 60
130 ON N GOTO 140,420,500
140 INPUT"ENTER THE INDUCTOR VALUE IN HENRIES. ";L
150 INPUT"ENTER THE CAPACITOR VALUE IN FARADS. ";C
160 FCO = .225 / SQR(L*C)
170 INPUT"ENTER THE LOAD RESISTANCE. ";RL
180 PRINT"THE SOURCE RESISTANCE IS ASSUMED TO BE ";RL;" OHMS."
200 PRINT"THE CUTOFF FREQUENCY IS ";FCO;" HZ."
210 PRINT""
220 INPUT"ENTER A TEST FREQUENCY. ";F
230 XL = 6.2832 * F * L
240 XC = 1/(6.2832 * F * C)
250 ZOUT = 1/SQR((1/RL)^2 + (1/XC)^2)
260 A1 = ATN(RL/ – XC) * 57.29578
270 R = COS(A1/57.29578) * ZOUT
280 X = SIN(A1/57.29578) * ZOUT
290 ZT = SQR((RL + R)^2 + (X + XL)^2)
300 A2 = ATN((X + XL)/(RL + R)) * 57.29578
310 A = A1 – A2
320 PRINT"THE OUTPUT VOLTAGE PHASE ANGLE = ";A;" DEGREES."
330 AF = 2 * ZOUT/ZT
340 PRINT"THE ATTENUATION FACTOR = ";AF
350 PRINT""
360 IF PP = 1 THEN GOTO 390
370 FF = AF: PP = 1
380 F = F * 10:PRINT"AT ";F;" Hz.":GOTO 230
390 DB = 20 * ((LOG(FF/AF))/LOG(10))
400 PRINT"THE ROLLOFF SLOPE = ";DB;" dB/DECADE."
410 PP = 0:PRINT"":GOTO 60
420 REM ** DESIGN SECTION **
430 INPUT"ENTER THE LOAD RESISTANCE. ";RL
440 INPUT"ENTER THE DESIRED CUTOFF FREQUENCY. ";FCO
450 C = .225/(FCO * RL)
460 L = .225 * RL/FCO
470 PRINT"THE REQUIRED INDUCTANCE IS ";L;" HENRIES."
480 PRINT"THE REQUIRED CAPACITANCE IS ";C;" FARADS."
490 GOTO 60
500 CLS:END
```

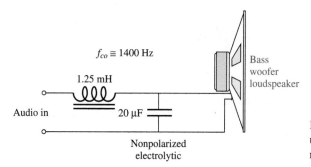

Figure 21-28   A lowpass filter used in an audio crossover network for a speaker system.

Lowpass Filter Applications

Lowpass filters are used for two general purposes: to pass DC while blocking AC, and to separate low frequencies from high frequencies. Figure 21-28 shows an extremely common application for the RCL inverted L lowpass filter. The filter shown is a lowpass crossover network for a high-fidelity speaker system. The cutoff frequency for this filter is about 1,400 Hz. This means all frequencies below 1,400 Hz are passed to the bass woofer with little or no attenuation. All frequencies above the 1,400 Hz cutoff frequency will be attenuated at the rate of 12 dB/octave, or 40 dB/decade.

Figure 21-29 shows how the lowpass filter is used to pass DC and block AC. A lowpass π filter such as this offers excellent filtering of the pulsating DC supplied from the bridge rectifier. In the United States, the fullwave rectified AC produces 120 Hz pulsating DC while in some other countries the pulsating DC is 100 Hz (50 Hz line frequency). This pulsating DC must be filtered to eliminate the pulsations. The DC output voltage should be as smooth as possible. Figure 21-29 lists two general design guidelines for such a filter: The capacitive reactance must be much less than the load resistance ($X_C < R_L/10$), and the inductive reactance must be much more than the load resistance ($X_L > 10R_L$) at the frequency of the pulsating DC. These design guidelines insure that the filter cutoff frequency is very low, so the 120 Hz or 100 Hz is attenuated a very large amount.

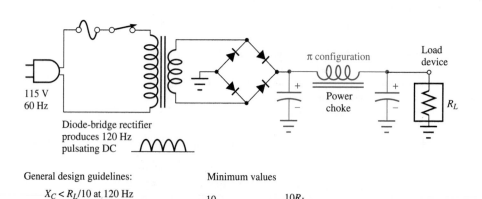

General design guidelines:

$X_C < R_L/10$ at 120 Hz

$X_L > 10R_L$ at 120 Hz

Minimum values

$$C = \frac{10}{2\pi f R_L} \qquad L = \frac{10R_L}{2\pi f}$$

Figure 21-29   The lowpass pi (π) filter used in some DC power supplies.

It is interesting to note that the lowpass filter is also an integrator circuit. The choke in the power supply integrates the pulsating current, while the capacitors integrate the pulsating voltage. Together they approach full integration (a perfectly smooth DC) (see Section 14-4).

Take time to go back and review these lowpass RL, RC, and RCL filters. Make a conscious effort to compare component placement and characteristics for the different filters. Then test your understanding by answering the questions of Self-Check 21-3.

1. Determine the cutoff frequency for an RL lowpass inverted L filter if $R_L = 3.3\ \text{k}\Omega$ and $L = 1\ \text{mH}$.

2. Design a 2 kHz RC lowpass inverted L filter using a 1 kΩ series resistor. ($C = ?$)

3. Is the output-voltage phase angle, for frequencies above cutoff, (a) greater than or (b) less than $-45°$ for an RC lowpass inverted L filter?

4. Describe or draw a lowpass filter that provides a rolloff slope of 40 dB/decade.

5. What happens to the output impedance of a lowpass RCL π filter as the applied frequency is increased above cutoff?

21-4 Bandpass Filters

In this section, we examine common configurations of bandpass filters. You will see how series resonant and parallel resonant circuits may be used to pass a band of frequencies and attenuate all frequencies outside the passband. It would be good for you to review Chapter 20, since these filters are designed using the principles and formulas presented in that chapter.

Series Resonant Bandpass Filters

Series Resonant Bandpass Frequency Response

Figure 21-30 illustrates the most common series resonant bandpass filter configuration, the inverted L. The signal provided by the source must pass through the series resonant circuit to the load. As you know from previous study, the series resonant circuit acts like a short at the resonant frequency (except for inductor winding resistance R_W). Thus, the resonant frequency is passed to the load resistor (R_L) with little or no attenuation. Above and below resonance, X_C and X_L become unbalanced and do not completely cancel. Below resonance the circuit acts capacitively, since $X_C > X_L$. Above resonance the circuit acts inductively, since $X_L > X_C$. Therefore, frequencies above and below resonance are faced with some reactive opposition ($X_L - X_C$) in series with the load resistance. This series reactance forms a voltage divider with the load resistance. The farther away from resonance a frequency is, the greater the voltage drop across the reactance, and the lower the output voltage (load voltage).

See color photo insert page D12, photograph 74.

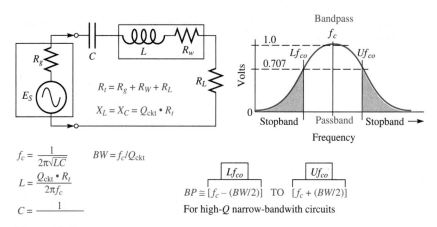

$f_c = \dfrac{1}{2\pi\sqrt{LC}}$ $BW = f_c/Q_{ckt}$

$L = \dfrac{Q_{ckt} \cdot R_t}{2\pi f_c}$

$C = \dfrac{1}{}$

$BP \cong [f_c - (BW/2)]$ TO $[f_c + (BW/2)]$

For high-Q narrow-bandwith circuits

Figure 21-30   Series resonant bandpass filters.

Determining Center Frequency and Bandpass

As shown in Figure 21-30, the series resonant bandpass filter is designed and analyzed using formulas presented in Chapter 20.

(20.1) $f_c = f_r = 1/(2\pi\sqrt{LC})$

(20.4) $Q_{ckt} = X_L/R_t$ (circuit Q for series resonance)

where $R_t = R_g + R_W + R_L$.

(20.8) $BW = f_c/Q_{ckt}$

(20.9) $BP \cong [f_c - (BW/2)] \rightarrow [f_c + (BW/2)]$ (approximation for high Q_{ckt})

Designing for a Specific f_c and BW

To design a series resonant bandpass filter for a specific frequency and bandwidth, the inductor and capacitor must be chosen so that X_L and X_C are Q_{ckt} times greater than R_t. Recall that $f_c/BW = Q_{ckt}$ and $Q_{ckt} = X_L/R_t$. Therefore, the required X_L and X_C will equal $Q_{ckt} \cdot R_t$. The following formulas include Q_{ckt} to meet this requirement:

$$L = \frac{Q_{ckt} \cdot R_t}{2\pi f_c} \tag{21.14}$$

$$C = 1/(2\pi f_c R_t Q_{ckt}) \tag{21.15}$$

where $Q_{ckt} = f_c/BW$ and $R_t = R_g + R_W + R_L$.

EXAMPLE   21-7

Let's suppose you want a filter that will pass frequencies from 88 MHz to 108 MHz. The signal source has a resistive impedance R_g of 300 Ω and the load is also 300 Ω (R_L). The total resistance R_t is 600 Ω. The center frequency is approximately 98 MHz. The total bandwidth is 20 MHz ($BW = 108$ MHz $- 88$ MHz). Thus, the circuit Q must be 4.9 ($Q_{ckt} = f_o/BW = 98$ MHz/20 MHz). Using Formulas 21.14 and 21.15:

$$L = \frac{4.9 \cdot 600 \ \Omega}{2 \cdot \pi \cdot 98 \ \text{MHz}} \cong 4.77 \ \mu\text{H}$$

$$C = \frac{1}{2 \cdot \pi \cdot 98 \ \text{MHz} \cdot 600 \cdot 4.9} \cong 0.55 \ \text{pF}$$

Series Resonant Bandpass Phase vs Frequency

Example 21-8 illustrates the analysis of a series resonant bandpass filter. Note how the phase angle ($\angle\theta$) of the output voltage changes with frequency. As shown in the phase plot, the output-voltage phase angle is $0°$ at the center frequency, $+45°$ at the lower cutoff frequency, and $-45°$ at the upper cutoff frequency. At the lower cutoff frequency, $X_C > X_L$ and the circuit acts capacitively. This means the total current will lead the source voltage (ICE). The leading current produces a voltage drop across the output resistor that leads the source voltage. At the upper cutoff frequency, $X_L > X_C$ and the circuit acts inductively. This means the total current will lag the source voltage (ELI). The lagging current produces a voltage drop across the output resistor that lags the source voltage. At any frequency, the output voltage phase angle is determined from this expression: $\angle\theta = \tan^{-1}[(X_C - X_L)/R_t]$, where X_L and X_C are determined for the frequency of interest and R_t is the sum of all resistance.

Series Resonant Bandpass Bode Plot Analysis

The Bode plot of Example 21-8 illustrates the characteristic double slope of a bandpass filter. The actual rolloff slope on each side depends on the circuit Q; the higher the Q, the steeper the slopes (high Q = steep slope, low Q = gradual slope). The slope may be a few dB/decade for a low-Q circuit to over 100 dB/decade for a very-high-Q bandpass filter circuit. For a series resonant bandpass filter, the voltage divider expression is: $V_{\text{out}}/V_{\text{in}} = R_L/\sqrt{R_t^2 + (X_L - X_C)^2}$, (where $R_t = R_g + R_W + R_L$). Design Note 21-3 summarizes the analysis and design of the series resonant bandpass filter. Take time to study it carefully before continuing.

EXAMPLE   21-8

INSERTION LOSS FORMULA

$$\text{dB}_{\text{atten.}} = 20 \log\left[\frac{R_L}{\sqrt{R_t^2 + (X_L - X_C)^2}}\right]$$

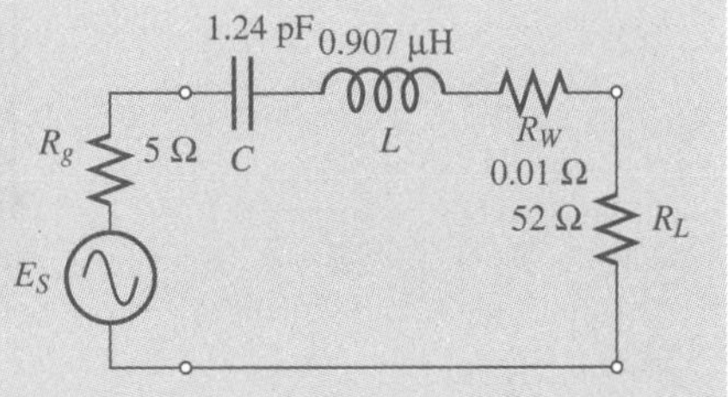

Figure 21-31

(Continued next page)

EXAMPLE 21-8 (Continued)

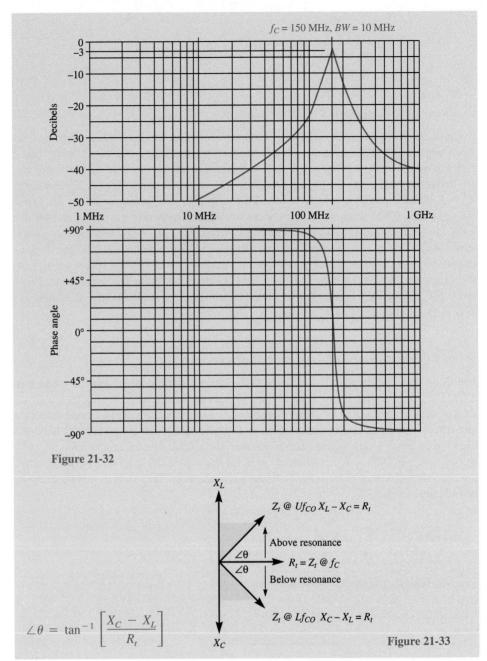

$f_C = 150$ MHz, $BW = 10$ MHz

Figure 21-32

X_L

Z_t @ Uf_{CO} $X_L - X_C = R_t$

Above resonance

$R_t = Z_t$ @ f_C

Below resonance

Z_t @ Lf_{CO} $X_C - X_L = R_t$

X_C

$$\angle\theta = \tan^{-1}\left[\frac{X_C - X_L}{R_t}\right]$$

Figure 21-33

D E S I G N   N O T E   2 1 - 3 :   Series Resonant Bandpass

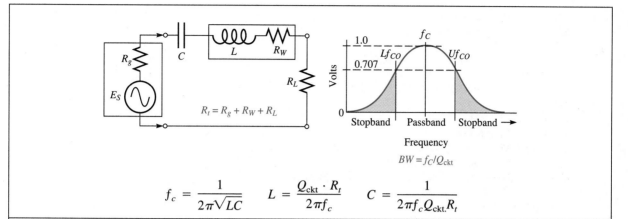

$$f_c = \frac{1}{2\pi\sqrt{LC}} \qquad L = \frac{Q_{ckt} \cdot R_t}{2\pi f_c} \qquad C = \frac{1}{2\pi f_c Q_{ckt.} R_t}$$

DESIGN AND ANALYSIS PROGRAM

```
10 CLS
20 PRINT"SERIES RESONANT BANDPASS FILTERS"
30 PRINT"":PRINT""
40 PRINT"THIS IS A DESIGN AND ANALYSIS PROGRAM FOR SERIES RESONANT BANDPASS FILTERS"
50 PRINT"** NOTE: XL = XC = Q * RT AT RESONANCE. **"
60 PRINT""
70 PRINT"(1) ANALYZE A SERIES RESONANT BANDPASS FILTER"
80 PRINT"(2) DESIGN A SERIES RESONANT BANDPASS FILTER"
90 PRINT"(3) QUIT"
100 PRINT""
110 INPUT"SELECT 1, 2, OR 3 - ";N
120 IF N<1 OR N>3 THEN GOTO 60
125 PRINT""
130 ON N GOTO 140,510,630
140 INPUT"ENTER THE INDUCTOR VALUE IN HENRIES. ";L
150 INPUT"ENTER THE CAPACITOR VALUE IN FARADS. ";C
160 INPUT"ENTER THE LOAD RESISTANCE. ";RL
170 INPUT"ENTER THE GENERATOR RESISTANCE. ";RG
180 INPUT"ENTER THE INDUCTOR WINDING RESISTANCE. ";RW
190 FC = .159155/SQR(L * C)
200 RT = RG + RW + RL
210 PRINT"THE CENTER FREQUENCY IS ";FC;" HZ."
220 XL = 6.2832 * FC * L
230 PRINT"THE CIRCUIT Q IS ";XL/RT
240 BW = FC/(XL/RT)
250 PRINT"THE BANDWIDTH IS ";BW;" HZ."
300 IF XL/RT < 10 THEN GOTO 330
310 PRINT"THE LOWER CUTOFF FREQUENCY IS APPROXIMATELY ";FC - (BW/2);" HZ."
320 PRINT"THE UPPER CUTOFF FREQUENCY IS APPROXIMATELY ";FC + (BW/2);" HZ."
330 PRINT""
340 INPUT"ENTER A TEST FREQUENCY. ";F
350 XL = 6.2832 * F * L
360 XC = 1 / (6.2832 * F * C)
370 ZT = SQR(RT^2 + (XL - XC)^2)
```

(Continued next page)

DESIGN NOTE 21-3 (Continued)

```
380 A = −ATN((XL − XC)/RT) * 57.29578
390 PRINT"THE PHASE ANGLE IS ";A;" DEGREES."
400 AF = RL/ZT
410 PRINT"THE ATTENUATION FACTOR = ";AF
420 PRINT""
430 IF PP = 1 THEN GOTO 480
440 FF = AF:PP = 1
450 IF F<FC THEN F = F/10
460 IF F>FC THEN F = 10 * F
470 PRINT"AT ";F;" HZ.":GOTO 350
480 DB = 20 * ((LOG(FF/AF))/LOG(10))
490 PRINT"THE ROLLOFF SLOPE = ";DB;" DB/DECADE."
500 PP = 0:PRINT"":GOTO 50
510 REM DESIGN SECTION
520 INPUT"ENTER THE LOAD RESISTANCE. ";RL
530 INPUT"ENTER THE GENERATOR RESISTANCE. ";RG
540 RT = RG + RL
550 INPUT"ENTER THE DESIRED CENTER FREQUENCY. ";FC
560 INPUT"ENTER THE DESIRED BANDWIDTH. ";BW
570 Q = FC/BW
580 C = .159155/(FC * RT * Q)
590 L = .159155 * RT * Q/FC
600 PRINT"":PRINT"THE REQUIRED INDUCTANCE IS ";L;" HENRIES."
610 PRINT"THE REQUIRED CAPACITANCE IS ";C;" FARADS."
620 GOTO 50
630 CLS:END
```

Parallel Resonant Bandpass Filters

Parallel Resonant Bandpass Frequency Response

Figure 21-34 illustrates the inverted L parallel resonant bandpass filter. Note that the parallel resonant circuit is in parallel with the load. The signal provided by the source must pass over the tank to the load. From previous study, you know the parallel resonant circuit acts like an open (ideally) at the resonant frequency. Thus, the resonant frequency is passed on to the load resistor R_L with little or no attenuation. Above and below resonance, I_C and I_L become unbalanced and do not cancel. This means the line current to the tank increases and the tank impedance decreases. Since the output voltage is developed across the tank and the load resistance, the decreased tank impedance causes a decrease in output voltage. The farther away from resonance a frequency is, the greater the decrease in output voltage, with most of the source voltage dropped across the series resistance.

Determining Center Frequency and Bandpass

The center frequency, for a parallel resonant bandpass filter, such as is shown in Figure 21-34, is found using the same basic resonant-frequency formula. This formula is reasonably accurate for tank circuits having a $Q > 10$. For low-Q tank circuits, the correction factor $\sqrt{Q^2/(Q^2 + 1)}$ must be used (see Section 20-7).

$$(20.1) \quad f_c = f_r = 1/(2\pi\sqrt{LC})$$

$$(20.25) \quad Q_{\text{ckt}} = (R_P \parallel Z_{\text{tank}})/X_L \quad \text{(circuit } Q \text{ for parallel resonance)}$$

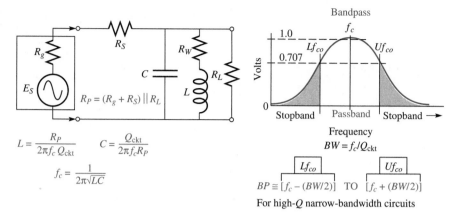

$$L = \frac{R_P}{2\pi f_c \, Q_{ckt}} \qquad C = \frac{Q_{ckt}}{2\pi f_c R_P}$$

$$f_c = \frac{1}{2\pi\sqrt{LC}}$$

$$BW = f_c / Q_{ckt}$$

$$BP \cong [f_c - (BW/2)] \quad \text{TO} \quad [f_c + (BW/2)]$$

For high-Q narrow-bandwidth circuits

Figure 21-34   Parallel resonant bandpass filters.

where $R_P = (R_g + R_S) \| R_L$ and $Z_{tank} = X_L^2/R_W$.

(20.8) $BW = f_c/Q_{ckt}$

(20.9) $BP \cong [f_c - (BW/2)] \rightarrow [f_c + (BW/2)]$ (approximation for high Q_{ckt})

Designing for a Specific f_c and BW

To design a parallel resonant bandpass filter for a specific frequency and bandwidth, the inductor and capacitor must be chosen so that X_L and X_C are Q_{ckt} times less than $(R_g + R_S) \| R_L$. Recall that $f_c/BW = Q_{ckt}$ and $Q_{ckt} \cong [(R_g + R_S) \| R_L]/X_L$. Therefore, the required X_L and X_C equal $[(R_g + R_S) \| R_L]/Q_{ckt}$. The following formulas include Q_{ckt} to meet this requirement: (A high-Q inductor must be used for these design formulas to be accurate. The effects of winding resistance (R_W) are ignored for simplicity.)

$$L = R_P/(2\pi f_c Q_{ckt}) \tag{21.16}$$

$$C = Q_{ckt}/(2\pi f_c R_P) \tag{21.17}$$

where $Q_{ckt} = f_c/BW$ and $R_P = (R_g + R_S) \| R_L$.

EXAMPLE   21-9

Suppose you want a filter that will pass frequencies from 7 MHz to 7.5 MHz. The signal source has a resistive impedance R_g of 50 Ω, a 100 kΩ series resistance is used (R_S), and the load is 1 MΩ (R_L). The total equivalent parallel resistance is 90,950 Ω ($R_P = 100,050\ \Omega \| 1\ M\Omega$). The center frequency is approximately 7.25 MHz. The total bandwidth is 500 kHz ($BW = 7.5$ MHz $-$ 7 MHz). The circuit Q must be approximately 14.5 ($Q_{ckt} = f_c/BW = 7.25$ MHz/0.5 MHz). By applying the values of R_P and Q_{ckt} to Formulas 21.16 and 21.17, we find L and C to be:

$$L = \frac{90,950\ \Omega}{2 \cdot \pi \cdot 7.25\ \text{MHz} \cdot 14.5} \cong 138\ \mu\text{H}$$

$$C = \frac{14.5}{2 \cdot \pi \cdot 7.25\ \text{MHz} \cdot 90,950} \cong 3.5\ \text{pF}$$

EXAMPLE 21-10

INSERTION LOSS FORMULA

$$dB_{atten.} = 20 \log \left[\frac{(-jX_C \parallel +jX_L \parallel R_L)}{(-jX_C \parallel +jX_L \parallel R_L) + (R_g + R_S)} \right]$$

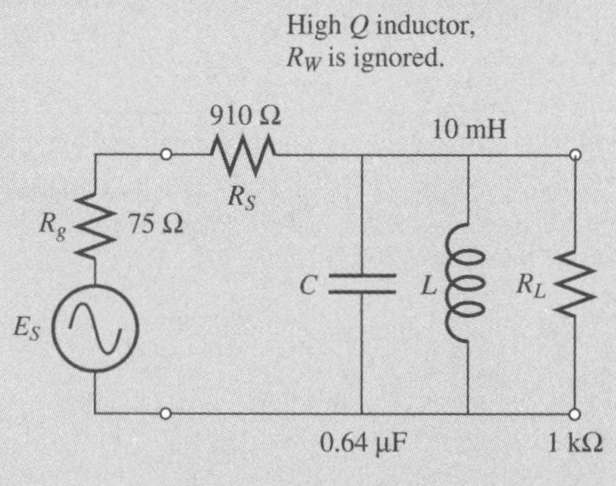

High Q inductor,
R_W is ignored.

Figure 21-35

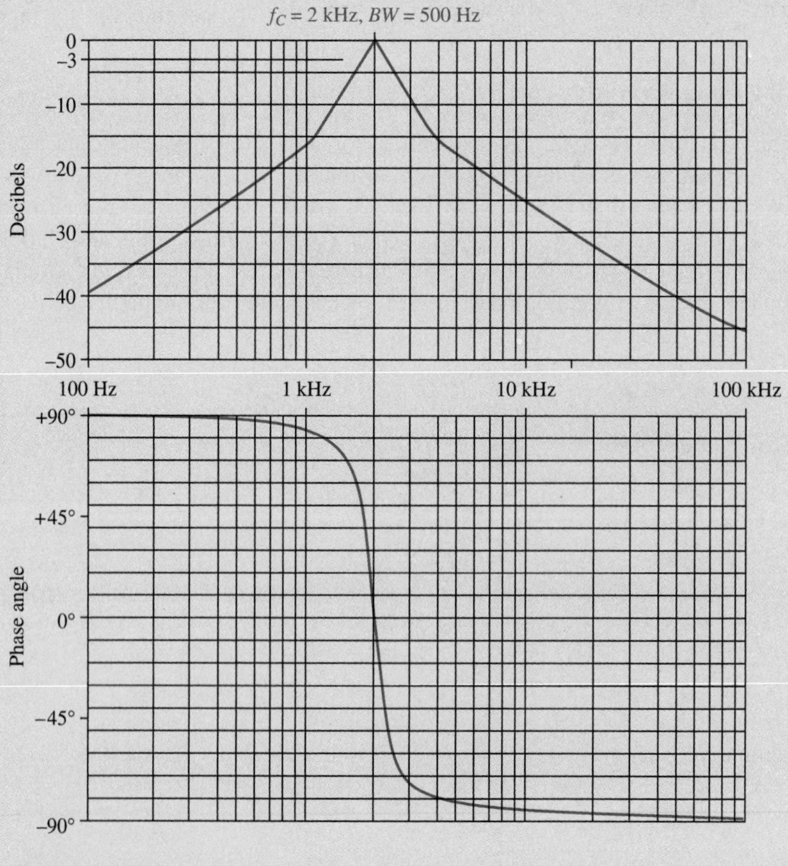

Figure 21-36

$$V_{out} \angle x = I_t \angle y \cdot Z_{out} \angle z$$

OUTPUT VOLTAGE PHASE ANGLE $= \angle x = \angle y + \angle z$

Parallel Resonant Bandpass Phase vs Frequency

Note in Example 21-10 that the output-voltage phase angle changes in the same manner as for the series resonant bandpass filter: $0°$ at f_c, $+45°$ at Lf_{co}, and $-45°$ at Uf_{co}. Above the center frequency, the circuit acts capacitively, since $I_C > I_L$. This causes the output voltage to lag the total current and source voltage (ICE). Below the center frequency, the circuit acts inductively, since $I_L > I_C$. This causes the output voltage to lead the total current and source voltage (ELI). For the parallel resonant bandpass filter, the output-voltage phase angle at any frequency is the sum of the phase angles for total current and output impedance ($V_{out} \angle x = I_t \angle y \cdot Z_{out} \angle z$ where $\angle x = \angle y + \angle z$.

Parallel Resonant Bandpass Bode Plot Analysis

Once again, the Bode plot of Example 21-10 illustrates the characteristic double slope of a bandpass filter. As with series resonant bandpass filters, the actual rolloff slope on each side depends on the circuit Q; the higher the Q, the steeper the slope. For a parallel resonant bandpass filter, the voltage divider expression is as shown in Example 21-10; it must be resolved following the rules for complex algebra. The voltage divider expression has also been simplified by ignoring the effects of the inductor winding resistance (R_W is assumed to be $0\ \Omega$). Thus, this formula is only accurate when high-Q inductors are used. Design Note 21-4 summarizes the analysis and design of the parallel resonant bandpass filter.

Series + Parallel Resonant Bandpass Filters

As shown in Figure 21-37, series resonant circuits and parallel resonant circuits can be combined to form very selective bandpass filters. Filters that employ more than one resonant circuit are able to achieve very steep rolloff slopes of well over 100 dB/decade (depending on circuit Q and number of tuned circuits employed). When series and parallel resonant circuits are combined, it becomes practical to consider different configurations such as the T and π.

The mathematical analysis and design of these series + parallel resonant filters is very complex and well beyond the intended purpose of this text. In practice, look-up tables and computer programs are used to aid in the design of these filters.

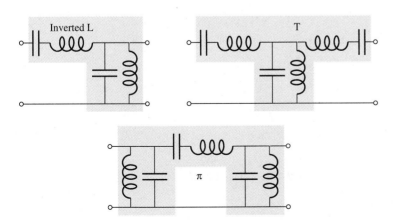

Figure 21-37 Series + parallel resonant bandpass filters.

DESIGN NOTE 21-4:   Parallel Resonant Bandpass

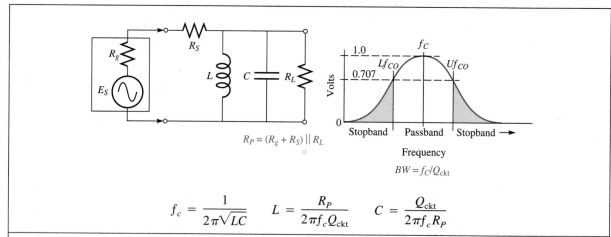

$$R_P = (R_g + R_S) \,\|\, R_L$$

$$BW = f_C/Q_{ckt}$$

$$f_c = \frac{1}{2\pi\sqrt{LC}} \qquad L = \frac{R_P}{2\pi f_c Q_{ckt}} \qquad C = \frac{Q_{ckt}}{2\pi f_c R_P}$$

DESIGN AND ANALYSIS PROGRAM

```
10 CLS
20 PRINT"PARALLEL RESONANT BANDPASS FILTERS"
30 PRINT"":PRINT""
40 PRINT"THIS IS A DESIGN AND ANALYSIS PROGRAM FOR PARALLEL RESONANT BANDPASS FILTERS"
60 PRINT""
70 PRINT"(1) ANALYZE A PARALLEL RESONANT BANDPASS FILTER"
80 PRINT"(2) DESIGN A PARALLEL RESONANT BANDPASS FILTER"
90 PRINT"(3) QUIT"
100 PRINT""
110 INPUT"SELECT 1, 2, OR 3 - ";N
120 IF N<1 OR N>3 THEN GOTO 60
125 PRINT""
130 ON N GOTO 140,590,720
140 INPUT"ENTER THE INDUCTOR VALUE IN HENRIES. ";L
150 INPUT"ENTER THE CAPACITOR VALUE IN FARADS. ";C
160 INPUT"ENTER THE SERIES RESISTANCE. ";RS
170 INPUT"ENTER THE GENERATOR RESISTANCE. ";RG
180 INPUT"ENTER THE LOAD RESISTANCE. ";RL
190 IF RL = 0 THEN RL = 1E-12
200 FC = .159155/SQR(L * C)
210 PRINT"THE CENTER FREQUENCY IS ";FC;" HZ."
220 XL = 6.2832 * FC * L
230 REM RP = EQUIVALENT PARALLEL RESISTANCE
240 RP = 1/((1/(RG + RS)) + (1/RL))
250 Q = RP/XL
260 PRINT"THE CIRCUIT Q IS ";Q
270 BW = FC/Q
280 PRINT"THE BANDWIDTH IS ";BW;" HZ."
330 IF Q<10 THEN GOTO 360
340 PRINT"THE LOWER CUTOFF FREQUENCY IS APPROXIMATELY ";FC - (BW/2);" HZ."
350 PRINT"THE UPPER CUTOFF FREQUENCY IS APPROXIMATELY ";FC + (BW/2);" HZ."
360 PRINT""
370 INPUT"ENTER A TEST FREQUENCY. ";F
380 XL = 6.2832 * F * L
390 XC = 1/(6.2832 * F * C)
```

```
400 ZOUT = 1 / SQR((1/RL)^2 + (1/XC − 1/XL)^2)
410 A1 = −ATN((1/XC − 1/XL)/(1/RL)) ∗ 57.29578
420 R = COS(A1/57.29578) ∗ ZOUT
430 X = SIN(A1/57.29578) ∗ ZOUT
440 ZT = SQR((RG + RS + R)^2 + X^2)
450 A2 = ATN(X/(RG + RS + R)) ∗ 57.29578
460 A = A1 − A2
470 AF = ZOUT/ZT
480 PRINT"THE PHASE ANGLE IS ";A;" DEGREES."
490 PRINT"THE ATTENUATION FACTOR = ";AF
500 PRINT""
510 IF PP = 1 THEN GOTO 560
520 FF = AF:PP = 1
530 IF F<FC THEN F = F/10
540 IF F>FC THEN F = 10 ∗ F
550 PRINT"AT ";F;" HZ.":GOTO 380
560 DB = 20 ∗ ((LOG(FF/AF))/LOG(10))
570 PRINT"THE ROLLOFF SLOPE = ";DB;" DB/DECADE."
580 PP = 0:PRINT"":GOTO 50
590 REM DESIGN SECTION
600 INPUT"ENTER THE SERIES RESISTANCE. ";RS
610 INPUT"ENTER THE LOAD RESISTANCE. ";RL
620 INPUT"ENTER THE GENERATOR RESISTANCE. ";RG
630 INPUT"ENTER THE DESIRED CENTER FREQUENCY. ";FC
640 INPUT"ENTER THE DESIRED BANDWIDTH. ";BW
650 RP = 1/((1/(RG + RS)) + (1/RL))
660 Q = FC/BW
670 C = Q/(6.283 ∗ FC ∗ RP)
680 L = RP/(6.283 ∗ FC ∗ Q)
690 PRINT"":PRINT"THE REQUIRED INDUCTANCE IS ";L;" HENRIES."
700 PRINT"THE REQUIRED CAPACITANCE IS ";C;" FARADS."
710 GOTO 50
720 CLS:END
```

Highpass/Lowpass Bandpass Filters

How can you design a bandpass filter to have a wide bandpass and at the same time have fairly steep rolloffs (skirts)? This is an interesting problem, since for a resonant filter, a wide bandwidth requires a low circuit Q, which means the rolloff is very gradual (20 dB/decade or less). However, we do *not* have to use a resonant filter to obtain a bandpass response. We can combine a highpass LC filter and a lowpass LC filter to get any desired bandwidth and at the same time have skirt slopes of 40 dB/decade. By using this approach, the bandpass width does not depend on circuit Q.

Figure 21-38 illustrates a very common application for the highpass/lowpass bandpass filter. This is a filter designed to pass a band of midrange audio frequencies (800 Hz to 6 kHz) to a midrange speaker. The highpass part of the filter is the inverted L section at the input. The cutoff frequency for this filter is about 800 Hz. All frequencies above 800 Hz pass through this highpass section with very little attenuation. The next section is the lowpass inverted L filter with a cutoff frequency of about 6,000 Hz. As you can see from the Bode plot, the overlapping responses of the highpass and lowpass filters produce the desired bandpass response. Thus, a bandwidth of 5,200 Hz is obtained with rolloff slopes of 40 dB/decade.

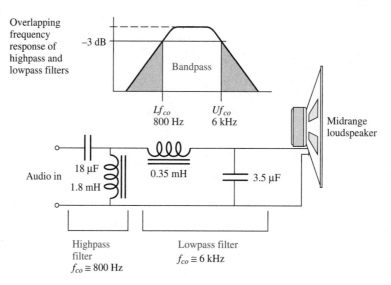

Overlapping
frequency
response of
highpass and
lowpass filters

−3 dB

Bandpass

Lf_{co}
800 Hz

Uf_{co}
6 kHz

Midrange
loudspeaker

Audio in

18 µF

1.8 mH

0.35 mH

3.5 µF

Highpass
filter
$f_{co} \cong 800$ Hz

Lowpass filter
$f_{co} \cong 6$ kHz

Figure 21-38   Highpass-lowpass bandpass filter.

SELF-CHECK 21-4

1. Briefly explain how a series resonant circuit that is in series with the signal path to the load can act as a bandpass filter.

2. What determines the bandwidth and cutoff frequencies for a series resonant bandpass filter?

3. Calculate the approximate center frequency for a lower cutoff frequency of 20 MHz and an upper cutoff frequency of 20.5 MHz.

4. Determine the proper value of L and C for a series resonant bandpass filter that has a center frequency of 150 kHz, a bandwidth of 10 kHz, and a load resistance of 200 Ω (R_g and $R_W = 0$ Ω).

5. What basically determines the slope of the skirts for the Bode plot of a resonant bandpass filter?

6. What kind of a bandpass filter has wide bandwidth, yet very steep rolloff?

21-5 Bandstop Filters

Series Resonant Bandstop Filters

Series Resonant Bandstop Frequency Response

See color photo insert page D12, photograph 75.

Figure 21-39 illustrates the inverted L series resonant bandstop filter. In this filter, the signal provided by the source must pass through the series resistor and across the series resonant circuit to the load. Since the series resonant circuit acts like a short at the resonant frequency (except for inductor winding resistance, R_W), there is no voltage available at the output. At resonance, the source voltage is dropped entirely across the series resistor.

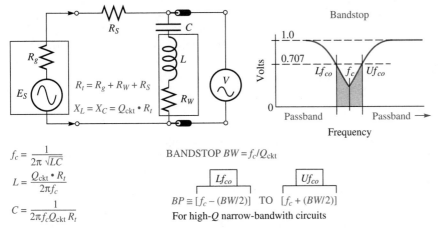

$$f_c = \frac{1}{2\pi \sqrt{LC}}$$

$$L = \frac{Q_{ckt} \cdot R_t}{2\pi f_c}$$

$$C = \frac{1}{2\pi f_c Q_{ckt} R_t}$$

BANDSTOP $BW = f_c / Q_{ckt}$

| Lf_{co} | | Uf_{co} |

$BP \cong [f_c - (BW/2)]$ TO $[f_c + (BW/2)]$
For high-Q narrow-bandwith circuits

Figure 21-39 Series resonant bandstop filters.

Above and below resonance, where X_C and X_L do not completely cancel, an output voltage is developed across the difference reactance $(X_L - X_C)$. The farther away from resonance a frequency is, the greater the voltage drop across the series resonant circuit and load. Since these filters reject a band of frequencies between two cutoff frequencies, they are often referred to as **bandreject filters**.

Determining Center Frequency and Bandstop

The design and analysis of the series resonant bandstop filter is accomplished using the same formulas as for bandpass filters.

E X A M P L E 21-11

Suppose you want to design a filter that will reject frequencies from 55 Hz to 65 Hz. We'll assume the signal source has a resistive impedance of 75 Ω (R_g) and the series resistance is 1 kΩ (R_S). The total resistance is 1,075 Ω (R_t) (we'll assume R_W is insignificant). The center frequency is 60 Hz. The total bandwidth is 10 Hz ($BW = 65 - 55$ Hz). Therefore, the circuit Q must be 6 ($Q_{ckt} = f_c/BW = 60$ Hz/10 Hz). Thus, using Formulas 21.14 and 21.15:

$$L = \frac{6 \cdot 1{,}075 \ \Omega}{2 \cdot \pi \cdot 60 \ \text{Hz}} = 17.1 \ \text{H}$$

$$C = \frac{1}{2 \cdot \pi \cdot 6 \cdot 60 \ \text{Hz} \cdot 1{,}075 \ \Omega} \cong 0.411 \ \mu\text{F}$$

Series Resonant Bandstop Phase vs Frequency

Example 21-12 illustrates the analysis of a series resonant bandstop filter. As shown in the phase plot, the output-voltage phase angle is 0° at the center frequency, $-45°$ at the lower cutoff frequency, and $+45°$ at the upper cutoff frequency. At the lower cutoff frequency, $X_C > X_L$ and the circuit acts capacitively. Since the output is taken across L and C, and it is acting capacitively, the phase angle of the output voltage must be negative

(ICE). At the upper cutoff frequency, $X_L > X_C$ and the circuit acts inductively. Thus, the output voltage has a positive phase angle (ELI). At any frequency, the output-voltage phase angle is determined as the sum of the output-impedance phase angle and the total-current phase angle ($V_{out} \angle x = I_t \angle y \cdot Z_{out} \angle z$, where $\angle x = \angle y + \angle z$).

Notice from the phase-angle graph how the output-voltage phase angle switches very quickly between $+90°$ and $-90°$ as the frequency is varied slightly above and below the center frequency. This represents a $180°$ phase change over a very small change in frequency. Why does this occur? This occurs because the output impedance changes very quickly between being inductive or capacitive on either side of f_c. At the same time, the current phase angle remains very close to $0°$ slightly above and below f_c, since R_S is very large compared to the slight difference between X_C and X_L. Thus, below f_c, $V_{out} \angle -90° = I_t \angle 0° \cdot Z_{out} \angle -90°$ and, above f_c, $V_{out} \angle +90° = I_t \angle 0° \cdot Z_{out} \angle +90°$.

Series Resonant Bandstop Bode Plot Analysis

The Bode plot of Example 21-12 illustrates the characteristic double slope of a bandstop filter. For bandstop filters, the double slope produces a notch in the Bode plot. It is for this reason that bandstop filters are also called **notch filters**. Like bandpass filters, the operating bandwidth of the notch filter is the frequency spacing between the lower and upper cutoff frequencies (between the lower and upper half-power points, the -3 dB points). The actual rolloff slope on each side will depend on the circuit Q and the inductor Q; the higher the circuit Q and inductor Q, the steeper the slope. If the slope is gradual, the depth of the notch will be shallow, reaching a maximum attenuation of, say, 10 dB, 20 dB, or 30 dB (notch at -10 dB, -20 dB, or -30 dB). If the slope is steep, due to high circuit and inductor Q, the depth of the notch will be great, reaching maximum attenuations in the neighborhood of 60 dB, 80 dB, or more (notch at -60 dB, -80 dB, or more). The slope and notch depth can be made steeper and deeper by decreasing the total circuit resistance, particularly the inductor-winding resistance. The winding resistance is the last remaining output impedance when X_C and X_L cancel at the center frequency. Thus, the output voltage is developed across this small resistance: the smaller the winding resistance, the lower the output voltage and the deeper the notch. For a series resonant bandstop filter, the voltage divider expression is: $V_{out}/V_{in} = \sqrt{R_W{}^2 + (X_L - X_C)^2}/\sqrt{R_t{}^2 + (X_L - X_C)^2}$, (where $R_t = R_g + R_W + R_S$). Maximum insertion loss occurs at the center notch frequency where X_C and X_L cancel, leaving the voltage divider expression as R_W/R_t. Design Note 21-5 summarizes the analysis and design of the series resonant bandstop filter.

Parallel Resonant Bandstop Filters

Parallel Resonant Bandstop Frequency Response

Figure 21-42 illustrates the inverted L parallel resonant bandstop filter. Note that the parallel resonant circuit is in series with the load. The signal provided by the source must pass through the tank to the load. As you know, the parallel resonant circuit acts like an open (ideally) at the resonant frequency. Thus, the resonant frequency is blocked from getting to the load resistor (R_L). In practice, the tank circuit is not an absolute open—just a very high impedance. Recall that the impedance of the tank depends on the X_L and the inductor's winding resistance ($Z_{tank} = X_L{}^2/R_W$). Thus, most of the source voltage at resonance is dropped across the impedance of the tank and not the load. Above and below resonance, the tank impedance decreases, dropping less voltage and leaving more voltage for the load resistance. The farther away from resonance a frequency is, the greater the output voltage.

EXAMPLE 21-12

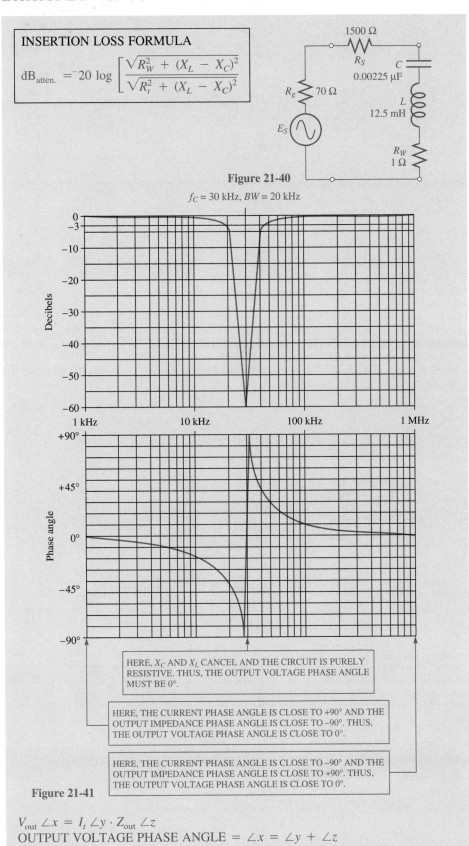

INSERTION LOSS FORMULA

$$dB_{atten.} = -20 \log \left[\frac{\sqrt{R_W^2 + (X_L - X_C)^2}}{\sqrt{R_t^2 + (X_L - X_C)^2}} \right]$$

1500 Ω
R_S
C
0.00225 μF

R_g 70 Ω

E_S

L
12.5 mH

R_W
1 Ω

Figure 21-40

$f_C = 30$ kHz, $BW = 20$ kHz

HERE, X_C AND X_L CANCEL AND THE CIRCUIT IS PURELY RESISTIVE. THUS, THE OUTPUT VOLTAGE PHASE ANGLE MUST BE 0°.

HERE, THE CURRENT PHASE ANGLE IS CLOSE TO +90° AND THE OUTPUT IMPEDANCE PHASE ANGLE IS CLOSE TO –90°. THUS, THE OUTPUT VOLTAGE PHASE ANGLE IS CLOSE TO 0°.

HERE, THE CURRENT PHASE ANGLE IS CLOSE TO –90° AND THE OUTPUT IMPEDANCE PHASE ANGLE IS CLOSE TO +90°. THUS, THE OUTPUT VOLTAGE PHASE ANGLE IS CLOSE TO 0°.

Figure 21-41

$V_{out} \angle x = I_t \angle y \cdot Z_{out} \angle z$

OUTPUT VOLTAGE PHASE ANGLE $= \angle x = \angle y + \angle z$

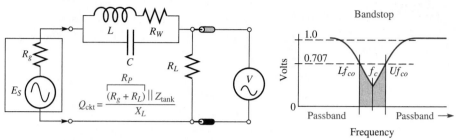

$Q_{ckt} \cong R_P/X_L$ when R_W is very low and Z_{tank} is very large. Therefore, at f_c, $X_L \cong R_P/Q_{ckt}$

$$f_c = \frac{1}{2\pi\sqrt{LC}}$$

$$L = \frac{R_P}{2\pi f_c Q_{ckt}}$$

$$C = \frac{Q_{ckt}}{2\pi f_c R_P}$$

BANDSTOP $BW = f_c/Q_{ckt}$

Bandstop $\cong [f_c - (BW/2)]$ TO $[f_c + (BW/2)]$

For high-Q narrow-bandwidth circuits

Figure 21-42 Parallel resonant bandstop filters.

Determining Center Frequency and Bandstop

The design and analysis of the series resonant bandstop filter is accomplished using the same formulas as for bandpass filters.

DESIGN NOTE 21-5: Series Resonant Bandstop

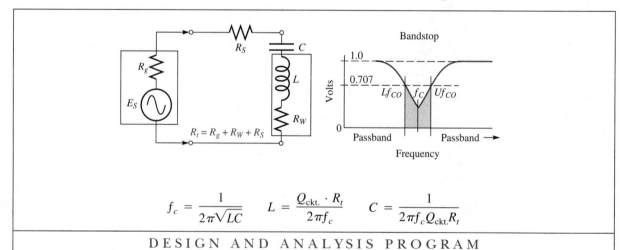

$$f_c = \frac{1}{2\pi\sqrt{LC}} \qquad L = \frac{Q_{ckt.} \cdot R_t}{2\pi f_c} \qquad C = \frac{1}{2\pi f_c Q_{ckt.} R_t}$$

DESIGN AND ANALYSIS PROGRAM

```
10 CLS
20 PRINT"SERIES RESONANT BANDSTOP FILTERS"
30 PRINT"":PRINT""
40 PRINT"THIS IS A DESIGN AND ANALYSIS PROGRAM FOR SERIES RESONANT BANDSTOP FILTERS"
50 PRINT"** NOTE: XL = XC = Q * RT AT RESONANCE. **"
60 PRINT""
70 PRINT"(1) ANALYZE A SERIES RESONANT BANDSTOP FILTER"
```

```
 80 PRINT"(2) DESIGN A SERIES RESONANT BANDSTOP FILTER"
 90 PRINT"(3) QUIT"
100 PRINT""
110 INPUT"SELECT 1, 2, OR 3 − ";N
120 IF N<1 OR N>3 THEN GOTO 60
125 PRINT""
130 ON N GOTO 140,480,600
140 INPUT"ENTER THE INDUCTOR VALUE IN HENRIES. ";L
150 INPUT"ENTER THE CAPACITOR VALUE IN FARADS. ";C
160 INPUT"ENTER THE SERIES RESISTANCE. ";RS
170 INPUT"ENTER THE GENERATOR RESISTANCE. ";RG
180 INPUT"ENTER THE INDUCTOR WINDING RESISTANCE. ";RW
190 FC = .159155/SQR(L * C)
200 RT = RG + RW + RS
210 PRINT"THE CENTER FREQUENCY IS ";FC;" HZ."
220 XL = 6.2832 * FC * L
230 PRINT"THE CIRCUIT Q IS ";XL/RT
240 BW = FC/(XL/RT)
250 PRINT"THE BANDWIDTH IS ";BW;" HZ."
300 IF XL/RT<10 THEN GOTO 330
310 PRINT"THE LOWER CUTOFF FREQUENCY IS APPROXIMATELY ";FC − (BW/2);" HZ."
320 PRINT"THE UPPER CUTOFF FREQUENCY IS APPROXIMATELY ";FC + (BW/2);" HZ."
330 PRINT""
340 INPUT"ENTER A TEST FREQUENCY. ";F
350 XL = 6.2832 * F * L
360 XC = 1/(6.2832 * F * C)
370 ZT = SQR(RT^2 + (XL − XC)^2)
380 A1 = ATN((XC − XL)/RT) * 57.29578
390 IF RW = 0 THEN RW = 1E − 12
400 A2 = ATN((XL − XC)/RW) * 57.29578
410 A = A1 + A2
420 PRINT"THE PHASE ANGLE IS ";A;" DEGREES."
430 AF = SQR(RW^2 + (XL − XC)^2) / ZT
440 PRINT"THE ATTENUATION FACTOR = ";AF
450 DB = 20 * (LOG(1/AF)/LOG(10))
460 PRINT"THE INSERTION LOSS AT ";F;" HZ IS ";DB;" DB."
470 GOTO 50
480 REM DESIGN SECTION
490 INPUT"ENTER THE SERIES RESISTANCE. ";RS
500 INPUT"ENTER THE GENERATOR RESISTANCE. ";RG
510 RT = RG + RS
520 INPUT"ENTER THE DESIRED CENTER FREQUENCY. ";FC
530 INPUT"ENTER THE DESIRED BANDWIDTH. ";BW
540 Q = FC/BW
550 C = .159155/(FC * RT * Q)
560 L = .159155 * RT * Q/FC
570 PRINT"":PRINT"THE REQUIRED INDUCTANCE IS ";L;" HENRIES."
580 PRINT"THE REQUIRED CAPACITANCE IS ";C;" FARADS."
590 GOTO 50
600 CLS:END
```

EXAMPLE 21-13

Let's suppose you want a filter that will reject frequencies from 12 kHz to 14 kHz. Let's say the signal source has a resistive impedance of 70 Ω (R_g), and the load resistance is 600 Ω (R_L). The center frequency is 13 kHz. The total bandwidth is 2 kHz ($BW = 14$ kHz $- 12$ kHz). The circuit Q must be 6.5 ($Q_{ckt} = f_c/BW =$ 13 kHz/2 kHz). $R_g + R_L$ is in parallel with the tank for the purpose of determining circuit Q ($Q_{ckt} = (R_g + R_L)/Z_L$). By applying the values of R_P and Q_{ckt} to Formulas 21.16 and 21.17, we find L and C to be: ($R_P = R_g + R_L = 670$ Ω; $Q_{ckt} = 6.5$)

$$L = \frac{670 \ \Omega}{2 \cdot \pi \cdot 13 \ \text{kHz} \cdot 6.5} \cong 1.26 \ \text{mH}$$

$$C = \frac{6.5}{2 \cdot \pi \cdot 13 \ \text{kHz} \cdot 670 \ \Omega} \cong 0.12 \ \mu\text{F}$$

Parallel Resonant Bandstop Phase vs Frequency

Note, in Example 21-14, the output-voltage phase angle changes in the same manner as for the series resonant bandstop filter: $0°$ at f_c, $-45°$ at Lf_{co}, and $+45°$ at Uf_{co}. Above the center frequency, the circuit acts capacitively, since $I_C > I_L$. This causes the output voltage to lead the source voltage (ICE). Below the center frequency, the circuit acts inductively, since $I_L > I_C$. This causes the output voltage to lag the source voltage (ELI). For the parallel resonant bandstop filter, the output-voltage phase angle at any frequency is the sum of the phase angles for total current and output impedance ($V_{out} \angle x = I_t \angle y \cdot Z_{out} \angle z$ where $\angle x = \angle y + \angle z$; when polar numbers are multiplied, their angles are added). In this case, $\angle z = 0°$ since the output impedance is a resistor. Therefore, the output-voltage phase angle is the same as the total-current phase angle ($V_{out} \angle x = I_t \angle y$). Like the series resonant bandstop filter, the output-voltage phase angle switches quickly between $+90°$ and $-90°$ with slight frequency changes above and below f_c. In this case, it is because the tank circuit changes abruptly from inductive to capacitive and vice versa as I_C and I_L are imbalanced slightly above and below f_c. The abrupt current phase shift from $+90°$ (capacitive) to $-90°$ (inductive) causes the output voltage to abruptly change phase.

EXAMPLE 21-14

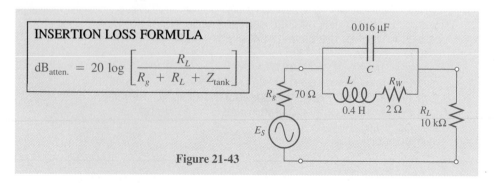

INSERTION LOSS FORMULA

$$\text{dB}_{\text{atten.}} = 20 \log \left[\frac{R_L}{R_g + R_L + Z_{\text{tank}}} \right]$$

Figure 21-43

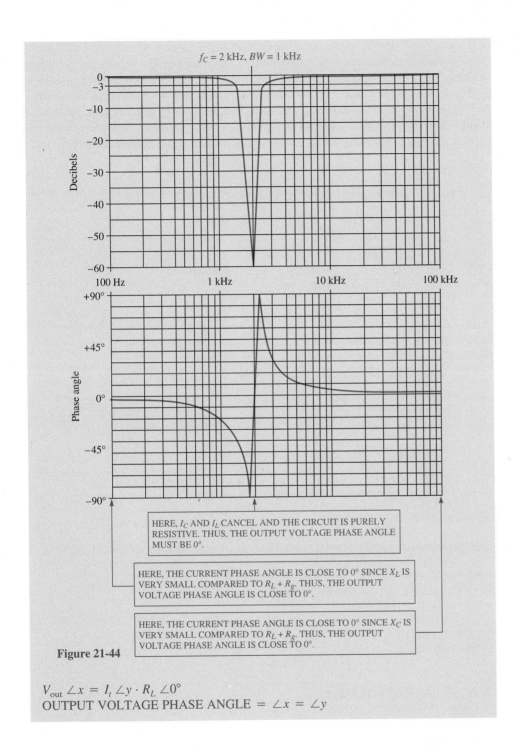

$f_C = 2$ kHz, $BW = 1$ kHz

HERE, I_C AND I_L CANCEL AND THE CIRCUIT IS PURELY RESISTIVE. THUS, THE OUTPUT VOLTAGE PHASE ANGLE MUST BE 0°.

HERE, THE CURRENT PHASE ANGLE IS CLOSE TO 0° SINCE X_L IS VERY SMALL COMPARED TO $R_L + R_g$. THUS, THE OUTPUT VOLTAGE PHASE ANGLE IS CLOSE TO 0°.

HERE, THE CURRENT PHASE ANGLE IS CLOSE TO 0° SINCE X_C IS VERY SMALL COMPARED TO $R_L + R_g$. THUS, THE OUTPUT VOLTAGE PHASE ANGLE IS CLOSE TO 0°.

Figure 21-44

$V_{\text{out}} \angle x = I_t \angle y \cdot R_L \angle 0°$
OUTPUT VOLTAGE PHASE ANGLE $= \angle x = \angle y$

Parallel Resonant Bandstop Bode Plot Analysis

Once again, the Bode plot of Example 21-14 illustrates the notch filters' characteristics. Again, the actual rolloff slope on each side of the notch will depend on the circuit and inductor Q; the higher the Q, the steeper the slopes and the deeper the notch. The depth of the notch will depend on the tank impedance, which in turn depends on the inductor Q. When tank impedance is high, tank voltage is high, leaving less voltage for the output (a deeper notch). For a parallel resonant bandstop filter, the voltage divider expression is as shown in Example 21-14. This insertion loss formula is in a simplified form and must be expanded and resolved following the rules for complex algebra. The tank impedance changes with frequency and consists of $R_W + jX_L \| -jX_C$. Design Note 21-6 includes the expanded formulas for the analysis and design of parallel resonant bandstop filters.

DESIGN NOTE 21-6:   Parallel Resonant Bandstop

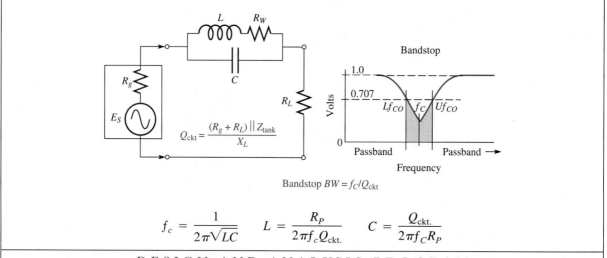

$$f_c = \frac{1}{2\pi\sqrt{LC}} \qquad L = \frac{R_P}{2\pi f_c Q_{ckt.}} \qquad C = \frac{Q_{ckt.}}{2\pi f_c R_P}$$

DESIGN AND ANALYSIS PROGRAM

```
10  CLS
20  PRINT"PARALLEL RESONANT BANDSTOP FILTERS"
30  PRINT"":PRINT""
40  PRINT"THIS IS A DESIGN AND ANALYSIS PROGRAM FOR PARALLEL RESONANT BANDSTOP FILTERS"
60  PRINT""
70  PRINT"(1) ANALYZE A PARALLEL RESONANT BANDSTOP FILTER"
80  PRINT"(2) DESIGN A PARALLEL RESONANT BANDSTOP FILTER"
90  PRINT"(3) QUIT"
100 PRINT""
110 INPUT"SELECT 1, 2, OR 3 – ";N
120 IF N<1 OR N>3 THEN GOTO 60
125 PRINT""
130 ON N GOTO 140,550,670
140 INPUT"ENTER THE INDUCTOR VALUE IN HENRIES. ";L
150 INPUT"ENTER THE CAPACITOR VALUE IN FARADS. ";C
160 INPUT"ENTER THE LOAD RESISTANCE. ";RL
170 INPUT"ENTER THE GENERATOR RESISTANCE. ";RG
180 INPUT"ENTER THE INDUCTOR RESISTANCE. ";RW
190 IF RW = 0 THEN RW = 1E−12
200 FC = .159155/SQR(L ∗ C)
210 PRINT"THE CENTER FREQUENCY IS ";FC;" HZ."
220 XL = 6.2832 ∗ FC ∗ L
```

```
230 REM RP = EQUIVALENT PARALLEL RESISTANCE
240 RP = 1/((1/(RG + RL)) + (1/(XL^2/RW)))
250 Q = RP/XL
260 PRINT"THE CIRCUIT Q IS ";Q
270 BW = FC/Q
280 PRINT"THE BANDWIDTH IS ";BW;" HZ."
330 IF Q<10 THEN GOTO 360
340 PRINT"THE LOWER CUTOFF FREQUENCY IS APPROXIMATELY ";FC - (BW/2);" HZ."
350 PRINT"THE UPPER CUTOFF FREQUENCY IS APPROXIMATELY ";FC + (BW/2);" HZ."
360 PRINT""
370 INPUT"ENTER A TEST FREQUENCY. ";F
380 XL = 6.2832 * F * L
390 XC = 1/(6.2832 * F * C)
400 Z1 = SQR((XC * XL)^2 + (XC * RW)^2) / SQR(RW^2 + (XL - XC)^2)
410 A1 = ATN((XC * RW)/(XC * XL)) * 57.29578
420 A2 = ATN((XL - XC)/RW) * 57.29578
430 A = A1 - A2
440 R = COS(A/57.29578) * Z1
450 X = SIN(A/57.29578) * Z1
460 ZT = SQR((RG + RL + R)^2 + X^2)
470 A3 = ATN(X/(RG + RL + R)) * 57.29578
480 A4 = -A3
490 AF = RL/ZT
500 PRINT"THE PHASE ANGLE = ";A4;" DEGREES."
510 PRINT"THE ATTENUATION FACTOR = ";AF
520 DB = 20 * ((LOG(1/AF))/LOG(10))
530 PRINT"THE INSERTION LOSS = ";DB;" DB."
540 GOTO 60
550 REM DESIGN SECTION
560 INPUT"ENTER THE LOAD RESISTANCE. ";RL
570 INPUT"ENTER THE GENERATOR RESISTANCE. ";RG
580 INPUT"ENTER THE DESIRED CENTER FREQUENCY. ";FC
590 INPUT"ENTER THE DESIRED BANDWIDTH. ";BW
600 RP = RG + RL
610 Q = FC/BW
620 C = Q/(6.2832 * FC * RP)
630 L = RP/(6.2832 * FC * Q)
640 PRINT"":PRINT"THE REQUIRED INDUCTANCE IS ";L;" HENRIES."
650 PRINT"THE REQUIRED CAPACITANCE IS ";C;" FARADS."
660 GOTO 60
670 CLS:END
```

Series + Parallel Resonant Bandstop Filters

As shown in Figure 21-45, series resonant circuits and parallel resonant circuits can be combined to form very selective bandstop filters. Once again, when series and parallel resonant circuits are combined, it becomes practical to consider different configurations such as the T and π.

The inverted L configuration employs a parallel resonant circuit in series with the signal path and a series resonant circuit in parallel with the output. Recall that at resonance the series resonant circuit ideally acts like a short and the tank circuit ideally acts like an

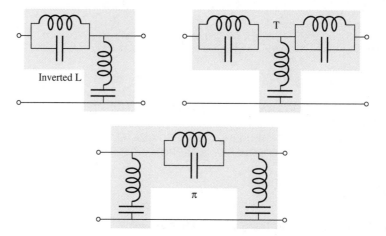

Figure 21-45   Series + parallel resonant bandstop filters.

open. Thus, the applied voltage at the resonant frequency is blocked by the tank, and anything that gets through is shorted to ground by the series resonant circuit. Above or below resonance, the series resonant circuit shows an increase in impedance and the parallel resonant circuit a decrease. As a result, less voltage is dropped across the tank circuit and more across the series resonant circuit at the output. Also, as frequency is increased or decreased above and below the center frequency, the input impedance to the filter decreases while the output impedance increases.

S E L F - C H E C K   2 1 - 5

1. Briefly explain how a series resonant circuit can act as a bandstop filter.

2. What determines the bandstop *BW* and cutoff frequencies for a series resonant bandstop filter?

3. Determine the proper values of *L* and *C* for a series resonant bandstop filter that has a center frequency of 15 kHz, a bandwidth of 1 kHz, and a series resistance of 100 Ω (R_g and R_W = 0 Ω).

4. What basically determines the slope of the skirts for the Bode plot of a resonant bandstop filter?

5. List the factors that determine the circuit *Q* of a parallel resonant bandstop filter.

Summary

FORMULAS

(21.1) $dB_{atten.} = 20 \cdot \log (V_{out}/V_{in})$ (insertion loss formula)

where V_{out}/V_{in} is a voltage divider expression for the particular filter.

High- and Lowpass *RL* and *RC* Filters

(21.2) $f_{co} = R/(2\pi L)$ (for *RL* inverted L filter analysis)

(21.3) $R = 2\pi L f_{co}$ (for design of *RL* inverted L filters)

(21.4) $L = R/(2\pi f_{co})$ (for design of *RL* inverted L filters)

(21.5) $f_{co} = 1/(2\pi CR_L)$ (for *RC* inverted L filter analysis)

(21.6) $R_L = 1/(2\pi C f_{co})$ (for design of *RC* inverted L filters)

(21.7) $C = 1/(2\pi R_L f_{co})$ (for design of *RC* inverted L filters)

Highpass *RCL* Filters

(21.8) $f_{co} = 1/(2.828\pi \sqrt{LC}) \cong 0.113/\sqrt{LC}$
    (for highpass *RCL* inverted L filters, where $R_g = R_L$)

(21.9) $L = R_L/(2.828\pi f_{co}) \cong 0.113\, R_L/f_{co}$

(21.10) $C = 1/(2.828\pi R_L f_{co}) \cong 0.113/(R_L f_{co})$
    (for highpass *RCL* inverted L filter design, where $R_g = R_L$)

Lowpass *RCL* Filters

(21.11) $f_{co} = 1/(1.414\pi \sqrt{LC}) \cong 0.225/\sqrt{LC}$
    (for lowpass *RCL* inverted L filters where $R_g = R_L$)

(21.12) $L = R_L/(1.414\pi f_{co}) \cong 0.225 R_L/f_{co}$

(21.13) $C = 1/(1.414\pi R_L f_{co}) \cong 0.225/(R_L f_{co})$
    (for lowpass *RCL* inverted L filter design, where $R_g = R_L$)

Bandpass and Bandstop Filters

(20.1) $f_c = f_r = 1/(2\pi\sqrt{LC})$ (series or parallel resonance)

(20.8) Bandpass or bandstop $BW = f_c/Q_{ckt}$
    (for any resonant circuit)

(20.9) Bandpass or bandstop $\cong [f_c - (BW/2)] \rightarrow [f_c + (BW/2)]$
    (an approximation for any resonant circuit with $Q_{ckt} > 10$)

Series Resonant Bandpass and Bandstop Filters

(20.4) $Q_{ckt} = X_L/R_t$ (circuit *Q* for series resonance)
    (for bandpass, where $R_t = R_g + R_W + R_L$; for bandstop, where $R_t = R_g + R_W + R_S$)

(21.14) $L = \dfrac{Q_{ckt} \cdot R_t}{2\pi f_c}$

(21.15) $C = 1/(2\pi f_c R_t Q_{ckt})$
    (for bandpass, where $Q_{ckt} = f_c/BW$ and $R_t = R_g + R_W + R_L$; for bandstop, where $Q_{ckt} = f_c/BW$ and $R_t = R_g + R_W + R_S$)

Parallel Resonant Bandpass and Bandstop Filters

(20.25) $Q_{ckt} = (R_P \parallel Z_{tank})/X_L$ (circuit Q for parallel resonance)

(for bandpass, where $R_P = (R_g + R_S) \parallel R_L$ and $Z_{tank} = X_L^2/R_W$; for bandstop, where $R_P = R_g + R_L$ and $Z_{tank} = X_L^2/R_W$)

(21.16) $L = R_P/(2\pi f_c Q_{ckt})$

(21.17) $C = Q_{ckt}/(2\pi f_c R_P)$

(for bandpass, where $Q_{ckt} = f_c/BW$ and $R_P = (R_g + R_S) \| R_L$); for bandstop, where $Q_{ckt} = f_c/BW$ and $R_P = R_g + R_L$)

CONCEPTS

- The four basic filter types are highpass, lowpass, bandpass, and bandstop.
- The three basic filter configurations are inverted L, T, and π.
- The cutoff frequency marks the dividing line between the passband and stopband.
- The amplitude of the cutoff frequency is 0.707 times the maximum voltage of the passband frequencies. (Also known as the half-power frequency and half-power point at -3 dB)
- Frequencies in the stopband roll off at a rate that is measured in dB/octave or dB/decade.
- The rolloff rate is called the slope.
- The Bode plot is a graphic representation of a filter's output voltage in response to changes in applied frequency.
- The insertion loss of a filter is the decibel measure of the amount of attenuation the filter produces at different frequencies.
- The insertion loss at a cutoff frequency is 3 dB.
- Generally, a highpass filter has a series capacitor, a parallel inductor, or both.
- Generally, a lowpass filter has a series inductor, a parallel capacitor, or both.
- A bandpass filter can be made by placing a series resonant circuit in series with a load, a parallel resonant circuit in parallel with a load, or both.
- A bandstop filter can be made by placing a series resonant circuit in parallel with a load or a parallel resonant circuit in series with a load, or both.
- Theoretical rolloff slope for highpass and lowpass RC and RL filters is 6 dB/octave, or 20 dB/decade.
- Theoretical rolloff slope for highpass and lowpass RCL filters is 12 dB/octave, or 40 dB/decade.
- The bandwidth of a bandpass or bandstop filter depends on circuit Q.
- The depth of the notch for bandstop filters is dependent on the inductor Q—specifically R_W.

SPECIAL TERMS

- Passive filters, active filters
- Filter types, filter configurations (inverted L, T, and π)
- Highpass, lowpass, bandpass, bandstop
- Frequency response, cutoff frequency (f_{co})
- Half-power frequency (HPF), half-power point (HPP)
- Rolloff, passband, stopband, skirts, slope

- Linear graph system, semilog graph system
- Log–log graph system
- Bode plot
- Phase response
- Insertion loss
- Bandreject filter, notch filter

Need to Know Solution

1. The alternator whine can be eliminated by placing a heavy-winding laminated-iron-core power choke in series with the + VDC supply line and by placing a large-value capacitor in parallel with the radio (between + VDC and ground).

2. The power-supply filter is inadequate for the load. The filter capacitors are being discharged by the heavy demand for current that occurs during transmit. Much larger capacitors must be added. The larger capacitors will hold a larger charge and will not suffer as great a loss in voltage between rectified DC pulses. The drop in filter capacitor voltage between DC pulses causes regular low-frequency fluctuations in the DC. This fluctuating DC is the transmitted hum.

3. A notch filter is needed to remove the hum while not eliminating too much of the desired audio spectrum. The frequency of the hum must first be identified with an oscilloscope. Then, the notch filter can be designed for a narrow bandstop bandwidth and deep notch.

4. A highpass filter must be designed to pass all frequencies above the cutoff frequency specified for the tweeter. An *RCL* inverted L filter will work well, providing a rolloff slope of 40 dB/decade below cutoff.

Questions

21-1 Introduction to Filters

1. What is a passive filter?
2. What is the purpose of a bandreject filter?
3. The cutoff frequency forms a dividing line between what two bands?
4. Why is the cutoff frequency referred to as a half-power frequency?
5. What is a "skirt"?
6. What is an octave of frequencies? What is a decade of frequencies?
7. What is a Bode plot?
8. What is filter insertion loss?

21-2 Highpass Filters
RL

9. What is the relationship between R and X_L at the cutoff frequency for a highpass *RL* filter?
10. Frequencies below the cutoff frequency of a highpass filter are attenuated more than how many decibels?

11. Explain why the output-voltage phase angle decreases as frequency is increased for a highpass *RL* filter.
12. What is the ideal rolloff slope for an *RL* filter?

RC

13. What is the relationship between *R* and X_C at the cutoff frequency for a highpass *RC* filter?
14. Explain why the output-voltage phase angle increases as frequency is decreased for a highpass *RC* filter.
15. How much insertion loss does a highpass *RC* filter produce at the cutoff frequency?

RCL

16. Draw an *RCL* highpass inverted L filter.
17. Briefly describe the operation of an *RCL* highpass inverted L filter in terms of frequency effects on X_C and X_L.
18. What is the ideal rolloff slope for an *RCL* filter?
19. What are the two general purposes for highpass filters?

21-3 Lowpass Filters
RL

20. What is the relationship between *R* and X_L at the cutoff frequency for a lowpass *RL* filter?
21. Frequencies above the cutoff frequency of a lowpass filter are attenuated more than how many decibels?
22. Explain why the output-voltage phase angle increases as frequency is increased for a lowpass *RL* filter.

RC

23. What is the relationship between *R* and X_C at the cutoff frequency for a lowpass *RC* filter?
24. Explain why the output-voltage phase angle increases as frequency is increased for a lowpass *RC* filter.

RCL

25. Briefly describe the operation of an *RCL* lowpass inverted L filter in terms of frequency effects on X_C and X_L.
26. What are the two general purposes for lowpass filters?
27. List at least two common applications for lowpass filters.

21-4 Bandpass Filters
Series Resonant

28. Describe the operation of a series resonant bandpass filter at the center frequency and above and below the center frequency.
29. What two factors determine the bandwidth for a series resonant bandpass filter?
30. What two factors determine the circuit *Q* for a series resonant bandpass filter?

31. What is the output-voltage phase angle at the center frequency for a series resonant bandpass filter?
32. What factor determines the slope of the skirts for a series resonant bandpass filter?

Parallel Resonant

33. Describe the operation of a parallel resonant bandpass filter at the center frequency and above and below the center frequency.
34. What factors determine the circuit Q and bandwidth for a parallel resonant bandpass filter?
35. What is the output-voltage phase angle at the center frequency for a parallel resonant bandpass filter?
36. What factor determines the slope of the skirts for a parallel resonant bandpass filter?

Series + Parallel Resonant

37. What is the advantage to combining series and parallel resonant circuits to form a bandpass filter?
38. Describe the cooperative interaction of the series resonant and parallel resonant circuits in a bandpass filter.

Highpass/Lowpass

39. Give two reasons for using a highpass filter and a lowpass filter to make a bandpass filter.
40. Draw a highpass/lowpass bandpass filter.

21-5 Bandstop Filters
Series Resonant

41. Explain how a series resonant circuit can be used as a bandstop filter.
42. Give two other names for bandstop filters.
43. What two factors determine the bandstop bandwidth for a series resonant bandstop filter?
44. What are the factors that determine the circuit Q for a series resonant bandstop filter?
45. What is the output-voltage phase angle (a) exactly at the center frequency, (b) slightly below the center frequency, and (c) slightly above the center frequency?
46. What determines the depth of the notch at the center frequency for a series resonant bandstop filter?

Parallel Resonant

47. Explain how a parallel resonant circuit can be used as a bandstop filter.
48. What two factors determine the bandstop bandwidth for a parallel resonant bandstop filter?
49. What are the factors that determine the circuit Q for a parallel resonant bandstop filter?
50. What is the output-voltage phase angle (a) exactly at the center frequency, (b) slightly below the center frequency, and (c) slightly above the center frequency?
51. What determines the depth of the notch at the center frequency for a parallel resonant bandstop filter?

Series + Parallel Resonant

52. What is the advantage to combining series and parallel resonant circuits to form a bandstop filter?
53. Draw a series + parallel resonant bandstop T filter.
54. Describe the cooperative interaction of the series resonant and parallel resonant circuits in a bandstop filter.

Problems

21-2 Highpass Filters
RL

1. What is the cutoff frequency for an *RL* highpass inverted L filter that has a 4.7 kΩ resistor and a 2 mH inductor?
2. Using a 1 kΩ resistor, determine the inductor value needed to make a 5 kHz highpass inverted L filter.
3. A 22 kΩ resistor and a 500 μH inductor form a highpass inverted L filter. Determine the cutoff frequency and the amount of insertion loss at the following frequencies: 1 MHz, 7 MHz, and 20 MHz.

RC

4. Using a 1 kΩ resistor, determine the capacitor value needed to make a 1 kHz highpass inverted L filter.
5. Using a 0.1 μF capacitor, determine the resistor value needed to make a 300 Hz highpass inverted L filter.
6. A 56 kΩ resistor and a 0.01 μF capacitor form a highpass inverted L filter. Determine the cutoff frequency and the amount of insertion loss at the following frequencies: 100 Hz, 250 Hz, and 500 Hz.

RCL

7. What is the cutoff frequency for an *RCL* highpass inverted L filter that has a 600 Ω load resistor, a 0.03 μF capacitor, and a 250 μH inductor?
8. Design an *RCL* highpass inverted L filter for a cutoff frequency of 10 kHz. The filter is to be inserted between a generator resistance and a load resistance of 600 Ω.

21-3 Lowpass Filters
RL

9. What is the cutoff frequency for an *RL* lowpass inverted L filter that has a 6.8 kΩ resistor and a 2 mH inductor?
10. Using a 2.7 kΩ resistor, determine the inductor value needed to make a 3 kHz lowpass inverted L filter.
11. A 33 kΩ resistor and a 1.5 mH inductor form a lowpass inverted L filter. Determine the cutoff frequency and the amount of insertion loss at the following frequencies: 300 kHz, 3 MHz, and 30 MHz.

RC

12. What is the cutoff frequency for an *RC* lowpass inverted L filter that has an 8.2 kΩ resistor and a 0.022 μF capacitor?
13. Using a 910 Ω resistor, determine the capacitor value needed to make a 3 kHz lowpass inverted L filter.
14. A 47 kΩ resistor and a 0.1 μF capacitor form a lowpass inverted L filter. Determine the cutoff frequency and the amount of insertion loss at the following frequencies: 3 Hz, 40 Hz, and 400 Hz.

RCL

15. What is the cutoff frequency for an *RCL* lowpass inverted L filter that has 0.05 μF capacitor, and a 400 μH inductor?
16. Design an *RCL* lowpass inverted L filter for a cutoff frequency of 3 kHz. The filter is to be inserted between a generator resistance and a load resistance of 1 kΩ.

21-4 Bandpass Filters
Series Resonant

17. What is the center frequency for a series resonant bandpass filter that has a 100 μH inductor and an 800 pF capacitor?
18. Design a series resonant bandpass filter for a center frequency of 1 kHz and a bandwidth of 50 Hz. The generator resistance is 70 Ω and the load resistance is 600 Ω. ($R_W \cong 0 \ \Omega$)

Parallel Resonant

19. What is the center frequency for a parallel resonant bandpass filter that has a 10 μH inductor and a 100 pF capacitor? (assume a high-Q inductor)
20. What is the center frequency for a parallel resonant bandpass filter that has a 15 mH inductor and a 0.01 μF capacitor? (assume a high-Q inductor)
21. Design a parallel resonant bandpass filter for a center frequency of 800 Hz and a bandwidth of 80 Hz. The generator resistance is 70 Ω, the series resistance is 10 kΩ, and the load resistance is 10 kΩ. ($R_W \cong 0 \ \Omega$)

21-5 Bandstop Filters
Series Resonant

22. Find the center frequency and bandstop bandwidth for a series resonant bandstop filter having the following parameters: L = 1 mH, C = 0.33 μF, R_S = 10 Ω, R_g = 1 Ω, and R_W = 1 Ω.
23. Design a series resonant bandstop filter with a bandstop from 10 kHz to 14 kHz. (assume the following: R_S = 1 kΩ, R_g = 75 Ω, and R_W = 2 Ω)

Parallel Resonant

24. Find the center frequency and bandstop bandwidth for a parallel resonant bandstop filter having the following parameters: L = 10 mH, C = 0.47 μF, R_L = 1 kΩ, R_g = 50 Ω, and R_W = 2 Ω.

25. Find the center frequency and bandstop bandwidth for a parallel resonant bandstop filter having the following parameters: $L = 5\ \mu H$, $C = 30$ pF, $R_L = 22$ kΩ, $R_g = 70$ Ω, and $R_W \cong 0$ Ω.
26. Design a parallel resonant bandstop filter with a bandstop from 150 kHz to 200 kHz. (assume $R_L = 10$ kΩ, $R_g = 70$ Ω, and $R_W \cong 0$ Ω)

Answers to Self-Checks

Self-Check 21-1

1. Resistors, capacitors, and inductors
2. Highpass, lowpass, bandpass, and bandstop
3. The filter components are arranged in the shape of a T.
4. A Bode plot is a filter's frequency response plotted on a semilog graph system. The horizontal axis is frequency and the vertical axis is scaled in decibels.
5. Insertion loss is the amount of drop of filter output voltage and power at a particular frequency. It is usually expressed in decibels. The Bode plot is created from the insertion loss formula for the filter.

Self-Check 21-2

1. >0.707 · input voltage
2. >3 dB
3. 42,972 Hz
4. 0.00339 μF
5. 6 dB/octave or 20 dB/decade
6. Less than 45°
7. An *RCL* highpass inverted L filter—capacitor in series and inductor in parallel with the load

Self-Check 21-3

1. 525,211 Hz
2. 0.0796 μF
3. Greater than $-45°$
4. An *RCL* lowpass inverted L filter—inductor in series and capacitor in parallel with the load
5. Decreases with X_C as frequency is increased

Self-Check 21-4

1. At the resonant, or center, frequency, the series resonant circuit is a near short passing nearly all voltage to the load. Above and below the center frequency, the impedance of the series resonant circuit increases, dropping voltage and leaving less for the load.
2. Bandwidth is determined by the center frequency and circuit Q, which is determined by X_L and total series resistance.
3. 20.3 MHz
4. $L = 3.18$ mH; $C = 354$ pF

5. Circuit Q
6. Highpass/lowpass bandpass filter

Self-Check 21-5

1. The output voltage is taken across the series resonant circuit. At resonance, the series resonant circuit is a near short having very little voltage drop. Above and below the center frequency, the output voltage increases with the impedance of the series resonant circuit.
2. Center frequency and circuit Q, which is determined by X_L and total series resistance
3. $L = 0.0159$ H; $C = 0.0071$ μF
4. Circuit Q and inductor Q
5. The equivalent parallel resistance ($R_P = R_g + R_L$), and the tank impedance (determined by inductor Q)

SUGGESTED PROJECTS

1. Add some of the main concepts and formulas from this chapter to your Electricity and Electronics Notebook.

2. Construct some of the circuits illustrated in the examples in this chapter. Use semilog graph paper and make a Bode plot for each filter. Compare your Bode plot to those shown in the examples. You will need a variable-frequency generator (bench signal generator) and an oscilloscope (an AC voltmeter can be used as long as the maximum frequency limit of the meter is not exceeded). Consider constructing Examples 21-1, 21-2, 21-4, 21-5, 21-11, and 21-13.

3. Design and build some filters of your own. Use the design and analysis formulas provided in this chapter. Use the BASIC computer programs if you have access to a computer.

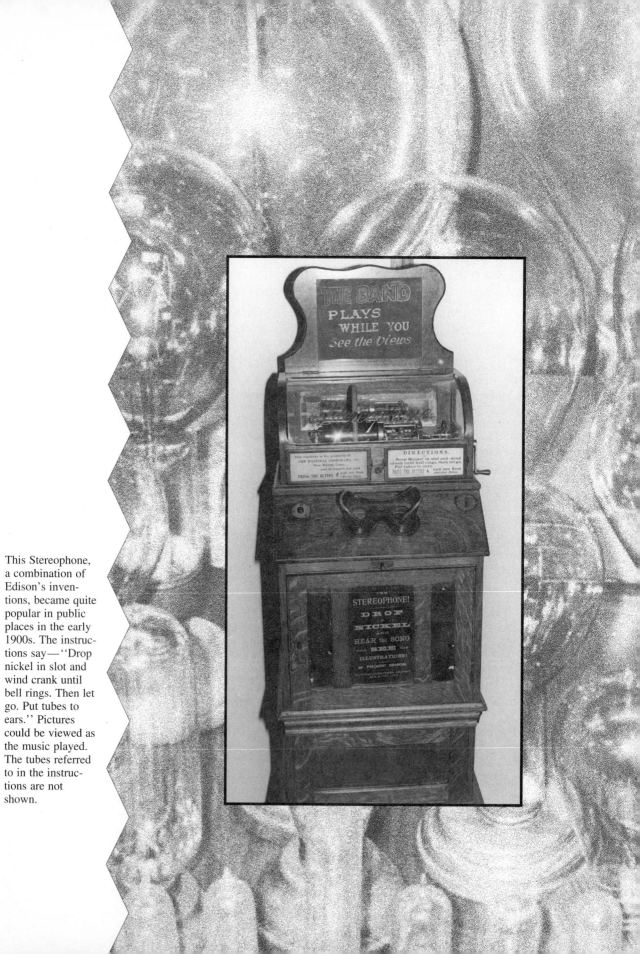

This Stereophone, a combination of Edison's inventions, became quite popular in public places in the early 1900s. The instructions say—''Drop nickel in slot and wind crank until bell rings. Then let go. Put tubes to ears.'' Pictures could be viewed as the music played. The tubes referred to in the instructions are not shown.

Chapter 22

Advanced AC
Circuit Analysis

OBJECTIVES

After studying this chapter, you will
understand and be able to use
• admittance in AC circuit analysis.
• basic circuit-analysis techniques to analyze
  complex AC circuits.
• the superposition theorem for AC circuit
  analysis.

• Thevenin's theorem for AC circuit analysis.
• Norton's theorem for AC circuit analysis.
• Millman's theorem for AC circuit analysis.
• the maximum power transfer theorem for AC
  circuit analysis.

INTRODUCTION

This final chapter covering advanced AC circuit analysis will be a challenge to your ability
to concentrate on circuit-analysis processes and the use of your scientific calculator. Here
you will use complex algebra more than ever before, yet the material will not seem particularly difficult because it is related to topics covered in previous chapters. Very soon,
you will find yourself working through what would be impossible for a beginner. You will
need to be systematic and cautious when working out solutions to problems. Use plenty
of scrap paper and write out each step as you solve a problem. Study all examples carefully.
I think you will enjoy your experience in this final chapter.

NEED TO KNOW

Here is a practical problem the solution to which you might actually determine before you read this chapter.

A certain transmitter has a known output impedance of 50 Ω − $j20$ Ω at the desired operating frequency. What must the impedance of the transmitting antenna be for maximum power transfer?

This problem may seem to be a riddle to you because you really have not covered enough AC theory to reach a solution. How you put together the information you have already learned will determine whether you can offer an early solution.

This is only one of the many topics of which you have a need to know in this final chapter of *Experiencing Electricity and Electronics*. Enjoy!

22-1 Conductance, Susceptance, and Admittance

Conductance (*G*)

Conductance (*G*) is simply the reciprocal of resistance. The SI unit of measure for conductance is the **siemen (S)**. Before the siemen was accepted as the standard unit, the **mho** was used (*ohm* spelled backwards). Conductance is often useful in simplifying calculations involved in AC and DC circuit analysis. For example, the total resistance of a dozen parallel resistors is readily obtained by adding the conductances for all of the resistors and taking the reciprocal of the total conductance:

$$G_t = (G_1 + G_2 + G_3 + \cdots) = 1/R_t = (1/R_1 + 1/R_2 + 1/R_3 + \cdots)$$
$$R_t = 1/G_t = 1/(G_1 + G_2 + G_3 + \cdots) = 1/(1/R_1 + 1/R_2 + 1/R_3 + \cdots)$$

(This is, in fact, the reciprocal formula for parallel components.)

$$G = 1/R \quad \text{(S), (1 S = 1/1 } \Omega) \tag{22.1}$$
$$R = 1/G \quad (\Omega), (1 \ \Omega = 1/1 \text{ S}) \tag{22.2}$$
$$R \cdot G = 1 \quad (1 \ \Omega \cdot 1 \text{ S} = 1) \tag{22.3}$$

Susceptance (*B*)

Susceptance (*B*) is simply the reciprocal of reactance (capacitive or inductive). As for conductance, the unit of measure for susceptance is the siemen (S). Again, it is sometimes convenient to convert a reactance to susceptance in the analysis process. This is illustrated in Figure 22-1 and Example 22-1. (*X* = reactance)

$$B = 1/X \quad \text{(S), (1 S = 1/1 } \Omega) \tag{22.4}$$

also $+jB = 1/-jX_C$ and $-jB = 1/+jX_L$

$$X = 1/B \quad (\Omega), (1 \ \Omega = 1/1 \text{ S}) \tag{22.5}$$

also $+jX_L = 1/-jB$ and $-jX_C = 1/+jB$

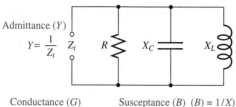

Conductance (G)

$$G = \frac{1}{R}$$

Susceptance (B) (B) = 1/X)

$$+jB = \frac{1}{-jX_C} \qquad -jB = \frac{1}{+jX_L}$$

The siemen is the unit of measure for Y, G and B.
One siemen = 1 S = 1/1Ω

$$Z_t = \cfrac{1}{\cfrac{1}{R} + \cfrac{1}{-jX_C} + \cfrac{1}{+jX_L}} \qquad \boxed{\begin{array}{l}\text{Reciprocal}\\\text{formula for}\\\text{parallel}\\\text{components}\end{array}}$$

$$Z_t = \frac{1}{Y} = \frac{1}{G + jB_C - jB_L} \quad (\Omega)$$

$$\frac{1}{Z_t} = \frac{1}{R} + \frac{1}{-jX_C} + \frac{1}{+jX_L} = Y = \overset{\text{Admittance}}{\underset{\text{Susceptance}}{G + \overbrace{jB_C - jB_L}}} \quad (S)$$

Figure 22-1 Admittance (Y), conductance
(G), and susceptance (B).

$$X \cdot B = 1 \qquad (1\ \Omega \cdot 1\ S = 1) \tag{22.6}$$

also $+jX_L \cdot -jB = 1$ and $-jX_C \cdot +jB = 1$

Admittance (Y)

Admittance (Y) is simply the reciprocal of impedance (Z). The unit of measure for admittance is also the siemen (S).

$$Y = 1/Z \qquad (S),\ (1\ S = 1/1\ \Omega) \tag{22.7}$$

also $Y\angle + \theta = 1/Z\angle - \theta$ and $Y\angle - \theta = 1/Z\angle + \theta$

$$Z = 1/Y \qquad (\Omega),\ (1\ \Omega = 1/1\ S) \tag{22.8}$$

also $Z\angle + \theta = 1/Y\angle - \theta$ and $Z\angle - \theta = 1/Y\angle + \theta$

$$Z \cdot Y = 1 \qquad (1\ \Omega \cdot 1\ S = 1) \tag{22.9}$$

also $Z\angle + \theta \cdot Y\angle - \theta = 1\angle 0°$ and $Z\angle - \theta \cdot Y\angle + \theta = 1\angle 0°$

Figure 22-1 illustrates the use of admittance in determining the total impedance of a parallel *RCL* circuit. Just as impedance is a combination of resistance and reactance ($Z = R + jX$), admittance is a combination of conductance and susceptance ($Y = G + jB$). Also, just as the term *impedance* is loosely used in place of resistance or reactance when referring to an AC circuit, the term *admittance* may loosely be used in place of conductance or susceptance.

Example 22-1 illustrates, in step-by-step fashion, how the total impedance of a parallel *RCL* circuit is calculated using conductance, susceptance, and admittance. First, the resistance is converted to conductance (Formula 22.1). Next, the reactive branches are converted to susceptances (Formula 22.4). The conductance and susceptances are then added together following the rules of addition for complex numbers. Finally, the total impedance is determined by taking the reciprocal of the admittance (Formula 22.8).

EXAMPLE   22-1

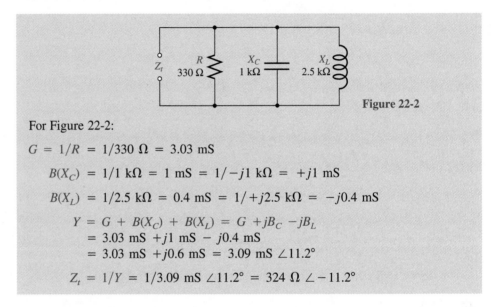

Figure 22-2

For Figure 22-2:

$G = 1/R = 1/330 \ \Omega = 3.03 \ \text{mS}$

$B(X_C) = 1/1 \ \text{k}\Omega = 1 \ \text{mS} = 1/-j1 \ \text{k}\Omega = +j1 \ \text{mS}$

$B(X_L) = 1/2.5 \ \text{k}\Omega = 0.4 \ \text{mS} = 1/+j2.5 \ \text{k}\Omega = -j0.4 \ \text{mS}$

$$Y = G + B(X_C) + B(X_L) = G + jB_C - jB_L$$
$$= 3.03 \ \text{mS} + j1 \ \text{mS} - j0.4 \ \text{mS}$$
$$= 3.03 \ \text{mS} + j0.6 \ \text{mS} = 3.09 \ \text{mS} \ \angle 11.2°$$

$$Z_t = 1/Y = 1/3.09 \ \text{mS} \ \angle 11.2° = 324 \ \Omega \ \angle -11.2°$$

We will make further use of admittance in the next sections, which cover advanced analysis. For now, take a few minutes to solve the problems of Self-Check 22-1. Verify your answers from those provided at the end of the chapter before continuing.

SELF-CHECK   22-1

1. What is the conductance of 470 Ω?

2. What is the susceptance of 1.5 kΩ of capacitive reactance?

3. What is the susceptance of 36 kΩ of inductive reactance?

4. What is the admittance for 220 Ω $+j400 \ \Omega$ $-j300 \ \Omega$?

5. What is the total admittance for a circuit consisting of a 1 kΩ resistor in parallel with 2 kΩ of capacitive reactance in parallel with 3 kΩ of inductive reactance? (all three components in parallel)

22-2 Complex AC Circuits

Finding Z_t for Complex Circuits

EXAMPLE 22-2

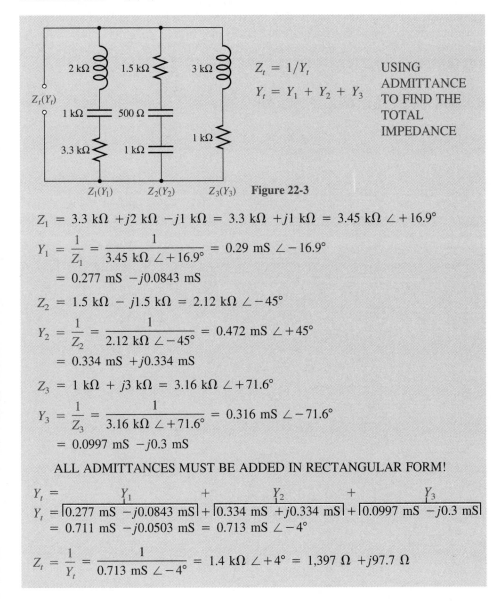

$Z_t = 1/Y_t$

$Y_t = Y_1 + Y_2 + Y_3$

USING ADMITTANCE TO FIND THE TOTAL IMPEDANCE

$Z_1(Y_1)$ $Z_2(Y_2)$ $Z_3(Y_3)$ **Figure 22-3**

$Z_1 = 3.3 \text{ k}\Omega + j2 \text{ k}\Omega - j1 \text{ k}\Omega = 3.3 \text{ k}\Omega + j1 \text{ k}\Omega = 3.45 \text{ k}\Omega \angle +16.9°$

$Y_1 = \dfrac{1}{Z_1} = \dfrac{1}{3.45 \text{ k}\Omega \angle +16.9°} = 0.29 \text{ mS} \angle -16.9°$

$\quad = 0.277 \text{ mS} - j0.0843 \text{ mS}$

$Z_2 = 1.5 \text{ k}\Omega - j1.5 \text{ k}\Omega = 2.12 \text{ k}\Omega \angle -45°$

$Y_2 = \dfrac{1}{Z_2} = \dfrac{1}{2.12 \text{ k}\Omega \angle -45°} = 0.472 \text{ mS} \angle +45°$

$\quad = 0.334 \text{ mS} + j0.334 \text{ mS}$

$Z_3 = 1 \text{ k}\Omega + j3 \text{ k}\Omega = 3.16 \text{ k}\Omega \angle +71.6°$

$Y_3 = \dfrac{1}{Z_3} = \dfrac{1}{3.16 \text{ k}\Omega \angle +71.6°} = 0.316 \text{ mS} \angle -71.6°$

$\quad = 0.0997 \text{ mS} - j0.3 \text{ mS}$

ALL ADMITTANCES MUST BE ADDED IN RECTANGULAR FORM!

$Y_t = \qquad \overline{Y_1} \qquad + \qquad \overline{Y_2} \qquad + \qquad \overline{Y_3}$

$Y_t = \overline{0.277 \text{ mS} - j0.0843 \text{ mS}} + \overline{0.334 \text{ mS} + j0.334 \text{ mS}} + \overline{0.0997 \text{ mS} - j0.3 \text{ mS}}$

$\quad = 0.711 \text{ mS} - j0.0503 \text{ mS} = 0.713 \text{ mS} \angle -4°$

$Z_t = \dfrac{1}{Y_t} = \dfrac{1}{0.713 \text{ mS} \angle -4°} = 1.4 \text{ k}\Omega \angle +4° = 1,397 \ \Omega + j97.7 \ \Omega$

Overall Parallel Circuits

The total impedance for an overall parallel circuit is found using one of the three standard parallel formulas. Perhaps the easiest, or most straightforward, approach to finding Z_t is to take the reciprocal of the sum of all branch admittances. This approach is illustrated in

Example 22-2. Notice that the total impedance is the reciprocal of the total admittance, and the total admittance is the sum of the individual branch admittances. The method used here is to find the impedance of each branch and convert it to admittance. Once all branch admittances are determined, they are added to form the total admittance. Finally, the reciprocal of the total admittance is taken to find the total circuit impedance. Notice that the complex numbers are divided in polar form and added in rectangular form. The total admittance must be in polar form before the reciprocal can be taken. The final total impedance can remain in polar form or it can be converted to rectangular form. In rectangular form, we see that the entire circuit is equivalent to 1,397 Ω of resistance in series with 97.7 Ω of inductive reactance (1,397 Ω + j97.7 Ω).

EXAMPLE   22-3

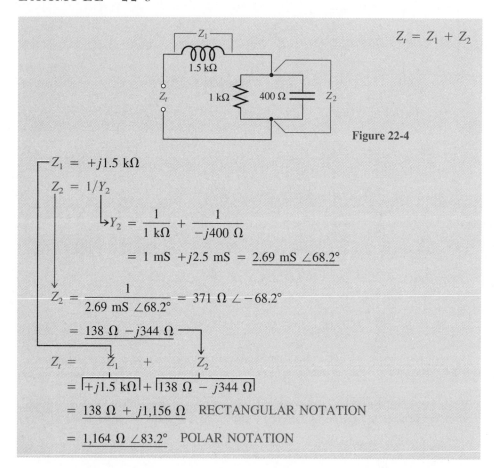

Figure 22-4

$$Z_1 = +j1.5 \text{ k}\Omega$$

$$Z_2 = 1/Y_2$$

$$Y_2 = \frac{1}{1 \text{ k}\Omega} + \frac{1}{-j400 \ \Omega}$$

$$= 1 \text{ mS} + j2.5 \text{ mS} = \underline{2.69 \text{ mS} \angle 68.2°}$$

$$Z_2 = \frac{1}{2.69 \text{ mS} \angle 68.2°} = 371 \ \Omega \angle -68.2°$$

$$= \underline{138 \ \Omega - j344 \ \Omega}$$

$$Z_t = Z_1 + Z_2$$

$$= \boxed{+j1.5 \text{ k}\Omega} + \boxed{138 \ \Omega - j344 \ \Omega}$$

$$= \underline{138 \ \Omega + j1{,}156 \ \Omega} \quad \text{RECTANGULAR NOTATION}$$

$$= \underline{1{,}164 \ \Omega \angle 83.2°} \quad \text{POLAR NOTATION}$$

Overall Series Circuits

Example 22-3 illustrates a fairly simple overall series circuit. The total impedance of an overall series circuit is simply the sum of all series impedances (each parallel section is converted to a series impedance). In Figure 22-4, there are two series impedances, labeled

Z_1 and Z_2. Z_2 must first be converted to a series impedance so it may be added to Z_1. To do this, we can find the total admittance for this parallel section (Y_2) and take its reciprocal to find Z_2 ($Z_2 = 1/Y_2$). As you can see, the total admittance for the parallel section is the sum of the two branch admittances ($Y_2 = 1/1\ k\Omega + 1/-j400\ \Omega$). Y_2 is found to be 2.69 mS $\angle 68.2°$, and the reciprocal of Y_2 is found to be 138 $\Omega - j344\ \Omega$. Thus, the parallel section is converted to a series impedance of 138 $\Omega - j344\ \Omega$. The final step is to add all series impedances to find the total impedance for the entire overall series circuit ($Z_t = Z_1 + Z_2 = +j1.5\ k\Omega + 138\ \Omega - j344\ \Omega = 138\ \Omega + j1,156\ \Omega = 1,164\ \Omega \angle 83.2°$).

EXAMPLE 22-4

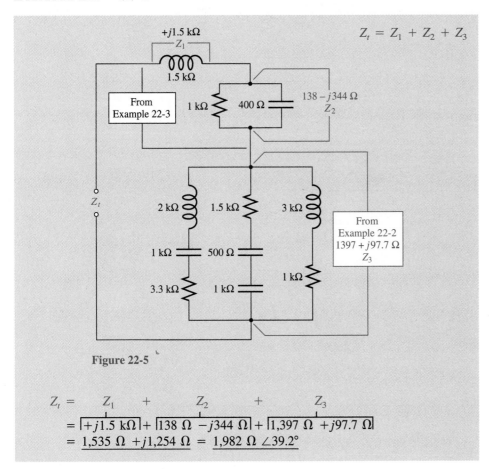

Figure 22-5

$$Z_t = \quad Z_1 \quad + \quad Z_2 \quad + \quad Z_3$$
$$= |+j1.5\ k\Omega| + |138\ \Omega - j344\ \Omega| + |1,397\ \Omega + j97.7\ \Omega|$$
$$= 1,535\ \Omega + j1,254\ \Omega = \underline{1,982\ \Omega \angle 39.2°}$$

Example 22-4 combines Examples 22-2 and 22-3. This example simply emphasizes the fact that each parallel section must be converted to an equivalent series impedance before the total overall impedance can be found. Again, the total impedance for any overall series circuit is simply the sum of all series impedances. That is why the parallel sections must first be converted to equivalent series impedances. Notice also that all series impedances are added in rectangular form.

Finding I_t for Complex Circuits

EXAMPLE   22-5

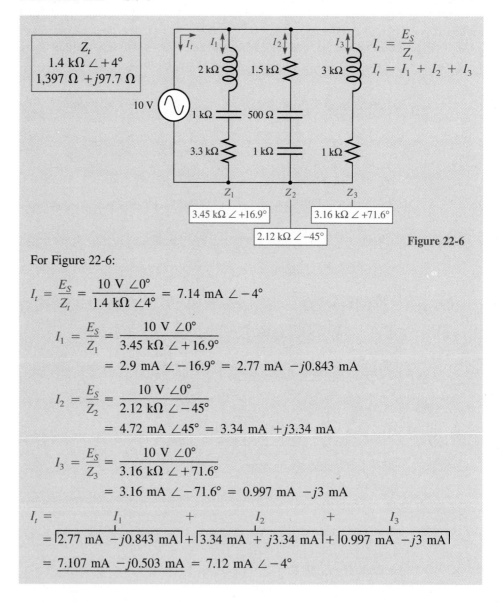

Figure 22-6

For Figure 22-6:

$$I_t = \frac{E_S}{Z_t} = \frac{10 \text{ V} \angle 0°}{1.4 \text{ k}\Omega \angle 4°} = 7.14 \text{ mA} \angle -4°$$

$$I_1 = \frac{E_S}{Z_1} = \frac{10 \text{ V} \angle 0°}{3.45 \text{ k}\Omega \angle +16.9°}$$

$$= 2.9 \text{ mA} \angle -16.9° = 2.77 \text{ mA} -j0.843 \text{ mA}$$

$$I_2 = \frac{E_S}{Z_2} = \frac{10 \text{ V} \angle 0°}{2.12 \text{ k}\Omega \angle -45°}$$

$$= 4.72 \text{ mA} \angle 45° = 3.34 \text{ mA} +j3.34 \text{ mA}$$

$$I_3 = \frac{E_S}{Z_3} = \frac{10 \text{ V} \angle 0°}{3.16 \text{ k}\Omega \angle +71.6°}$$

$$= 3.16 \text{ mA} \angle -71.6° = 0.997 \text{ mA} -j3 \text{ mA}$$

$$I_t = \qquad I_1 \qquad + \qquad I_2 \qquad + \qquad I_3$$

$$= \boxed{2.77 \text{ mA} -j0.843 \text{ mA}} + \boxed{3.34 \text{ mA} + j3.34 \text{ mA}} + \boxed{0.997 \text{ mA} -j3 \text{ mA}}$$

$$= \underline{7.107 \text{ mA} -j0.503 \text{ mA}} = 7.12 \text{ mA} \angle -4°$$

Using Ohm's Law

The total current for any circuit can be found by simply using Ohm's Law as long as the source voltage and total circuit impedance is known. The circuit of Example 22-5 is borrowed from Example 22-2. The total impedance for this circuit is already known from previous calculation ($Z_t = 1.4 \text{ k}\Omega \angle +4°$). The complex impedance is divided into the 10 V source to reveal a total current of 7.14 mA at an angle of $-4°$. If we wish to know the phase angle for the current, the rules for division of polar numbers must be followed. Otherwise, the total current is simply stated as 7.14 mA.

Adding Branch Currents

The total current for Figure 22-6 may also be determined by adding all branch currents. If this method is used, the phase angle for each branch current cannot be ignored in the calculations. Each branch current must be calculated using Ohm's Law and complex algebra. Notice the branch currents must be added in rectangular form. Thus, when a branch current is calculated, it must immediately be converted to rectangular form. The slight difference in answers between using Ohm's Law and adding branch currents is simply due to rounding error (7.14 mA as compared to 7.12 mA).

Finding Voltage Drops for Complex Circuits

Using Ohm's Law

Voltage drops in an overall series circuit can be calculated using Ohm's Law. The total current of the circuit and the impedance of each series section must be known or predetermined. Example 22-6 demonstrates how this is done. The impedance of each of the three series sections was determined in Example 22-4. The total current is determined using Ohm's Law. It is important to note that the total current and the impedance for each series section must be expressed in polar form in order to determine the voltage drop across each series section (complex numbers should be divided in polar form). Once each voltage drop is determined, they should be totaled using phasor addition to see if their sum equals the source voltage (10 V $\angle 0°$). Note that the voltage drops are added in rectangular form according to the rules for addition of complex numbers.

EXAMPLE   22-6

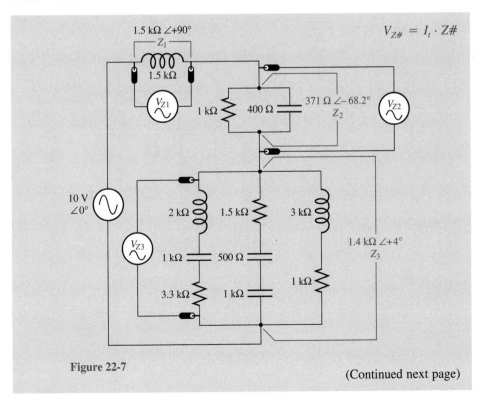

Figure 22-7

(Continued next page)

EXAMPLE 22-6 (Continued)

USING OHM'S LAW TO FIND AC VOLTAGE DROPS FOR FIGURE 22-7

$$Z_t = \underline{1{,}535 \ \Omega \ +j1{,}254 \ \Omega} = \underline{1{,}982 \ \Omega \ \angle 39.2°}$$

$$I_t = \frac{E_S}{Z_t} = \frac{10 \text{ V } \angle 0°}{1{,}982 \ \Omega \ \angle 39.2°} = \underline{5.05 \text{ mA } \angle -39.2°}$$

$$V_{Z1} = I_t \cdot Z_1 = 5.05 \text{ mA } \angle -39.2° \cdot 1.5 \text{ k}\Omega \ \angle +90$$
$$= \underline{7.58 \text{ V } \angle +50.8°} = \underline{4.79 \text{ V } +j5.87 \text{ V}}$$

$$V_{Z2} = I_t \cdot Z_2 = 5.05 \text{ mA } \angle -39.2° \cdot 371 \ \Omega \ \angle -68.2°$$
$$= \underline{1.87 \text{ V } \angle -107.4°} = \underline{-0.56 \text{ V } -j1.78 \text{ V}}$$

$$V_{Z3} = I_t \cdot Z_3 = 5.05 \text{ mA } \angle -39.2° \cdot 1.4 \text{ k}\Omega \ \angle +4°$$
$$= \underline{7.07 \text{ V } \angle -35.2°} = \underline{5.78 \text{ V } -j4.08 \text{ V}}$$

KIRCHOFF'S VOLTAGE LAW FOR AC CIRCUITS

$$E_S = \qquad V_{Z1} \qquad + \qquad V_{Z2} \qquad + \qquad V_{Z3}$$

$$= \boxed{4.79 \text{ V } +j5.87 \text{ V}} + \boxed{-0.56 \text{ V } -j1.78 \text{ V}} + \boxed{5.78 \text{ V } -j4.08 \text{ V}}$$

$$= 10 \text{ V } \angle 0° \quad \text{GOOD!}$$

EXAMPLE 22-7

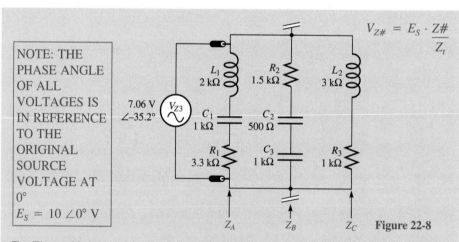

NOTE: THE PHASE ANGLE OF ALL VOLTAGES IS IN REFERENCE TO THE ORIGINAL SOURCE VOLTAGE AT 0°

$E_S = 10 \ \angle 0° \text{ V}$

$$V_{Z\#} = E_S \cdot \frac{Z\#}{Z_t}$$

Figure 22-8

For Figure 22-8:

$$Z_A = 3.3 \text{ k}\Omega \ -j1 \text{ k}\Omega \ +j2 \text{ k}\Omega = 3.3 \text{ k}\Omega \ +j1 \text{ k}\Omega = 3.45 \text{ k}\Omega \ \angle +16.9°$$
$$Z_B = 1.5 \text{ k}\Omega \ -j1 \text{ k}\Omega \ -j0.5 \text{ k}\Omega = 1.5 \text{ k}\Omega \ -j1.5 \text{ k}\Omega = 2.12 \text{ k}\Omega \ \angle -45°$$
$$Z_C = 1 \text{ k}\Omega \ +j3 \text{ k}\Omega = 3.16 \text{ k}\Omega \ \angle +71.6°$$

NOTE: V_{Z3} acts as an E_S at $\angle -35.2°$ in the following calculations.

$$V_{L1} = E_S \cdot Z_{L1}/Z_A = 7.06 \text{ V } \angle -35.2° \cdot 2 \text{ k}\Omega \ \angle +90°/3.45 \text{ k}\Omega \ \angle +16.9°$$
$$= 4.09 \text{ V } \angle +37.9°$$

$$V_{C1} = E_S \cdot Z_{C1}/Z_A = 7.06 \text{ V} \angle -35.2° \cdot 1 \text{ k}\Omega \angle -90°/3.45 \text{ k}\Omega \angle +16.9°$$
$$= 2.05 \text{ V} \angle -142°$$

$$V_{R1} = E_S \cdot R_1/Z_A = 7.06 \text{ V} \angle -35.2° \cdot 3.3 \text{ k}\Omega \angle 0°/3.45 \text{ k}\Omega \angle +16.9°$$
$$= 6.75 \text{ V} \angle -52.1°$$

$$V_{R2} = E_S \cdot R_2/Z_B = 7.06 \text{ V} \angle -35.2° \cdot 1.5 \text{ k}\Omega \angle 0°/2.12 \text{ k}\Omega \angle -45°$$
$$= 5 \text{ V} \angle 9.8°$$

$$V_{C2} = E_S \cdot Z_{C2}/Z_B = 7.06 \text{ V} \angle -35.2° \cdot 500 \text{ }\Omega \angle -90°/2.12 \text{ k}\Omega \angle -45°$$
$$= 1.67 \text{ V} \angle -80.2°$$

$$V_{C3} = E_S \cdot Z_{C3}/Z_B = 7.06 \text{ V} \angle -35.2° \cdot 1 \text{ k}\Omega \angle -90° \text{ }\Omega/2.12 \text{ k}\Omega \angle -45°$$
$$= 3.33 \text{ V} \angle -80.2°$$

$$V_{L2} = E_S \cdot Z_{L2}/Z_C = 7.06 \text{ V} \angle -35.2° \cdot 3 \text{ k}\Omega \angle +90° \text{ }\Omega/3.16 \text{ k}\Omega \angle +71.6°$$
$$= 6.7 \text{ }\Omega \angle -16.8°$$

$$V_{R3} = E_S \cdot R_3/Z_C = 7.06 \text{ V} \angle -35.2° \cdot 1 \text{ k}\Omega \angle 0° \text{ }\Omega/3.16 \text{ k}\Omega \angle +71.6°$$
$$= 2.23 \text{ V} \angle -106.8°$$

Using the Voltage Divider Rule (Proportion Formula)

Voltage drops may also be determined according to the voltage divider rule (proportion formula). Figure 22-8 is a close-up look at the large parallel section of Figure 22-7. The voltage across this section was calculated to be 7.06 V $\angle -35.2°$. Thus, each of the three series branches has an applied source voltage of 7.06 V $\angle -35.2°$. The voltage across each component of each branch can be determined using the proportion formula as follows:

$$V_{Z\#} = E_S \cdot Z\#/Z_t \qquad \text{(general voltage divider rule)} \qquad (22.10)$$

The formula indicates that the voltage across any series impedance is in proportion to the ratio of its impedance to the total series impedance. In Example 22-7, the impedance for each of the branches (Z_A, Z_B, Z_C) is calculated first. As you can see, the individual voltage drops for each component are calculated for each branch from left to right. In each case, the parallel-section voltage (7.06 V $\angle -35.2°$) is multiplied by the individual component's impedance divided by the total impedance for the branch. The total impedance for each branch must be in polar form, since division is being accomplished in the proportion formula. One final note: The phase angle of voltage across the parallel section (7.06 V $\angle -35.2°$) is in reference to the source voltage of Example 22-6 (10 V $\angle 0°$). Thus, all phase angles for all voltage drops in each branch of the large parallel section are in reference to the original source voltage as well (10 V $\angle 0°$). For example, V_{R3} is at an angle of $-106.8°$ with respect to the source voltage at 0°.

Try your skills by solving for all impedances, currents, and voltages drops in the following Self-Check.

SELF-CHECK 22-2

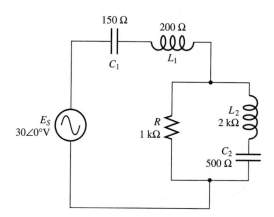

Figure 22-9

Calculate the following for Figure 22-9:

$Z_t =$ _____ $I_{L2} =$ _____ $V_R =$ _____

$I_t =$ _____ $V_{C1} =$ _____ $V_{C2} =$ _____

$I_R =$ _____ $V_{L1} =$ _____ $V_{L2} =$ _____

Include the phase angle in each answer.
Show all answers in polar form.

22-3 The Superposition Theorem for AC Circuits

You probably recall the superposition theorem from Chapter 8, where you added it to your DC circuit-analysis tool kit. Here, we pull it out once again and apply it to AC circuit analysis. You'll notice that the statement of the theorem here is nearly the same as before:

Superposition Theorem for AC Circuits

The actual currents and voltages in a circuit containing multiple voltage sources are the phasor sum of currents and voltages produced by each individual voltage source.

Notice we are using the "phasor sum" of currents and voltages, not just the "algebraic sum." Study Examples 22-8 and 22-9 carefully to see how superposition is applied using complex algebra.

EXAMPLE   22-8

Use superposition to determine the voltage across R_L in Figure 22-10.

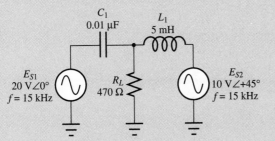

Figure 22-10

Step 1: Determine V_L with each voltage source. The other source is replaced with its internal impedance. Here, we consider the voltage sources to be ideal and replace them with a short (WIRE).

At 15 kHz, $X_{C1} = \dfrac{1}{2 \cdot \pi \cdot 15 \text{ kHz} \cdot 0.01 \ \mu\text{F}} = 1{,}061 \ \Omega$

$X_{L1} = 2 \cdot \pi \cdot 15 \text{ kHz} \cdot 5 \text{ mH} = 471 \ \Omega$

Using E_{S1} alone, the circuit appears as Figure 22-11.

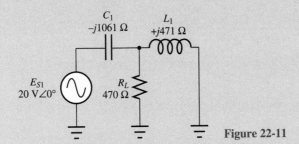

Figure 22-11

$$Z_t = -j1{,}061 \ \Omega + \cfrac{1}{\cfrac{1}{470 \ \Omega} + \cfrac{1}{+j471 \ \Omega}}$$

$$= -j1{,}061 \ \Omega + \frac{1}{2.13 \text{ mS} - j2.12 \text{ mS}}$$

$$= -j1{,}061 \ \Omega + \frac{1}{3 \text{ mS} \ \angle 44.9°}$$

$$= -j1{,}061 \ \Omega + 333 \ \Omega \ \angle -44.9°$$

$$= -j1{,}061 \ \Omega + 236 \ \Omega - j235 \ \Omega$$

$$= 236 \ \Omega - j1{,}296 \ \Omega = 1{,}317 \ \Omega \ \angle -80°$$

(Continued next page)

EXAMPLE 22-8 (Continued)

$$V_{RL1} = E_{S1} \cdot (R_L \| L_1)/Z_t$$

$$= 20 \text{ V } \angle 0° \cdot \frac{333 \ \Omega \ \angle -44.9°}{1{,}317 \ \Omega \ \angle -80°}$$

$$= 20 \text{ V } \angle 0° \cdot 0.253 \ \angle +35.1°$$

$$= \underline{5.06 \text{ V } \angle +35.1°}$$

Using E_{S2} alone, the circuit appears as Figure 22-12.

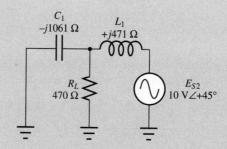

Figure 22-12

$$Z_t = +j471 \ \Omega + \cfrac{1}{\cfrac{1}{470 \ \Omega} + \cfrac{1}{-j1{,}061 \ \Omega}}$$

$$= +j471 \ \Omega + \frac{1}{2.13 \text{ mS} + j0.943 \text{ mS}}$$

$$= +j471 \ \Omega + \frac{1}{2.33 \text{ mS} \ \angle +23.9°}$$

$$= +j471 \ \Omega + 429 \ \Omega \ \angle -23.9°$$

$$= +j471 \ \Omega + 392 \ \Omega -j174 \ \Omega$$

$$= 392 \ \Omega + j297 \ \Omega = 492 \ \Omega \ \angle +37.1°$$

$$V_{RL2} = E_{S2} \cdot (R_L \| C_1)/Z_t$$

$$= 10 \text{ V } \angle +45° \cdot \frac{429 \ \Omega \ \angle -23.9°}{492 \ \Omega \ \angle +37.1°}$$

$$= 10 \text{ V } \angle +45° \cdot 0.872 \ \angle -61°$$

$$= \underline{8.72 \text{ V } \angle -16°}$$

Step 2: Use phasor addition to determine the actual voltage across R_L with both sources.

$$V_{RL} = V_{RL1} + V_{RL2}$$

$$= 5.06 \text{ V } \angle +35.1° + 8.72 \text{ V } \angle -16°$$

$$= 4.14 \text{ V } + j2.91 \text{ V } + 8.38 \text{ V } - j2.40 \text{ V}$$

$$= 12.5 \text{ V } + j0.51 \text{ V } = \underline{12.5 \text{ V } \angle +2.34°}$$

In Example 22-9, we apply the Superposition Theorem to current sources.

EXAMPLE   22-9

Use superposition to find the current through R_L in Figure 22-13.

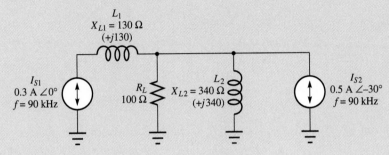

Figure 22-13

Step 1: Determine I_{RL} for each separate current source. The ideal current source has an infinite internal impedance. Therefore, each current source is replaced with an open. Using I_{S1} alone, the circuit looks like Figure 22-14.

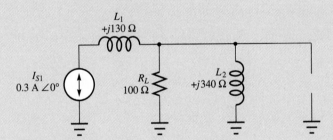

Figure 22-14

I_{S1} proceeds through L_1 and must be divided between R_L and L_2. Therefore, we must use the current divider rule to find I_{RL1} (Formula 8.2 from Chapter 8).

$$I_{RL1} = I_{S1} \cdot \frac{X_{L2}}{R + jX_{L2}}$$

$$= 0.3 \text{ A } \angle 0° \cdot \frac{+j340 \ \Omega}{100 \ \Omega + j340 \ \Omega}$$

$$= 0.3 \text{ A } \angle 0° \cdot \frac{340 \ \Omega \angle +90°}{354 \ \Omega \angle +73.6°}$$

$$= 0.3 \text{ A } \angle 0° \cdot 0.960 \ \angle +16.4°$$

$$= 0.288 \text{ A } \angle +16.4° = \underline{0.276 \text{ A } + j0.0813 \text{ A}}$$

Using I_{S2} alone, the circuit looks like Figure 22-15.

(Continued next page)

EXAMPLE   22-9   (Continued)

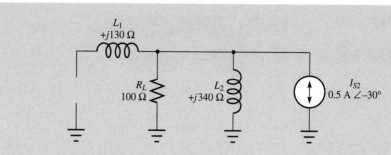

Figure 22-15

As you can see, I_{S2} is divided between R_L and L_2. L_1 receives no current.

$$I_{RL2} = I_{S2} \cdot \frac{X_{L2}}{R + jX_{L2}}$$

$$= 0.5 \text{ A} \angle -30° \cdot \frac{340 \ \Omega \ \angle +90°}{354 \ \Omega \ \angle +73.6°}$$

$$= 0.5 \text{ A} \angle -30° \cdot 0.960 \ \angle +16.4°$$

$$= 0.48 \text{ A} \angle -13.6° = \underline{0.467 \text{ A} - j0.113 \text{ A}}$$

Step 2: Add the currents through R_L using the rectangular complex forms.

$$I_{RL} = I_{RL1} + I_{RL2}$$

$$= 0.276 \text{ A} + j0.0813 \text{ A} + 0.467 \text{ A} - j0.113 \text{ A}$$

$$= 0.743 \text{ A} - j0.0317 \text{ A} = \underline{0.744 \text{ A} \angle -2.44°}$$

Now try your skills on Self-Check 22-3.

SELF-CHECK 22-3

1. Use superposition to determine V_{RL} in Figure 22-16.

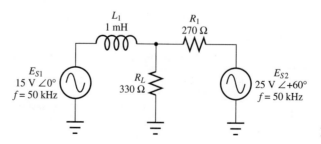

Figure 22-16

2. Use superposition to determine I_{RL} in Figure 22-17.

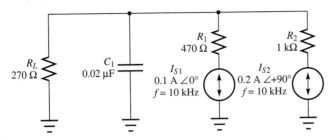

Figure 22-17

22-4 Thevenin's Theorem for AC Circuits

Thevenin's Theorem for AC Circuits

Any linear circuit consisting of any combination of resistance, inductance, and capacitance, and having one or more voltage sources and two output terminals, can be replaced with a single voltage source (V_{Th}) and an equivalent series impedance (Z_{Th}) for the purpose of circuit analysis.

This statement of Thevenin's theorem should sound familiar to you. Yet, there are some important changes here from the statement of the theorem in Chapter 8. Here we include inductors and capacitors and are dealing with complex impedances. Nevertheless, the overall concept and procedure is the same. Examples 22-10 and 22-11 will demonstrate Thevenin's theorem for AC circuits.

EXAMPLE   22-10

Convert the circuit of Figure 22-18 to a Thevenin AC model and determine the output voltage when R_L is (a) 100 Ω, (b) 500 Ω, and (c) 1 kΩ.

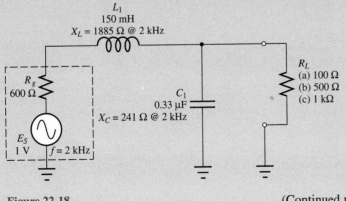

Figure 22-18   (Continued next page)

EXAMPLE   22-10   (Continued)

Step 1: Remove the load and determine the Thevenin voltage at the output terminals.

$$
\begin{aligned}
V_{Th} = V_{C1} &= 1 \text{ V } \angle 0° \cdot \frac{-j241 \ \Omega}{600 \ \Omega + j1{,}885 \ \Omega - j241 \ \Omega} \\
&= 1 \text{ V } \angle 0° \cdot \frac{-j241 \ \Omega}{600 \ \Omega + j1{,}644 \ \Omega} \\
&= 1 \text{ V } \angle 0° \cdot \frac{241 \ \Omega \ \angle -90°}{1{,}750 \ \Omega \ \angle +70°} \\
&= 1 \text{ V } \angle 0° \cdot 0.138 \ \angle -160° \\
&= \underline{0.138 \text{ V } \angle -160°}
\end{aligned}
$$

Step 2: Replace the voltage source with its internal impedance and determine the Thevenin circuit impedance at the output terminals with the load removed as shown in Figure 22-19.

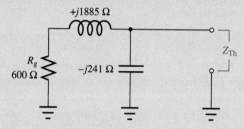

Figure 22-19

$$
\begin{aligned}
Z_{Th} &= -j241 \ \Omega \| (600 + j1{,}885 \ \Omega) \\[4pt]
&= \cfrac{1}{\cfrac{1}{-j241 \ \Omega} + \cfrac{1}{600 + j1{,}885 \ \Omega}} \\[4pt]
&= \cfrac{1}{+j0.00415 \text{ S} + \cfrac{1}{1{,}978 \ \Omega \ \angle +72.3°}} = \frac{1}{+j4{,}150 \ \mu S + 506 \ \mu S \ \angle -72.3°} \\[4pt]
&= \frac{1}{+j4{,}150 \ \mu S + 154 \ \mu S - j482 \ \mu S} = \frac{1}{154 \ \mu S + j3{,}668 \ \mu S} \\[4pt]
&= \frac{1}{3{,}671 \ \mu S \ \angle +87.6°} = 272.4 \ \Omega \ \angle -87.6°
\end{aligned}
$$

Step 3: Draw the Thevenin model. Connect the load and determine the output voltage. Study Figure 22-20.

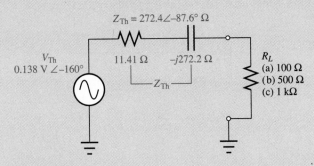

Figure 22-20

(a) $R_L = 100 \ \Omega$

$$V_{RL} = 0.138 \text{ V} \angle -160°\cdot \frac{100 \ \Omega}{100 \ \Omega + 11.41 \ \Omega - j272.2 \ \Omega}$$

$$= 0.138 \text{ V} \angle -160° \cdot \frac{100 \ \Omega \angle 0°}{294 \ \Omega \angle -67.7°}$$

$$= 0.138 \text{ V} \angle -160° \cdot 0.340 \angle +67.7°$$

$$= \underline{0.047 \text{ V} \angle -92.3°}$$

(b) $R_L = 500 \ \Omega$

$$V_{RL} = 0.138 \text{ V} \angle -160° \cdot \frac{500 \ \Omega}{500 \ \Omega + 11.41 \ \Omega - j272.2 \ \Omega}$$

$$= 0.138 \text{ V} \angle -160° \cdot \frac{500 \ \Omega \angle 0°}{579.3 \ \Omega \angle -28°}$$

$$= 0.138 \text{ V} \angle -160° \cdot 0.863 \angle +28°$$

$$= \underline{0.119 \text{ V} \angle -132°}$$

(c) $R_L = 1 \ k\Omega$

$$V_{RL} = 0.138 \text{ V} \angle -160° \cdot \frac{1,000 \ \Omega}{1,000 \ \Omega + 11.41 \ \Omega - j272.2 \ \Omega}$$

$$= 0.138 \text{ V} \angle -160° \cdot \frac{1,000 \ \Omega \angle 0°}{1,047 \ \Omega \angle -15.1°}$$

$$= 0.138 \text{ V} \angle -160° \cdot 0.955 \angle +15.1°$$

$$= \underline{0.132 \text{ V} \angle -145°}$$

Note: The Thevenin model and all calculations for different loads are valid for only one frequency. The model is frequency dependent.

EXAMPLE   22-11

Convert the circuit of Figure 22-21 to a Thevenin AC model and determine the output voltage when the load is (a) 400 Ω and (b) $+j250$ Ω.

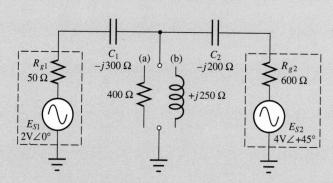

Figure 22-21

Step 1: With the load removed, determine the Thevenin voltage at the output terminals. In this case, the output voltage is determined by both E_{S1} and E_{S2}. We will use superposition to find V_{Th}, where

$$V_{out(no\ load)} = V_{Th} = V_{out1} + V_{out2}$$

V_{out1} is the output voltage due to E_{S1} and V_{out2} is the output voltage due to E_{S2}. Figure 22-22 is used to determine V_{out1}.

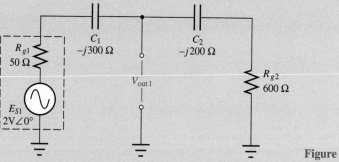

Figure 22-22

Notice in Figure 22-22, E_{S2} is replaced with its internal impedance (600 Ω).

$$V_{out1} = 2\ V \angle 0° \cdot \frac{600\ \Omega - j200\ \Omega}{50\ \Omega - j300\ \Omega - j200\ \Omega + 600\ \Omega}$$

$$= 2\ V \angle 0° \cdot \frac{600\ \Omega - j200\ \Omega}{650\ \Omega - j500\ \Omega}$$

$$= 2\ V \angle 0° \cdot \frac{632.5\ \Omega \angle -18.4°}{820\ \Omega \angle -37.6°}$$

$$= 2\ V \angle 0° \cdot 0.771 \angle +19.2°$$

$$= \underline{1.54\ V \angle +19.2°}$$

Next, Figure 22-23 is used to determine V_{out2}. Notice E_{S1} is replaced with its internal impedance (50 Ω).

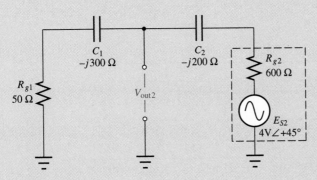

Figure 22-23

$$V_{out2} = 4\text{ V} \angle +45° \cdot \frac{50\ \Omega - j300\ \Omega}{600\ \Omega - j200\ \Omega - j300\ \Omega + 50\ \Omega}$$

$$= 4\text{ V} \angle +45° \cdot \frac{50 - j300\ \Omega}{650 - j500\ \Omega}$$

$$= 4\text{ V} \angle +45° \cdot \frac{304\ \Omega \angle -80.5°}{820\ \Omega \angle -37.6°}$$

$$= 4\text{ V} \angle +45° \cdot 0.371 \angle -42.9°$$

$$= \underline{1.48\text{ V} \angle +2.1°}$$

$$V_{Th} = V_{out1} + V_{out2}$$

$$= 1.54\text{ V} \angle +19.2° + 1.48\text{ V} \angle +2.1°$$

$$= 1.45\text{ V} + j0.506\text{ V} + 1.48\text{ V} + j0.054\text{ V}$$

$$= 2.93\text{ V} + j0.56\text{ V} = \underline{2.98\text{ V} \angle +10.8°}$$

Step 2: Replace both sources with their internal impedance and determine the Thevenin impedance at the output terminals as shown in Figure 22-24.

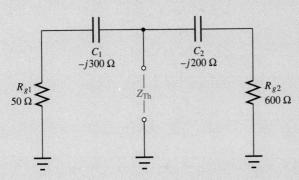

Figure 22-24

(Continued next page)

EXAMPLE 22-11 (Continued)

$$Z_{Th} = \cfrac{1}{\cfrac{1}{50 \ \Omega - j300 \ \Omega} + \cfrac{1}{600 \ \Omega - j200 \ \Omega}}$$

$$= \cfrac{1}{\cfrac{1}{304 \ \Omega \ \angle -80.5°} + \cfrac{1}{632 \ \Omega \ \angle -18.4°}}$$

$$= \cfrac{1}{3.29 \ \text{mS} \ \angle 80.5° + 1.58 \ \text{mS} \ \angle 18.4°}$$

$$= \cfrac{1}{0.543 \ \text{mS} + j3.24 \ \text{mS} + 1.5 \ \text{mS} + j0.499 \ \text{mS}}$$

$$= \cfrac{1}{2.04 \ \text{mS} + j3.74 \ \text{mS}} = \cfrac{1}{4.26 \ \text{mS} \ \angle +61.4°}$$

$$= 235 \ \Omega \ \angle -61.4° = 112 \ \Omega - j206 \ \Omega$$

Step 3: Draw the Thevenin model as shown in Figure 22-25. Determine the load voltage for (a) the 400 Ω resistor, and (b) the $+j250 \ \Omega$ inductor.

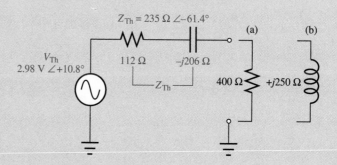

Figure 22-25

(a) $R_L = 400 \ \Omega$

$$V_{RL} = 2.98 \ \text{V} \ \angle +10.8° \cdot \frac{400 \ \Omega}{400 \ \Omega - j206 \ \Omega + 112 \ \Omega}$$

$$= 2.98 \ \text{V} \ \angle +10.8° \cdot \frac{400 \ \Omega \ \angle 0°}{552 \ \Omega \ \angle -21.9°}$$

$$= 2.98 \ \text{V} \ \angle +10.8° \cdot 0.725 \ \angle +21.9°$$

$$= \underline{2.16 \ \text{V} \ \angle +32.7°}$$

(b) $X_L = +j250 \ \Omega$

$$V_L = 2.98 \ \text{V} \ \angle +10.8° \cdot \frac{j250 \ \Omega}{j250 \ \Omega - j206 \ \Omega + 112 \ \Omega}$$

$$= 2.98 \ \text{V} \ \angle +10.8° \cdot \frac{250 \ \Omega \ \angle +90°}{120 \ \Omega \ \angle +21.4°}$$

$$= 2.98 \text{ V} \angle +10.8° \cdot 2.08 \angle 68.6°$$

$$= \underline{6.2 \text{ V} \angle +79.4°}$$

Why is this load voltage greater than the source voltage? The circuit is near resonance where $X_C = X_L$. Thus, we are seeing a Q-rise in voltage across the inductor. See Chapter 20.

Now that you have seen Thevenin's theorem for AC circuits in use, try your skills with Self-Check 22-4.

SELF-CHECK   22-4

Thevenize Figure 22-26 and determine the load voltage.

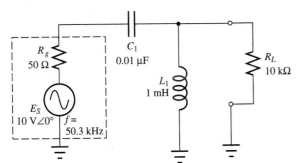

R_g 50 Ω

C_1 0.01 μF

L_1 1 mH

R_L 10 kΩ

E_S 10 V∠0° $f = $ 50.3 kHz

Figure 22-26

22-5   Norton's Theorem for AC Circuits

Norton's Theorem for AC Circuits

Any linear circuit consisting of any combination of resistance, inductance, and capacitance, and having one or more voltage sources and two output terminals, can be replaced with a single constant-current source (I_N) and a parallel complex impedance (Z_N) for the purpose of circuit analysis.

Here we include inductors and capacitors in the use of Norton's theorem just as we did with Thevenin's theorem. Naturally, the circuit becomes frequency dependent. How the circuit acts depends on the frequency of the applied AC. This makes the computer almost essential for circuit analysis over a wide range of frequencies. Our investigation here is limited to single-frequency analysis because of the tremendous volume of computation required. However, there are many software packages available for IBM® and Macintosh® computers that permit rapid and thorough circuit analysis over any range of frequencies. You will most likely be introduced to this type of computer analysis later on.

For now, it is important for you to gain a basic understanding of AC circuit analysis by working through the procedures using your scientific or engineering calculator. Examples 22-12, 22-13, and 22-14 will help you understand the use of Norton's theorem in AC circuit analysis.

EXAMPLE 22-12

Nortonize the circuit of Figure 22-27 and determine the current through R_L.

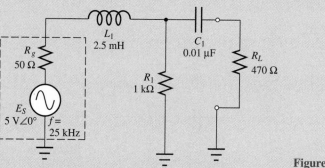

Figure 22-27

Before we begin the Nortonizing procedure, X_{L1} and X_{C1} must be determined at 25 kHz.

$$X_{L1} = 2\pi f L = 2 \cdot \pi \cdot 25 \text{ kHz} \cdot 2.5 \text{ mH} = 393 \ \Omega$$

$$X_{C1} = \frac{1}{2\pi f C} = \frac{1}{2 \cdot \pi \cdot 25 \text{ kHz} \cdot 0.01 \ \mu\text{F}} = 637 \ \Omega$$

Step 1: Replace the load with a short and determine the Norton current (I_N) that flows through the short. As shown in Figure 22-28, we're looking for I_{C1}.

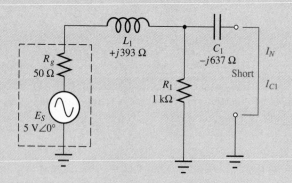

Figure 22-28

First, we must find Z_t, then I_t. I_t is divided between R_1 and C_1 in Figure 22-28.

$$Z_t = 50 \ \Omega + j393 \ \Omega + (R_1 \| C_1)$$

$$(R_1 \| C_1) = \cfrac{1}{\cfrac{1}{1 \text{ k}\Omega} + \cfrac{1}{-j637 \ \Omega}} = \frac{1}{1 \text{ mS} + j1.57 \text{ mS}}$$

$$= \frac{1}{1.86 \text{ mS} \angle +57.5°} = 537 \text{ }\Omega \angle -57.5°$$

$$537 \text{ }\Omega \angle -57.5° = 289 \text{ }\Omega - j453 \text{ }\Omega$$

$$Z_t = 50 \text{ }\Omega + j393 \text{ }\Omega + 289 \text{ }\Omega - j453 \text{ }\Omega$$

$$= 339 \text{ }\Omega - j60 \text{ }\Omega = 344 \text{ }\Omega \angle -10°$$

$$I_t = E_S/Z_t = \frac{5 \text{ V} \angle 0°}{344 \text{ }\Omega \angle -10°} = 0.0145 \text{ A} \angle +10°$$

$$I_N = I_{C1} = 0.0145 \text{ A} \angle +10° \cdot \frac{1,000 \text{ }\Omega \angle 0°}{1,000 \text{ }\Omega - j637 \text{ }\Omega}$$

$$= 0.0145 \text{ A} \angle +10° \cdot \frac{1,000 \text{ }\Omega \angle 0°}{1,186 \text{ }\Omega \angle -32.5°}$$

$$= 0.0145 \text{ A} \angle +10° \cdot 0.843 \angle +32.5°$$

$$= \underline{0.0122 \text{ A} \angle +42.5°}$$

Step 2: Replace the voltage source with its internal impedance and calculate the Norton impedance (Z_N) at the output terminals as shown in Figure 22-29.

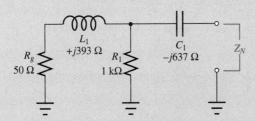

Figure 22-29

$$Z_N = -j637 \text{ }\Omega + [1 \text{ k}\Omega \| (50 \text{ }\Omega + j393 \text{ }\Omega)]$$

$$= -j637 \text{ }\Omega + \frac{1}{\dfrac{1}{1 \text{ k}\Omega} + \dfrac{1}{50 \text{ }\Omega + j393 \text{ }\Omega}}$$

$$= -j637 \text{ }\Omega + \frac{1}{1 \text{ mS} + \dfrac{1}{396 \text{ }\Omega \angle +82.7°}}$$

$$= -j637 \text{ }\Omega + \frac{1}{1 \text{ mS} + 2.53 \text{ mS} \angle -82.7°}$$

$$= -j637 \text{ }\Omega + \frac{1}{1 \text{ mS} + 0.321 \text{ mS} - j2.51 \text{ mS}}$$

$$= -j637 \text{ }\Omega + \frac{1}{1.321 \text{ mS} - j2.51 \text{ mS}}$$

$$= -j637 \text{ }\Omega + \frac{1}{2.84 \text{ mS} \angle -62.2°}$$

$$= -j637 \text{ }\Omega + 352 \text{ }\Omega \angle +62.2°$$

$$= -j637 \text{ }\Omega + 164 \text{ }\Omega + j311 \text{ }\Omega$$

$$= \underline{164 \text{ }\Omega - j326 \text{ }\Omega} = \underline{365 \text{ }\Omega \angle -63.3°}$$

(Continued next page)

EXAMPLE   22-12   (Continued)

Step 3: Draw the Norton model as shown in Figure 22-30. Calculate the current through R_L.

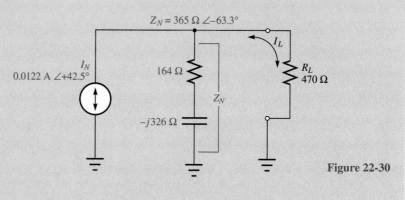

Figure 22-30

$$I_L = 0.0122 \text{ A} \angle +42.5° \cdot \frac{365 \ \Omega \ \angle -63.3°}{634 \ \Omega \ - \ j326 \ \Omega}$$

$$= 0.0122 \text{ A} \angle +42.5° \cdot \frac{365 \ \Omega \ \angle -63.3°}{713 \ \Omega \ \angle -27.2°}$$

$$= 0.0122 \text{ A} \angle +42.5° \cdot 0.512 \ \angle -36.1°$$

$$= \underline{0.00625 \text{ A} \angle +6.4°}$$

In Example 22-13, we convert a Thevenin model to a Norton model.

EXAMPLE   22-13

Convert the circuit of Figure 22-31 to a Norton model and calculate load current and voltage.

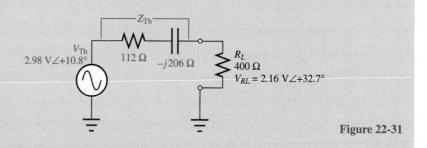

Figure 22-31

Step 1: Determine the Norton impedance.

$$Z_N = Z_{Th} = 112 - j206 \ \Omega = 235 \ \Omega \ \angle -61.4°$$

Step 2: Determine the Norton current.

$$I_N = V_{Th}/Z_{Th} = \frac{2.98 \text{ V} \angle +10.8°}{235 \text{ } \Omega \angle -61.4°}$$

$$= 0.0127 \text{ A} \angle +72.2°$$

Step 3: Draw the Norton model as shown in Figure 22-32. Calculate I_L and V_L.

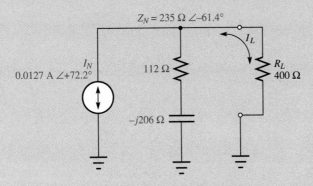

$Z_N = 235 \text{ } \Omega \angle -61.4°$

I_L

I_N
0.0127 A $\angle +72.2°$

112 Ω

R_L
400 Ω

$-j206 \text{ } \Omega$

Figure 22-32

$$I_L = 0.0127 \text{ A} \angle +72.2° \cdot \frac{235 \text{ } \Omega \angle -61.4°}{512 \text{ } \Omega - j206 \text{ } \Omega}$$

$$= 0.0127 \text{ A} \angle +72.2° \cdot \frac{235 \text{ } \Omega \angle -61.4°}{552 \text{ } \Omega \angle -21.9°}$$

$$= 0.0127 \text{ A} \angle +72.2° \cdot 0.426 \angle -39.5°$$

$$= \underline{0.00541 \text{ A} \angle +32.7°}$$

$$V_L = 0.00541 \text{ A} \angle +32.7° \cdot 400 \text{ } \Omega = 2.16 \text{ V} \angle +32.7°$$

Example 22-14 makes use of the Thevenin to Norton conversion process.

EXAMPLE 22-14

Convert the circuit of Figure 22-33 to the Norton model and calculate load current.

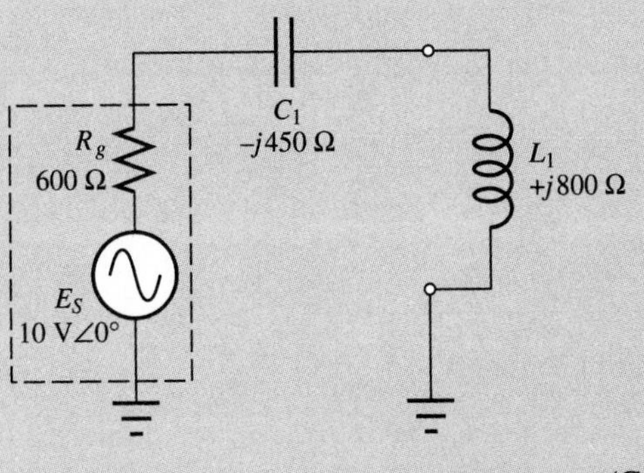

R_g
600 Ω

C_1
$-j450 \text{ } \Omega$

L_1
$+j800 \text{ } \Omega$

E_S
10 V$\angle 0°$

Figure 22-33

(Continued next page)

EXAMPLE   22-14   (Continued)

Step 1: Determine Z_N.

$Z_N = R_g + X_{C1} = 600\ \Omega - j450\ \Omega$

$\quad = 750\ \Omega\ \angle -36.9°$

Step 2: Determine I_N.

$I_N = E_S/Z_N = \dfrac{10\ \text{V}\ \angle 0°}{750\ \Omega\ \angle -36.9°} = 0.0133\ \text{A}\ \angle +36.9°$

Step 3: Draw the Norton model as shown in Figure 22-34 and calculate the current through the inductive load.

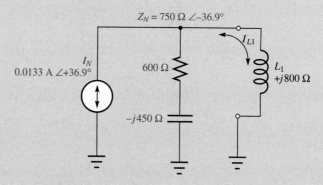

Figure 22-34

$I_{L1} = 0.0133\ \text{A}\ \angle +36.9° \cdot \dfrac{750\ \Omega\ \angle -36.9°}{600\ \Omega - j450\ \Omega + j800\ \Omega}$

$\quad = 0.0133\ \text{A}\ \angle +36.9° \cdot \dfrac{750\ \Omega\ \angle -36.9°}{695\ \Omega\ \angle +30.3°}$

$\quad = 0.0133\ \text{A}\ \angle +36.9° \cdot 1.08\ \angle -67.2°$

$\quad = \underline{0.0144\ \text{A}\ \angle -30.3°}$

Take time now to test your skills with Self-Check 22-5.

S E L F - C H E C K   2 2 - 5

1. Nortonize the circuit of Figure 22-35.

2. Calculate the load current through a 560 Ω resistor connected to the output terminals of Figure 22-35.

3. Calculate the load current through an impedance of 400 Ω + $j300$ Ω connected to the output terminals of Figure 22-35.

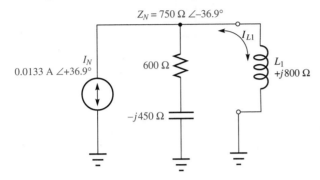

Figure 22-35

22-6   Millman's Theorem for AC Circuits

Millman's Theorem for AC Circuits

Any circuit composed of any number of parallel voltage sources can be reduced to a single equivalent voltage source by (1) converting each voltage source to a current source, (2) combining all parallel current sources and impedances into one current source, and (3) converting the equivalent current source to a voltage source.

Millman's theorem for AC circuits is illustrated in Figure 22-36. Figure 22-36a illustrates the type of circuit to which Millman's theorem applies, parallel voltage sources. The equivalent current sources are shown in Figure 22-36b. As you can see, this requires Thevenin to Norton conversion, where $Z_{Th} = Z_N$ and $I_N = E_S/Z_{Th}$. Thus, Z_1 and Z_1 and $I_{S1} = E_{S1}/Z_1$, and so on. In Figure 22-36c all current sources are combined, currents are added, and parallel impedances are combined using complex algebra. Finally, the Norton model is converted to the Thevenin model where $Z_{Th} = Z_N = Z_{eq}$ and $E_{Th} = E_{eq} = I_N \cdot Z_N = I_{eq} \cdot Z_{eq}$.

Millman's formulas, adapted from Chapter 8, provide a means of quick conversion from Figure 22-36a to 22-36d.

$$Z_{eq} = \cfrac{1}{\cfrac{1}{Z_1} + \cfrac{1}{Z_2} + \cfrac{1}{Z_3} + \cdots + \cfrac{1}{Z_\#}} \tag{22.11}$$

$$E_{eq} = \cfrac{\cfrac{E_{S1}}{Z_1} + \cfrac{E_{S2}}{Z_2} + \cfrac{E_{S3}}{Z_3} + \cdots + \cfrac{E_{S\#}}{Z_\#}}{\cfrac{1}{Z_1} + \cfrac{1}{Z_2} + \cfrac{1}{Z_3} + \cdots + \cfrac{1}{Z_\#}} \tag{22.12}$$

Carefully study Example 22-15.

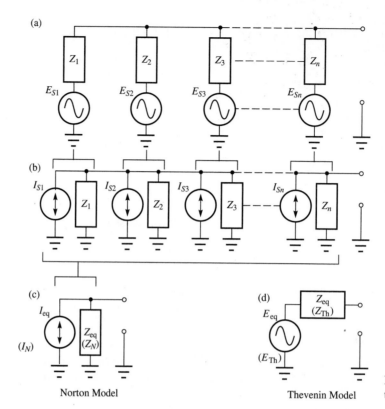

(a)

(b)

(c) Norton Model

(d) Thevenin Model

Figure 22-36 Millman's theorem for AC circuits.

EXAMPLE 22-15

Convert the circuit of Figure 22-37 to its Thevenin equivalent using Millman's theorem. Determine the output voltage across a 300 Ω + j150 Ω load connected to the output terminals.

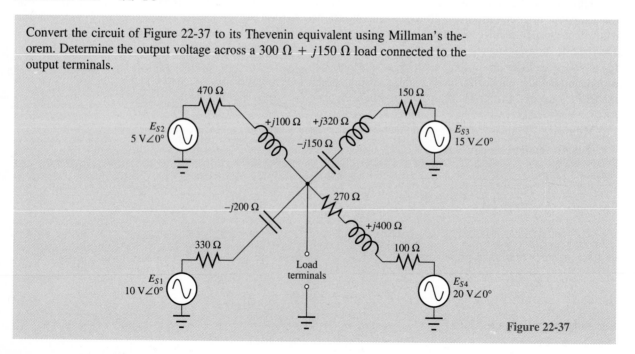

Figure 22-37

Step 1: Determine Z_{eq} for the circuit.

$$Z_1 = 330 \ \Omega - j200 \ \Omega = 386 \ \Omega \ \angle -31.2°$$

$$Z_2 = 470 \ \Omega + j100 \ \Omega = 481 \ \Omega \ \angle +12°$$

$$Z_3 = 150 \ \Omega + j320 \ \Omega - j150 \ \Omega = 227 \ \Omega \ \angle +48.6°$$

$$Z_4 = 370 \ \Omega + j400 \ \Omega = 545 \ \Omega \ \angle +47.2°$$

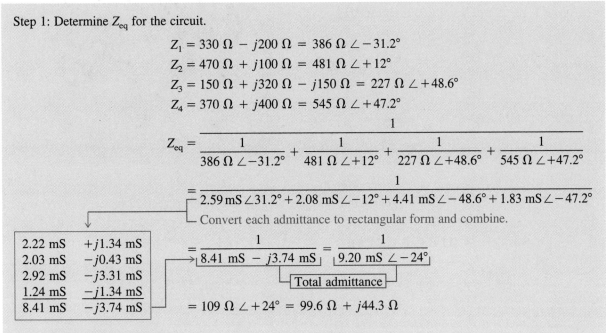

$$Z_{eq} = \cfrac{1}{\cfrac{1}{386 \ \Omega \ \angle -31.2°} + \cfrac{1}{481 \ \Omega \ \angle +12°} + \cfrac{1}{227 \ \Omega \ \angle +48.6°} + \cfrac{1}{545 \ \Omega \ \angle +47.2°}}$$

$$= \cfrac{1}{2.59 \ \text{mS} \ \angle 31.2° + 2.08 \ \text{mS} \ \angle -12° + 4.41 \ \text{mS} \ \angle -48.6° + 1.83 \ \text{mS} \ \angle -47.2°}$$

Convert each admittance to rectangular form and combine.

2.22 mS	$+j1.34$ mS
2.03 mS	$-j0.43$ mS
2.92 mS	$-j3.31$ mS
1.24 mS	$-j1.34$ mS
8.41 mS	$-j3.74$ mS

$$= \frac{1}{8.41 \ \text{mS} - j3.74 \ \text{mS}} = \frac{1}{9.20 \ \text{mS} \ \angle -24°}$$

Total admittance

$$= 109 \ \Omega \ \angle +24° = 99.6 \ \Omega + j44.3 \ \Omega$$

Step 2: Calculate E_{eq} for the circuit.

$$E_{eq} = \frac{\dfrac{10 \ \text{V} \ \angle 0°}{386 \ \Omega \ \angle -31.2°} + \dfrac{5 \ \text{V} \ \angle 0°}{481 \ \Omega \ \angle +12°} + \dfrac{15 \ \text{V} \ \angle 0°}{227 \ \Omega \ \angle +48.6°} + \dfrac{20 \ \text{V} \ \angle 0°}{545 \ \Omega \ \angle +47.2°}}{9.20 \ \text{mS} \ \angle -24°}$$

Total admittance from Step 1

22.2 mA	$+j13.4$ mA
10.2 mA	$-j2.16$ mA
43.7 mA	$-j49.6$ mA
24.9 mA	$-j26.9$ mA
101 mA	$-j76.5$ mA

$$= \frac{25.9 \ \text{mA} \ \angle +31.2° + 10.4 \ \text{mA} \ \angle -12° + 66.1 \ \text{mA} \ \angle -48.6° + 36.7 \ \text{mA} \ \angle -47.2°}{9.20 \ \text{mS} \ \angle -24°}$$

$$= \frac{127 \ \text{mA} \ \angle -37.1°}{9.20 \ \text{mS} \ \angle -24°} = 13.8 \ \text{V} \ \angle -13.1°$$

Step 3: Draw the equivalent Thevenin model. Connect the load and calculate output voltage as illustrated in Figure 22-38.

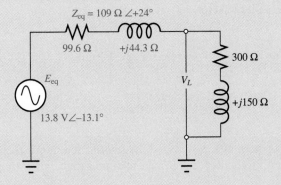

Figure 22-38

(Continued next page)

EXAMPLE 22-15 (Continued)

$$V_L = E_{eq} \cdot Z_{out}/Z_{total}$$

$$V_L = 13.8 \text{ V} \angle -13.1° \cdot \frac{300 \ \Omega + j150 \ \Omega}{399.6 \ \Omega + j194.3 \ \Omega}$$

$$= 13.8 \text{ V} \angle -13.1° \cdot \frac{335 \ \Omega \angle +26.6°}{444 \ \Omega \angle +25.9°}$$

$$= 13.8 \text{ V} \angle -13.1° \cdot 0.755 \angle +0.7° = \underline{10.4 \text{ V} \angle -12.4°}$$

Now try your skill by solving the problem of Self-Check 22-6.

SELF-CHECK 22-6

 Determine the load voltage for the circuit of Figure 22-39.

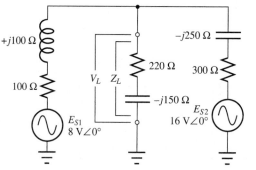

Figure 22-39

22-7 The Maximum Power Transfer Theorem for AC Circuits

Maximum Power Transfer Theorem for AC Circuits

The maximum amount of power will be delivered to a load when the load impedance is the conjugate of the source impedance.

Believe it or not, we have already covered the information in this section. Why cover it again? Cement! That's right, look at the concept of maximum power transfer from another angle and solidify it.

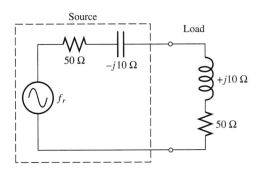

Figure 22-40   Maximum power load.

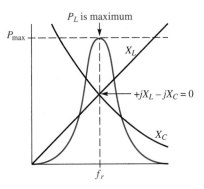

Figure 22-41   The power curve for an AC circuit.

Now, here's what you already know. From Chapter 8 you know that the load resistance must match the resistance of the source for maximum power transfer to the load. Second, from Chapter 18, you know that the power efficiency of a series RCL circuit is maximum at resonance. Recall that this is so because $X_C = X_L$ and they cancel ($-jX_C + jX_L = 0$). At resonance, X_C and X_L are said to be conjugates. Look at Figure 22-40.

In Figure 22-40, the load impedance is the conjugate of the source impedance.

$$Z_S = 50\ \Omega - j10\ \Omega \qquad \text{and} \qquad Z_L = 50\ \Omega + j10\ \Omega$$

You may recall that the concept of conjugate was presented in Chapter 18.

The circuit of Figure 22-40 has been matched for maximum power transfer. Note that at the resonant frequency (f_r) $+j10\ \Omega$ cancels $-j10\ \Omega$, leaving the source and load purely resistive where the 50 Ω load matches the 50 Ω source for maximum power transfer. The $+j10\ \Omega$ inductor performs power-factor correction for us to ensure maximum power efficiency. Now, doesn't all of this sound familiar?

Figure 22-41 demonstrates maximum power transfer for AC circuits with a bell curve. Note that this graph is a typical bandpass curve.

Study Examples 22-16 and 22-17.

EXAMPLE   22-16

The internal impedance of an AC source at a desired frequency is known to be 600 Ω − $j100$ Ω. What must the load impedance be for maximum power transfer?

The conjugate of 600 Ω − $j100$ Ω is simply 600 Ω + $j100$ Ω. Thus, Z_L = 600 Ω + $j100$ Ω.

If the unloaded source voltage is 20 V $\angle 0°$, what is the maximum power delivered to the load?

$$P_{L(max)} = \frac{(E_S/2)^2}{R_L} \qquad \text{(From Chapter 8)}$$

$$= \frac{(20\ \text{V}\ \angle 0°/2)^2}{600\ \Omega} = \frac{100}{600} = \underline{167\ \text{mW}}$$

EXAMPLE 22-17

At what frequency will the source of Figure 22-42 transfer maximum power to the load?

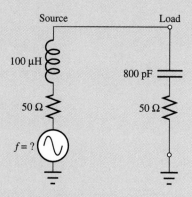

Figure 22-42

The frequency is the resonant frequency.

$$f_r = \frac{1}{2\pi\sqrt{LC}} \quad \text{(From Chapter 20)}$$

$$= \frac{1}{2\pi\sqrt{100\ \mu H \cdot 800\ pF}} = 562.7\ kHz$$

What are the source and load impedances at this resonant frequency?

$$X_L = 2\pi f_r L = 2 \cdot \pi \cdot 562.7\ kHz \cdot 100\ \mu H = 354\ \Omega$$

Thus, $Z_S = 50\ \Omega + j354\ \Omega$ and $Z_L = 50\ \Omega - j354\ \Omega$.

Time for the final Self-Check.

SELF-CHECK 22-7

1. The impedance of a source is known to be $100\ \Omega - j15\ \Omega$. What must the impedance of the load be for maximum power transfer?

2. Is the application of the maximum power transfer theorem for AC circuits frequency dependent? Explain.

3. For Figure 22-43, determine the proper load impedance for maximum power transfer and the actual maximum load power.

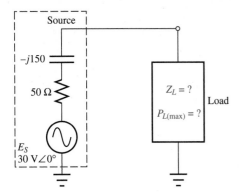

Figure 22-43

Summary

FORMULAS

Conductance

(22.1) $G = 1/R$ (S), (1 S = 1/1 Ω)

(22.2) $R = 1/G$ (Ω), (1 Ω = 1/1 S)

(22.3) $R \cdot G = 1$ (1 $\Omega \cdot$ 1 S = 1)

Susceptance

(22.4) $B = 1/X$ (S), (1 S = 1/1 Ω)

   also $+jB = 1/-jX_C$ and $-jB = 1/+jX_L$

(22.5) $X = 1/B$ (Ω), (1 Ω = 1/1 S)

   also $+jX_L = 1/-jB$ and $-jX_C = 1/+jB$

(22.6) $X \cdot B = 1$ (1 $\Omega \cdot$ 1 S = 1)

   also $+jX_L \cdot -jB = 1$ and $-jX_C \cdot jB = 1$

Admittance

(22.7) $Y = 1/Z$ (S), (1 S = 1/1 Ω)

   also $Y \angle +\theta = 1/Z \angle -\theta$ and $Y \angle -\theta = 1/Z \angle +\theta$

(22.8) $Z = 1/Y$ (Ω), (1 Ω = 1/1 S)

   also $Z \angle +\theta = 1/Y \angle -\theta$ and $Z \angle -\theta = 1/Y \angle +\theta$

(22.9) $Z \cdot Y = 1$ (1 $\Omega \cdot$ 1 S = 1)

   also $Z \angle +\theta \cdot Y \angle -\theta = 1 \angle 0°$ and $Z \angle -\theta \cdot Y \angle +\theta = 1 \angle 0°$

General Formula for Series Voltage Drops

(22.10) $V_{Z\#} = E_S \cdot Z^{\#}/Z_t$ (general voltage divider rule)

Millman's Theorem for AC Circuits

$$(22.11)\quad Z_{eq} = \cfrac{1}{\cfrac{1}{Z_1} + \cfrac{1}{Z_2} + \cfrac{1}{Z_3} + \cdots + \cfrac{1}{Z_\#}}$$

$$(22.12)\quad Z_{eq} = \cfrac{\dfrac{E_{S1}}{Z_1} + \dfrac{E_{S2}}{Z_2} + \dfrac{E_{S3}}{Z_3} + \cdots + \dfrac{E_{S\#}}{Z_\#}}{\dfrac{1}{Z_1} + \dfrac{1}{Z_2} + \dfrac{1}{Z_3} + \cdots + \dfrac{1}{Z_\#}}$$

CONCEPTS

- Conductance (G) is the reciprocal of resistance and is measured in siemens (S).
- Susceptance (B) is the reciprocal of reactance and is measured in siemens (S).
- Admittance (Y) is the reciprocal of impedance and is measured in siemens (S).
- The total impedance for a complex overall parallel AC circuit may be determined using Ohm's Law, if E_S and I_t are known, or the reciprocal formula using all branch impedances.
- The total impedance for a complex overall series AC circuit may be determined using Ohm's Law, if E_S and I_t are known, or by adding all series impedances in rectangular form.
- The total current for a complex AC circuit may be determined using Ohm's Law if E_S and Z_t are known.
- The total current for complex overall parallel AC circuits may be determined by adding all branch currents in rectangular form.
- The voltage drops for series impedances of a complex AC circuit may be determined using Ohm's Law or the proportion formula (voltage divider rule).
- AC circuits are frequency dependent because of changes in X_C and X_L with frequency.
- Circuit-analysis theorems are applied to AC circuits following the same basic procedures as for DC circuits except complex algebra must be used to handle complex impedances.
- When a source has an internal complex impedance at a desired frequency, the load must have an impedance that is the conjugate of the source impedance for maximum power to be transferred.

PROCEDURES

The Superposition Theorem

1. Perform a complete analysis of the AC circuit using only one source at a time. Each other source is replaced with its internal impedance.
2. Use phasor addition to determine the actual voltage across and current through circuit components.

Thevenin's Theorem

1. Remove the load and determine the Thevenin voltage (V_{Th}) at the output terminals.
2. Replace each voltage source with its internal impedance and calculate the Thevenin circuit impedance at the output terminals with the load removed.

Norton's Theorem

1. Replace the load with a short and determine the Norton current (I_N) that flows through the short.
2. Determine the Norton impedance (Z_N) the same as Thevenin's. $Z_{Th} = Z_N$

Millman's Theorem

1. Each parallel voltage source is converted to an equivalent current source using Thevenin to Norton conversion.
2. Parallel current sources are combined, as are the parallel impedances, using complex algebra.
3. The resulting equivalent current source is converted to an equivalent voltage source using Norton to Thevenin conversion.

SPECIAL TERMS

- admittance $\left(Y = \dfrac{1}{Z} \right)$

- susceptance $\left(B = \dfrac{1}{X} \right)$

- conductance $\left(G = \dfrac{1}{R} \right)$

Need to Know Solution

For maximum power transfer, which is what you want, the impedance of the antenna must be the conjugate of the transmitter's output impedance. Therefore, $Z_{ant.} = 50\ \Omega + j20\ \Omega$. The $+j20\ \Omega$ of the antenna cancels the $-j20\ \Omega$ of the transmitter, leaving the output circuit purely resistive. This means the radiation resistance of the antenna is 50 Ω and the maximum amount of power is converted to radiated electromagnetic energy.

Problems

22-1 Conductance, Susceptance, and Admittance

1. What is the conductance of a 51 Ω resistor?
2. What is the susceptance of 2,300 Ω of capacitive reactance? Express it in rectangular and polar form.
3. What is the susceptance of 150 kΩ of inductive reactance? Express it in rectangular and polar form.
4. What is the admittance of an impedance of 1.5 MΩ $\angle 45°$?
5. What is the admittance of an impedance of 400 Ω − $j300$ Ω?
6. What is the total admittance for a parallel *RCL* circuit that has 2.7 kΩ of resistance in the first branch, 5 kΩ of capacitive reactance in the second branch, and 3 kΩ of inductive reactance in the third branch? Use this total admittance to find the total impedance.

22-2 Complex AC Circuits

7.

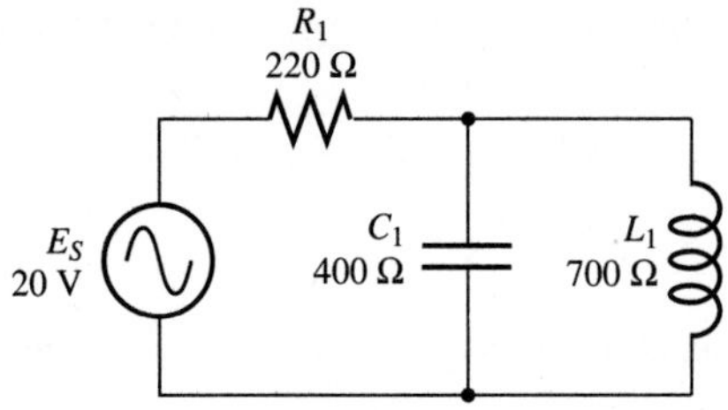

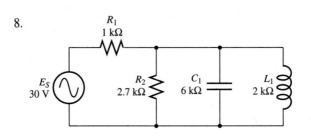

Figure 22-44

Show phase angles for all voltages and currents in Figure 22-44.

a. $Z_t =$ d. $V_{C1} =$
b. $I_t =$ e. $I_{C1} =$
c. $V_{R1} =$ f. $I_{L1} =$

8.

Figure 22-45

Show phase angles for all voltages and currents in Figure 22-45.

a. $Z_t =$ e. $I_{C1} =$
b. $I_t =$ f. $i_{L1} =$
c. $V_{R1} =$ g. $I_{R2} =$
d. $V_{C1} =$

9.

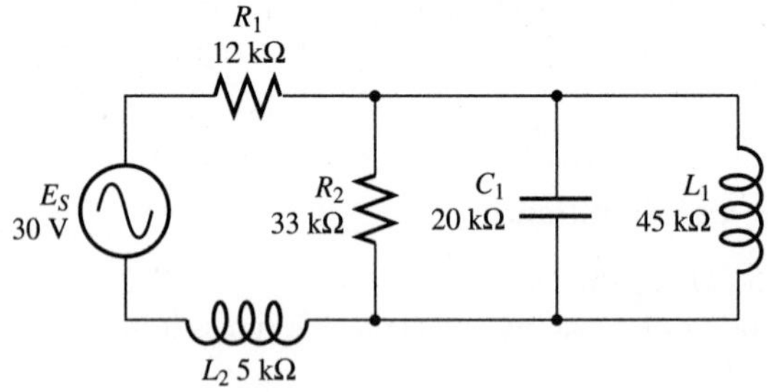

Figure 22-46

L_2 5 kΩ

Show phase angles for all voltages and currents in Figure 22-46.

a. $Z_t =$ e. $V_{L2} =$
b. $I_t =$ f. $I_{L1} =$
c. $V_{R1} =$ g. $I_{R2} =$
d. $V_{C1} =$

10.

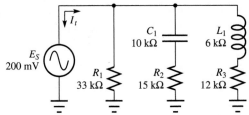

Figure 22-47

Show phase angles for all voltages and currents in Figure 22-47.
a. $Z_t =$
b. $I_t =$
c. $V_{R2} =$
d. $V_{C1} =$
e. $V_{R3} =$

11.

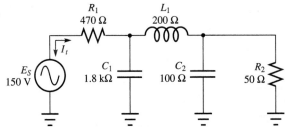

Figure 22-48

Show phase angles for all voltages and currents in Figure 22-48.
a. $Z_t =$
b. $I_t =$
c. $V_{R1} =$

12.

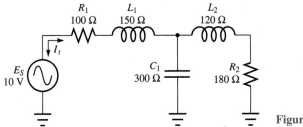

Figure 22-49

Show phase angles for all voltages and currents in Figure 22-49.
a. $Z_t =$
b. $I_t =$
c. $V_{R1} =$

22-3 The Superposition Theorem for AC Circuits

13. Use superposition to determine the voltage across R_L in Figure 22-50. Show all work in steps.

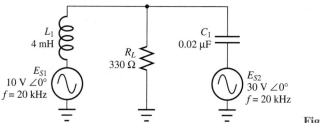

Figure 22-50

14. Use superposition to determine the voltage across R_L in Figure 22-51. Show all work in steps.

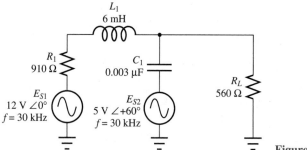

Figure 22-51

15. Use superposition to find the current through R_L in Figure 22-52. Show all work in steps.

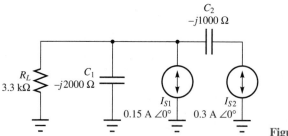

Figure 22-52

16. Use superposition to find the current through R_L in Figure 22-53. Show all work in steps.

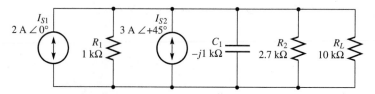

Figure 22-53

22-4 Thevenin's Theorem for AC Circuits

17. Convert Figure 22-54 to a Thevenin AC model and determine the voltage across R_L. Show all work in steps.

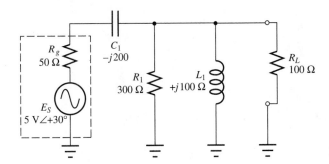

Figure 22-54

18. Convert Figure 22-55 to a Thevenin AC model and determine the voltage across R_L. Show all work in steps.

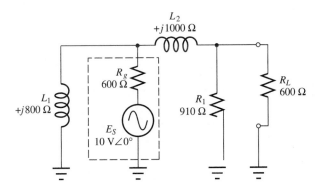

Figure 22-55

19. Convert Figure 22-56 to a Thevenin AC model and determine the voltage across a load impedance of $400 \, \Omega - j300 \, \Omega$. Show all work.

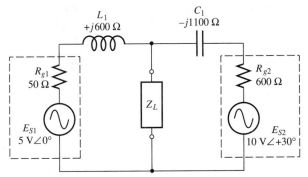

Figure 22-56

20. Convert Figure 22-56 to a Thevenin AC model and determine the voltage across a load impedance of $100 \, \Omega + j200 \, \Omega$. Show all work.

22-5 Norton's Theorem for AC Circuits

21. Convert Figure 22-57 to a Norton AC model and determine the current through R_L. Show work.

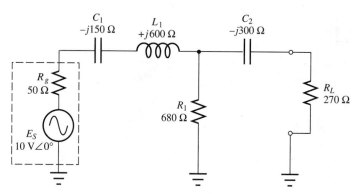

Figure 22-57

22. Convert Figure 22-58 to a Norton AC model and determine the current through the load impedance. Show work.

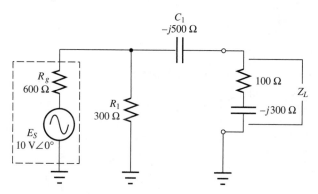

Figure 22-58

23. Convert Figure 22-54 to a Norton AC model and determine load current. Show work.
24. Convert Figure 22-55 to a Norton AC model and determine load current. Show work.

22-6 Millman's Theorem for AC Circuits

25. Convert Figure 22-59 to a Thevenin AC model using Millman's theorem. Show all work in steps.

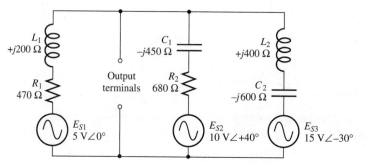

Figure 22-59

26. Determine the voltage at the output terminals of Figure 22-59 with a 200 Ω + $j100$ Ω load connected. Show work.
27. Convert Figure 22-60 to a Thevenin AC model using Millman's theorem. Show all work in steps.

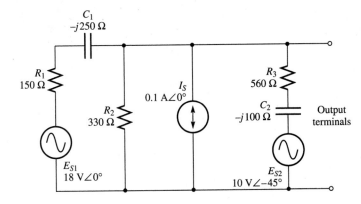

Figure 22-60

28. Determine the voltage across the output terminals of Figure 22-60 with a 400 Ω + $j200$ Ω load connected. Show all work.

22-7 The Maximum Power Transfer Theorem for AC Circuits

29. What should the load impedance of Figure 22-54 be to get maximum output power (maximum load power)?
30. What should the load impedance of Figure 22-56 be to get maximum load power?
31. Calculate the maximum possible load power for Figure 22-54.
32. Calculate the maximum possible load power for Figure 22-56.

Answers to Self-Checks

Self-Check 22-1

1. 2.128 mS
2. $+j666$ μS
3. $-j27.8$ μS
4. 4.14 mS $\angle -24.4°$
5. 1 mS $+ j0.167$ mS $= 1.014$ mS $\angle +9.5°$

Self-Check 22-2

$Z_t = 860.7$ Ω $\angle 36.5°$ $I_{L2} = 19.3$ mA $\angle -92.8°$ $V_R = 29$ V $\angle -2.8°$

$I_t = 34.9$ mA $\angle -36.5°$ $V_{C1} = 5.24$ V $\angle -126.5°$ $V_{C2} = 9.7$ V $\angle -182.8°$

$I_R = 29$ mA $\angle -2.8°$ $V_{L1} = 6.98$ V $\angle +53.5°$ $V_{L2} = 38.7$ V $\angle -2.8°$

Self-Check 22-3

1. $V_{RL1} = 2.744$ V $- j5.804$ V $= 6.42$ V $\angle -64.7°$
   $V_{RL2} = 1.02$ V $+ j12.4$ V $= 12.4$ V $\angle +85.3°$
   $V_{RL} = 3.764$ V $+ j6.596$ V $= 7.59$ V $\angle +60.3°$

2. $I_{RL1} = 0.0896\ \text{A} - j0.0303\ \text{A} = 0.0946\ \text{A} \angle -18.7°$
   $I_{RL2} = 0.00606\ \text{A} - j0.0179\ \text{A} = 0.0189\ \text{A} \angle -71.3°$
   $I_{RL} = 0.0957\ \text{A} - j0.0124\ \text{A} = 0.0965\ \text{A} \angle -7.38°$

Self-Check 22-4

$X_L = 316\ \Omega,\ X_C = 316\ \Omega$
$V_{\text{Th}} = 63.2\ \text{V} \angle +90°,\ Z_{\text{Th}} = 2{,}020\ \Omega \angle +8.13°$
$Z_{\text{Th}} = 2{,}000\ \Omega + j286\ \Omega$
$V_{RL} = 52.6\ \text{V} \angle +88.6°$

Self-Check 22-5

1. $Z_N = 552\ \Omega \angle -74.8°$
   $I_N = 0.0178\ \text{A} \angle -57.7°$
2. $I_L = 0.0111\ \text{A} \angle -95.4°$
3. $I_L = 0.0166\ \text{A} \angle -109.4°$

Self-Check 22-6

$Z_{\text{eq}} = 129\ \Omega \angle +25.8° = 116\ \Omega + j56.1\ \Omega$
$E_{\text{eq}} = 9.39\ \text{V} \angle +14.8°,\ V_L = 7.16\ \text{V} \angle -3.9°$

Self-Check 22-7

1. $Z_L = 100\ \Omega + j15\ \Omega$
2. Yes! The resonant frequency where $X_C = X_L$
3. $Z_L = 50\ \Omega + j150\ \Omega,\ P_{L(\text{max})} = 4.5\ \text{W}$

SUGGESTED PROJECTS

1. Add some of the main concepts, formulas, and procedures from this chapter to your Electricity and Electronics Notebook.

POSTSCRIPT

Well, you have just finished the entire book and you have gained much valuable knowledge and experience in electricity and electronics. Together, we have built a strong foundation upon which you will continue to build. I trust you have found this book to be of value and that you will want to keep it as a reference. Perhaps I will have a part in your continued journey through the world of electronic devices and integrated circuits via my new book, *Exploring Electronic Devices*. Until next time, I wish the best for you in your career!—*Mark E. Hazen*

Appendix A

English/Metric Conversion (cgs/mks, SI)

English	Conversion Factor (to multiply by)		Metric
angstrom (Å)	× =	$1 \cdot 10^{-10}$ $1 \cdot 10^{10}$	= ×	meters (m)
inches (in.)	× =	2.54 0.3937	= ×	centimeters (cm)
feet (ft)	× =	0.3048 3.281	= ×	meters (m)
yards (yd)	× =	0.9144 1.094	= ×	meters (m)
miles (mi)	× =	1.60934 0.6214	= ×	kilometers (km)
square yards (yd²)	× =	0.8361 1.196	= ×	square meters (m²)
circular mil	× =	$5.067 \cdot 10^{-10}$ $1.9736 \cdot 10^{9}$	= ×	square meters (m²)
cubic yards (yd³)	× =	0.7646 1.308	= ×	cubic meters (m³)
ounces (oz)	× =	28.35 0.0353	= ×	grams (g)

English		Conversion Factor (to multiply by)		Metric
pounds (lb)	× =	0.4536 2.205	= ×	kilograms (kg)
Fahrenheit (F)	× =	5/9 (after subtracting 32) 9/5 (then add 32)	= ×	Celsius (C)
gauss (G)	× =	$1 \cdot 10^{-4}$ $1 \cdot 10^{4}$	= ×	tesla (T)
maxwell (Mx)	× =	$1 \cdot 10^{-8}$ $1 \cdot 10^{8}$	= ×	weber (wb)
gilbert (Gb)	× =	0.794 1.26	= ×	ampere turns (At)
oersted (Oe)	× =	79.4 0.0126	= ×	ampere turns per meter (At/m)
British Thermal Unit (BTU)	× =	1,056 $9.47 \cdot 10^{-4}$	= ×	joule (J)
calorie (c)	× =	4.19 0.239	= ×	joule (J)
foot pound (ft · lb)	× =	1.356 0.7375	= ×	joule (J)
horsepower (hp)	× =	746 $1.34 \cdot 10^{-3}$	= ×	watt (W)
feet per second (ft/s)	× =	0.3048 3.2808	= ×	meters per second (m/s)
miles per hour (mi/h)	× =	1.609 0.622	= ×	kilometers per hour (km/h)

Appendix B

The Periodic Table

IA																	VIIA	O
1 **H** 1.0079																	1 **H** 1.0079	2 **He** 4.0026
	IIA											IIIA	IVA	VA	VIA			
3 **Li** 6.941	4 **Be** 9.0122											5 **B** 10.811	6 **C** 12.011	7 **N** 14.0067	8 **O** 15.9994	9 **F** 18.9984	10 **Ne** 20.1797
11 **Na** 22.9898	12 **Mg** 24.3050	IIIB	IVB	VB	VIB	VIIB	VIII			IB	IIB	13 **Al** 26.9815	14 **Si** 28.0855	15 **P** 30.9738	16 **S** 32.066	17 **Cl** 35.4527	18 **Ar** 39.948
19 **K** 39.0983	20 **Ca** 40.078	21 **Sc** 44.9559	22 **Ti** 47.88	23 **V** 50.9415	24 **Cr** 51.9961	25 **Mn** 54.9380	26 **Fe** 55.847	27 **Co** 58.9332	28 **Ni** 58.69	29 **Cu** 63.39	30 **Zn** 65.39	31 **Ga** 69.723	32 **Ge** 72.61	33 **As** 74.9216	34 **Se** 78.96	35 **Br** 79.904	36 **Kr** 83.80
37 **Rb** 85.4678	38 **Sr** 87.62	39 **Y** 88.9059	40 **Zr** 91.224	41 **Nb** 92.9064	42 **Mo** 95.94	43 **Tc** (98)	44 **Ru** 101.07	45 **Rh** 102.9055	46 **Pd** 106.42	47 **Ag** 107.8682	48 **Cd** 112.411	49 **In** 114.82	50 **Sn** 118.710	51 **Sb** 121.75	52 **Te** 127.60	53 **I** 126.9045	54 **Xe** 131.29
55 **Cs** 132.9054	56 **Ba** 137.327	57 ***La** 138.9055	72 **Hf** 178.49	73 **Ta** 180.9479	74 **W** 183.85	75 **Re** 186.207	76 **Os** 190.2	77 **Ir** 192.22	78 **Pt** 195.08	79 **Au** 196.9665	80 **Hg** 200.59	81 **Tl** 204.3833	82 **Pb** 207.2	83 **Bi** 208.9804	84 **Po** (209)	85 **At** (210)	86 **Rn** (222)
87 **Fr** (223)	88 **Ra** (226)	89 ****Ac** (227)	104 **Unq** (261)	105 **Unp** (262)	106 **Unh** (263)	107 **Uns** (262)	108	109									

* Lanthanide Series

58 **Ce** 140.115	59 **Pr** 140.9076	60 **Nd** 144.24	61 **Pm** (145)	62 **Sm** 150.36	63 **Eu** 151.965	64 **Gd** 157.25	65 **Tb** 158.9523	66 **Dy** 162.50	67 **Ho** 164.9303	68 **Er** 167.26	69 **Tm** 168.9342	70 **Yb** 173.04	71 **Lu** 174.967

** Actinide Series

90 **Th** 232.0381	91 **Pa** 231.0359	92 **U** 238.0289	93 **Np** (237)	94 **Pu** (244)	95 **Am** (243)	96 **Cm** (247)	97 **Bk** (247)	98 **Cf** (251)	99 **Es** (252)	100 **Fm** (257)	101 **Md** (258)	102 **No** (259)	103 **Lr** (260)

Note: Atomic masses are 1985 IUPAC values (up to four decimal places).

Appendix C

Algebraic Operations

The Golden Rule for Solving Algebraic Equations

"Any operation performed on one side of an equation must also be performed on the opposite side of the equation."

Using Addition and Subtraction in Solving Equations

> **Rule**
>
> What is added to or subtracted from one side of the equation must also be added to or subtracted from the other side of the equation.

EXAMPLE Solve for I_1: $I_t = I_1 + 10$
10 must be subtracted from each side to leave I_1 by itself.

$$I_t - 10 = I_1 + 10 - 10$$
$$I_t - 10 = I_1 + \cancel{10} - \cancel{10}$$
$$I_t - 10 = I_1$$
$$\boxed{I_1 = I_t - 10}$$

EXAMPLE Solve for I_t: $I_t - I_1 - I_2 - I_3 = 0$
I_1, I_2, & I_3 must be added to both sides to leave I_t by itself.

$$I_t - I_1 - I_2 - I_3 + I_1 + I_2 + I_3 = 0 + I_1 + I_2 + I_3$$
$$I_t - \cancel{I_1 - I_2 - I_3} + \cancel{I_1 + I_2 + I_3} = 0 + I_1 + I_2 + I_3$$
$$I_t \qquad\qquad\qquad = 0 + I_1 + I_2 + I_3$$
$$\boxed{I_t = I_1 + I_2 + I_3}$$

EXAMPLE Solve for V_{R3}: $E_s - V_{R1} = V_{R2} + V_{R3}$
V_{R2} must be subtracted from both sides to leave V_{R3} by itself.

$$E_s - V_{R1} - V_{R2} = V_{R2} + V_{R3} - V_{R2}$$
$$E_s - V_{R1} - V_{R2} = \cancel{V_{R2}} + V_{R3} - \cancel{V_{R2}}$$
$$E_s - V_{R1} - V_{R2} = \qquad V_{R3}$$
$$\boxed{V_{R3} = E_s - V_{R1} - V_{R2}}$$

Using Multiplication and Division in Solving Equations

> **Rule**
>
> Both sides of the equation must be multiplied or divided by the same factor, constant, or variable when the equation is simplified or rearranged.

Solve for E:

$$I = \frac{E}{R}$$

Both sides must be multiplied by R to leave E by itself.

$$I \cdot R = \frac{E}{R} \cdot R$$

$$I \cdot R = E$$

$$\boxed{E = I \cdot R}$$

EXAMPLE

Solve for f: $X_L = 2\pi f L$
Both sides must be divided by $2\pi L$ to leave f by itself.

$$\frac{X_L}{2\pi L} = \frac{2\pi f L}{2\pi L}$$

$$\frac{X_L}{2\pi L} = f$$

$$\boxed{f = \frac{X_L}{2\pi L}}$$

EXAMPLE

Solve for C:

$$X_C = \frac{1}{2\pi f C}$$

Both sides must be multiplied by C then divided by X_C.

$$X_C \cdot C = \frac{1}{2\pi f C} \cdot C$$

$$\frac{X_C \cdot C}{X_C} = \frac{\frac{1}{2\pi f}}{X_C} = \frac{1}{2\pi f X_C}$$

$$\boxed{C = \frac{1}{2\pi f X_C}}$$

EXAMPLE

Using Squares and Square Roots in Solving Equations

Rule

If it is necessary to square one side of the equation, the other side must also be squared. If it is necessary to take the square root of one side of the equation, the square root must also be taken of the other side.

EXAMPLE   Solve for E:

$$P = \frac{E^2}{R}$$

Multiply both sides of the equation by R, then take the square root of both sides to solve for E.

$$P \cdot R = \frac{E^2}{\cancel{R}} \cdot \cancel{R}$$
$$\sqrt{P \cdot R} = \sqrt{E^2}$$
$$\sqrt{P \cdot R} = E$$
$$\boxed{E = \sqrt{P \cdot R}}$$

EXAMPLE   Solve for X_L: $Z_t = \sqrt{R^2 + X_L^2}$

Both sides must be squared, R^2 must be subtracted from both sides, and the square root must then be taken from both sides to solve for X_L by itself.

$$Z_t^2 = \sqrt{R^2 + X_L^2}^{\,2}$$
$$Z_t^2 = R^2 + X_L^2$$
$$Z_t^2 - R^2 = \cancel{R^2} + X_L^2 - \cancel{R^2}$$
$$Z_t^2 - R^2 = X_L^2$$
$$\sqrt{Z_t^2 - R^2} = \sqrt{X_L^2}$$
$$\sqrt{Z_t^2 - R^2} = X_L$$
$$\boxed{X_L = \sqrt{Z_t^2 - R^2}}$$

Dividing One Fraction by Another

Rule

Invert the fraction in the denominator and multiply the two fractions numerator to numerator and denominator to denominator.

EXAMPLE

$$\frac{\dfrac{A}{B}}{\dfrac{C}{D}} = \frac{A}{B} \cdot \frac{D}{C} = \frac{AD}{BC}$$

Using Cross Multiplication

EXAMPLE

$$\frac{A}{B} = \frac{C}{D}$$

$$\frac{A}{B} \diagdown \frac{C}{D} \rightarrow AD = BC$$

Now the equation can easily be solved for any one of the variables.

Adding and Subtracting Fractions

> **Rule**
>
> Fractions cannot be added or subtracted unless they have the same value denominator (common denominator).

EXAMPLE

$$\frac{3}{4} + \frac{5}{6} = ?$$

An easily obtained common denominator is simply the product of the two denominators. The numerator of each fraction must also be multiplied in the conversion process as follows:

$$\frac{3}{4} = \frac{3 \cdot 6}{4 \cdot 6} = \frac{18}{24} \quad \text{and} \quad \frac{5}{6} = \frac{5 \cdot 4}{6 \cdot 4} = \frac{20}{24}$$

$$\frac{18}{24} + \frac{20}{24} = \frac{18 + 20}{24} = \frac{38}{24} = 1\frac{14}{24} = 1\frac{7}{12}$$

Appendix D

Basic Trigonometry

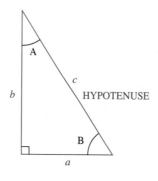

Pythagorean Theorem
$$a^2 + b^2 = c^2$$
$$c = \sqrt{a^2 + b^2}$$
$$a = \sqrt{c^2 - b^2}$$
$$b = \sqrt{c^2 - a^2}$$

Sine Function: (side opposite the angle over the hypotenuse)

$\sin \angle A = a/c$, and $\angle A = \arcsin (a/c)$
$\qquad\qquad\qquad\qquad = \sin^{-1} (a/c)$

$\sin \angle B = b/c$, and $\angle B = \arcsin (b/c)$
$\qquad\qquad\qquad\qquad = \sin^{-1} (b/c)$

Cosine Function: (side adjacent to the angle over the hypotenuse)

$\cos \angle A = b/c$, and $\angle A = \arccos (b/c)$
$\qquad\qquad\qquad\qquad = \cos^{-1} (b/c)$

$\cos \angle B = a/c$, and $\angle B = \arccos (a/c)$
$\qquad\qquad\qquad\qquad = \cos^{-1} (a/c)$

Tangent Function: (side opposite the angle over the side adjacent to the angle)

$\tan \angle A = a/b$, and $\angle A = \arctan (a/b)$
$\qquad\qquad\qquad\qquad = \tan^{-1} (a/b)$

$\tan \angle B = b/a$, and $\angle B = \arctan (b/a)$
$\qquad\qquad\qquad\qquad = \tan^{-1} (b/a)$

Appendix E

Simultaneous Equations and Determinants

Second Order Determinants

Second order determinants are used to solve for the common variables of two simultaneous equations.

Basic Format:

Equation #1: $aV_1 + bV_2 = c$

Equation #2: $dV_1 + eV_2 = f$

(where a, b, d, & e are coefficients of the variables V_1 & V_2, and c & f are known quantities)

$$V_1 = \frac{\begin{vmatrix} c & b \\ f & e \end{vmatrix}}{\begin{vmatrix} a & b \\ d & e \end{vmatrix}} \qquad V_2 = \frac{\begin{vmatrix} a & c \\ d & f \end{vmatrix}}{\begin{vmatrix} a & b \\ d & e \end{vmatrix}}$$

The barred configurations in the numerators and denominators are referred to as *determinants*.

Solving for Each Determinant:

$$\begin{vmatrix} a & b \\ d & e \end{vmatrix} = \begin{matrix} ae - db \\ \begin{vmatrix} a & b \\ d & e \end{vmatrix} \end{matrix} = ae - db$$

Multiply diagonally and subtract the products.

Solving for Each Variable:

$$V_1 = \frac{\begin{vmatrix} c & b \\ f & e \end{vmatrix}}{\begin{vmatrix} a & b \\ d & e \end{vmatrix}} = \frac{ce - fb}{ae - db} = \frac{ce - fb}{ae - db}$$

$$V_2 = \frac{\begin{vmatrix} a & c \\ d & f \end{vmatrix}}{\begin{vmatrix} a & b \\ d & e \end{vmatrix}} = \frac{af - dc}{ae - db} = \frac{af - dc}{ae - db}$$

Third Order Determinants

Third order determinants are used to solve for the common variables of three simultaneous equations.

Basic Format:

Equation #1: $a_1V_1 + b_1V_2 + c_1V_3 = d_1$

Equation #2: $a_2V_1 + b_2V_2 + c_2V_3 = d_2$

Equation #3: $a_3V_1 + b_3V_2 + c_3V_3 = d_3$

$$V_1 = \frac{\begin{vmatrix} d_1 & b_1 & c_1 \\ d_2 & b_2 & c_2 \\ d_3 & b_3 & c_3 \end{vmatrix}}{\begin{vmatrix} a_1 & b_1 & c_1 \\ a_2 & b_2 & c_2 \\ a_3 & b_3 & c_3 \end{vmatrix}} \qquad V_2 = \frac{\begin{vmatrix} a_1 & d_1 & c_1 \\ a_2 & d_2 & c_2 \\ a_3 & d_3 & c_3 \end{vmatrix}}{\begin{vmatrix} a_1 & b_1 & c_1 \\ a_2 & b_2 & c_2 \\ a_3 & b_3 & c_3 \end{vmatrix}} \qquad V_3 = \frac{\begin{vmatrix} a_1 & b_1 & d_1 \\ a_2 & b_2 & d_2 \\ a_3 & b_3 & d_3 \end{vmatrix}}{\begin{vmatrix} a_1 & b_1 & c_1 \\ a_2 & b_2 & c_2 \\ a_3 & b_3 & c_3 \end{vmatrix}}$$

SOLVING FOR EACH DETERMINANT:

$\boxed{a_3b_2c_1 + b_3c_2a_1 + c_3a_2b_1}$

expression #2

$\boxed{c_3a_2b_1}$

$\boxed{b_3c_2a_1}$

$\boxed{a_3b_2c_1}$

$$\begin{vmatrix} a_1 & b_1 & c_1 \\ a_2 & b_2 & c_2 \\ a_3 & b_3 & c_3 \end{vmatrix} \rightarrow \begin{vmatrix} a_1 & b_1 & c_1 \\ a_2 & b_2 & c_2 \\ a_3 & b_3 & c_3 \end{vmatrix} \begin{matrix} a_1 & b_1 \\ a_2 & b_2 \\ a_3 & b_3 \end{matrix} \rightarrow$$

first two columns repeated

$\boxed{a_1b_2c_3}$

$\boxed{a_1b_2c_3 + b_1c_2a_3 + c_1a_2b_3}$

expression #1

$\boxed{b_1c_2a_3}$

$\boxed{c_1a_2b_3}$

$$\begin{vmatrix} a_1 & b_1 & c_1 \\ a_2 & b_2 & c_2 \\ a_3 & b_3 & c_3 \end{vmatrix} = \text{expression \#1} - \text{expression \#2}$$

$$= \boxed{a_1b_2c_3 + b_1c_2a_3 + c_1a_2b_3} - \boxed{a_3b_2c_1 + b_3c_2a_1 + c_3a_2b_1}$$

expression #1   expression #2

Third Order Determinants Continued

Equation #1: $a_1V_1 + b_1V_2 + c_1V_3 = d_1$

Equation #2: $a_2V_1 + b_2V_2 + c_2V_3 = d_2$

Equation #3: $a_3V_1 + b_3V_2 + c_3V_3 = d_3$

$$V_1 = \frac{\begin{vmatrix} d_1 & b_1 & c_1 \\ d_2 & b_2 & c_2 \\ d_3 & b_3 & c_3 \end{vmatrix} \;\text{This determinant must be resolved the same way the common denominator determinant was resolved.}}{\boxed{a_1b_2c_3 + b_1c_2a_3 + c_1a_2b_3} - \boxed{a_3b_2c_1 + b_3c_2a_1 + c_3a_2b_1}}$$

$$V_2 = \frac{\begin{vmatrix} a_1 & d_1 & c_1 \\ a_2 & d_2 & c_2 \\ a_3 & d_3 & c_3 \end{vmatrix} \;\text{This determinant must be resolved the same way the common denominator determinant was resolved.}}{\boxed{a_1b_2c_3 + b_1c_2a_3 + c_1a_2b_3} - \boxed{a_3b_2c_1 + b_3c_2a_1 + c_3a_2b_1}}$$

$$V_3 = \frac{\begin{vmatrix} a_1 & b_1 & d_1 \\ a_2 & b_2 & d_2 \\ a_3 & b_3 & d_3 \end{vmatrix} \;\text{This determinant must be resolved the same way the common denominator determinant was resolved.}}{\boxed{a_1b_2c_3 + b_1c_2a_3 + c_1a_2b_3} - \boxed{a_3b_2c_1 + b_3c_2a_1 + c_3a_2b_1}}$$

EXAMPLE

Equation #1: $11V_1 - 3V_2 - 8V_3 = 15$

Equation #2: $-3V_1 + 10V_2 - 5V_3 = 0$

Equation #3: $-8V_1 - 5V_2 + 23V_3 = 0$

$$V_1 = \frac{\begin{vmatrix} 15 & -3 & -8 & 15 & -3 \\ 0 & 10 & -5 & 0 & 10 \\ 0 & -5 & 23 & 0 & -5 \end{vmatrix}}{\begin{vmatrix} 11 & -3 & -8 & 11 & -3 \\ -3 & 10 & -5 & -3 & 10 \\ -8 & -5 & 23 & -8 & -5 \end{vmatrix}}$$

$= \dfrac{[(15)(10)(23) + (-3)(-5)(0) + (-8)(0)(-5)]\; -[(0)(10)(-8) + (-5)(-5)(15) + (23)(0)(-3)]}{[(11)(10)(23) + (-3)(-5)(-8) + (-8)(-3)(-5)]\; -[(-8)(10)(-8) + (-5)(-5)(11) + (23)(-3)(-3)]}$

$= \dfrac{3{,}450 - 375}{2{,}290 - 1{,}122} = \dfrac{3{,}075}{1{,}168} = 2.633$

Third Order Determinant Example Continued:

Equation #1: $11V_1 - 3V_2 - 8V_3 = 15$

Equation #2: $-3V_1 + 10V_2 - 5V_3 = 0$

Equation #3: $-8V_1 - 5V_2 + 23V_3 = 0$

$$V_2 = \frac{\begin{vmatrix} 11 & 15 & -8 \\ -3 & 0 & -5 \\ -8 & 0 & 23 \end{vmatrix}}{\begin{vmatrix} 11 & -3 & -8 \\ -3 & 10 & -5 \\ -8 & -5 & 23 \end{vmatrix}}$$

$$= \frac{[(11)(0)(23) + (15)(-5)(-8) + (-8)(-3)(0)] - [(-8)(0)(-8) + (0)(-5)(11) + (23)(-3)(15)]}{[(11)(10)(23) + (-3)(-5)(-8) + (-8)(-3)(-5)] - [(-8)(10)(-8) + (-5)(-5)(11) + (23)(-3)(-3)]}$$

$$= \frac{600 - (-1{,}035)}{2{,}290 - 1{,}122} = \frac{1{,}635}{1{,}168} = 1.4$$

$$V_3 = \frac{\begin{vmatrix} 11 & -3 & 15 \\ -3 & 10 & 0 \\ -8 & -5 & 0 \end{vmatrix}}{\begin{vmatrix} 11 & -3 & -8 \\ -3 & 10 & -5 \\ -8 & -5 & 23 \end{vmatrix}}$$

$$= \frac{[(11)(10)(0) + (-3)(0)(-8) + (15)(-3)(-5)] - [(-8)(10)(15) + (-5)(0)(11) + (0)(-3)(-3)]}{[(11)(10)(23) + (-3)(-5)(-8) + (-8)(-3)(-5)] - [(-8)(10)(-8) + (-5)(-5)(11) + (23)(-3)(-3)]}$$

$$= \frac{225 - (-1{,}200)}{2{,}290 - 1{,}122} = \frac{1{,}425}{1{,}168} = 1.22$$

CHECK ANSWERS: (PLACE THE VALUE OF EACH VARIABLE IN THE EQUATIONS)

EQUATION #1 V_1 V_2 V_3

$11(2.633) - 3(1.4) - 8(1.22) = 28.963 - 4.2 - 9.76 = 15$

EQUATION #2

$-3(2.633) + 10(1.4) - 5(1.22) = -7.899 + 14 - 6.1 = 0$

EQUATION #3

$-8(2.633) - 5(1.4) + 23(1.22) = -21.064 - 7 + 28.06 = 0$

VARIABLE VALUES CHECK OK!

Appendix F

Standard Resistor Values

Standard 5% and 10% Resistor Values

Bands 1 / 2	Ohms			kΩ			MΩ	
	Standard Colors of the 3rd Band							
	gold	blk	brn	red	org	yel	grn	blu
brn/blk		**10**	**100**	**1.0**	**10**	**100**	**1.0**	**10**
brn/brn		11	110	1.1	11	110	1.1	11
brn/red		**12**	**120**	**1.2**	**12**	**120**	**1.2**	**12**
brn/org		13	130	1.3	13	130	1.3	13
brn/grn		**15**	**150**	**1.5**	**15**	**150**	**1.5**	**15**
brn/blu		16	160	1.6	16	160	1.6	16
brn/gry		**18**	**180**	**1.8**	**18**	**180**	**1.8**	**18**
red/blk		20	200	2.0	20	200	2.0	20
red/red		**22**	**220**	**2.2**	**22**	**220**	**2.2**	**22**
red/yel		24	240	2.4	24	240	2.4	
red/vio	**2.7**	**27**	**270**	**2.7**	**27**	**270**	**2.7**	
org/blk	3.0	30	300	3.0	30	300	3.0	
org/org	**3.3**	**33**	**330**	**3.3**	**33**	**330**	**3.3**	
org/blu	3.6	36	360	3.6	36	360	3.6	
org/wht	**3.9**	**39**	**390**	**3.9**	**39**	**390**	**3.9**	
yel/org	4.3	43	430	4.3	43	430	4.3	
yel/vio	**4.7**	**47**	**470**	**4.7**	**47**	**470**	**4.7**	
grn/brn	5.1	51	510	5.1	51	510	5.1	
grn/blu	**5.6**	**56**	**560**	**5.6**	**56**	**560**	**5.6**	
blu/red	6.2	62	620	6.2	62	620	6.2	
blu/gry	**6.8**	**68**	**680**	**6.8**	**68**	**680**	**6.8**	
vio/grn	7.5	75	750	7.5	75	750	7.5	
gry/red	**8.2**	**82**	**820**	**8.2**	**82**	**820**	**8.2**	
whit/brn	9.1	91	910	9.1	91	910	9.1	

Note: Boldfaced values are available only in 10% tolerance. All values are available in 5% tolerance. Carbon-film and carbon-composition resistors.

Standard 1% and 2% Resistor Values

1%	2%	1%	2%	1%	2%	1%	2%
100	100	182		332	332	604	
102		187	187	340		619	619
105	105	191		348	348	634	
107		196	196	357		649	649
110	110	200		365	365	665	
113		205	205	374		681	681
115	115	210		383	383	698	
118		215	215	392		715	715
121	121	221		407	407	732	
124		226	226	412		750	750
127	127	232		422	422	765	
130		237	237	432		787	787
133	133	243		442	442	806	
137		249	249	453		825	825
140	140	255		464	464	845	
143		261	261	475		866	866
147	147	267		487	487	887	
150		274	274	499		909	909
154	154	280		511	511	931	
158		287	287	523		953	953
162	162	294		536	536	976	
165		301	301	549			
169	169	309		562	562		
174		316	316	576			
178	178	324		590	590		

Metal and carbon film resistors.

Answers to Odd-Numbered Problems

CHAPTER 1

Section 2

1. $1.55 \cdot 10^5 = 155 \cdot 10^3 = 0.155 \cdot 10^6$
3. $4.5 \cdot 10^{-6}$
5. $9.465 \cdot 10^2 = 0.9465 \cdot 10^3$

Section 3

7. $2.1801674 \cdot 10^{13} = 2.1801674 \text{ E}13$
9. $3.3165 \cdot 10^9 = 3.3165 \text{ E}9$
11. $7.54 \cdot 10^6 = 7.54 \text{ E}6$
13. 2.998
15. $1.79333 \cdot 10^7 = 1.79333 \text{ E}7$
17. $2 \cdot 10^{-11} = 2 \text{ E}-11$
19. $1.8315 \cdot 10^7 = 1.8315 \text{ E}7$
21. 6.684
23. $2 \cdot 10^4 = 2 \text{ E}4$
25. 2.019
27. $5.76 \cdot 10^{-10} = 5.76 \text{ E}-10$
29. 4.9

Section 4

31. $2\ \mu\text{F}$
33. 4 pF
35. $150 \text{ E}6\ \Omega = 150\ \text{M}\Omega$
37. $12.5 \text{ E}9\ \text{Hz} = 12.5\ \text{GHz}$
39. $47 \text{ E}3\ \Omega = 47\ \text{k}\Omega$
41. 10,000 pF
43. $23\ \mu\text{A}$

CHAPTER 2

Section 2

1. 0.0979 N of repulsion

Section 3

3. 3.33 A
5. 3.413 A

Section 5

7. 6.67 mS
9. 2 S

CHAPTER 3

Section 2

1. 10 h
3. 9 V with a capacity of 4 AH
5. 12 cells

Section 3

7. $5{,}100\ \Omega$
9. $19.2\ \Omega$

Section 6

11. $12\ \text{k}\Omega$ 11.4 to 12.6 kΩ
13. $91\ \text{k}\Omega$ 81.9 to 100.1 kΩ
15. $3\ \text{M}\Omega$ 2.7 to 3.3 MΩ
17. $1{,}400\ \Omega$ 1,386 to 1,414 Ω
19. $511\ \Omega$ 505.9 to 516.1 Ω

CHAPTER 4

Section 1

	Resistance Ω	Applied Voltage	Current A	Conductance S
1.	**500 Ω**	50 mV	100 μA	**2 mS**
3.	**240 Ω**	24 V	100 mA	**4.17 mS**
5.	130 kΩ	**6.5 V**	50 μA	**7.69 μS**
7.	**1.92 kΩ**	23 kV	12 A	**521 μS**
9.	**47 Ω**	100 mV	**2.13 mA**	21.3 mS

Section 2

11. 2.5 W

	Power W	Resistance Ω	Current A	Voltage V
13.	25 W	2.5 kΩ	100 mA	250 V
15.	26.2 mW	22 kΩ	1.09 mA	24 V
17.	90 μW	1.5 MΩ	7.75 μA	11.62 V
19.	2.45 mW	33 kΩ	273 μA	9 V
21.	190 mW	328.9 kΩ	760 μA	250 V

23. 288 W
25. 9.82 mW
27. Yes
29. $35° \cdot 60$ mW/°C $= 2.1$ W, 10 W $- 2.1$ W $=$ 7.9 W
31. 46.1 W
33. #kWh $= 12.5$ kWh
35. cost $= \$86.40$

CHAPTER 5

Section 1

1. 130.7 kΩ
3. Increase

Section 2

5. 1.273 mA

Section 3

7. 33 V

Section 4

9. $V_R = 4$ V
11. -1.26 V $- 4.17$ V $- 5.94$ V $- 2.78$ V $-$ 10.36 V $- 9.48$ V $+ 34$ V $= 0$ V

Section 5

13.

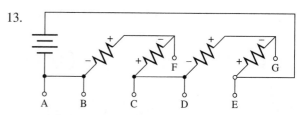

Section 5

15. relatively negative: A, B
    relatively positive: C, D, E, G

Section 7

17. $V_A = -11.12$ V, $V_B = -20.34$ V, $V_C = -24$ V, $I_t = 135.6$ mA, $P_{R1} = 496$ mW, $P_{R2} = 1.25$ W, $P_{R3} = 1.51$ W
19. $R_1 = R_2 = R_3 = 20$ k$\Omega \cdot 4$ V/12 V $= 6.7$ kΩ
    $P_{R1} = P_{R2} = P_{R3} = 4^2/6{,}700 = 2.39$ mW, use $\frac{1}{8}$ W resistor

CHAPTER 6

Section 1

1. $N = R/R_t$, $N = 100$ $\Omega/20$ $\Omega = 5$
3. $R_t = 2{,}628$ Ω

Section 2

5. $G_t = 2.363$ μS, $R_t = 423$ kΩ, $I_t = 236$ μA, $I_{B1} = 100$ μA, $I_{B2} = 66.7$ μA, $I_{B3} = 19.6$ μA, $I_{B4} = 50$ μA

Section 3

7. $I_{R1} = 23.7$ mA, $I_{R2} = 92.6$ mA, $I_{R3} = 33.9$ mA
9. $R_{\text{shunt}} = 33.3$ Ω, $P_{\text{shunt}} = 750$ mW

CHAPTER 7

Section 1

1. $R_t = 7.89$ kΩ
3. $R_t = 4{,}040$ Ω

Section 2

5. $R_t = 14{,}139\ \Omega$, $I_t = 1.7$ mA
   $P_t = 40.8$ mW, $I_{R1} = 1.7$ mA, $V_{R1} = 3.7$ V,
   $V_{R2} = V_{R3} = 3.3$ V, $V_{R4} = 17$ V, $I_{R2} = 1$ mA,
   $I_{R3} = 0.7$ mA, $P_{R1} = 6.3$ mW, $P_{R2} = 3.3$ mW,
   $P_{R3} = 2.3$ mW, $P_{R4} = 28.9$ mW
7. $R_t = 268.3$ kΩ, $I_t = 205\ \mu$A, $E_s = 55$ V,
   $P_t = 11.3$ mW
   $V_{R1} = V_{R2} = 27.68$ V, $V_{R4} = V_{R5} = 6.82$ V,
   $V_{R3} = 20.5$ V
   $I_{R1} = I_{R2} = 102.5\ \mu$A, $I_{R4} = 83.2\ \mu$A,
   $I_{R5} = 121.8\ \mu$A
   $P_{R1} = P_{R2} = 2.84$ mW, $P_{R3} = 4.20$ mW,
   $P_{R4} = 0.57$ mW, $P_{R5} = 0.83$ mW

Section 3

9. $R_t = 1.13$ kΩ

Section 4

11. $I_{R1} = 219.5$ mA, $V_{R1} = 4.39$ V, $V_{R2} = V_{R3} = 5.61$ V
    $I_{R2} = 119.4$ mA, $I_{R3} = 100.2$ mA, $I_{R4} = 109$ mA
    $V_{R4} = 1.63$ V, $V_{R5} = V_{R6} = 8.36$ V
    $I_{R5} = 83.6$ mA, $I_{R6} = 25.3$ mA, $I_t = I_{R1} + I_{R4} = 328.5$ mA, $R_t = 30.4\ \Omega$

Section 5

13. $R_x = 535.7\ \Omega$
15. $V_{\text{meter}} = 3.75$ V

Section 6

17.   OPEN
    V_{R1} 24 V $\cdot$ 100 Ω/320 Ω = 7.5 V
    V_{R2} 24 V $\cdot$ 220 Ω/320 Ω = 16.5 V
    P_{R1} (7.5 V)^2/100 Ω = 0.563 W
    P_{R2} (16.5 V)^2/220 Ω = 1.238 W

                                CLOSED
    V_{R1} 24 V $\cdot$ 100 Ω/250 Ω = 9.6 V
    V_{R2} 24 V $\cdot$ 150 Ω/250 Ω = 14.4 V
    P_{R1} (9.6 V)^2/100 Ω = 0.922 W
    P_{R2} (14.4 V)^2/220 Ω = 0.943 W

19. $V_{R1} = 26$ V, $V_{R2} = 16.4$ V, $V_{R3} = 17.6$ V,
    $V_{R4} = 17.6$ V, $V_{R5} = 34$ V
    $P_{R1} = 9.94$ mW, $P_{R2} = 5.72$ mW,
    $P_{R3} = 5.53$ mW

Section 7

21. R_3 is OPEN.
23. R_1 is SHORTED.

CHAPTER 8

Section 1

1.

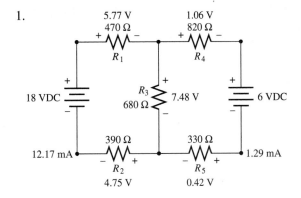

Section 2

3. $V_{\text{Th}} = 10$ V $\cdot$ 220 Ω/1,220 Ω = 1.8 V,
   $R_{\text{Th}} = 1{,}000\ \Omega\ \|\ 220\ \Omega = 180\ \Omega$
5. $V_{\text{Th}} = (-16$ V $\cdot$ 750 Ω/1,570 Ω) +
   (24 V $\cdot$ 820 Ω/1,570 Ω) = +4.9 V
   $R_{\text{Th}} = 820\ \Omega\ \|\ 750\ \Omega = 392\ \Omega$
7. $V_{470\Omega} = 2.67$ V

Section 3

9. $I_N = 12.5$ mA, $R_{\text{Th}} = R_N = 392\ \Omega$
11. $I_N = 1.43$ mA, $R_N = 12.1$ kΩ
13. $V_{\text{Th}} = 17.3$ V, $R_{\text{Th}} = 12.1$ kΩ

Section 4

15. $V_{\text{eq}} = 11.6$ V, $R_{\text{eq}} = 13.2$ kΩ, $V_L = 5$ V

Section 5

17. $I_N = 4.32$ mA, $R_N = 223\ \Omega$
19. $V_L = 0.298$ V, $I_L = 2.98$ mA

Section 6

21. $R_L = 100$ kΩ, $P_{\text{max}} = 563\ \mu$W

Section 7

23. $V_{R1} = 7.39$ V, $I_{R1} = 15.72$ mA, $V_{R2} = 6.13$ V,
$I_{R2} = 15.72$ mA, $V_{R3} = 4.48$ V, $I_{R3} = 6.59$ mA,
$V_{R4} = 7.47$ V, $I_{R4} = 9.11$ mA, $V_{R5} = 3.01$ V,
$I_{R5} = 9.11$ mA

Section 8

25. loop#1: $40{,}000\,I_1 - 12{,}000\,I_2 - 0\,I_3 = +3$ V
loop#2: $-12{,}000\,I_1 + 55{,}800\,I_2 - 10{,}000\,I_3 = 0$ V
loop#3: $0\,I_1 - 10{,}000\,I_2 + 55{,}000\,I_3 = +6$ V

Section 9

27. $R_A = 18.73$ kΩ, $R_B = 11.95$ kΩ, $R_C = 11.02$ kΩ
29. $V_{AB} = 2.74$ V

CHAPTER 9

Section 1

1. 255 µWb
3. 500,000,000 Φ
5. 0.3 T
7. $B = 1.25$ mT

Section 4

9. 131.25 mA
11. $H = 6{,}000$ At/m
13. $H = 24{,}000$ At/m
15. $L = 0.1257$ m, $H = 1{,}791$ At/m
17. 2,779 At/m
19. $\mu378$ µT/At/m, $H = 7{,}500$ At/m, $B = 2.835$ T
21. 0.4679 T

Section 9

23. $V_{\text{ind}} = 250$ V

CHAPTER 11

Section 2

1. $V_M = 0.1$ V, $I_{\text{shunt}} = 499$ mA, $R_{\text{shunt}} = 0.2$ Ω
3. $R_{\text{SH1}} = 600$ Ω, $R_{\text{SH2}} = 30$ Ω, $R_{\text{SH3}} = 3$ Ω,
$R_1 = 570$ Ω, $R_2 = 27$ Ω, $R_3 = 3$ Ω

Section 3

5. The meter resistance is significant since the total circuit resistance is 5 kΩ.
7. $R_{x1} = 39.5$ kΩ
$R_{x2} = 960$ kΩ
$R_{x3} = 3$ MΩ
$R_{x4} = 6$ MΩ
9. Sensitivity $= 1/25$ µA $= 40$ kΩ/V
11. $R_{\text{total}} = 5$ MΩ, $I_{\text{meter}} = 20$ µA

Section 4

13. $R_t = 60$ kΩ, $R_L = 54$ kΩ, $R_Z = 10.8$ kΩ
15. $R_t = 3$ V/25 µA $= 120$ kΩ
$R_L = (120$ kΩ $- 750$ Ω$)/1.1 = 108{,}409$ Ω
$R_Z = 0.2 \cdot 108{,}409$ Ω $= 21{,}681$ Ω

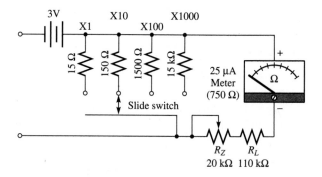

CHAPTER 12

Section 1

1. $L = 10$ µH
3. $V_{\text{ind}} = 1.25$ mV
5. $L = 4$ H
7. $E = 0.0156$ J

Section 2

9. $L = 1.58$ µH
11. 50 µH, since the number of turns and coil length are cut in half.

Section 4

13. $L_t = 155.5$ mH
15. $L_t = 469$ µH
17. $L_t = 5$ H
19. $L_t = 12.5$ mH

CHAPTER 13

Section 1

1. $\mathscr{E} = 5{,}000$ V/m
3. $C = 10$ μF
5. $Q = 40{,}000$ μC $= 0.04$ C
7. $V = 2{,}400$ V
9. $E = 9$ mJ

Section 2

11. $\epsilon = 61.95$ μF/m
13. A $= 0.002$ m^2, Air $\epsilon = 8.85$ pF/m, C $= 177$ pF

Section 4

15. 100 MFD $= 100$ μF
17. 4,700 pF $= 0.0047$ μF, 100 V
19. 3,000 pF $= 0.003$ μF, $\pm 5\%$
21. P750 $=$ positive temperature coefficient of 750 parts per million for every degree centigrade change from 25° C.

Section 5

23. $C_t = 0.207$ μF
25. $C_t = 60$ μF

Section 6

27. $Q_t = Q_1 = Q_2 = Q_3 = 3$ mC
29. $Q_1 = 7.05$ mC, $Q_2 = 15$ mC, $Q_3 = 3$ mC, $Q_t = 25$ mC
31. $Q_t = C_t \cdot V$; $C_t = 28.57$ pF, $Q_t = 28.57$ pF $\cdot$ 100 V $= 2.857$ nC
33. $Q_1 = 0.1$ μC, $Q_2 = 0.2$ μC, $Q_3 = 0.47$ μC, $Q_t = 0.77$ μC
35. $V_{C1} = 3.93$ V, $V_{C2} = 1.85$ V, $V_{C3} = 9.23$ V

CHAPTER 14

Section 1

1. Fully charged after 5 TC, 1 TC $= RC$, 5 TC $= 1.35$ s
3. 1 TC $= 63\%$ of full charge. 1 TC $= 1$ s
5. $\dfrac{25 \text{ ms}}{5 \cdot 0.1 \ \mu\text{F}} = 50$ kΩ

Section 2

7. Fully decayed after 5 TC, 1 TC $= L/R$, 5 TC $= 125$ μs
9. $\dfrac{5 \cdot 2 \text{ mH}}{1 \ \mu\text{s}} = 10$ kΩ

Section 3

11. 1 TC $= RC = 1$ ms, # TC $= 1$ ms/1 ms $= 1$ TC maximum charge current $= 10$ V/10 k$\Omega = 1$ mA instantaneous charge current at 1 TC $=$ 1 mA $\cdot$ 0.37 $= 370$ μA
13. 1 TC $= L/R = 1$ ns, # TC $= 1.5$ ns/1 ns $= 1.5$ TC initial decay voltage $= 500$ μA $\cdot$ 100 k$\Omega = 50$ V instantaneous decay voltage at 1.5 TC $=$ 50 V $\cdot$ 0.223 $= 11.2$ V
15. 5 V/24 V $= 0.208 = 20.8\% = 1.57$ TC, 1 TC $=$ 7.5 s; therefore, 1.57 TC $= 7.5$ s $\cdot$ 1.57 $= 11.78$ s
17. maximum initial current $= 13.64$ mA, 1 TC $=$ 45.45 μs
$i = I_F + (I_I - I_F) \cdot e^{-(t/1\text{TC})} = 0$ A $+$ (13.64 mA $- 0$ A) $\cdot e^{-(100\mu s/45.45\mu s)}$
$= 13.64$ mA $\cdot e^{-(100\mu s/45.45\mu s)} = 1.51$ mA
19. 1 TC $= RC = 47$ k$\Omega \cdot 250$ μF $=$
11.75 s $- 1$ TC $\cdot \ln\left(\dfrac{V - V_F}{V_I - V_F}\right) = t$, -11.75 s $\cdot$
$\ln\left(\dfrac{6.5 \text{ V} - 12 \text{ V}}{0 - 12 \text{ V}}\right) = t = 9.17$ s

Section 5

21. maximum build-up current $= 10$ V/1 k$\Omega =$ 10 mA
initial decay voltage $= 10$ mA $\cdot$ 100 k$\Omega =$ 1,000 V

CHAPTER 15

Section 1

1. $45°/360° = 1/X$, $X = 360°/45° = 8$
3. $T - 1/f = 1/10$ kcps $= 0.1$ ms
5. $e_i = E_p \cdot \sin \theta = 20$ V $\cdot \sin 60° = 17.32$ V
7. $v = 360° \cdot f = 360° \cdot 10$ MHz $= 3.6 \cdot 10^9$ °/s
9. $v = 360° \cdot f = 360° \cdot 3$ kHz $= 1.08 \cdot 10^{6°}$/s
$e_i = 8$ V $\cdot \sin (1{,}650°) = 8$ V $\cdot -0.5 = -4$ V
11. $v = 18{,}000$ °/s, $i_i = -30$ A

13. $\omega = 2\pi f = 6.283 \cdot 600$ kHz $= 3,769,800$ rad/s
15. $e_i = E_p \cdot \sin(\omega \cdot t + 0) = 50$ V $\cdot \sin[6,000$
    rad/s $\cdot 50 \mu s + (0°/57.3°)]$
    $= 50$ V$_p \cdot \sin(0.3$ rad$) = 50$ V$_p \cdot 0.296 =$
    14.8 V

Section 2

17. 16 A$_{p\text{-}p}$
19. $I_{avg} = 318.5$ mA$_{avg}$
21. 4.71 A$_p$
23. $A_{RMS} = 35.4$ mA
25. 169.7 V$_p$
27. 4 A $= 4$ A$_{RMS}$, 20 V$_{p\text{-}p} = 7.07$ V$_{RMS}$,
    $P = 28.28$ W

Section 3

29. 15 V$_{p\text{-}p} = 7.5$ V$_p = 5.3$ V, $R_t = 1210$ Ω,
    $I_t = 4.38$ mA
    $V_{220\Omega} = 0.964$ V, $P_{220\Omega} = 4.22$ mW,
    $V_{430\Omega} = 1.88$ V, $P_{430\Omega} = 8.23$ mW
    $V_{560\Omega} = 2.45$ V, $P_{560\Omega} = 10.73$ mW
31. 36 V$_p = 25.45$ V, $I_{1M\Omega} = 25.45 \mu$A,
    $P_{1M\Omega} = 648 \mu$W
    $I_{470k\Omega} = 54.1 \mu$A, $P_{470k\Omega} = 1.38$ mW,
    $I_{820k\Omega} = 31 \mu$A, $P_{820k\Omega} = 790 \mu$W
    $I_t = 110.55 \mu$A

Section 4

33.
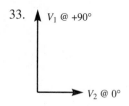

35. $360° = 200° = 160°$
37.

I_R V_R

In phase phasors

Section 5

39. 50 dB $= 5$ Bels
41. dB $= 20 \log(V_{out}/V_{in}) = 20 \log(0.5$ V$/20$ V$) =$
    -32 dB
43. also 14 dB power gain
45. $P_{out} = 5$ W
47. dB $= 10 \log(100$ W$/100$ kW$) =$
    -30 dB

Section 7

49. $11 \cdot 5$ kHz $= 55$ kHz
51. $T = 1/f = 1/400$ pps $= 2.5$ ms
    % duty cycle $= (PW/T) \cdot 100\% =$
    $(0.3$ ms$/2.5$ ms$) \cdot 100\% = 12\%$

CHAPTER 16

Section 3

1. $X_L = 400$ Ω
3. $X_L = 0.06283$ Ω, $X_L = 62.83$ Ω, $X_L = 62.83$ kΩ
5. $L = 7.32$ mH
7. $f = 239$ kHz

Section 4

9. $X_{Lt} = 820$ Ω, $I_t = 12.2$ mA, $V_{L1} = 4.88$ V,
   $V_{L2} = 3.9$ V, $V_{L3} = 1.22$ V
11. $L_t = 900 \mu$H, $V_{L1} = 4.44$ V, $V_{L2} = 8.89$ V,
    $V_{L3} = 6.67$ V
13. $X_{Lt} = 550$ Ω, $I_t = 54.5$ mA, $I_{L1} = I_{L2} =$
    13.64 mA, $I_{L3} = 27.27$ mA

Section 5

15. $X_C = 500$ Ω
17. $X_C = 1,592$ Ω, 15.92 Ω, 0.1592 Ω
19. $C = 0.0032 \mu$F $= 3,200$ pF
21. $f = 796$ MHz

Section 6

23. $X_{Ct} = 110$ kΩ, $I_t = 1.36$ mA, $V_{C1} = 95.5$ V,
    $V_{C2} = 41$ V, $V_{C3} = 13.6$ V
25. $I_{C1} = 12.5$ mA, $I_{C2} = 6.25$ mA, $I_{C3} = 25$ mA,
    $I_t = 43.75$ mA, $X_{Ct} = 571$ Ω

CHAPTER 17

Section 1

1. $k = \Phi_M/\Phi_t = 70 \mu$Wb$/120 \mu$Wb $= 0.583$
3. $L_M = 8.5$ mH
5. $L_M = 0.105$ mH, $L_t = L_1 + L_2 + 2L_M =$
   1.1 mH
7. $L_M = 52.5 \mu$H, $k = 0.238$

Section 2

9. $E_s = 23$ V
11. $E_P/E_s = N_P/N_S = I_S/I_P = 120$ V$/24$ V $= 5:1$,
    therefore: $I_P/I_S = 1:5$
13. $I_P = 700$ mA
15. $I_S = 232$ mA

Section 3

17. $Z_P/Z_S = 9:4 = 2.25:1$
19. $1:69.4$
21. $Z_P = Z_S \cdot (N_P/N_S)^2 = 600 \ \Omega \cdot (3.5/1)^2 =$
    $7,350 \ \Omega$
23. $Z_S = 3,038 \ \Omega$

Section 4

25. $P_{S1} = 45$ VA, $P_{S2} = 25.2$ VA, $P_{S(\text{total})} =$
    70.2 VA $= P_P$
27. The transformer is rated in VA, since very little
    electrical energy is converted to heat or
    mechanical motion in the transformer itself. The
    energy, or power, is transferred from the primary
    circuit to the secondary circuit.

CHAPTER 18

Section 1

1. a. 5 ft. b. 11.2 in. c. 10 cm
3. $j^2 = -1 = 180°$, $j^3 = -j = 270° = -90°$,
   $j^4 = 1 = 0°$, $j^5 = +j = 90°$
5. a. $671 \ \Omega \ \angle 26.6°$ b. $292 \ \Omega \ \angle 59°$
   c. $142 \ \Omega \ \angle 32°$ d. $112 \ \Omega \ \angle -63.4°$

Section 2

7. a. $-260 - j165$ b. $-9 - j6$
9. a. $4,502.4 \ \angle 36.8°$
   b. $40.96 \ \angle 0°$
   c. $1,700 \ \angle -152°$
11. a. $2 \ \angle 53.2°$
    b. $0.0316 \ \angle 18.4°$
13. a. $320 \ \angle 0°$ b. $6,000 \ \angle -150°$

Section 3

15. $T = 50 \ \mu s$, $12.5 \ \mu s$, $6.25 \ \mu s$, $8.33 \ \mu s$
17. a. $V_R = 141.4$ V, $V_L = 141.4$ V
    b. $V_R = 35$ V, $V_L = 60.6$ V
19. a. $\angle 63.4°$ b. $\angle 33.7°$

21. $X_L = 6,283 \ \Omega$, $Z_t = 6,362 \ \Omega$, $I_t = 1.57 \ \mu A$,
    $V_R = 1.57$ mV, $V_L = 9.87$ mV, $\angle \theta = 81°$

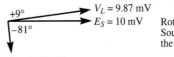

Rotated diagram.
Source voltage is
the reference at 0°.

23. $f = 63,662$ Hz, $Z_t = 11.05$ kΩ, $I_t = 1.1$ mA
    $V_R = 5.1$ V, $V_L = 11$ V, $\angle \theta = 64.8°$

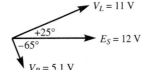

Rotated diagram.
Source voltage is
the reference at 0°.

Section 4

25. $X_C = 159$ kΩ @ 1 kHz, $3,183 \ \Omega$ @ 50 kHz,
    $20 \ \Omega$ @ 8 MHz, $4 \ \Omega$ @ 40 MHz
27. a. $V_R = 15.3$ V, $V_C = 12.9$ V
    b. $V_R = 20$ V, $V_C = 34.6$ V
    c. $V_R = 23$ mV, $V_C = 19.3$ mV
29. a. $\angle -71.6°$ b. $\angle -32°$
    c. $\angle -73.3°$ d. $\angle -23.1°$
31. $X_C = 177 \ \Omega$, $Z_t = 232 \ \Omega$, $I_t = 60 \ \mu A$,
    $V_R = 9$ mV, $V_C = 10.6$ mV
    $\angle \theta = \tan^{-1}(-X_C/R) = -49.7°$

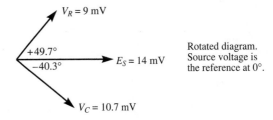

Rotated diagram.
Source voltage is
the reference at 0°.

33. $C = 354$ pF, $Z_t = 316.23 \ \Omega$, $I_t = 94.87 \ \mu A$,
    $V_R = 9.49$ mV, $V_C = 28.46$ mV
    $\angle \theta = \tan^{-1}(-X_C/R) = -71.6°$

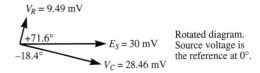

Rotated diagram.
Source voltage is
the reference at 0°.

Section 5

35. a. $E_S = 12.17$ V b. $E_S = 13.45$ V
    c. $E_S = 1.8$ V

37. problem 35a.

39. $T = 12.5\ \mu s$, $E_S = 20.025$ V $\angle 2.86°$, V_R lags by
2.86°: $T_D = 99.3$ ns
V_L leads by 87.14°: $T_D = 3.03\ \mu s$, V_C lags by
92.86°: $T_D = 3.224\ \mu s$

41. $L = 3.82$ mH

43. $X_C = 177\ \Omega$, $X_L = 942.5\ \Omega$, $Z_t =$
833.6 Ω $\angle 66.7°$, $I_t = 24$ mA
$V_R = 7.92$ V, $V_C = 4.25$ V, $V_L = 22.62$ V
$\angle \theta = \tan^{-1}[(X_L - X_C)/R] = 66.7°$

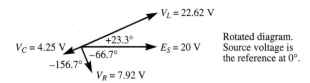

45. $E_S = 31.6$ mV $\angle 18.4°$, $R = 30\ \Omega$, $X_C = 40\ \Omega$,
$X_L = 50\ \Omega$, $Z_t = 31.6\ \Omega$
$\angle \theta \times \tan^{-1}[(X_L - X_C)/R] = 18.4°$, $C =$
796 pF, $L = 1.59\ \mu H$

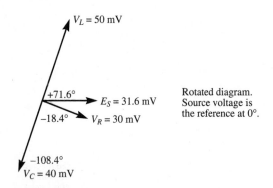

47. $R = 100\ \Omega$, $V_C = 20$ V, $f = 159.2$ Hz,
$X_L = 500\ \Omega$, $V_L = 50$ V
$E_s = 31.6$ V, $Z_T = 100\ \Omega +j300\ \Omega = 316.2\ \Omega$
$\angle \theta = \tan^{-1}[(V_L - V_C)/V_R] = 71.6°$

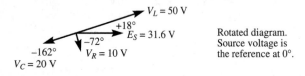

Section 6

49. $P_R = 0.0064$ pW, $P_L = 0.0018$ pVAR,
$P_C = 0.00225$ pVAR
$P_A = 0.0064$ pVA, $PF = P_R/P_A \cong 1$

51. $P_R = 3.5$ W, $P_L = 2.5$ VAR, $P_C = 7.5$ VAR,
$P_A = 6.1$ VA, $PF = 0.5738$

CHAPTER 19

Section 1

1. $X_L = 188.5\ \Omega$, $I_L = 212.2\ \mu A$, $I_R = 40\ \mu A$,
$I_t = 216\ \mu A$
$Z_t = 185.2\ \Omega$, $\angle \theta = \tan^{-1}(-I_L/I_R) = -79.3°$

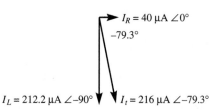

3. $f = 60$ Hz, $I_L = 153$ mA, $I_R = 63.9$ mA,
$I_t = 63.9$ mA $-j153$ mA $= 165.8$ mA
$Z_t = 693.6\ \Omega$, $\angle \theta = \tan^{-1}(-I_L/I_R) = -67.3°$

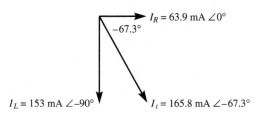

5. $I_L = \sin \angle 28° \cdot 1.13$ mA $= 0.531$ mA
$I_R = \cos \angle 28° \cdot 1.13$ mA $= 0.998$ mA
$E_s = 99.8$ mV, $X_L = 188\ \Omega$, $L = 30\ \mu H$,
$Z_t = 88.3\ \Omega$

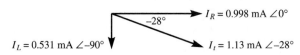

Section 2

7. $X_C = 199\ \Omega$, $I_C = 176\ \mu A$, $I_R = 106\ \mu A$,
$I_t = 106\ \mu A +j176\ \mu A = 205.5\ \mu A$
$Z_t = 170\ \Omega$, $\angle \theta = \tan^{-1}(I_C/I_R) = 58.9°$

$I_C = 176\ \mu A \angle +90°$ $I_t = 205.5\ \mu A \angle +58.9°$
+58.9°
$I_R = 106\ \mu A \angle 0°$

9. $X_C = 398\ \Omega$, $R = 110\ \Omega$, $f = 400$ Hz,
$Z_t = 106\ \Omega$, $\angle \theta = \tan^{-1}(I_C/I_R) = 15.5°$

$I_C = 0.352$ A $\angle +90°$ $I_t = 1.32$ A $\angle +15.5°$
+15.5°
$I_R = 1.27$ A $\angle 0°$

Section 3

11. $X_C = 265 \ \Omega$, $X_L = 377 \ \Omega$, $I_R = 16.7$ mA,
    $I_C = 18.9$ mA, $I_L = 13.3$ mA
    $I_t = 16.7$ mA $+ j5.6$ mA $= 17.6$ mA, $\angle \theta =$
    $\tan^{-1}[(I_C - I_L)/I_R] = 18.5°$, $Z_t = 284 \ \Omega$

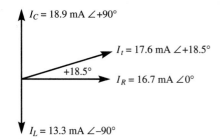

13. $X_C = 797 \ \Omega$, $X_L = 126 \ \Omega$, $R = 50 \ \Omega$, $C =$
    20 pF, $L = 2 \ \mu$H
    $I_t = 4 \ \mu$A $-j1.339 \ \mu$A $= 4.22 \ \mu$A, $\angle \theta =$
    $\tan^{-1}[(I_C - I_L)/I_R] = -18.5°$, $Z_t = 47.4 \ \Omega$

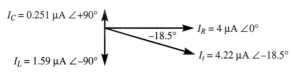

15. $R = 120 \ \Omega$, $f = 50$ Hz, $X_L = 942.5 \ \Omega$,
    $I_R = 667$ mA, $I_C = 503$ mA, $I_L = 85$ mA
    $I_t = 667$ mA $+j418$ mA $= 787$ mA,
    $\angle \theta = \tan^{-1}[(I_C - I_L)/I_R] = +32.1°$
    $Z_t = 102 \ \Omega$

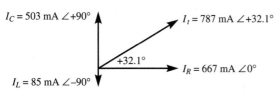

Section 4

17. $P_R = 53.4$ W, $P_C = 40.24$ VAR, $P_L = 6.8$ VAR,
    $P_A = 63$ VA
    $PF = P_R/P_A = 53.4$ W$/63$ VA $= 0.85$

CHAPTER 20

Section 3

1. $f_r = 1.84$ MHz
3. $f_r = 50.28$ kHz

5. $C = 4.76$ pF
7. $C = 3.38 \ \mu$F
9. $L = 96$ mH
11. $f_r = 386$ kHz, $X_L = 412 \ \Omega$, $Q_{ckt} = 2.05$
13. $Q = 1/PF = 1/0.0005 = 2,000$; capacitor Q is
    the reciprocal of its PF
15. $V_L = 3.5$ V
17. $f_r = 162$ MHz, $X_L = 245 \ \Omega$, $Q_{ckt} = 12.3$,
    $BW = 13.2$ MHz
19. $f_r = 96.8$ kHz, $X_L = 547 \ \Omega$, $Q_{ckt} = 10.94$,
    $BW = 8.85$ kHz
    $BP \cong [f_r - (BW/2)]$ to $[f_r + (BW/2)] \cong$
    92,375 Hz to 101,225 Hz
21. $X_L = 9.4 \ \Omega$, $Q_{ckt} = 15$, $C = 5.64 \ \mu$F, $R_t =$
    0.63 Ω

Section 6

23. $Z_{tank} = X_C{}^2/R_W = 600^2/2.5 = 144$ kΩ
25. $Z_{tank} = X_L{}^2/R_W = 750^2/5 = 112.5$ kΩ,
    $I_{tank} = 26.7$ mA
27. $I_{tank} = 1.88$ mA
29. $Z_{tank} = 31.25$ kΩ, $I_{line} = 320 \ \mu$A, $I_R = 100 \ \mu$A,
    $I_t = 420 \ \mu$A, $Z_t = 23.8$ kΩ
31. $Z_{tank} = 76.9$ kΩ, $Z_t = 123.9$ kΩ, $I_t = 0.807 \ \mu$A,
    $V_{tank} = 62.07$ mV, $V_{Rs} = 37.93$ mV

Section 7

33. $Q_{tank} = 124$, $Z_{tank} = 76.88$ kΩ
    $Q_{ckt} = (R_P \| Z_{tank})/X_L$ (where $R_P = R_P \| R_g =$
    100 k$\Omega \|$ 250 k$\Omega = 71.4$ kΩ)
    $= (71.4$ k$\Omega \| 76.88$ k$\Omega)/620 \ \Omega = 37$ k$\Omega/620 \ \Omega$
    $= 59.7$
35. $Q_{tank} = X_C/R_W = 65 \ \Omega/1 \ \Omega = 65$,
    $Z_{tank} = X_C{}^2/R_W = 65^2/1 = 4,225 \ \Omega$
    $R_P = R_g + R_S = 600 \ \Omega + 10$ k$\Omega = 10.6$ kΩ
    $Q_{ckt} = (R_P \| Z_{tank})/X_L = (10.6$ k$\Omega \| 4,225 \ \Omega)/65 \ \Omega$
    $= 3,021 \ \Omega/65 \ \Omega = 46.5$
37. $f_r = 6.625$ MHz; correction factor not needed.
39. $f_r = 318$ kHz; correction factor is needed.
    $f_r = 314.93$ kHz
41. $f_r = 107.2$ kHz
    $X_L = 67.4 \ \Omega$
    $Z_{tank} = 3,489 \ \Omega$
    $R_P = R_g + R_S = 3$ k$\Omega + 27$ k$\Omega \cong 30$ kΩ
    $Q_{ckt} = (R_P \| Z_{tank})/X_L = (30$ k$\Omega \| 3,489 \ \Omega)/67.4 \ \Omega$
    $= 3,126 \ \Omega/67.4 \ \Omega = 46.4$
    $BW = f_r/Q_{ckt} = 2,310$ Hz
    $BP \cong [f_r - (BW/2)]$ to $[f_r + (BW/2)] \cong$
    105.613 kHz to 108.787 kHz

43. Use a higher-Q inductor or increase the parallel resistance.

45. $f_{r(\text{max})} = 64.9$ MHz,
    $f_{r(\text{min})} = 50.3$ MHz

47. $C_{(\text{min})} = 501$ pF
    $C_{(\text{max})} = 804$ pF

CHAPTER 21

Section 2

1. $f_{co} = 374$ kHz

3. $f_{co} = 7$ MHz
   X_L @ 1 MHz $= 3{,}142$ Ω, X_L @ 7 MHz $= 22$ kΩ,
   X_L @ 20 MHz $= 62.8$ kΩ
   $dB_{\text{atten}} = 20 \log[X_L/\sqrt{(R^2 + X_L{}^2)}]$
   @ 1 MHz: $dB_{\text{atten}} = 20$
   $\log[3{,}142/\sqrt{22k^2 + 3{,}142^2}] = -17$ dB
   @ 7 MHz: $dB_{\text{atten}} = 20$
   $\log[22k/\sqrt{22k^2 + 22k^2}] = -3$ dB
   @ 20 MHz: $dB_{\text{atten}} = 20$
   $\log[62.8k/\sqrt{22k^2 + 62.8k^2}] = -0.5$ dB

5. $R = 5{,}305$ Ω

7. $f_{co} = 41.1$ kHz

Section 3

9. $f_{co} = 541.4$ kHz

11. $f_{co} = 3.5$ MHz
    X_L @ 300 kHz $= 2{,}827$ Ω, X_L @ 3 MHz $=$
    28.27 kΩ, X_L @ 30 MHz $= 282.7$ kΩ
    $dB_{\text{atten}} = 20 \log[R/\sqrt{(R^2 + X_L{}^2)}]$
    @ 300 kHz: $dB_{\text{atten}} = -0.03$ dB,
    @ 3 MHz: $dB_{\text{atten}} = -2.4$ dB
    @ 30 MHz: $dB_{\text{atten}} = -18.7$ dB

13. $C = 0.0583$ μF (0.06 μF)

15. $f_{co} = 50.3$ kHz

Section 4

17. $f_C = 563$ kHz

19. $f_C = 5$ MHz

21. $R_P = (R_g + R_S) \| R_L = (70$ Ω $+ 10$ kΩ$) \| 10$ kΩ
    $= 5{,}017$ Ω
    $Q_{\text{ckt}} = f_C/BW = 800/80 = 10$, $L = 99.8$ mH,
    $C = 0.397$ μF

Section 5

23. Bandstop $BW = 14$ kHz $- 10$ kHz $= 4$ kHz,
    $f_C = 12$ kHz
    $Q_{\text{ckt}} = 3$, $R_t = 1{,}077$ Ω, $L = 42.9$ mH,
    $C = 0.0041$ μF

25. $f_C = 13$ MHz, $X_L = 408.4$ Ω, $Q_{\text{ckt}} = 54$,
    $BW = 240.74$ kHz

CHAPTER 22

Section 1

1. 19.23 mS

3. $B = 1/X = 1/150$ kΩ $= 6.67$ μS $=$
   $-j6.67$ μS $= 6.67$ μS $\angle -90°$

5. $Z = 400 - j300$ Ω $= 500$ Ω $\angle -36.9°$
   $Y = 1/Z = 1/500$ Ω $\angle -36.9° =$
   2 mS $\angle +36.9°$

Section 2

7. $Z_1 = (400 \angle -90° \cdot 700 \angle +90°)/(+j700$
   $-j400)$
   $= 280{,}000 \angle 0°/300 \angle +90° =$
   933 Ω $\angle -90°$
   a. $Z_t = 220$ Ω $- j933$ Ω $= 959$ Ω $\angle -76.7°$
   b. $I_t = 20$ V $\angle 0°/959$ Ω $\angle -76.7° =$
   20.86 mA $\angle +76.7°$
   c. $V_{R1} = 20.86$ mA $\angle +76.7° \cdot 220$ Ω $\angle 0° =$
   4.59 V $\angle +76.7°$
   d. $V_{C1} = 20.86$ mA $\angle +76.7° \cdot 933$ Ω $\angle -90° =$
   19.46 V $\angle -13.3°$
   e. $I_{C1} = 19.46$ V $\angle -13.3°/400$ Ω $\angle -90° =$
   48.7 mA $\angle +76.7°$
   f. $I_{L1} = 19.46$ V $\angle -13.3°/700$ Ω $\angle +90° =$
   27.8 mA $\angle -103.3°$

9. $Y_1 = 1/33$ kΩ $\angle 0° + 1/20$ kΩ $\angle -90° +$
   $1/45$ kΩ $\angle +90° = 41.1$ μS $\angle +42.5°$
   $Z_1 = 1/Y_1 = 1/41.1$ μS $\angle +42.5° =$
   24.33 kΩ $\angle -42.5° = 17.94$ kΩ $- j16.43$ kΩ
   a. $Z_t = 32$ kΩ $\angle -21°$
   b. $I_t = 937.5$ μA $\angle +21°$
   c. $V_{R1} = 11.25$ V $\angle +21°$
   d. $V_{C1} = 22.81$ V $\angle -21.5°$
   e. $V_{L2} = 4.69$ V $\angle +111°$
   f. $I_{L1} = 506.9$ μA $\angle -111.5°$
   g. $I_{R2} = 691.2$ μA $\angle -21.5°$

11. $Z_1 = 44.6 \ \Omega \ \angle -26.6° = 40 \ \Omega \ -j20 \ \Omega$
$Z_2 = 40 \ \Omega \ +j180 \ \Omega = 184.4 \ \Omega \ \angle +77.5°$
$Z_3 = 205 \ \Omega \ \angle +76.5° = 48 \ \Omega \ +j200 \ \Omega$
a. $Z_t = 555 \ \Omega \ \angle +21°$
b. $I_t = 270 \ \text{mA} \ \angle -21°$
c. $V_{R1} = 127 \ \text{V} \ \angle -21°$

Section 3

13. $X_{L1} = +j503 \ \Omega, X_{C1} = -j398 \ \Omega$
$V_{RL1} = 6.46 \ \text{V} \ \angle -99.7°, V_{RL2} = $
24.4 V $\angle +80.3°$
$V_{RL} = 6.46 \ \text{V} \ \angle -99.7° + 24.4 \ \text{V} \ \angle +80.3°$
$\quad = -1.09 \ \text{V} - j6.37 \ \text{V} + 4.11 \ \text{V} + j24.1 \ \text{V}$
$\quad = 3.02 \ \text{V} + j17.73 \ \text{V} = 18 \ \text{V} \ \angle +80.3°$

15. $I_{RL1} = 0.078 \ \text{A} \ \angle -58.8° = 0.040 \ \text{A} - j0.067 \ \text{A}$
$I_{RL2} = 0.155 \ \text{A} \ \angle -58.8° = 0.080 \ \text{A} - j0.133 \ \text{A}$
$I_{RL} = I_{RL1} + I_{RL2}$
$\quad = 0.040 \ \text{A} - j0.067 \ \text{A} + 0.080 \ \text{A} - j0.133 \ \text{A}$
$\quad = 0.120 \ \text{A} - j0.2 \ \text{A} = 0.233 \ \text{A} \ \angle -59°$

Section 4

17. $V_{\text{Th}} = 3.5 \ \text{V} \ \angle +155.6°$
$Z_{\text{Th}} = 144 \ \Omega \ \angle +49.6° = 93.3 \ \Omega + j110 \ \Omega$
$V_L = 1.58 \ \text{V} \ \angle +126°$

19. $V_{\text{Th1}} = 7.65 \ \text{V} \ \angle -23.8°, V_{\text{Th2}} = $
7.34 V $\angle +153°$
$V_{\text{Th}} = V_{\text{Th1}} + V_{\text{Th2}} = 0.519 \ \text{V} \ \angle +27.6°$
$Z_{\text{Th}} = 926 \ \Omega \ \angle +61.2° = 446 \ \Omega + j811 \ \Omega$
$V_L = 0.263 \ \text{V} \ \angle -40.4°$

Section 5

21. $I_N = I_{C2}$ with output shorted $= 0.036 \ \text{A} \ \angle -27°$
$Z_N = 194 \ \Omega - j191 \ \Omega = 272 \ \Omega \ \angle -44.6°$
$I_{RL} = I_N \cdot \dfrac{Z_N}{Z_N + R_L} = 0.0195 \ \text{A} \ \angle -49.2°$

23. $I_N = 0.0243 \ \text{A} \ \angle +106°$
$Z_N = Z_{\text{Th}} = 144 \ \Omega \ \angle +49.6° = 93.3 \ \Omega + $
$j110 \ \Omega$
$I_{RL} = 0.0158 \ \Omega \ \angle +126°$

Section 6

25. $Z_{\text{eq}} = Z_{\text{Th}} = 176 \ \Omega \ \angle -60° = 88 \ \Omega - j152 \ \Omega$
$E_{\text{eq}} = E_{\text{Th}} = 15.6 \ \text{V} \ \angle -4.4°$

27. $Z_{\text{eq}} = Z_{\text{Th}} = 137 \ \Omega \ \angle -26.4° = 123 - j60.9 \ \Omega$
$E_{\text{eq}} = E_{\text{Th}} = 21 \ \text{V} \ \angle -10.1°$

Section 7

29. $Z_L = 93.3 \ \Omega - j110 \ \Omega$ (the conjugate of Z_{Th})

31. $P_{L(\text{max})} = \dfrac{(3.5 \ \text{V}/2)^2}{93.3 \ \Omega} = 32.8 \ \text{mW}$

Glossary

absolute permeability (μ) actual permeability measured in units of gauss per oersted (G/Oe, cgs unit) or teslas per ampere-turn per meter (T/At/m, SI unit)

active filter a filter that includes an active device such as a transistor, FET, or integrated circuit

adjustable component value of the component may be changed manually by the user—usually requires a selection of taps (connecting terminals) or a repositionable metallic collar—as in tapped inductors or adjustable wirewound resistors (not a rheostat or potentiometer)

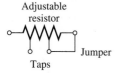

Adjustable resistor

Taps   Jumper

admittance (Y) the reciprocal of impedance—$Y = 1/Z$

air gap the air space between the south and north pole of a permanent or electromagnet—area of flux leakage between poles

Air gap

N   S

ALNICO permanent magnet alloy composed of aluminum, nickel, cobalt, and iron

alternating current (AC) electrical current that changes its direction of flow at some rate or frequency—house current

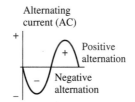

Alternating current (AC)

+   Positive alternation

−   Negative alternation

alternation a half cycle—all positive or all negative

alternator an AC generator

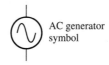

AC generator symbol

ammeter an instrument used to measure electrical current—placed in-line with the device whose current is being measured—a very-low-impedance instrument

A   Ammeter symbol

ampere (A) electrical unit for current—equals one coulomb of electrons flowing past a given point every second of time—1 A = 1 C/s

Ampère, André Marie (1775–1836)—French mathematician and physicist—discovered electromagnetism—invented first electromagnet and ammeter to measure current

ampere hours (AH)   unit of measure for charge capacity

ampere-turn   standard SI unit of measure for magnetomotive force (MMF)—current times number of turns

amplitude   amount or level or magnitude as in the *amplitude of an audio signal*

angular velocity (v or ω)   the rate at which a waveform cycle changes amplitude—related to angular change—measured in degrees per second or radians per second—$v = 360° \cdot f$ and $\omega = 2\pi f$

anode   the electrode, or terminal, of a device from which electrons are exiting—examples: the plate of a vacuum tube—the positive terminal of a component or device

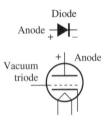

apparent power (P_A)   total power applied to an AC circuit—includes reactive and real power—measured in voltamperes (VA)

armature   the moving part of an electromechanical system or device—as in a relay armature or the armature of a motor or generator

atom   the smallest part of an element that retains the characteristics of that element

atomic dipole   ferromagnetic atom having magnetic polarity

atomic number   the number of protons in the nucleus of an atom

attenuate   to reduce in amplitude or strength

audio frequencies   frequencies that fall within the range of human hearing—frequencies below 20 kHz

autotransformer   a single-tapped winding transformer

Autotransformer

available power   the amount of power that is capable of being supplied by a power source—supply voltage times the maximum supply current

average value   as in average voltage or current—the arithmetic average of all instantaneous values of voltage or current for a single alternation of a sine wave

Ayrton shunt   universal shunt—"daisy-chained" resistances used as meter shunts for multirange ammeters

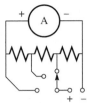

Ayrton shunt circuit

BASIC   Beginner's All-purpose Symbolic Instruction Code—a very common computer language

bandpass (*BP*)   the actual range of frequency that is passed or rejected by a tuned circuit—measured in hertz (Hz)—ranges from a lower cutoff frequency to an upper cutoff frequency

bandpass filter   a filter that passes all frequencies within a specified band—all frequencies between a lower and upper cutoff frequency are passed

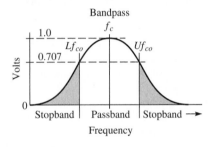

bandstop filter   bandreject filter—notch filter—trap—a filter that rejects all frequencies within a specified band—all frequencies between a lower and upper cutoff frequency are rejected

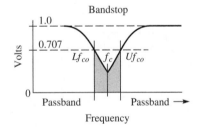

bandwidth (*BW*)   the size of the frequency range that is passed or rejected by a tuned circuit—measured in hertz (Hz)—$BW = f/Q_{ckt}$

Bel (B)   a logarithmic function or relationship—a power ratio of $10/1$—$1 \text{ B} = \log(10)$

Bell, Alexander Graham   (1847–1922)—Scottish-born American—first scientist to develop apparatus capable of transmitting and receiving voice

Bode plot   a frequency response graph—vertical scale in decibels and horizontal scale in hertz—linear vertical scale and logarithmic horizontal scale

brushes   carbon contacts that rub against the commutator or slip rings of a motor or generator—used to transfer electrical power to or from a rotating armature

build-up   refers to the increase in current and magnetic field of an inductor after an EMF is first applied

bus (bus wire)   a common current and voltage distribution wire—can be positive, negative, "hot," neutral, or common ground

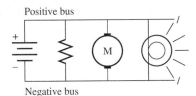

Positive bus

Negative bus

capacitance (C)   measure of a capacitor's ability to store an electric charge on its plates—unit of measure is the farad (F)

capacitor   a component that has the ability to store an electrical charge—blocks DC, passes AC—made of two metal plates separated by an insulator

Capacitor
symbols

carbon composition resistor   resistance is determined by the material composition of a carbon-based mixture used to form the resistor slug

carbon film resistor   a thin carbon film deposited on a tubular insulator

cathode   the electrode, or terminal, of a device at which electrons are arriving—examples: cathode of a vacuum tube—the negative terminal of a component or device

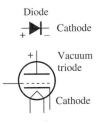

Diode

Cathode

Vacuum
triode

Cathode

capacity   a measure of the amount of service you can expect to get from a fully charged cell or battery—measured in ampere hours (AH)

Celsius temperature scale   centigrade—0° is the freezing point as compared to 32° of the Fahrenheit system—used for most scientific applications

ceramic   very high dielectric-constant insulator—used in capacitors and for general high voltage insulation—very high compression strength—very weak stress strength

cgs   centimeter-gram-second system of units—not the international standard—English standard

charged   having a positive or negative electrical characteristic—having an imbalance in the number of electrons and protons

charging rate   the amount of reverse current used to recharge a battery (secondary cells)

chemical energy   energy contained in the molecular structure of a substance

chemical energy storage   energy stored in a chemical reaction such as a charged battery

choke   an inductor used to pass DC and block AC—used as a component in a lowpass filter

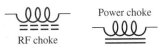

RF choke   Power choke

circuit breaker   a resettable circuit protection device

circular mil (CM)   the cross-sectional area of a circular wire that is 0.001 (1 mil) in diameter—1 CM = 1 mil^2

clamp-on probe   used to measure AC current—operates by mutual induction

closed circuit voltage   actual source voltage delivered to a load device

coefficient of coupling (k)   a ratio or decimal that expresses the relationship between the number of flux linking two coils and the total flux—no units

color code   system of color rings or dots used to identify the value of a resistor, capacitor, or inductor

commutator   that portion of a generator that mechanically creates DC ripple from AC—segmented, curved, conductive plates used to transfer DC ripple from the rotating armature to carbon brushes and terminals

complex impedance   consists of resistance and reactance—can be expressed in rectangular or polar form—see *complex number*

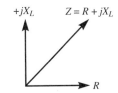

$+jX_L$ $Z = R + jX_L$

R

complex number   a number that has real and imaginary parts—a number used to express an impedance composed of resistance and reactance—complex numbers may appear in rectangular or polar form—example: $50 + j30 \; \Omega = 58.3 \; \Omega \; \angle 31°$

complex plane   a rectangular coordinate system consisting of real numbers on the horizontal axis and imaginary numbers on the vertical axis

compound motor   DC motor that has a series and a shunt wired field coil

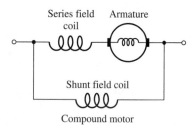

Compound motor

conductance (G)   the reciprocal of resistance—$G = 1/R$—measured in mhos or siemens (S)—1 S = 1/1 Ω

conduction   a method of delivering a charge by direct physical contact

conductors   elements that have many free electrons and valence electrons that are loosely bonded to the atom—metals such as silver, copper, gold, aluminum, and iron

constant-current source   a current source that has a very high internal resistance, so changes in load resistance have no effect on the load current

constant-voltage source   a voltage source that has a very low internal resistance, so changes in load resistance have no effect on load voltage

contactor   a heavy-duty solenoid-operated device used for high-power circuit control

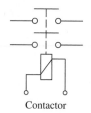

Contactor

continuity   having an unbroken, complete, path for current—not having an OPEN

coulomb (C)   a quantity of charge equal to $6.25 \cdot 10^{18}$ electrons

Coulomb, Charles   (1736–1806)—French scientist who worked with static electricity and magnetism—derived mathematical formula to calculate force between two charged bodies

counterelectromotive force (CEMF)   a voltage that is induced across the windings of a motor armature as it rotates—opposes incoming current and external EMF

countermagnetomotive force (CMMF)   creates an opposing magnetic field—field opposes the external field that induced the current in the first place

CRT   cathode ray tube—used in oscilloscopes and televisions

Currie temperature   the temperature at which a permanent magnet loses all of its magnetism and ferrous materials lose their magnetic properties

current   the organized flow of electrical charges

current divider   a circuit that divides current—a parallel circuit

cycle life   number of times a secondary cell can be recharged—number of charge and discharge cycles

damped waveform   a waveform that decreases in amplitude with each cycle—caused by circuit resistance

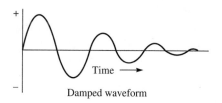

Damped waveform

damping   the steady decay of a ringing waveform—caused by circuit resistance

decade   a jump in magnitude—ten times or 1/10

decay   refers to the decrease in current and magnetic field when an EMF is removed from an inductor

decibel (dB)   one-tenth of a Bel—1 dB = 1/10 Bel—see *Bel*

degaussing   a process of demagnetization accomplished by applying an alternating magnetic field that gradually decreases in strength

delta configuration   components arranged in the shape of a triangle (Δ)

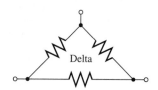

Delta

diamagnetic having a very low permeability and very high reluctance to magnetic flux—takes on an opposing magnetic polarity in the presence of a very strong magnetic field

dielectric an insulator—the name given to an insulator used between the plates of a capacitor

dielectric resistance resistance of the dielectric in a capacitor

dielectric strength voltage rating of dielectric materials—maximum voltage before breakdown occurs

difference of potential potential difference—electromotive force measured in volts

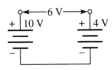

differentiator a circuit that produces positive and negative pulses from an applied pulsating DC square wave—high-pass filter

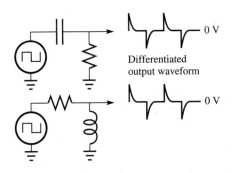

diode a semiconductor device that permits current in only one direction

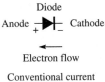

DIP dual in-line package—an integrated-circuit-type package

DIP switch switch or switches designed into a dual in-line package

direct current (DC) electrical current that constantly flows in one direction—as from a battery

dissipated power (P_D) power that is converted to heat or mechanical motion, or to some other form of energy

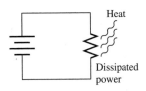

DMM digital multimeter

documentation written description of a circuit or piece of equipment including a schematic—a written description of work performed on a circuit or piece of equipment

domain magnetic domain—a magnetically organized cluster of atomic dipoles

DPDT double pole, double throw—switch nomenclature

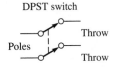

DPST double pole, single throw—switch nomenclature

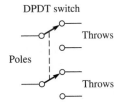

dry cell a voltaic cell that contains an electrolyte that is not a liquid (more like a paste)

duty cycle indicates the relative amount of time a pulse is "on" compared to the total cycle time—% duty cycle = $100\% \cdot PW/T$

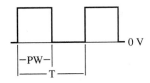

eddy currents induced currents in the core of electromagnetic induction devices—account for an undesirable I^2R power loss

effective value   as in effective voltage or current—same as RMS value—the amount of DC required to produce the same heating effect as a given amount of AC—RMS = 0.707 · peak value

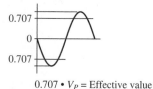

0.707 · V_P = Effective value

electric current   see *electron flow*

electric field   an energy field composed of electrostatic lines of force (flux)

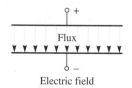

Electric field

electricity   the presence of a difference in electrical potential, which may result in the flow of current when a conductive path is provided

electrochemical   a source of EMF in which a chemical reaction takes place that produces free electrons as a result of molecular bonding

electrodes   conductive elements that perform a point of contact function in devices such as batteries, vacuum tubes, capacitors, etc.—often designated as having a specific polarity

electrolysis   a process by which a direct current is passed through a conductive liquid, causing a chemical reaction in which certain elements are removed from a compound or solution—example: the separation of hydrogen and oxygen from water

electrolyte   a substance that permits ionic conduction, or current, within a voltaic cell or electrolytic capacitor

electrolytic   having an electrolyte—large-value capacitor types—aluminum and tantalum electrolytics

electromagnetic   electrical and magnetic, one the cause and the other is the result—as relating to the electromagnetic production of electricity: mechanical energy is converted to electrical energy by the relative physical motion of a conductor through a magnetic field

electromagnetic induction   a means by which current is forced to flow in a conductor under the influence of an external magnetic field—there must be relative motion between the conductor and the magnetic field

electromagnetic trip mechanism   circuit breaker mechanism that responds to the electromagnetic field created by excessive current—solenoid trip

Electromagnetic
trip circuit breaker

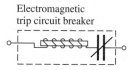

electromotive force (EMF or E)   the force or difference of electrical potential that causes electrical current—measured in volts

electron   a small subatomic particle that orbits the nucleus of an atom—negatively charged particle

electron flow   the organized migration of electrons from atom to atom in a conductor, or semiconductor, under the influence of some source of energy

electroscope   a small instrument used to demonstrate the repulsion effect of like charges

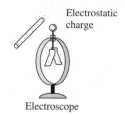

Electrostatic
charge

Electroscope

electrostatic   having a static electrical charge

electrostatic flux density (D)   number of flux per unit area—$D = \Phi/m^2$ = flux/square meter

electrostatic induction   charge by repulsion—no physical contact with charging source

elements   the simplest and most basic substances in the universe of which all things are made—examples: hydrogen, oxygen, iron, copper, argon

E notation   E stands for exponent—a notation similar to scientific notation that permits very large and very small numbers to be expressed in abbreviated form—this notation is used with most scientific calculators and computers—1,255 = 1.255 E + 3 (or E3) and 0.0047 = 4.7 E − 3

exponent   the power of 10—a superscripted number indicating magnitude—in 10^{-2}, the -2 is the exponent

farad (F)   the SI unit of measure for capacitance—the amount of capacitance needed by a capacitor to store one coulomb of electrons under the influence of an applied EMF of one volt

Faraday, Michael   (1791–1867)—British lecturer and scientist—discovered electromagnetic induction

Faraday's Law   the amount of voltage induced across a coil is dependent upon three factors: (1) the amount of flux cutting across the coil, (2) the number of turns of the coil, and (3) the rate at which the flux cuts across the turns of the coil

fault condition current   a fuse rating—the amount of current the fuse will sustain without failure

Fermi, Enrico   (1901–1954)—Italian-born American physicist—in 1942, performed the world's first controlled nuclear fission chain reaction

ferromagnetic   materials having magnetic properties similar to iron—iron alloys

field intensity (ℰ)   the strength of the electric field—intensity of the total flux produced by the EMF—measured in volts per meter (V/m)

field intensity (*H*)   strength of the magnetic field—intensity of the total flux produced by MMF—measured in At/m (SI unit) and oersteds (cgs unit)

fixed component   value determined by the manufacturer—value cannot be changed by the user—as in fixed carbon or metal film resistors or fixed capacitors and inductors

Fixed value resistor

Fixed value capacitor

Fixed value inductor

flux (Φ)   magnetic or electrostatic lines of force

flux density (*B*)   concentration of flux per unit area—measured in gauss (G, cgs unit) or tesla (T, SI unit)

flywheel effect   see *ringing*

free electrons   electrons that are normally in the valence shell of an atom but are able to break free under the influence of an external source of energy

frequency (*f*)   the rate at which AC cycles are repeated—measured in cycles per second (cps) or hertz (Hz)

$f = 1/T$

full differentiation   an applied pulsating DC square wave is converted to a perfect AC square wave—requires a very long time constant

full integration   an applied pulsating DC square wave is converted to a pure DC—a very long time constant is required

full-scale deflection (FSD)   the full-scale movement of the pointer of an analog meter

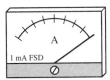

Full scale deflection

fundamental   as in *fundamental frequency*—the first frequency in a harmonic series—the first harmonic

fuse   a circuit protection device—prevents additional circuit damage when a component becomes shorted—contains a thin current-sensitive element that burns open when its current rating is exceeded

Fuse symbol

fuse response time   the length of time required for a fuse to open in response to a 200% overload

fuse voltage rating   the highest voltage for which the fuse is designed to safely interrupt

galvanic cell   see *voltaic cell*

galvanometer   an analog meter whose pointer is normally at rest at center scale

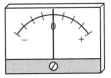

Galvanometer

gang tuning   as in variable capacitors or inductors that are changed in value at the same time—one control knob connected to many components through a common mechanical system

Gang tuning

gauss (G)   cgs system unit of measure for flux density—one flux line passing through a unit area of one square centimeter

Gauss, Karl Friedrich   (1777–1855)—German mathematician and experimenter

generator action   relative motion of a conductor and magnetic field that results in the production of electrical current

gilbert (Gb)   cgs system unit of measure for magnetomotive force—1 Gb = 0.794 ampere-turns

ground   common circuit connection—may be an earth ground, common return for AC line voltage—metal chassis, copper foil, common wire

Ground symbols

half-power frequency (HPF)   a frequency at which the power level drops to half—any cutoff frequency—frequency that corresponds to the half-power point (−3 dB point)—voltage amplitude is $0.707 \cdot V_{max}$

half-power point (HPP)   see *half-power frequency*

Hall effect   if a current-carrying conductor is placed in a magnetic field perpendicular to the flux lines, a small voltage develops across the width of the conductor perpendicular to the flux lines and the flow of current

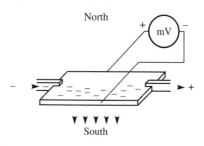

harmonic series   a group of frequencies that are harmonically related—the spacing between the frequencies is equal to the first frequency of the harmonic series—example: 3 kHz, 6 kHz, 9 kHz, 12 kHz, 15 kHz, etc.

henry (H)   the unit of measure for inductance—one henry equals one induced volt per one ampere per second change in current (1 H = 1 V/1 A/s)

Henry, Joseph   (1797–1878)—American—taught mathematics and physics—was a professor of philosophy at Princeton University—first director of the Smithsonian Institution

hermetically sealed   total airtight encapsulation

hertz (Hz)   standard SI unit for frequency—1 Hz = 1 cps

Hertz, Heinrich Rudolph   (1857–1894)—German physicist—discovered electromagnetic waves—demonstrated the first spark-gap transmitter

highpass filter   a filter that passes all frequencies above a specified cutoff frequency—a differentiator circuit

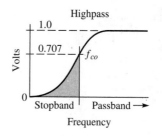

hypotenuse   side opposite the right angle of a right triangle

hysteresis   to lag behind—magnetization level and polarity ($\pm B$) lagging behind the magnetizing force ($\pm H$)

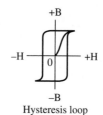

Hysteresis loop

IF transformer   intermediate frequency transformer—used in radio receivers—tuned to a specific frequency such as 455 kHz (for AM radios)

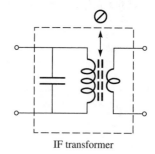

IF transformer

impedance (Z)   a circuit's or component's opposition to AC—usually includes a combination of resistance and reactance—measured in ohms

inductance (L)   self-inductance—the measure of a coil's ability to induce a voltage across itself as a result of a change in current in its windings

induction   the process of delivering an electrical charge without physical contact—repulsion of like charges is employed

inductor a component that is able to store energy in the form of a magnetic field—passes DC and opposes AC—a coil

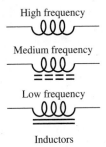

High frequency

Medium frequency

Low frequency

Inductors

insertion loss the loss in voltage and power that results from a filter being inserted into a circuit—insertion loss varies with frequency and depends on filter design

instantaneous value the amount of voltage or current at a specific instant in time—the voltage or current at a specific point in an AC waveform

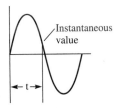

Instantaneous value

insulator a material that has very tightly bonded valence electrons—for most practical purposes, does not conduct electricity—examples: ceramic, glass, rubber, resins

integrator a circuit that creates an average DC level from an applied pulsating DC—lowpass filter

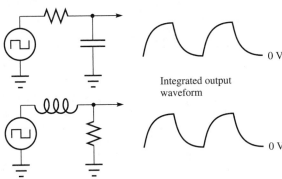

Integrated output waveform

0 V

0 V

internal resistance (r_i) the resistance inside a voltage source such as a battery—limits the amount of current a source can deliver

interwinding capacitance capacitance that exists between the windings of an inductor

inverter a DC to AC power converter

ion an atom or molecule that is not neutral in polarity

ionization the process of creating ions

joule (J) a standard SI unit for energy—1 J = 1 Ws (watt second)

keeper a ferromagnetic material used as a bridge between poles of a magnet—helps maintain the strength of the magnet

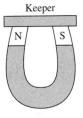

Keeper

Kelvin temperature scale (K) Celsius scale plus 273°—water freezes at 273°K = 0°C = 32°F

kilowatthour (kWh) unit of measure for accumulated power usage or energy usage—power companies charge customers on the basis of the number of kWh

Kirchhoff, Gustav Robert (1824–1887)—famous for voltage and current circuit-analysis laws

Kirchhoff's Current Law the algebraic sum of all currents entering and leaving any point in a circuit must equal zero

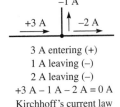

−1 A

+3 A   −2 A

3 A entering (+)
1 A leaving (−)
2 A leaving (−)
+3 A − 1 A − 2 A = 0 A
Kirchhoff's current law

Kirchhoff's Voltage Law the algebraic sum of all voltages in a closed loop must equal zero

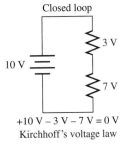

Closed loop

3 V

10 V

7 V

+10 V − 3 V − 7 V = 0 V
Kirchhoff's voltage law

laminated iron core iron core made of many layers of thin, insulated iron sheets—used to minimize eddy currents

leakage current current through the dielectric of a capacitor—undesirable

leakage flux magnetic flux that passes through the air instead of through a ferromagnetic medium

Leakage flux
in the air gap

Lenz, Heinrich F. (1804–1865)—German-born Russian scientist—discovered the principle of countermagnetomotive force (CMMF)

Lenz's Law induced current is produced in such a direction as to produce a magnetic field that opposes the external magnetic field that generated it—explains why it takes work to crank a generator to produce electricity

load the amount of current demanded by a load device

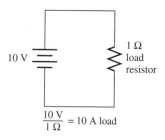

10 V

1 Ω
load
resistor

$$\frac{10\ V}{1\ \Omega} = 10\ A\ load$$

load device a device that is connected to a power supply, or circuit, that demands current from the supply—examples: resistors, lamps, motors, etc.

loading effect the effect a meter or any device has when connected to a voltage source or circuit—the degree to which a voltage decreases due to an increased load (due to an increased demand for current)

long time constant the time required is much greater than the time allowed, where the time required is 5 TC and the time allowed is the applied pulse width—long TC is when 5 TC > PW

lowpass filter a filter that passes all frequencies below a specified cutoff frequency—an integrator circuit

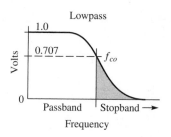

Lowpass

1.0

0.707 f_{co}

Volts

0

Passband | Stopband →

Frequency

Maximum Power Transfer theorem the amount of power a source can deliver to a load is limited by the internal resistance of the source and will be at a maximum value when the load resistance is equal to the internal resistance of the source

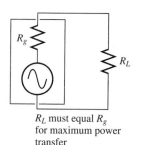

R_g

R_L

R_L must equal R_g
for maximum power
transfer

maxwell (Mx) one flux line

Maxwell, James Clerk (1831–1879)—British scientist—in 1864, he proposed that energy traveled through space as electromagnetic waves—extensive work with electricity and magnetism

magnetic polarity having two poles that are of opposite magnetic polarity—north or south

Magnetic polarity

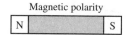

N | S

magnetohydrodynamic power generation (MHD) a means of generating DC by pumping a conductive fluid through a duct at high speeds—the fluid flows through a stationary magnetic field—the relative motion of fluid through the magnetic field induces a voltage perpendicular to the magnetic field and direction of fluid flow

magnetomotive force (MMF) the magnetic pressure that determines the number of flux lines through a medium—the amount of current times the number of turns—measured in ampere-turns (At)

metal film resistor thin metal film deposited on a tubular insulator—precision, close tolerance resistors ($\leq 1\%$)

Metal film resistor

meter sensitivity amount of current needed to cause a meter to indicate full scale—current needed for full-scale deflection

mho older unit for conductance—1 mho = 1/1 Ω = 1 siemen (S)

micro switch a short-travel, light-pressure switch

Millman's theorem any circuit composed of any number of parallel voltage sources can be reduced to a single equivalent voltage source by: (1) converting each voltage source to a current source, (2) combining all current sources and resistors, then (3) converting back to a single voltage source from the equivalent current source

mksa meter, kilogram, second, ampere—standard units of the international system of units (SI)—SI is often referred to as the MKSA system

molecule small substance building block composed of two or more atoms—tight bonding with shared valence electrons

monolithic of one mass or block

motor action the mechanical motion created by the interaction of a current-carrying conductor and a neighboring magnetic field

multimeter one meter with many functions—analog or digital

mumetal very-high-permeability iron alloy—used for magnetic shielding

mutual inductance (L_M) the property of one inductor inducing voltage across a neighboring inductor—measured in henries (H)

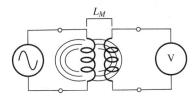

natural logarithm (ln, e^x) logarithms using 2.718 as the base—example: $3{,}562 = e^{8.178} = 2.718^{8.178}$

NC normally closed—refers to a closed switch condition

negative ion an atom or molecule that has one or more extra electrons—negatively charged

negative temperature coefficient (NTC) component's value changes inversely with temperature, as with an NTC thermistor whose resistance decreases as temperature increases

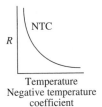
Negative temperature coefficient

neutron a subatomic particle found in the nucleus of all atoms except hydrogen atoms—has a mass slightly greater than a proton

NO normally open—refers to an open switch condition

nodal analysis circuit-analysis technique that makes use of nodal voltage or current

node a connecting point or junction for two or more components

nonpolarized having no polarity—as in *nonpolarized electrolytic capacitors*

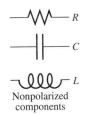
Nonpolarized components

Norton's theorem any linear circuit, consisting of resistances and one or more voltage sources and having two output terminals, can be replaced with a single constant current source and a single parallel resistance for the purpose of circuit analysis

Ohmmeter symbol

nucleus the central part of an atom—contains neutrons and protons (+ charged)

null to balance out—to bring to a minimum level—as in nulling the difference of potential of a Wheatstone bridge

octave an increase by a factor of 2 or a decrease by ½—a doubling of frequency or cutting a frequency in half—there is one octave from 400 Hz to 800 Hz

oersted (Oe) cgs system unit of measure for field intensity—1 Oe = 1 Gb/cm = 79.4 At/m

Oersted, Hans Christian (1777–1851)—Denmark—discovered that a current-carrying conductor produces a magnetic field

ohm (Ω) the unit of electrical resistance—see *resistance*

Ohm, Georg Simon (1787–1854)—German scientist who discovered the mathematical relationship between current, voltage, and resistance ($I = E/R$)

ohmmeter an instrument used to measure resistance—number of ohms of resistance is indicated on an analog meter (pointer and scale) or a digital display—circuit power is first removed, component is isolated, then measured

OPEN a break in a circuit—no continuity—current is disrupted—0 A of current—maximum voltage measured across an open

open circuit voltage available source voltage when no load device is connected

operator the angle associated with a magnitude in a polar number system

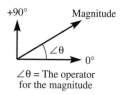

∠θ = The operator for the magnitude

operator *j* the *j* is used to signify 90° increments—$+j = +90°$ and $-j = -90°$—also, $j = \sqrt{-1}$—example: $+j10\ \Omega = 10\ \Omega$ at an angle of $+90° = 10\ \Omega$ of inductive reactance

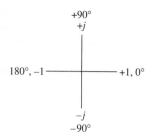

oscillator an electronic device (circuit) used to create AC at some specific frequency—a signal source

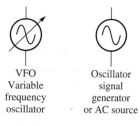

VFO Variable frequency oscillator

Oscillator signal generator or AC source

overall parallel circuit a parallel/series circuit—the circuit can be reduced to a simple parallel circuit

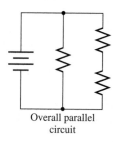

Overall parallel circuit

overall series circuit a series/parallel circuit—the circuit can be reduced to a simple series circuit

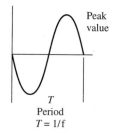

Overall series circuit

parallel circuit a circuit in which components are placed in parallel—more than one path for current—current is divided among parallel branches

paramagnetic material having very low permeability and high reluctance—may become only slightly magnetized under the influence of an extremely strong magnetic field

passive filter a filter that is made only of components such as resistors, capacitors, and inductors—no active devices

peak value as in *peak voltage or current*—the maximum value reached during a cycle

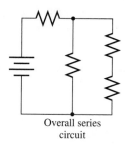

Peak value

T Period T = 1/f

period (*T*) time for one cycle—$T = 1/f$

periodic table a table of all known elements—elements arranged by the number of protons in the nucleus (atomic number)

periodic waveform a waveform that is repeated at regular intervals

permalloy iron/nickel permanent magnetic material

permanent magnet (PM) ferromagnetic material whose domains remain significantly aligned after an external magnetic field is removed

permeability (μ) the measure of the ease with which magnetic flux can be established in a material

permittivity (ϵ) absolute permittivity—the measure of the ease with which a dielectric accepts electrostatic flux—measured in farads per meter (F/m)

phase response the phase shift in output voltage created by a filter as a result of a change in applied frequency

phasor a ray or arrow that is used to represent magnitude and direction related to time—represents magnitude and phase angle

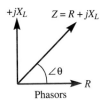

Phasors

photoelectric as in *a photoelectric source of EMF*—photon bombardment from the sun produces electron flow in semiconductor materials—photovoltaic cells (PV)

photoresistor resistor whose resistance changes with light intensity

Photoresistor

photovoltaic cell (PV) a thin semiconductor plate that yields an organized DC under the influence of light intensity—see *photoelectric*

Photovoltaic cell

piezoelectric effect pertains to crystalline minerals such as rochelle salt, quartz, and tourmaline—applying a voltage across the crystal produces physical stress—applying physical stress to the crystal produces a voltage

planetary an organized system revolving around a central focal point (central object)

polarity the sign designating one of two opposite states—positive or negative

polarized of two opposite states or positions—plus and minus

Polarized

Electrolytic capacitor

polar notation a notation used to express complex numbers using magnitude and phase angle—example 150 V $\angle 45°$

pole one of two opposite states or positions—positive or negative—in a switch, it is a single common terminal that is switched to one or more other terminals

positive ion an atom or molecule having a deficiency of electrons—positively charged

positive temperature coefficient (PTC) component's value changes directly with temperature as with a PTC thermistor whose resistance increases as temperature increases

Positive temperature coefficient

potential energy relational energy—energy held as a stationary mass that is being acted upon by a force in relationship to a certain point—as in gravity acting on water stored in an elevated reservoir

potentiometer ("pot") a variable resistor used to vary DC and AC levels—a variable voltage divider—knob or screwdriver controlled—trimmer pots, rotary pots, slide pots

Potentiometer

Wiper

power (P) the rate at which work is done and energy is expended—a quantity of work per unit of time—power = work/time

power derating lowering the safe upper limit of power dissipation for a component or device because of unfavorable ambient conditions (increased surrounding temperature, poor air circulation, etc.)

power factor (*PF*) a decimal that expresses the relationship between real power and apparent power in an AC circuit—$PF = P_R/P_A$—no units

power loss   power that is intended for a system or circuit but is lost in the resistance of conductors or inefficiency of an electromechanical machine

power of 10   used in scientific notation—it is the magnitude of a number—the exponent to the base 10—examples: 10^3, 10^{-6}, 10^{12}, 10^{-9}

power rating   the maximum amount of power that a component or device can safely dissipate under certain conditions of temperature and humidity

prefix   a three- or four-letter syllable added to the beginning of a word to indicate magnitude or dimension—examples: *kilo* as in *kilo*gram, *micro* as in *micro*farad, *mega* as in *mega*hertz

primary cell   a single-use, nonrechargeable, voltaic cell (battery)—examples: carbon zinc cells, alkaline cells

primary node   the junction of three or more components

primary winding   the input winding of a transformer

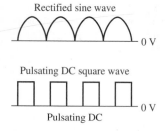

Transformer

printed circuit (PC)   a circuit layout etched on a copper-clad board

proton   a small but heavy subatomic particle contained in the nucleus of an atom—has a mass 1,836 times an electron

pulsating DC   direct current that does not remain constant in amplitude—returns to 0 V at regular intervals—should not be confused with DC ripple voltage

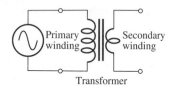

Pulsating DC

pulse duration (*PD*)   same as pulse width (*PW*)—length of time the pulse is "on"

Pythagorean theorem   in a right triangle, the sum of the squares of the two sides adjacent to the right angle (*a* and *b*) is equal to the square of the side opposite the right angle (*c*)—formula: $a^2 + b^2 = c^2$

quadrature   of a 90° phase relationship

quality factor (*Q*)   a number that represents the ratio of reactance to resistance (X/R)—a figure of merit for tuned circuits and inductors and capacitors—used to determine the bandwidth of a tuned circuit—has no units

radian (rad)   an angular distance defined by a circumferal arc equal in length to the radius of a circle—1 rad = 57.3°

Circle contains 2 π radians

reactance (*X*)   as in *inductive and capacitive reactance* (X_L and X_C)—a reactive component's reaction to AC—measured in ohms—opposition to AC

reactive power (*P_r*)   power that is stored in reactive components—measured in voltamperes (VA) or voltamperes reactive (VAR)

real power (*P_R*)   true power or average power—power dissipated by resistance in an AC circuit—power that is actually converted to mechanical motion or heat—measured in watts (W)

rectangular coordinate system   Cartesian coordinate system—an *X,Y* graph system—horizontal and vertical scales provide *X,Y* coordinates

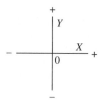

Rectangular coordinate system

rectangular notation   a notation used to express a complex number in terms of its real and imaginary parts—example: $3 - j5$ A

reflected impedance   the impedance of the primary winding of a transformer as a result of a load on a secondary winding—measured in ohms—primary voltage divided by primary current

relative permeability (*μ_r*)   the permeability of a material relative to air where air = 1

relative permittivity (*ε_r*)   permittivity of a dielectric compared to air or a vacuum—has no units

relay   a solenoid-operated switch used for circuit control

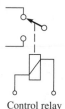

Control relay

reluctance ($\mathcal{R}$ or R_m)   the measure of opposition to the establishment of magnetic flux in a material

residual magnetism   magnetism that remains in a material after a magnetizing force is removed

resistance (R)   opposition to electrical current—measured in ohms (Ω)—a resistance of 1 Ω will allow 1 A of current to flow under a pressure of 1 V

resistivity (ρ)   see *specific resistance*

resistor   a component that is used to limit current and provide a desired voltage drop—types: carbon composition, carbon and metal film, wirewound

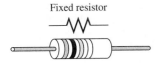

Fixed resistor

resonant   having the ability to resonate

resonant frequency (f_r)   the frequency at which a circuit is designed to resonate

resonate   to respond in such a way as to reproduce or echo stimulus

retentivity   the ability of a material to retain a quantity of residual magnetism

rheostat   a variable resistor used to control the amount of current in a circuit—often wirewound

Rheostat

ringing   the exchange of energy between an inductor and a capacitor—sinusoidal in nature

ripple voltage   DC ripple—rapid and continuous variations in a DC voltage and current

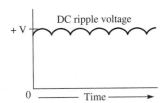

RMS   root mean square—see *effective value*

rolloff   the part of a frequency response graph that drops off—skirt—its slope is measured in dB/octave or dB/decade of frequency change

rotary switch   sometimes called ''wafer switch''—switching is obtained with a rotary motion of a mechanical knob and shaft

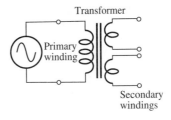

Rotary switch

rotor   the rotating part of a generator or motor—armature

sawtooth waveform   a waveform with a linear rise and fall—gradual rise and very sharp fall

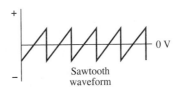

schematic   a circuit diagram made of symbols that represent circuit components and devices

schematic symbols   those symbols used to create a schematic—icons that represent components and devices

scientific notation   a notation used to express very large and very small numbers as a number between 1 and 10 times a power of ten: $5{,}285 = 5.285 \cdot 10^3$ and $0.0022 = 2.2 \cdot 10^{-3}$

secondary cell   a rechargeable voltaic cell (battery)—examples: nickel-cadmium cells

secondary winding   an output winding of a transformer

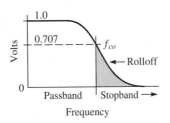

semiconductors   elements whose valence electrons are more tightly bonded than that of conductors—usually have four valence electrons—examples: carbon, germanium, silicon, gallium

series circuit a circuit in which components are connected end to end—only one path for current—voltage is divided among series components

series motor DC motor whose field coil is wired in series with the armature

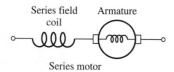

Series motor

shaded pole motor low-power single-phase AC induction motor—uses single loop copper strap to warp or offset the changing magnetic field, which creates a sweeping action of the field

shelf life the length of time a battery can sit idle before it is considered unusable

shells definite orbits and energy levels surrounding the nucleus of an atom

shield a metal enclosure used to block or trap magnetic flux

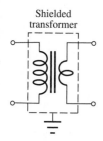

Shielded transformer

short a condition that causes the resistance of a component or device to become zero ohms (or close to it)—zero volts is measured across a "dead" short

short circuit rating a fuse rating—the maximum current a fuse can safely interrupt at a rated voltage

short time constant the time required is much less than the time allowed where the time required is 5 TC and the time allowed is the applied pulse width—short TC is when $5\,TC < PW$

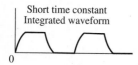

Short time constant
Integrated waveform

shunt to place in parallel—a parallel component

shunt motor DC motor whose field coil is wired in parallel with the armature

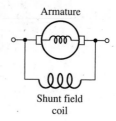

Armature

Shunt field coil

SI Système International d'Unités—the International System of units

siemen (S) standard unit for conductance—1 S = 1/1 Ω

simultaneous equations equations pertaining to a single system that each contain the same variables—these equations can be solved together to determine the value of each variable

sine wave an AC waveform the amplitude of which varies with the sine of the angle from 0 to 360 degrees through each cycle

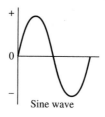

Sine wave

SIP single in-line package—all package terminals in a single row—used for multiple resistor packaging and various integrated circuits

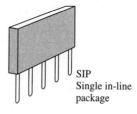

SIP
Single in-line package

slip rings used on the armature of some AC motors and generators to transfer electrical energy via brushes

slope a rate of change—can be rate of change in voltage or current (V/s or A/s)—can be rate of change in amplitude as for filters (dB/octave or dB/decade)—the steepness of a graph or ramp—does not have to be time related—negative slope is heading forward—positive slope is heading upward

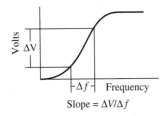

Slope = $\Delta V/\Delta f$

SMT   surface mount technology—components, devices, and ICs are solder mounted on the copper side of a printed circuit—no lead holes needed—for higher density circuit design

solder   an alloy of tin and lead used to make permanent electrical connections or to join metal surfaces

solenoid   an electromagnetic coil—electromagnet

source voltage (E_S)   voltage applied to a circuit source of electromotive force (EMF)

Source voltage

SPDT   single pole, double throw—switch nomenclature

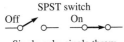

Single pole, double throw

specific gravity   the ratio of a liquid's weight to the weight of an equal volume of distilled water

specific resistance (ρ)   the amount of resistance of a one-foot length of wire with a cross-sectional area of one circular mil (1 CM)—different for different conductive materials—measured in circular mil ohms per foot ($CM \cdot \Omega/ft$)

SPST   single pole, single throw—switch nomenclature

SPST switch
Off   On

Single pole, single throw

squirrel-cage rotor   a brushless armature found in many AC induction motors

static   stationary—no movement or change

static charge   the buildup or stripping away of electrons from an object—the positive or negative charge is stationary

stator   the stationary part of an electromechanical system or device—as in the *stator windings* of a motor or generator and the *stator plates* of a variable capacitor

stray capacitance   circuit capacitance that exists between conductors—not shown in a schematic

stray inductance   circuit inductance of every conductor—exists in component leads, cables, copper foils, etc.—not shown in a schematic

superposition   a method of circuit analysis in which a multiple-voltage-source circuit can be solved for currents and voltage drops—currents and voltage drops produced by each voltage source are superimposed (combined)

supersonic   frequencies above the range of human hearing

susceptance (B)   the reciprocal of reactance—$B = 1/X$

switch   a circuit control device—used to apply and remove power or to control special functions—operates in one of two conditions, OPEN or CLOSED

synchronous motor   an AC motor whose armature stays in step with the rotating magnetic field of the stator windings

tank   not as in military equipment—a parallel resonant circuit—capacitor and inductor in parallel

tank current (I_{tank})   current that is ringing in the tank—$I_{tank} = E_{tank}/X_C = E_{tank}/X_L$

tank impedance (Z_{tank})   the overall impedance created by the tank circuit—$Z_{tank} = X_L^2/R_W$—measured in ohms—purely resistive at resonance

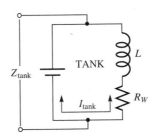

tank Q (Q_{tank})   the quality factor of the tank itself—$Q_{tank} = X_L/R_W$—no units

tantalum   a metal used in very small high-value electrolytic capacitors

taut-band movement   meter movement whose armature is suspended by a spring-loaded band—generally more expensive and more sensitive than jewel movements

temporary magnet   a ferromagnetic material that is only magnetized under the influence of a magnetizing force—has very low retentivity

tesla (T)   a standard SI unit of measure for flux density—1 T = 10,000 gauss—equals one weber of flux passing through a unit area of one square meter

Tesla, Nikola   (1856–1943)—Yugoslavian-born American inventor—invented the AC induction motor—famous for the Tesla coil

thermal trip mechanism circuit breaker mechanism that responds to the heating effect of excessive current—bimetal strip or solder pool

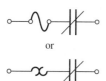

or

Thermal trip circuit breaker

thermistor thermal resistor—changes resistance with temperature—available with positive or negative temperature coefficients

Thermister

thermocouple used to generate an EMF in response to applied heat—formed of two dissimilar conductors

Thermocouple

thermoelectric a thermally generated source of EMF—heat used to create electrical current as in a thermocouple

thermoelectric generator the combination of a heat source and a thermopile for the production of electricity

thermopile a battery of thermocouples—series/parallel arrangement of thermocouples to obtain a higher power output

Thevenin's theorem any linear circuit, consisting of resistances and one or more voltage sources and having two output terminals, can be replaced by a single voltage source and a single series resistor for the purpose of circuit analysis—used to evaluate the effects of different loads

time constant (TC or T) resistance times capacitance (RC), or inductance divided by resistance (L/R)

toggle switch a lever action type switch

toroid as in *toroidal inductor* or *toroidal coil* or *toroidal core*—donut shaped—ring shaped

Toroid

total resistance (R_t) the total resistance offered to a voltage source by a DC circuit—the equivalent resistance of a circuit

transducer a device used to convert one form of energy to another—examples: a microphone (acoustical to electrical), a loudspeaker (electrical to mechanical), thermocouple (heat to electrical and electrical to heat plus cold on opposite ends of the thermocouple)

transformer an electromagnetic device that makes use of mutual inductance—voltage and current can be stepped up or down—has a primary winding and one or more secondary windings—windings wound on same core—an AC component

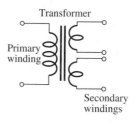

Transformer

Primary winding

Secondary windings

transient in the process of changing level or amplitude—transient voltage and transient current

trap a resonant circuit used to stop, block, or reject a frequency

triangle waveform a ramp waveform that has rising and falling edges of equal, yet opposite, slope

0 V

Triangle waveform

trimmer a low-value component used for fine adjustments—as in *trimmer potentiometers* and *trimmer capacitors*

Trimmer capacitor

tuneable circuit a circuit that may be tuned to resonate at any frequency within a given range of frequencies

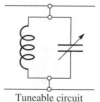

Tuneable circuit

tuned circuit   a circuit that is tuned to or resonant at a particular frequency

turns ratio   the ratio of input turns to output turns of a transformer—determines current and voltage step-up or step-down—determines transformation ratio

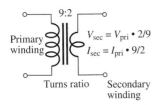

$V_{sec} = V_{pri} \cdot 2/9$
$I_{sec} = I_{pri} \cdot 9/2$

Turns ratio   Secondary winding

valence   a positive or negative number that indicates the number of electrons contained in the outer shell of an atom (outer shell = valence shell)—examples: $+2$ means there are 2 electrons in the valence shell, -2 means the valence shell is 2 electrons short of being full (8 is full), a valence of $+2$ = a valence of -6

variable component   value may be changed very easily by the user—knob or tool variable—as in *variable tuning capacitors, potentiometers, and rheostats*

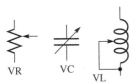

Variable components

variac   a variable AC autotransformer

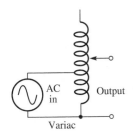

Variac

varistor   a device that exhibits a sudden drop in resistance when an applied voltage exceeds its voltage threshold—used to protect equipment from power line voltage spikes

Varistor

vector   a ray or arrow that indicates magnitude and physical direction

volt   unit of measure for electromotive force—the potential difference that exists when one joule of energy is needed to move one coulomb of electrons between two points—$1 \text{ V} = 1 \text{ J}/1 \text{ C}$

Volta, Count Alessandro   (1745–1827)—Italian scientist—invented the battery and the capacitor—experimented with electrolysis

voltage divider   a series circuit that divides the source voltage

voltage drop   the amount of voltage developed across an impedance in a circuit

voltaic cell   a single chemical unit that produces a certain amount of voltage during a chemical reaction—amount of voltage determined by substances involved in the reaction

Voltaic cell

voltmeter   an instrument used to measure an electrical difference of potential—placed in parallel with the component, device, or source—a high-impedance instrument

Voltmeter

voltmeter sensitivity   gives an indication of the possible loading effect of the voltmeter on a circuit that is to be measured—measured in ohms per volt—the greater the Ω/V, the higher the voltmeter sensitivity—inexpensive analog voltmeters typically have a sensitivity of 20 kΩ/V

VOM   volt-ohm-milliammeter—multimeter

watt (W)   standard SI unit for electrical power—a rate at which energy is used—$1 \text{ W} = 1 \text{ J/s}$ (1 joule of energy per second of time)—also, $1 \text{ W} = 1 \text{ VA}$ (1 volt ampere)

wattmeter   an instrument used to measure power in watts

Wattmeter

weber (Wb)   a standard SI unit for number of flux lines—$1 \text{ Wb} = 100 \text{ M}\Phi$ (flux)—usually expressed as microwebers (μWb)

Weber, Wilhelm   (1804–1890)—German physicist who experimented with electromagnetism

wet cell   a voltaic cell containing a liquid electrolyte

Wheatstone bridge   a circuit used to establish a balance in difference of potential between two points—used in servo motor control—used in test equipment to measure unknown impedance

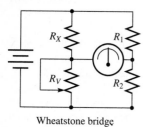

Wheatstone bridge

Wheatstone, Sir Charles   (1802–1875)—British physicist and inventor—ironically, he did not invent the Wheatstone bridge

wirewound resistor   constantan or nichrome wire wound on an insulator base—high power ratings and close tolerance

working volts DC (WVDC)   the maximum voltage that may safely be applied to a capacitor—maximum DC or peak AC

wye configuration   a group of components connected in a ''Y'' shape arrangement—also referred to as a ''T'' configuration

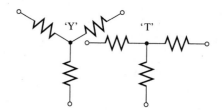

Wye and Tee configuration

zero-adjust   the rheostat fine adjustment on an ohmmeter used to set the pointer to exactly full scale before measurements are made

Index

Edison is shown here in one of his laboratories holding an invention that resembles a light bulb but is actually one of the earliest vacuum tubes. Edison discovered that electricity would flow from a heated filament to a metal plate to which a positive potential was applied. This became known as the "Edison effect." He did not pursue this concept further and others such as John Flemming and Lee DeForest received acclaim for the invention of the vacuum tube diode and vacuum tube amplifier.